Mechanics of Solids

Third Edition

Why are competent engineers so vital?

Engineering is among the most important of all professions. It is the authors' opinion that engineers save more lives than medical doctors (physicians). For example, poor water or the lack of it is the second largest cause of human death in the world, and if engineers are given the 'tools', they can solve this problem. The largest cause of human death is the malarial mosquito, and even death due to malaria can be decreased by engineers – by providing helicopters for spraying areas infected by the mosquito and making and designing medical syringes and pills to protect people against catching all sorts of diseases, including malaria. Most medicines are produced by engineers! How does the engineer put 1 mg of 'medicine' precisely and individually into millions of individual pills, at an affordable price?

Moreover, one of the biggest contributions by humankind was the design of the agricultural tractor, which was designed and built by engineers to increase food production many-fold, for a human population which more or less quadruples every century! It is also interesting to note that the richest countries in the world are very heavily industrialised. Engineers create wealth! Most other professions don't!

Even in blue-sky projects, engineers play a major role. For example, most rocket scientists are chartered engineers or their equivalents and Americans call their chartered engineers (and their equivalents) scientists. Astronomers are space scientists and not rocket scientists; they could not design a rocket to conquer outer space. Even modern theoretical physicists are mainly interested in astronomy and cosmology and also nuclear science. In general a theoretical physicist cannot, without special training, design a submarine structure to dive to the bottom of the Mariana Trench, which is 11.52 km or 7.16 miles deep, or design a very long bridge, a tall city skyscraper or a rocket to conquer outer space. It may be shown that the load on a submarine pressure hull of diameter 10 m and length 100 m is equivalent to carrying the total weight of about 7 million London double-decker buses!

This book presents a solid foundation for the reader in mechanics of solids, on which s/he can safely build tall buildings and long bridges that may last for a thousand years or more. It is the authors' experience that it is most unwise to attempt to build such structures on shaky foundations; they may come tumbling down – with disastrous consequences.

Some quotes from ***Albert Einstein*** (14 March 1879–18 April 1955)

'Scientists investigate that which already is; Engineers create that which has never been'

'Imagination is more important than knowledge. For knowledge is limited to all we now know and understand, while imagination embraces the entire world, and all there ever will be to know and understand'

'Everybody is a genius. But if you judge a fish by its ability to climb a tree, it will live its whole life believing that it is stupid'

'To stimulate creativity, one must develop the childlike inclination for play'

Mechanics of Solids

Third Edition

Carl Ross *BSc(Hons), PhD, DSc, CEng, FRINA, MSNAME*

John Bird *BSc(Hons), CEng, CMath, CSci, FIMA, FIET, FCollT*

and Andrew Little *BSc, PhD, CEng, FIMechE, FIED, FHEA*

Routledge
Taylor & Francis Group
LONDON AND NEW YORK

Third edition published 2022
by Routledge
2 Park Square, Milton Park, Abingdon, Oxon, OX14 4RN

and by Routledge
605 Third Avenue, New York, NY 10158

Routledge is an imprint of the Taylor & Francis Group, an informa business

First edition published by Prentice Hall PTR 1996
Second edition published by Routledge 2016

British Library Cataloguing-in-Publication Data
A catalogue record for this book is available from the British Library

Library of Congress Cataloging-in-Publication Data
Names: Ross, C. T. F., 1935- author. | Bird, John, author. | Little,
Andrew, author.
Title: Mechanics of solids / Carl Ross, John Bird, Andrew Little.
Description: Third edition. | Abingdon, Oxon ; New York, NY : Routledge,
2022. | Includes bibliographical references and index.
Identifiers: LCCN 2021019550 (print) | LCCN 2021019551 (ebook) | ISBN
9780367651411 (hardback) | ISBN 9780367651404 (paperback) | ISBN
9781003128021 (ebook)
Subjects: LCSH: Structural analysis (Engineering)
Classification: LCC TA645 .R682 2022 (print) | LCC TA645 (ebook) | DDC
624.1/71—dc23
LC record available at https://lccn.loc.gov/2021019550
LC ebook record available at https://lccn.loc.gov/2021019551

ISBN: 978-0-367-65141-1 (hbk)
ISBN: 978-0-367-65140-4 (pbk)
ISBN: 978-1-003-12802-1 (ebk)

Typeset in Times New Roman
by KnowledgeWorks Global Ltd.

Access the companion website: www.routledge.com/bird

Mechanics of Solids

Mechanics of Solids provides an introduction to the behaviour of solid materials under various loading conditions, focusing upon the fundamental concepts and principles of statics and stress analysis. As the primary recommended text of the Council of Engineering Institutions for university undergraduates studying mechanics of solids it is essential reading for mechanical engineering undergraduates and also students on many civil, structural, aeronautical and other engineering courses. The mathematics in this book has been kept as straightforward as possible and worked examples are used to reinforce key concepts. Practical stress and strain scenarios are covered, including simple stress and strain, torsion, bending, elastic failure and buckling. Many examples are given of thin-walled structures, beams, struts and composite structures. This third edition includes new chapters on matrix algebra, linear elastic fracture mechanics, material property considerations and more on strain energy methods. The companion website www.routledge.com/cw/bird provides full solutions to all 575 further problems in the text, multiple-choice tests, a list of essential formulae, resources for adopting course instructors, together with several practical demonstrations by Professor Ross.

Carl Ross gained his first degree in Naval Architecture from King's College, Durham University, his PhD in Structural Engineering from the Victoria University of Manchester, and was awarded his DSc in Ocean Engineering from the CNAA, London. His research in the field of engineering led to advances in the design of submarine pressure hulls. His publications and guest lectures to date exceed some 290 papers and books, and he was Professor Structural Dynamics at the University of Portsmouth, UK. On the website www.routledge.com/cw/bird are several of Carl Ross's practical demonstrations.

John Bird is the former Head of Applied Electronics in the Faculty of Technology at Highbury College, Portsmouth, UK. More recently, he has combined freelance lecturing at the University of Portsmouth with examiner responsibilities for Advanced Mathematics with City and Guilds and examining for the International Baccalaureate Organisation. He has over 45 years' experience of successfully teaching, lecturing, instructing, training, educating and planning of trainee engineers study programmes. He is the author of 146 textbooks on engineering, science and mathematical subjects, with worldwide sales of over one million copies. He is a chartered engineer, a chartered mathematician, a chartered scientist and a Fellow of three professional institutions. He has recently retired from lecturing at the Royal Navy's Defence College of Marine Engineering in the Defence College of Technical Training at H.M.S. Sultan, Gosport, Hampshire, UK.

Andrew Little completed an undergraduate apprenticeship with Rolls-Royce Aero Division at the City University, London. He subsequently worked as a Stress Engineer for Rolls-Royce and then for companies such as Ferranti Computer Systems and Plessey Aerospace, designing equipment for high stress and vibration environments. After joining the University of Portsmouth, he became involved with pressure vessel research and completed his PhD whilst lecturing full-time. He has taught subjects such as Solid Mechanics, Dynamics, Design and Computer Aided Design over his 30 years at Portsmouth. He is a chartered engineer, a Fellow of the Institution of Mechanical Engineers and has published 80 academic papers. Now semi-retired, Andrew is still an external examiner and an online tutor.

Practical Demonstrations by Professor Carl Ross

may be viewed at: www.routledge.com/cw/bird or at carltfross YouTube

Tensile Test Experiment

Torsion of Circular Section Shafts

The Poisson's Ratio Experiment

Thin-Walled Circular Cylinders Under Internal pressure

Combined Bending and Torsion of Circular Section Shafts

Buckling of Axially Loaded Struts in Compression

Asymmetrical or Unsymmetrical Bending of Beams

Collapse of Submarine Pressure Hulls

Buckling of a Rail Car Tank Under External Pressure

Bottle Buckling – Simulating Submarine Pressure Hull Collapse

Recent Advances in Submarine Pressure Hull Design

Contents

Preface

Failure of most components, be they electrical or mechanical devices, can be attributed to stress values. These components vary from small electrical switches to supertankers, or from automobile components to aircraft structures or bridges or submarines, and so on. Thus, some knowledge of stress analysis is of great importance to all engineers, be they electrical, mechanical, civil or structural engineers, naval architects or others.

Although, today, there is much emphasis on designing structures by methods dependent on computers, it is very necessary for the engineer to be capable of interpreting computer output, and also to know how to use the computer program correctly. For this reason most of the earlier chapters in this text thoroughly emphasise fundamental principles, and when these are clearly understood, the reader can move on to using computer methods. Moreover, in recent years, the SmartPhone has become ubiquitous – in that some 70 per cent of Americans and 94 per cent of South Koreans possess one, together with about 880 million Chinese, at the present time of writing! These figures will grow in future years, and computer methods will become even more important. Ross has shown that by using an app, called DosBox, he has incorporated about 100 compiled finite element programs, which he gives away free, for use on a SmartPhone/iPad, etc. These programs which are written in compiled DOS BASIC and Visual BASIC, lend themselves to be easily run on a SmartPhone, or iPad or laptop, etc. Ross believes that the SmartPhone will revolutionise teaching in colleges and universities in the next few years.

The importance of this book, therefore, is that it introduces the fundamental concepts and principles of statics and stress analysis, and then applies these concepts and principles to a large number of practical problems which do not necessarily require computers. The authors consider that it is essential for all engineers who are involved in structural design, whether they use computer methods or not, to be at least familiar with the fundamental concepts and principles that are discussed and demonstrated in this text, so they can analyse the structure via a computational device.

The third edition of this book incorporates new material on linear elastic fracture mechanics, material property considerations, matrix algebra revision and further work on energy methods.

The book should appeal to undergraduates in all branches of engineering, construction and architecture, at all levels, and also to other students on BTEC and similar courses in engineering and construction. It contains a large number of worked examples, which are presented in great detail, so that many readers will be able to grasp the concepts by working under their own initiative. Most of the chapters contain a section on examples for practice, where readers can test their newly acquired skills.

Chapter 1 is on revisionary mathematics, which the student must be familiar with. This is taught in a step-by-step method, which is easily understood.

Chapter 2 is on further revisionary mathematics, and introduces a novel method of teaching integral and differential calculus, together with an introduction of elementary vector algebra. However, in this textbook, a considerable amount of higher mathematical knowledge is needed in order to understand the theory. If this is an area where help is needed, then help is available, in the form of recommended textbooks by John Bird (see pages 49 and 50).

Chapter 3 is on statics, and after an introduction of the principles of statics, the method is applied to plane pin-jointed trusses, and to the calculation of bending moments and shearing forces in beams. The stress analysis of cables, supporting distributed and point loads, is also investigated.

Chapter 4 is on simple stress and strain, and after some fundamental definitions are given, application is made to a number of practical problems, including compound bars and problems involving stresses induced by temperature change.

Chapter 5 is on geometrical properties, and shows the reader how to calculate second moments of area and the positions of centroids for a number of different two-dimensional shapes. These principles are then extended to 'built-up' sections, such as the cross-sections of 'I' beams, tees, etc.

Chapter 6 is on stresses due to the bending of symmetrical sections, and after proving the well-known formulae for bending stresses, the method is applied to a number of practical problems involving bending stresses, and also the problem on combined and direct stress. Composite beams are also considered.

Chapter 7 is on beam deflections due to bending, and after deriving the differential equation relating deflection and bending moment, based on small deflection elastic theory, this equation is then applied to a number of statically determinate and statically indeterminate beams. The area-moment theorem is also derived, and then applied to the deflection of a cantilever with a varying cross-section.

Chapter 8 is on torsion. After deriving the well-known torsion formulae, these are then applied to a number of circular-section shafts, including compound shafts. The method is also applied to the stress analysis of close-coiled helical springs, and then extended to the stress analysis of thin-walled open and closed non-circular sections. The elastic-plastic stress analysis of circular-section shafts is also considered.

Chapter 9 is on complex stress and strain, and commences with the derivation of the equation for direct stress and shear stress at any angle to the co-ordinate stresses. These relationships are later extended to determine the principal stresses in terms of the co-ordinate stresses. The method is also applied to two-dimensional strains, and the equations for both two-dimensional stress and two-dimensional strain are applied to a number of practical problems, including the analysis of strains recorded from shear pairs and strain rosettes. Mohr's circle for stress and strain is also introduced. Compliance and stiffness matrices are derived for the application to composite structures.

Chapter 10 is on membrane theory for thin-walled circular cylinders and spheres, and commences with a derivation of the elementary formulae for hoop and longitudinal stress in a thin-walled circular cylinder under uniform internal pressure. A similar process is also used for determining the membrane stress in a thin-walled spherical shell under uniform internal pressure. These formulae are then applied to a number of practical examples.

Chapter 11 is on energy methods, and commences by stating the principles of the most popular energy theorems in stress analysis. Expressions are then derived for the strain energy of rods, beams and torque bars, and these expressions are then applied to a number of problems involving thin curved bars and rigid-jointed frames. The method is also used for investigating stresses and bending moments in rods and beams under impact, and an application to a beam supported by a wire is also considered. The unit load method is also introduced. The elastic-plastic bending of beams is considered, together with the plastic design of beams.

Chapter 12 is on the theories of elastic failure. The five major theories of elastic failure are introduced and, through the use of two worked examples, some of the differences between the five major theories are demonstrated.

Chapter 13 is on thick cylinders and spheres, and commences by determining the equations for the hoop and radial stresses of thick cylinders under pressure. The Lamé line is introduced and applied to a number of cases, including thick compound tubes with interference fits. The plastic theory of thick tubes is discussed, as are the theories for thick spherical shells and for rotating rings and discs.

Chapter 14 is on the buckling of struts, and commences by discussing the Euler theory for axially loaded struts. The theory is extended to investigate the inelastic instability of axially loaded struts through the use of the Rankine-Gordon formula, together with BS449. The chapter also considers eccentrically loaded and initially curved struts, and derives the Perry-Robertson formula.

Chapter 15 is on the asymmetrical bending of beams. This chapter covers the asymmetrical bending of symmetrical section beams and the bending and deflections of beams of asymmetrical cross-sections. It also shows how to calculate the second moments of area of asymmetrical cross-sections.

Chapter 16 is on shear stresses in bending and shear deflections. This chapter shows how both vertical and horizontal shearing stresses can occur owing to bending in the vertical plane. The theory is extended to determine shear stresses in open and closed thin-walled curved tubes and to calculate shear centre positions for these structures. Shear deflections due to bending are also discussed.

Chapter 17 is on experimental strain analysis, and discusses a number of different methods in experimental strain analysis, in particular those in electrical resistance strain gauges and photoelasticity, together with the use of stress lacquers.

Chapter 18 is on the introduction to matrix algebra, the terminology used, the addition, subtraction, and multiplication of matrices and calculating the determinants and inverses of 2 by 2 and 3 by 3 matrices. The main purpose for using matrices is in solving simultaneous equations. This chapter is included for those whose knowledge of this topic may need a little revision.

Chapter 19 is on composites, where the stress analyses of new materials are discussed. The equations for these orthotropic materials are very different to those of isotropic materials, and they are derived from fundamental principles. Many of these materials have a much larger strength: weight ratio than materials used in the past, and they will prove to be very important for futuristic engineering projects.

Chapter 20 is on the matrix displacement method, and commences by introducing this method, together with the finite element method. A stiffness matrix is obtained for a rod element and applied to a statically indeterminate plane pin-jointed truss, with the aid of a worked example. Similarly, another stiffness matrix is derived for a beam element and applied to a continuous beam. The chapter also shows how to use a SmartPhone/iPad to analyse more complicated structures, together with the analysis of a rigid-jointed plane frame.

Chapter 21 is on the finite element proper, and derives stiffness matrices for an in-plane plate element, together with a specialist rod element. The chapter then shows how to analyse a gear tooth, under a concentrated load.

Chapter 22 is on fracture mechanics, where the existence of cracks within engineering components is acknowledged, together with some of the ways of quantifying their effects. Energy considerations, along with energy release during crack propagation is considered which gives rise to methods of measuring key material parameters from which others can be derived. Stress intensity factors are introduced, which enable the prediction of acceptable stress levels. The issue of plasticity due to the high stresses around the crack tip are then discussed and the effect of cracks on the cyclic life of components that undergo cyclic loading is also addressed.

Chapter 23 is on material property considerations and addresses how the materials from which engineering components are made respond to cyclical loading and high temperature environments. The effects of cyclic loading on acceptable stress levels are examined along with the issues that arise from different loading patterns (e.g. using Goodman diagrams and Miner's rule). The effects of surface condition and corrosive environments are also discussed. The concept of creep at high temperature is introduced, together with its significance; graphs are used to show how data can be cross-plotted and presented for best use. Creep parameters are derived and their benefits for design purposes shown, and the topic of relaxation is also covered.

Mechanics of Solids, 3rd Edition contains some **275 worked problems**, followed by over **575 further problems** (all **with answers**). The further problems are contained within some **81 Exercises**; each Exercise follows on directly from the relevant section of work, every few pages. In addition, the text contains **100 multiple-choice questions** (all **with answers**). Where at all possible, the problems mirror practical situations found in mechanical/civil/structural engineering. **Over 700 line diagrams** enhance the understanding of the theory.

At regular intervals throughout the text are some **5 Revision Tests** to check understanding. For example, Revision Test 1 covers material contained in Chapter 1, Test 2 covers the material in Chapter 2, Test 3 covers the material in Chapters 3 to 8, and so on. No answers are given for the questions in the Revision Tests, but an **Instructor's guide** has been produced giving full solutions and suggested marking scheme. The guide is offered online free to lecturers/instructors at www.routledge.com/cw/bird

At the end of the text, a list of relevant **formulae** is included for easy reference. A number of famous engineers/mathematicians are featured in the text, some of whom have their image and a short biography included, with more available on the website.

'**Learning by Example**' is at the heart of *Mechanics of Solids, 3rd Edition.*

Carl Ross, John Bird & Andrew Little

Free web downloads
The following support material is available from *www.routledge.com/cw/bird*

For Students:

1. Full worked solutions to all 575 further questions contained in the 81 Practice Exercises
2. A list of essential formulae
3. A list of the notation used in *Mechanics of Solids*
4. Multiple-choice test questions
5. Information on 27 famous engineers mentioned in the text
6. Video links to practical demonstrations by Professor Carl Ross

For Lecturers/Instructors:

1–6. As per students 1–6 above
7. Full solutions and marking scheme for each of the 5 Revision Tests; also, each test may be downloaded for distribution to students
8. All illustrations used in the text may be downloaded for use in PowerPoint presentations

Chapter 1

Revisionary mathematics

Why it is important to understand: **Revisionary mathematics**

Mathematics is a vital tool for professional and chartered engineers. It is used in mechanical and manufacturing engineering, in electrical and electronic engineering, in civil and structural engineering, in naval architecture and marine engineering and in aeronautical and rocket engineering. In these various branches of engineering, it is very often much cheaper and safer to design your artefact with the aid of mathematics – rather than through guesswork. 'Guesswork' may be reasonably satisfactory if you are designing an exactly similar artefact as one that has already proven satisfactory; however, the classification societies will usually require you to provide the calculations proving that the artefact is safe and sound. Moreover, these calculations may not be readily available to you and you may have to provide fresh calculations, to prove that your artefact is 'roadworthy'. For example, if you design a tall building or a long bridge by 'guesswork', and the building or bridge does not prove to be structurally reliable, it could cost you a fortune to rectify the deficiencies. This cost may dwarf the initial estimate you made to construct these artefacts, and cause you to go bankrupt. Thus, without mathematics, the prospective professional or chartered engineer is very severely handicapped.

At the end of this chapter you should be able to:

- convert radians to degrees
- convert degrees to radians
- calculate sine, cosine and tangent for large and small angles
- calculate the sides of a right-angled triangle
- use Pythagoras' theorem
- use the sine and cosine rules for acute angled triangles
- understand trigonometric identities
- expand equations containing brackets
- be familiar with summing vulgar fractions
- understand and perform calculations with percentages
- understand and use the laws of indices
- solve simple simultaneous equations

1.1 Introduction

As highlighted above, it is not possible to understand aspects of mechanical engineering without a good knowledge of mathematics. This chapter highlights some areas of mathematics which will make the understanding of the engineering in the following chapters a little easier. For more advanced mathematics, see the references on pages 49 and 50.

1.2 Radians and degrees

There are 2π radians or 360° in a complete circle, thus:

$$\boldsymbol{\pi \text{ radians} = 180°} \qquad \text{from which}$$

$$\mathbf{1\ rad} = \frac{\mathbf{180°}}{\boldsymbol{\pi}} \text{ or } \mathbf{1°} = \frac{\boldsymbol{\pi}}{\mathbf{180}}\ \mathbf{rad}$$

where $\pi = 3.14159265358979323846\ldots$ to 20 decimal places!

Problem 1. Convert the following angles to degrees correct to 3 decimal places:
(a) 0.1 rad (b) 0.2 rad (c) 0.3 rad

(a) $0.1\text{ rad} = 0.1\text{ rad} \times \frac{180°}{\pi\text{ rad}} = \mathbf{5.730°}$

(b) $0.2\text{ rad} = 0.2\text{ rad} \times \frac{180°}{\pi\text{ rad}} = \mathbf{11.459°}$

(c) $0.3\text{ rad} = 0.3\text{ rad} \times \frac{180°}{\pi\text{ rad}} = \mathbf{17.189°}$

Problem 2. Convert the following angles to radians correct to 4 decimal places:
(a) 5° (b) 10° (c) 30°

(a) $5° = 5° \times \frac{\pi\text{ rad}}{180°} = \frac{\pi}{36}\text{rad} = \mathbf{0.0873\ rad}$

(b) $10° = 10° \times \frac{\pi\text{ rad}}{180°} = \frac{\pi}{18}\text{rad} = \mathbf{0.1745\ rad}$

(c) $30° = 30° \times \frac{\pi\text{ rad}}{180°} = \frac{\pi}{6}\text{rad} = \mathbf{0.5236\ rad}$

Now try the following Practice Exercise

Practice Exercise 1. Radians and degrees

1. Convert the following angles to degrees correct to 3 decimal places (where necessary):
   (a) 0.6 rad (b) 0.8 rad
   (c) 2 rad (d) 3.14159 rad
   [(a) 34.377° (b) 45.837°
   (c) 114.592° (d) 180°]
2. Convert the following angles to radians correct to 4 decimal places:
   (a) 45° (b) 90°
   (c) 120° (d) 180°
   [(a) $\frac{\pi}{4}$ rad or 0.7854 rad
   (b) $\frac{\pi}{2}$ rad or 1.5708 rad
   (c) $\frac{2\pi}{3}$ rad or 2.0944 rad
   (d) π rad or 3.1416 rad]

1.3 Measurement of angles

Angles are measured starting from the horizontal 'x' axis, in an **anticlockwise direction**, as shown by θ_1 to θ_4 in Figure 1.1. An angle can also be measured in a **clockwise direction**, as shown by θ_5 in Figure 1.1, but in this case the angle has a negative sign before it. If, for example, $\theta_4 = 320°$ then $\theta_5 = -40°$

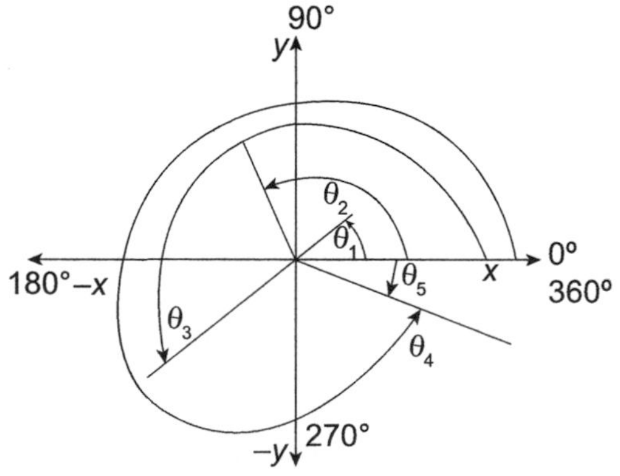

Figure 1.1

Problem 3. Use a calculator to determine the cosine, sine and tangent of the following angles, each measured anticlockwise from the horizontal 'x' axis, each correct to 4 decimal places:

(a) 30° (b) 120° (c) 250°
(d) 320° (e) 390° (f) 480°

(a) cos 30° = **0.8660** sin 30° = **0.5000**
tan 30° = **0.5774**

(b) cos 120° = **– 0.5000** sin 120° = **0.8660**
tan 120° = **– 1.7321**

(c) cos 250° = **– 0.3420** sin 250° = **– 0.9397**
tan 250° = **2.7475**

(d) cos 320° = **0.7660** sin 320° = **– 0.6428**
tan 320° = **– 0.8391**

(e) cos 390° = **0.8660** sin 390° = **0.5000**
tan 390° = **0.5774**

(f) cos 480° = **– 0.5000** sin 480° = **0.8660**
tan 480° = **– 1.7321**

These angles are now drawn in Figure 1.2. Note that cosine and sine are always between – 1 and + 1 but that tangent can be > 1 and < 1 to > – ∞ and < + ∞.

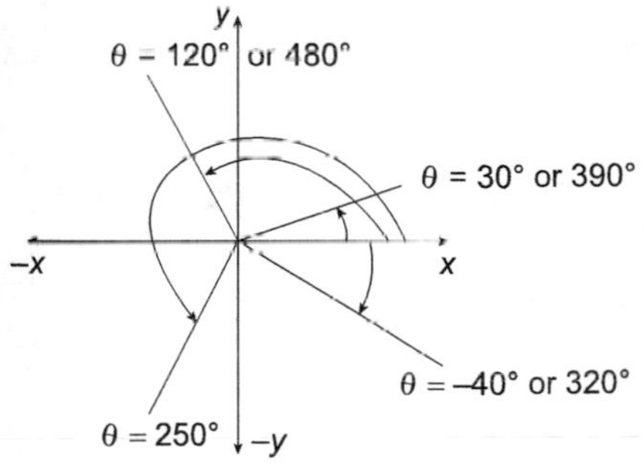

Figure 1.2

Note from Figure 1.2 that θ = 30° is the same as θ = 390° and so are their cosines, sines and tangents. Similarly, note that θ = 120° is the same as θ = 480° and so are their cosines, sines and tangents. Also, note that θ = – 40° is the same as θ = + 320° and so are their cosines, sines and tangents.

It is noted from above that

- in the **first quadrant**, i.e. where θ varies from 0° to 90°, all (A) values of cosine, sine and tangent are positive
- in the **second quadrant**, i.e. where θ varies from 90° to 180°, only values of sine (S) are positive
- in the **third quadrant**, i.e. where θ varies from 180° to 270°, only values of tangent (T) are positive
- in the **fourth quadrant**, i.e. where θ varies from 270° to 360°, only values of cosine (C) are positive.

These positive signs, A, S, T and C are shown in Figure 1.3, spelling the word CAST when moving anti-clockwise starting in the 4th quadrant.

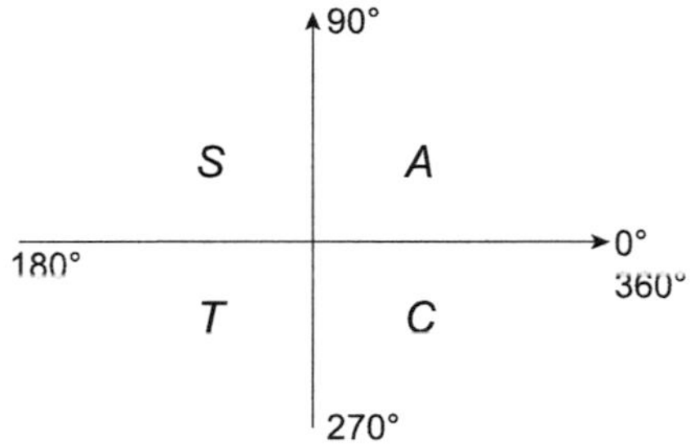

Figure 1.3

Now try the following Practice Exercise

Practice Exercise 2. Measurement of angles

1. Find the cosine, sine and tangent of the following angles, where appropriate each correct to 4 decimal places:
(a) 60° (b) 90° (c) 150°
(d) 180° (e) 210° (f) 270°
(g) 330° (h) – 30° (i) 420°
(j) 450° (k) 510°
[(a) 0.5, 0.8660, 1.7321
(b) 0, 1, ∞
(c) – 0.8660, 0.5, – 0.5774
(d) –1, 0, 0
(e) – 0.8660, – 0.5, 0.5774
(f) 0, –1, – ∞
(g) 0.8660, – 0.5000, – 0.5774
(h) 0.8660, – 0.5000, – 0.5774
(i) 0.5, 0.8660, 1.7321
(j) 0, 1, ∞
(k) – 0.8660, 0.5, – 0.5774]

1.4 Trigonometry revision

(a) Sine, cosine and tangent

From Figure 1.4, $\sin\theta = \dfrac{bc}{ac}$ $\cos\theta = \dfrac{ab}{ac}$ $\tan\theta = \dfrac{bc}{ab}$

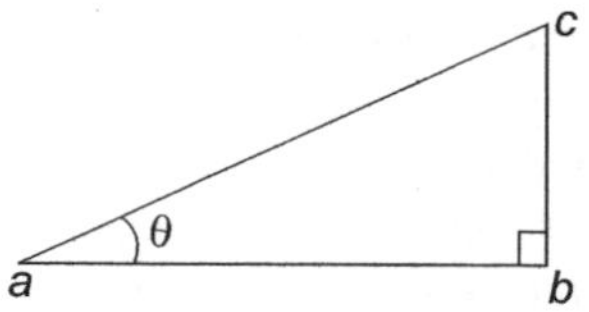

Figure 1.4

Problem 4. In Figure 1.4, if $ab = 2$ and $ac = 3$, determine the angle θ.

It is convenient to use the expression for cos θ, since 'ab' and 'ac' are given.

Hence $\quad \cos\theta = \dfrac{ab}{ac} = \dfrac{2}{3} = 0.66667$

from which $\quad \theta = \cos^{-1}(0.66667) = \mathbf{48.19^\circ}$

Problem 5. In Figure 1.4, if $bc = 1.5$ and $ac = 2.2$, determine the angle θ.

It is convenient to use the expression for sin θ, since 'bc' and 'ac' are given.

Hence $\quad \sin\theta = \dfrac{bc}{ac} = \dfrac{1.5}{2.2} = 0.68182$

from which $\quad \theta = \sin^{-1}(0.68182) = \mathbf{42.99^\circ}$

Problem 6. In Figure 1.4, if $bc = 8$ and $ab = 1.3$, determine the angle θ.

It is convenient to use the expression for tan θ, since 'bc' and 'ab' are given.

Hence $\quad \tan\theta = \dfrac{bc}{ab} = \dfrac{8}{1.3} = 6.1538$

from which $\quad \theta = \tan^{-1}(6.1538) = \mathbf{80.77^\circ}$

(b) Pythagoras' theorem

Pythagoras' theorem* states that
$(\text{hypotenuse})^2 = (\text{adjacent side})^2 + (\text{opposite side})^2$
i.e. in the triangle of Figure 1.5

$$\boldsymbol{ac^2 = ab^2 + bc^2}$$

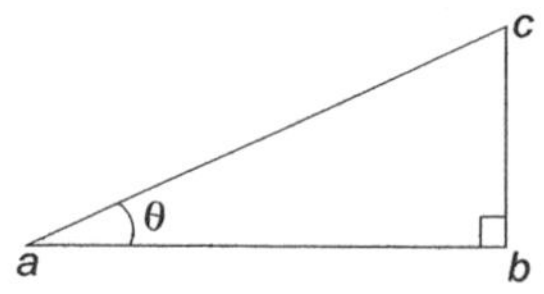

Figure 1.5

Problem 7. In Figure 1.5, if $ab = 5.1$ m and $bc = 6.7$ m, determine the length of the hypotenuse, ac.

From Pythagoras $\quad ac^2 = ab^2 + bc^2$

$$= 5.1^2 + 6.7^2 = 26.01 + 44.89$$
$$= 70.90$$

from which $\quad \boldsymbol{ac} = \sqrt{70.90} = \mathbf{8.42\ m}$

Now try the following Practice Exercise

Practice Exercise 3. Sines, cosines and tangents and Pythagoras' theorem

In problems 1 to 5, refer to Figure 1.5.

1. If $ab = 2.1$ m and $bc = 1.5$ m, determine angle θ. [35.54°]
2. If $ab = 2.3$ m and $ac = 5.0$ m, determine angle θ. [62.61°]

*Pythagoras of Samos (Born about 570 BC and died about 495 BC) was an Ionian Greek philosopher and mathematician. He is best known for the Pythagorean theorem, which states that in a right-angled triangle $a^2 + b^2 = c^2$. To find out more about Pythagoras go to www.routledge.com/cw/bird

3. If bc = 3.1 m and ac = 6.4 m, determine angle θ. [28.97°]

4. If ab = 5.7 cm and bc = 4.2 cm, determine the length ac. [7.08 cm]

5. If ab = 4.1 m and ac = 6.2 m, determine length bc. [4.65 m]

(c) The sine and cosine rules

For the triangle ABC shown in Figure 1.6

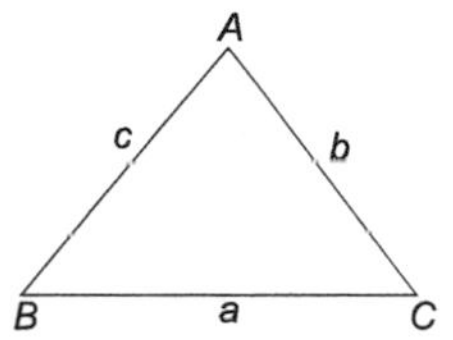

Figure 1.6

the sine rule states $$\frac{\boldsymbol{a}}{\sin \boldsymbol{A}} = \frac{\boldsymbol{b}}{\sin \boldsymbol{B}} = \frac{\boldsymbol{c}}{\sin \boldsymbol{C}}$$

and the cosine rule states $\boldsymbol{a^2 = b^2 + c^2 - 2bc\cos A}$

Problem 8. In Figure 1.6, if $a = 3$ m, $A = 20°$ and $B = 120°$, determine lengths b, c and angle C.

Using the sine rule $\frac{a}{\sin A} = \frac{b}{\sin B}$

i.e. $\frac{3}{\sin 20°} = \frac{b}{\sin 120°}$

from which $b = \frac{3\sin 120°}{\sin 20°} = \frac{3 \times 0.8660}{0.3420}$

$= \mathbf{7.596\ m}$

Angle, $\boldsymbol{C} = 180° - 20° - 120° = \mathbf{40°}$

Using the sine rule again gives $\frac{c}{\sin C} = \frac{a}{\sin A}$

i.e. $c = \frac{a\sin C}{\sin A} = \frac{3 \times \sin 40°}{\sin 20°}$

$= \mathbf{5.638\ m}$

Problem 9. In Figure 1.6, if $b = 8.2$ cm. $c = 5.1$ cm and $A = 70°$, determine the length a and angles B and C.

From the cosine rule

$a^2 = b^2 + c^2 - 2bc\cos A$

$= 8.2^2 + 5.1^2 - 2 \times 8.2 \times 5.1 \times \cos 70°$

$= 67.24 + 26.01 - 2(8.2)(5.1)\cos 70°$

$= 64.643$

Hence **length,** $\boldsymbol{a} = \sqrt{64.643} = \mathbf{8.04\ cm}$

Using the sine rule $\frac{a}{\sin A} = \frac{b}{\sin B}$

i.e. $\frac{8.04}{\sin 70°} = \frac{8.2}{\sin B}$

from which $8.04 \sin B = 8.2 \sin 70°$

and $\sin B = \frac{8.2\sin 70°}{8.04} = 0.95839$

and $\boldsymbol{B} = \sin^{-1}(0.95839) = \mathbf{73.41°}$

Since $A + B + C = 180°$, then

$C = 180° - A - B = 180° - 70° - 73.41° = \mathbf{36.59°}$

Now try the following Practice Exercise

Practice Exercise 4. Sine and cosine rules

In problems 1 to 4, refer to Figure 1.6.

1. If $b = 6$ m, $c = 4$ m and $B = 100°$, determine angles A and C and length a. [$A = 38.96°$, $C = 41.04°$, $a = 3.83$ m]

2. If $a = 15$ m, $c = 23$ m and $B = 67°$, determine length b and angles A and C. [$b = 22.01$ m, $A = 38.86°$, $C = 74.14°$]

3. If $a = 4$ m, $b = 8$ m and $c = 6$ m, determine angle A. [28.96°]

4. If $a = 10.0$ cm, $b = 8.0$ cm and $c = 7.0$ cm, determine angles A, B and C. [$A = 83.33°$, $B = 52.62°$, $C = 44.05°$]

5. In Figure 1.7, PR represents the inclined jib of a crane and is 10.0 m long. PQ is 4.0 m long. Determine the inclination of the jib to the vertical (i.e. angle P) and the length of tie QR.

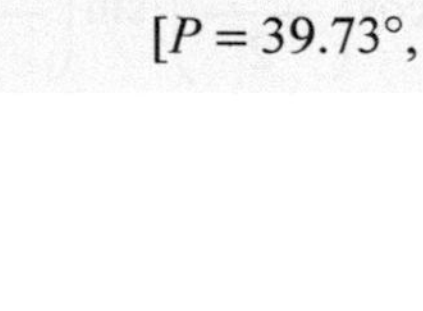

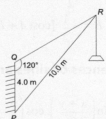

Figure 1.7

[$P = 39.73°$, $QR = 7.38$ m]

(d) Trigonometric identities

A **trigonometric identity** is a relationship that is true for all values of the unknown variable. From time to time such identities are needed in engineering. Below is a list (without proofs) of some of the main trigonometric identities.

Identities involving trigonometric ratios

$$\sec\theta = \frac{1}{\cos\theta} \quad \operatorname{cosec}\theta = \frac{1}{\sin\theta} \quad \cot\theta = \frac{1}{\tan\theta}$$

$$\tan\theta = \frac{\sin\theta}{\cos\theta}$$

$$\cos^2\theta + \sin^2\theta = 1 \quad 1 + \tan^2\theta = \sec^2\theta$$

$$\cot^2\theta + 1 = \operatorname{cosec}^2\theta$$

Compound angle formulae

$$\sin(A \pm B) = \sin A \cos B \pm \cos A \sin B$$

$$\cos(A \pm B) = \cos A \cos B \mp \sin A \sin B$$

$$\tan(A \pm B) = \frac{\tan A \pm \tan B}{1 \mp \tan A \tan B}$$

If **R** $\mathbf{\sin(\omega t + \alpha)} = \boldsymbol{a} \mathbf{\sin \omega t} + \boldsymbol{b} \mathbf{\cos \omega t}$

then $a = R\cos\alpha$, $b = R\sin\alpha$, $R = \sqrt{(a^2 + b^2)}$ and $a = \tan^{-1}\frac{b}{a}$

Double angles

$$\sin 2A = 2\sin A\cos A$$

$$\cos 2A = \cos^2 A - \sin^2 A = 2\cos^2 A - 1 = 1 - 2\sin^2 A$$

$$\tan 2A = \frac{2\tan A}{1 - \tan^2 A}$$

Products of sines and cosines into sums or differences

$$\sin A\cos B = \frac{1}{2}[\sin(A+B) + \sin(A-B)]$$

$$\cos A\sin B = \frac{1}{2}[\sin(A+B) - \sin(A-B)]$$

$$\cos A\cos B = \frac{1}{2}[\cos(A+B) + \cos(A-B)]$$

$$\sin A\sin B = -\frac{1}{2}[\cos(A+B) - \cos(A-B)]$$

Sums or differences of sines and cosines into products

$$\sin x + \sin y = 2\sin\left(\frac{x+y}{2}\right)\cos\left(\frac{x-y}{2}\right)$$

$$\sin x - \sin y = 2\cos\left(\frac{x+y}{2}\right)\sin\left(\frac{x-y}{2}\right)$$

$$\cos x + \cos y = 2\cos\left(\frac{x+y}{2}\right)\cos\left(\frac{x-y}{2}\right)$$

$$\cos x - \cos y = -2\sin\left(\frac{x+y}{2}\right)\sin\left(\frac{x-y}{2}\right)$$

Now try the following Practice Exercise

Practice Exercise 5. Trigonometric identities

1. Show that: (a) $\sin\left(x + \frac{\pi}{3}\right) + \sin\left(x + \frac{2\pi}{3}\right) = \sqrt{3}\cos x$
   and (b) $-\sin\left(\frac{3\pi}{2} - \phi\right) = \cos\phi$
2. Prove that: (a) $\sin\left(\theta + \frac{\pi}{4}\right) - \sin\left(\theta - \frac{3\pi}{4}\right) = \sqrt{2}(\sin\theta + \cos\theta)$
   and (b) $\frac{\cos(270° + \theta)}{\cos(360° - \theta)} = \tan\theta$
3. Prove the following identities:
   (a) $1 - \frac{\cos 2\phi}{\cos^2\phi} = \tan^2\phi$
   (b) $\frac{1 + \cos 2t}{\sin^2 t} = 2\cot^2 t$
   (c) $\frac{(\tan 2x)(1 + \tan x)}{\tan x} = \frac{2}{1 - \tan x}$
   (d) $2\operatorname{cosec} 2\theta\cos 2\theta = \cot\theta - \tan\theta$

In Problems 4 to 7, express as sums or differences:

4. $2\sin 7t\cos 2t$ [$(\sin 9t + \sin 5t)$]
5. $4\cos 8x\sin 2x$ [$2(\sin 10x - \sin 6x)$]
6. $2\sin 7t\sin 3t$ [$\cos 4t - \cos 10t$]
7. $6\cos 3\theta\cos\theta$ [$3(\cos 4\theta + \cos 2\theta)$]

In Problems 8 to 10, express as products:

8. $\sin 3x + \sin x$ [$2\sin 2x\cos x$]
9. $\frac{1}{2}(\sin 9\theta - \sin 7\theta)$ [$\cos 8\theta\sin\theta$]
10. $\cos 5t + \cos 3t$ [$2\cos 4t\cos t$]

1.5 Brackets

The use of brackets, which are used in many engineering equations, is explained through the following worked problems.

Problem 10. Expand the bracket to determine A, given $A = a(b + c + d)$

Multiplying each term in the bracket by 'a' gives:

$$A = a(b + c + d) = \mathbf{ab + ac + ad}$$

Problem 11. Expand the brackets to determine A, given $A = a[b(c + d) - e(f - g)]$

When there is more than one set of brackets the innermost brackets are multiplied out first. Hence

$$A = a[b(c + d) - e(f - g)] = a[bc + bd - ef + eg]$$

Note that $- e \times - g = + eg$

Now multiplying each term in the square brackets by 'a' gives:

$$A = \mathbf{abc + abd - aef + aeg}$$

Problem 12. Expand the brackets to determine A, given $A = a[b(c + d - e) - f(g - h\{j - k\})]$

The inner brackets are determined first, hence

$$A = a[b(c + d - e) - f(g - h\{j - k\})]$$
$$= a[b(c + d - e) - f(g - hj + hk)]$$
$$= a[bc + bd - be - fg + fhj - fhk]$$

i.e. $A = \mathbf{abc + abd - abe - afg + afhj - afhk}$

Problem 13. Evaluate A, given $A = 2[3(6 - 1) - 4(7\{2 + 5\} - 6)]$

$$A = 2[3(6 - 1) - 4(7\{2 + 5\} - 6)]$$
$$= 2[3(6 - 1) - 4(7 \times 7 - 6)]$$
$$= 2[3 \times 5 - 4 \times 43]$$
$$= 2[15 - 172] = 2[- 157] = \mathbf{- 314}$$

Now try the following Practice Exercise

Practice Exercise 6. Brackets

In problems 1 to 2, evaluate A

1. $A = 3(2 + 1 + 4)$ [21]
2. $A = 4[5(2 + 1) - 3(6 - 7)]$ [72]

Expand the brackets in problems 3 to 7.

3. $2(x - 2y + 3)$ $[2x - 4y + 6]$
4. $(3x - 4y) + 3(y - z) - (z - 4x)$ $[7x - y - 4z]$
5. $2x + [y - (2x + y)]$ [0]
6. $24a - [2\{3(5a - b) - 2(a + 2b)\} + 3b]$ $[11b - 2a]$
7. $ab[c + d - e(f - g + h\{i + j\})]$ $[abc + abd - abef + abeg - abehi - abehj]$

1.6 Fractions

An example of a fraction is $\frac{2}{3}$ where the top line, i.e. the 2, is referred to as the **numerator** and the bottom line, i.e. the 3, is referred to as the **denominator**.

A **proper fraction** is one where the numerator is smaller than the denominator, examples being $\frac{2}{3}, \frac{1}{2}, \frac{3}{8}, \frac{5}{16}$, and so on.

An **improper fraction** is one where the denominator is smaller than the numerator, examples being $\frac{3}{2}, \frac{2}{1}, \frac{8}{3}, \frac{16}{5}$, and so on.

Addition of fractions is demonstrated in the following worked problems.

Problem 14. Evaluate A, given $A = \frac{1}{2} + \frac{1}{3}$

The lowest common denominator of the two denominators 2 and 3 is 6, i.e. 6 is the lowest number that both 2 and 3 will divide into.

Then $\frac{1}{2} = \frac{3}{6}$ and $\frac{1}{3} = \frac{2}{6}$ i.e. both $\frac{1}{2}$ and $\frac{1}{3}$ have the common denominator, namely 6.

The two fractions can therefore be added as

$$\mathbf{A} = \frac{1}{2} + \frac{1}{3} = \frac{3}{6} + \frac{2}{6} = \frac{3+2}{6} = \mathbf{\frac{5}{6}}$$

Problem 15. Evaluate A, given $A = \frac{2}{3} + \frac{3}{4}$

A common denominator can be obtained by multiplying the two denominators together, i.e. the common denominator is $3 \times 4 = 12$

The two fractions can now be made equivalent, i.e. $\frac{2}{3} = \frac{8}{12}$ and $\frac{3}{4} = \frac{9}{12}$

so that they can be easily added together, as follows

$$A = \frac{2}{3} + \frac{3}{4} = \frac{8}{12} + \frac{9}{12} = \frac{8+9}{12} = \frac{17}{12}$$

i.e. $\mathbf{A = \frac{2}{3} + \frac{3}{4} = 1\frac{5}{12}}$

Problem 16. Evaluate A, given $A = \frac{1}{6} + \frac{2}{7} + \frac{3}{2}$

A suitable common denominator can be obtained by multiplying $6 \times 7 = 42$, because all three denominators divide exactly into 42.

Thus $\frac{1}{6} = \frac{7}{42}$ $\quad \frac{2}{7} = \frac{12}{42}$ $\quad$ and $\quad \frac{3}{2} = \frac{63}{42}$

Hence $A = \frac{1}{6} + \frac{2}{7} + \frac{3}{2}$

$$= \frac{7}{42} + \frac{12}{42} + \frac{63}{42} = \frac{7+12+63}{42} = \frac{82}{42} = \frac{41}{21}$$

i.e. $\quad \mathbf{A = \frac{1}{6} + \frac{2}{7} + \frac{3}{2} = 1\frac{20}{21}}$

Problem 17. Determine A as a single fraction, given $A = \frac{1}{x} + \frac{2}{y}$

A common denominator can be obtained by multiplying the two denominators together, i.e. xy

Thus $\frac{1}{x} = \frac{y}{xy}$ $\quad$ and $\quad \frac{2}{y} = \frac{2x}{xy}$

Hence $\quad A = \frac{1}{x} + \frac{2}{y} = \frac{y}{xy} + \frac{2x}{xy}$

i.e. $\quad \mathbf{A = \frac{y + 2x}{xy}}$

Note that addition, subtraction, multiplication and division of fractions may be determined using a **calculator** (for example, the CASIO fx-991ES PLUS).

Locate the $\frac{\square}{\square}$ and $\square\frac{\square}{\square}$ functions on your calculator (the latter function is a shift function found above the $\frac{\square}{\square}$ function) and then check the following worked problems.

Problem 18. Evaluate $\frac{1}{4} + \frac{2}{3}$

(i) Press $\frac{\square}{\square}$ function

(ii) Type in 1

(iii) Press ↓ on the cursor key and type in 4

(iv) $\frac{1}{4}$ appears on the screen

(v) Press → on the cursor key and type in +

(vi) Press $\frac{\square}{\square}$ function

(vii) Type in 2

(viii) Press ↓ on the cursor key and type in 3

(ix) Press → on the cursor key

(x) Press = and the answer $\frac{11}{12}$ appears

(xi) Press $S \Leftrightarrow D$ function and the fraction changes to a decimal 0.9166666 …

Thus, $\mathbf{\frac{1}{4} + \frac{2}{3} = \frac{11}{12}}$ = **0.9167** as a decimal, correct to 4 decimal places.

It is also possible to deal with **mixed numbers** on the calculator.

Press Shift then the $\frac{\square}{\square}$ function and $\square\frac{\square}{\square}$ appears.

Problem 19. Evaluate $5\frac{1}{5} - 3\frac{3}{4}$

(i) Press Shift then the $\frac{\square}{\square}$ function and $\square\frac{\square}{\square}$ appears on the screen

(ii) Type in 5 then → on the cursor key

(iii) Type in 1 and ↓ on the cursor key

(iv) Type in 5 and $5\frac{1}{5}$ appears on the screen

(v) Press → on the cursor key

(vi) Type in – and then press Shift then the $\frac{\square}{\square}$ function and $5\frac{1}{5} - \square\frac{\square}{\square}$ appears on the screen

(vii) Type in 3 then → on the cursor key

(viii) Type in 3 and ↓ on the cursor key

(ix) Type in 4 and $5\frac{1}{5} - 3\frac{3}{4}$ appears on the screen

(x) Press = and the answer $\frac{29}{20}$ appears

(xi) Press $S \Leftrightarrow D$ function and the fraction changes to a decimal 1.45.

Thus $\mathbf{5\frac{1}{5} - 3\frac{3}{4} = \frac{29}{20} = 1\frac{9}{20}}$ = **1.45** as a decimal.

Now try the following Practice Exercise

Practice Exercise 7. Fractions

In problems 1 to 3, evaluate the given fractions

1. $\frac{1}{3}+\frac{1}{4}$ $\left[\frac{7}{12}\right]$
2. $\frac{1}{5}+\frac{1}{4}$ $\left[\frac{9}{20}\right]$
3. $\frac{1}{6}+\frac{1}{2}-\frac{1}{5}$ $\left[\frac{7}{15}\right]$

In problems 4 and 5, use a calculator to evaluate the given expressions

4. $\frac{1}{3}-\frac{3}{4}\times\frac{8}{21}$ $\left[\frac{1}{21}\right]$
5. $\frac{3}{4}\times\frac{4}{5}-\frac{2}{3}\div\frac{4}{9}$ $\left[-\frac{9}{10}\right]$
6. Evaluate $\frac{3}{8}+\frac{5}{6}-\frac{1}{2}$ as a decimal, correct to 4 decimal places. $\left[\frac{17}{24}=0.7083\right]$
7. Evaluate $8\frac{8}{9}\div 2\frac{2}{3}$ as a mixed number. $\left[3\frac{1}{3}\right]$
8. Evaluate $3\frac{1}{5}\times 1\frac{1}{3}-1\frac{7}{10}$ as a decimal, correct to 3 decimal places. [2.567]
9. Determine $\frac{2}{x}+\frac{3}{y}$ as a single fraction. $\left[\frac{3x+2y}{xy}\right]$

1.7 Percentages

Percentages are used to give a common standard. The use of percentages is very common in many aspects of commercial life, as well as in engineering. Interest rates, sale reductions, pay rises, exams and VAT are all examples where percentages are used.

Percentages are fractions having 100 as their denominator.

For example, the fraction $\frac{40}{100}$ is written as 40% and is read as 'forty per cent'.

The easiest way to understand percentages is to go through some worked examples.

Problem 20. Express 0.275 as a percentage

$$0.275 = 0.275 \times 100\% = \mathbf{27.5\%}$$

Problem 21. Express 17.5% as a decimal number

$$17.5\% = \frac{17.5}{100} = \mathbf{0.175}$$

Problem 22. Express $\frac{5}{8}$ as a percentage

$$\frac{5}{8} = \frac{5}{8}\times 100\% = \frac{500}{8}\% = \mathbf{62.5\%}$$

Problem 23. In two successive tests a student gains marks of 57/79 and 49/67. Is the second mark better or worse than the first?

$$57/79 = \frac{57}{79} = \frac{57}{79}\times 100\% = \frac{5700}{79}\%$$
$$= \mathbf{72.15\%} \text{ correct to 2 decimal places}$$

$$49/67 = \frac{49}{67} = \frac{49}{67}\times 100\% = \frac{4900}{67}\%$$
$$= \mathbf{73.13\%} \text{ correct to 2 decimal places}$$

Hence, **the second test is marginally better than the first test.**

This question demonstrates how much easier it is to compare two fractions when they are expressed as percentages.

Problem 24. Express 75% as a fraction

$$75\% = \frac{75}{100} = \frac{3}{4}$$

The fraction $\frac{75}{100}$ is reduced to its simplest form by cancelling, i.e. dividing numerator and denominator by 25.

Problem 25. Express 37.5% as a fraction

$$37.5\% = \frac{37.5}{100}$$

$= \dfrac{375}{1000}$ by multiplying numerator and denominator by 10

$= \dfrac{15}{40}$ by dividing numerator and denominator by 25

$= \dfrac{3}{8}$ by dividing numerator and denominator by 5

Problem 26. Find 27% of £65

27% of £65 $= \dfrac{27}{100} \times 65$ = **£17.55** by calculator

Problem 27. A 160 GB iPod is advertised as costing £190 excluding VAT. If VAT is added at 20%, what will be the total cost of the iPod?

VAT = 20% of £190 $= \dfrac{20}{100} \times 190$ = £38

Total cost of iPod = £190 + £38 = **£228**

A quicker method to determine the total cost is 1.20 × £190 = **£228**

Problem 28. Express 23 cm as a percentage of 72 cm, correct to the nearest 1%

23 cm as a percentage of 72 cm $= \dfrac{23}{72} \times 100\%$

$= 31.94444\ldots\%$

= **32%** correct to the nearest 1%

Problem 29. A box of screws increases in price from £45 to £52. Calculate the percentage change in cost, correct to 3 significant figures.

$$\% \text{ change} = \frac{\text{new value} - \text{original value}}{\text{original value}} \times 100\%$$

$$= \frac{52-45}{45} \times 100\,\% = \frac{7}{45} \times 100$$

= **15.6% = percentage change in cost**

Problem 30. A drilling speed should be set to 400 rev/min. The nearest speed available on the machine is 412 rev/min. Calculate the percentage over-speed.

% over-speed

$$= \frac{\text{available speed} - \text{correct speed}}{\text{correct speed}} \times 100\%$$

$$= \frac{412-400}{400} \times 100\%$$

$$= \frac{12}{400} \times 100\% = \mathbf{3\%}$$

Now try the following Practice Exercise

Practice Exercise 8. Percentages

In problems 1 and 2, express the given numbers as percentages.

1. 0.057 [5.7%]
2. 0.374 [37.4%]
3. Express 20% as a decimal number [0.20]
4. Express $\dfrac{11}{16}$ as a percentage [68.75%]
5. Express $\dfrac{5}{13}$ as a percentage, correct to 3 decimal places [38.462%]
6. Place the following in order of size, the smallest first, expressing each as percentages, correct to 1 decimal place:
   (a) $\dfrac{12}{21}$ (b) $\dfrac{9}{17}$ (c) $\dfrac{5}{9}$ (d) $\dfrac{6}{11}$
   [(b), (d), (c), (a)]
7. Express 65% as a fraction in its simplest form $\left[\dfrac{13}{20}\right]$
8. Calculate 43.6% of 50 kg [21.8 kg]
9. Determine 36% of 27 m [9.72 m]
10. Calculate correct to 4 significant figures:
    (a) 18% of 2758 tonnes
    (b) 47% of 18.42 grams
    (c) 147% of 14.1 seconds
    [(a) 496.4 t (b) 8.657 g (c) 20.73 s]

11. Express: (a) 140 kg as a percentage of 1 t (b) 47 s as a percentage of 5 min (c) 13.4 cm as a percentage of 2.5 m [(a) 14% (b) 15.67% (c) 5.36%]

12. A computer is advertised on the internet at £520, exclusive of VAT. If VAT is payable at 20%, what is the total cost of the computer? [£624]

13. Express 325 mm as a percentage of 867 mm, correct to 2 decimal places. [37.49%]

14. When signing a new contract, a Premiership footballer's pay increases from £15,500 to £21,500 per week. Calculate the percentage pay increase, correct 3 significant figures. [38.7%]

15. A metal rod 1.80 m long is heated and its length expands by 48.6 mm. Calculate the percentage increase in length. [2.7%]

1.8 Laws of indices

The manipulation of indices, powers and roots is a crucial underlying skill needed in algebra.

Law 1: When multiplying two or more numbers having the same base, the indices are added.

For example $2^2 \times 2^3 = 2^{2+3} = 2^5$

and $5^4 \times 5^2 \times 5^3 = 5^{4+2+3} = 5^9$

More generally $\boldsymbol{a^m \times a^n = a^{m+n}}$

For example $a^3 \times a^4 = a^{3+4} = a^7$

Law 2: When dividing two numbers having the same base, the index in the denominator is subtracted from the index in the numerator.

For example $\frac{2^5}{2^3} = 2^{5-3} = 2^2$

and $\frac{7^8}{7^5} = 7^{8-5} = 7^3$

More generally $\boldsymbol{\frac{a^m}{a^n} = a^{m-n}}$

For example $\frac{c^5}{c^2} = c^{5-2} = c^3$

Law 3: When a number which is raised to a power is raised to a further power, the indices are multiplied.

For example $\left(2^2\right)^3 = 2^{2\times3} = 2^6$

and $\left(3^4\right)^2 = 3^{4\times2} = 3^8$

More generally $\boldsymbol{(a^m)^n = a^{mn}}$

For example $\left(d^2\right)^3 = d^{2\times3} = d^6$

Law 4: When a number has an index of 0, its value is 1.

For example, $3^0 = 1$ (check with your calculator)

and $17^0 = 1$

More generally $\boldsymbol{a^0 = 1}$

Law 5: A number raised to a negative power is the reciprocal of that number raised to a positive power.

For example $3^{-4} = \frac{1}{3^4}$ and $\frac{1}{2^{-3}} = 2^3$

More generally $\boldsymbol{a^{-n} = \frac{1}{a^n}}$

For example $a^{-2} = \frac{1}{a^2}$

Law 6: When a number is raised to a fractional power the denominator of the fraction is the root of the number and the numerator is the power

For example $8^{\frac{2}{3}} = \sqrt[3]{8^2} = (2)^2 = 4$

and $25^{\frac{1}{2}} = \sqrt[2]{25^1} = \sqrt{25^1}$

$= \pm 5$ (Note that $\sqrt{\ } \equiv \sqrt[2]{\ }$)

More generally $\boldsymbol{a^{\frac{m}{n}} = \sqrt[n]{a^m}}$

For example $x^{\frac{4}{3}} = \sqrt[3]{x^4}$

Problem 31. Evaluate in index form $5^3 \times 5 \times 5^2$

$5^3 \times 5 \times 5^2 = 5^3 \times 5^1 \times 5^2$ (Note that 5 means 5^1)

$= 5^{3+1+2} = \boldsymbol{5^6}$ from law 1

Problem 32. Evaluate $\frac{3^5}{3^4}$

From law 2 $\frac{3^5}{3^4} = 3^{5-4} = 3^1 = \mathbf{3}$

Problem 33. Evaluate $\frac{2^4}{2^4}$

$\frac{2^4}{2^4} = 2^{4-4}$ from law 2

$= 2^0 = \mathbf{1}$ from law 4

Any number raised to the power of zero equals 1

Problem 34. Evaluate $\frac{3 \times 3^2}{3^4}$

$\frac{3 \times 3^2}{3^4} = \frac{3^1 \times 3^2}{3^4} = \frac{3^{1+2}}{3^4} = \frac{3^3}{3^4}$

$= 3^{3-4} = 3^{-1}$ from laws 1 and 2

$= \mathbf{\frac{1}{3}}$ from law 5

Problem 35. Evaluate $\frac{10^3 \times 10^2}{10^8}$

$\frac{10^3 \times 10^2}{10^8} = \frac{10^{3+2}}{10^8} = \frac{10^5}{10^8}$ from law 1

$= 10^{5-8} = 10^{-3}$ from law 2

$= \frac{1}{10^{+3}} = \frac{1}{1000}$ from law 5

Hence $\frac{10^3 \times 10^2}{10^8} = \mathbf{10^{-3}} = \mathbf{\frac{1}{1000}} = \mathbf{0.001}$

Problem 36. Simplify: (a) $(2^3)^4$ (b) $(3^2)^5$ expressing the answers in index form.

From law 3 (a) $(2^3)^4 = 2^{3\times 4} = \mathbf{2^{12}}$

(b) $(3^2)^5 = 3^{2\times 5} = \mathbf{3^{10}}$

Problem 37. Evaluate: $\frac{(10^2)^3}{10^4 \times 10^2}$

From laws 1, 2 and 3 $\frac{(10^2)^3}{10^4 \times 10^2} = \frac{10^{(2\times 3)}}{10^{(4+2)}}$

$= \frac{10^6}{10^6} = 10^{6-6}$

$= 10^0 = \mathbf{1}$

Problem 38. Evaluate (a) $4^{1/2}$ (b) $16^{3/4}$ (c) $27^{2/3}$ (d) $9^{-1/2}$

(a) $4^{1/2} = \sqrt{4} = \pm\mathbf{2}$

(b) $16^{3/4} = \sqrt[4]{16^3} = (2)^3 = \mathbf{8}$

(Note that it does not matter whether the 4th root of 16 is found first or whether 16 cubed is found first – the same answer will result.)

(c) $27^{2/3} = \sqrt[3]{27^2} = (3)^2 = \mathbf{9}$

(d) $9^{-1/2} = \frac{1}{9^{1/2}} = \frac{1}{\sqrt{9}} = \frac{1}{\pm 3} = \pm\mathbf{\frac{1}{3}}$

Problem 39. Simplify $a^2b^3c \times ab^2c^5$

$a^2b^3c \times ab^2c^5 = a^2 \times b^3 \times c \times a \times b^2 \times c^5$

$= a^2 \times b^3 \times c^1 \times a^1 \times b^2 \times c^5$

Grouping together like terms gives

$a^2 \times a^1 \times b^3 \times b^2 \times c^1 \times c^5$

Using law 1 of indices gives

$a^{2+1} \times b^{3+2} \times c^{1+5} = a^3 \times b^5 \times c^6$

i.e. $\mathbf{a^2b^3c \times ab^2c^5 = a^3\, b^5\, c^6}$

Problem 40. Simplify $\frac{x^5y^2z}{x^2yz^3}$

$\frac{x^5y^2z}{x^2yz^3} = \frac{x^5 \times y^2 \times z}{x^2 \times y \times z^3} = \frac{x^5}{x^2} \times \frac{y^2}{y^1} \times \frac{z}{z^3}$

$= x^{5-2} \times y^{2-1} \times z^{1-3}$ by law 2

$= x^3 \times y^1 \times z^{-2} = \mathbf{x^3\, y\, z^{-2}}$ or $\mathbf{\frac{x^3\, y}{z^2}}$

Now try the following Practice Exercise

Practice Exercise 9. Laws of indices

In questions 1 to 18, evaluate without the aid of a calculator.

1. Evaluate $2^2 \times 2 \times 2^4$ [$2^7 = 128$]
2. Evaluate $3^5 \times 3^3 \times 3$ in index form [3^9]
3. Evaluate $\frac{2^7}{2^3}$ [$2^4 = 16$]
4. Evaluate $\frac{3^3}{3^5}$ $\left[3^{-2} = \frac{1}{3^2} = \frac{1}{9}\right]$
5. Evaluate 7^0 [1]

6. Evaluate $\dfrac{2^3 \times 2 \times 2^6}{2^7}$ $[2^3 = 8]$

7. Evaluate $\dfrac{10 \times 10^6}{10^5}$ $[10^2 = 100]$

8. Evaluate $10^4 \div 10$ $[10^3 = 1000]$

9. Evaluate $\dfrac{10^3 \times 10^4}{10^9}$

$$\left[10^{-2} = \frac{1}{10^2} = \frac{1}{100} = 0.01\right]$$

10. Evaluate $5^6 \times 5^2 \div 5^7$ $[5]$

11. Evaluate $(7^2)^3$ in index form $[7^6]$

12. Evaluate $(3^3)^2$ $[3^6 = 729]$

13. Evaluate $\dfrac{3^7 \times 3^4}{3^5}$ in index form $[3^6]$

14. Evaluate $\dfrac{(9 \times 3^2)^3}{(3 \times 27)^2}$ in index form $[3^4]$

15. Evaluate $\dfrac{(16 \times 4)^2}{(2 \times 8)^3}$ $[1]$

16. Evaluate $\dfrac{5^{-2}}{5^{-4}}$ $[5^2 = 25]$

17. Evaluate $\dfrac{3^2 \times 3^{-4}}{3^3}$ $\left[3^{-5} = \dfrac{1}{3^5} = \dfrac{1}{243}\right]$

18. Evaluate $\dfrac{7^2 \times 7^{-3}}{7 \times 7^{-4}}$ $[7^2 = 49]$

In problems 19 to 36, simplify the following, giving each answer as a power:

19. $z^2 \times z^6$ $[z^8]$

20. $a \times a^2 \times a^5$ $[a^8]$

21. $n^8 \times n^{-5}$ $[n^3]$

22. $b^4 \times b^7$ $[b^{11}]$

23. $b^2 \div b^5$ $\left[b^{-3} \text{ or } \dfrac{1}{b^3}\right]$

24. $c^5 \times c^3 \div c^4$ $[c^4]$

25. $\dfrac{m^5 \times m^6}{m^4 \times m^3}$ $[m^4]$

26. $\dfrac{(x^2)(x)}{x^6}$ $\left[x^{-3} \text{ or } \dfrac{1}{x^3}\right]$

27. $\left(x^3\right)^4$ $[x^{12}]$

28. $\left(y^2\right)^{-3}$ $\left[y^{-6} \text{ or } \dfrac{1}{y^6}\right]$

29. $\left(t \times t^3\right)^2$ $[t^8]$

30. $\left(c^{-7}\right)^{-2}$ $[c^{14}]$

31. $\left(\dfrac{a^2}{a^5}\right)^3$ $\left[a^{-9} \text{ or } \dfrac{1}{a^9}\right]$

32. $\left(\dfrac{1}{b^3}\right)^4$ $\left[\dfrac{1}{b^{12}} \text{ or } b^{-12}\right]$

33. $\left(\dfrac{b^2}{b^7}\right)^{-2}$ $[b^{10}]$

34. $\dfrac{1}{\left(s^3\right)^3}$ $\left[\dfrac{1}{s^9} \text{ or } s^{-9}\right]$

35. $p^3qr^2 \times p^2q^5r \times pqr^2$ $[p^6q^7r^5]$

36. $\dfrac{x^3y^2z}{x^5yz^3}$ $\left[x^{-2}yz^{-2} \text{ or } \dfrac{y}{x^2z^2}\right]$

1.9 Simultaneous equations

The solution of simultaneous equations is demonstrated in the following worked problems.

Problem 41. If 6 apples and 2 pears cost £1.80 and 8 apples and 6 pears cost £2.90, calculate how much an apple and a pear each cost.

Let an apple = A and a pear = P, then:

$$6A + 2P = 180 \quad (1)$$

$$8A + 6P = 290 \quad (2)$$

From equation (1), $6A = 180 - 2P$

and $A = \dfrac{180 - 2P}{6} = 30 - 0.3333P$ (3)

From equation (2), $8A = 290 - 6P$

and $A = \dfrac{290 - 6P}{8} = 36.25 - 0.75P$ (4)

Equating (3) and (4) gives

$$30 - 0.3333P = 36.25 - 0.75P$$

i.e. $0.75P - 0.3333P = 36.25 - 30$

and $0.4167P = 6.25$

and $P = \dfrac{6.25}{0.4167} = 15$

Substituting in (3) gives $A = 30 - 0.3333(15)$
$= 30 - 5 = 25$

Hence, **an apple costs 25p and a pear costs 15p.** The above method of solving simultaneous equations is called the **substitution method.**

Problem 42. If 6 bananas and 5 peaches cost £3.45 and 4 bananas and 8 peaches cost £4.40, calculate how much a banana and a peach each cost.

Let a banana = B and a peach = P, then:

$$6B + 5P = 345 \quad (1)$$

$$4B + 8P = 440 \quad (2)$$

Multiplying equation (1) by 2 gives

$$12B + 10P = 690 \quad (3)$$

Multiplying equation (2) by 3 gives

$$12B + 24P = 1320 \quad (4)$$

Equation (4) – equation (3) gives $14P = 630$

from which $P = \dfrac{630}{14} = 45$

Substituting in (1) gives $6B + 5(45) = 345$

i.e. $6B = 345 - 5(45)$

i.e. $6B = 120$

and $B = \dfrac{120}{6} = 20$

Hence, **a banana costs 20p and a peach costs 45p.** The above method of solving simultaneous equations is called the **elimination method.**

Problem 43. If 20 bolts and 2 spanners cost £10, and 6 spanners and 12 bolts cost £18, how much do a spanner and a bolt each cost?

Let s = a spanner and b = a bolt.

Therefore $2s + 20b = 10 \quad (1)$

and $6s + 12b = 18 \quad (2)$

Multiplying equation (1) by 3 gives

$$6s + 60b = 30 \quad (3)$$

Equation (3) – equations (2) gives $48b = 12$

from which $b = \dfrac{12}{48} = 0.25$

Substituting in (1) gives $2s + 20(0.25) = 10$

i.e. $2s = 10 - 20(0.25)$

i.e. $2s = 5$

and $s = \dfrac{5}{2} = 2.5$

Therefore, **a spanner costs £2.50 and a bolt costs £0.25** or **25p**

Now try the following Practice Exercise

Practice Exercise 10. Simultaneous equations

1. If 5 apples and 3 bananas cost £1.45 and 4 apples and 6 bananas cost £2.42, determine how much an apple and a banana each cost. [apple = 8p, banana = 35p]
2. If 7 apples and 4 oranges cost £2.64 and 3 apples and 3 oranges cost £1.35, determine how much an apple and a banana each cost. [apple = 28p, orange = 17p]
3. Three new cars and four new vans supplied to a dealer together cost £93000, and five new cars and two new vans of the same models cost £99000. Find the respective costs of a car and a van. [car = £15000, van = £12000]
4. In a system of forces, the relationship between two forces F_1 and F_2 is given by:

$$5F_1 + 3F_2 = -6$$
$$3F_1 + 5F_2 = -18$$

Solve for F_1 and F_2 [$F_1 = 1.5$, $F_2 = -4.5$]

5. Solve the simultaneous equations:
$a+b=7$
$a-b=3$ $[a=5, b=2]$

6. Solve the simultaneous equations:
$8a-3b=51$
$3a+4b=14$ $[a=6, b=-1]$

There is some more revisionary mathematics in Chapter 2 following. For the more advanced mathematics that will be needed within the theory of this text, some mathematics references are listed on pages 49 and 50. The only reason for studying engineering mathematics is so that it can be used in the engineering. Help is at hand!

For fully worked solutions to each of the problems in Exercises 1 to 10 in this chapter, go to the website:
www.routledge.com/cw/bird

Revision Test 1 *Revisionary mathematics*

This Revision Test covers the material contained in Chapter 1. *The marks for each question are shown in brackets at the end of each question.*

1. Convert, correct to 2 decimal places:
   (a) 76.8° to radians.
   (b) 1.724 radians to degrees. (4)

2. In triangle JKL in Figure RT1.1, find:
   (a) length KJ correct to 3 significant figures.
   (b) $\sin L$ and $\tan K$, each correct to 3 decimal places.

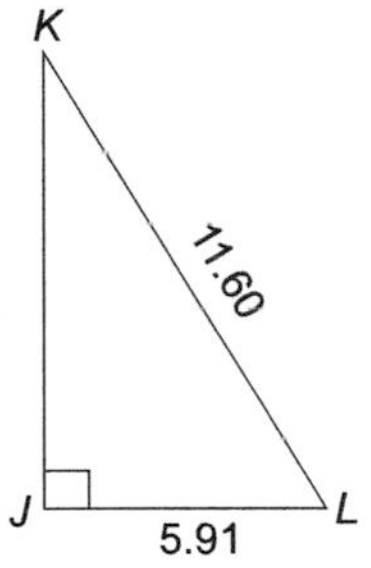

Figure RT1.1

(4)

3. In triangle PQR in Figure RT1.2, find angle P in decimal form, correct to 2 decimal places.

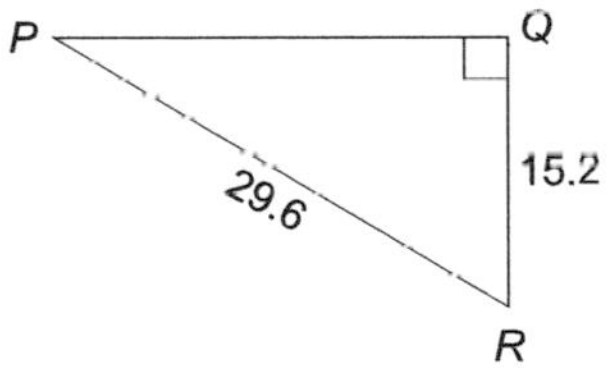

Figure RT1.2

(2)

4. In triangle ABC in Figure RT1.3, find lengths AB and AC, correct to 2 decimal places.

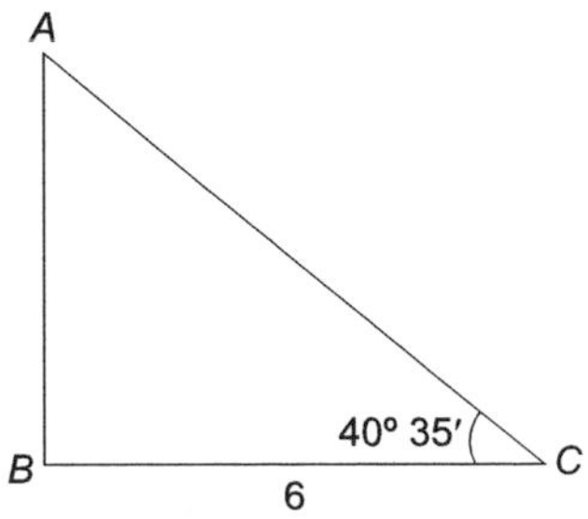

Figure RT1.3

(4)

5. A triangular plot of land ABC is shown in Figure RT1.4. Solve the triangle and determine its area.

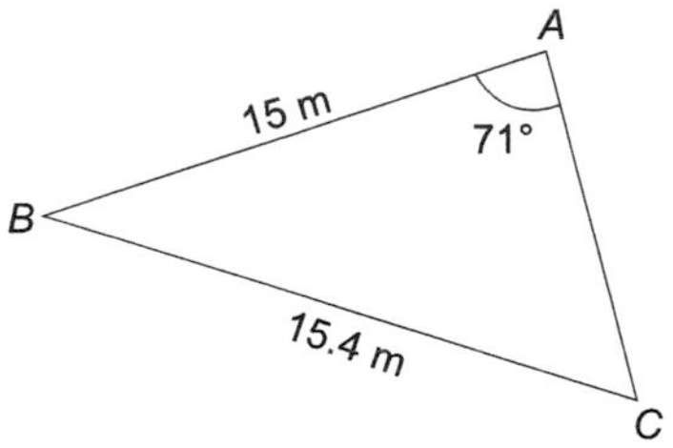

Figure RT1.4

(9)

6. Figure RT1.5 shows a roof truss PQR with rafter $PQ = 3$ m. Calculate the length of (a) the roof rise PP', (b) rafter PR, and (c) the roof span QR. Find also (d) the cross-sectional area of the roof truss.

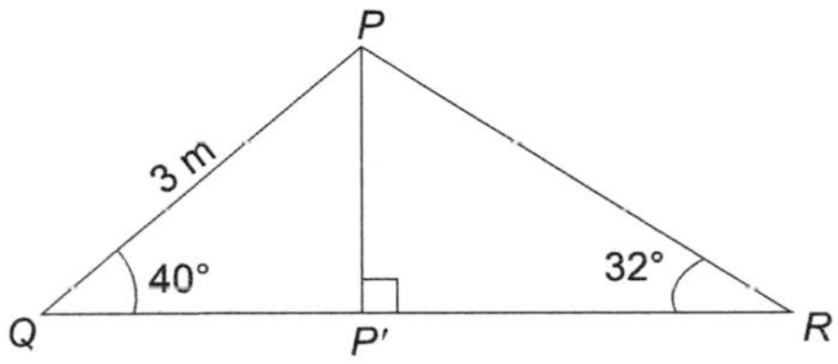

Figure RT1.5

(10)

7. Solve triangle ABC given $b = 10$ cm, $c = 15$ cm and $\angle A = 60°$. (7)

8. Remove the brackets and simplify $2(3x - 2y) - (4y - 3x)$ (3)

9. Remove the brackets and simplify $10a - [3(2a - b) - 4(b - a) + 5b]$ (4)

10. Determine, correct to 1 decimal place, 57% of 17.64 g (2)

11. Express 54.7 mm as a percentage of 1.15 m, correct to 3 significant figures. (3)

12. Simplify:

    (a) $\frac{3}{4} - \frac{7}{15}$ (b) $1\frac{5}{8} - 2\frac{1}{3} + 3\frac{5}{6}$ (8)

13. Use a calculator to evaluate:

(a) $1\frac{7}{9}\times\frac{3}{8}\times3\frac{3}{5}$

(b) $6\frac{2}{3}\div1\frac{1}{3}$

(c) $1\frac{1}{3}\times2\frac{1}{5}\div\frac{2}{5}$ (10)

14. Evaluate:

(a) $3\times2^3\times2^2$

(b) $49^{\frac{1}{2}}$ (4)

15. Evaluate:

(a) $\frac{2^7}{2^2}$ (b) $\frac{10^4\times10\times10^5}{10^6\times10^2}$ (4)

16. Evaluate:

(a) $\frac{2^3\times2\times2^2}{2^4}$

(b) $\frac{\left(2^3\times16\right)^2}{\left(8\times2\right)^3}$

(c) $\left(\frac{1}{4^2}\right)^{-1}$ (7)

17. Evaluate:

(a) $(27)^{-\frac{1}{3}}$ (b) $\frac{\left(\frac{3}{2}\right)^{-2}-\frac{2}{9}}{\left(\frac{2}{3}\right)^2}$ (5)

18. Solve the simultaneous equations:

(a) $2x+y=6$
$5x-y=22$

(b) $4x-3y=11$
$3x+5y=30$ (10)

For lecturers/instructors/teachers, fully worked solutions to each of the problems in Revision Test 1, together with a full marking scheme, are available at the website:
www.routledge.com/cw/bird

Multiple-Choice Questions Test 1

This test covers the material in Chapter 1 on revisionary mathematics. All questions have only one correct answer (answers on page 496).

1. 73° is equivalent to:
(a) 23.24 rad (b) 1.274 rad
(c) 0.406 rad (d) 4183 rad

2. 0.52 radians is equivalent to:
(a) 93.6° (b) 0.0091°
(c) 1.63° (d) 29.79°

3. $3\pi/4$ radians is equivalent to:
(a) 135° (b) 270°
(c) 45° (d) 67.5°

4. In the right-angled triangle *ABC* shown in Figure M1.1, sine *A* is given by:
(a) b/a (b) c/b
(c) b/c (d) a/b

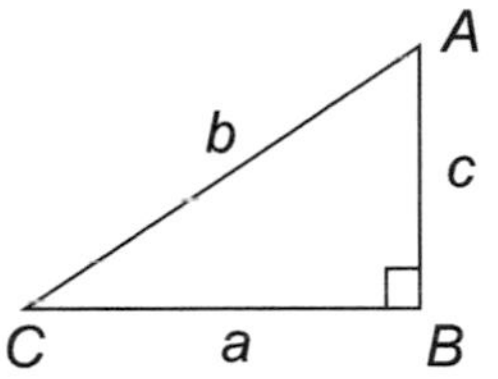

Figure M1.1

5. In the right-angled triangle *ABC* shown in Figure M1.1, cosine *C* is given by:
(a) a/b (b) c/b
(c) a/c (d) b/a

6. In the right-angled triangle *ABC* shown in Figure M1.1, tangent *A* is given by:
(a) b/c (b) a/c
(c) a/b (d) c/a

7. In the right-angled triangle *PQR* shown in Figure M1.2, angle *R* is equal to:
(a) 41.41° (b) 48.59°
(c) 36.87° (d) 53.13°

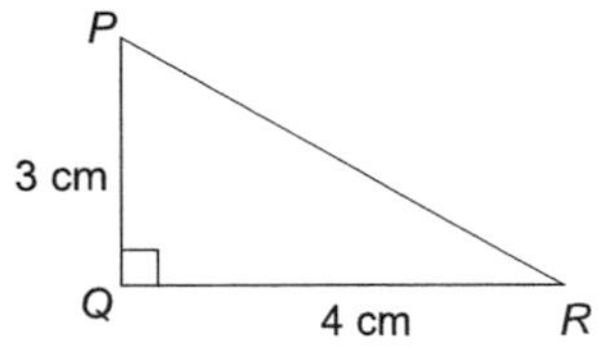

Figure M1.2

8. In the triangle *ABC* shown in Figure M1.3, side '*a*' is equal to:
(a) 61.27 mm (b) 86.58 mm
(c) 96.41 mm (d) 54.58 mm

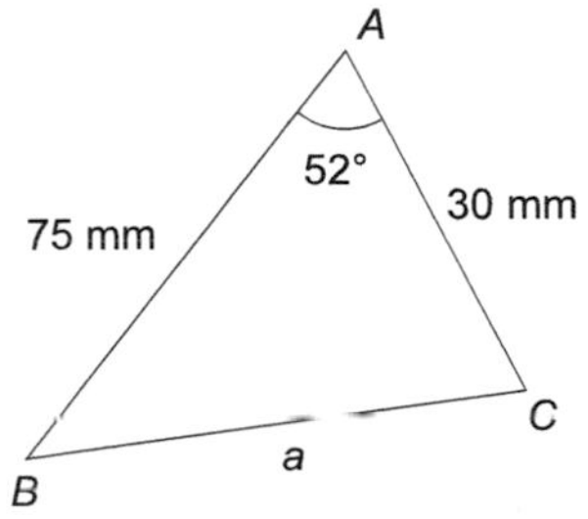

Figure M1.3

9. In the triangle *ABC* shown in Figure M1.3, angle *B* is equal to:
(a) 0.386° (b) 22.69°
(c) 74.71° (d) 23.58°

10. Removing the brackets from the expression $a[b+2c-d\{(e-f)-g(m-n)\}]$ gives:
(a) $ab+2ac-ade-adf+adgm-adgn$
(b) $ab+2ac-ade-adf-adgm-adgn$
(c) $ab+2ac-ade+adf+adgm-adgn$
(d) $ab+2ac-ade-adf+adgm+adgn$

11. $\frac{5}{6}+\frac{1}{5}-\frac{2}{3}$ is equal to:
(a) $\frac{1}{2}$ (b) $\frac{11}{30}$
(c) $-\frac{1}{2}$ (d) $1\frac{7}{10}$

12. $1\frac{1}{3}+1\frac{2}{3}\div 2\frac{2}{3}-\frac{1}{3}$ is equal to:
(a) $1\frac{2}{7}$ (b) $\frac{19}{24}$
(c) $2\frac{1}{21}$ (d) $1\frac{5}{8}$

13. $\frac{3}{4}\div 1\frac{3}{4}$ is equal to:
(a) $\frac{3}{7}$ (b) $1\frac{9}{16}$
(c) $1\frac{5}{16}$ (d) $2\frac{1}{2}$

14. 11 mm expressed as a percentage of 41 mm is:

(a) 2.68, correct to 3 significant figures
(b) 2.6, correct to 2 significant figures
(c) 26.83, correct to 2 decimal places
(d) 0.2682, correct to 4 decimal places

15. The value of $\frac{2^{-3}}{2^{-4}} - 1$ is equal to:

(a) 1 (b) 2
(c) $-\frac{1}{2}$ (d) $\frac{1}{2}$

16. In an engineering equation $\frac{3^4}{3^r} = \frac{1}{9}$. The value of r is:

(a) − 6 (b) 2
(c) 6 (d) − 2

17. $16^{-\frac{3}{4}}$ is equal to:

(a) 8 (b) $-\frac{1}{2^3}$
(c) 4 (d) $\frac{1}{8}$

18. The engineering expression $\frac{(16 \times 4)^2}{(8 \times 2)^4}$ is equal to:

(a) 4 (b) 2^{-4}
(c) $\frac{1}{2^2}$ (d) 1

19. $(16^{-\frac{1}{4}} - 27^{-\frac{2}{3}})$ is equal to:

(a) $\frac{7}{18}$ (b) − 7
(c) $1\frac{8}{9}$ (d) $-8\frac{1}{2}$

20. The solution of the simultaneous equations 3a − 2b = 13 and 2a + 5b = − 4 is:

(a) a = − 2, b = 3 (b) a = 1, b = − 5
(c) a = 3, b = − 2 (d) a = − 7, b = 2

Chapter 2

Further revisionary mathematics

Why it is important to understand: **Further revisionary mathematics**

In engineering there are many different quantities to get used to, and hence many units to become familiar with. For example, force is measured in newtons, electric current is measured in amperes and pressure is measured in pascals. Sometimes the units of these quantities are either very large or very small and hence prefixes are used. For example, 1000 pascals may be written as 10^3 Pa which is written as 1 kPa in prefix form, the k being accepted as a symbol to represent 1000 or 10^3. Studying, or working, in an engineering discipline, you very quickly become familiar with the standard units of measurement, the prefixes used and engineering notation. An electronic calculator is extremely helpful with engineering notation.

Most countries have used the metric system of units for many years; however, there are other countries, such as the USA, who still use the imperial system. Hence, metric to imperial unit conversions, and vice versa, are internationally important and are contained in this chapter.

Graphs have a wide range of applications in engineering and in physical sciences because of their inherent simplicity. A graph can be used to represent almost any physical situation involving discrete objects and the relationship among them. If two quantities are directly proportional and one is plotted against the other, a straight line is produced. Examples of this include an applied force on the end of a spring plotted against spring extension, the speed of a flywheel plotted against time, and strain in a wire plotted against stress (Hooke's law). In engineering, the straight line graph is the most basic graph to draw and evaluate.

There are many practical situations engineers have to analyse which involve quantities that are varying. Typical examples include the stress in a loaded beam, the temperature of an industrial chemical, the rate at which the speed of a vehicle is increasing or decreasing, the current in an electrical circuit or the torque on a turbine blade. Differential calculus, or differentiation, is a mathematical technique for analysing the way in which functions change. This chapter explains how to differentiate the five most common functions. Engineering is all about problem solving and many problems in engineering can be solved using calculus. Physicists, chemists, engineers and many other scientific and technical specialists use calculus in their everyday work; it is a technique of fundamental importance. Integration has numerous applications in engineering and science and some typical examples include determining areas, mean and r.m.s. values, volumes of solids of revolution, centroids, second moments of area and differential equations. Standard integrals are covered in this chapter, and for any further studies in engineering, differential and integral calculus are unavoidable.

Vectors are an important part of the language of science, mathematics and engineering. They are used to discuss multivariable calculus, electrical circuits with oscillating currents, stress and strain in structures and materials, and flows of atmospheres and fluids, and they have many other applications. Resolving a vector into components is a precursor to computing things with or about a vector quantity. Because position, velocity, acceleration, force, momentum and angular momentum are all vector quantities, resolving vectors into components is a most important skill required in any engineering studies.

At the end of this chapter, you should be able to:

- state the seven SI units
- understand derived units
- recognise common engineering units
- understand common prefixes used in engineering
- use engineering notation and prefix form with engineering units
- understand and calculate metric to imperial conversions and vice versa
- understand rectangular axes, scales and co-ordinates
- plot given co-ordinates and draw the best straight line graph
- determine the gradient and vertical-axis intercept of a straight line graph
- state the equation of a straight line graph
- plot straight line graphs involving practical engineering examples
- state that calculus comprises two parts – differential and integral calculus
- differentiate $y = ax^n$ by the general rule
- differentiate sine, cosine, exponential and logarithmic functions
- understand that integration is the reverse process of differentiation
- determine integrals of the form ax^n where n is fractional, zero or a positive or negative integer
- integrate standard functions – cos ax, sin ax, e^{ax}, $\frac{1}{x}$
- evaluate definite integrals
- evaluate 2 by 2 and 3 by 3 determinants
- determine scalar (and dot) products of two vectors
- determine vector (or cross) products of two products

2.1 Units, prefixes and engineering notation

Of considerable importance in engineering is knowledge of units of engineering quantities, the prefixes used with units, and engineering notation.
We need to know, for example, that

80 kN = 80×10^3 N which means 80,000 newtons
25 mJ = 25×10^{-3}J which means 0.025 joules
50 nF = 50×10^{-9}F which means 0.000000050 farads

This is explained in this chapter.

SI units

The system of units used in engineering and science is the *Système Internationale d'Unités* (**International system of units**) usually abbreviated to SI units, and is based on the metric system. This was introduced in 1960 and is now adopted by the majority of countries as the official system of measurement.

The basic seven units used in the SI system are listed below with their symbols.

There are, of course, many other units than the seven shown below. These other units are called **derived units** and are defined in terms of the standard units listed below.

Quantity	Unit	Symbol	
Length	metre	m	(1 m = 100 cm = 1000 mm)
Mass	kilogram	kg	(1 kg = 1000 g)
Time	second	s	
Electric current	ampere	A	
Thermodynamic temperature	Kelvin	K	(K = °C + 273)
Luminous intensity	candela	cd	
Amount of substance	mole	mol	

For example, speed is measured in metres per second, therefore using two of the standard units, i.e. length and time.

Some derived units are given **special names**.

For example, force = mass × acceleration, has units of kilogram metre per second squared, which uses three of the base units, i.e. kilograms, metres and seconds. The unit of kg m/s^2 is given the special name of a **newton.***

Below is a list of some quantities and their units that are common in engineering.

Quantity	Unit	Symbol
Length	metre	m
Area	square metre	m^2
Volume	cubic metre	m^3
Mass	kilogram	kg
Time	second	s
Electric current	ampere	A
Speed, velocity	metre per second	m/s
Acceleration	metre per second squared	m/s^2
Density	kilogram per cubic metre	kg/m^3
Temperature	Kelvin or Celsius	K or °C
Angle	radian or degree	rad or °
Angular velocity	radian per second	rad/s
Frequency	hertz	Hz
Force	newton	N
Pressure	pascal	Pa
Energy, work	joule	J
Power	watt	W
Electric potential	volt	V
Capacitance	farad	F
Electrical resistance	ohm	Ω
Inductance	henry	H
Moment of force	newton metre	N m
Stress	pascal	Pa
Torque	newton metre	N m
Momentum	kilogram metre per second	kg m/s

Common prefixes

SI units may be made larger or smaller by using prefixes which denote multiplication or division by a particular amount.

*Sir Isaac Newton (25 December 1642 – 20 March 1727) was an English polymath. Newton showed that the motions of objects are governed by the same set of natural laws, by demonstrating the consistency between Kepler's laws of planetary motion and his theory of gravitation. The SI unit of force is the newton, named in his honour. To find out more about Newton go to www.routledge.com/cw/bird

The most common multiples are listed below. Knowledge of indices is needed since all of the prefixes are powers of 10 with indices that are a multiple of 3.

Prefix	Name	Meaning	
T	tera	multiply by 10^{12}	i.e. × 1,000,000,000,000
G	giga	multiply by 10^9	i.e. × 1,000,000,000
M	mega	multiply by 10^6	i.e. × 1,000,000
k	kilo	multiply by 10^3	i.e. × 1,000
m	milli	multiply by 10^{-3}	i.e. $\times \frac{1}{10^3} = \frac{1}{1000} = 0.001$
μ	micro	multiply by 10^{-6}	i.e. $\times \frac{1}{10^6} = \frac{1}{1,000,000} = 0.000\ 001$
n	nano	multiply by 10^{-9}	i.e. $\times \frac{1}{10^9} = \frac{1}{1,000,000,000} = 0.000\ 000\ 001$
p	pico	multiply by 10^{-12}	i.e. $\times \frac{1}{10^{12}} = \frac{1}{1,000,000,000,000} = 0.000\ 000\ 000\ 001$

Here are some examples of prefixes used with engineering units.

A **frequency of 15 GHz** means 15×10^9 Hz, which is 15,000,000,000 hertz,* i.e. 15 gigahertz is written as 15 GHz and is equal to 15 thousand million hertz. Instead of writing 15,000,000,000 hertz, it is much neater, takes up less space and prevents errors caused by having so many zeros, to write the frequency as 15 GHz.

A **voltage of 40 MV** means 40×10^6 V, which is 40,000,000 volts, i.e. 40 megavolts is written as 40 MV and is equal to 40 million volts.

Energy of **12 mJ** means 12×10^{-3} J or $\frac{12}{10^3}$ J or $\frac{12}{1000}$ J, which is 0.012 J, i.e. 12 millijoules is written as 12 mJ and is equal to 12 thousandths of a joule.†

A **time of 150 ns** means 150×10^{-9}s or $\frac{150}{10^9}$ s, which is 0.000 000 150 s, i.e. 150 nanoseconds is written as 150 ns and is equal to 150 thousand millionths of a second.

A **force of 20 kN** means 20×10^3N, which is 20,000 Newtons, i.e. 20 kilonewtons is written as 20 kN and is equal to 20 thousand Newtons.

Engineering notation

Engineering notation is a number multiplied by a power of 10 that **is always a multiple of 3**.

For example,

$43,645 = 43.645 \times 10^3$ in engineering notation and

$0.0534 = 53.4 \times 10^{-3}$ in engineering notation

*Heinrich Rudolf Hertz (22 February 1857 – 1 January 1894) was the first person to conclusively prove the existence of electromagnetic waves. The scientific unit of frequency was named the hertz in his honour. To find out more about Hertz go to www.routledge.com/cw/bird

In the list of engineering prefixes on this page it is apparent that all prefixes involve powers of 10 that are multiples of 3.

[†]James Prescott Joule (24 December 1818 – 11 October 1889) was an English physicist and brewer. He studied the nature of heat, and discovered its relationship to mechanical work. This led to the theory of conservation of energy, which in turn led to the development of the first law of thermodynamics. The SI derived unit of energy, the joule, is named after him. To find out more about Joule go to www.routledge.com/cw/bird

For example, a force of 43,645 N can be rewritten as 43.645×10^3N and from the list of prefixes can then be expressed as 43.645 kN.

Thus **43,645 N ≡ 43.645 kN**

To help further, on your calculator is an 'ENG' button. Enter the number 43,645 into your calculator and then press '='.

Now press the 'ENG' button and the answer is 43.645×10^3.

We then have to appreciate that 10^3 is the prefix 'kilo' giving **43,645 N ≡ 43.645 kN.**

In another example, let a current be 0.0745 A.

Enter 0.0745 into your calculator. Press '='.

Now press 'ENG' and the answer is 74.5×10^{-3}.

We then have to appreciate that 10^{-3} is the prefix 'milli' giving **0.0745 A ≡ 74.5 mA.**

Problem 1. Express the following in engineering notation and in prefix form:

(a) 300,000 W (b) 0.000068 H

(a) Enter 300,000 into the calculator. Press '='.
Now press 'ENG' and the answer is 300×10^3.
From the table of prefixes on page 24, 10^3 corresponds to kilo.
Hence, 300,000 W = 300×10^3W in engineering notation
= **300 kW** in prefix form

(b) Enter 0.000068 into the calculator. Press '='
Now press 'ENG' and the answer is 68×10^{-6}
From the table of prefixes on page 24, 10^{-6}corresponds to micro.
Hence, 0.000068 H =68×10^{-6}H in engineering notation
= **68 μH** in prefix form

Problem 2. Express the following in engineering notation and in prefix form:

(a) $42\times10^5\ \Omega$ (b) 47×10^{-10} F

(a) Enter 42 × 10^5into the calculator. Press '='
Now press 'ENG' and the answer is 4.2×10^6
From the table of prefixes on page 24, 10^6 corresponds to mega.
Hence, $42\times10^5\ \Omega = 4.2\times10^6\ \Omega$ in engineering notation
= **4.2 MΩ** in prefix form

(b) Enter $47\div10^{10} = \dfrac{47}{10{,}000{,}000{,}000}$ into the calculator. Press '='
Now press 'ENG' and the answer is 4.7×10^{-9}
From the table of prefixes on page 24, 10^{-9} corresponds to nano.
Hence, $47\div10^{10}$F =4.7×10^{-9}F in engineering notation
= **4.7 nF** in prefix form

Problem 3. Rewrite (a) 14,700 mm in metres (b) 276 cm in metres (c) 3.375 kg in grams

(a) $1\text{ m}=1000\text{ mm hence, }1\text{ mm}=\dfrac{1}{1000}=\dfrac{1}{10^3}=10^{-3}\text{m}$

Hence, 14,700 mm = 14,700 × 10^{-3}m = **14.7 m**

(b) $1\text{ m} = 100\text{ cm hence }1\text{ cm} = \dfrac{1}{100}=\dfrac{1}{10^2}=10^{-2}\text{m}$

Hence, 276 cm = 276 × 10^{-2}m = **2.76 m**

(c) 1 kg = 1000 g = 10^3g

Hence, 3.375 kg = 3.375 × 10^3g = **3375 g**

Now try the following Practice Exercise

Practice Exercise 11. Engineering notation

In problems 1 to 12, express in engineering notation in prefix form:

1. 60,000 Pa [60 kPa]
2. 0.00015 W [150 μW or 0.15 mW]
3. 5×10^7 V [50 MV]
4. 5.5×10^{-8} F [55 nF]
5. 100,000 N [100 kN]
6. 0.00054 A [0.54 mA or 540 μA]
7. $15\times10^5\,\Omega$ [1.5 MΩ]
8. 225×10^{-4} V [22.5 mV]
9. 35,000,000,000 Hz [35 GHz]
10. 1.5×10^{-11} F [15 pF]
11. 0.000017 A [17 μA]
12. 46200 Ω [46.2 kΩ]
13. Rewrite 0.003 mA in μA [3 μA]
14. Rewrite 2025 kHz as MHz [2.025 MHz]
15. Rewrite 5×10^4N in kN [50 kN]
16. Rewrite 300 pF in nF [0.3 nF]
17. Rewrite 6250 cm in metres [62.50 m]
18. Rewrite 34.6 g in kg [0.0346 kg]
19. The tensile stress acting on a rod is 5600000 Pa. Write this value in engineering notation. [5.6×10^6 Pa = 5.6 MPa]
20. The expansion of a rod is 0.0043 m. Write this in engineering notation. [4.3×10^{-3} m = 4.3 mm]

2.2 Metric – US/Imperial conversions

The Imperial System (which uses yards, feet, inches, etc. to measure length) was developed over hundreds of years in the UK, then the French developed the Metric System (metres) in 1670, which soon spread through Europe, even to England itself in 1960. But the USA and a few other countries still prefer feet and inches.

When converting from metric to imperial units, or vice versa, one of the following tables (2.1 to 2.8) should help.

Table 2.1 Metric to imperial length

Metric	US or Imperial
1 millimetre, mm	0.03937 inch
1 centimetre, cm = 10 mm	0.3937 inch
1 metre, m = 100 cm	1.0936 yard
1 kilometre, km = 1000 m	0.6214 mile

Problem 4. Calculate the number of inches in 350 mm, correct to 2 decimal places

350 mm = 350 × 0.03937 inches = **13.78 inches** from Table 2.1

Problem 5. Calculate the number of inches in 52 cm, correct to 4 significant figures

52 cm = 52 × 0.3937 inches = **20.47 inches** from Table 2.1

Problem 6. Calculate the number of yards in 74 m, correct to 2 decimal places

74 m = 74 × 1.0936 yards = **80.93 yds** from Table 2.1

Problem 7. Calculate the number of miles in 12.5 km, correct to 3 significant figures

12.5 km = 12.5 × 0.6214 miles = **7.77 miles** from Table 2.1

Table 2.2 Imperial to metric length

US or Imperial	Metric
1 inch, in	2.54 cm
1 foot, ft = 12 in	0.3048 m
1 yard, yd = 3 ft	0.9144 m
1 mile = 1760 yd	1.6093 km
1 nautical mile = 2025.4 yd	1.853 km

Problem 8. Calculate the number of centimetres in 35 inches, correct to 1 decimal places

35 inches = 35 × 2.54 cm = **88.9 cm** from Table 2.2

Problem 9. Calculate the number of metres in 66 inches, correct to 2 decimal places

66 inches = $\frac{66}{12}$ feet = $\frac{66}{12}$ × 0.3048 m = **1.68 m** from Table 2.2

Problem 10. Calculate the number of metres in 50 yards, correct to 2 decimal places

50 yards = 50 × 0.9144 m = **45.72 m** from Table 2.2

Problem 11. Calculate the number of kilometres in 7.2 miles, correct to 2 decimal places

7.2 miles = 7.2 × 1.6093 km = **11.59 km** from Table 2.2

Problem 12. Calculate the number of (a) yards (b) kilometres in 5.2 nautical miles

(a) 5.2 nautical miles = 5.2 × 2025.4 yards
= **10532 yards** from Table 2.2
(b) 5.2 nautical miles = 5.2 × 1.853 km
= **9.636 km** from Table 2.2

Table 2.3 Metric to Imperial area

Metric	US or Imperial
1 cm^2 = 100 mm^2	0.1550 in^2
1 m^2 = 10,000 cm^2	1.1960 yd^2
1 hectare, ha = 10,000 m^2	2.4711 acres
1 km^2 = 100 ha	0.3861 $mile^2$

Problem 13. Calculate the number of square inches in 47 cm^2, correct to 4 significant figures

47 cm^2 = 47 × 0.1550 in^2 = **7.285 in^2** from Table 2.3

Problem 14. Calculate the number of square yards in 20 m^2, correct to 2 decimal places

20 m^2 = 20 × 1.1960 yd^2 = **23.92 yd^2** from Table 2.3

Problem 15. Calculate the number of acres in 23 hectares of land, correct to 2 decimal places

23 hectares = 23 × 2.4711 acres = **56.84 acres** from Table 2.3

Problem 16. Calculate the number of square miles in a field of 15 km^2 area, correct to 2 decimal places

15 km^2 = 15 × 0.3861 $mile^2$ = **5.79 $mile^2$** from Table 2.3

Table 2.4 Imperial to metric area

US or Imperial	Metric
1 in^2	6.4516 cm^2
1 ft^2 = 144 in^2	0.0929 m^2
1yd^2 = 9 ft^2	0.8361 m^2
1 acre = 4840 yd^2	4046.9 m^2
1 $mile^2$ = 640 acres	2.59 km^2

Problem 17. Calculate the number of square centimetres in 17.5 in^2, correct to the nearest square centimetre

17.5 in^2 = 17.5 × 6.4516 cm^2 = **113 cm^2** from Table 2.4

Problem 18. Calculate the number of square metres in 205 ft^2, correct to 2 decimal places

205 ft^2 = 205 × 0.0929 m^2 = **19.04 m^2** from Table 2.4

Problem 19. Calculate the number of square metres in 11.2 acres, correct to the nearest square metre

11.2 acres = 11.2 × 4046.9 m^2 = **45325 m^2** from Table 2.4

Problem 20. Calculate the number of square kilometres in 12.6 $mile^2$, correct to 2 decimal places

12.6 $mile^2$ = 12.6 × 2.59 km^2 = **32.63 km^2** from Table 2.4

Table 2.5 Metric to imperial volume/capacity

Metric	US or Imperial
1 cm^3	0.0610 in^3
1 dm^3 = 1000 cm^3	0.0353 ft^3
1 m^3 = 1000 dm^3	1.3080 yd^3
1 litre = 1 dm^3 = 1000 cm^3	2.113 fluid pt = 1.7598 pt

Problem 21. Calculate the number of cubic inches in 123.5 cm^3, correct to 2 decimal places

123.5 cm^3 = 123.5 × 0.0610 cm^3 = **7.53 cm^3** from Table 2.5

Problem 22. Calculate the number of cubic feet in 144 dm^3, correct to 3 decimal places

144 dm^3 = 144 × 0.0353 ft^3 = **5.083 ft^3** from Table 2.5

Problem 23. Calculate the number of cubic yards in 5.75 m^3, correct to 4 significant figures

5.75 m^3 = 5.75 × 1.3080 yd^3 = **7.521 yd^3** from Table 2.5

Problem 24. Calculate the number of US fluid pints in 6.34 litres of oil, correct to 1 decimal place

6.34 litre = 6.34 × 2.113 US fluid pints = **13.4 US fluid pints** from Table 2.5

Table 2.6 Imperial to metric volume/capacity

US or Imperial	Metric
1 in^3	16.387 cm^3
1 ft^3	0.02832 m^3
1 US fl oz = 1.0408 UK fl oz	0.0296 litres
1 US pint (16 fl oz) = 0.8327 UK pt	0.4732 litre
1 US gal (231 in^3) = 0.8327 UK gal	3.7854 litre

Problem 25. Calculate the number of cubic centimetres in 3.75 in^3, correct to 2 decimal places

3.75 in^3 = 3.75 × 16.387 cm^3 = **61.45 cm^3** from Table 2.6

Problem 26. Calculate the number of cubic metres in 210 ft^3, correct to 3 significant figures

210 ft^3 = 210 × 0.02832 m^3 = **5.95 m^3** from Table 2.6

Problem 27. Calculate the number of litres in 4.32 US pints, correct to 3 decimal places

4.32 US pints = 4.32 × 0.4732 litres = **2.044 litres** from Table 2.6

Problem 28. Calculate the number of litres in 8.62 US gallons, correct to 2 decimal places

8.62 US gallons = 8.62 × 3.7854 litre = **32.63 litre** from Table 2.6

Table 2.7 Metric to imperial mass

Metric	US or Imperial
1 g = 1000 mg	0.0353 oz
1 kg = 1000 g	2.2046 lb
1 tonne, t, = 1000 kg	1.1023 short ton
1 tonne, t, = 1000 kg	0.9842 long ton

The British ton is the long ton, which is 2240 pounds, and the US ton is the short ton which is 2000 pounds.

Problem 29. Calculate the number of ounces in a mass of 1346 g, correct to 2 decimal places

1346 g = 1346 × 0.0353 oz = **47.51 oz** from Table 2.7

Problem 30. Calculate the mass, in pounds, in a 210.4 kg mass, correct to 4 significant figures

210.4 kg = 210.4 × 2.2046 lb = **463.8 lb** from Table 2.7

Problem 31. Calculate the number of short tons in 5000 kg, correct to 2 decimal places

5000 kg = 5 t = 5 × 1.1023 short tons = **5.51 short tons** from Table 2.7

Table 2.8 Imperial to metric mass

US or Imperial	Metric
1 oz = 437.5 grain	28.35 g
1 lb = 16 oz	0.4536 kg
1 stone = 14 lb	6.3503 kg
1 hundredweight, cwt = 112 lb	50.802 kg
1 short ton	0.9072 tonne
1 long ton	1.0160 tonne

Problem 32. Calculate the number of grams in 5.63 oz, correct to 4 significant figures

5.63 oz = 5.63 × 28.35 g = **159.6 g** from Table 2.8

Problem 33. Calculate the number of kilograms in 75 oz, correct to 3 decimal places

$75\text{ oz} = \frac{75}{16}\text{ lb} = \frac{75}{16} \times 0.4536\text{ kg} = \mathbf{2.126\text{ kg}}$ from Table 2.8

Problem 34. Convert 3.25 cwt into (a) pounds (b) kilograms

(a) 3.25 cwt = 3.25 × 112 lb = **364 lb** from Table 2.8

(b) 3.25 cwt = 3.25 × 50.802 kg = **165.1 kg** from Table 2.8

Temperature

To convert from Celsius‡ to Fahrenheit, first multiply by 9/5, then add 32.
To convert from Fahrenheit to Celsius, first subtract 32, then multiply by 5/9

Problem 35. Convert 35°C to degrees Fahrenheit

$$F = \frac{9}{5}C + 32 \text{ hence } 35°C = \frac{9}{5}(35) + 32 = 63 + 32 = \mathbf{95°F}$$

Problem 36. Convert 113°F to degrees Celsius

$$C = \frac{5}{9}(F - 32) \text{ hence } 113°F = \frac{5}{9}(113 - 32) = \frac{5}{9}(81) = \mathbf{45°C}$$

‡**Anders Celsius** (27 November 1701 – 25 April 1744) proposed the Celsius temperature scale which takes his name. His thermometer was calibrated with a value of 100° for the freezing point of water and 0° for the boiling point. To find out more about Celsius go to www.routledge.com/cw/bird

Now try the following Practice Exercise

Practice Exercise 12. Metric/imperial conversions

In the following problems, use the metric/imperial conversions in Tables 2.1 to 2.8.

1. Calculate the number of inches in 476 mm, correct to 2 decimal places. [18.74 in]
2. Calculate the number of inches in 209 cm, correct to 4 significant figures. [82.28 in]
3. Calculate the number of yards in 34.7 m, correct to 2 decimal places. [37.95 yd]
4. Calculate the number of miles in 29.55 km, correct to 2 decimal places. [18.36 mile]
5. Calculate the number of centimetres in 16.4 inches, correct to 2 decimal places. [41.66 cm]
6. Calculate the number of metres in 78 inches, correct to 2 decimal places. [1.98 m]

7. Calculate the number of metres in 15.7 yards, correct to 2 decimal places. [14.36 m]
8. Calculate the number of kilometres in 3.67 miles, correct to 2 decimal places. [5.91 km]
9. Calculate the number of (a) yards (b) kilometres in 11.23 nautical miles. [(a) 22745 yd (b) 20.81 km]
10. Calculate the number of square inches in 62.5 cm^2, correct to 4 significant figures. [9.688 in^2]
11. Calculate the number of square yards in 15.2 m^2, correct to 2 decimal places. [18.18 yd^2]
12. Calculate the number of acres in 12.5 hectares, correct to 2 decimal places. [30.89 acres]
13. Calculate the number of square miles in 56.7 km^2, correct to 2 decimal places. [21.89 $mile^2$]
14. Calculate the number of square centimetres in 6.37 in^2, correct to the nearest square centimetre. [41 cm^2]
15. Calculate the number of square metres in 308.6 ft^2, correct to 2 decimal places. [28.67 m^2]
16. Calculate the number of square metres in 2.5 acres, correct to the nearest square metre. [10117 m^2]
17. Calculate the number of square kilometres in 21.3 $mile^2$, correct to 2 decimal places. [55.17 km^2]
18. Calculate the number of cubic inches in 200.7 cm^3, correct to 2 decimal places. [12.24 in^3]
19. Calculate the number of cubic feet in 214.5 dm^3, correct to 3 decimal places. [7.572 ft^3]
20. Calculate the number of cubic yards in 13.45 m^3, correct to 4 significant figures. [17.59 yd^3]
21. Calculate the number of US fluid pints in 15 litres, correct to 1 decimal place. [31.7 fluid pints]
22. Calculate the number of cubic centimetres in 2.15 in^3, correct to 2 decimal places. [35.23 cm^3]
23. Calculate the number of cubic metres in 175 ft^3, correct to 4 significant figures. [4.956 m^3]
24. Calculate the number of litres in 7.75 US pints, correct to 3 decimal places. [3.667 litre]
25. Calculate the number of litres in 12.5 US gallons, correct to 2 decimal places. [47.32 litre]
26. Calculate the number of ounces in 980 g, correct to 2 decimal places. [34.59 oz]
27. Calculate the mass, in pounds, in 55 kg, correct to 4 significant figures. [121.3 lb]
28. Calculate the number of short tons in 4000 kg, correct to 3 decimal places. [4.409 short tons]
29. Calculate the number of grams in 7.78 oz, correct to 4 significant figures. [220.6 g]
30. Calculate the number of kilograms in 57.5 oz, correct to 3 decimal places. [1.630 kg]
31. Convert 2.5 cwt into (a) pounds (b) kilograms. [(a) 280 lb (b) 127.2 kg]
32. Convert 55°C to degrees Fahrenheit. [131°F]
33. Convert 167°F to degrees Celsius. [75 °C]

2.3 Straight line graphs

A graph is a visual representation of information, showing how one quantity varies with another related quantity.

The most common method of showing a relationship between two sets of data is to use a pair of reference axes – these are two lines drawn at right angles to each other (often called **Cartesian** or **rectangular axes**), as shown in Figure 2.1.

The horizontal axis is labelled the x-axis, and the vertical axis is labelled the y-axis.

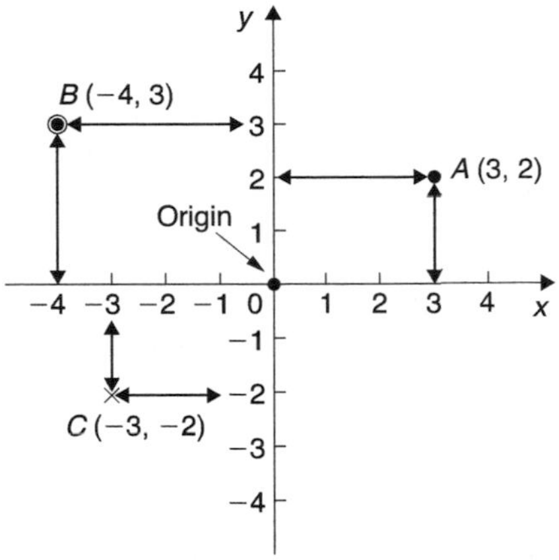

Figure 2.1

The point where x is 0 and y is 0 is called the **origin**.
x values have **scales** that are positive to the right of the origin and negative to the left.
y values have scales that are positive up from the origin and negative down from the origin.
Co-ordinates are written with brackets and a comma in between two numbers.
For example, point A is shown with co-ordinates (3, 2) and is located by starting at the origin and moving 3 units in the positive x direction (i.e. to the right) and then 2 units in the positive y direction (i.e. up).
When co-ordinates are stated, the first number is always the x value, and the second number is always the y value.
Also in Figure 2.1, point B has co-ordinates (− 4, 3) and point C has co-ordinates (− 3, − 2).
The following table gives the force F newtons which, when applied to a lifting machine, overcomes a corresponding load of L newtons.

F (newtons)	19	35	50	93	125	147
L (newtons)	40	120	230	410	540	680

1. Plot L horizontally and F vertically.
2. Scales are normally chosen such that the graph occupies as much space as possible on the graph paper. So in this case, the following scales are chosen:
   Horizontal axis (i.e. L): 1 cm = 50 N
   Vertical axis (i.e. F): 1 cm = 10 N
3. Draw the axes and label them L (newtons) for the horizontal axis and F (newtons) for the vertical axis.
4. Label the origin as 0.
5. Write on the horizontal scaling at 100, 200, 300, and so on, every 2 cm.
6. Write on the vertical scaling at 10, 20, 30, and so on, every 1 cm.
7. Plot on the graph the co-ordinates (40, 19), (120, 35), (230, 50), (410, 93), (540, 125) and (680, 147) marking each with a cross or a dot.
8. Using a ruler, draw the best straight line through the points. You will notice that not all of the points lie exactly on a straight line. This is quite normal with experimental values. In a practical situation it would be surprising if all of the points lay exactly on a straight line.
9. Extend the straight line at each end.
10. From the graph, determine the force applied when the load is 325 N. It should be close to 75 N. This process of finding an equivalent value within the given data is called **interpolation**. Similarly, determine the load that a force of 45 N will overcome. It should be close to 170 N.
11. From the graph, determine the force needed to overcome a 750 N load. It should be close to 161 N. This process of finding an equivalent value outside the given data is called **extrapolation**. To extrapolate we need to have extended the straight line drawn. Similarly, determine the force applied when the load is zero. It should be close to 11 N. Where the straight line crosses the vertical axis is called the **vertical-axis intercept**. So in this case, the vertical-axis intercept = 11 N at co-ordinates (0, 11).

The graph you have drawn should look something like Figure 2.2.

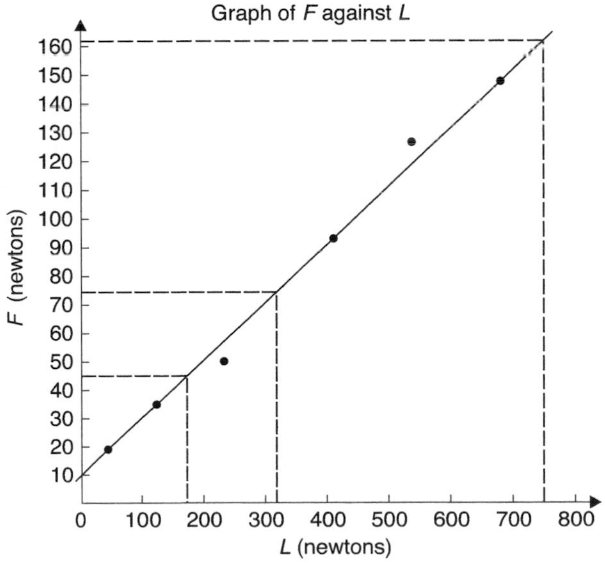

Figure 2.2

In another example, let the relationship between two variables x and y be $y = 3x + 2$

When $x = 0$, $y = 3 \times 0 + 2 = 0 + 2 = 2$
When $x = 1$, $y = 3 \times 1 + 2 = 3 + 2 = 5$
When $x = 2$, $y = 3 \times 2 + 2 = 6 + 2 = 8$, and so on.

The co-ordinates (0, 2), (1, 5) and (2, 8) have been produced and are plotted as shown in Figure 2.3.

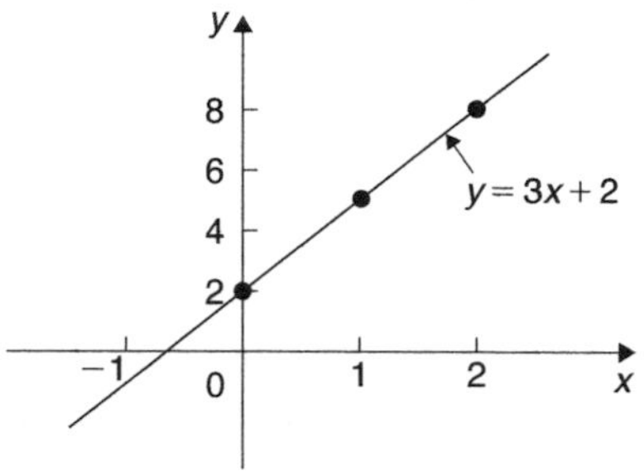

Figure 2.3

When the points are joined together **a straight line graph results**, i.e. $y = 3x + 2$ is a straight line graph.

Now try the following Practice Exercise

Practice Exercise 13. Straight line graphs

1. Corresponding values obtained experimentally for two quantities are:

x	− 5	− 3	− 1	0	2	4
y	− 13	− 9	− 5	− 3	1	5

Plot a graph of y (vertically) against x (horizontally) to scales of 2 cm = 1 for the horizontal x-axis and 1 cm = 1 for the vertical y-axis. (This graph will need the whole of the graph paper with the origin somewhere in the centre of the paper). From the graph find:

(a) the value of y when $x = 1$
(b) the value of y when $x = -2.5$
(c) the value of x when $y = -6$
(d) the value of x when $y = 5$

[(a) − 1 (b) − 8 (c) − 1.5 (d) 4]

2. Corresponding values obtained experimentally for two quantities are:

x	− 2.0	− 0.5	0	1.0	2.5	3.0	5.0
y	−13.0	− 5.5	− 3.0	2.0	9.5	12.0	22.0

Use a horizontal scale for x of 1 cm = $\frac{1}{2}$ unit and a vertical scale for y of 1 cm = 2 units and draw a graph of x against y. Label the graph and each of its axes. By interpolation, find from the graph the value of y when x is 3.5

[14.5]

3. Draw a graph of $y - 3x + 5 = 0$ over a range of $x = -3$ to $x = 4$. Hence determine (a) the value of y when $x = 1.3$ and (b) the value of x when $y = -9.2$

[(a) − 1.1 (b) − 1.4]

4. The speed n rev/min of a motor changes when the voltage V across the armature is varied. The results are shown in the following table:

n (rev/min)	560	720	900	1010	1240	1410
V (volts)	80	100	120	140	160	180

It is suspected that one of the readings taken of the speed is inaccurate. Plot a graph of speed (horizontally) against voltage (vertically) and find this value. Find also (a) the speed at a voltage of 132 V, and (b) the voltage at a speed of 1300 rev/min.

[1010 rev/min should be 1070 rev/min; (a) 1000 rev/min (b) 167 V]

2.4 Gradients, intercepts and equation of a graph

Gradient

The **gradient or slope** of a straight line is the ratio of the change in the value of y to the change in the value of x between any two points on the line. If, as x increases, (→), y also increases, (↑), then the gradient is positive.
In Figure 2.4(a), a straight line graph $y = 2x + 1$ is shown. To find the gradient of this straight line, choose two points on the straight line graph, such as A and C.
Then construct a right angled triangle, such as ABC, where BC is vertical and AB is horizontal.

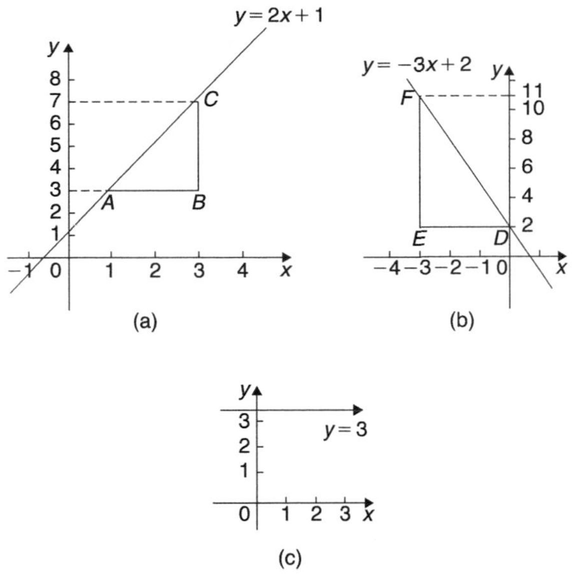

Figure 2.4

Then **gradient of** $\boldsymbol{AC} = \dfrac{\text{change in } y}{\text{change in } x}$

$$= \frac{CB}{BA} = \frac{7-3}{3-1} = \frac{4}{2} = \mathbf{2}$$

In Figure 2.4(b), a straight line graph $y = -3x + 2$ is shown. To find the gradient of this straight line, choose two points on the straight line graph, such as D and F. Then construct a right angled triangle, such as DEF, where EF is vertical and DE is horizontal.

Then **gradient of** $\boldsymbol{DF} = \dfrac{\text{change in } y}{\text{change in } x}$

$$= \frac{FE}{ED} = \frac{11-2}{-3-0} = \frac{9}{-3} = \mathbf{-3}$$

Figure 2.4(c) shows a straight line graph $y = 3$. **Since the straight line is horizontal the gradient is zero.**

y-axis intercept

The value of y when $x = 0$ is called the ***y*-axis intercept**. In Figure 2.5(a) the y-axis intercept is 1 and in Figure 2.4(b) the y-axis intercept is 2.

Equation of a straight line graph

The general equation of a straight line graph is:

$$y = mx + c$$

where m is the gradient or slope, and c is the y-axis intercept

Thus, as we have found in Figure 2.4(a), $y = 2x + 1$ represents a straight line of gradient 2 and y-axis intercept 1. So, given an equation $y = 2x + 1$, we are able to state, on sight, that the gradient = 2 and the y-axis intercept = 1, without the need for any analysis.

Similarly, in Figure 2.4(b), $y = -3x + 2$ represents a straight line of gradient -3 and y-axis intercept 2.

In Figure 2.4(c), $y = 3$ may be rewritten as $y = 0x + 3$ and therefore represents a straight line of gradient 0 and y-axis intercept 3.

Here are some worked problems to help understanding of gradients, intercepts and equation of a graph.

Problem 37. Determine for the straight line shown in Figure 2.5 (a) the gradient and (b) the equation of the graph

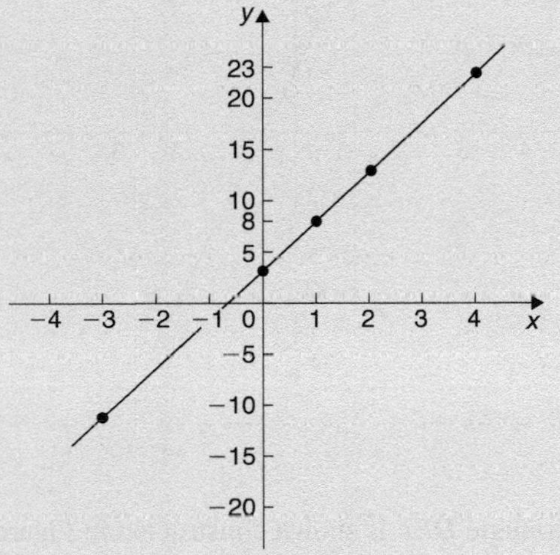

Figure 2.5

(a) A right angled triangle ABC is constructed on the graph as shown in Figure 2.6.

$$\textbf{Gradient} = \frac{AC}{CB} = \frac{23-8}{4-1} = \frac{15}{3} = \mathbf{5}$$

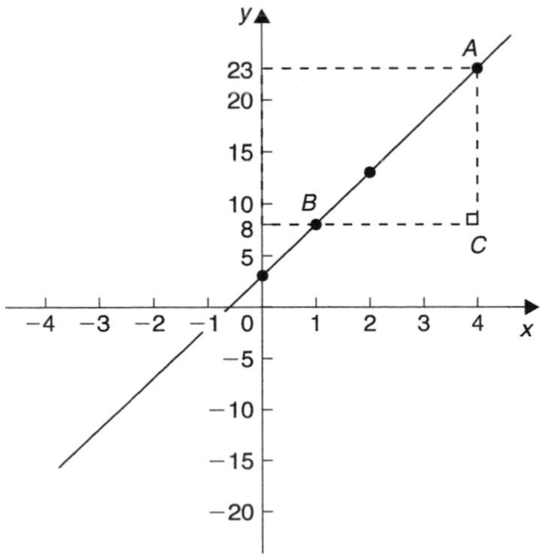

Figure 2.6

(b) The y-axis intercept at $x = 0$ is seen to be at $y = 3$

$y = mx + c$ is a straight line graph where m = gradient and c = y-axis intercept.
From above, $m = 5$ and $c = 3$.
Hence, equation of graph is: $\mathbf{y = 5x + 3}$

Problem 38. Determine the equation of the straight line shown in Figure 2.7.

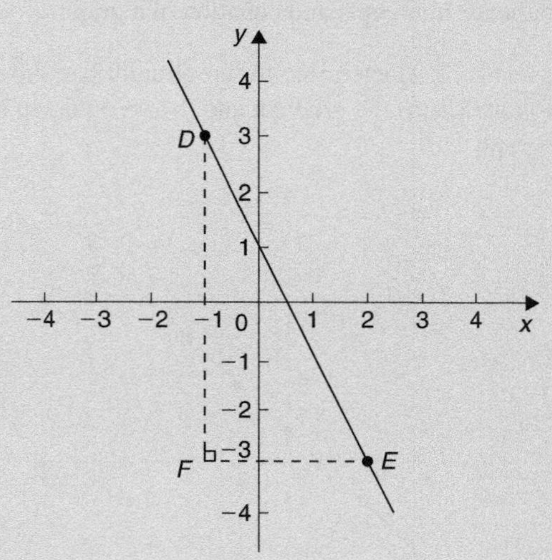

Figure 2.7

The triangle DEF is shown constructed in Figure 2.7.

$$\text{Gradient of } DE = \frac{DF}{FE} = \frac{3-(-3)}{-1-2} = \frac{6}{-3} = -2$$

and the y-axis intercept = 1

Hence, **the equation of the straight line is:** $y = mx + c$
i.e. $\mathbf{y = -2x + 1}$

Now try the following Practice Exercise

Practice Exercise 14. Gradients, intercepts and equation of a graph

1. The equation of a line is $4y = 2x + 5$. A table of corresponding values is produced and is shown below. Complete the table and plot a graph of y against x. Find the gradient of the graph.

x	−4	−3	−2	−1	0	1	2	3	4
y		−0.25			1.25				3.25

[Missing values: − 0.75, 0.25, 0.75, 1.75, 2.25, 2.75 Gradient = $\frac{1}{2}$]

2. Determine the gradient and intercept on the y-axis for each of the following equations:
(a) $y = 4x - 2$ (b) $y = -x$
(c) $y = -3x - 4$ (d) $y = 4$

[(a) 4, − 2 (b) − 1, 0 (c) − 3, − 4 (d) 0, 4]

3. Draw on the same axes the graphs of $y = 3x - 5$ and $3y + 2x = 7$. Find the co-ordinates of the point of intersection.

[(2, 1)]

4. A piece of elastic is tied to a support so that it hangs vertically, and a pan, on which weights can be placed, is attached to the free end. The length of the elastic is measured as various weights are added to the pan and the results obtained are as follows:

Load, W (N)	5	10	15	20	25
Length, l (cm)	60	72	84	96	108

Plot a graph of load (horizontally) against length (vertically) and determine: (a) the value of length when the load is 17 N, (b) the value of load when the length is 74 cm, (c) its gradient, and (d) the equation of the graph.

[(a) 89 cm (b) 11 N (c) 2.4 (d) $l = 2.4W + 48$]

5. The following table gives the effort P to lift a load W with a small lifting machine:

W (N)	10	20	30	40	50	60
P (N)	5.1	6.4	8.1	9.6	10.9	12.4

Plot W horizontally against P vertically and show that the values lie approximately on a straight line. Determine the probable relationship connecting P and W in the form $P = aW + b$.

[$P = 0.15W + 3.5$]

2.5 Practical straight line graphs

When a set of co-ordinate values are given or are obtained experimentally and it is believed that they follow a law of the form $y = mx + c$, then if a straight line can be drawn reasonably close to most of the co-ordinate values when plotted, this verifies that a law of the form $y = mx + c$ exists. From the graph, constants m (i.e. gradient) and c (i.e. y-axis intercept) can be determined. Here is a worked practical problem.

Problem 39. In an experiment demonstrating Hooke's law, the strain in an aluminium wire was measured for various stresses. The results were:

Stress N/mm^2	4.9	8.7	15.0
Strain	0.00007	0.00013	0.00021
Stress N/mm^2	18.4	24.2	27.3
Strain	0.00027	0.00034	0.00039

Plot a graph of stress (vertically) against strain (horizontally). Find:

(a) Young's Modulus of Elasticity for aluminium which is given by the gradient of the graph

(b) the value of the strain at a stress of 20 N/mm^2

(c) the value of the stress when the strain is 0.00020

The co-ordinates (0.00007, 4.9), (0.00013, 8.7), and so on, are plotted as shown in Figure 2.8. The graph produced is the best straight line which can be drawn corresponding to these points. (With experimental results it is unlikely that all the points will lie exactly on a straight line.) The graph, and each of its axes, are labelled. Since the straight line passes through the origin, then stress is directly proportional to strain for the given range of values.

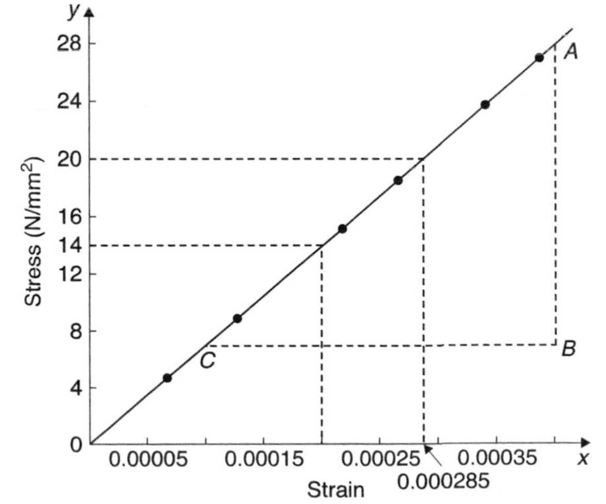

Figure 2.8

(a) The gradient of the straight line AC is given by

$$\frac{AB}{BC} = \frac{28-7}{0.00040-0.00010} = \frac{21}{0.00030}$$

$$= \frac{21}{3\times10^{-4}} = \frac{7}{10^{-4}} = 7\times10^4 = 70000 \text{ N/mm}^2$$

Thus, **Young's Modulus of Elasticity§ for aluminium is 70000 N/mm^2.**

§Thomas Young (13 June 1773 – 10 May 1829) was an English polymath. He is perhaps best known for his work on Egyptian hieroglyphics and the Rosetta Stone, but Young also made notable scientific contributions to the fields of vision, light, solid mechanics, energy, physiology, language and musical harmony. Young's modulus relates the stress in a body to its associated strain. To find out more about Young go to www.routledge.com/cw/bird

Since 1 $m^2 = 10^6$ mm^2, 70000 N/mm^2 is equivalent to 70000 × 10^6 N/m^2, i.e. **70 × 10^9 N/m^2 (or pascals)**

From Figure 2.8:

(b) the value of the strain at a stress of 20 N/mm^2 is **0.000285**, and

(c) the value of the stress when the strain is 0.00020 is **14 N/mm^2**

NB 1 N/mm^2 = 1 MN/m^2 = 1 MPa

Now try the following Practice Exercise

Practice Exercise 15. Practical problems involving straight line graphs

1. The following table gives the force F newtons which, when applied to a lifting machine, overcomes a corresponding load of L newtons.

Force F newtons	25	47	64	120	149	187
Load L newtons	50	140	210	430	550	700

Choose suitable scales and plot a graph of F (vertically) against L (horizontally). Draw the best straight line through the points. Determine from the graph (a) the gradient, (b) the F-axis intercept, (c) the equation of the graph, (d) the force applied when the load is 310 N, and (e) the load that a force of 160 N will overcome. (f) If the graph were to continue in the same manner, what value of force will be needed to overcome a 800 N load?

[(a) 0.25 (b) 12 (c) F = 0.25L + 12 (d) 89.5 N (e) 592 N (f) 212 N]

2. The following table gives the results of tests carried out to determine the breaking stress σ of rolled copper at various temperatures, t:

Stress σ (N/cm^2)	8.51	8.07	7.80	7.47	7.23	6.78
Temperature t (°C)	75	220	310	420	500	650

Plot a graph of stress (vertically) against temperature (horizontally). Draw the best straight line through the plotted co-ordinates. Determine the slope of the graph and the vertical axis intercept.

[− 0.003, 8.73 N/cm^2]

3. The velocity v of a body after varying time intervals t was measured as follows:

t (seconds)	2	5	8	11	15	18
v (m/s)	16.9	19.0	21.1	23.2	26.0	28.1

Plot v vertically and t horizontally and draw a graph of velocity against time. Determine from the graph (a) the velocity after 10 s, (b) the time at 20 m/s and (c) the equation of the graph.

[(a) 22.5 m/s (b) 6.5 s (c) $v = 0.7t + 15.5$]

4. An experiment with a set of pulley blocks gave the following results:

Effort, E (newtons)	9.0	11.0	13.6	17.4	20.8	23.6
Load, L (newtons)	15	25	38	57	74	88

Plot a graph of effort (vertically) against load (horizontally) and determine (a) the gradient, (b) the vertical axis intercept, (c) the law of the graph, (d) the effort when the load is 30 N and (e) the load when the effort is 19 N.

[(a) $\frac{1}{5}$ or 0.2 (b) 6 (c) $E = 0.2L + 6$ (d) 12 N (e) 65 N]

2.6 Introduction to calculus

Calculus is a branch of mathematics involving or leading to calculations dealing with continuously varying functions – such as velocity and acceleration, rates of change and maximum and minimum values of curves.

Calculus has widespread applications in science and engineering and is used to solve complicated problems for which algebra alone is insufficient.

Calculus is a subject that falls into two parts:

(i) **differential calculus** (or **differentiation**) and

(ii) **integral calculus** (or **integration**)

2.7 Basic differentiation revision

Differentiation of common functions

The **standard derivatives** summarised in Table 2.9 and are true for all real values of x.

Table 2.9 Standard differentials

y or $f(x)$	$\frac{dy}{dx}$ or $f'(x)$
ax^n	anx^{n-1}
$\sin ax$	$a\cos ax$
$\cos ax$	$-a\sin ax$
e^{ax}	$a\,e^{ax}$
$\ln ax$	$\frac{1}{x}$

In Table 2.9, $\frac{dy}{dx}$ is the gradient, or slope, of 'y' with respect to 'x'.

The **differential coefficient of a sum or difference** is the sum or difference of the differential coefficients of the separate terms.

Thus, if $f(x) = p(x) + q(x) - r(x)$, (where f, p, q and r are functions)

then $\quad f'(x) = p'(x) + q'(x) - r'(x)$

Differentiation of common functions is demonstrated in the following worked problems.

Problem 40. Find the differential coefficients of:

(a) $y = 12x^3$ (b) $y = \frac{12}{x^3}$

If $y = ax^n$ then $\frac{dy}{dx} = anx^{n-1}$

(a) Since $y = 12x^3$, $a = 12$ and $n = 3$

thus $\frac{dy}{dx} = (12)(3)x^{3-1} = \mathbf{36x^2}$

(b) $y = \frac{12}{x^3}$ is rewritten in the standard ax^n form as $y = 12x^{-3}$ and in the general rule $a = 12$ and $n = -3$

$$\text{Thus } \frac{dy}{dx} = (12)(-3)x^{-3-1} = -36x^{-4} = -\frac{\mathbf{36}}{\mathbf{x^4}}$$

Problem 41. Differentiate: (a) $y = 6$ (b) $y = 6x$

(a) $y = 6$ may be written as $y = 6x^0$, i.e. in the general rule $a = 6$ and $n = 0$

$$\text{Hence } \frac{dy}{dx} = (6)(0)x^{0-1} = \mathbf{0}$$

In general, **the differential coefficient of a constant is always zero.**

(b) Since $y = 6x$, in the general rule $a = 6$ and $n = 1$

$$\text{Hence } \frac{dy}{dx} = (6)(1)x^{1-1} = 6x^0 = \mathbf{6}$$

In general, the differential coefficient of kx, where k is a constant, is always k.

Problem 42. Find the derivatives of: (a) $y = 3\sqrt{x}$ (b) $y = \frac{5}{\sqrt[3]{x^4}}$

(a) $y = 3\sqrt{x}$ is rewritten in the standard differential form as $y = 3x^{1/2}$

In the general rule, $a = 3$ and $n = \frac{1}{2}$

$$\text{Thus } \frac{dy}{dx} = (3)\left(\frac{1}{2}\right)x^{\frac{1}{2}-1} = \frac{3}{2}x^{-\frac{1}{2}} = \frac{3}{2x^{1/2}} = \frac{\mathbf{3}}{\mathbf{2\sqrt{x}}}$$

(b) $y = \frac{5}{\sqrt[3]{x^4}} = \frac{5}{x^{4/3}} = 5x^{-4/3}$ in the standard differential form.

In the general rule, $a = 5$ and $n = -\frac{4}{3}$

$$\text{Thus } \frac{dy}{dx} = (5)\left(-\frac{4}{3}\right)x^{(-4/3)-1} = \frac{-20}{3}x^{-7/3} = \frac{-20}{3x^{7/3}} = \frac{\mathbf{-20}}{\mathbf{3\sqrt[3]{x^7}}}$$

Problem 43. Find the differential coefficients of: (a) $y = 3\sin 4x$ (b) $f(t) = 2\cos 3t$ with respect to the variable

(a) When $y = 3\sin 4x$, $a = 4$, so that $\frac{dy}{dx} = (3)(4\cos 4x)$
$= \mathbf{12\cos 4x}$ from Table 2.9

(b) When $f(t) = 2\cos 3t$, $a = 3$, so that
$f'(t) = (2)(-3\sin 3t) = \mathbf{-6\sin 3t}$

Problem 44. Determine the derivatives of:

(a) $y = 3e^{5x}$ (b) $f(\theta) = \frac{2}{e^{3\theta}}$ (c) $y = 6\ln 2x$

(a) When $y = 3e^{5x}$, $a = 5$, so that $\frac{dy}{dx} = (3)(5)e^{5x} = \mathbf{15e^{5x}}$ from Table 2.9

(b) $f(\theta) = \frac{2}{e^{3\theta}} = 2e^{-3\theta}$, a = − 3, so that $f'(\theta) = (2)(-3)e^{-3\theta} = -6e^{-3\theta} = \frac{\mathbf{-6}}{\mathbf{e^{3\theta}}}$

(c) When $y = 6\ln 2x$ then $\frac{dy}{dx} = 6\left(\frac{1}{x}\right) = \frac{\mathbf{6}}{\mathbf{x}}$

Problem 45. Find the gradient of the curve $y = 3x^4 - 2x^2 + 5x - 2$ at the points $(0, -2)$ and $(1, 4)$

The gradient of a curve at a given point is given by the corresponding value of the derivative. Thus, since $y = 3x^4 - 2x^2 + 5x - 2$ then

the gradient $= \frac{dy}{dx} = 12x^3 - 4x + 5$

At the point (0, −2), $x = 0$.
Thus the gradient $= 12(0)^3 - 4(0) + 5 = \mathbf{5}$
At the point (1, 4), $x = 1$.
Thus the gradient $= 12(1)^3 - 4(1) + 5 = \mathbf{13}$

Now try the following Practice Exercise

Practice exercise 16. Further problems on differentiating common functions

In Problems 1 to 6 find the differential coefficients of the given functions with respect to the variable.

1. (a) $5x^5$ (b) $\frac{1}{x}$ [(a) $25x^4$ (b) $-\frac{1}{x^2}$]
2. (a) $\frac{-4}{x^2}$ (b) 6 [(a) $\frac{8}{x^3}$ (b) 0]
3. (a) $2x$ (b) $2\sqrt{x}$ [(a) 2 (b) $\frac{1}{\sqrt{x}}$]
4. (a) $3\sqrt[3]{x^5}$ (b) $\frac{4}{\sqrt{x}}$ [(a) $5\sqrt[3]{x^2}$ (b) $-\frac{2}{\sqrt{x^3}}$]
5. (a) $2 \sin 3x$ (b) $-4 \cos 2x$ [(a) $6 \cos 3x$ (b) $8 \sin 2x$]
6. (a) $2e^{6x}$ (b) $4 \ln 9x$ [(a) $12e^{6x}$ (b) $\frac{4}{x}$]
7. Find the gradient of the curve $y = 2t^4 + 3t^3 - t + 4$ at the points (0, 4) and (1, 8) [−1, 16]
8. (a) Differentiate $y = \frac{2}{\theta^2} + 2 \ln 2\theta - 2(\cos 5\theta + 3 \sin 2\theta) - \frac{2}{e^{3\theta}}$

   (b) Evaluate $\frac{dy}{d\theta}$ when $\theta = \frac{\pi}{2}$, correct to 4 significant figures.

   [(a) $\frac{-4}{\theta^3} + \frac{2}{\theta} + 10 \sin 5\theta - 12 \cos 2\theta + \frac{6}{e^{3\theta}}$ (b) 22.30]

2.8 Revision of integration

The general solution of integrals of the form ax^n

The general solution of integrals of the form $\int ax^n dx$, where 'a' and 'n' are constants and $n \neq -1$, is given by:

$$\int ax^n\, dx = \frac{ax^{n+1}}{n+1} + c$$

Note that $\int ax^n dx$ actually represents the area under the curve $y = ax^n$ and c is an arbitrary constant.

Integrals of this type are called **indefinite integrals**. (Section 2.9 explains **definite integrals** where limits are applied to the integral, for example, $\int_a^b ax^n dx$)

Using the above rule gives:

(i) For $\int 3x^4 dx$, $a = 3$ and $n = 4$, so that

$$\int 3x^4 dx = \frac{3x^{4+1}}{4+1} + c = \frac{3}{5}x^5 + c$$

(ii) For $\int \frac{4}{9}t^3\, dt$, $a = \frac{4}{9}$, $n = 3$ and $t = x$, so that

$$\int \frac{4}{9}t^3\, dt = \frac{4}{9}\left(\frac{t^{3+1}}{3+1}\right) + c = \frac{4}{9}\left(\frac{t^4}{4}\right) + c = \frac{1}{9}t^4 + c$$

(iii) For $\int \frac{2}{x^2} dx = \int 2x^{-2} dx$, $a = 2$ and $n = -2$, so that

$$\int 2x^{-2}\, dx = (2)\left(\frac{x^{-2+1}}{-2+1}\right) + c = \frac{2x^{-1}}{-1} + c$$

$$= -\frac{2}{x} + c$$

(iv) For $\int \sqrt{x}\, dx = \int x^{\frac{1}{2}}\, dx$, a = 1 and n = $\frac{1}{2}$, so that

$$\int x^{\frac{1}{2}}\, dx = (1)\frac{x^{\frac{1}{2}+1}}{\frac{1}{2}+1} + c = \frac{x^{\frac{3}{2}}}{\frac{3}{2}} + c = \frac{2}{3}\sqrt{x^3} + c$$

Each of these results may be checked by differentiation.

(a) The integral of a constant k is $kx + c$.

For example, $\int 8\, dx = 8x + c$ and $\int 5\, dt = 5t + c$

(b) When a sum of several terms is integrated the result is the sum of the integrals of the separate terms. For example,

$$\int (3x + 2x^2 - 5)dx = \int 3x\, dx + \int 2x^2\, dx - \int 5\, dx$$

$$= \frac{3x^2}{2} + \frac{2x^3}{3} - 5x + c$$

Standard integrals

Since integration is the reverse process of differentiation the standard integrals listed in Table 2.10 may be deduced and readily checked by differentiation.

Table 2.10 Standard integrals

y	$\int y\,dx$
1. $\int ax^n$	$\frac{ax^{n+1}}{n+1}+c$ (except when $n=-1$)
2. $\int \cos ax\,dx$	$\frac{1}{a}\sin ax+c$
3. $\int \sin ax\,dx$	$-\frac{1}{a}\cos ax+c$
4. $\int e^{ax}\,dx$	$\frac{1}{a}e^{ax}+c$
5. $\int \frac{1}{x}dx$	$\ln x+c$

Problem 46. Determine: $\int 7x^2\,dx$

The standard integral, $\int ax^n\,dx=\frac{ax^{n+1}}{n+1}+c$

When $a=7$ and $n=2$,

$$\int 7x^2\,dx=\frac{7x^{2+1}}{2+1}+c=\frac{\mathbf{7x^3}}{\mathbf{3}}+\mathbf{c}\text{ or }\frac{\mathbf{7}}{\mathbf{3}}\mathbf{x^3}+\mathbf{c}$$

Problem 47. Determine: $\int 2t^3\,dt$

When $a=2$ and $n=3$,

$$\int 2t^3\,dt=\frac{2t^{3+1}}{3+1}+c=\frac{2t^4}{4}+c=\frac{\mathbf{1}}{\mathbf{2}}\mathbf{t^4}+\mathbf{c}$$

Note that each of the results in worked examples 1 and 2 may be checked by differentiating them.

Problem 48. Determine: $\int 8\,dx$

$\int 8\,dx$ is the same as $\int 8x^0\,dx$ and, using the general rule when $a=8$ and n = 0 gives:

$$\int 8x^0\,dx=\frac{8x^{0+1}}{0+1}+c=\mathbf{8x+c}$$

In general, if k is a constant then $\int \mathbf{k\,dx=kx+c}$

Problem 49. Determine: $\int 2x\,dx$

When $a=2$ and $n=1$, then

$$\int 2x\,dx=\int 2x^1\,dx=\frac{2x^{1+1}}{1+1}+c=\frac{2x^2}{2}+c=\mathbf{x^2+c}$$

Problem 50. Determine: $\int \frac{5}{x^2}dx$

$$\int \frac{5}{x^2}dx=\int 5x^{-2}\,dx$$

Using the standard integral, $\int ax^n\,dx$, when $a=5$ and $n=-2$ gives:

$$\int 5x^{-2}\,dx=\frac{5x^{-2+1}}{-2+1}+c=\frac{5x^{-1}}{-1}+c=-5x^{-1}+c=-\frac{\mathbf{5}}{\mathbf{x}}+\mathbf{c}$$

Problem 51. Determine: $\int 3\sqrt{x}\,dx$

For fractional powers it is necessary to appreciate $\sqrt[n]{a^m}=a^{\frac{m}{n}}$ – see Chapter 1, page 11.

$$\int 3\sqrt{x}\,dx=\int 3x^{\frac{1}{2}}\,dx=\frac{3x^{\frac{1}{2}+1}}{\frac{1}{2}+1}+c=\frac{3x^{\frac{3}{2}}}{\frac{3}{2}}+c$$

$$=2x^{\frac{3}{2}}+c=\mathbf{2\sqrt{x^3}+c}$$

Problem 52. Determine: $\int 4\cos 3x\,dx$

From 2 of Table 2.10,

$$\int 4\cos 3x\,dx=(4)\left(\frac{1}{3}\right)\sin 3x+c=\frac{\mathbf{4}}{\mathbf{3}}\mathbf{\sin 3x+c}$$

Problem 53. Determine: $\int 5\sin 2\theta\,d\theta$

From 3 of Table 2.10,

$$\int 5\sin 2\theta\,d\theta=(5)\left(-\frac{1}{2}\right)\cos 2\theta+c=-\frac{\mathbf{5}}{\mathbf{2}}\mathbf{\cos 2\theta+c}$$

Problem 54. Determine: $\int 5e^{3x}\,dx$

From 4 of Table 2.10, $\int 5e^{3x}\,dx=(5)\left(\frac{1}{3}\right)e^{3x}+c$

$$=\frac{\mathbf{5}}{\mathbf{3}}\mathbf{e^{3x}+c}$$

Problem 55. Determine: $\int \frac{3}{5x} dx$

From 5 of Table 2.10,

$$\int \frac{3}{5x} dx = \int \left(\frac{3}{5}\right)\left(\frac{1}{x}\right) dx = \mathbf{\frac{5}{3} \ln x + c}$$

Now try the following Practice Exercise

Practice Exercise 17. Standard integrals

Determine the following integrals:

1. (a) $\int 4\,dx$ (b) $\int 7x\,dx$

[(a) $4x + c$ (b) $\frac{7}{2}x^2 + c$]

2. (a) $\int 5x^3\,dx$ (b) $\int 3t^7\,dt$

[(a) $\frac{5}{4}x^4 + c$ (b) $\frac{3}{8}t^8 + c$]

3. (a) $\int \frac{2}{5}x^2\,dx$ (b) $\int \frac{5}{6}x^3\,dx$

[(a) $\frac{2}{15}x^3 + c$ (b) $\frac{5}{24}x^4 + c$]

4. (a) $\int \left(2x^4 - 3x\right)dx$ (b) $\int \left(2 - 3t^3\right)dt$

[(a) $\frac{2}{5}x^5 - \frac{3}{2}x^2 + c$ (b) $2t - \frac{3}{4}t^4 + c$]

5. (a) $\int \frac{4}{3x^2}\,dx$ (b) $\int \frac{3}{4x^4}\,dx$

[(a) $-\frac{4}{3x} + c$ (b) $-\frac{1}{4x^3} + c$]

6. (a) $2\int \sqrt{x^3}\,dx$ (b) $\int \frac{1}{4}\sqrt[4]{x^5}\,dx$

[(a) $\frac{4}{5}\sqrt{x^5} + c$ (b) $\frac{1}{9}\sqrt[4]{x^9} + c$]

7. (a) $\int 3\cos 2x\,dx$ (b) $\int 7\sin 3\theta\,d\theta$

[(a) $\frac{3}{2}\sin 2x + c$ (b) $-\frac{7}{3}\cos 3\theta + c$]

8. (a) $\int 3\sin\frac{1}{2}x\,dx$ (b) $\int 6\cos\frac{1}{3}x\,dx$

[(a) $-6\cos\frac{1}{2}x + c$ (b) $18\sin\frac{1}{3}x + c$]

9. (a) $\int \frac{3}{4}e^{2x}\,dx$ (b) $\int \frac{2}{3x}\,dx$

[(a) $\frac{3}{8}e^{2x} + c$ (b) $\frac{2}{3}\ln x + c$]

2.9 Definite integrals

Integrals containing an arbitrary constant c in their results are called **indefinite integrals** since their precise value cannot be determined without further information.

Definite integrals are those in which limits are applied. If an expression is written as $\left[x\right]_a^b$ 'b' is called the **upper limit** and 'a' the **lower limit**.

The operation of applying the limits is defined as: $\left[x\right]_a^b = (b) - (a)$

For example, the increase in the value of the integral x^2 as x increases from 1 to 3 is written as $\int_1^3 x^2\,dx$, where the lower limit is 1, and the upper limit is 3.

Applying the limits gives:

$$\int_1^3 x^2\,dx = \left[\frac{x^3}{3} + c\right]_1^3 = \left(\frac{3^3}{3} + c\right) - \left(\frac{1^3}{3} + c\right)$$

$$= (9 + c) - \left(\frac{1}{3} + c\right) = \mathbf{8\frac{2}{3}}$$

Note that the 'c' term always cancels out when limits are applied and it need not be shown with definite integrals.

Problem 56. Evaluate: $\int_1^2 3x\,dx$

$$\int_1^2 3x\,dx = \left[\frac{3x^2}{2}\right]_1^2 = \left\{\frac{3}{2}(2)^2\right\} - \left\{\frac{3}{2}(1)^2\right\}$$

$$= 6 - 1\frac{1}{2} = \mathbf{4\frac{1}{2}}$$

Problem 57. Evaluate: $\int_{-2}^3 (4 - x^2)\,dx$

$$\int_{-2}^3 (4 - x^2)\,dx = \left[4x - \frac{x^3}{3}\right]_{-2}^3$$

$$= \left\{4(3) - \frac{(3)^3}{3}\right\} - \left\{4(-2) - \frac{(-2)^3}{3}\right\}$$

$$= \left\{12 - 9\right\} - \left\{-8 - \frac{-8}{3}\right\}$$

$$= \left\{3\right\} - \left\{-5\frac{1}{3}\right\} = \mathbf{8\frac{1}{3}}$$

Problem 58. Evaluate: $\int_0^2 x(3 + 2x)\,dx$

$$\int_0^2 x(3+2x)\,dx = \int_0^2 (3x + 2x^2)\,dx = \left[\frac{3x^2}{2} + \frac{2x^3}{3}\right]_0^2$$

$$= \left\{\frac{3(2)^2}{2} + \frac{2(2)^3}{3}\right\} - \left\{0 + 0\right\}$$

$$= 6 + \frac{16}{3} = \mathbf{11\frac{1}{3}} \text{ or } \mathbf{11.33}$$

Problem 59. Evaluate: $\int_0^{\pi/2} 3\sin 2x\,dx$

$$\int_0^{\pi/2} 3\sin 2x\,dx = \left[(3)\left(-\frac{1}{2}\right)\cos 2x\right]_0^{\pi/2} = \left[-\frac{3}{2}\cos 2x\right]_0^{\pi/2}$$

$$= \left\{-\frac{3}{2}\cos 2\left(\frac{\pi}{2}\right)\right\} - \left\{-\frac{3}{2}\cos 2(0)\right\}$$

$$= \left\{-\frac{3}{2}\cos \pi\right\} - \left\{-\frac{3}{2}\cos 0\right\}$$

$$= \left\{-\frac{3}{2}(-1)\right\} - \left\{-\frac{3}{2}(1)\right\}$$

$$= \frac{3}{2} + \frac{3}{2} = \mathbf{3}$$

Problem 60. Evaluate: $\int_1^2 4\cos 3t\,dt$

$$\int_1^2 4\cos 3t\,dt = \left[(4)\left(\frac{1}{3}\right)\sin 3t\right]_1^2 = \left[\frac{4}{3}\sin 3t\right]_1^2$$

$$= \left\{\frac{4}{3}\sin 6\right\} - \left\{\frac{4}{3}\sin 3\right\}$$

Note that limits of trigonometric functions are always expressed in **radians** – thus, for example, sin6 means the sine of 6 radians = – 0.279415.... where π radians = 180°

Hence $\int_1^2 4\cos 3t\,dt = \left\{\frac{4}{3}(-0.279415..)\right\}$

$$- \left\{\frac{4}{3}(0.141120..)\right\}$$

$$= (-0.37255) - (0.18816)$$

$$= \mathbf{-0.5607}$$

Problem 61. Evaluate: $\int_1^2 4e^{2x}\,dx$ correct to 4 significant figures.

$$\int_1^2 4e^{2x}\,dx = \left[\frac{4}{2}e^{2x}\right]_1^2 = 2\left[e^{2x}\right]_1^2 = 2[e^4 - e^2]$$

$$= 2[54.5982 - 7.3891]$$

$$= \mathbf{94.42}$$

Problem 62. Evaluate: $\int_1^4 \frac{3}{4u}\,du$ correct to 4 significant figures

$$\int_1^4 \frac{3}{4u}\,du = \left[\frac{3}{4}\ln u\right]_1^4 = \frac{3}{4}[\ln 4 - \ln 1]$$

$$= \frac{3}{4}[1.3863 - 0] = \mathbf{1.040}$$

Now try the following Practice Exercise

Practice Exercise 18. Definite integrals

In Problems 1 to 7, evaluate the definite integrals (where necessary, correct to 4 significant figures).

1. (a) $\int_1^2 x\,dx$ (b) $\int_1^2 (x-1)\,dx$

 [(a) 1.5 (b) 0.5]

2. (a) $\int_1^4 5x^2\,dx$ (b) $\int_{-1}^1 -\frac{3}{4}t^2\,dt$

 [(a) 105 (b) – 0.5]

3. (a) $\int_{-1}^2 (3 - x^2)\,dx$ (b) $\int_1^3 (x^2 - 4x + 3)\,dx$

 [(a) 6 (b) – 1.333]

4. (a) $\int_0^4 2\sqrt{x}\,dx$ (b) $\int_2^3 \frac{1}{x^2}\,dx$

 [(a) 10.67 (b) 0.1667]

5. (a) $\int_0^\pi \frac{3}{2}\cos\theta\,d\theta$ (b) $\int_0^{\pi/2} 4\cos\theta\,d\theta$

 [(a) 0 (b) 4]

6. (a) $\int_{\pi/6}^{\pi/3} 2\sin 2\theta\,d\theta$ (b) $\int_0^2 3\sin t\,dt$

 [(a) 1 (b) 4.248]

7. (a) $\int_0^1 3e^{3t}\,dt$ (b) $\int_2^3 \frac{2}{3x}\,dx$

 [(a) 19.09 (b) 0.2703]

2.10 Simple vector analysis

Determinants

For a **2 by 2** determinant: $\begin{vmatrix} a & b \\ c & d \end{vmatrix} = ad - bc$

For example $\begin{vmatrix} 5 & -3 \\ 4 & 2 \end{vmatrix} = (5)(2) - (-3)(4) = 10 - -12$

$= \mathbf{22}$

For a **3 by 3** determinant $\begin{vmatrix} a & b & c \\ d & e & f \\ g & h & i \end{vmatrix}$

$$= a\begin{vmatrix} e & f \\ h & i \end{vmatrix} - b\begin{vmatrix} d & f \\ g & i \end{vmatrix} + c\begin{vmatrix} d & e \\ g & h \end{vmatrix}$$

$$= a(e \times i - f \times h) - b(d \times i - f \times g) + c(d \times h - e \times g)$$

For example $\begin{vmatrix} 3 & 1 & -4 \\ -5 & 2 & 1 \\ 6 & -3 & -1 \end{vmatrix}$

$$= 3\begin{vmatrix} 2 & 1 \\ -3 & -1 \end{vmatrix} - 1\begin{vmatrix} -5 & 1 \\ 6 & -1 \end{vmatrix} + (-4)\begin{vmatrix} -5 & 2 \\ 6 & -3 \end{vmatrix}$$

$$= 3(-2 - -3) - 1(5 - 6) - 4(15 - 12)$$

$$= 3(1) - 1(-1) - 4(3) = 3 + 1 - 12 = \mathbf{-8}$$

i, j, k unit vectors

A method of completely specifying the direction of a vector in space relative to some reference point is to use three unit vectors, *i*, *j* and *k*, mutually at right angles to each other, as shown in Figure 2.9.

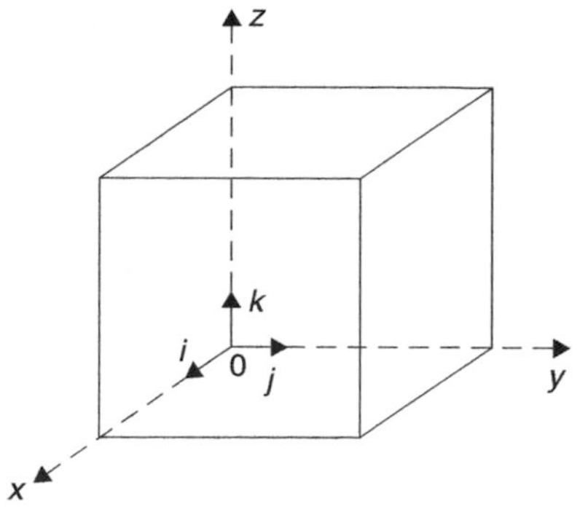

Figure 2.9

The unit vectors are defined by: $\boldsymbol{i} = (1, 0, 0)$ $\boldsymbol{j} = (0, 1, 0)$ $\boldsymbol{k} = (0, 0, 1)$

The scalar or dot product of two vectors

Let $\boldsymbol{a} = a_1\boldsymbol{i} + a_2\boldsymbol{j} + a_3\boldsymbol{k}$ and $\boldsymbol{b} = b_1\boldsymbol{i} + b_2\boldsymbol{j} + b_3\boldsymbol{k}$
then $\boldsymbol{a} \bullet \boldsymbol{b} = (a_1\boldsymbol{i} + a_2\boldsymbol{j} + a_3\boldsymbol{k}) \bullet (b_1\boldsymbol{i} + b_2\boldsymbol{j} + b_3\boldsymbol{k})$
Multiplying out the brackets gives:
$\boldsymbol{a} \bullet \boldsymbol{b}$

$$= a_1b_1\boldsymbol{i} \bullet \boldsymbol{i} + a_1b_2\boldsymbol{i} \bullet \boldsymbol{j} + a_1b_3\mathrm{i} \bullet \boldsymbol{k} + a_2b_1\boldsymbol{j} \bullet \boldsymbol{i} + a_2b_2\boldsymbol{j} \bullet \boldsymbol{j} + a_2b_3\mathrm{j} \bullet \boldsymbol{k} + a_3b_1\boldsymbol{k} \bullet \boldsymbol{i} + a_3b_2\boldsymbol{k} \bullet \boldsymbol{j} + a_3b_3\boldsymbol{k} \bullet \boldsymbol{k}$$

However, it may be shown that

$\boldsymbol{i} \bullet \boldsymbol{i} = \boldsymbol{j} \bullet \boldsymbol{j} = \boldsymbol{k} \bullet \boldsymbol{k} = 1$ and

$\boldsymbol{i} \bullet \boldsymbol{j} = \boldsymbol{i} \bullet \boldsymbol{k} = \boldsymbol{j} \bullet \boldsymbol{i} = \boldsymbol{j} \bullet \boldsymbol{k} = \boldsymbol{k} \bullet \boldsymbol{i} = \boldsymbol{k} \bullet \boldsymbol{j} = 0$

Thus, the **scalar or dot product**
$\boldsymbol{a} \bullet \boldsymbol{b} = a_1b_1 + a_2b_2 + a_3b_3$
For example, if $\boldsymbol{a} = 2\boldsymbol{i} + \boldsymbol{j} - \boldsymbol{k}$ and $\boldsymbol{b} = \boldsymbol{i} - 3\boldsymbol{j} + 2\boldsymbol{k}$ then
$\boldsymbol{a} \bullet \boldsymbol{b} = (2)(1) + (1)(-3) + (-1)(2) = \mathbf{-3}$
A further example: calculate the work done by a force $\boldsymbol{F} = (-5\boldsymbol{i} + \boldsymbol{j} + 7\boldsymbol{k})$ N when its point of application moves from point $(-2\boldsymbol{i} - 6\boldsymbol{j} + \boldsymbol{k})$ m to the point $(\boldsymbol{i} - \boldsymbol{j} + 10\boldsymbol{k})$ m
Work done = $\boldsymbol{F} \bullet \boldsymbol{d}$
where $\boldsymbol{d} = (\boldsymbol{i} - \boldsymbol{j} + 10\boldsymbol{k}) - (-2\boldsymbol{i} - 6\boldsymbol{j} + \boldsymbol{k})$
$= 3\boldsymbol{i} + 5\boldsymbol{j} + 9\boldsymbol{k}$
Hence **work done** $= (-5\boldsymbol{i} + \boldsymbol{j} + 7\boldsymbol{k}) \bullet (3\boldsymbol{i} + 5\boldsymbol{j} + 9\boldsymbol{k})$
$= -15 + 5 + 63 = \mathbf{53\ N\ m}$

The vector or cross product of two vectors

When $\boldsymbol{a} = a_1\boldsymbol{i} + a_2\boldsymbol{j} + a_3\boldsymbol{k}$ and $\boldsymbol{b} = b_1\boldsymbol{i} + b_2\boldsymbol{j} + b_3\boldsymbol{k}$ then it may be shown that

the **vector or cross product,** $\boldsymbol{a} \times \boldsymbol{b} = \begin{vmatrix} i & j & k \\ a_1 & a_2 & a_3 \\ b_1 & b_2 & b_3 \end{vmatrix}$

For example, if $\boldsymbol{a} = \boldsymbol{i} + 4\boldsymbol{j} - 2\boldsymbol{k}$ and $\boldsymbol{b} = 2\boldsymbol{i} - \boldsymbol{j} + 3\boldsymbol{k}$

then $\boldsymbol{a} \times \boldsymbol{b} = \begin{vmatrix} i & j & k \\ 1 & 4 & -2 \\ 2 & -1 & 3 \end{vmatrix}$

$$= \boldsymbol{i}\begin{vmatrix} 4 & -2 \\ -1 & 3 \end{vmatrix} - \boldsymbol{j}\begin{vmatrix} 1 & -2 \\ 2 & 3 \end{vmatrix} + \boldsymbol{k}\begin{vmatrix} 1 & 4 \\ 2 & -1 \end{vmatrix}$$

$$= \boldsymbol{i}(12 - 2) - \boldsymbol{j}(3 - -4) + \boldsymbol{k}(-1 - 8)$$

$$= \mathbf{10\boldsymbol{i} - 7\boldsymbol{j} - 9\boldsymbol{k}}$$

A further example: *a* force of $(2\boldsymbol{i} - \boldsymbol{j} + \boldsymbol{k})$ newtons acts on *a* line through point *P* having co-ordinates

(0, 3, 1) metres. Determine the moment vector and its magnitude about point Q having co-ordinates (4, 0, – 1) metres.

Position vector, $\boldsymbol{r} = (0\boldsymbol{i} + 3\boldsymbol{j} + \boldsymbol{k}) - (4\boldsymbol{i} + 0\boldsymbol{j} - \boldsymbol{k})$
$= -4\boldsymbol{i} + 3\boldsymbol{j} + 2\boldsymbol{k}$

Moment, $\boldsymbol{M} = \boldsymbol{r} \times \boldsymbol{f}$ where

$$\boldsymbol{M} = \begin{vmatrix} i & j & k \\ -4 & 3 & 2 \\ 2 & -1 & 1 \end{vmatrix}$$

$$= (3+2)\boldsymbol{i} - (-4-4)\boldsymbol{j} + (4-6)\boldsymbol{k}$$

$$= \mathbf{(5\boldsymbol{i} + 8\boldsymbol{j} - 2\boldsymbol{k})\ N\ m}$$

Now try the following Practice Exercises

Practice Exercise 19. Simple vector analysis

1. Calculate the value of $\begin{vmatrix} 2 & -3 \\ 5 & 1 \end{vmatrix}$ [17]

2. Evaluate $\begin{vmatrix} 1 & 2 \\ -5 & -7 \end{vmatrix}$ [3]

3. Find the value of $\begin{vmatrix} 2 & 3 & -1 \\ -5 & 0 & 4 \\ 1 & -4 & -2 \end{vmatrix}$ [– 6]

4. Find the value of $\begin{vmatrix} -3 & 2 & 5 \\ 7 & 1 & -2 \\ 4 & 0 & -1 \end{vmatrix}$ [– 19]

5. Find the scalar product $\boldsymbol{a} \bullet \boldsymbol{b}$ when $\boldsymbol{a} = 3\boldsymbol{i} + 2\boldsymbol{j} - \boldsymbol{k}$ and $\boldsymbol{b} = 2\boldsymbol{i} + 5\boldsymbol{j} + \boldsymbol{k}$ [15]

6. Find the scalar product $\boldsymbol{p} \bullet \boldsymbol{q}$ when $\boldsymbol{p} = 2\boldsymbol{i} - 3\boldsymbol{j} + 4\boldsymbol{k}$ and $\boldsymbol{q} = \boldsymbol{i} + 2\boldsymbol{j} + 5\boldsymbol{k}$ [16]

7. Given $\boldsymbol{p} = 2\boldsymbol{i} - 3\boldsymbol{j}$, $\boldsymbol{q} = 4\boldsymbol{j} - \boldsymbol{k}$ and $\boldsymbol{r} = \boldsymbol{i} + 2\boldsymbol{j} - 3\mathbf{k}$, determine the quantities (a) $\boldsymbol{p} \bullet \boldsymbol{q}$ (b) $\boldsymbol{p} \bullet \boldsymbol{r}$ (c) $\boldsymbol{q} \bullet \boldsymbol{r}$

   [(a) – 12 (b) – 4 (c) 11]

8. Calculate the work done by a force $\boldsymbol{F} = (-2\boldsymbol{i} + 3\boldsymbol{j} + 5\boldsymbol{k})$ N when its point of application moves from point $(-\boldsymbol{i} - 5\boldsymbol{j} + 2\boldsymbol{k})$ m to the point $(4\boldsymbol{i} - \boldsymbol{j} + 8\mathbf{k})$ m

   [32 N m]

9. Given that $\boldsymbol{p} = 3\boldsymbol{i} + 2\boldsymbol{k}$, $\boldsymbol{q} = \boldsymbol{i} - 2\boldsymbol{j} + 3\boldsymbol{k}$ and $\boldsymbol{r} = -4\boldsymbol{i} + 3\boldsymbol{j} - \boldsymbol{k}$ determine (a) $\boldsymbol{p} \times \boldsymbol{q}$ (b) $\boldsymbol{q} \times \boldsymbol{r}$

   [(a) $4\boldsymbol{i} - 7\boldsymbol{j} - 6\boldsymbol{k}$ (b) $-7\boldsymbol{i} - 11\boldsymbol{j} - 5\boldsymbol{k}$]

10. A force of $(4\boldsymbol{i} - 3\boldsymbol{j} + 2\boldsymbol{k})$ newtons acts on a line through point P having co-ordinates (2, 1, 3) metres. Determine the moment vector about point Q having co-ordinates (5, 0, – 2) metres.

   [$\boldsymbol{M} = (17\boldsymbol{i} + 26\boldsymbol{j} + 5\boldsymbol{k})$ N m]

For the more advanced mathematics that will be needed within the theory of Chapters 3 to 23 in this text, some help is at hand – some mathematics textbook references are listed on pages 49 and 50. Mathematics is a tool that is needed to understand the theory behind the engineering. Think of mathematics as a 'toolbox' – you never know when you might need it to solve equations, transpose formulae, use differential or integral calculus, trigonometry or matrices. All of these topics – and more – are needed to understand *Mechanics of Solids*; it is hoped that the references given will be of help.

For fully worked solutions to each of the problems in Exercises 11 to 19 in this chapter, go to the website:
www.routledge.com/cw/bird

Revision Test 2 Further revisionary mathematics

This Revision Test covers the material in Chapter 2. *The marks for each question are shown in brackets at the end of each question.*

1. If 12 inches = 30.48 cm, find the number of millimetres in 23 inches. (3)

2. State the SI unit of: (a) force (b) pressure (c) work (3)

3. State the quantity that has an SI unit of: (a) kilograms (b) rad/s (c) hertz (d) m^3 (4)

4. Express the following in engineering notation in prefix form:

   (a) 250,000 kg (b) 0.00005 *s* (c) 2×10^8 W

   (d) 750×10^{-8} J (4)

5. Rewrite: (a) 0.0067 mA in μA (b) 40×10^4 kN as MN (2)

6. Determine the value of *P* in the following table of values

x	0	1	4
$y = 3x - 5$	-5	-2	P

   (2)

7. Corresponding values obtained experimentally for two quantities are:

x	-5	-3	-1	0	2	4
y	-17	-11	-5	-2	4	10

   Plot a graph of *y* (vertically) against *x* (horizontally) to scales of 1 cm = 1 for the horizontal *x*-axis and 1 cm = 2 for the vertical *y*-axis.

   From the graph find:

   (a) the value of *y* when $x = 3$

   (b) the value of *y* when $x = -4$

   (c) the value of *x* when $y = 1$

   (d) the value of *x* when $y = -20$ (8)

8. If graphs of *y* against *x* were to be plotted for each of the following, state (i) the gradient, and (ii) the *y*-axis intercept.

   (a) $y = -5x + 3$ (b) $y = 7x$ (c) $2y + 4 = 5x$ (d) $5x + 2y = 6$ (8)

9. The resistance *R* ohms of a copper winding is measured at various temperatures *t*°C and the results are as follows:

R (Ω)	38	47	55	62	72
t (°C)	16	34	50	64	84

   Plot a graph of *R* (vertically) against *t* (horizontally) and find from it

   (a) the temperature when the resistance is 50 Ω

   (b) the resistance when the temperature is 72°C

   (c) the gradient

   (d) the equation of the graph. (10)

10. Differentiate the following functions with respect to *x*:

    (a) $y = 5x^2 - 4x + 9$ (b) $y = x^4 - 3x^2 - 2$ (4)

11. If $y = \frac{3}{x}$ determine $\frac{dy}{dx}$ (2)

12. Given $f(t) = \sqrt{t^5}$, find $f'(t)$ (2)

13. Calculate the gradient of the curve $y = 3\cos\frac{x}{3}$ at $x = \frac{\pi}{4}$ correct to 3 decimal places. (4)

14. Find the gradient of the curve $f(x) = 7x^2 - 4x + 2$ at the point (1, 5) (3)

15. If $y = 5 \sin 3x - 2 \cos 4x$ find $\frac{dy}{dx}$ (2)

16. Determine the value of the differential coefficient of $y = 5 \ln 2x - \frac{3}{e^{2x}}$ when $x = 0.8$, correct to 3 significant figures. (4)

In Problems 17 to 19, determine the indefinite integrals.

17. (a) $\int \left(x^2 + 4\right) dx$ (b) $\int \frac{1}{x^3} dx$ (4)

18. (a) $\int \frac{2}{\sqrt[3]{x^2}} dx$ (b) $\int \left(e^{0.5x} + \frac{1}{3x} - 2\right) dx$ (6)

19. (a) $\int (2+\theta)^2\, d\theta$ (b) $\int \left(\cos\frac{1}{2}x + \frac{3}{x} - e^{2x}\right) dx$ (6)

Evaluate the integrals in problems 20 to 22, each, where necessary, correct to 4 significant figures:

20. (a) $\int_1^3 \left(t^2 - 2t\right) dt$ (b) $\int_{-1}^2 \left(2x^3 - 3x^2 + 2\right) dx$ (6)

21. (a) $\int_0^{\pi/3} 3\sin 2t\, dt$ (b) $\int_{\pi/4}^{3\pi/4} \cos\frac{1}{3}x\, dx$ (7)

22. (a) $\int_0^1 \left(\sqrt{x} + 2e^x\right) dx$ (b) $\int_1^2 \left(r^3 - \frac{1}{r}\right) dr$ (6)

23. Evaluate (a) $\begin{vmatrix} 5 & -3 \\ 4 & 2 \end{vmatrix}$ (b) $\begin{vmatrix} 3 & 1 & -4 \\ -5 & 2 & 1 \\ 6 & -3 & -1 \end{vmatrix}$

(7)

24. If $\boldsymbol{a} = 2\boldsymbol{i} + 4\boldsymbol{j} - 5\boldsymbol{k}$ and $\boldsymbol{b} = 3\boldsymbol{i} - 2\boldsymbol{j} + 6\boldsymbol{k}$ determine:
(i) $\boldsymbol{a} \bullet \boldsymbol{b}$ (ii) $\boldsymbol{a} \times \boldsymbol{b}$ (7)

25. Determine the work done by a force of F newtons acting at a point A on a body, when A is displaced to point B, the co-ordinates of A and B being (2, 5, – 3) and (1, – 3, 0) metres respectively, and when $\boldsymbol{F} = 2\boldsymbol{i} - 5\boldsymbol{j} + 4\boldsymbol{k}$ newtons. (5)

26. A force $\boldsymbol{F} = 3\boldsymbol{i} - 4\boldsymbol{j} + \boldsymbol{k}$ newtons act on a line passing through a point P. Determine moment $\boldsymbol{M}$ of the force $\boldsymbol{F}$ about a point Q when P has co-ordinates (4, – 1, 5) metres and Q has co-ordinates (4, 0, – 3) metres. (6)

For lecturers/instructors/teachers, fully worked solutions to each of the problems in Revision Test 2, together with a full marking scheme, are available at the website:
www.routledge.com/cw/bird

Multiple-Choice Questions Test 2

This test covers the material in Chapter 2 on Further Revisionary Mathematics. All questions have only one correct answer (answers on page 496).

1. Differentiating $y = 4x^5$ gives:

   (a) $\frac{dy}{dx} = \frac{2}{3}x^6$ (b) $\frac{dy}{dx} = 20x^4$

   (c) $\frac{dy}{dx} = 4x^6$ (d) $\frac{dy}{dx} = 5x^4$

2. $\int (5 - 3t^2)\, dt$ is equal to:

   (a) $5 - t^3 + c$ (b) $-3t^3 + c$

   (c) $-6t + c$ (d) $5t - t^3 + c$

3. A graph of resistance against voltage for an electrical circuit is shown in Figure M2.1. The equation relating resistance R and voltage V is:

   (a) $R = 1.45\ V + 40$ (b) $R = 0.8\ V + 20$

   (c) $R = 1.45\ V + 20$ (d) $R = 1.25\ V + 20$

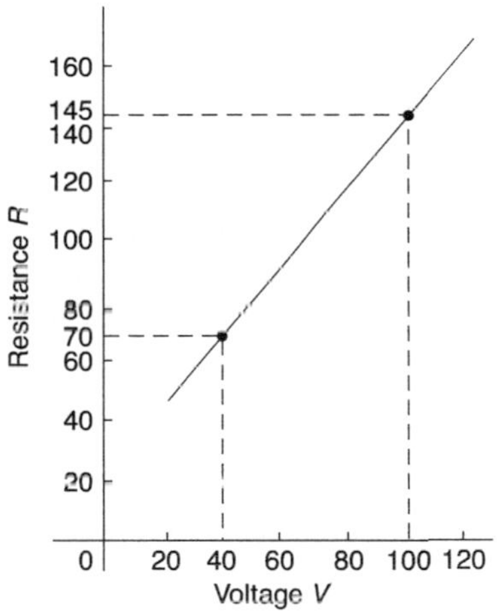

Figure M2.1

4. The gradient of the curve $y = -2x^3 + 3x + 5$ at $x = 2$ is:

   (a) -21 (b) 27 (c) -16 (d) -5

5. The value of $\int_0^1 (3\sin 2\theta - 4\cos\theta)\, d\theta$, correct to 4 significant figures, is:

   (a) -0.06890 (b) -1.242
   (c) -2.742 (d) -1.569

6. A graph of y against x, two engineering quantities, produces a straight line.

   A table of values is shown below:

x	2	-1	p
y	9	3	5

   The value of p is:

   (a) $-\frac{1}{2}$ (b) -2 (c) 3 (d) 0

7. If $y = 5\sqrt{x^3} - 2$, $\frac{dy}{dx}$ is equal to:

   (a) $\frac{15}{2}\sqrt{x}$ (b) $2\sqrt{x^5} - 2x + c$

   (c) $\frac{5}{2}\sqrt{x} - 2$ (d) $5\sqrt{x} - 2x$

8. $\int \frac{2}{9}t^3\, dt$ is equal to:

   (a) $\frac{t^4}{18} + c$ (b) $\frac{2}{3}t^2 + c$

   (c) $\frac{2}{9}t^4 + c$ (d) $\frac{2}{9}t^3 + c$

9. A water tank is in the shape of a rectangular prism having length 1.5 m, breadth 60 cm and height 300 mm. If 1 litre = 1000 cm^3, the capacity of the tank is:

   (a) 27 litre (b) 2.7 litre
   (c) 2700 litre (d) 270 litre

10. Evaluating $\int_0^{\pi/3} 3\sin 3x\, dx$ gives:

    (a) 1.503 (b) -18 (c) 2 (d) 6

11. Here are four equations in x and y. When x is plotted against y, in each case a straight line results.

    (i) $y + 3 = 3x$ (ii) $y + 3x = 3$

    (iii) $\frac{y}{2} - \frac{3}{2} = x$ (iv) $\frac{y}{3} = x + \frac{2}{3}$

Which of these equations are parallel to each other?

(a) (i) and (ii) (b) (i) and (iv)
(c) (ii) and (iii) (d) (ii) and (iv)

12. Differentiating $i = 3\sin 2t - 2\cos 3t$ with respect to t gives:

(a) $3\cos 2t + 2\sin 3t$

(b) $6(\sin 2t - \cos 3t)$

(c) $\frac{3}{2}\cos 2t + \frac{3}{2}\sin 3t$

(d) $6(\cos 2t + \sin 3t)$

13. $\int(5\sin 3t - 3\cos 5t)\,d$ is equal to:

(a) $-5\cos 3t + 3\sin 5t + c$

(b) $15(\cos 3t + \sin 3t) + c$

(c) $-\frac{5}{3}\cos 3t - \frac{3}{5}\sin 5t + c$

(d) $\frac{3}{5}\cos 3t - \frac{5}{3}\sin 5t + c$

14. Which of the straight lines shown in Figure M2.2 has the equation $y + 4 = 2x$?

(a) (iv) (b) (ii) (c) (iii) (d) (i)

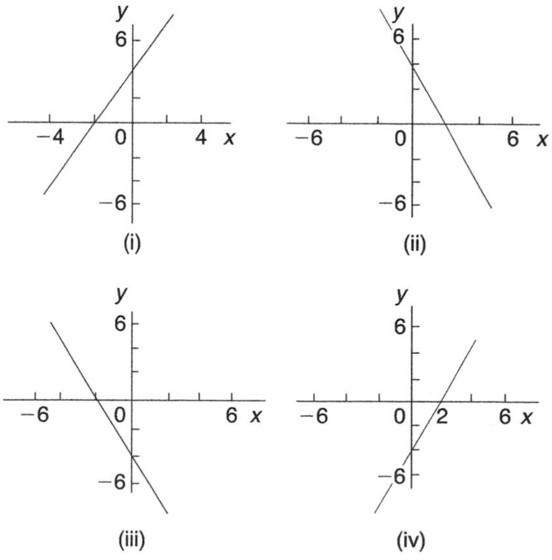

Figure M2.2

15. Given $y = 3e^x + 2\ln 3x$, $\frac{dy}{dx}$ is equal to:

(a) $6e^x + \frac{2}{3x}$ (b) $3e^x + \frac{2}{x}$

(c) $6e^x + \frac{2}{x}$ (d) $3e^x + \frac{2}{3}$

16. Given $f(t) = 3t^4 - 2$, $f'(t)$ is equal to:

(a) $12t^3 - 2$ (b) $\frac{3}{4}t^5 - 2t + c$

(c) $12t^3$ (d) $3t^5 - 2$

17. $\int 1 + \frac{4}{e^{2x}}\,dx$ is equal to:

(a) $\frac{8}{e^{2x}} + c$ (b) $x + \frac{4}{e^{2x}}$

(c) $x - \frac{2}{e^{2x}} + c$ (d) $x - \frac{8}{e^{2x}} + c$

18. In an experiment demonstrating Hooke's law, the strain in a copper wire was measured for various stresses. The results included

Stress (megapascals)	18.24	24.00	39.36
Strain	0.00019	0.00025	0.00041

When stress is plotted vertically against strain horizontally a straight line graph results. Young's modulus of elasticity for copper, which is given by the gradient of the graph, is:

(a) 96000 Pa (b) 1.04×10^{-11} Pa
(c) 96 Pa (d) 96×10^9 Pa

19. The gradient of the curve $y = 4x^2 - 7x + 3$ at the point (1, 0) is

(a) 3 (b) 0 (c) 1 (d) −7

20. Evaluating $\int_1^2 2e^{3t}\,dt$, correct to 4 significant figures, gives:

(a) 2300 (b) 255.6 (c) 766.7 (d) 282.3

Mathematics help – some references

In Chapters 1 and 2, some basic mathematics revision involving algebra, trigonometry, units, graphs, introduction to calculus and vectors is briefly revised. However, in this textbook, a considerable amount of higher mathematical knowledge is needed in order to understand the theory. There is no way around this – any studies in engineering are based on applied mathematics and the topics comprising *Mechanics of Solids* assumes a solid mathematical background. If this is an area where you need help, then help is available, in particular in the following two texts:

***Bird's Higher Engineering Mathematics 9th Edition* 2021 John Bird (Routledge) ≡ HEM**

***Bird's Comprehensive Engineering Mathematics* 2018 John Bird (Routledge) ≡ CEM**

Below is a table outlining the mathematics needed in chapters 3 to 23 of this textbook *Mechanics of Solids*, together with the chapters within the two recommended books that explain the topics.

Chapter	Maths Required	Suggested References	
3. Statics	Algebra/Transposition	HEM Chap 1	CEM Chaps 9/10/12
	Trigonometry	HEM Chap 8	CEM Chap 38
	Quadratic formula	HEM Chap 1	CEM Chap 14
	Differentiation	HEM Chap 25	CEM Chap 53
	Integration	HEM Chap 35	CEM Chap 63
	Sine waves	HEM Chap 11	CEM Chap 39
	Hyperbolic functions	HEM Chaps 12/30	CEM Chaps 24/65
4. Stress and strain	Algebra/Transposition	HEM Chap 1	CEM Chaps 9/10/12
	Integration	HEM Chap 35	CEM Chap 63
	Laws of logarithms	HEM Chaps 3/4	CEM Chaps 15/16
5. Geometrical properties of symmetrical sections	Integration	HEM Chap 35	CEM Chap 63
	Quadratic equations	HEM Chap 1	CEM Chap 14
	Trigonometric identities	HEM Chap 13	CEM Chap 42
	Double angles	HEM Chap 15	CEM Chap 44
	Simpson's rule	HEM Chap 17	CEM Chap 30
6. Bending stresses in beams	Algebra/Transposition	HEM Chap 1	CEM Chaps 9/10/12
	Integration	HEM Chap 35	CEM Chap 63
	Quadratic equations	HEM Chap 1	CEM Chap 14
7. Beam deflections due to bending	Algebra/Transposition	HEM Chap 1	CEM Chaps 9/10/12
	Integration	HEM Chap 35	CEM Chap 63
	Integration by parts	HEM Chap 42	CEM Chap 68
8. Torsion	Algebra/Transposition	HEM Chap 1	CEM Chaps 9/10/12
	Integration	HEM Chap 35	CEM Chap 63
	Newton-Raphson method	HEM Chap 26	CEM Chap 23
9. Complex stress and strain	Trigonometry	HEM Chap 8	CEM Chap 38
	Trigonometric identities	HEM Chap 13	CEM Chap 42
	Double angles	HEM Chap 15	CEM Chap 44
	Partial differentiation	HEM Chaps 32	CEM Chap 60
	Matrices	HEM Chaps 20/21	CEM Chaps 47/48
10. Membrane theory for thin-walled circular cylinders and spheres	Integration	HEM Chap 35	CEM Chap 63
	Double angles	HEM Chap 15	CEM Chap 44

Chapter	Maths Required	Suggested References	
11. Energy methods	Differentiation	HEM Chap 25	CEM Chap 53
	Partial differentiation	HEM Chaps 32	CEM Chap 60
	Integration	HEM Chap 35	CEM Chap 63
	Double angles	HEM Chap 15	CEM Chap 44
	Quadratic equations	HEM Chap 1	CEM Chap 14
12. Theories of elastic failure	Algebra/Transposition	HEM Chap 1	CEM Chaps 9/10/12
	Solving equations	HEM Chap 1	CEM Chap 11
13. Thick cylinders and spheres	Algebra/Transposition	HEM Chap 1	CEM Chaps 9–12
	Differentiation	HEM Chap 25	CEM Chap 53
	Straight line graphs	HEM Chap 16	CEM Chap 31
	Trigonometry	HEM Chap 8	CEM Chaps 37/38
	Integration	HEM Chap 35	CEM Chap 63
	Integration by parts	HEM Chap 42	CEM Chap 68
14. The buckling of struts	Differentiation	HEM Chap 25	CEM Chap 53
	Second order differential equations	HEM Chaps 50/51	CEM Chaps 81/82
	Trigonometry	HEM Chap 8	CEM Chap 38
	Double angles	HEM Chap 15	CEM Chap 44
	Quadratic equations	HEM Chap 1	CEM Chap 14
15. Asymmetrical bending of beams	Integration	HEM Chap 35	CEM Chap 63
	Double integrals	HEM Chap 44	CEM Chap 70
	Trigonometric identities	HEM Chap 13	CEM Chap 42
	Double angles	HEM Chap 15	CEM Chap 44
16. Shear stresses in bending and shear deflections	Integration	HEM Chap 35	CEM Chap 63
	Integration – algebraic substitution	HEM Chap 38	CEM Chap 64
	Trigonometry	HEM Chap 8	CEM Chap 38
	Double angles	HEM Chap 15	CEM Chap 44
	Differentiation	HEM Chap 25	CEM Chap 53
	Newton's method	HEM Chap 26	CEM Chap 23
18. An introduction to matrix algebra	Matrices	HEM Chap 20	CEM Chap 47
19. Composites	Matrices	HEM Chaps 20/21	CEM Chaps 47/48
	Integration	HEM Chap 35	CEM Chap 63
20. The matrix displacement method	Matrices	HEM Chaps 20/21	CEM Chaps 47/48
21. The finite element method	Matrices	HEM Chaps 20/21	CEM Chaps 47/48
	Partial differentiation	HEM Chaps 32	CEM Chap 60
22. An introduction to linear elastic fracture mechanics	Algebra	HEM Chap 1	CEM Chap 9/10/12
	Differentiation	HEM Chap 25	CEM Chap 53
	Integration	HEM Chap 35	CEM Chap 63
	Partial differentiation	HEM Chap 32	CEM Chap 60
	Differential equations	HEM Chaps 46/47	CEM Chap 52/53
23. Material property considerations	Algebra	HEM Chap 1	CEM Chap 9/10/12
	Graphs	HEM Chap 16	CEM Chap 31
	Logarithms/Exponentials	HEM Chaps 3/4	CEM Chap 15/16
	Integration	HEM Chap 35	CEM Chap 63

Notation used in Mechanics of Solids

Unless otherwise stated, the following symbols are used in Chapters 3 to 23 in this text:

E	Young's modulus of elasticity
F	shearing force (SF)
G	shear of rigidity modulus
g	acceleration due to gravity
I	second moment of area
J	polar second moment of area
K	bulk modulus
l	length
M	bending moment (BM)
P	load or pressure
R, r	radius
T	torque or temperature change
t	thickness
W	concentrated load
ω	load/unit length
$\hat{x}$	maximum value of x
α	coefficient of linear expansion
γ	shear strain
ε	direct or normal strain
θ	angle of twist
λ	load factor
ν	Poisson's ratio
ρ	density
σ	direct or normal stress
τ	shear stress
P_e	Euler buckling load
P_E	Rankine buckling load
σ_{yp}	yield stress
W_e	plastic collapse load
τ_{yp}	yield stress in shear
(k)	= elemental stiffness matrix in local co-ordinates
(k°)	= elemental stiffness matrix in global co-ordinates
(p_i)	= a vector of internal nodal forces
(q°)	= a vector of external nodal forces in global co-ordinates
(u_i)	= a vector of nodal displacements in local co-ordinates
(u_i°)	= a vector of nodal displacements in global co-ordinates
(K_{11})	= that part of the system stiffness matrix that corresponds to the 'free' displacements
(Ξ)	= a matrix of directional cosines
(I)	= identity matrix
$\begin{pmatrix} & \\ & \end{pmatrix}$	= a square or rectangular matrix
$\begin{pmatrix} \\ \end{pmatrix}$	= a column vector
( )	= a row vector
(0)	= a null matrix
NA	neutral axis
KE	kinetic energy
PE	potential energy
UDL	uniformly distributed load
WD	work done
⇒	vector defining the direction of rotation, according to the *right-hand screw rule*. The direction of rotation, according to the right-hand screw rule, can be obtained by pointing the right hand in the direction of the open arrow, and rotating it *clockwise*.

Some SI units used in stress analysis

s	second (time)
m	metre
kg	kilogram (mass)
N	newton (force)
Pa	pascal (pressure) = 1 N/m^2
MPa	megapascal (10^6 pascals)
bar	pressure, where 1 bar = 10^5 N/m^2 = 14.5 lbf/in^2
kg/m^3	kilograms/cubic metre (density)
W	watt (power), where 1 watt = 1 ampere × 1 volt = 1 N m/s = 1 joule/s
hp	horsepower (power), where 1 hp = 745.7 W

Author's note on the SI system

Is it not interesting to note that an *apple* weighs approximately 1 *newton?*

Greek Alphabet		
Letter	Upper Case	Lower Case
Alpha	Α	α
Beta	Β	β
Gamma	Γ	γ
Delta	Δ	δ
Epsilon	Ε	ε
Zeta	Ζ	ζ
Eta	Η	η
Theta	Θ	θ
Iota	Ι	ι
Kappa	Κ	κ
Lambda	L	λ
Mu	Μ	μ
Nu	Ν	ν
Xi	Ξ	ξ
Omicron	Ο	ο
Pi	Π	π
Rho	Ρ	ρ
Sigma	Σ	σ
Tau	Τ	τ
Upsilon	Υ	υ
Phi	Φ	φ
Chi	Χ	χ
Psi	Ψ	ψ
Omega	Ω	ω

Chapter 3

Statics

***Why it is important to understand:* Statics**

All the structures that are analysed in the present chapter are assumed to be in *equilibrium*. This is a fundamental assumption that is made in the structural design of most structural components, some of which can then be designed quite satisfactorily from statical considerations alone. Such structures are said to be *statically determinate*.

Most modern structures, however, cannot be solved by considerations of statics alone, as there are more unknown 'forces' than there are simultaneous equations obtained from observations of statical equilibrium. Such structures are said to be *statically indeterminate*, and analysis of this class of structure will not be carried out in this chapter.

Examples of statically determinate and statically indeterminate pin-jointed trusses are given in Figures 3.1 and 3.2 in Section 3.1, where the applied concentrated loads and support reactions (*R* and *H*) are shown by arrows.

At the end of this chapter you should be able to:

- understand pin-jointed trusses and the terms rod, tie, strut and mechanism
- appreciate the criteria for sufficiency of bracing
- use trigonometry in statics calculations
- state the three equilibrium considerations in statics
- calculate the reactions on horizontal beams and forces in the members of pin-jointed trusses
- use the method of joints and the method of sections for analysing statically determinate plane pin-jointed trusses
- define bending moment and shearing force
- understand concentrated, uniformly distributed, hydrostatic and varying distributed loads
- understand simply supported beams and cantilevers
- produce bending moment and shearing force diagrams
- appreciate the term point of contraflexure
- understand the relationship between bending moment, shearing force and intensity of load
- appreciate different forms of cables
- undertake calculations on cables under self-weight and under concentrated loads
- appreciate the use of cables in suspension bridges

Mathematical references

In order to understand the theory involved in this chapter on **statics**, knowledge of the following mathematical topics is required: ***algebraic manipulation/transposition, trigonometry, quadratic formulae, standard differentiation and integration, sine waves and hyperbolic functions.*** If help/revision is needed in these areas, see page 49 for some textbook references.

3.1 Plane pin-jointed trusses

The structures of Figures 3.1 and 3.2 are called pin-jointed trusses because the joints are assumed to be held together by smooth frictionless pins. The reason for this assumption is that the solution of trusses with pin joints is considerably simpler than if the joints were assumed welded (i.e. rigid joints). It should be noted that the external loads in Figures 3.1 and 3.2 are assumed to be applied at the joints, and providing this assumption is closely adhered to, the differences between the internal member forces determined from a pin-jointed truss calculation and those obtained from a rigid-jointed 'truss' calculation will be small as shown in Problem 5 on page 62.

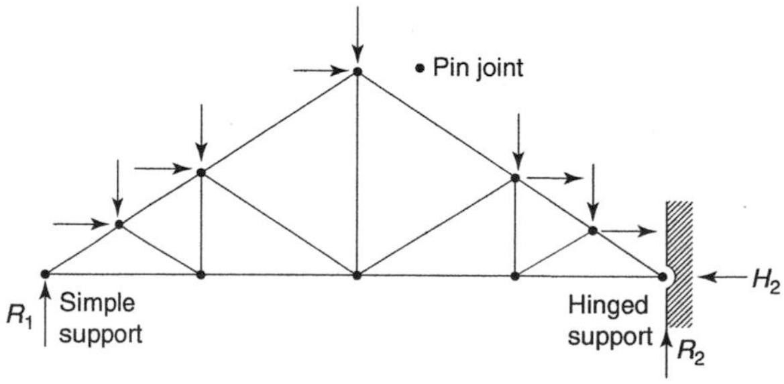

Figure 3.1 Statically determinate pin-jointed truss

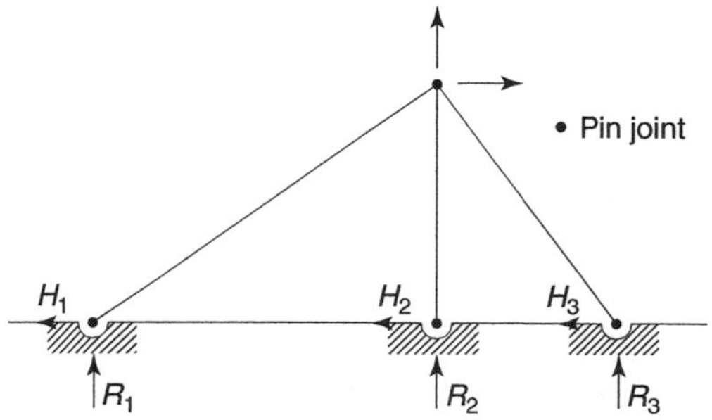

Figure 3.2 Statically indeterminate pin-jointed truss

The members of a pin-jointed truss are called *rods,* and these elements are assumed to withstand loads axially, so that they are in tension or in compression or in a state of zero load. When a rod is subjected to tension it is called a *tie* and when it is in compression, it is called a *strut,* as shown in Figure 3.3. The internal resisting forces inside ties and struts will act in an opposite direction to the externally applied forces, as shown in Figure 3.4. Ties, which are in tension, are said to have internal resisting forces of *positive magnitude,* and struts, which are in compression, are said to have internal resisting forces of *negative magnitude.*

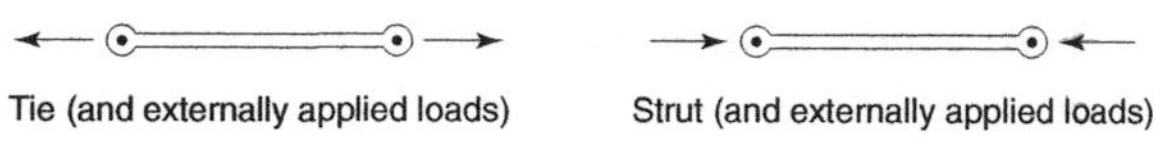

Figure 3.3 External loads acting on ties and struts

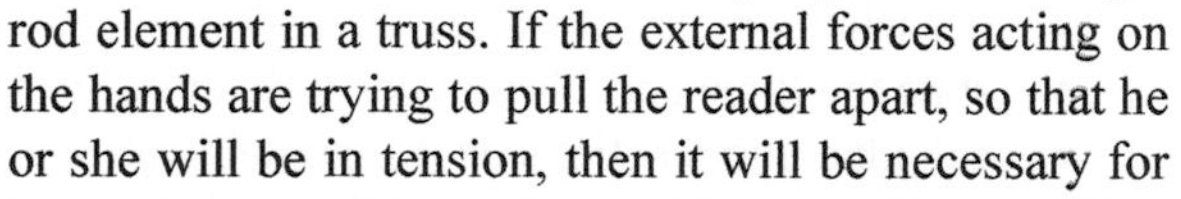

Figure 3.4 Sign conventions for ties and struts

In order to help the reader understand the direction of the internal forces acting in a rod, the reader can imagine his or her hands and arms acting together as a single rod element in a truss. If the external forces acting on the hands are trying to pull the reader apart, so that he or she will be in tension, then it will be necessary for the reader to pull inwards to achieve equilibrium. That is, if the reader is in tension, then he or she will have to pull inwards as shown by the diagram for the tie in Figure 3.4. Alternatively, if the reader is in compression, then he or she will have to push outwards to achieve equilibrium, as shown by the strut in Figure 3.4.
The structure of Figure 3.2 is said to be statically indeterminate to the first degree, or that it has one *redundant member;* this means that for the structure to be classified as a structure, one rod may be removed. If two rods were removed from Figure 3.2 or one from

Figure 3.1, the structure (or part of it) would become a *mechanism,* and collapse. It should be noted, however, that if a rod were removed from Figure 3.2, although the structure can still be classified as a structure and not a mechanism, it may become too weak and collapse in any case.

3.2 Criterion for sufficiency of bracing

A simple formula, which will be given without proof, to test whether or not a pin-jointed truss is statically determinate, is as follows:

$$2j = r + R \tag{3.1}$$

where j = number of pin joints
r = number of rods
R = minimum number of reacting forces

To test equation (3.1), consider the pin-jointed truss of Figure 3.1:

$j = 10 \quad r = 17 \quad R = 3$

[$R = 3$ is obtained from counting the three reactions – namely R_1, R_2 and H_2]

Therefore $2 \times 10 = 17 + 3$
i.e. $20 = 20$
i.e. **statically determinate.**

Now consider the truss of Figure 3.2:

$j = 4 \quad r = 3 \quad R = 6$

[$R = 6$ is obtained by counting the number of reactions – namely R_1, H_1, R_2, H_2, R_3 and H_3]

$$\text{or } 2 \times 4 = 3 + 6 \tag{3.2}$$

i.e. $8 \neq 9$
(i.e. there is **one redundancy**).

NB If $2j > r + R$, then the structure (or part of it) is a mechanism and will collapse under load. Similarly, from equation (3.2), it can be seen that if $2j < r + R$, the structure may be statically indeterminate.

Prior to analysing statically determinate structures, it will be necessary to cover some elementary mathematics, and show how this can be applied to some simple equilibrium problems.

3.3 Mathematics used in statics

Consider the right-angled triangle of Figure 3.5.
From elementary trigonometry:

$$\sin\theta = \frac{bc}{ac} \quad \cos\theta = \frac{ab}{ac} \quad \tan\theta = \frac{bc}{ab} \tag{3.3}$$

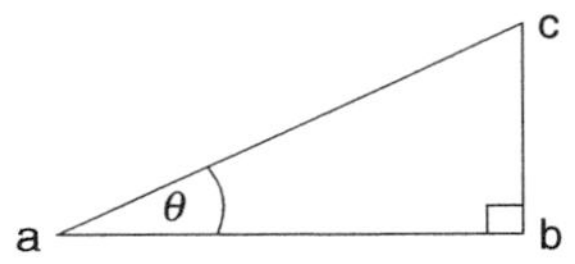

Figure 3.5 Right-angled triangle

It will now be shown how these simple trigonometric formulae can be used for elementary statics.

A scalar is a quantity which only has magnitude. Typical scalar quantities are length, time, mass and temperature; that is, they have magnitude, but there is no direction of these quantities.

A vector is a quantity which has both magnitude and direction. Typical vector quantities are force, weight, velocity, acceleration, etc. A vector quantity, say $\boldsymbol{F}$, can be displayed by a scaled drawing, as shown by Figure 3.6.

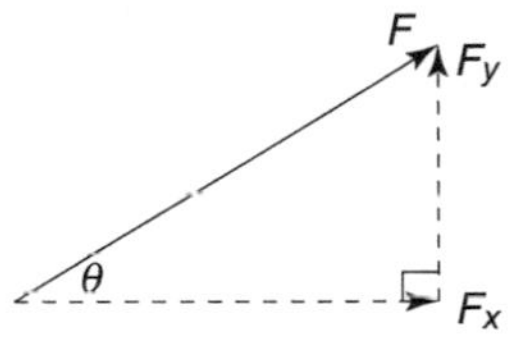

Figure 3.6 A vector quantity

The same vector quantity $\boldsymbol{F}$ can be represented by the two vectors $\boldsymbol{F}_x$ and $\boldsymbol{F}_y$ shown in Figure 3.6, where $\boldsymbol{F}_x$ is the horizontal component of $\boldsymbol{F}$ and $\boldsymbol{F}_y$ is the vertical component of $\boldsymbol{F}$. Both $\boldsymbol{F}_x$ and $\boldsymbol{F}_y$ can be related to $\boldsymbol{F}$ and the angle that $\boldsymbol{F}$ acts with respect to the horizontal, namely θ, as follows.

From equation (3.3)

$$\sin\theta = \frac{F_y}{F} \quad \text{or} \quad F_y = F\sin\theta \tag{3.4}$$

From equation (3.3)

$$\cos\theta = \frac{F_x}{F} \quad \text{or} \quad F_x = F\cos\theta \tag{3.5}$$

That is, the horizontal component of $\boldsymbol{F}$ is $\boldsymbol{F}\cos\theta$ and the vertical component of $\boldsymbol{F}$ is $\boldsymbol{F}\sin\theta$.

It will now be shown how expressions such as equations (3.4) and (3.5) can be used for elementary statics.

3.4 Equilibrium considerations

In *two dimensions,* the following three equilibrium considerations apply:

(a) Vertical equilibrium must be satisfied, i.e.

Upward forces = downward forces

(b) Horizontal equilibrium must be satisfied, i.e.

Forces to the left = forces to the right

(c) Rotational equilibrium must be satisfied at all points, i.e.

Clockwise couples = counter-clockwise (or anti-clockwise) couples

Use of the above vertical and horizontal equilibrium considerations is self-explanatory, but use of the rotational equilibrium consideration is not quite as self-evident; this deficiency will now be clarified through the use of a worked example.

Problem 1. Determine the reactions R_A and R_B acting on the ends of the horizontal beams of Figures 3.7 and 3.8. Neglect the self-weight of the beam.

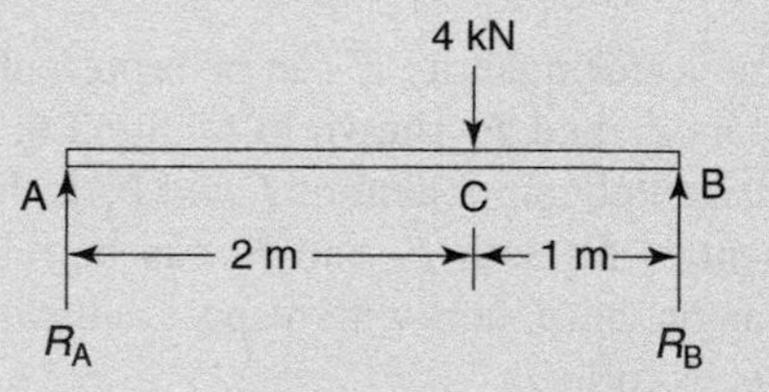

Figure 3.7 Beam with a point load at C

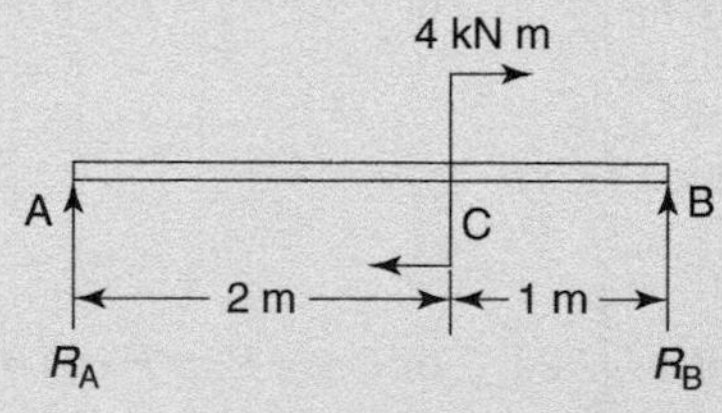

Figure 3.8 Beam with a clockwise couple at C

(a) *Beam with a point load at C*

Take moments about the point B in Figure 3.7. In practice, moments could be taken anywhere, but by taking moments about B, the reaction R_B would not exert a couple about B, and this would leave only one unknown, namely R_A. It would have been just as convenient to have taken moments about A.

By taking moments about B, it is meant that

clockwise couples about B = counter-clockwise (or anti-clockwise) couples about B

That is, for rotational equilibrium about B,

$$R_A \times (2 + 1) = 4 \times 1$$

or $$\boldsymbol{R_A} = \frac{4}{3} = \mathbf{1.333\ kN}$$

Resolving forces vertically (i.e. seeking vertical equilibrium),

Upward forces = downward forces

i.e. $$R_A + R_B = 4\ \text{kN}$$

from above $$R_A = 1.333\ \text{kN}$$

therefore $$\boldsymbol{R_B} = 4 - 1.333 = \mathbf{2.667\ kN}$$

(b) *Beam with a clockwise couple about C*

Taking moments about B (i.e. seeking rotational equilibrium about B)

clockwise couples about B = counter-clockwise couples about B

$$R_A \times (2 + 1) + 4 = 0$$

or $$\boldsymbol{R_A} = -\frac{4}{3} = \mathbf{-1.333\ kN}$$

That is, R_A is acting downwards.

Resolving forces vertically (i.e. seeking vertical equilibrium),

upward forces = downward forces

$$R_A + R_B = 0$$

Hence $$\boldsymbol{R_B} = -R_A = -(-1.333) = \mathbf{1.333\ kN}$$

To demonstrate the static analysis of pin-jointed trusses, the following two examples will be considered.

Problem 2. Determine the internal forces in the members of the pin-jointed truss of Figure 3.9.

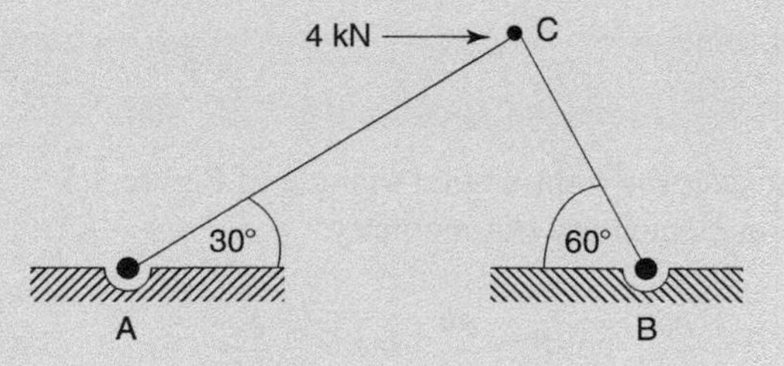

Figure 3.9 Simple pin-jointed truss

Assume all unknown forces are in tension, as shown by the arrows in Figure 3.10. If this assumption is incorrect, the sign for the force in the rod will be negative, indicating that the member is in compression.

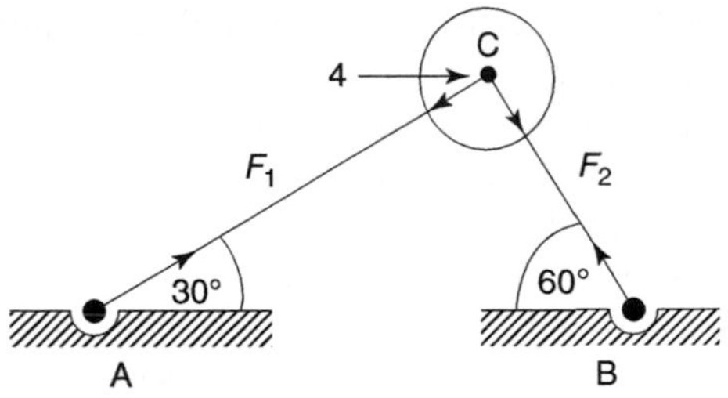

Figure 3.10 Assumed internal forces in simple truss

By drawing an imaginary circle around joint C, equilibrium can be sought around the joint with the aid of the **free-body diagram** of Figure 3.11.

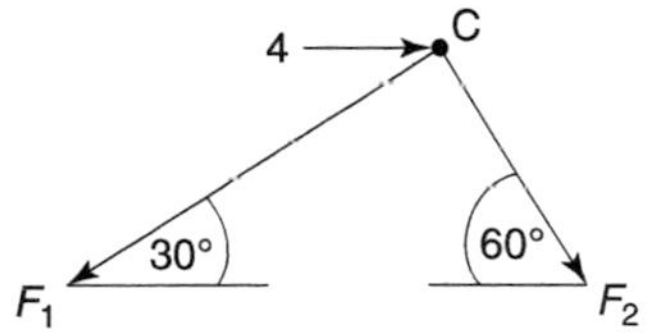

Figure 3.11 Equilibrium around joint C (free-body diagram)

Resolving forces vertically

upward forces = downward forces

$$0 = F_1 \sin 30° + F_2 \sin 60°$$

Therefore $\quad F_1 \sin 30° = -F_2 \sin 60°$

and $\quad F_1 = \dfrac{-F_2 \sin 60°}{\sin 30°}$

i.e. $\quad F_1 = -1.732\, F_2 \qquad (3.6)$

Resolving forces horizontally

forces to the left = forces to the right

$$F_1 \cos 30° = 4 + F_2 \cos 60° \qquad (3.7)$$

Substituting equation (3.6) into equation (3.7)

$$-1.732\, F_2 \cos 30° = 4 + F_2 \cos 60°$$

$$-1.5\, F_2 = 4 + 0.5\, F_2$$

$$-1.5\, F_2 - 0.5\, F_2 = 4$$

$$-2\, F_2 = 4$$

$$\mathbf{F_2 = -2\ kN}$$

(i.e. this member is in compression)

Hence, from equation (3.6) $\quad F_1 = -1.732 \times -2$ kN

i.e. $\quad \mathbf{F_1 = 3.464\ kN}$

The forces in the members of this truss are shown in Figure 3.12.

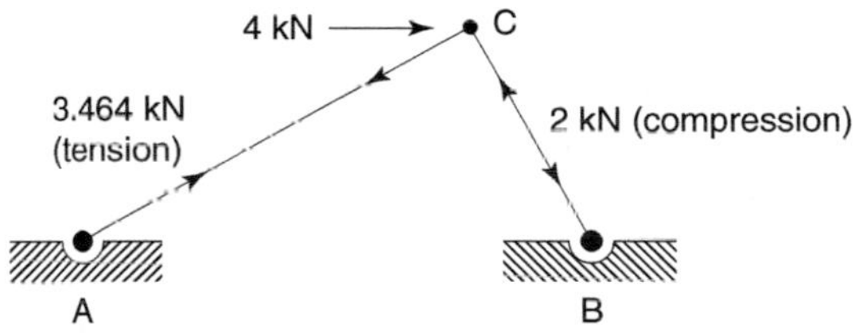

Figure 3.12 Internal forces in truss

Problem 3. Determine the internal forces in the members of the statically determinate pin-jointed truss shown in Figure 3.13.

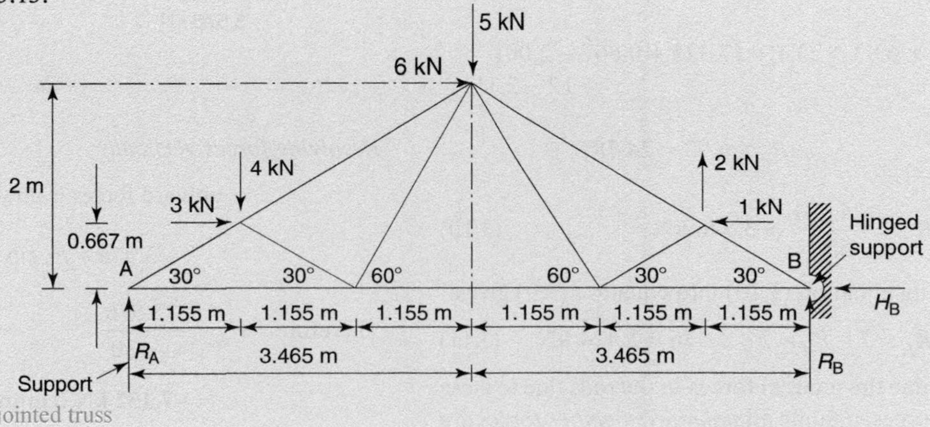

Figure 3.13 Pin-jointed truss

From Figure 3.13 it can be seen that to achieve equilibrium, it will be necessary for the three reactions R_A, R_B and H_B to act. These reactions balance vertical and horizontal forces and rotational couples.

To determine the unknown reactions, it will be necessary to obtain three simultaneous equations from the three equilibrium considerations described in Section 3.5.

Consider *horizontal equilibrium*

forces to the left = forces to the right

$$H_B = 3 + 6 - 1$$

i.e. $$\mathbf{H_B = 8\ kN} \tag{3.8}$$

Consider *vertical equilibrium*

upward forces = downward forces

$$R_A + R_B + 2 = 4 + 5$$

i.e. $$R_B = 7 - R_A \tag{3.9}$$

Consider *rotational equilibrium.* Now rotational equilibrium can be considered at any point in the plane of the truss, but to simplify arithmetic, it is better to take moments about a point through which an unknown force acts, so that this unknown force has no moment about this point. In this case, it will be convenient to *take moments* about A or B.

Taking moments about B means that

clockwise couples about B = counter-clockwise couples about B

$$R_A \times (3.465 + 3.465) + 3 \times 0.667 + 6 \times 2 + 2 \times 1.155 = 4 \times (3.465 + 1.155 + 1.155) + 5 \times 3.465 + 1 \times 0.667$$

i.e. $$R_A \times 6.93 + 2.001 + 12 + 2.31 = 23.1 + 17.325 + 0.667$$

i.e. $$R_A \times 6.93 = 23.1 + 17.325 + 0.667 - 2.001 - 12 - 2.31$$

i.e. $$R_A \times 6.93 = 24.781$$

i.e. $$\mathbf{R_A} = \frac{24.781}{6.93} = \mathbf{3.576\ kN} \tag{3.10}$$

Substituting equation (3.10) into equation (3.9) gives:

$$\mathbf{R_B} = 7 - R_A = 7 - 3.576 = \mathbf{3.424\ kN} \tag{3.11}$$

To determine the internal forces in the rods due to these external forces, assume all *unknown member forces are in tension.* This assumption will be found to be convenient, because if a member is in compression, the sign of its member force will be negative, which is the correct sign for a compressive force.

The method adopted in this section for determining member forces is called the **method of joints.** It consists of isolating each joint in turn by making an imaginary cut through the members around that joint and then considering vertical and horizontal equilibrium between the internal member forces and the external forces at that joint (i.e. a free-body diagram of the joint). As only vertical and horizontal equilibrium is considered at each joint, it is necessary to start the analysis at a joint where there are only *two unknown member forces.* In this case, consideration must first be made at either joint 1 or joint 7 (see Figure 3.14).

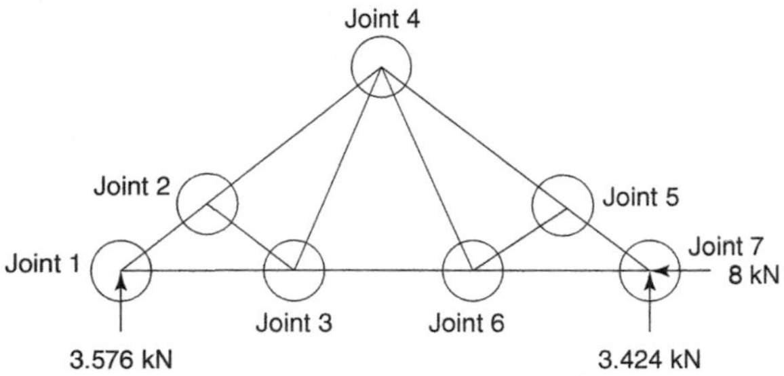

Figure 3.14 Joint numbers for pin-jointed truss

Joint 1 (see the free-body diagram of Figure 3.15)

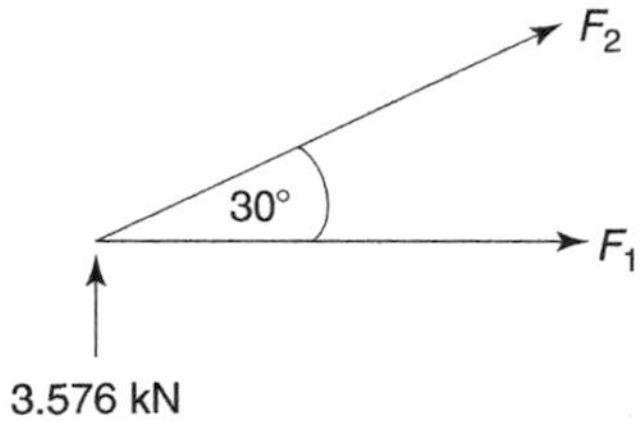

Figure 3.15 Joint 1 (free-body diagram)

Resolving forces vertically

upward forces = downward forces

$$3.576 + F_2 \sin 30° = 0$$

Therefore $$F_2 = -\frac{3.576}{\sin 30°}$$

$$= \mathbf{-7.152\ kN\ (compression)} \tag{3.12}$$

Resolving forces horizontally

forces to the left = forces to the right

$$0 = F_1 \cos 30° + F_1 \text{ or } F_1 = -F_2 \cos 30°$$

Therefore $F_1 = -(-7.152) \cos 30°$
$= \mathbf{6.194\ kN\ (tension)}$ (3.13)

Joint 2 (see the free-body diagram of Figure 3.16)

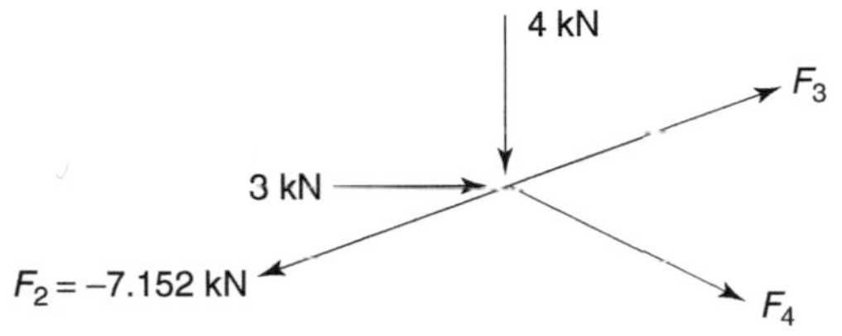

Figure 3.16 Joint 2 (free-body diagram)

Resolving horizontally

$$F_3 \cos 30° + F_4 \cos 30° + 3 = F_2 \cos 30°$$

Therefore $F_3(\cos 30°) + F_4(\cos 30°) + 3$
$= (-7.152)(\cos 30°)$ from (3.12)

$$F_3(\cos 30°) + F_4(\cos 30°) + 3 = -6.194$$

$$F_3(\cos 30°) = -6.194 - F_4(\cos 30°) - 3$$

$$F_3 = \frac{-9.194 - F_4(\cos 30°)}{\cos 30°} = \frac{-9.194}{\cos 30°} - \frac{F_4(\cos 30°)}{\cos 30°}$$

i.e. $F_3 = -10.616 - F_4$ (3.14)

Resolving vertically

$$F_3 \sin 30° = F_4 \sin 30° + F_2 \sin 30° + 4$$

Therefore $F_3(0.5) = F_4(0.5) + (-7.152)(0.5) + 4$ from (3.12)

i.e. $F_3(0.5) = F_4(0.5) + 0.424$

i.e. $F_3 = \dfrac{F_4(0.5) + 0.424}{0.5} = F_4 + 0.848$ (3.15)

Substituting equation (3.14) into (3.15) gives:

$$-10.616 - F_4 = F_4 + 0.848$$

i.e. $-10.616 - 0.848 = 2F_4$

i.e. $F_4 = \dfrac{-10.616 - 0.848}{2}$

$= \mathbf{-5.732\ kN\ (compression)}$ (3.16)

From equation (3.14)

$$F_3 = -10.616 - F_4 = -10.616 - (-5.732)$$

$\mathbf{F_3 = -4.884\ kN\ (compression)}$ (3.17)

Joint 3 (see the free-body diagram of Figure 3.17)

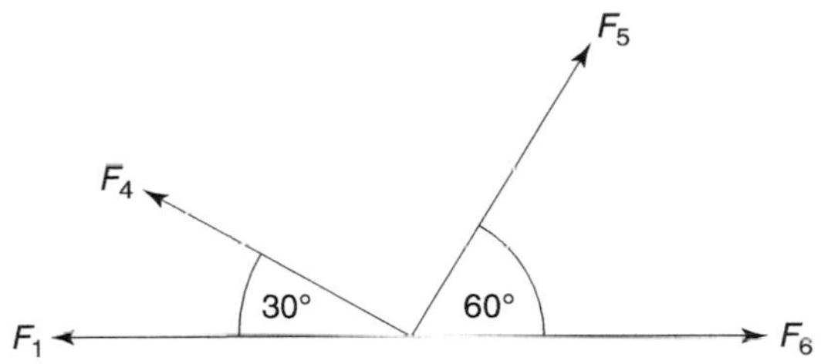

Figure 3.17 Joint 3 (free-body diagram)

Resolving vertically

$$F_5 \sin 60° + F_4 \sin 30° = 0$$

i.e. $F_5 \sin 60° + (-5.732)\sin 30° = 0$

i.e. $F_5 \sin 60° - 2.866 = 0$

i.e. $F_5 = \dfrac{2.866}{\sin 60°} = \mathbf{3.309\ kN\ (tensile)}$ (3.18)

Resolving horizontally

$$F_1 + F_4 \cos 30° = F_6 + F_5 \cos 60°$$

i.e. $6.194 + (-5.732)\cos 30° = F_6 + (3.309)\cos 60°$ from (3.13), (3.16) and (3.18)

i.e. $F_6 = 6.194 + (-5.732)\cos 30° - (3.309)\cos 60°$

i.e. $\mathbf{F_6 = -0.425\ kN\ (compression)}$ (3.19)

Joint 4 (see the free-body diagram of Figure 3.18)

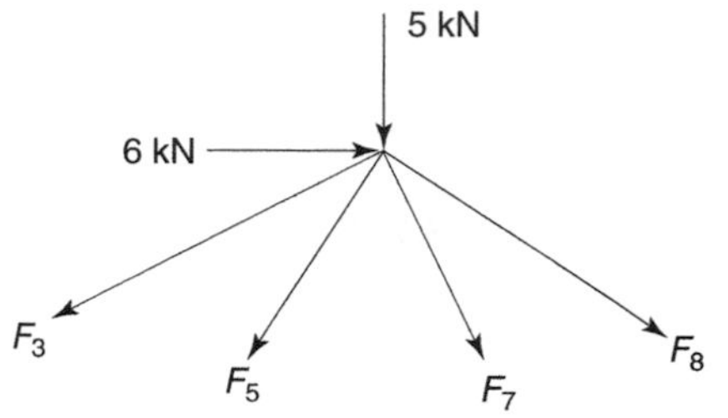

Figure 3.18 Joint 4 (free-body diagram)

Resolving vertically

$$0 = 5 + F_3 \sin 30° + F_5 \sin 60° + F_7 \sin 60° + F_8 \sin 30°$$

i.e. $$0 = 5 + (-4.884) \sin 30° + (3.309) \sin 60° + F_7 \sin 60° + F_8 \sin 30°$$

i.e. $$0 = 5 + -2.442 + 2.866 + F_7 (0.866) + F_8 (0.5)$$ from (3.13), (3.17) and (3.18)

i.e. $$0.866\, F_7 + 0.5\, F_8 = -5.424 \qquad (3.20)$$

Resolving horizontally

$$6 + F_8 \cos 30° + F_7 \cos 60° = F_3 \cos 30° + F_5 \cos 60°$$

i.e. $$6 + F_8 (0.866) + F_7 (0.5) = (-4.884)(0.866) + (3.309)(0.5)$$ from (3.17) and (3.18)

i.e. $$F_8 (0.866) + F_7 (0.5) = (-4.884)(0.866) + (3.309)(0.5) - 6$$

i.e. $$F_8 (0.866) + F_7 (0.5) = -8.575$$

and $$F_8 = \frac{-0.5F_7 - 8.575}{0.866} = -0.577\, F_7 - 9.902 \qquad (3.21)$$

Substituting equation (3.21) into (3.20) gives:

$$0.866\, F_7 + 0.5(-0.577\, F_7 - 9.902) = -5.424$$

i.e. $$0.866\, F_7 - 0.2885\, F_7 - 4.951 = -5.424$$

i.e. $$0.5775\, F_7 = 4.951 - 5.424 = -0.473$$

and $$\boldsymbol{F_7} = \frac{-0.473}{0.5775} = \mathbf{-0.82\ kN\ (compression)} \qquad (3.22)$$

From (3.21) $$F_8 = -0.577\, F_7 - 9.902 = (-0.577)(-0.82) - 9.902$$

i.e. $$\boldsymbol{F_8} = \mathbf{-9.43\ kN\ (compression)} \qquad (3.23)$$

Joint 5 (see the free-body diagram of Figure 3.19)

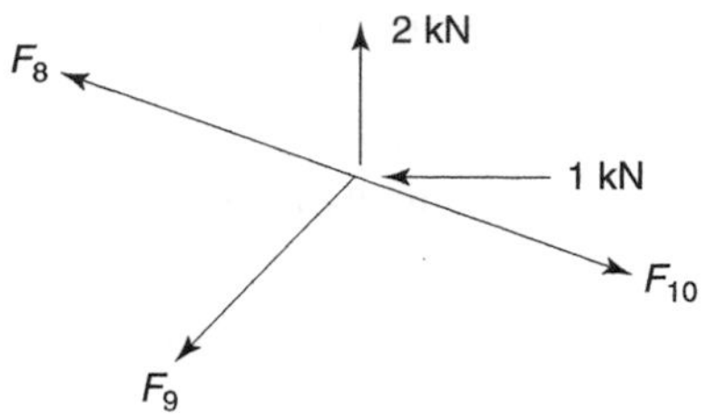

Figure 3.19 Joint 5 (free-body diagram)

Resolving vertically

$$2 + F_8 \sin 30° = F_{10} \sin 30° + F_9 \sin 30°$$

i.e. $$2 + (-9.43)(0.5) = 0.5\, F_{10} + 0.5\, F_9$$ from (3.23)

i.e. $$-2.715 = 0.5\, F_{10} + 0.5\, F_9$$

or $$F_{10} + F_9 = \frac{-2.715}{0.5} = -5.43 \text{ kN} \qquad (3.24)$$

Resolving horizontally

$$F_8 \cos 30° + F_9 \cos 30° + 1 = F_{10} \cos 30°$$

i.e. $$(-9.43)(0.866) + 0.866\, F_9 + 1 = 0.866\, F_{10}$$

i.e. $$-8.1664 + 0.866\, F_9 + 1 = 0.866\, F_{10}$$

i.e. $$-7.1664 = 0.866\, F_{10} - 0.866\, F_9$$

or $$F_{10} - F_9 = \frac{-7.1664}{0.866} = -8.28 \qquad (3.25)$$

Adding (3.24) and (3.25) gives:

$$2\, F_{10} = -13.71$$

Therefore $$\boldsymbol{F_{10}} = \frac{-13.71}{2} = \mathbf{-6.86\ kN\ (compression)} \qquad (3.26)$$

Substituting (3.26) into (3.25) gives:

$$-6.86 - F_9 = -8.28$$

i.e. $$\boldsymbol{F_9} = -6.86 + 8.28 = \mathbf{1.42\ kN\ (tensile)} \qquad (3.27)$$

Joint 7 (see the free-body diagram of Figure 3.20)

Only one unknown force is required, namely F_{11} therefore it will be easier to consider joint 7, rather than joint 6.

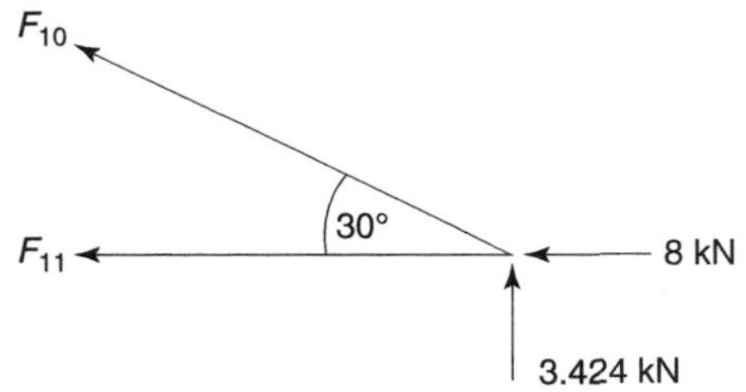

Figure 3.20 Joint 7 (free-body diagram)

Resolving horizontally

$$F_{11} + 8 + F_{10} \cos 30° = 0$$

Therefore $F_{11} + 8 + (-6.86) \cos 30° = 0$ from (3.26)

i.e. $\boldsymbol{F_{11}} = -8 + (6.86) \cos 30°$

$= \mathbf{-2.06\ kN\ (compression)}$ (3.28)

Another method for analysing statically determinate plane pin-jointed trusses is called the **method of sections.** In this method, an imaginary cut is made through the truss, so that there are no more than three unknown internal forces to be determined, across that section. By considering equilibrium, these unknown internal forces can be calculated. Prior to applying the method, however, it is usually necessary first to determine the support reactions acting on the framework. To demonstrate the method, the following example will be considered.

Problem 4. Using the method of sections, determine the forces in the three members, namely a, b and c, of Figure 3.21.

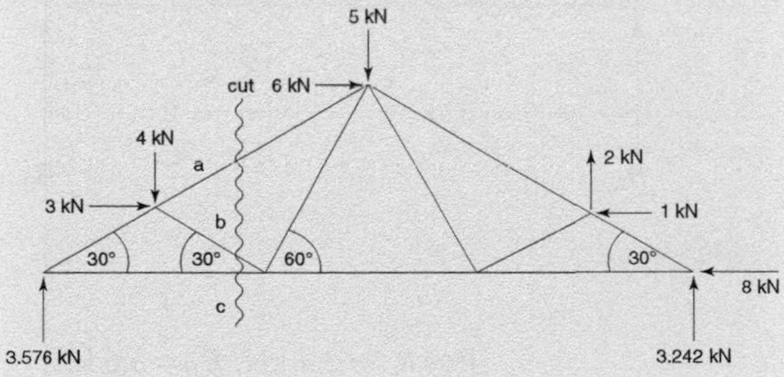

Figure 3.21 Pin-jointed truss

It can be seen that this problem is the same as Problem 3.
Assume all unknown internal forces are in tension and make an imaginary cut through the frame, as shown in Figure 3.21. Consider equilibrium of the section to the left of this cut, as shown by the free-body diagram of Figure 3.22.

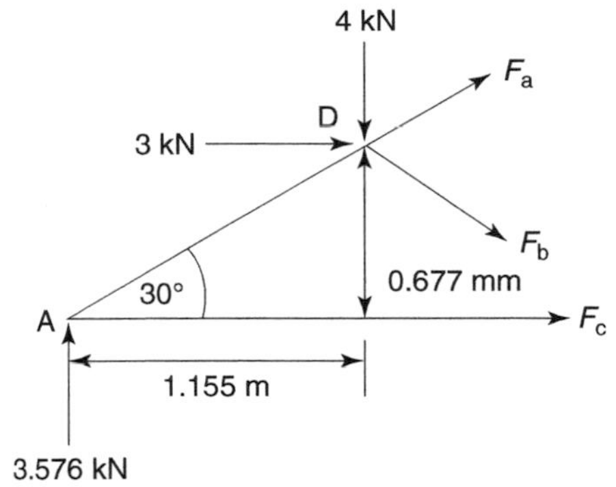

Figure 3.22 Section of truss (free-body diagram)

Taking moments about D

clockwise couples about D = counter-clockwise couples about D

$$3.576 \times 1.155 = F_c \times 0.667$$

from which $\boldsymbol{F_c} = \dfrac{3.576 \times 1.155}{0.667} = \mathbf{6.192\ kN}$

Resolving forces vertically

upward forces = downward forces

$$3.576 + F_a \sin 30° = 4 + F_b \sin 30°$$

$$(F_a - F_b) \sin 30° = 0.424$$

$$(F_a - F_b) = \frac{0.424}{\sin 30°} = 0.848$$

i.e. $F_a = 0.848 + F_b$ (3.29)

Resolving forces horizontally

forces to the left = forces to the right

$$3 + F_c + F_a \cos 30° + F_b \cos 30° = 0$$

i.e. $3 + 6.192 + F_a \cos 30° + F_b \cos 30° = 0$

i.e. $(F_a + F_b) \cos 30° = -9.192$ (3.30)

Substituting equation (3.29) into (3.30) gives:

$$(0.848 + F_b + F_b)(0.866) = -9.192$$

i.e. $0.734 + 1.732\ F_b = -9.192$

i.e. $1.732\,F_b = -9.192 - 0.734 = -9.926$

and $F_b = \dfrac{-9.926}{1.732} = \mathbf{-5.731\ kN}$ (3.31)

Substituting equation (3.31) into (3.29) gives:

$$F_a = 0.848 + F_b = 0.848 - 5.731$$

i.e. $\mathbf{F_a = -4.884\ kN}$

Thus, $F_a = -4.884$ kN, $F_b = -5.731$ kN and $F_c = 6.192$ kN similar to Problem 3.

Problem 5. Using the computer programs TRUSS and PLANEFRAME of reference 1, and Chapter 18, determine the member forces in the plane pin-jointed truss of Figure 3.13. The program TRUSS assumes that the frame has pin joints, and the program PLANEFRAME assumes that the frame has rigid (or welded) joints; the results are given in Table 3.1, where for PLANEFRAME the cross-sectional areas of the members were assumed to be 1000 times greater than their second moments of area.

As can be seen from Table 3.1, the assumption that the joints are pinned gives similar results to the much more difficult problem of assuming that the joints are rigid (or welded), providing, of course, that the externally applied loads are at the joints. It the applied loads are applied between the joints, then bending can occur, which can be catered for by PLANEFRAME, but not by TRUSS; thus TRUSS and PLANEFRAME will only agree if the applied loads are placed at the joints.

Table 3.1 Member forces in framework

Force	Pin-jointed truss (kN)	Rigid-jointed truss (kN)
F_1	6.194	6.043
F_2	−7.152	−7.021
F_3	−4.884	−4.843
F_4	−5.730	−5.601
F_5	3.309	3.258
F_6	−0.424	−0.417
F_7	−0.823	−0.868
F_8	−9.426	−9.384
F_9	1.426	1.421
F_{10}	−6.847	−6.829
F_{11}	−2.071	−2.086

Now try the following Practice Exercise

Practice Exercise 20. Pin-jointed trusses

For all problems, neglect self-weight.

1. Determine the reactions R_A and R_B for the simply supported beams of Figures 3.23(a) and (b)

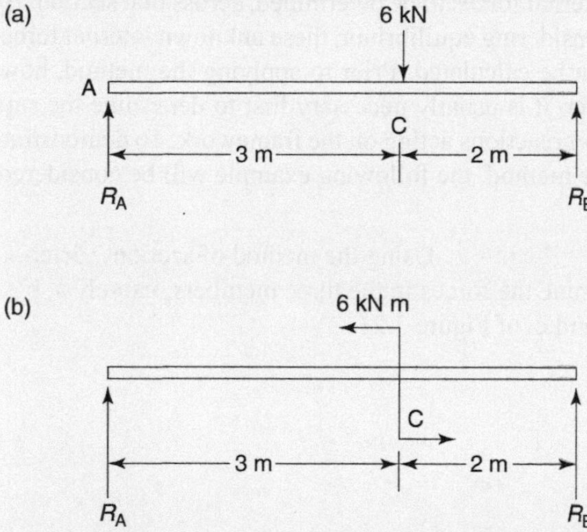

Figure 3.23

[(a) $R_A = 2.4$ kN, $R_B = 3.6$ kN; (b) $R_A = 1.2$ kN, $R_B = -1.2$ kN]

2. Determine the reactions R_A and R_B for the simply supported beams of Figures 3.24(a) and (b)

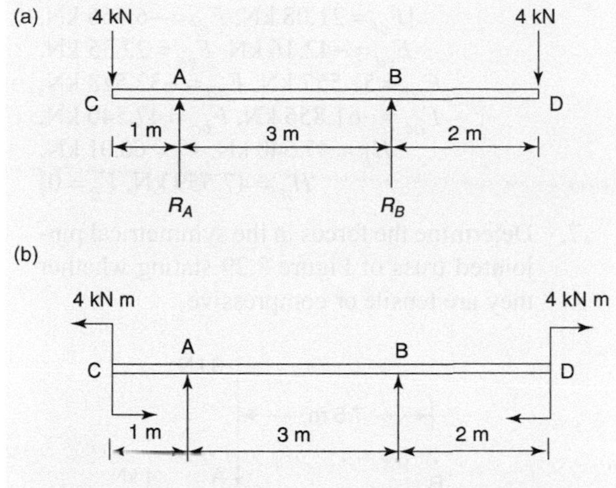

Figure 3.24

[(a) $R_A = 2.667$ kN, $R_B = 5.333$ kN; (b) $R_A = R_B = 0$]

3. Determine the internal forces in the members of the plane pin-jointed truss of Figures 3.25(a) and (b).

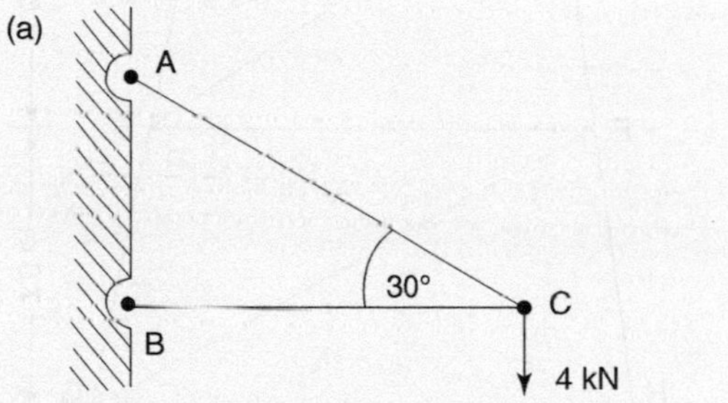

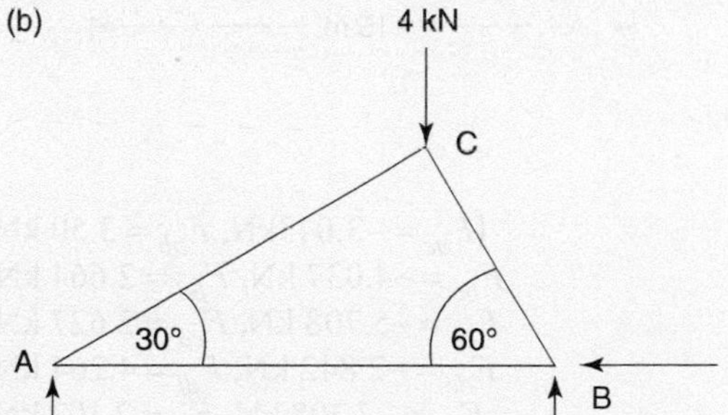

Figure 3.25

[(a) $F_{ac} = 8$ kN, $F_{bc} = -6.928$ kN; (b) $F_{bc} = -3.464$ kN, $F_{ac} = -2$ kN, $F_{ab} = 1.732$ kN]

4. A plane pin-jointed truss is firmly pinned at its base, as shown in Figure 3.26. Determine the forces in the members of this truss, stating whether they are in tension or compression.

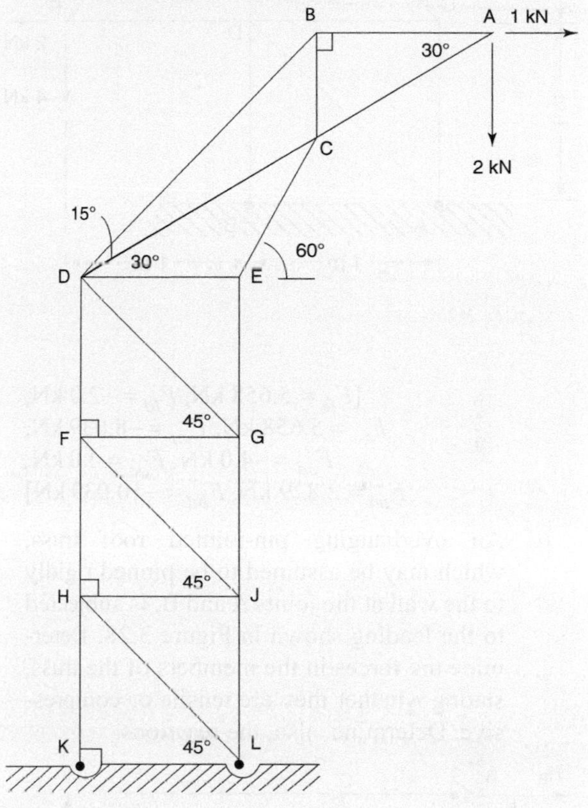

Figure 3.26

[$F_{ac} = -4$ kN, $F_{ab} = 4.464$ kN, $F_{bd} = 6.313$ kN, $F_{bc} = -4.464$ kN, $F_{ce} = -7.733$ kN, $F_{cd} = 0.465$ kN, $F_{de} = -3.867$ kN, $F_{eg} = -6.70$ kN, $F_{dg} = -1.413$ kN, $F_{df} = 5.695$ kN, $F_{gf} = 1.0$ kN, $F_{gj} = -7.699$ kN, $F_{fj} = -1.414$ kN, $F_{fh} = 6.695$ kN, $F_{jh} = 1.0$ kN, $F_{jl} = -8.699$ kN, $F_{hl} = -1.414$ kN, $F_{hk} = 7.695$ kN]

5. The plane pin-jointed truss of Figure 3.27 is firmly pinned at A and B and subjected to two point loads at point F. Determine the forces in the members, stating whether they are tensile or compressive.

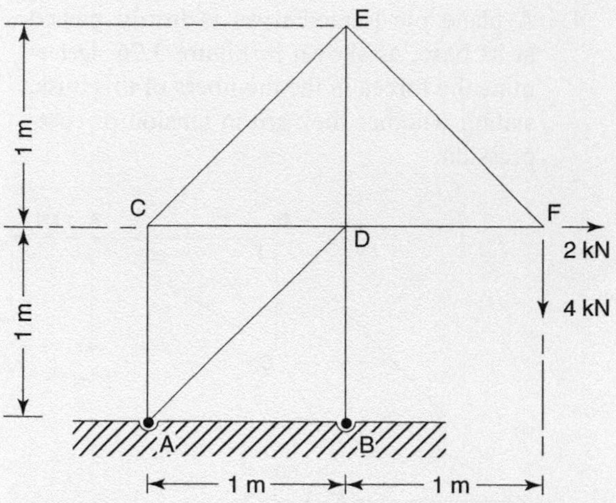

Figure 3.27

[F_{fe} = 5.658 kN, F_{fd} = −2.0 kN, F_{ec} = 5.658 kN, F_{ed} = −8.039 kN, F_{cd} = −4.0 kN, F_{ac} = 4.0 kN, F_{ad} = 2.829 kN, F_{bd} = −10.039 kN]

6. An overhanging pin-jointed roof truss, which may be assumed to be pinned rigidly to the wall at the joints A and B, is subjected to the loading shown in Figure 3.28. Determine the forces in the members of the truss, stating whether they are tensile or compressive. Determine, also, the reactions.

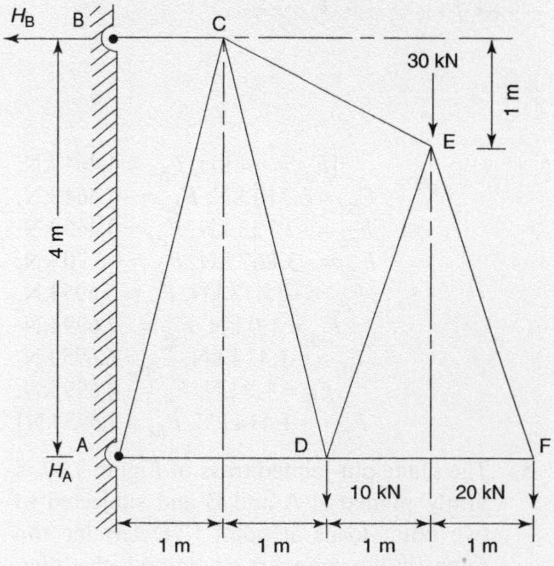

Figure 3.28

[F_{ef} = 21.08 kN, F_{df} = −6.666 kN, F_{ed} = −42.16 kN, F_{ec} = 22.35 kN, F_{cd} = 51.557 kN, F_{ad} = −32.528 kN, F_{ac} = −61.856 kN, F_{bc} = 47.540 kN, H_B = 47.540 kN, V_A = 60.01 kN, H_A = 47.559 kN, V_B = 0]

7. Determine the forces in the symmetrical pin-jointed truss of Figure 3.29 stating whether they are tensile or compressive.

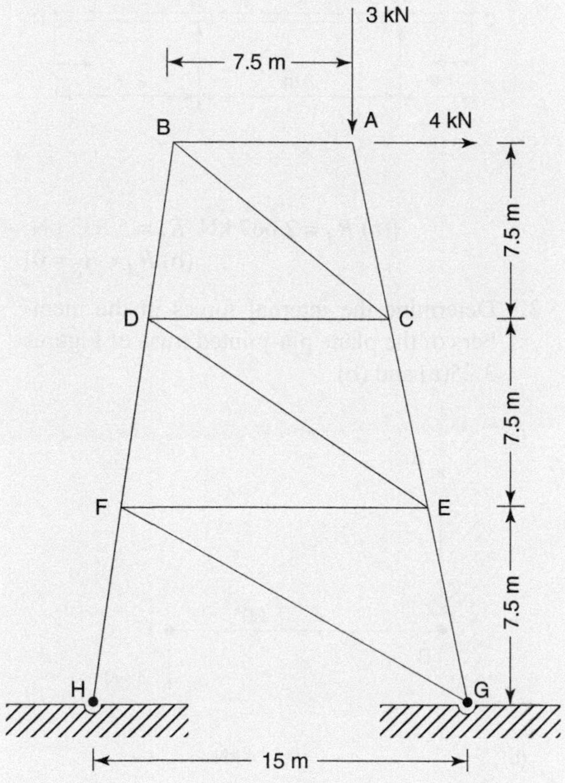

Figure 3.29

[F_{ac} = −3.043kN, F_{ab} = 3.50 kN, F_{bc} = −4.037 kN, F_{bd} = 2.664 kN, F_{ce} = −5.708 kN, F_{cd} = 2.627 kN, F_{de} = −2.842 kN, F_{df} = 4.264 kN, F_{eg} = −7.308 kN, F_{ef} = 2.102 kN, F_{fg} = −2.194 kN, F_{fh} = 5.330 kN]

8. Determine the reactions R_A and R_B for the simply supported beams of Figures 3.30(a) and (b).

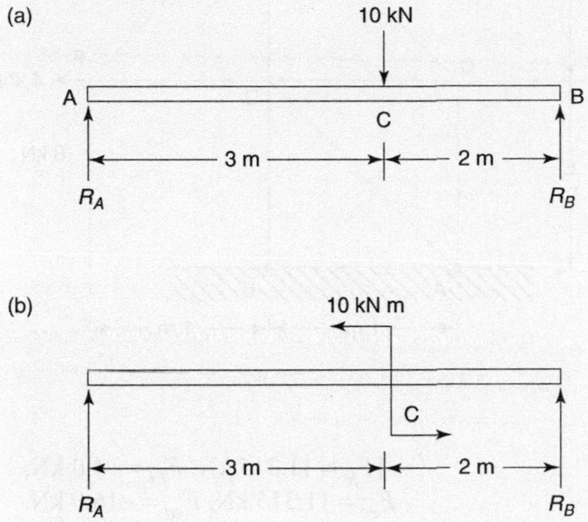

Figure 3.30

[(a) $R_A = 4$ kN, $R_B = 6$ kN
(b) $R_A = 2$ kN, $R_B = -2$ kN]

9. Determine the reactions R_A and R_B for the simply supported beams of Figures 3.31(a) and (b).

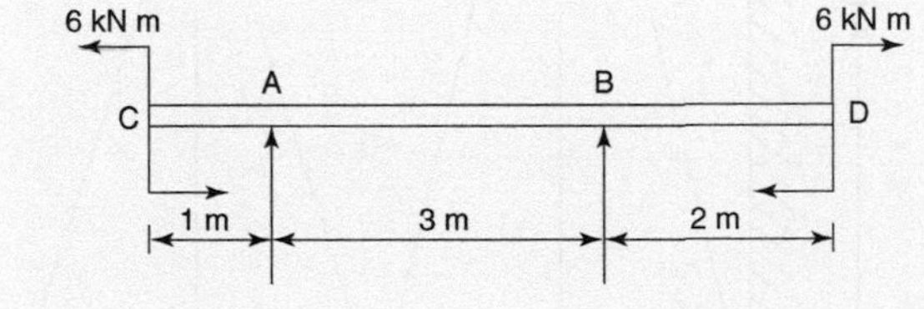

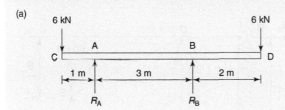

Figure 3.31

[(a) $R_A = 4$ kN, $R_B = 8$ kN (b) $R_A = R_B = 0$]

10. Determine the internal forces in the members of the plane pin-jointed truss of Figures 3.32(a) and (b).

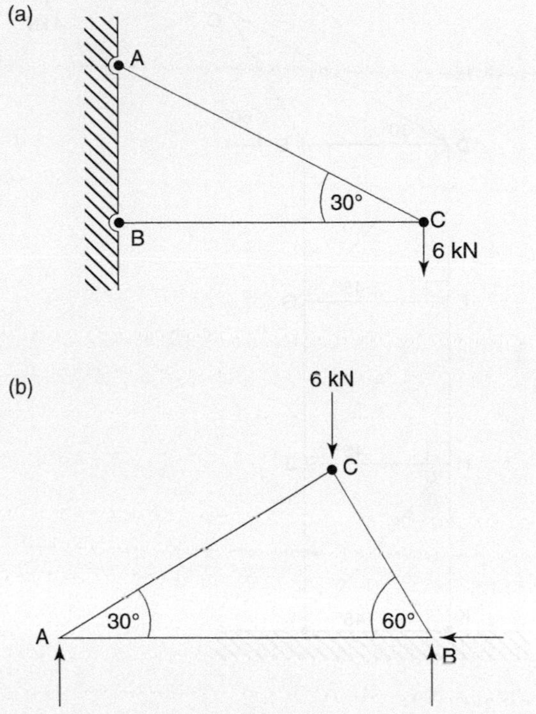

Figure 3.32

[(a) $F_{ac} = 12$ kN, $F_{bc} = -10.392$ kN
(b) $F_{bc} = -5.196$ kN, $F_{ac} = -3$ kN,
$F_{ab} = 2.598$ kN]

11. A plane pin-jointed truss is firmly pinned at its base, as shown in Figure 3.33. Determine the forces in the members of this truss, stating whether they are in tension or compression.

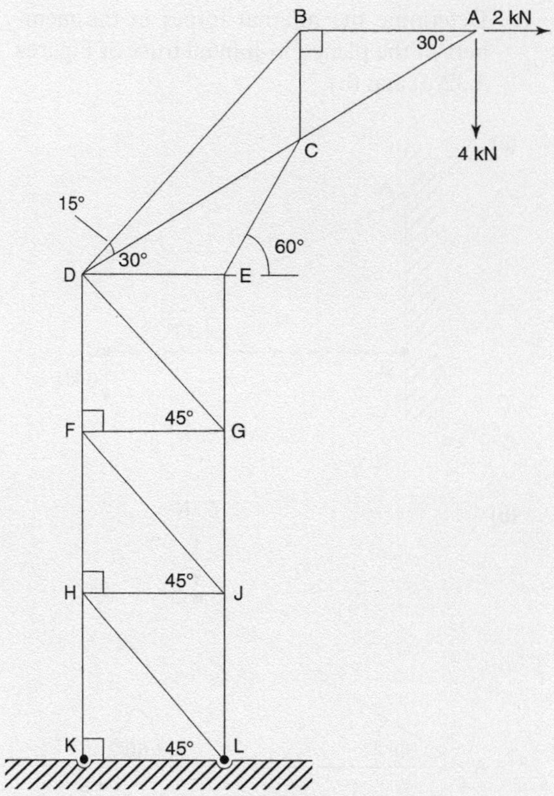

Figure 3.33

$[F_{ac} = -8$ kN, $F_{ab} = 8.928$ kN,
$F_{bd} = 12.626$ kN, $F_{bc} = -8.928$ kN,
$F_{ce} = -15.462$ kN, $F_{cd} = 0.928$ kN,
$F_{de} = -7.731$ kN, $F_{eg} = -13.390$ kN,
$F_{dg} = -2.828$ kN, $F_{df} = 11.390$ kN,
$F_{gf} = 2.0$ kN, $F_{gj} = -15.390$ kN,
$F_{fj} = -2.828$ kN, $F_{fh} = 13.390$ kN,
$F_{jh} = 2.0$ kN, $F_{jl} = -17.390$ kN,
$F_{hl} = -2.828$ kN, $F_{hk} = 15.390$ kN$]$

12. The plane pin-jointed truss of Figure 3.34 is firmly pinned at A and B and subjected to two point loads at point F. Determine the forces in the members, stating whether they are tensile or compressive.

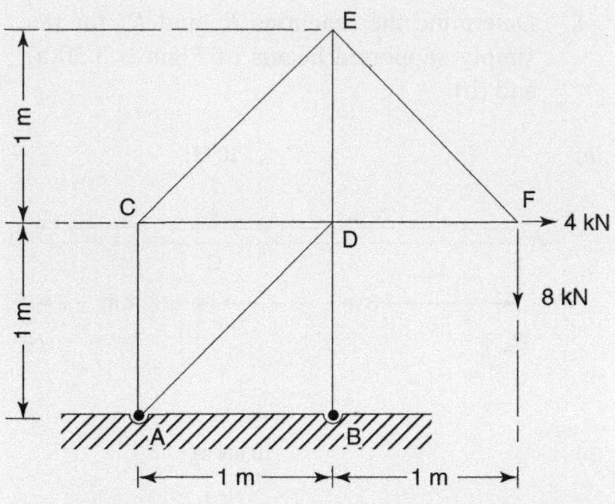

Figure 3.34

$[F_{fe} = 11.315$ kN, $F_{fd} = -4.0$ kN,
$F_{ec} = 11.315$ kN, $F_{ed} = -16.0$ kN,
$F_{cd} = -8.0$ kN, $F_{ac} = 8.0$ kN,
$F_{ad} = 5.658$ kN, $F_{bd} = -20.0$ kN$]$

13. An overhanging pin-jointed roof truss, which may be assumed to be pinned rigidly to the wall at the joints A and B, is subjected to the loading shown in Figure 3.35. Determine the forces in the members of the truss, stating whether they are tensile or compressive. Determine also the reactions.

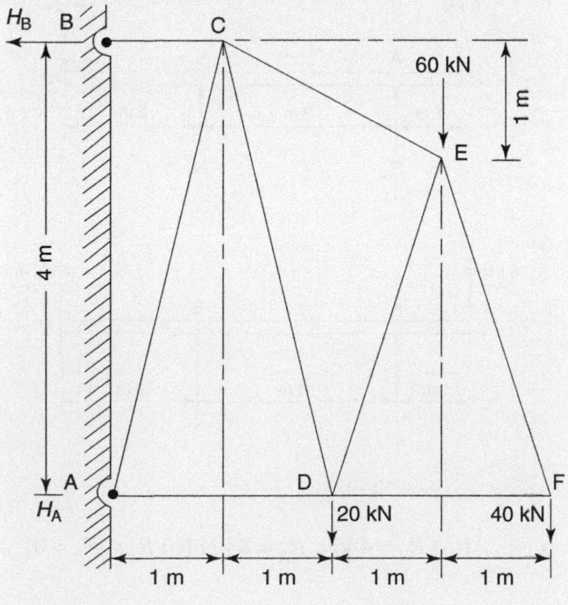

Figure 3.35

$[F_{ef} = 42.16$ kN, $F_{df} = -13.332$ kN,
$F_{ed} = -88.35$ kN, $F_{ec} = 46.14$ kN,
$F_{cd} = 107.06$ kN, $F_{ad} = -67.27$ kN,
$F_{ac} = -128.32$ kN, $F_{bc} = 98.45$ kN,
$H_B = 98.45$ kN, $V_A = 124.49$ kN,
$H_A = 98.45$ kN, $V_B = 0]$

14. Determine the forces in the symmetrical pin-jointed truss of Figure 3.36 stating whether they are tensile or compressive.

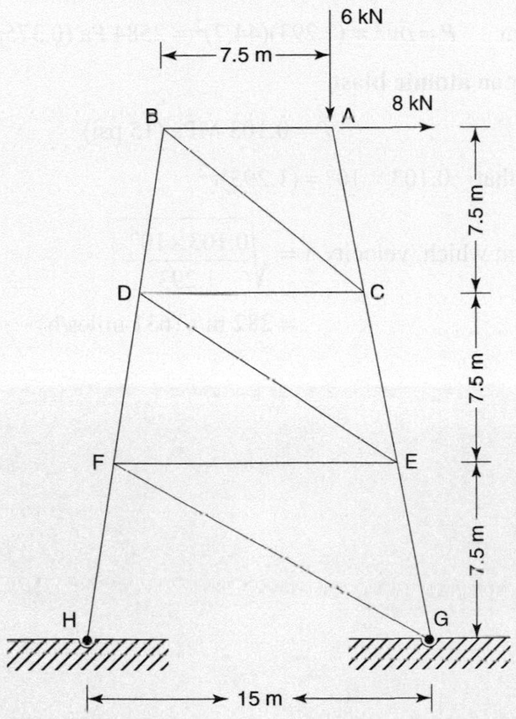

Figure 3.36

$[F_{ac} = -6.085$kN, $F_{ab} = 7.0$ kN,
$F_{bc} = -8.074$ kN, $F_{bd} = 5.329$ kN,
$F_{ce} = -11.416$ kN, $F_{cd} = 5.254$ kN,
$F_{de} = -5.684$ kN, $F_{df} = 8.529$ kN,
$F_{eg} = -14.615$ kN, $F_{ef} = 4.204$ kN,
$F_{fg} = -4.390$ kN, $F_{fh} = 10.663$ kN]

3.5 Bending moment and shearing force

These are very important in the analysis of beams and rigid-jointed frameworks, but the latter structures will not be considered in the present chapter, because they are much more suitable for computer analysis (see reference 1 on page 497 and Chapter 20). Once again, as in Section 3.1, all the beams will be assumed to be statically determinate and in equilibrium.

Definition of bending moment (*M*)

A *bending moment, M,* acting at any particular section on a beam, in equilibrium, can be defined as the resultant of all the couples acting on one side of the beam at that particular section. The resultant of the couples acting on either side of the appropriate section can be considered, as the beam is in equilibrium, so that the beam will be either in a *sagging* condition or in a *hogging* one, at that section, as shown in Figure 3.37.

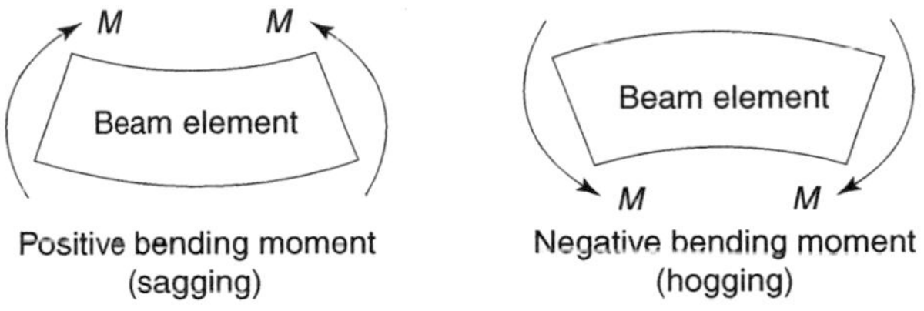

Figure 3.37 Sagging and hogging bending action

The sign convention for bending moments will be that a sagging moment is assumed to be *positive* and a hogging one is assumed to be *negative.* The units for bending moment are N m, kN m, etc.

Definition of shearing force (*F*)

A *shearing force, F,* acting at any particular section on a *horizontal beam,* in equilibrium, can be defined as the resultant of the *vertical forces* acting on one side of the beam at that particular section. The resultant of the vertical forces acting on either side of the appropriate section can be considered, as the beam is in equilibrium, as shown in Figure 3.38 which also shows the sign conventions for positive and negative shearing forces. The units for shearing force are N, kN, MN, etc. Prior to analysing beams it will be necessary to describe the symbols used for loads and supports and the various types of beam.

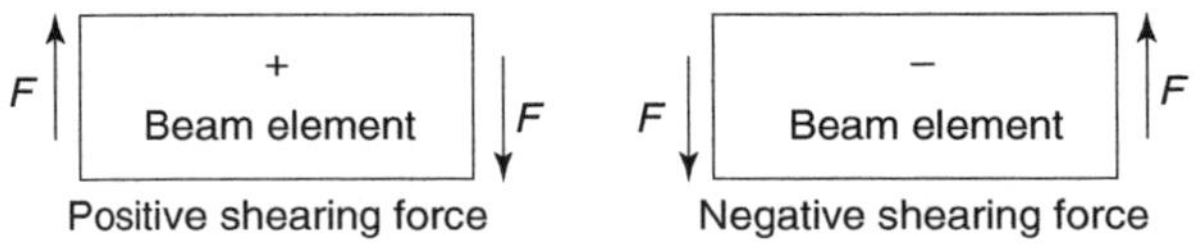

Figure 3.38 Positive and negative shearing forces

3.6 Loads

These can take various forms, including concentrated loads and couples, which are shown as arrows in Figure 3.39, and distributed loads, as shown in Figures 3.40 to 3.42.

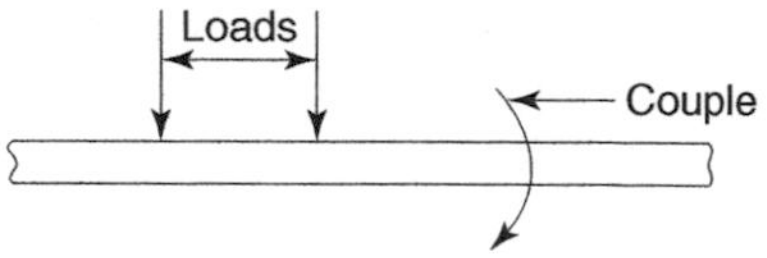

Figure 3.39 Concentrated loads and couples

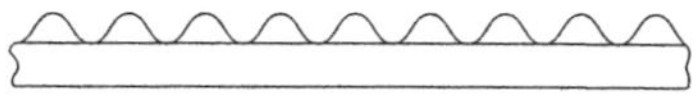

Figure 3.40 Uniformly distributed load (acting downwards)

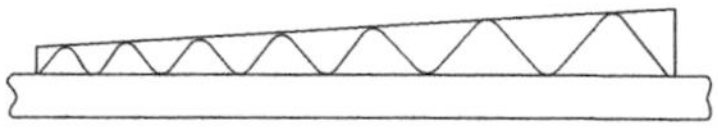

Figure 3.41 Hydrostatic load (acting downloads)

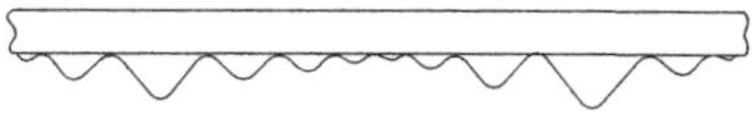

Figure 3.42 A varying distributed load (acting upwards)

Concentrated loads are assumed to act at points. This is, in general, a pessimistic assumption. Typical concentrated loads appear in the form of loads transmitted through wheels, hanging weights, etc. The units for concentrated loads are N, kN, MN, etc.

Uniformly distributed loads are assumed to be distributed uniformly over the length or part of the length of beam. Typical uniformly distributed loads are due to wind load, snow load, self-weight, etc. The units for uniformly distributed loads are N/m, kN/m, etc.

Hydrostatic (or trapezoidal) loads are assumed to increase or decrease linearly with length, as shown in Figure 3.41.They usually appear in containers carrying liquids or in marine structures, etc., which are attempting to contain the water, etc. The units of hydrostatic loads are N/m, kN/m, etc.

Varying distributed loads do not have the simpler shapes of uniformly distributed and hydrostatic loads described earlier. Typical cases of varying distributed loads are those due to the self-weight of a ship, together with the buoyant forces acting on its hull.

Wind loads can be calculated from momentum considerations, as follows:

$$\text{Wind pressure} = \rho v^2 = P$$

where ρ = density of air (kg/m^3)
v = velocity of air (m/s)
P = pressure in pascals† (N/m^2)

This simple expression assumes that the wind acts on a flat surface and that the wind is turned through 90° from its original direction, so that the calculation for P is, in general, overestimated, and on the so-called 'safe' side.

For *air* at standard temperature and pressure (s.t.p.)

$$\rho = 1.293 \text{ kg/m}^3$$

so that, when $v = 44.7$ m/s (100 miles/h)

then $P = \rho v^2 = (1.293)(44.7)^2 = 2584$ Pa (0.375psi)

For an **atomic blast**

$$P = 0.103 \text{ MPa (15 psi)}$$

so that $0.103 \times 10^6 = (1.293)v^2$

from which, velocity, $v = \sqrt{\dfrac{0.103 \times 10^6}{1.293}}$

$= 282$ m/s (631 miles/h)

† Blaise Pascal (19 June 1623 – 19 August 1662) was a French polymath who made important contributions to the study of fluids and clarified the concepts of pressure and vacuum. He corresponded with Pierre de Fermat on probability theory, strongly influencing the development of modern economics and social science. The unit of pressure, the *pascal*, is named in his honour. To find out more about Pascal go to www.routledge.com/cw/bird

NB For a standard living room of dimensions, say, 5 m × 5 m × 2.5 m, the air contained in the room will weigh about 80.8 kg (178.1 lbf), and for a large hall of dimensions, say, 30 m × 30 m × 15 m, the air contained in the hall will weigh about 17456 kg (17.1 tons)!
Structural engineers usually calculate the pressure due to wind acting on buildings as $\rho v^2/2$.
In general, the pressure due to water increases linearly with the depth of the water, according to the expression

$$P = \rho g h$$

where P = pressure in pascals (N/m²)
h = depth of water (m)
g = acceleration due to gravity (m/s²)
ρ = density of water (kg/m³), which for pure water, at normal temperature and pressure, is 1000 kg/m³, and for sea water is 1020 kg/m³

At a *depth* of 1000 m (in sea water),

$$\text{pressure, } P = \rho g h = (1020)(9.81)(1000)$$
$$= 10 \text{ MPa } (1451 \text{ psi})$$

so that, for a submarine of average diameter 10 m and of length 100 m, the total load due to water pressure will be about 37,700 MN (3.77×10^6 tons)!
NB 1 MN ≈ 100 tons.

The **Mariana Trench** in the Pacific Ocean is about 11 km deep where the hydrostatic pressure will be 1020 × 9.81 × 11000 = 110 MPa = 1591 psi = 7.12 ton/in²

3.7 Types of beam

The simplest types of beam are those shown in Figures 3.43 and 3.44. The beam of Figure 3.43, which is statically determinate, is supported on two knife edges or **simple supports.** In practice, however, a true knife-edge support is not possible, as such edges will not have zero area.

The beam of Figure 3.44, which is also statically determinate, is rigidly fixed (encastré) at its right end and is called a **cantilever.**

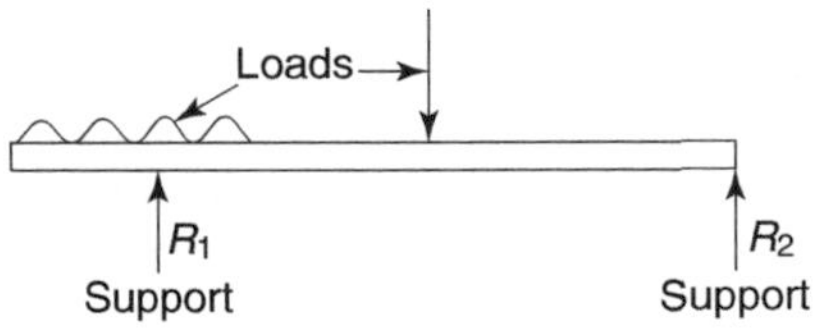

Figure 3.43 Statically determinate beam on simple supports

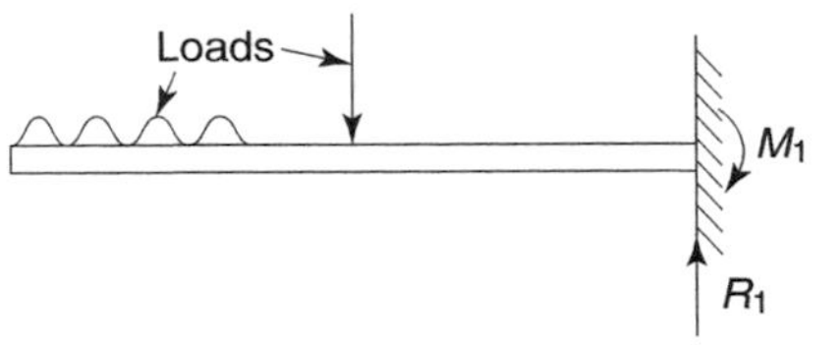

Figure 3.44 Cantilever

It can be seen from Figures 3.43 and 3.44 that in both cases, there are two reacting 'forces', which in the case of Figure 3.43 are R_1 and R_2 and in the case of Figure 3.44 are the vertical reaction R_1 and the restraining couple M_1. These beams are said to be statically determinate, because the two unknown reactions can be found from two simultaneous equations, which can be obtained by resolving forces vertically and taking moments.
Statically indeterminate beams cannot be analysed through simple statical considerations alone, because for such beams, the number of unknown 'reactions' is more than the equations that can be derived from statical observations. For example, the **propped cantilever** of Figure 3.45 has one redundant reacting 'force' (i.e. either R_1 or M_2) and the beam of Figure 3.46, which is encastré at both ends, has two redundant reacting 'forces' (i.e. either M_1 and M_2 or R_1 and M_1 or R_2 and M_2). These 'forces' are said to be redundant, because if they did not exist, the structure could still be classified as a structure, as distinct from a mechanism.

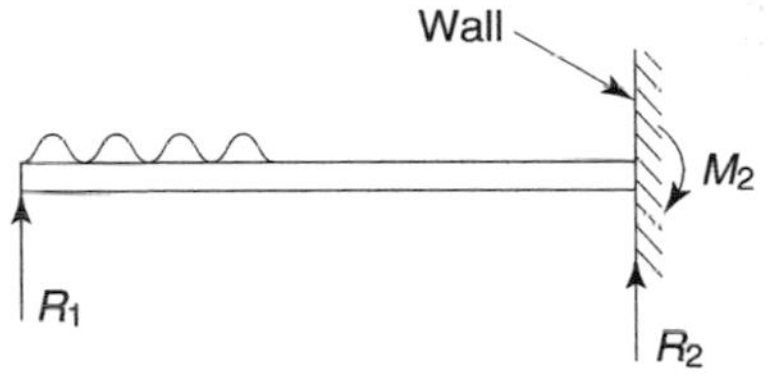

Figure 3.45 Propped cantilever

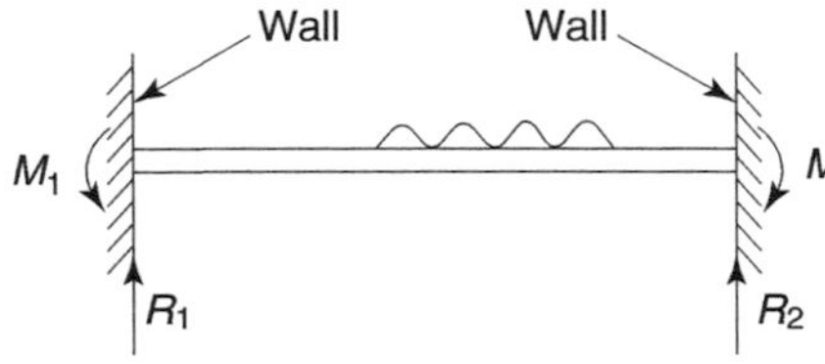

Figure 3.46 Beam with encastré ends

Other statically indeterminate beams are like the continuous beam of Figure 3.47, which has three redundant reacting forces (i.e. it is statically indeterminate to the third degree).

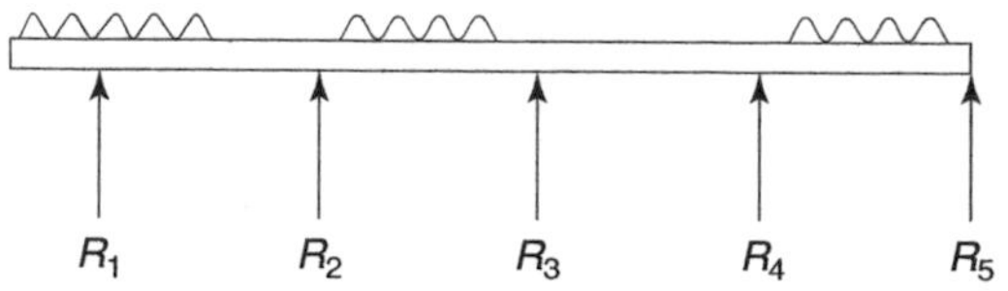

Figure 3.47 Continuous beam with several supports

NB By an 'encastré end', it is meant that all movement, including rotation, is prevented at that end.

3.8 Bending moment and shearing force diagrams

To demonstrate bending moment and shearing force action, Problem 6 below will first be considered, but on this occasion, the shearing forces and bending moments acting on both sides of any particular section will be calculated. The reason for doing this is to demonstrate the nature of bending moment and shearing force.

Problem 6. Determine expressions for the bending moment and shearing force distributions along the length of the beam in Figure 3.48. Hence, or otherwise, plot these bending moment and shearing force distributions.

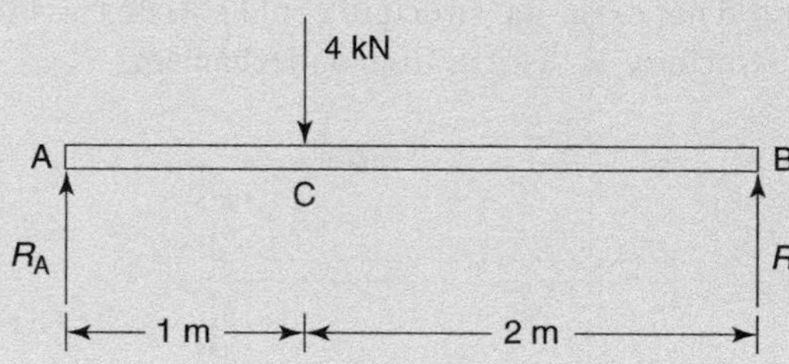

Figure 3.48 Beam with end supports

First, it will be necessary to calculate the reactions R_A and R_B and this can be achieved by taking moments about a suitable point and resolving forces vertically.

Take moments about B

It is convenient to take moments about either A or B. as this will eliminate the moment due to either R_A or R_B respectively.

clockwise moments about B
= counter-clockwise moments about B

$$R_A \times (2+1)\text{ m} = 4\text{ kN} \times 2\text{ m}$$

from which, $\boldsymbol{R_A} = \frac{8}{3} = \mathbf{2.667\ kN}$

Resolving forces vertically

upward forces = downward forces

$$R_A + R_B = 4\text{ kN}$$

i.e. $2.667 + R_B = 4\text{ kN}$

and $\boldsymbol{R_B} = 4 - 2.667 = \mathbf{1.333\ kN}$

To determine the bending moment between A and C

Consider *any* distance x between A and C, as shown in Figure 3.49. From the figure, it can be seen that at *any* distance x, between A and C, the reaction R_A causes a bending moment equal to $R_A x$, which is sagging (i.e, positive). It can also be seen from Figure 3.49 that the forces to the right of the beam cause an equal and opposite moment, so that the beam bends at this point in the manner shown in Figure 3.49(b). Therefore

Bending moment = M = 2.667 x (3.32)

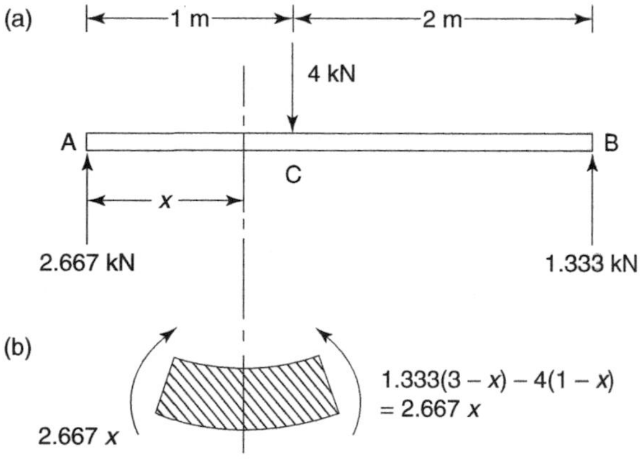

Figure 3.49 Bending moment between A and C

Equation (3.32) can be seen to be a straight line, which increases linearly from zero at A to a maximum value of 2.667 kN m at C.

To determine the shearing force between A and C

Consider any distance x between A and C as shown in Figure 3.50. From Figure 3.50(b), it can be seen that the vertical forces on the left of the beam tend to cause the left part of the beam at x to 'slide' upwards, whilst the vertical forces on the right of the beam tend to cause the right part of the beam at x to 'slide' downwards, i.e.

Shearing force at $x = F = 2.667$ kN (3.33)

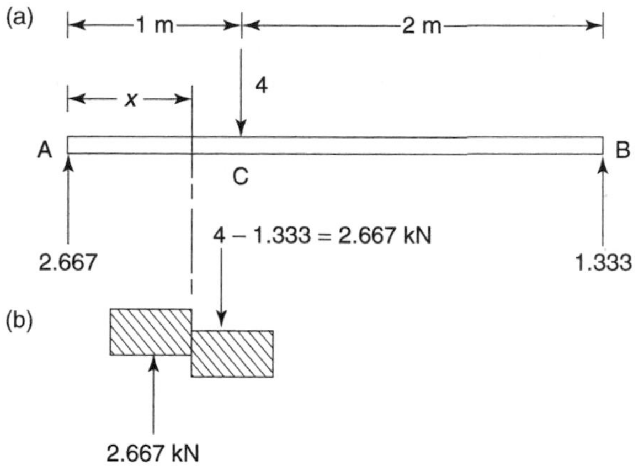

Figure 3.50 Shearing force between A and C

Equation (3.33) shows the shearing force to be constant between A and C, and is said to be positive, because the right side tends to move downwards, as shown in Figure 3.50(b).

To determine the bending moment between C and B

Consider any distance x between C and B, as shown in Figure 3.51. From the figure, it can be seen that at any distance x

Bending moment = $M = 4 - 1.333x$ (sagging) (3.34)

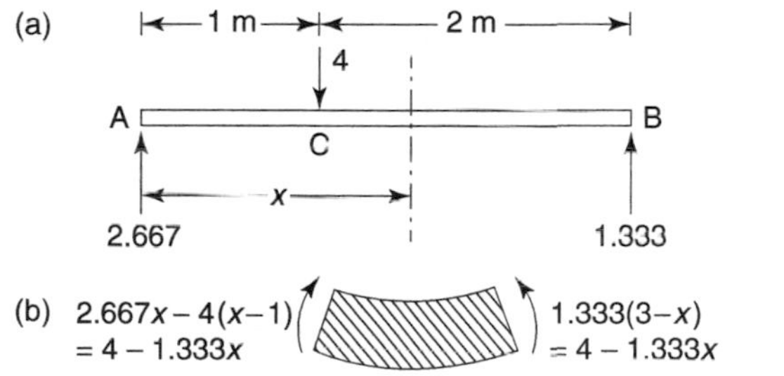

Figure 3.51 Bending moment and shearing force diagrams

Equation (3.34) shows the bending moment distribution between C and B to be decreasing linearly from 2.667 kN m at C to zero at B.

To determine the shearing force between C and B

Consider any distance x between C and B, as shown in Figure 3.52. From the figure, it can be seen that

Shearing force = $F = -1.333$ kN (3.35)

The shearing force F is constant between C and B, and it is negative because the right side of any section tends to 'slide' upwards and the left side to 'slide' downwards, as shown in Figure 3.52(b).

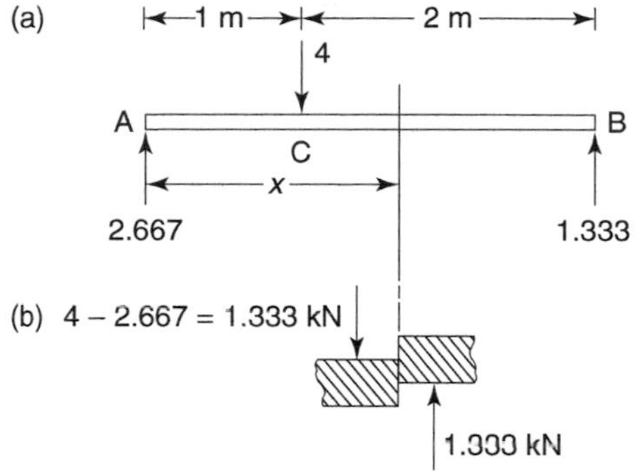

Figure 3.52 Shearing force between C and B

Bending moment and shearing force diagrams

To obtain the bending moment diagram, it is necessary to plot equations (3.32) and (3.34) in the manner shown in Figure 3.53(b), and to obtain the shearing force diagram, it is necessary to plot equations (3.33) and (3.35) in the manner shown in Figure 3.53(c).

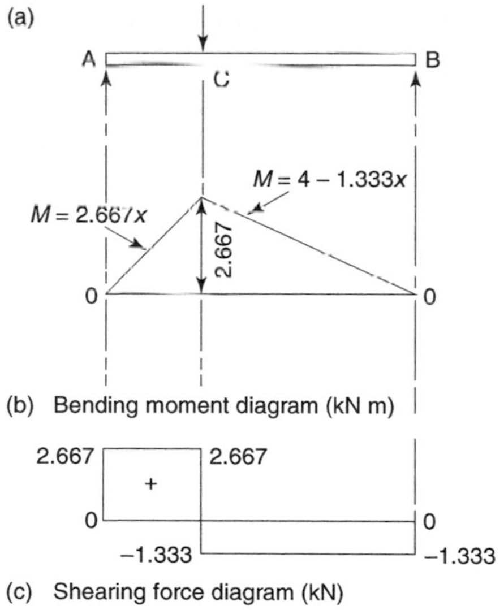

Figure 3.53 Bending moment and shearing force diagrams

From the calculations carried out above, it can be seen that to determine the bending moment or shearing force at any particular section, it is only necessary to consider the resultant of the forces on *one side of the section.* Either side of the appropriate section can be considered, as the beam is in equilibrium.

Problem 7. Determine the bending moment and shearing force distributions for the cantilever of Figure 3.54. Hence, or otherwise, plot the bending moment and shearing force diagrams.

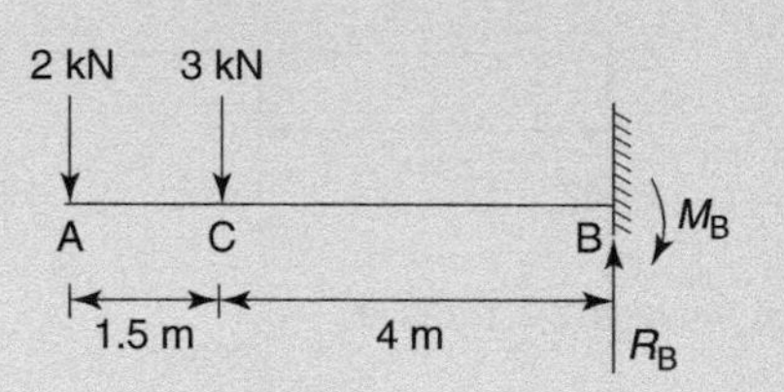

Figure 3.54 Cantilever

To determine R_B and M_B

Resolving forces vertically

$$\boldsymbol{R_B = 2 + 3 = 5\ \text{kN}}$$

Taking moments about B

$$M_B = 2 \times 5.5 + 3 \times 4$$

i.e. $$\boldsymbol{M_B = 23\ \text{kN m}}$$

To determine bending moment distributions

Consider span AC

At any given distance x from A, the forces to the left cause a hogging bending moment of $2x$, as shown in Figure 3.55, i.e.

$$M = -2x \qquad (3.36)$$

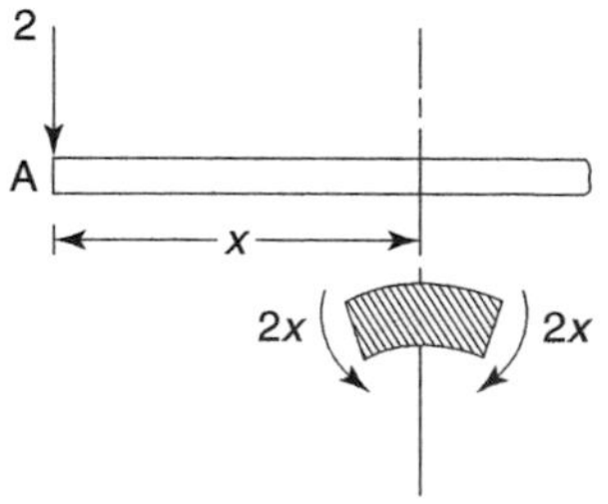

Figure 3.55 Bending moment between A and C

Equation (3.36) can be seen to increase linearly in magnitude from zero at the free end to 3 kN m at C.

Consider span CB

At any distance x from A, the forces to the left cause a hogging bending moment of $2x + 3(x - 1.5)$, as shown in Figure 3.56, i.e.

$$M = -2x - 3(x - 1.5)$$

i.e. $$M = -5x + 4.5 \qquad (3.37)$$

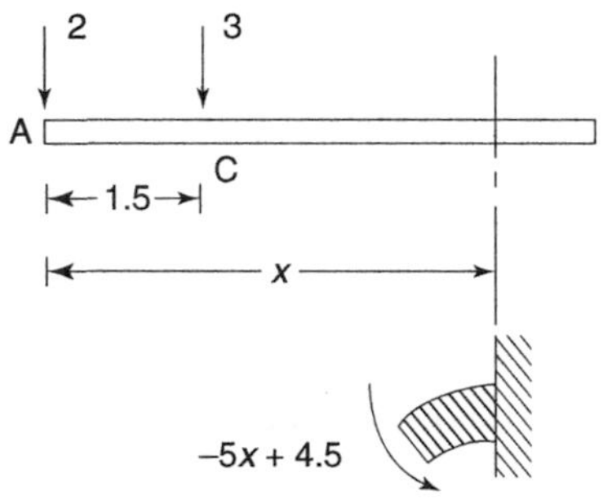

Figure 3.56 Bending moment between C and B

Equation (3.37) can be seen to increase linearly in magnitude from 3 kN m at C to 23 kN m at B.

To determine the shearing force distributions

Consider span AC

At any distance x, the resultant of the vertical forces to the left of this section causes a shearing

$$F = -2\ \text{kN} \qquad (3.38)$$

as shown in Figure 3.57.

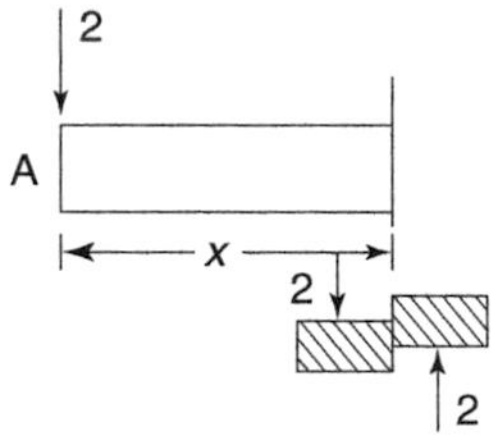

Figure 3.57 Shearing force between A and C

Consider span CB

At any given distance x, the resultant of the vertical forces to the left of this section causes a shearing force equal to

$$\boldsymbol{F = -2 - 3 = -5\ \text{kN}} \qquad (3.39)$$

as shown in Figure 3.58.

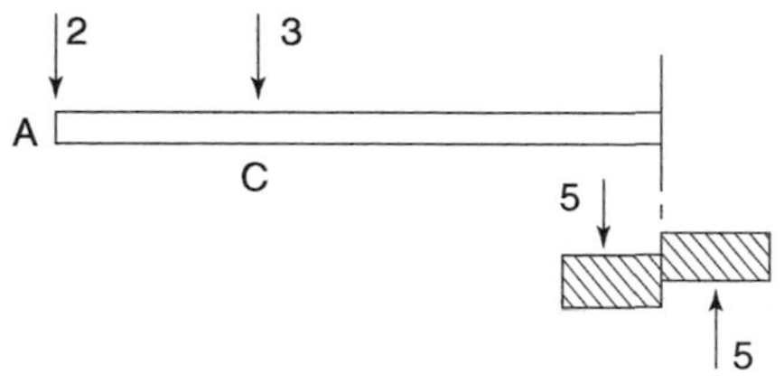

Figure 3.58 Shearing force between C and B

Bending moment and shearing force diagrams

Plots of the bending moment and shearing force distributions, along the length of the cantilever, are shown in Figure 3.59.

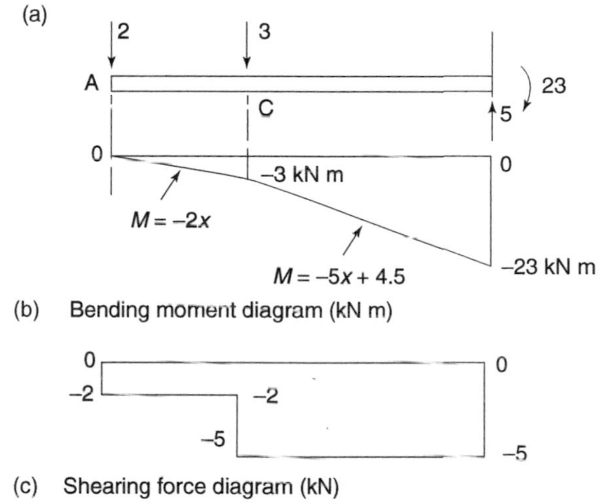

Figure 3.59 Bending moment and shearing force diagrams

Problem 8. Determine the bending moment and shearing force distributions for the beam shown in Figure 3.60 and plot the diagrams. Determine also the position of the point of contraflexure. It is convenient to divide the beam into spans AB, BC and CD, because there are discontinuities at B and C.

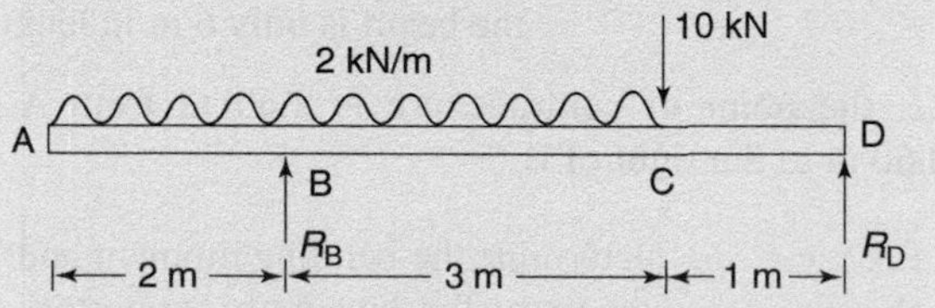

Figure 3.60 Beam ABCD

To determine R_B and R_D

Taking moments about D

$$R_B \times 4 = 10 \times 1 + 5 \times 2 \times 3.5$$

i.e. $$R_B = \frac{45}{4} = \mathbf{11.25\ kN}$$

Resolving forces vertically

$$R_B + R_D = 10 + 5 \times 2$$

i.e. $$11.25 + R_D = 20 \text{ from above}$$

i.e. $$R_D = 20 - 11.25 = \mathbf{8.75\ kN}$$

Consider span AB

From Figure 3.61, it can be seen that at any distance x from the left end, the bending moment M, which is hogging, is given by

$$M = -2 \times x \times \frac{x}{2} = -x^2 \qquad (3.40)$$

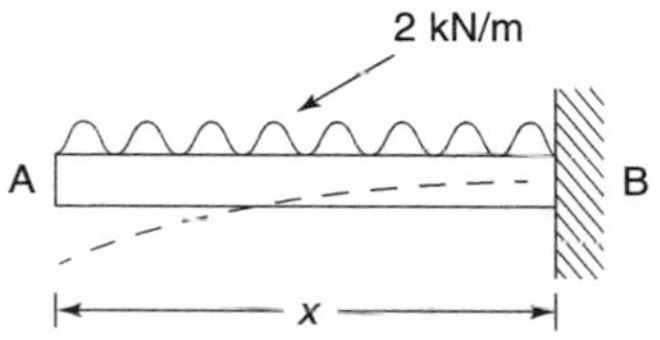

Figure 3.61

Equation (3.40) can be seen to be parabolic, which increases in magnitude from zero at A to −4 kN m at B. The equation is obtained by multiplying the weight of the load, which is $2x$, by the lever, which is $x/2$. It is negative, because the beam is hogging at this section. Furthermore, from Figure 3.61, it can be seen that the resultant of the vertical forces to the left of the section is $2x$, causing the left to 'slide' downwards, and the right to 'slide' upwards, i.e.

$$\text{Shearing force} = F = -2x \qquad (3.41)$$

Equation (3.41) is linear, increasing in magnitude from zero at A to 4 kN at B.

Consider span BC

At any distance x in Figure 3.62

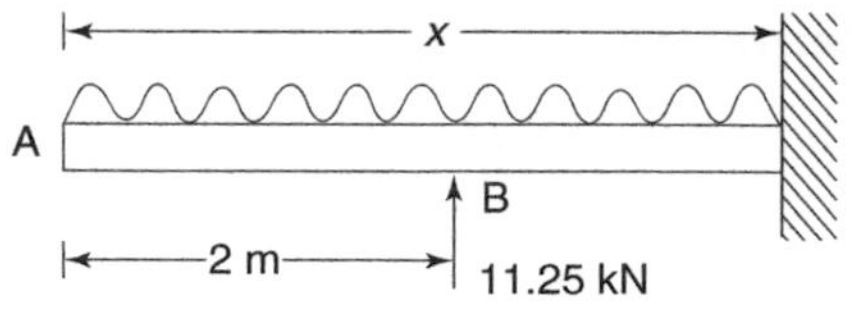

Figure 3.62

$$M = -2 \times x \times \frac{x}{2} + 11.25(x - 2)$$

i.e. $M = -x^2 + 11.25x - 22.5$ (a quadratic equation and hence parabolic) (3.42)

At $x = 2$, $M_B = -4$ kN m

At $x = 5$, $M_C = 8.75$ kN m

At any distance x

$$F = 11.25 - 2 \times x$$

i.e. $F = 11.25 - 2x$ (linear) (3.43)

At $x = 2$ $F_B = 7.25$ kN

At $x = 5$ $F_C = 1.25$ kN

Consider span CD

At any distance x in Figure 3.63

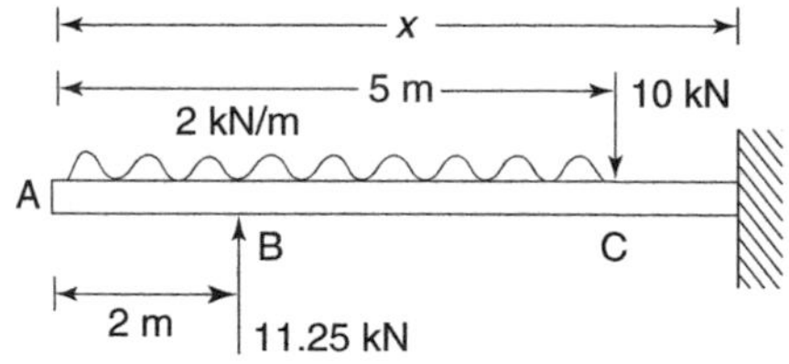

Figure 3.63

$$M = 11.25(x - 2) - 2 \times 5 \times (x - 2.5) - 10(x - 5)$$

i.e. $M = 52.5 - 8.75x$ (linear) (3.44)

At $x = 5$ $M_C = 8.75$ kN m

At $x = 6$ $M_D = 0$ (as required)

At any distance x in Figure 3.63,

$$F = 11.25 - 2 \times 5 - 10$$

i.e. $F = -\,8.75$ kN (constant) (3.45)

Bending moment and shearing force diagrams

From equations (3.40) to (3.45), the bending moment and shearing force diagrams can be plotted, as shown in Figure 3.64.

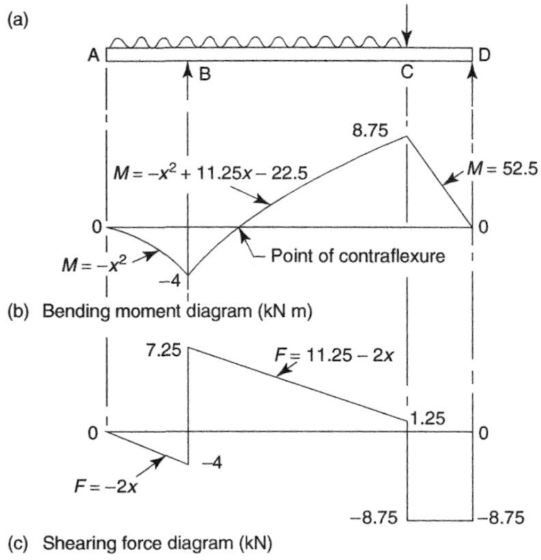

Figure 3.64

3.9 Point of contraflexure

The **point of contraflexure** is the point on a beam where the bending moment changes sign from a positive value to a negative one, or vice versa, and $M = 0$, as shown in Figure 3.64(b).

For this case, M must be zero between B and C, i.e. equation (3.42) must be used, so that

$$-x^2 + 11.25x - 22.5 = 0$$

from which $$x = \frac{-11.25 \pm \sqrt{11.25^2 - 4(-1)(-22.5)}}{2(-1)}$$

and $\mathbf{x = 2.60}$ **m** (or 8.65 m, which is ignored since the beam is only 6 m in length)

i.e. the point of contraflexure is 2.60 m from A or 0.60 m to the right of B.

Problem 9. Determine the bending moment and shearing force diagrams for the simply supported beam of Figure 3.65, which is acted upon by a clockwise couple of 3 kN m at B and a counter-clockwise couple of 5 kN m at C.

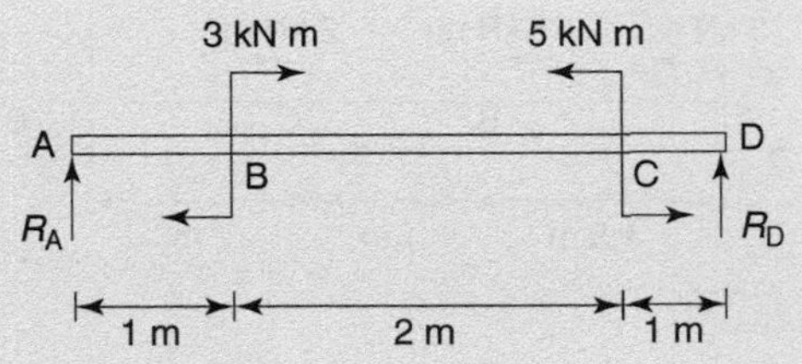

Figure 3.65 Beam with couples

To determine R_A and R_D

Take moments about D

$$R_A \times 4 + 3 = 5$$

i.e. $R_A \times 4 = 5 - 3 = 2$

from which $\boldsymbol{R_A} = \frac{2}{4} = \mathbf{0.5\ kN}$

Resolving vertically

$$R_A + R_D = 0$$

Therefore $\boldsymbol{R_D} = -R_A = \mathbf{-\ 0.5\ kN}$

i.e. R_D acts vertically downwards.

To determine the bending moment and shearing force distributions

Consider span AB
At any distance x in Figure 3.66

$$\boldsymbol{M} = \mathbf{0.5}\boldsymbol{x} \quad \text{(sagging)} \qquad (3.46)$$

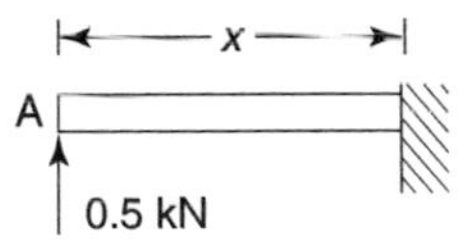

Figure 3.66

At A $M_A = 0$

At B $M_B = 0.5$ kN m

Similarly, considering vertical forces only,

$$\boldsymbol{F} = \mathbf{0.5\ kN} \text{ (constant)} \qquad (3.47)$$

This is positive, because the right hand is down.

Consider span BC
At any distance x in Figure 3.67

$$\boldsymbol{M} = \mathbf{0.5}\boldsymbol{x} + \mathbf{3} \quad \text{(sagging)} \qquad (3.48)$$

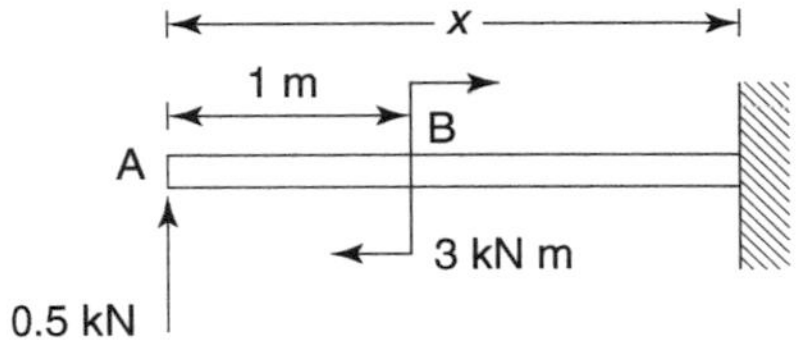

Figure 3.67

At B $M_B = 3.5$ kN m

At C $M_C = 4.5$ kN m

Similarly, considering vertical forces only,

$$\boldsymbol{F} = \mathbf{0.5\ kN} \text{ (constant)} \qquad (3.49)$$

Consider span CD
At any distance x in Figure 3.68

$$M = 0.5x + 3 - 5$$

i.e. $\boldsymbol{M} = \mathbf{0.5}\boldsymbol{x} - \mathbf{2}$ (hogging) (3.50)

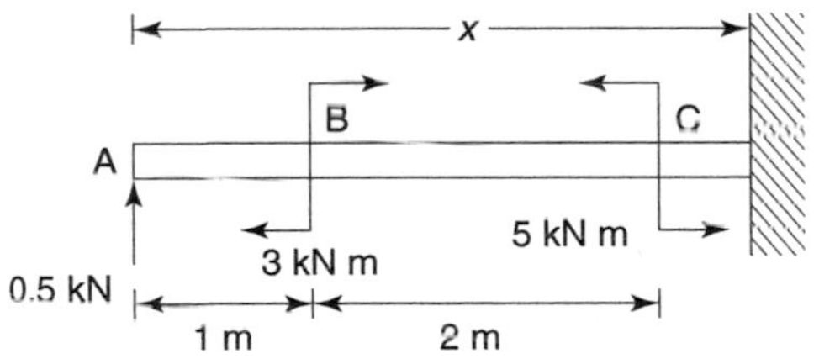

Figure 3.68

At C $M_C = -0.5$ kN m

At 0 $M_D = 0$

Similarly, considering vertical forces only,

$$\boldsymbol{F} = \mathbf{0.5\ kN} \text{ (constant)} \qquad (3.51)$$

The bending moment and shearing force distributions can be obtained by plotting equations (3.46) to (3.51), as shown in Figure 3.69.

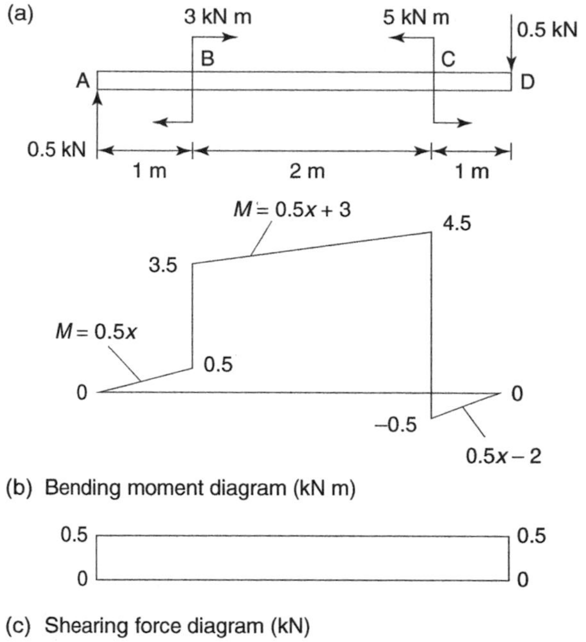

Figure 3.69 Bending moment and shearing force diagrams

Now try the following Practice Exercise

Practice Exercise 21. Bending moment and shear force diagrams

1. Determine the bending moments and shearing forces at the points A, B, C, D and E for the simply supported beam of Figure 3.70. Determine, also, the position of the point of contraflexure.

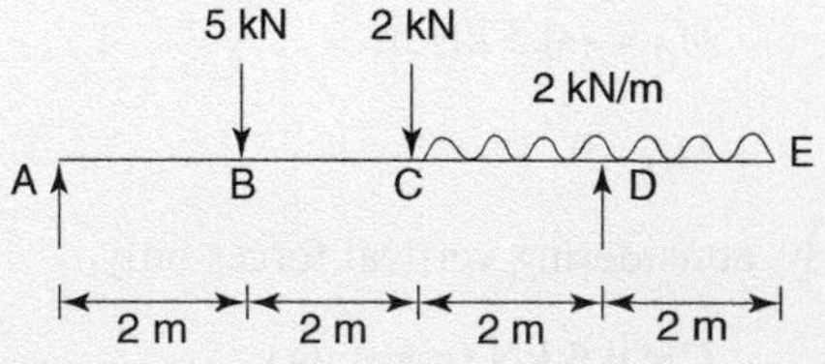

Figure 3.70

[*M* (kN m) → 0, 8, 6, – 4, 0; *F* (kN) → 4, 4/–1, – 1/–3, – 7/4, 0; 5.37 m to the right of A]

2. Determine the bending moments and shearing forces at the points A, B, C and D on the cantilever of Figure 3.71.

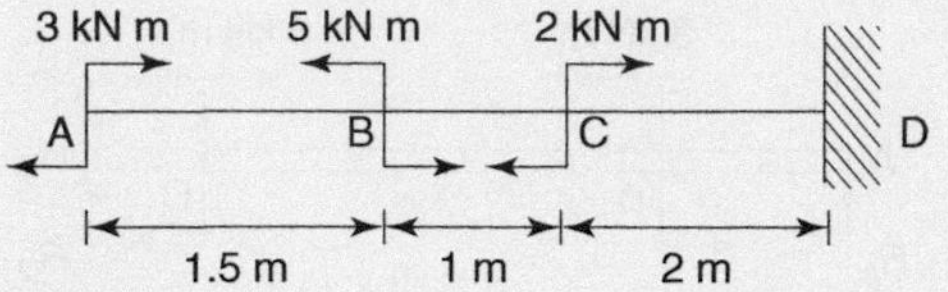

Figure 3.71

[*M* (kN m) → 3, 3/–2, – 2/0, 0;
F (kN) → 0, 0, 0, 0]

3. Determine the bending moments and shearing forces at the points A, B, C and D on the beam of Figure 3.72.

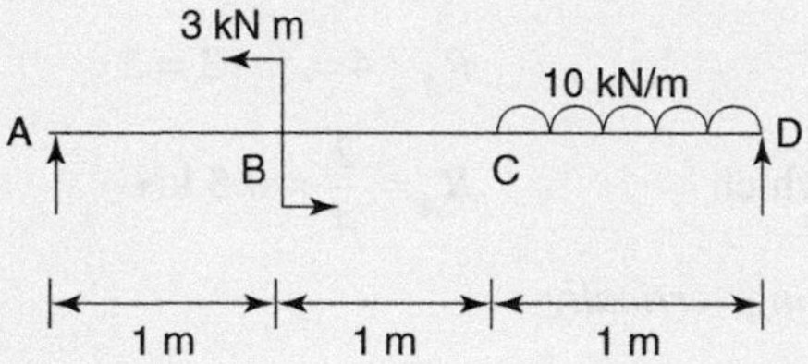

Figure 3.72

[*M* (kN m) → 0, 2.667/–0.333, 2.333, 0;
F (kN) → 2.667, 2.667, 2.667, – 7.333]

4. A uniform-section beam is simply supported at A and B, as shown in Figure 3.73. Determine the bending moments and shearing forces at the points C, A, D, E and B.

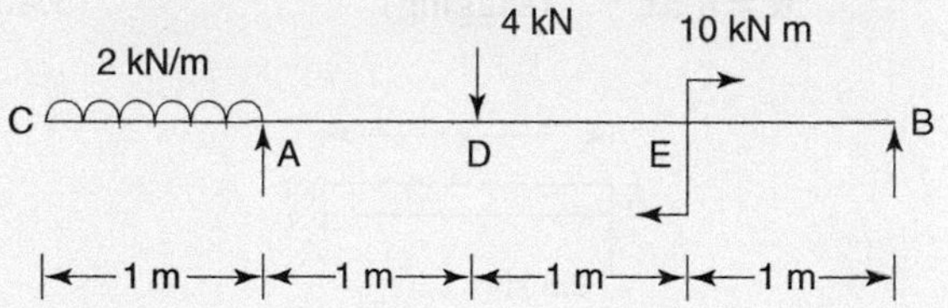

Figure 3.73

[*M* (kN m) → 0, –1, – 1.333,
– 5.667/4.333, 0;
F (kN) → 0, – 2/–0.333, –0.333/–4.333,
– 4.333, – 4.333]

5. A simply supported beam supports a distributed load, as shown in Figure 3.74. Obtain an expression for the value of *a*, so that the bending moment at the support will be of the same magnitude as that at mid-span.

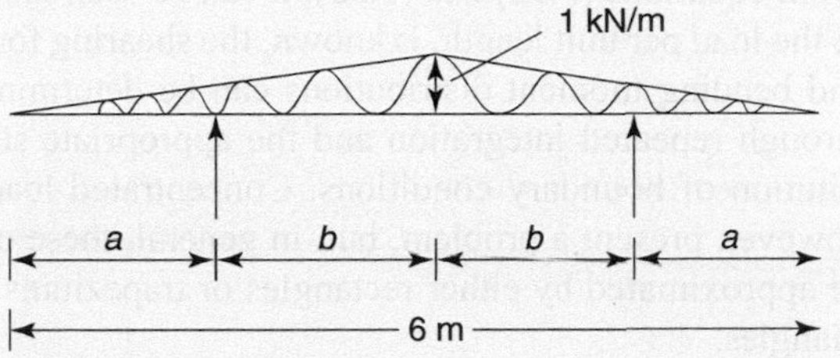

Figure 3.74

[$a^3/18 + 1.5a = 3$, $a = 1.788$ m]

6. Determine the bending moments and shearing forces at the points A, B, C, D and E for the simply supported beam of Figure 3.75. Determine also the position of the point of contraflexure.

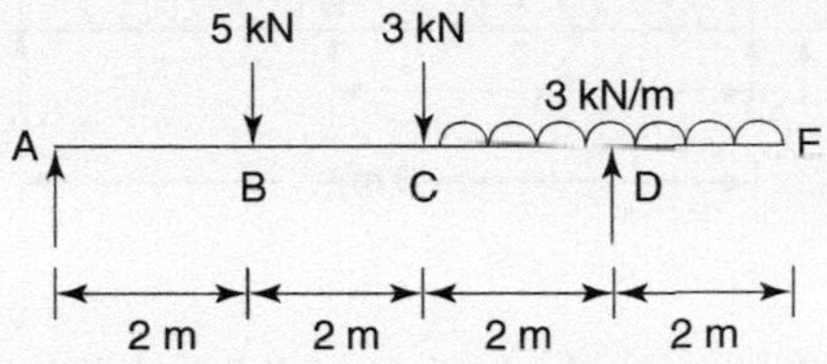

Figure 3.75

[M (kN m) → 0, 8.666, 7.332, – 6.0, 0;
F (kN) → 4.333, 4.333/–0.667, – 0.667/–3.667,
– 9.667/6.0, 0; 5.30 m to the right of A]

7. Determine the bending moments and shearing forces at the points A, B, C and D on the cantilever of Figure 3.76.

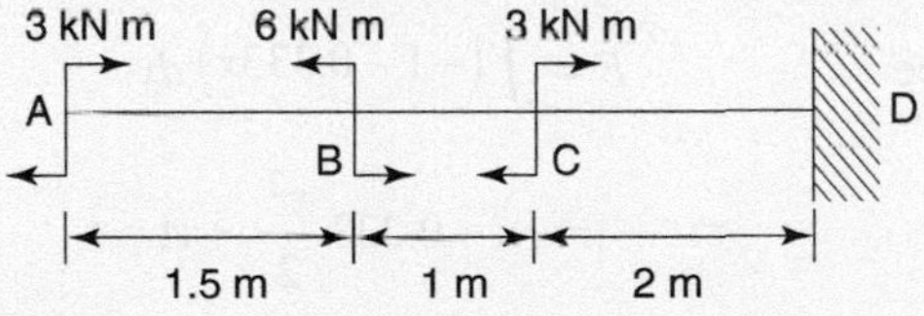

Figure 3.76

[M (kN m) → 3, 3/–3, –3/0; F (kN) → 0, 0, 0, 0]

8. Determine the bending moments and shearing forces at the points A, B, C and D on the beam of Figure 3.77.

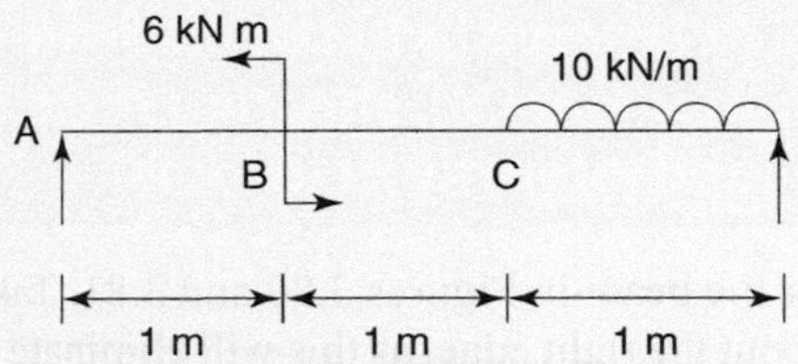

Figure 3.77

[M (kN m) → 0, 3.667/–2.333, 1.333, 0;
F (kN) → 3.667, 3.667, 3.667, – 6.333]

9. A uniform-section beam is simply supported at A and B, as shown in Figure 3.78. Determine the bending moments and shearing forces at points C, A, D, E and B

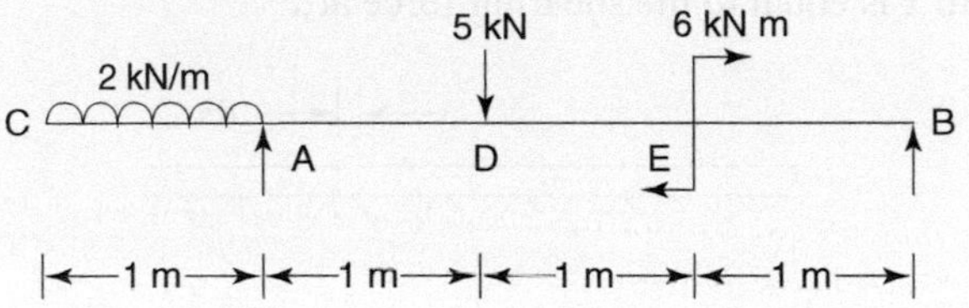

Figure 3.78

[M (kN m) → 0, – 1, – 0.667, – 2.666/3.333, 0;
F (kN) → 0, – 2/–1.667, –1.667/–6.667,
– 6.667, – 3.333]

10. A simply supported beam supports a distributed load, as shown in Figure 3.79. Obtain an expression for the value of a, so that the bending moment at the support will be of the same magnitude as that at mid-span.

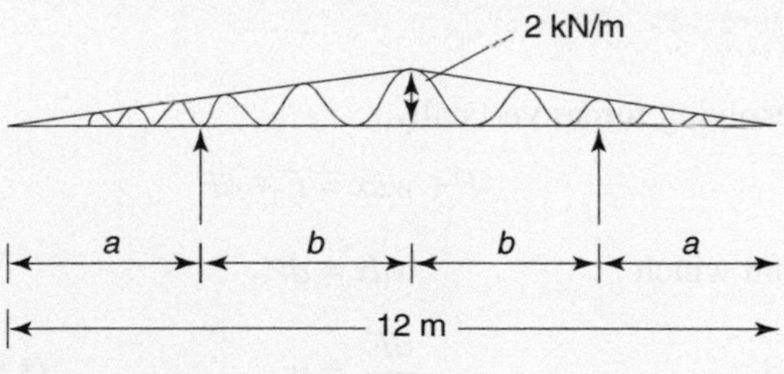

Figure 3.79

[$a^3 + 108a - 432 = 0$, $a = 3.862$ m]

3.10 Relationship between bending moment (*M*), shearing force (*F*) and intensity of load (*w*)

Consider the beam in Figures 3.80 and 3.81. Take moments about the right edge, as this will eliminate dF because the lever arm of dF is zero about the right edge; additionally, neglect the term $w\,dx\,dx/2$:

$$M + Fdx = M + dM$$

from which $$Fdx = dM$$

and $$\frac{dM}{dx} = F \qquad (3.52)$$

i.e. the derivative of the bending moment with respect to x is equal to the shearing force at x.

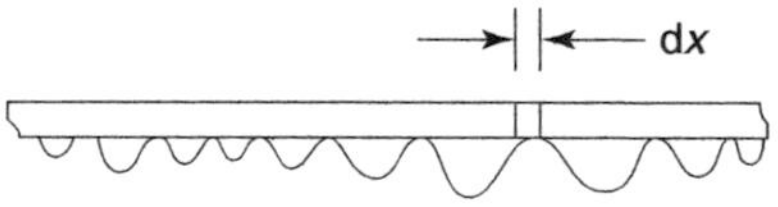

Figure 3.80 Beam with (positive) distribuited load

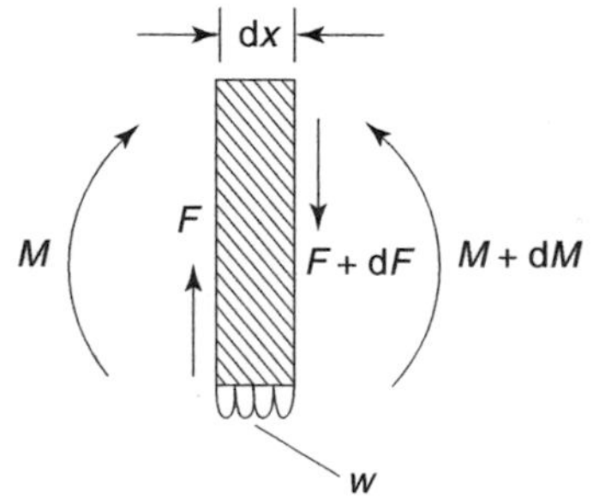

Figure 3.81 Beam element

Resolving forces vertically,

$$F + wdx = F + dF$$

from which $$wdx = dF$$

and $$\frac{dF}{dx} = w \qquad (3.53)$$

i.e. the derivative of the shearing force with respect to x is equal to w, the load per unit length, at x.
From equations (3.52) and (3.53),

$$\frac{d^2M}{dx^2} = w$$

From equations (3.52) and (3.53), it can be seen that if w, the load per unit length, is known, the shearing force and bending moment distributions can be determined through repeated integration and the appropriate substitution of boundary conditions. Concentrated loads, however, present a problem, but, in general, these can be approximated by either rectangles or trapeziums or triangles.

Problem 10. Determine the shearing force and bending moment distributions for the hydrostatically loaded beam of Figure 3.82, which is simply supported at its ends. Find also the position and value of the maximum value of bending moment, and plot the bending moment and shearing force diagrams.

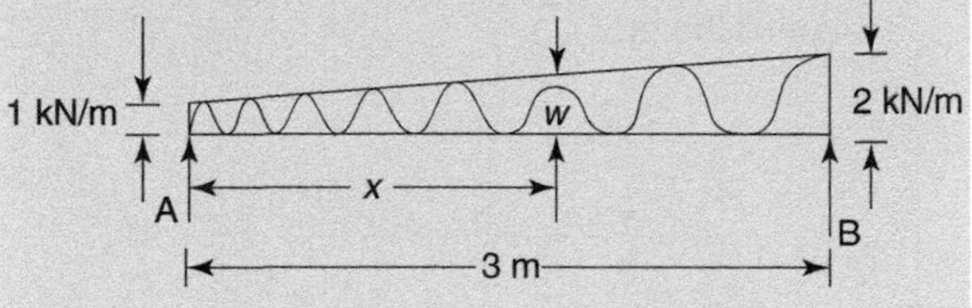

Figure 3.82 Hydrostatically loaded beam

At any distance x, the load per unit length is

$$w = -1 - 0.333x \qquad (3.54)$$

From (3.53) $$\frac{dF}{dx} = w$$

hence $$\frac{dF}{dx} = -1 - 0.333x$$

If $\frac{dF}{dx} = f(x)$, then $F = \int f(x)\,dx$

Hence $$F = \int\left(-1 - 0.333x\right)dx$$

$$= -x - 0.333\frac{x^2}{2} + A$$

from Chapter 2, page 38.

i.e. $$\boldsymbol{F = -x - 0.16667\,x^2 + A} \qquad (3.55)$$

From (3.52) $\frac{dM}{dx} = F$ from which, $M = \int F\,dx$

Hence $$M = \int\left(-x - 0.16667x^2 + A\right)dx$$

$$= -\frac{x^2}{2} - 0.16667\frac{x^3}{3} + Ax + B$$

i.e. $$\boldsymbol{M = -0.5x^2 - 0.05556x^3 + Ax + B} \qquad (3.56)$$

As there are two unknowns, two *boundary conditions* will be required to obtain the two simultaneous equations:

At $x = 0$ $\qquad M = 0$

Therefore from equation (3.56)

$$B = 0$$

At $x = 3$ $\qquad M = 0$

Therefore from equation (3.56)

$$A = 2$$

i.e. $\quad \mathbf{F = -x - 0.16667x^2 + 2} \qquad (3.57)$

and $\quad \mathbf{M = -0.5x^2 - 0.05556x^3 + 2x} \qquad (3.58)$

To obtain $\widehat{M}$ (the maximum bending moment)

$\widehat{M}$ occurs at the point where $\dfrac{dM}{dx} = 0$

i.e. from (3.58) $\quad -0.5(2x) - 0.05556(3x^2) + 2 = 0$

i.e. $\quad -x - 0.16668x^2 + 2 = 0$

or $\quad -0.16668x^2 - x + 2 = 0$

Solving the quadratic equation gives

$\mathbf{x = 1.583\ m}$ (to the right of A) $\qquad (3.59)$

or $x = -7.582$ m which is ignored

Substituting equation (3.59) into (3.58):

$$M = -0.5(1.583)^2 - 0.05556(1.583)^3 + 2(1.583)$$

i.e. $\quad \mathbf{M = 1.693\ kN\ m}$

Bending moment and shearing force diagrams

The bending moment and shearing force distributions can be obtained from equations (3.57) and (3.58), as shown in Figure 3.83.

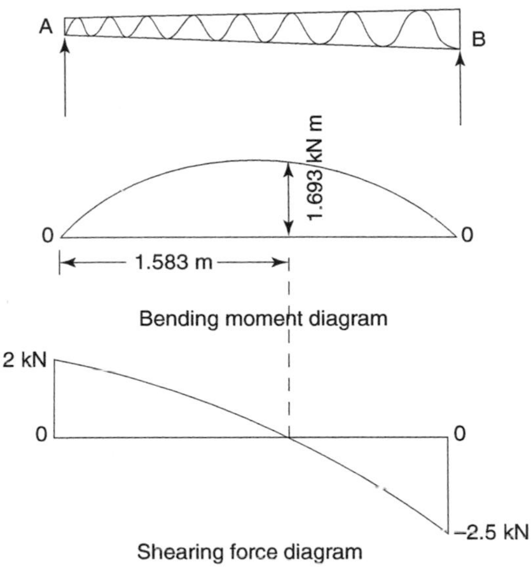

Figure 3.83 Bending moment and shearing force diagrams

Problem 11. A barge of uniform width 10 m and length 100 m can be assumed to be of weight 40,000 kN, which is uniformly distributed over its entire length.

Assuming that the barge is horizontal and is in equilibrium, determine the bending moment and shearing force distributions when the barge is subjected to upward buoyant forces from a wave, which is of sinusoidal shape as shown in Figure 3.84. The wave, whose height between peaks and trough is 3 m, may be assumed to have its peaks at the ends of the barge and its trough at the mid-length of the barge (amidships). Assume that ρ = density of water = 1020 kg/m^3 and $g = 9.81$ m/s^2

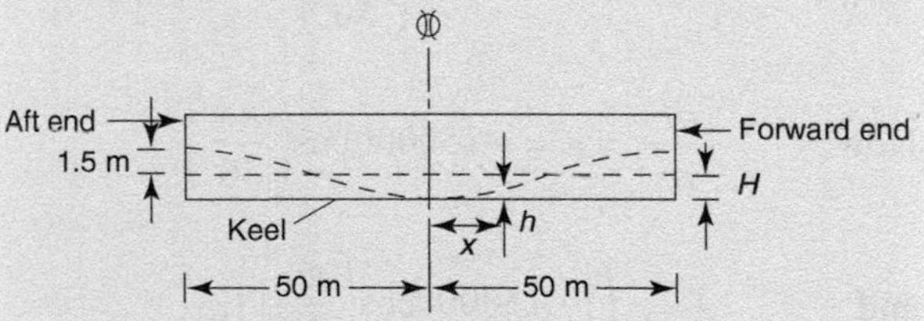

Figure 3.84 Barge subjected to a sinusoidal wave

To determine *H*, the depth of the still waterline

Weight of barge $= \rho \times g \times$ volume of water displaced

i.e. $\quad 40{,}000{,}000 = 1020 \times 9.81 \times (10 \times H \times 100)$

from which $$H = \frac{40{,}000{,}000}{1020 \times 9.81 \times 10 \times 100}$$

$= 3.998$ m, say, $\boldsymbol{H = 4}$ **m**

At any distance x from **amidships**, the height of the water above the keel is

$$h = 4 - 1.5\cos\left(\frac{\pi x}{50}\right) \tag{3.60}$$

The *upward load* per unit length acting on the barge will be due to the buoyancy, i.e. the buoyant load/unit length at x is

$$w_b = 1020 \times 9.81 \times 10 \times \left[4 - 1.5\cos\left(\frac{\pi x}{50}\right)\right]$$

i.e. $$w_b = 100062\left[4 - 1.5\cos\left(\frac{\pi x}{50}\right)\right]$$

i.e. $$w_b = 400248 - 150093\cos\left(\frac{\pi x}{50}\right)$$

or $$w_b = 400000 - 150000\cos\left(\frac{\pi x}{50}\right) \tag{3.61}$$

Downward forces, due to weight = upward forces due to buoyancy.
Now *downward load* per unit length is due to the self-weight of the barge, or 400,000 N/m.

Therefore $w = w_b - 400000$

$$w = 400000 - 150000\cos\left(\frac{\pi x}{50}\right) - 400000$$

i.e. $$\boldsymbol{w = -150000\cos\left(\frac{\pi x}{50}\right)}$$

Now $$\frac{dF}{dx} = w = -150000\cos\left(\frac{\pi x}{50}\right)$$

and $$F = \int\left(-150000\cos\left(\frac{\pi x}{50}\right)\right)dx$$

$$= -150000\left(\frac{1}{\frac{\pi}{50}}\sin\left(\frac{\pi x}{50}\right)\right) + A$$

$$= -\frac{50}{\pi} \times 150000\sin\left(\frac{\pi x}{50}\right) + A$$

At $x = 0$, $F = 0$ hence $\boldsymbol{A = 0}$ since $\sin 0 = 0$

Now $$\frac{dM}{dx} = \mathrm{F}$$

hence $$M = \int F\,dx = \int\left(-\frac{50}{\pi} \times 150000\sin\left(\frac{\pi x}{50}\right) + A\right)dx$$

$$= \frac{50}{\pi} \times 150000\left[\frac{1}{\frac{\pi}{50}}\cos\left(\frac{\pi x}{50}\right)\right] + Ax + B$$

$$= \left(\frac{50}{\pi}\right)^2 \times 150000\cos\left(\frac{\pi x}{50}\right) + Ax + B$$

At $x = 50$, M = 0 hence $0 = -\left(\frac{50}{\pi}\right)^2 \times 150000 + Ax + B$

since $\cos\pi = -1$

Therefore, with $A = 0$ from above,

$$B = \left(\frac{50}{\pi}\right)^2 \times 150000 = 38 \times 10^6$$

and $$M = \left(\frac{50}{\pi}\right)^2 \times 150000\cos\left(\frac{\pi x}{50}\right) + Ax + B$$

$$= 38 \times 10^6\cos\left(\frac{\pi x}{50}\right) + 0 + 38 \times 10^6$$

i.e. $$\mathbf{M = 38 \times 10^6\left[1 + \cos\left(\frac{\pi x}{50}\right)\right]} \tag{3.62}$$

and hence $$F = -\frac{50}{\pi} \times 150000\sin\left(\frac{\pi x}{50}\right) + A$$

i.e. $$\boldsymbol{F = -2387324\sin\left(\frac{\pi x}{50}\right)} \tag{3.63}$$

$\widehat{M}$ occurs amidships and when $x = 0$,

$$\widehat{M} = 38 \times 10^6\left[1 + \cos\left(\frac{\pi x}{50}\right)\right]$$

$$= 38 \times 10^6\left[1 + \cos 0\right] = \mathbf{76\ MN\ m}$$

Bending moment and shearing force diagrams

The bending moment and shearing force diagrams are shown in Figure 3.85.

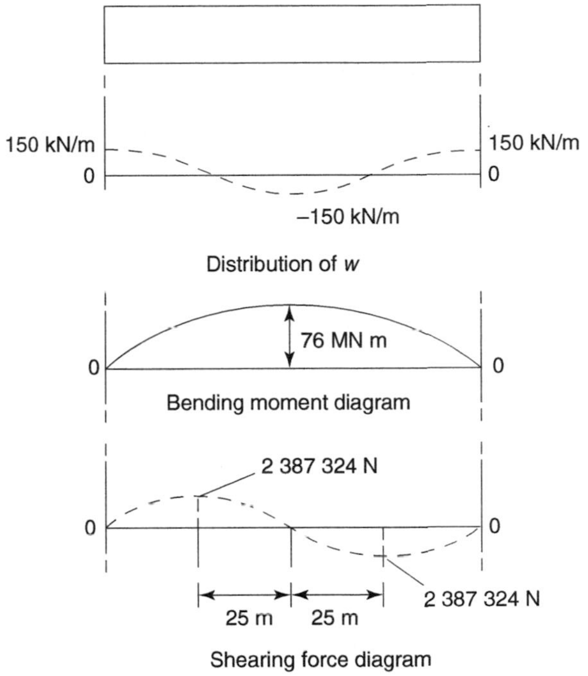

Figure 3.85 Bending moment and shearing force diagrams

Now try the following Practice Exercise

Practice Exercise 22. Relationship between bending moment (*M*), shearing force (*F*) and intensity of load (*w*)

1. Determine expressions for the shearing force and bending moment distributions for the hydrostatically loaded beam of Figure 3.86, which is simply supported at its ends. Find also the position and value of the maximum bending moment.

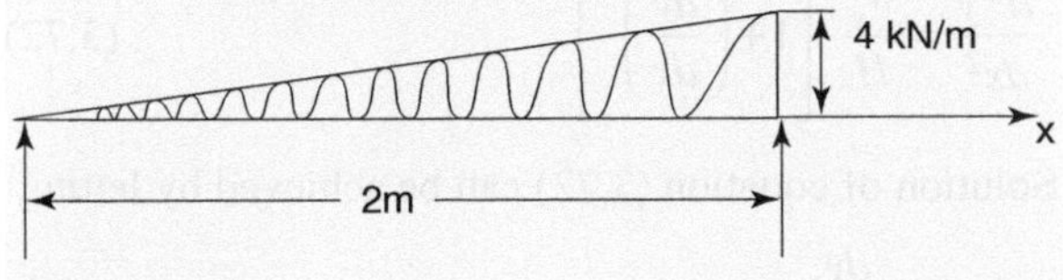

Figure 3.86

$[F = -x^2 + 1.333, M = -0.333x^3 + 1.333x,$

$1.155 \text{ m}, \widehat{M} = 1.026 \text{ kN m}]$

2. Determine expressions for the shearing force and bending moment diagrams for the hydrostatically loaded beam of Figure 3.87, which is simply supported at its ends. Hence, or otherwise, determine the position and value of the maximum bending moment.

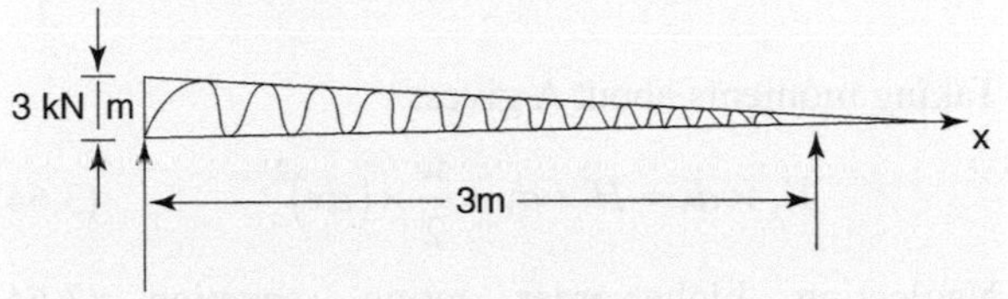

Figure 3.87

$[F = 3 - 3x + 0.5x^2, M = 3x - 1.5x^2 + \frac{x^3}{6},$

$1.268 \text{ m}, \widehat{M} = 1.732 \text{ kN m}]$

3.11 Cables

Cables, when acting as load-carrying members, appear in a number of different forms varying from power lines to cables used in suspension bridges, and from rods used in pre-stressed concrete to cables used in air-supported structures. When cables are used for pre-stressed concrete and for air-supported structures, the cables or rods are initially placed under tension, where, providing their stress values are within the elastic limit, their bending stiffness will increase with tension These problems, however, which are non-linear, are beyond the scope of this book and will not be discussed further.

Prior to analysis, it will be necessary to obtain the appropriate differential equation that governs the deflection of cables.

Consider an element of cable, loaded with a distributed load, *w*, which is uniform with respect to the *x* axis, as shown in Figure 3.88.

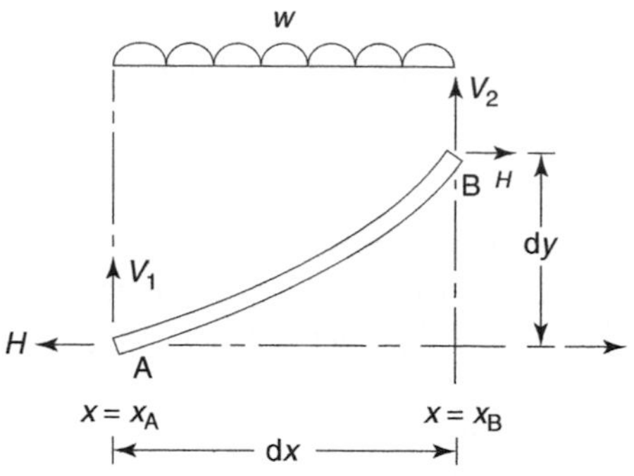

Figure 3.88 Cable element

Taking moments about A gives:

$$V_2 \times dx = H \times dy + \frac{w}{2} \times (dx)^2 \tag{3.64}$$

Neglecting higher-order terms, equation (3.64) becomes

$$V_2 = H \times \left.\frac{dy}{dx}\right|_{x = x_B} \tag{3.65}$$

Similarly, by taking moments about B,

$$H \times dy + V_1 \times dx = \frac{w}{2} \times (dx)^2$$

or $$V_1 = -H \times \left.\frac{dy}{dx}\right|_{x = x_A} \tag{3.66}$$

Resolving vertically gives:

$$V_1 + V_2 = w \times dx$$

or $$w = \frac{V_1 + V_2}{dx} \tag{3.67}$$

Substituting equations (3.65) and (3.66) into (3.67) gives:

$$w = \frac{-H \times (dy/dx)_{x=x_A} + H \times (dy/dx)_{x=x_B}}{dx}$$

i.e. $$w = \frac{H\left[(dy/dx)_{x=x_B} - (dy/dx)_{x=x_A}\right]}{dx}$$

i.e. $$w = \mathrm{H} \times \frac{d^2y}{dx^2}$$

or $$\frac{d^2y}{dx^2} = \frac{w}{H} \tag{3.68}$$

In general, when w varies with x, equation (3.68) becomes

$$\frac{d^2y}{dx^2} = \frac{w(x)}{H} \tag{3.69}$$

where $w(x)$ is the value of the load/unit length at any distance x.

Cable under self-weight

Consider an infinitesimally small length of cable, under its own weight, as shown in Figure 3.89.

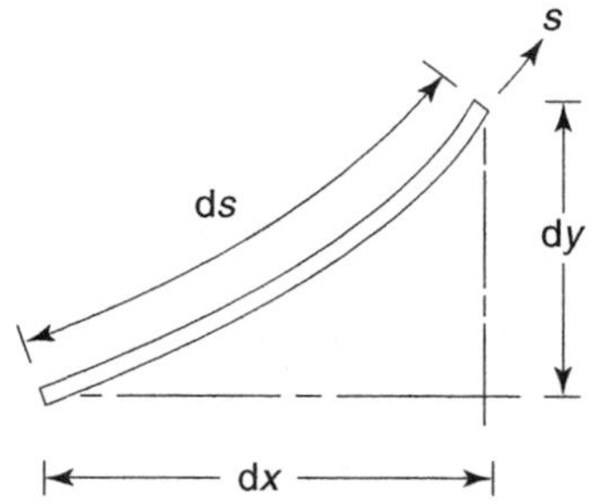

Figure 3.89 Infinitesimal length of cable under self-weight

As the element is infinitesimal

$$(ds)^2 = (dx)^2 + (dy)^2$$

or $$ds = \sqrt{(dx)^2 + (dy)^2} = \sqrt{(dx)^2\left[1 + \left(\frac{dy}{dx}\right)^2\right]}$$

i.e. $$ds = dx\left[1 + \left(\frac{dy}{dx}\right)^2\right]^{1/2} \tag{3.70}$$

Let w_s = weight/unit length of cable in the s direction,

$w(x)$ = weight/unit length of cable in the x direction, at any distance x

Then $$w(x) = w_s\left[1 + \left(\frac{dy}{dx}\right)^2\right]^{1/2} \tag{3.71}$$

Substituting equation (3.71) into equation (3.69)

$$\frac{d^2y}{dx^2} = \frac{w_s}{H}\left[1 + \left(\frac{dy}{dx}\right)^2\right]^{1/2} \tag{3.72}$$

Solution of equation (3.72) can be achieved by letting

$$\frac{dy}{dx} = Y \tag{3.73}$$

and $$\frac{d^2y}{dx^2} = \frac{dY}{dx} \tag{3.74}$$

Substituting equation (3.73) into (3.72)

$$\frac{dY}{dx} = \frac{w_s}{H}\left[1+Y^2\right]^{1/2}$$

or $$\frac{dY}{\left(1+Y^2\right)^{1/2}} = \frac{w_s}{H} \times dx$$

which by inspection, yields the following solution:

$$\sinh^{-1}(Y) = \frac{w_s}{H} \times x + C_1$$

or $$Y = \sinh\left(\frac{w_s}{H}x + C_1\right) \tag{3.75}$$

Substituting equation (3.73) into (3.75),

$$\frac{dy}{dx} = \sinh\left(\frac{w_s}{H}x + C_1\right)$$

or $$dy = \sinh\left(\frac{w_s}{H}x + C_1\right)dx$$

Hence $$y = \int \sinh\left(\frac{w_s}{H}x + C_1\right)dx$$

i.e. $$y = \frac{1}{\frac{w_s}{H}}\cosh\left(\frac{w_s}{H}x + C_1\right) + C_2$$

i.e. $$y = \frac{H}{w_s}\cosh\left(\frac{w_s}{H}x + C_1\right) + C_2 \tag{3.76}$$

where C_1 and C_2 are arbitrary constants which can be obtained from boundary value considerations.

Equation (3.76) can be seen to be the equation for a **catenary**, which is how a cable deforms naturally under its own weight.

The solution of equation (3.76) for practical cases is very difficult, but a good approximation for the small-deflection theory of cables can be obtained by assuming a parabolic variation for y, as in equation (3.77):

$$y = \frac{w}{2H}x^2 + C_1x + C_2 \tag{3.77}$$

where y is the deflection of the cable at any distance x (see Figure 3.90). C_1 and C_2 are arbitrary constants which can be determined from boundary value considerations, and w is load/unit length in the x direction.

To illustrate the solution of equation (3.77), consider the cable of Figure 3.90, which is supported at the same level at its ends. Let T be the tension in the cable at any distance x.

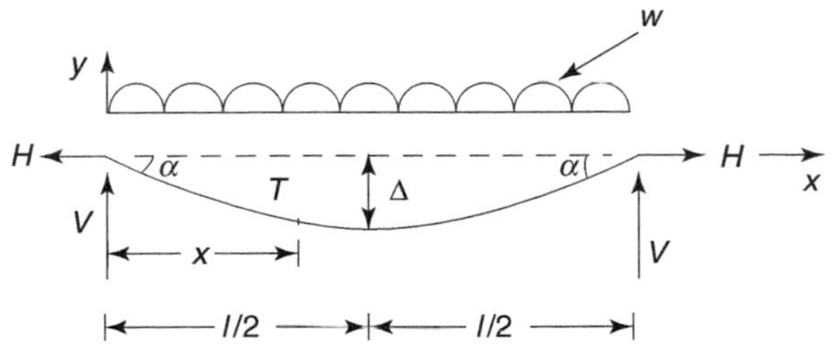

Figure 3.90 Cable with a uniformly distributed load

To determine C_1 and C_2

In equation (3.77), at $x = 0$, $y = 0$

Hence $$C_2 = 0 \tag{3.78}$$

Now $$\frac{dy}{dx} = \frac{w}{H}x + C_1$$

At $x = l/2$ $$\frac{dy}{dx} = 0$$

Hence $$0 = \frac{wl}{2H} + C_1$$

and $$C_1 = -\frac{wl}{2H} \tag{3.79}$$

Substituting equations (3.78) and (3.79) into (3.77) gives:

$$y = \frac{wx^2}{2H} - \frac{wlx}{2H} = \frac{w}{2H}\left(x^2 - lx\right) \tag{3.80}$$

To determine the sag Δ, substitute $x = l/2$ in equation (3.80), i.e.

$$\Delta = \frac{wl^2}{8H} - \frac{wl^2}{4H}$$

i.e. $$\textbf{sag} = \Delta = -\frac{wl^2}{8H} \tag{3.81}$$

From equation (3.81), it can be seen that if the sag is known, H can be found, and, hence, from equilibrium considerations, Δ can be calculated.

To determine T

Let T = tension in the cable at any distance x, as shown by Figures 3.90 and 3.91, and θ = angle of cable with the x axis, at x

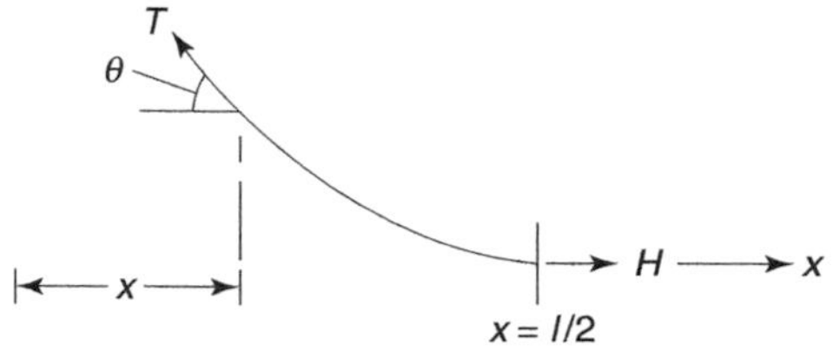

Figure 3.91 Element of cable

Resolving horizontally,

$$T\cos\theta = H \tag{3.82}$$

From equation (3.82), it can be seen that the maximum tension in the cable, namely $\hat{T}$, will occur at its steepest gradient, so that in this case

$$\hat{T} = \frac{H}{\cos\alpha} = H\sec\alpha \tag{3.83}$$

where $\hat{T}$ = maximum tension in the cable, which is at the ends of the cable in Figure 3.64, and α = slope of this cable at its ends

Problem 12. Determine the maximum tension in the cable of Figure 3.92, where w = 120 N/m and H = 20 kN.

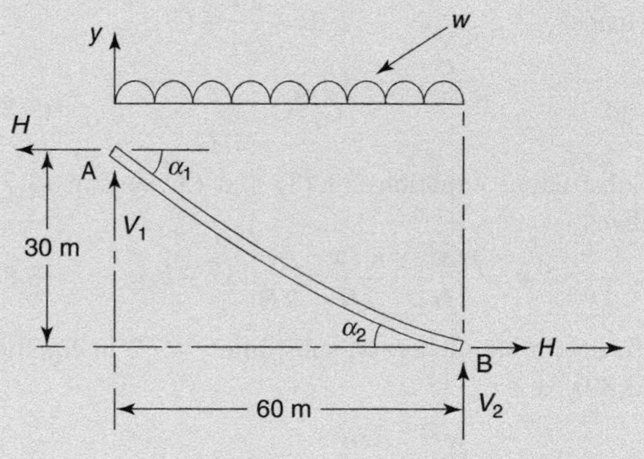

Figure 3.92 Cable supported at different levels

From equation (3.77)

$$y = \frac{w}{2H}x^2 + C_1x + C_2 \tag{3.84}$$

At $x = 0$, $y = 30$, hence $30 = 0 + 0 + C_2$

hence $\quad C_2 = \mathbf{30}$ (3.85)

At $x = 60$, $y = 0$, hence $0 = \dfrac{w}{2H}x^2 + C_1x + C_2$

i.e. $$0 = \frac{120}{2(20000)}60^2 + C_1(60) + 30 \tag{3.86}$$

since w = 120 N/m and H = 20000 N

i.e. $\quad 0 = 10.8 + 60C_1 + 30$

from which $\quad -40.8 = 60C_1$

and $$C_1 = \frac{-40.8}{60} = \mathbf{-0.68} \tag{3.87}$$

Therefore $\quad y = \dfrac{w}{2H}x^2 + C_1x + C_2$

$$= \frac{120x^2}{2\times 20000} - 0.68x + 30$$

i.e. $\quad y = 0.003x^2 - 0.68x + 30$

and $$\frac{dy}{dx} = \mathbf{0.006x - 0.68} \tag{3.88}$$

By inspection, the maximum slope, namely α_1 occurs at $x = 0$

$$\alpha_1 = \tan^{-1}(-0.68) = -34.22°$$

From equation (3.83)

$$\hat{T} = \frac{H}{\cos\alpha_1} = \frac{20000}{\cos(-34.22°)}$$

i.e. **maximum tension in the cable, $\hat{T}$ = 24.19 kN**

Cables under concentrated loads

Cables in this category are assumed to deform, as shown in Figures 3.93 to 3.95.

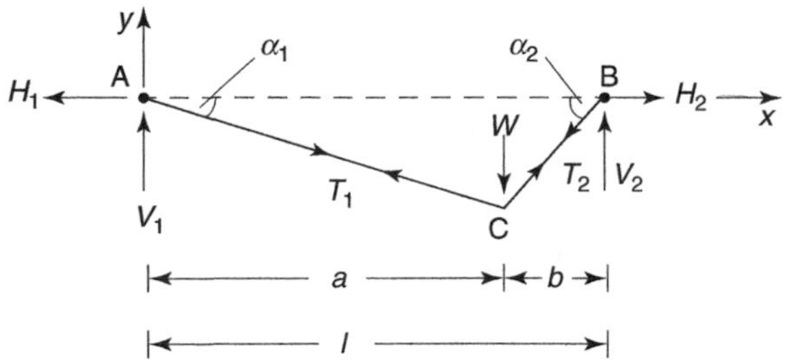

Figure 3.93 Cable with a single concentrated load

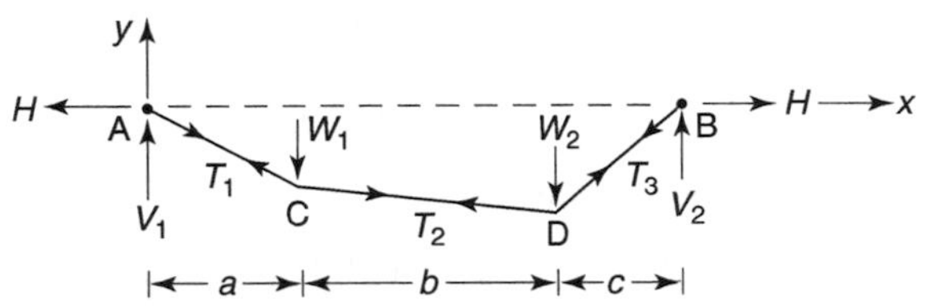

Figure 3.94 Cable with two concentrated loads, but with the end supports at the same level

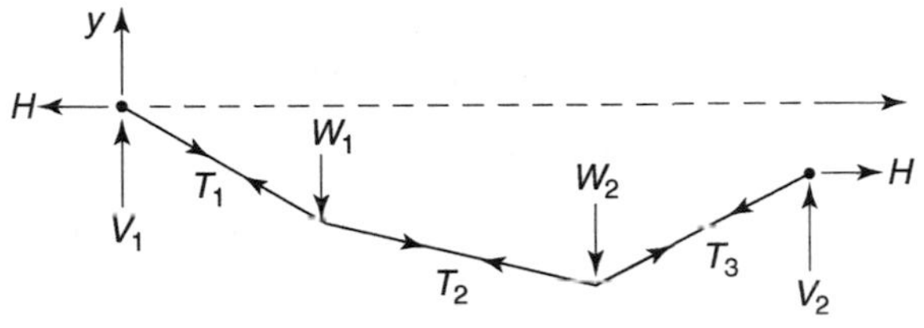

Figure 3.95 Cable with two concentrated loads, but with the end supports at the different levels

Problem 13. Determine expressions for T_1, T_2, H_1, H_2, V_1 and V_2 in terms of α_1 and α_2 for the cable of Figure 3.93.

Resolving vertically gives

$$V_1 + V_2 = W \qquad (3.89)$$

Taking moments about A

$$V_2 \times l = W \times a$$

Therefore $V_2 = \dfrac{W\,a}{l}$

Taking moments about B

$$V_1 \times l = W \times b$$

Therefore $V_1 = \dfrac{W\,b}{l}$

Consider the point C

Resolving horizontally at the point C gives:

$$T_1 \cos\alpha_1 = T_2 \cos\alpha_2$$

from which $T_1 = \dfrac{T_2 \cos\alpha_2}{\cos\alpha_1}$ (3.90)

Resolving vertically at the point C gives:

$$T_1 \sin\alpha_1 + T_2 \sin\alpha_2 = W$$

from which $T_1 = \dfrac{W}{\sin\alpha_1} - T_2 \dfrac{\sin\alpha_2}{\sin\alpha_1}$ (3.91)

Equating (3.90) and (3.91) gives:

$$\frac{W}{\sin\alpha_1} - T_2 \frac{\sin\alpha_2}{\sin\alpha_1} = \frac{T_2 \cos\alpha_2}{\cos\alpha_1}$$

i.e. $\dfrac{W}{\sin\alpha_1} = T_2\left(\dfrac{\cos\alpha_2}{\cos\alpha_1} + \dfrac{\sin\alpha_2}{\sin\alpha_1}\right)$

from which $T_2 = \dfrac{\dfrac{W}{\sin\alpha_1}}{\dfrac{\cos\alpha_2}{\cos\alpha_1} + \dfrac{\sin\alpha_2}{\sin\alpha_1}}$ (3.92)

Substituting equation (3.92) into (3.91), T_1 can be determined.

H_1 and H_2 can be determined by resolution, as follows.

Resolving horizontally gives

$$H_1 = H_2 = T_1 \cos\alpha_1 = T_2 \cos\alpha_2 \qquad (3.93)$$

Problem 14. Determine the tensions T_1, T_2 and T_3 that act in the cable of Figure 3.96. Hence, or otherwise, determine the end forces H_1, V_1, H_2 and V_2

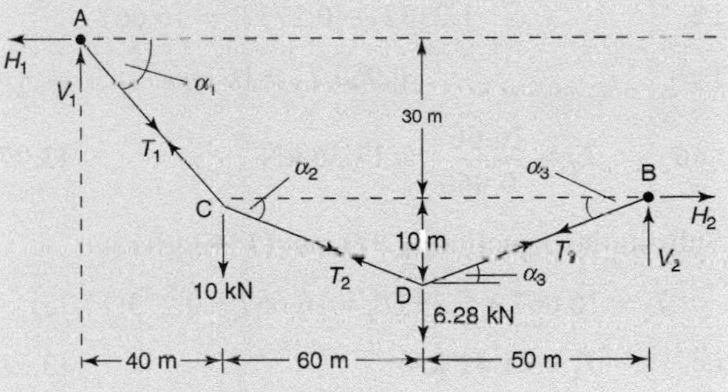

Figure 3.96 Cable with concentrated loads

Resolving vertically

$$V_1 + V_2 = 30 \qquad (3.94)$$

$$\boldsymbol{\alpha_1 = \tan^{-1}\left(\frac{30}{40}\right) = 36.87^\circ}$$

$$\boldsymbol{\alpha_2 = \tan^{-1}\left(\frac{10}{60}\right) = 9.46^\circ}$$

$$\boldsymbol{\alpha_3 = \tan^{-1}\left(\frac{10}{50}\right) = 11.31^\circ}$$

To determine T_1, T_2 and T_3

Resolving horizontally at C

$$T_1 \cos\alpha_1 = T_2 \cos\alpha_2$$

i.e. $$T_1 \cos 36.87° = T_2 \cos 9.46°$$

i.e. $$0.8\, T_1 = 0.986\, T_2$$

Therefore $$T_1 = \left(\frac{0.986}{0.8}\right) T_2$$

$$\mathbf{T_1 = 1.233 T_2} \quad (3.95)$$

Resolving vertically at C

$$T_1 \sin\alpha_1 = 10 + T_2 \sin\alpha_2$$

i.e. $$T_1 \sin 36.87° = 10 + T_2 \sin 9.86°$$

i.e. $$0.6\, T_1 = 10 + 0.164\, T_2$$

and $$T_1 = \frac{10}{0.6} + \frac{0.164}{0.6} T_2$$

i.e. $$\mathbf{T_1 = 16.667 + 0.273 T_1} \quad (3.96)$$

Equating (3.95) and (3.96) gives:

$$1.233 T_2 = 16.667 + 0.273 T_2$$

i.e. $$1.233 T_2 - 0.273 T_2 = 16.667$$

i.e. $$0.960\, T_2 = 16.667$$

and $$T_2 = \frac{16.667}{0.960} = \mathbf{17.36\ kN} \quad (3.97)$$

Substituting equation (3.97) into (3.96) gives:

$$T_1 = 16.667 + 0.273 T_2 = 16.667 + 0.273(17.36)$$

i.e. $$\mathbf{T_1 = 21.41\ kN} \quad (3.98)$$

Resolving horizontally at D

$$T_3 \cos\alpha_3 = T_2 \cos\alpha_2$$

from which $$T_3 = \frac{T_2 \cos\alpha_2}{\cos\alpha_3} = \frac{17.38(\cos 9.46°)}{\cos 11.31°}$$

i.e. $$\mathbf{T_3 = 17.48\ kN} \quad (3.99)$$

To determine H_1, V_1, H_2 and V_2

Resolving vertically at A

$$V_1 = T_1 \sin\alpha_1 = (21.41)(\sin 36.87°)$$

i.e. $$\mathbf{V_1 = 12.85\ kN}$$

Resolving horizontally at A

$$H_1 = T_1 \cos\alpha_1 = (21.41)(\cos 36.87°)$$

i.e. $$\mathbf{H_1 = 17.13\ kN}$$

Resolving vertically at B

$$V_2 = T_3 \sin\alpha_3 = (17.48)(\sin 11.31°)$$

i.e. $$\mathbf{V_2 = 3.43\ kN}$$

Resolving horizontally at B

$$H_2 = T_3 \cos\alpha_3 = (17.48)(\cos 11.31°)$$

i.e. $$\mathbf{H_2 = 17.14\ kN}$$

Note that $\mathbf{H_1 = H_2}$ (as required)

Now try the following Practice Exercise

Practice Exercise 23. Cables

1. Determine the maximum tensile force in the cable of Figure 3.97, and the vertical reactions at its ends, given the following: $w = 200$ N/m and $H = 30$ kN

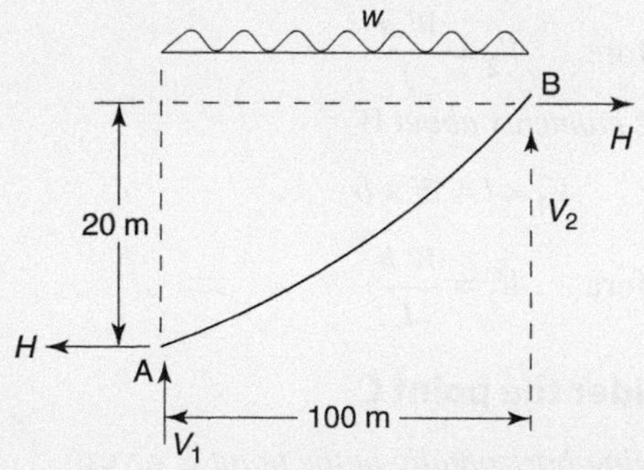

Figure 3.97 Cable with uniformly distributed load

[$V_1 = 4$ kN, $V_2 = 16$ kN, F(at B) =34.0 kN]

2. Determine the tensile forces in the cable of Figure 3.98, together with the end reactions.

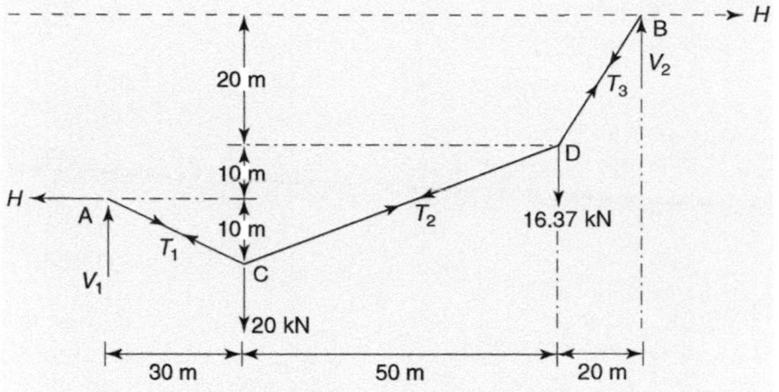

Figure 3.98 Cable with concertrated load

[$T_1 = 28.79$ kN, $T_2 = 29.41$ kN, $T_3 = 38.62$ kN; $H = 27.31$ kN, $V_1 = 9.10$ kN, $V_2 = 27.31$ kN]

3.12 Suspension bridges

The use of cables to improve the structural efficiency of bridges is widely adopted throughout the world, especially for suspension bridges. In this case, the cables are placed under tension between towers, so that the cables exert upward forces to support the main structure of the bridge, via vertical tie-bars, as shown in Figure 3.99. Most of the world's longest bridges are, in fact, suspension bridges.

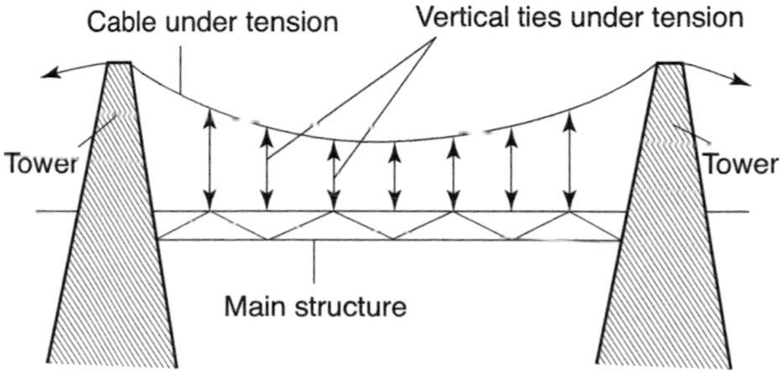

Figure 3.99 Suspension bridge

For fully worked solutions to each of the problems in Exercises 20 to 23 in this chapter, go to the website:
www.routledge.com/cw/bird

Chapter 4

Stress and strain

Why it is important to understand: **Stress and strain**

A good knowledge of some of the constants used in the study of the properties of materials is vital in most branches of engineering, especially in mechanical, manufacturing, aeronautical, civil and structural engineering. For example, most steels look the same, but steels used for the pressure hull of a submarine are about five times stronger than those used in the construction of a small building, and it is very important for the professional and chartered engineer to know what steel to use for what construction; this is because the cost of the high-tensile steel used to construct a submarine pressure hull is considerably higher than the cost of the mild steel, or similar material, used to construct a small building. The engineer must not only take into consideration the ability of the chosen material of construction to do the job, but also its cost. Similar arguments lie in manufacturing engineering, where the engineer must be able to estimate the ability of his/her machines to bend, cut or shape the artefact s/he is trying to produce, and at a competitive price! This chapter provides explanations of the different terms that are used in determining the properties of various materials and a description is given of the standard tensile test used to obtain the strength of various materials, especially the strength and material stiffness of metals. This is aided by a standard tensile test, where the relationship of axial load on a specimen and its axial deflection are described. The engineer can tell a lot from the results of this experiment. For example, the relationship between load and deflection is very different for steel and aluminium alloy and most other materials, including copper, zinc, brass, titanium, composites and so on. By carrying out this test, the engineer can determine the required properties of different materials, and this method, together with its interpretation, is discussed in this chapter. Proof stress, ductility, shear stress and shear strain, Poisson's ratio, hydrostatic stress, composite materials, thermal strain, compound bars, failure by fatigue and failure due to creep are all considered in this chapter.

At the end of this chapter you should be able to:

- recognise tensile and compressive stresses
- state, and appreciate the importance of, Hooke's law in structural design
- understand load-extension relationships – limit of proportionality, elastic limit, yield point, strain hardening and peak load
- define nominal stress and strain and Young's modulus
- appreciate proof stress

- define ductility
- define shear stress and strain
- define Poisson's ratio
- define hydrostatic stress
- appreciate the relationship between material constants E, G, K and v
- appreciate three-dimensional stress
- appreciate the advantages of composite materials
- define thermal strain
- understand the importance of compound bars
- appreciate failure by fatigue and due to creep

Mathematical references

In order to understand the theory involved in this chapter on **stress and strain**, knowledge of the following mathematical topics is required: ***algebraic manipulation/transposition, integration and laws of logarithms***. If help/revision is needed in these areas, see page 49 for some textbook references.

4.1 Introduction

The most elementary definition of stress is that it is the *load per unit area* acting on a surface, rather similar to pressure, except that it can be either tensile or compressive and it does not necessarily act normal to the surface, i.e.

$$\text{Stress} = \frac{\text{load}}{\text{area}} \tag{4.1}$$

In its simplest form, stress acts at an angle to the surface, as shown in Figure 4.1. However, in the form shown in Figure 4.1, it is difficult to apply stress analysis to practical problems, and because of this, the resultant stress is represented by a **normal** or **direct stress** σ, together with a **shear stress**, τ, as shown in Figure 4.2. The stress, σ, in Figure 4.2 is called a normal or direct stress because it acts perpendicularly to the surface under consideration, and the stress, τ, is called a shearing stress because it acts tangentially to the surface, causing shearing action, as shown in Figure 4.3. Thus, if a flat surface is subjected to a force, R, acting at an angle to the surface, it is convenient to represent this resultant force by its two perpendicular components, namely P and F, where P acts normal to the surface and F acts tangentially to the surface, as shown in Figure 4.4.

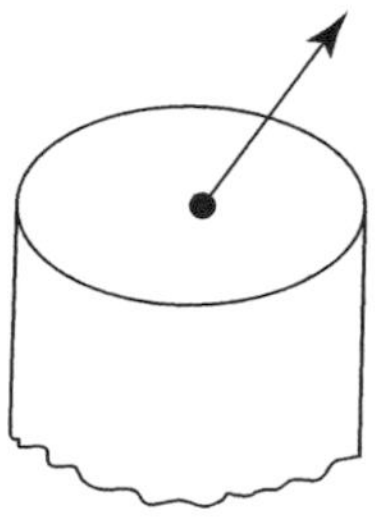

Figure 4.1

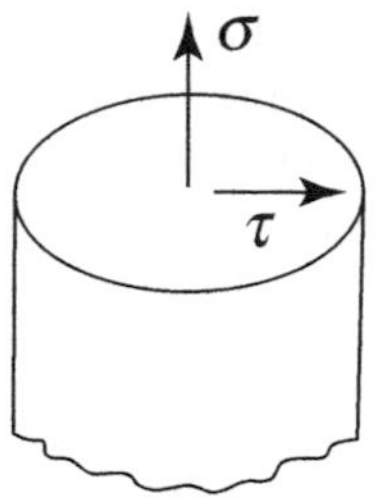

Figure 4.2 Normal and shear stress

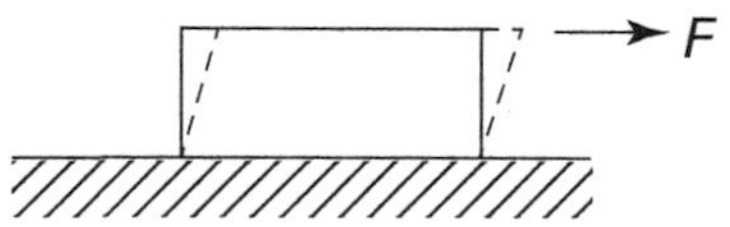

Figure 4.3 Shearing action of F

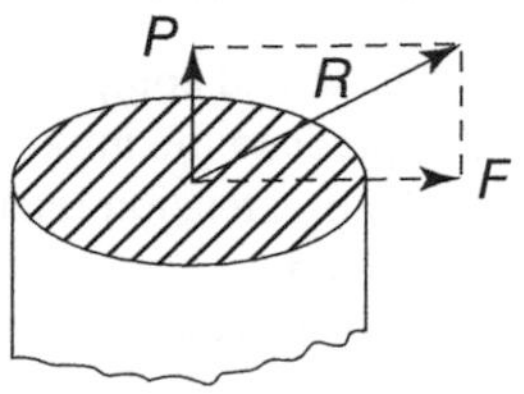

Figure 4.4 Components of R

The effect of P will be to increase the length of the structural component and to cause a normal or direct stress, σ, where

$$\sigma = \frac{P}{A} \quad (4.2)$$

and A is the cross-sectional area. Similarly, the effect of F will be to cause the component to suffer shear deformation, as shown in Figure 4.3, and to cause a shear stress, τ, where

$$\tau = \frac{F}{A} \quad (4.3)$$

The *sign convention for direct stress* is as follows:
Tensile stresses are positive
Compressive stresses are negative

4.2 Hooke's Law

If a length of wire, made from steel or aluminium alloy, is tested in tension, the wire will be found to increase its length linearly with increase in load, for 'smaller' values of load, so that

$$\text{load} \ \alpha \ \text{extension} \quad (4.4)$$

Expression (4.4) was discovered by Robert Hooke,* and it applies to many materials up to the *limit of proportionality* of the material.
In structural design, Hooke's law is very important for the following reasons:

(a) In general, it is not satisfactory to allow the stress in a structural component to exceed the limit of proportionality. This is because, if the stress exceeds this value, it is likely that certain parts of the structural component will suffer permanent deformation.
(b) If the stress in a structure does not exceed the material's limit of proportionality, the structure will return to its undeformed shape on removal of the loading.
(c) For many structures it is undesirable to allow them to suffer large deformations under normal loading.

* Robert Hooke (28 July 1635 – 3 March 1703) was an English natural philosopher, architect and polymath who, amongst other things, discovered the law of elasticity. To find out more about Hooke go to www.routledge.com/cw/bird

4.3 Load-extension relationships

A typical load-extension curve for mild steel, in tension, is shown in Figure 4.5.

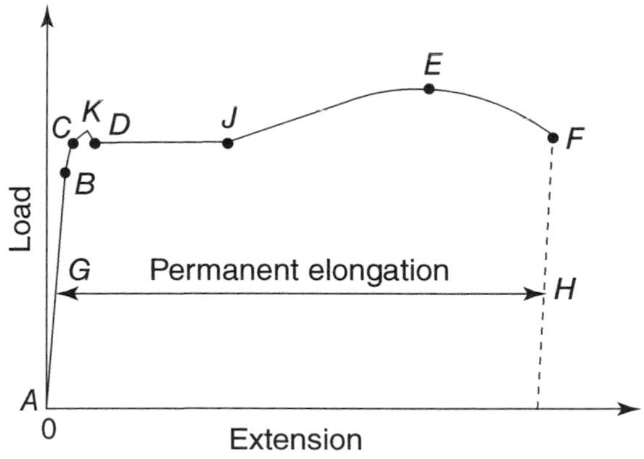

Figure 4.5 Load–extension curve for mild steel

Some important points on Figure 4.5

Limit of proportionality (B). Up to this point, the load-extension curve is linear and elastic.

Elastic limit (C). Up to this point, the material will recover its original shape on removal of the load. The section of the load-extension curve between the limit of proportionality and the elastic limit can be described as non-linear elastic.

Yield point (C/D/J). This is the point of the load-extension curve where the material suffers permanent deformation, i.e. the material behaves plastically beyond this point, and Poisson's ratio is approximately equal to 0.5 (For more on Poisson's ratio see Section 4.7 on page 98). The extension of the specimen from C to J is approximately 40 times greater than the extension of the specimen up to B.

Strain hardening (J/E). After the point J, the material strain hardens, where the slope of the load-extension curve, just above J, is about 1/50th of the slope between the origin and B.

Peak load (E). This load is used for calculating the *ultimate tensile stress* or *tensile strength* of the material. After this point, the specimen 'necks' and eventually fractures at F. ('Necking' means that a certain section of the specimen suffers a local decrease in its cross-sectional area).

Stress-strain curve

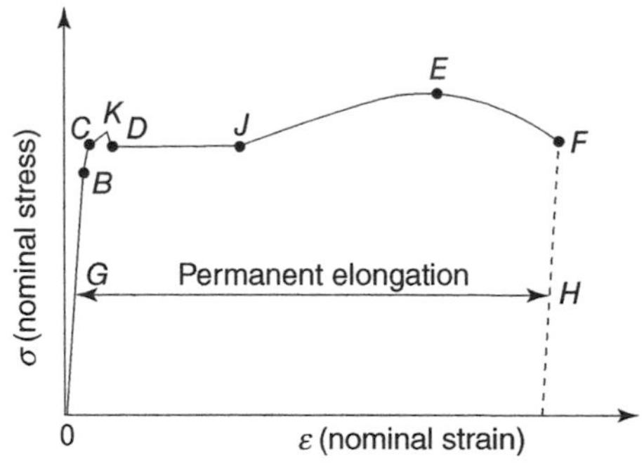

Figure 4.6 Nominal stress–nominal strain relationship

Normally, Figure 4.6 is preferred to Figure 4.5 where

$$\textbf{Nominal stress, } \sigma = \frac{P}{A}$$

$$\textbf{Nominal strain, } \varepsilon = \frac{\delta}{\textbf{original length}}$$

where A = original cross-sectional area

δ = deflection due to P

In Figure 4.6, σ_{YP} = yield stress, and for most structural designs the stress-strain relationship is assumed to be linear up to this point

σ_{UTS} = ultimate tensile stress or nominal peak stress = peak load/A

In the design of structures, the stress is normally not allowed to exceed the limit of proportionality, where the relationship between *0* and *B* can be put in the form

$$\frac{\text{stress}(\sigma)}{\text{strain}(\varepsilon)} = \text{E} \tag{4.5}$$

where E = Young's (or the elastic) modulus

ε = nominal strain

= extension per unit length (unitless)

To determine E for construction materials, such as mild steel, aluminium alloy, etc., it is normal to make a suitable specimen from the appropriate material and to load it in tension in a universal testing machine, as shown in Figure 4.7. Prior to loading the specimen, a small extensometer is connected to the specimen, and measurements are made of the extension of the specimen over the **gauge length *l*** of the extensometer against increasing load, up to the limit of proportionality. The cross-section of the specimen is usually circular and is sensibly constant over the gauge length. In general, E is approximately the same in tension as it is in compression for most structural materials, and some typical values are given in Table 4.1.

Figure 4.7 Specimen undergoing tensile test

Table 4.1 Young's modulus E (N/m^2)

Steel	Aluminium	Copper alloy	Concrete 'new'	Concrete 'old'	Oak (with grain)
2.1×10^{11}	7×10^{10}	1.2×10^{11}	1.9×10^{10}	3.6×10^{10}	1.2×10^{10}

For a practical video demonstration of a tensile test go to www.routledge.com/cw/bird and click on the menu for 'Mechanics of Solids 3rd Edition'

4.4 Proof stress

It should be noted that certain materials, such as aluminium alloy and high tensile steel, do not exhibit a definite yield point, such as that shown in Figure 4.8. For such cases, a 0.1% or 0.2% proof stress is used instead of a yield stress, as shown in Figure 4.8.

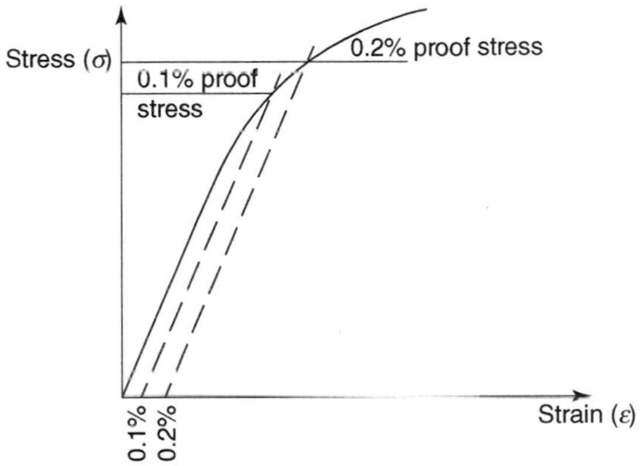

Figure 4.8 Stress–strain curve for aluminium alloy or a typical high tensile steel

To determine the 0.1% proof stress, a strain of 0.1% is set off along the horizontal axis of Figure 4.8, and a straight line is drawn from this point, parallel to the bottom section of the straight line part of the stress-strain relationship. The 0.1% proof stress is measured where the straight line intersects the stress-strain curve, as shown in Figure 4.8. A similar process is used to determine the 0.2% proof stress.

4.5 Ductility

Another important material property is ductility. This can be described as the ability of a material to suffer plastic deformation while still resisting increasing load. The more the material can suffer plastic deformation, the more ductile it is said to be.

Ductility can be measured either by the percentage reduction in area or by the percentage elongation of a measured length of the specimen namely its *gauge length:*

$$\text{Percentage reduction in area} = \frac{A_I - A_F}{A_I}\times 100\%$$

$$\text{Percentage elongation} = \frac{L_I - L_F}{L_I}\times 100\%$$

where A_I = initial cross-sectional area of the tensile specimen
A_F = final cross-sectional area of the tensile specimen
L_I = initial gauge length of the tensile specimen
L_F = final gauge length of the tensile specimen

It should be emphasised that comparisons between various values of percentage elongation will be dependent on whether or not the tensile specimens are geometrically similar.

Table 4.2 Circular cylindrical tensile specimen

Place	L_I	L_I/D_I
UK	$4\sqrt{\text{area}}$	3.54
USA	$4.51\sqrt{\text{area}}$	4.0
Europe	$5.65\sqrt{\text{area}}$	5.0

D_I = initial diameter of tensile specimen

Typical values used in various parts of the world for circular cylindrical specimens are shown in Table 4.2, where L_I is usually taken as 50 mm.

Brittle materials, such as cast iron, have low ductility. A stress-strain curve for a typical brittle material is shown in Figure 4.9.

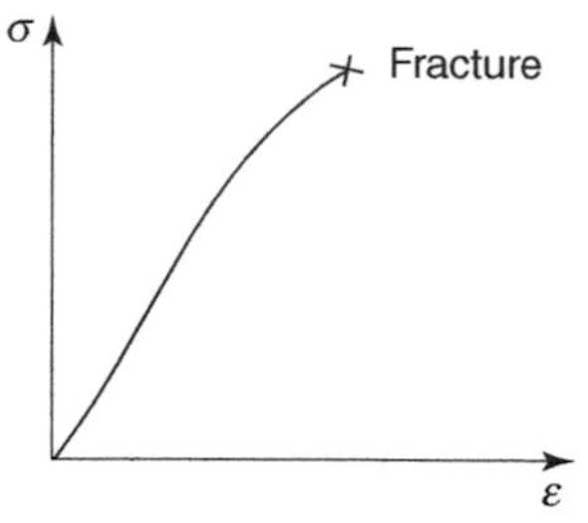

Figure 4.9 Stress–strain curve for a brittle material

Problem 1. Two structural components are joined together by a steel bolt with a screw thread, which has a diameter of 12.5 mm and a pitch of 1 mm, as shown in Figure 4.10. Assuming that the bolt is initially stress free and that the structural components are inextensible, determine the stress in the bolt; if it is tightened by rotating it clockwise by one-eighth of a turn, the bolt will increase its length by an amount δ over the length 50 mm,

where $\delta = \frac{1}{8} \times 1\text{ mm} = 1.25 \times 10^{-4}\text{ m}$. Assume that

Young's modulus for steel, $E = 2 \times 10^{11}\text{N/m}^2$

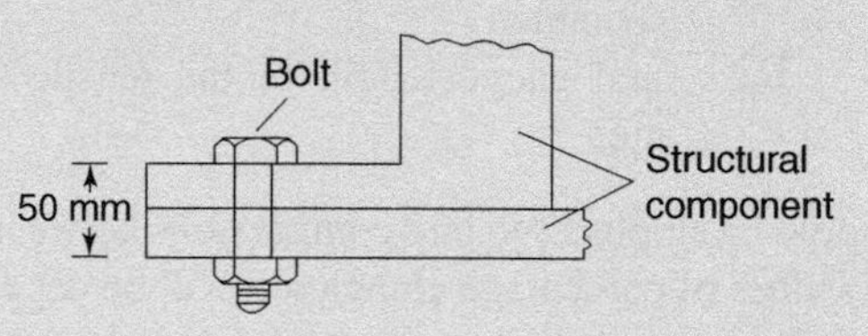

Figure 4.10 Bolt under tensile stress

From equation (4.5), Young's modulus, $E = \frac{\text{stress}, \sigma}{\text{strain}, \varepsilon}$

from which stress, $\sigma = E \times \varepsilon$

$$\text{Strain} = \varepsilon = \frac{\delta}{l} = \frac{1.25 \times 10^{-4}}{50 \times 10^{-3}} = 2.5 \times 10^{-3}$$

Hence, stress $= \sigma = E \times \varepsilon$

$$= 2 \times 10^{11}\text{ N/m}^2 \times 2.5 \times 10^{-3}$$

$$= 500 \times 10^6\text{ N/m}^2$$

i.e. **stress, σ = 500 MN/m² (tensile)**

Problem 2. A copper cable hangs down a vertical mineshaft. Determine the maximum permissible length of the cable if its maximum permissible stress, due to self-weight, must not exceed 10 MN/m². Hence, or otherwise, determine the maximum vertical deflection of this cable due to self-weight. Assume Young's modulus, $E = 1 \times 10^{11}\text{ N/m}^2$, density, $\rho = 8900\text{ kg/m}^3$ and acceleration due to gravity, $g = 9.81\text{m/s}^2$.

The maximum stress in the cable will be at the top. Let $\hat{\sigma}$ = maximum stress = 10 MN/m².

$$\text{Weight of cable} = \rho \times g \times \text{volume}$$

i.e. $\text{weight} = \rho \times g \times (A \times l)$

where A = cross-sectional area and l = length of cable. Maximum stress in the cable,

$$\hat{\sigma} = \frac{\text{weight of cable}}{\text{cross-sectional area}} = \frac{\rho \times g \times A \times l}{A}$$

$$= \rho \times g \times l$$

i.e. $$10 \times 10^6 \frac{\text{N}}{\text{m}^2} = 8900 \frac{\text{kg}}{\text{m}^3} \times 9.81 \frac{\text{m}}{\text{s}^2} \times l$$

from which $$l = \frac{10 \times 10^6 \frac{\text{kg m/s}^2}{\text{m}^2}}{8900 \frac{\text{kg}}{\text{m}^3} \times 9.81 \frac{\text{m}}{\text{s}^2}}$$

= 114.5 m (since 1 N = 1 kg m/s²)

i.e. **the maximum permissible length of the cable = 114.5 m.**

Now to determine δ, the maximum deflection of the cable due to self-weight. The stress in the cable varies *linearly* from zero at the bottom to 10 MN/m² at the top; the strain varies in a similar manner. Therefore

$$\text{average stress} = 5\text{ MN/m}^2$$

From equation (4.5), Young's modulus, $E = \frac{\text{stress}, \sigma}{\text{strain}, \varepsilon}$

from which, $\varepsilon = \frac{\sigma}{E}$

Hence, average strain

$$= \frac{\text{average stress}}{E} = \frac{5 \times 10^6}{1 \times 10^{11}} = 5 \times 10^{-5}$$

Now strain $= \frac{\delta}{l}$ i.e. $5 \times 10^{-5} = \frac{\delta}{114.5}$

and $\delta = 5 \times 10^{-5} \times 114.5 = 5.73 \times 10^{-3}\text{m}$

Hence, **the maximum deflection of the cable = 5.73 mm.**

Problem 3. Determine the profile of a vertical pillar which is to have a constant normal stress due to self-weight

Consider an element of the bar at any distance x from the bottom, as shown in Figure 4.11, where the cross-sectional area is A and the stress is σ.

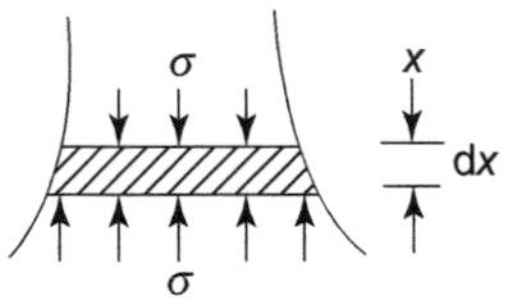

Figure 4.11 Constant strength pillar in compression

Resolving vertically, $\sigma(A + dA) = \sigma A + \rho g A\, dx$ (4.6)

where ρ = density of material, and g = acceleration due to gravity

and $\rho g A\, dx$ = weight of element (shaded area)

From equation (4.6) $\sigma(A + dA) = \sigma A + \rho g A\, dx$

Dividing throughout by σ gives

$$(A + dA) = A + \frac{\rho g A}{\sigma} dx$$

from which $$dA = \frac{\rho g A}{\sigma} dx$$

and $$\frac{dA}{A} = \frac{\rho g}{\sigma} dx$$

Integrating both sides gives $$\int \frac{dA}{A} = \int \frac{\rho g}{\sigma} dx$$

from which $$\ln A = \frac{\rho g x}{\sigma} + c$$

Let arbitrary constant $c = \ln C$, where C is another arbitrary constant.

Then $$\ln A = \frac{\rho g x}{\sigma} + \ln C$$

Rearranging gives $$\ln A - \ln C = \frac{\rho g x}{\sigma}$$

and from the laws of logarithms $$\ln \frac{A}{C} = \frac{\rho g x}{\sigma} \quad (4.7)$$

Taking the antilogarithm of both sides gives

$$\frac{A}{C} = e^{\frac{\rho g x}{\sigma}}$$

and $$A = C e^{\frac{\rho g x}{\sigma}} \quad (4.8)$$

i.e. if A is known for any value of x, C can be determined.

It can be seen from equation (4.8) that the cross-sectional area A is of exponential form, as shown by Figure 4.11; this is necessary for uniform stress to occur in a pillar under self-weight.

Now try the following Practice Exercise

Practice Exercise 24. Stress and strain

1. If a solid stone is dropped into the sea and comes to rest at a depth of 5000 m below the surface of the sea, what will be the stress in the stone? Take the density of seawater = 1020 kg/m^3 and g = 9.81 m/s^2.
   [−50 MN/m^2]

2. A solid bar of length 1 m consists of three shorter sections firmly joined together. Assuming the following apply, determine the change in length of the bar when it is subjected to an axial pull of 50 kN. Assume Young's modulus, E = 2×10^{11} N/m^2.

Section	Length (m)	Diameter (mm)
1	0.2	15
2	0.3	20
3	0.5	30

   [0.698 mm]

3. If the bar of Problem 2 were made from three different materials with the following elastic moduli, determine the change in length of the bar:

Section	E (N/m^2)
1	2×10^{11}
2	7×10^{10}
3	1×10^{11}

   [1.318 mm]

4. A circular-section solid bar of linear taper is subjected to an axial pull of 0.1 MN, as shown in Figure 4.12. If E = 2×10^{11} N/m^2 , by how much will the bar extend?

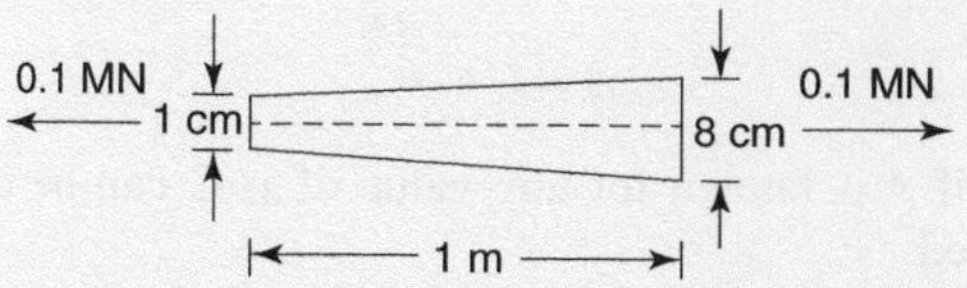

Figure 4.12 [0.796 mm]

5. If a solid stone is dropped into the sea and comes to rest at a depth of 11000 m below the surface of the sea, what will be the stress in the stone? Assume that density of sea water = 1020 kg/m^3 and $g = 9.81$ m/s^2.

[−110 MPa]

6. A solid bar of length 0.7 m consists of three shorter sections firmly joined together. Assuming the following apply, determine the change in length of the bar when it is subjected to an axial pull of 30 kN. Assume that Young's modulus, $E = 2 \times 10^{11}$ N/m^2.

Section	Length (m)	Diameter (mm)
1	0.1	10
2	0.2	15
3	0.4	20

[0.552 mm]

7. If the bar of question 6 were made from three different materials with the following elastic moduli, determine the change in length of the bar.

Section	$E\left(N/m^2\right)$
1	2×10^{11}
2	7×10^{10}
3	1×10^{11}

[1.058 mm]

8. A circular-section solid bar of linear taper is subjected to an axial pull of 0.2 MN, as shown in Figure 4.13. If $E = 2 \times 10^{11}$ N/m^2, by how much will the bar expand?

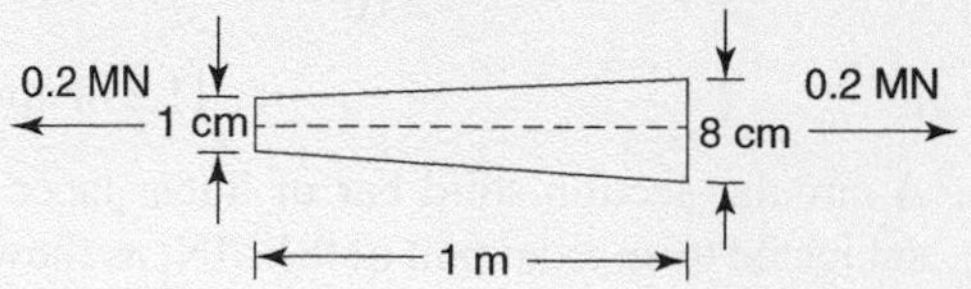

Figure 4.13 [1.592 mm]

4.6 Shear stress and shear strain

From Figure 4.2, it can be seen that shear stress acts tangentially to the surface and that this shear stress (τ) causes the shape of a body to deform, as shown in Figure 4.14. Although shear stress causes a change of shape (or shear strain γ), it does not cause a change in volume.

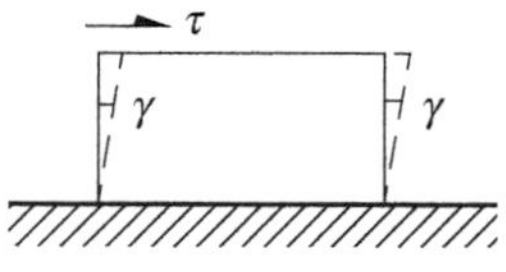

Figure 4.14 Shear stress τ and shear strain γ

By experiment, it has been found that for many materials, the relationship between shear stress and shear strain is given by the equation

$$\frac{\tau}{\gamma} = G \tag{4.9}$$

where G = modulus of rigidity or shear modulus (N/m^2, N/mm^2, MN/m^2, and so on) and γ = shear strain (unitless).

Complementary shear stress

The effect of shearing action on an element of material, as shown in Figure 4.15, will be to cause the system of shearing stresses (τ) in Figure 4.16.

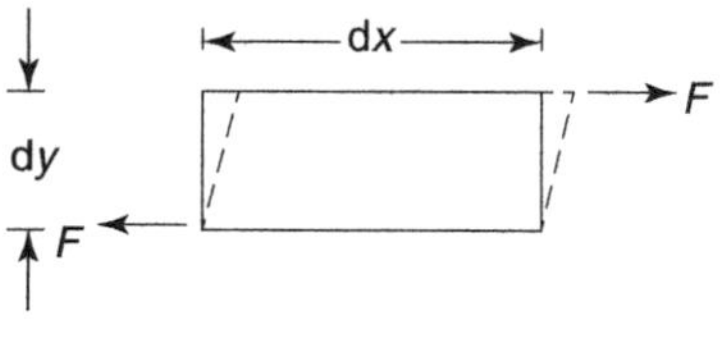

Figure 4.15 Shearing action

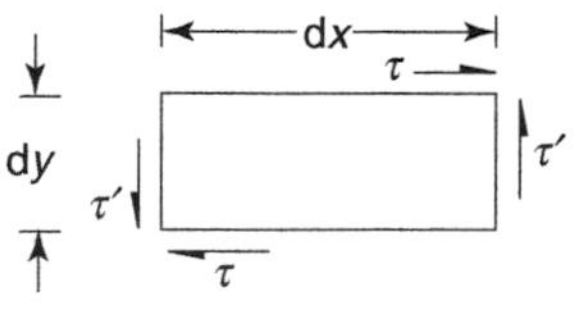

Figure 4.16 Shearing stresses acting on an element

Let t be the thickness of elemental lamina. By considerations of horizontal equilibrium, it is evident that τ will act on the top and bottom surfaces in the manner shown. The effect of these shearing stresses will be to cause a clockwise couple of $\tau \times t \times dx \times dy$, and from equilibrium considerations, the system of shearing stresses τ' must act in the direction shown. Hence, by taking moments about the bottom left-hand corner of the elemental lamina

$$\tau \times t \times dx \times dy = \tau' \times t \times dy \times dx$$

or $$\tau = \tau'$$

That is, the systems of shearing stresses are complementary and equal. Positive and negative shearing stresses are shown in Figure 4.17. Negative shearing stresses are said to cause a counter-clockwise couple, as shown on the vertical surfaces of Figure 4.16, and positive shearing stresses are said to cause a clockwise couple, as shown on the horizontal surfaces of Figure 4.17.

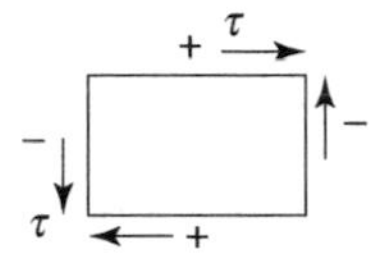

Figure 4.17 Positive and negative complementary shearing stresses

Failure due to exceeding the shear stress

Failure of a component can also take place if the shear stress of the material exceeds the shear stress in 'yield'.

For *ductile materials*

yield shear stress ≈ 0.577 × tensile yield stress (σ_{yp}) (see Section 12.6 on page 287)

For *brittle materials*

failure shear stress ≈ 0.5 × 0.1% proof stress

Problem 4. Two lengths of a material are connected together by a single rivet, as shown in Figure 4.18. If failure of this joint is due to shearing of the rivet, determine the force T to cause failure (in terms of yield stress and diameter of the rivet). The material for the rivet may be assumed to be ductile.

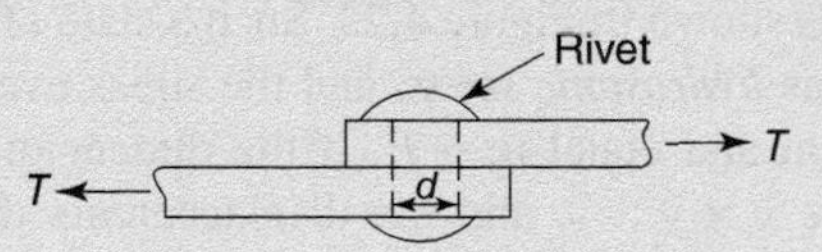

Figure 4.18 Joint using one rivet (elevation)

If the rivet fails in shear, it will fail as shown in Figure 4.19.

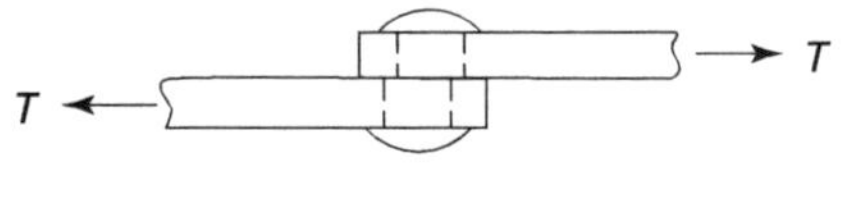

Figure 4.19 Failure mode of rivet (elevation)

Shear stress $\tau_{yield} = 0.577\,\sigma_{yp}$

$$= \frac{\text{force T}}{\text{cross-sectional area of rivet}}$$

$$= \frac{T}{\pi r^2} = \frac{T}{\frac{\pi d^2}{4}} = \frac{4T}{\pi d^2}$$

where d = diameter of rivet

from which **force,** $$\boldsymbol{T = \frac{0.577\,\sigma_{yp} \times \pi d^2}{4}} \quad (4.10)$$

Problem 5. Determine the downward force P required to punch a hole in the plate of Figure 4.20 (in terms of yield shear stress, diameter d and plate thickness t)

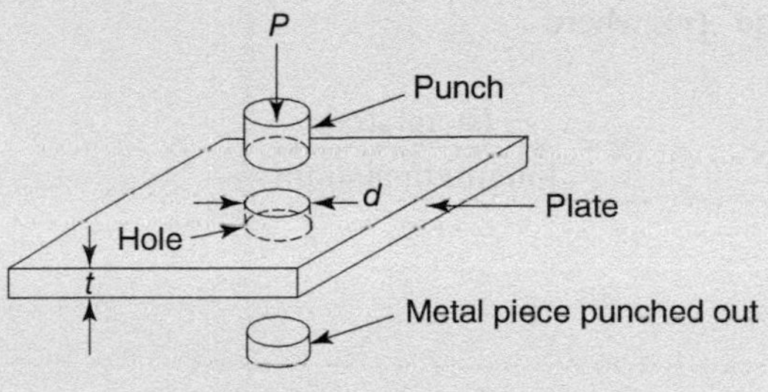

Figure 4.20 Hole punched in metal

The shear stress τ_{yp} that acts on the circular cylindrical metal piece, as the piece is being punched out, is shown in Figure 4.21.

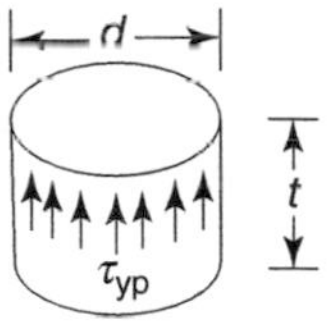

Figure 4.21 Shear stress acting on the metal piece

From equation (4.3) shear stress, $\tau = \frac{\text{force}}{\text{area}}$

from which, force = shear stress × area

Hence, force, $P = \tau_{yp} \times$ the area that the shear stress acts on.

The area that the shear stress acts on is the cylindrical curved surface (which is the circumference × the height of the cylinder).

Hence from Figure 4.20 $$\boldsymbol{P = \tau_{yp} \times \pi d \times t} \quad (4.11)$$

where τ_{yp} is the yield shear stress.

4.7 Poisson's ratio (ν)

If a length of wire or rubber or similar material is subjected to axial tension, as shown in Figure 4.22, then in addition to its length increasing, its lateral dimension will decrease owing to this axial stress.

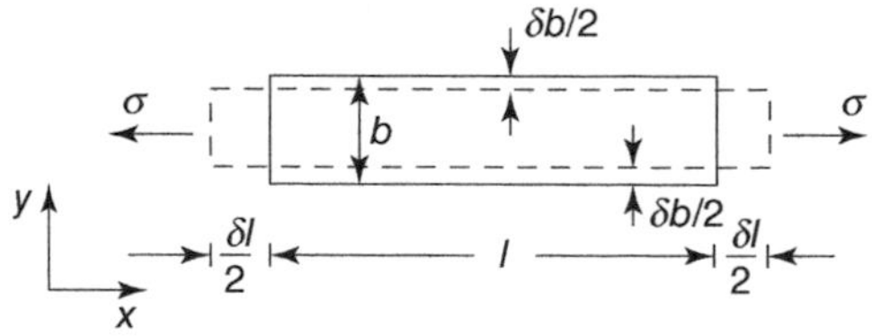

Figure 4.22 Axially loaded element

The relationship between the lateral strain and the axial strain, due to the axial stress, is known as **Poisson's ratio*** **(ν)**, where

$$\nu = \frac{-\text{ lateral strain}}{\text{longitudinal strain}}$$

*Siméon Denis Poisson (21 June 1781 – 25 April 1840), was a French mathematician, geometer, and physicist. The Poisson distribution in probability theory is named after him. To find out more about Poisson go to www.routledge.com/cw/bird

$$\text{Lateral strain} = \frac{-\delta b}{b} \quad \text{and longitudinal strain} = \frac{\delta l}{l}$$

where b = breadth, δb = increment of b, l = length and δl = increment of l

From Figure 4.20, it can be seen that although there is lateral strain, there is no lateral stress; thus, the relationship $\sigma/\varepsilon = E$ only applies for uniaxial stress in the direction of the uniaxial stress.

Now $\quad \varepsilon_x$ = longitudinal strain = σ/E

From above $\nu = \dfrac{-\text{ lateral strain}}{\text{longitudinal strain}}$ from which

$$\text{lateral strain} = -\nu \times \text{longitudinal strain}$$

i.e.
$$\varepsilon_y = \text{lateral strain} = -\nu\,\sigma/E \qquad (4.12)$$

Thus, for two- and three-dimensional systems of stress, equation (4.5) does not apply, and for such cases, the *Poisson effect* of equation (4.12) must also be included (see Chapter 9).

Typical values of Poisson's ratio are 0.3 for steel, 0.33 for aluminium alloy, 0.1 for concrete. It will be shown in the next section that ν cannot exceed 0.5.

For a practical video demonstration of Poisson's ratio experiment go to www.routledge.com/cw/bird and click on the menu for 'Mechanics of Solids 3rd Edition'

4.8 Hydrostatic stress

If a solid piece of material were dropped into the ocean, it would be subjected to a uniform external pressure P, caused by the weight of the water above it. If an internal elemental cube from this piece of material were examined, it would be found to be subjected to a three-dimensional system of stresses, where there is no shear stress, as shown in Figure 4.23. Such a state of stress is known as ***hydrostatic stress,*** and the stress everywhere is normal and equal to $-P$. If the dimensions of the cube are $x \times y \times z$, and the displacements due to P, corresponding to these dimensions, are δx, δy and δz, respectively, then from equation (4.12)

$$\left.\begin{aligned} \varepsilon_x &= \frac{\delta x}{x} = -\frac{P}{E}(1-\nu-\nu) \\ \varepsilon_y &= \frac{\delta y}{y} = -\frac{P}{E}(1-\nu-\nu) \\ \varepsilon_z &= \frac{\delta z}{z} = -\frac{P}{E}(1-\nu-\nu) \end{aligned}\right\} \qquad (4.13)$$

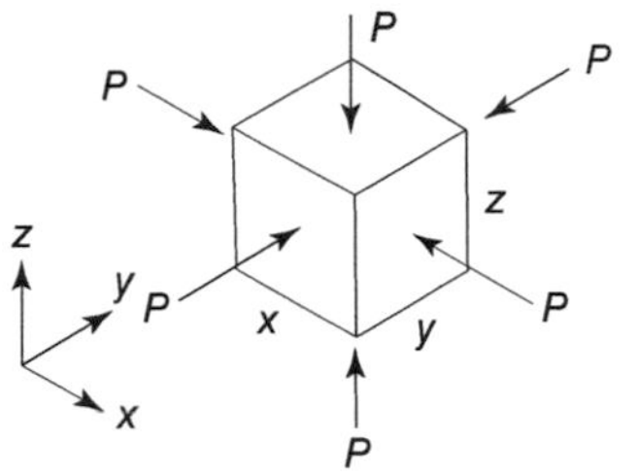

Figure 4.23 Elemental cube under hydrostatic stress

From equations (4.13), it can be seen that if $x = y = z$, and $v > 0.5$, then δx, δy and δz will be positive, which is impossible, i.e. *v cannot be greater than* 0.5. In fact, the stress system of Figure 4.1 (on page 90) will cause the volume of the element to decrease by an amount δV, so that

$$\text{volumetric strain} = \frac{\text{change in volume } (\delta V)}{\text{original volume } (V)} = \frac{\delta V}{V}$$

If the material obeys Hooke's law, then

$$\frac{\text{volumetric stress}}{\text{volumetric strain}} = K \tag{4.14}$$

where K is the **bulk modulus**. In this case, volumetric stress equals $-P$.

To determine volumetric strain

Now the original volume V is

$$V = x \times y \times z$$

and the new volume is

$$V + \delta V = x(1+\varepsilon_x) \times y(1+\varepsilon_y) \times z(1+\varepsilon_z)$$

Assuming the deflections are small and neglecting higher-order terms

$$V + \delta V = xyz(1+\varepsilon_x+\varepsilon_y+\varepsilon_z)$$

from which $\delta V = xyz(1+\varepsilon_x+\varepsilon_y+\varepsilon_z) - V$

$$= xyz(1+\varepsilon_x+\varepsilon_y+\varepsilon_z) - xyz$$

$$= xyz(\varepsilon_x+\varepsilon_y+\varepsilon_z) = V(\varepsilon_x+\varepsilon_y+\varepsilon_z)$$

and volumetric strain $= \dfrac{\delta V}{V} = \dfrac{V\left(\varepsilon_x+\varepsilon_y+\varepsilon_z\right)}{V}$

$$= (\varepsilon_x+\varepsilon_y+\varepsilon_z)$$

i.e. the volumetric strain is the sum of three co-ordinate strains ε_x, ε_y and ε_z.

4.9 Relationship between the material constants *E*, *G*, *K* and *v*

It will be shown in Chapter 9 that the relationships between the **elastic constants** are given by

$$G = \frac{E}{2(1+v)} \tag{4.15}$$

and

$$K = \frac{E}{3(1+2v)} \tag{4.16}$$

Some typical values of G and K (N/m^2) are shown in Table 4.3.

Table 4.3

Material	Steel	Aluminium alloy	Copper
G	8×10^{10}	2.6×10^{10}	4.4×10^{10}
K	1.67×10^{11}	6.68×10^{10}	1.33×10^{11}

4.10 Three-dimensional stress

Stress is a tensor, which can be represented diagrammatically as in Figure 4.24. If the front bottom-left corner of the cubic element of Figure 4.24 is sliced off, to yield the tetrahedral sub-element of Figure 4.25, it can be seen that equilibrium is achieved by the stress tensor σ_{ij} acting at an angle to the generic plane *abc*. It is evident, therefore, that even for a tensile specimen undergoing uniaxial stress, a three-dimensional system of stress can be obtained by examining an elemental cube, tilted at an angle to the axis, as shown in Figure 4.26.

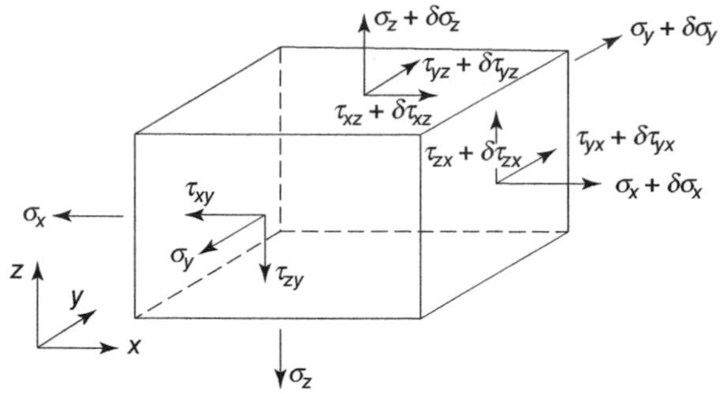

Figure 4.24 Three-dimensional stress system

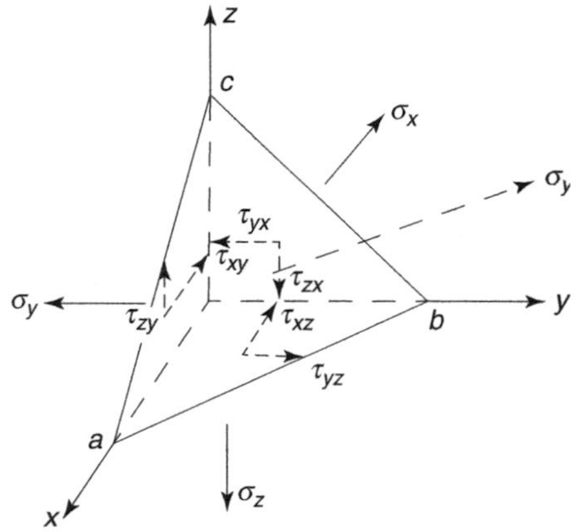

Figure 4.25 Three-dimensional stress

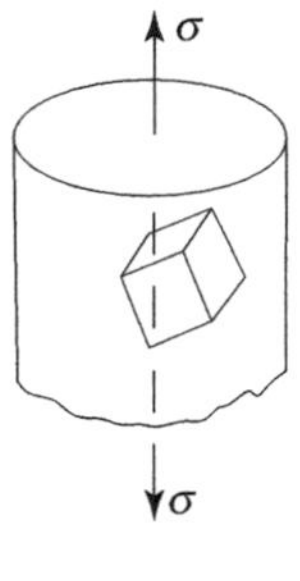

Figure 4.26 Elemental cube

Stress-strain relationship in three dimensions

In three dimensions, the stress-strain relationships are given by

$$\left.\begin{aligned} \varepsilon_x &= \frac{\sigma_x}{E} - \frac{\nu\sigma_y}{E} - \frac{\nu\sigma_z}{E} \\ \varepsilon_y &= \frac{\sigma_y}{E} - \frac{\nu\sigma_x}{E} - \frac{\nu\sigma_z}{E} \\ \varepsilon_z &= \frac{\sigma_z}{E} - \frac{\nu\sigma_x}{E} - \frac{\nu\sigma_y}{E} \end{aligned}\right\} \tag{4.17}$$

where ε_x, ε_y and ε_z are the co-ordinate strains in the *x, y* and *z* directions.

NB The stress-strain relationships quoted in equations (4.5), (4.9), (4.13), etc., are known as the *constitutive laws.*

4.11 Composite materials

So far we have only discussed traditional structural materials, such as metals, etc. In recent years, man-made materials have proved more popular than traditional materials for a number of applications. It is likely that in the future, these man-made materials will probably prove even more popular, especially when they become relatively cheaper to produce (see Chapter 19).

One reason for the popularity of man-made materials is that they prove more resistant to certain forms of corrosion. Another reason for their increasing popularity is that they have a better strength to weight ratio, as can be seen from the figures in Table 4.4.

This improved strength to weight ratio is of much importance when the weight of a structure is at a premium, such as in the cases of aircraft and submarine structures. It should also be noted from Table 4.4 that for composites, *tensile modulus* is used in place of Young's modulus.

Table 4.4 Material properties of man-made fibres

Material	Density (kg/m^3)	Tensile strength (GPa)	Tensile modulus (GPa)
E-glass fibre	2550	3.4	72
Boron fibre	2600	3.8	380
Carbon fibre (UHS)	1750	5.2	270
Piano wire	7860	3.0	210

UHS – Ultra High Strength

For two-dimensional structures, most composite materials are assumed to have orthogonal properties. That is, the material properties of the two-dimensional composite are different in directions perpendicular to each other. Some typical material properties of composites, made from a combination of fibres and resins, are shown in Table 4.5. For these composites, the fibres were laid in the x direction in a resin.

NB $\quad \nu_x E_y = \nu_y E_x \qquad (4.18)$

Table 4.5 Material properties of fibre composites

Material	Density (kg/m^3)	E_x (GPa)	E_y(GPa)	ν_x	G_{xy} (GPa)
Glass fibre/resin	1800	40	8.5	0.25	4.1
Carbon fibre/resin	1600	180	10	0.28	7.2
Kevlar/resin	1430	75	5.5	0.34	2.3
Mild steel	7860	200	200	0.30	7.7

E_x = tensile modulus in the x direction; E_y = tensile modulus in the y direction; G_{xy} = shear modulus in the x-y plane; ν_x = Poisson's ratio due to σ_x; ν_y = Poisson's ratio due to σ_y

To achieve more even values for the material properties of composites in all directions (see Chapter 19), it is normal to lay the fibres in several layers, as shown in Figure 4.27, the whole being set in a resin.

Figure 4.27 Five layers of fibre reinforcement

For two-dimensional orthotropic materials, the stress-strain relationships are as follows:

$$\left.\begin{aligned}\varepsilon_x &= \frac{\sigma_x}{E_x} - \frac{\nu_y \sigma_y}{E_y}\\ \varepsilon_y &= \frac{\sigma_y}{E_y} - \frac{\nu_x \sigma_x}{E_x}\end{aligned}\right\} \qquad (4.19)$$

$$\gamma_{xy} = \frac{\tau_{xy}}{G_{xy}} \qquad (4.20)$$

where ε_x = direct strain in the x direction

ε_y = direct strain in the y direction

and γ_{xy} = shear strain in the x-y plane

Equations (4.19) can be rearranged as shown below, which is more convenient for calculating stresses from measured strains.

From equation (4.19) $\varepsilon_x = \dfrac{\sigma_x}{E_x} - \dfrac{\nu_y \sigma_y}{E_y}$

from which $\quad \dfrac{\sigma_x}{E_x} = \varepsilon_x + \dfrac{\nu_y \sigma_y}{E_y}$

and $$\sigma_x = E_x \varepsilon_x + \frac{E_x \nu_y \sigma_y}{E_y} \qquad \text{(i)}$$

Also, from (4.19) $\quad \varepsilon_y = \dfrac{\sigma_y}{E_y} - \dfrac{\nu_x \sigma_x}{E_x}$

from which $\quad \dfrac{\sigma_y}{E_y} = \varepsilon_y + \dfrac{\nu_x \sigma_x}{E_x}$

and $$\sigma_y = E_y \varepsilon_y + \frac{E_y \nu_x \sigma_x}{E_x} \qquad \text{(ii)}$$

Substituting (ii) into (i) gives

$$\sigma_x = E_x \varepsilon_x + \frac{E_x \nu_y \left(E_y \varepsilon_y + \dfrac{E_y \nu_x \sigma_x}{E_x} \right)}{E_y}$$

$$= E_x \varepsilon_x + \frac{E_x \nu_y E_y \varepsilon_y + \nu_y E_y \nu_x \sigma_x}{E_y}$$

i.e. $\quad \sigma_x = E_x \varepsilon_x + E_x \nu_y \varepsilon_y + \nu_y \nu_x \sigma_x$

and $\quad \sigma_x - \nu_y \nu_x \sigma_x = E_x \varepsilon_x + E_x \nu_y \varepsilon_y$

Thus $\sigma_x(1-\nu_y\nu_x)=E_x(\varepsilon_x+\nu_y\varepsilon_y)$

from which $$\sigma_x=\frac{E_x}{(1-\nu_x\nu_y)}(\varepsilon_x-\nu_y\varepsilon_y) \quad (4.21)$$

Similarly, it may be shown that

$$\sigma_y=\frac{E_y}{(1-\nu_x\nu_y)}(\varepsilon_y-\nu_x\varepsilon_x) \quad (4.21)$$

4.12 Thermal strain

If a bar of length l and coefficient of linear expansion α is subjected to a temperature rise T, its length will increase by an amount αlT, as shown in Figure 4.28. Thus, at this temperature, the natural length of the bar is $l(1+\alpha T)$. In this condition, although the bar has a thermal strain of αT, it has no thermal stress, but if the free expansion αlT is prevented from taking place, compressive thermal stresses will occur.

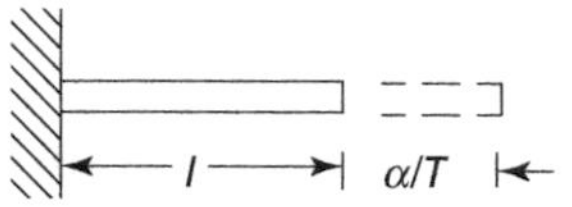

Figure 4.28 Free thermal expansion of a bar

Such problems are of great importance in a number of practical situations, including railway lines and pipe systems, and the following two examples will be used to demonstrate the problem.

Problem 6. A steel prop is used to stabilise a building, as shown in Figure 4.29. If the compressive stress in the prop at a temperature of 20°C is 50 MN/m^2, what will the stress be in the prop if the temperature is raised to 30°C? At what temperature will the prop cease to be effective? It may be assumed that the floor, the building or the ends of the prop do not move. Assume that Young's modulus, $E=2\times10^{11}\,N/m^2$ and the coefficient of linear expansion, $\alpha=15\times10^{-6}/°C$

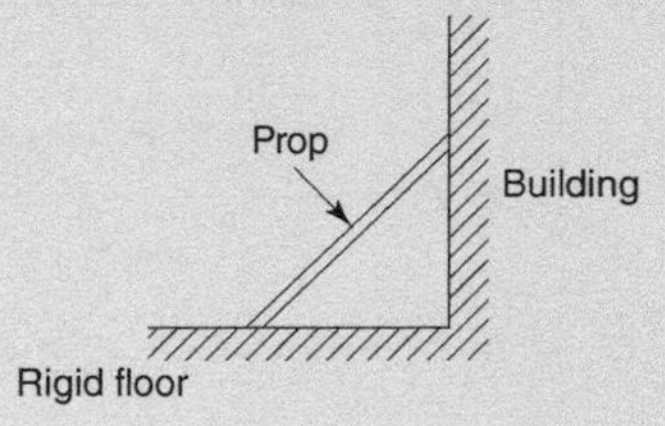

Figure 4.29 Prop

$$\text{Additional thermal strain} = -\frac{\text{extension}}{\text{original length}}$$

$$=-\frac{\alpha\times l\times T}{l}=-\alpha\times T$$

$$=-\,15\times10^{-6}\times(30-20)=-\,-150\times10^{-6}$$

Thermal stress,

$$\sigma_T=E\varepsilon_T=2\times10^{11}\times-150\times10^{-6}$$

$$=-\,30\times10^{6}\,N/m^2$$

$$=-\,30\ MN/m^2$$

Hence stress at 30°C = initial stress + σ_T

$= -\,50-30$

$= -\,80$ MN/m^2

For the prop to be ineffective, it will be necessary for the temperature to drop, so that the initial compressive stress of 50 MN/m^2 is nullified, i.e.

Thermal stress = 50 MN/m^2

$$\text{Thermal strain}=\frac{\sigma}{E}=\frac{50\times10^6}{2\times10^{11}}=2.5\times10^{-4}$$

i.e. $2.5\times10^{-4}=-\alpha T$

where T is the temperature fall.

Hence, $$T=\frac{2.5\times10^{-4}}{-\alpha}=\frac{2.5\times10^{-4}}{-15\times10^{-6}}=-16.67°C$$

i.e. **the temperature for the prop to be ineffective** = 20 − 16.67 = **3.33°C**

Problem 7. A steel rail may be assumed to be in a stress-free condition at 10°C. If the stress required to cause buckling is − 75 MN/m^2, what temperature rise will cause the rail to buckle, assuming that the rail is rigidly restrained at its ends and that its material properties are as in Problem 6.

$$\text{Thermal strain, } \varepsilon=\frac{\sigma}{E}=aT \quad (4.22)$$

i.e. $$\frac{75\times10^6}{2\times10^{11}}=15\times10^{-6}\times T$$

from which, $$T=\frac{75\times10^6}{2\times10^{11}\times15\times10^{-6}}=25$$

i.e. temperature rise = T = 25°C

and **the temperature to cause buckling** = 10°C + 25°C = **35°C**

Now try the following Practice Exercise

Practice Exercise 25. Thermal strain

1. If the bar of Problem 4 in Practice Exercise 24, page 95, were prevented from moving axially by two rigid walls and subjected to a temperature rise of 10°C, what would be the maximum stress in the bar? Assume the 0.1 MN load is not acting. Assume that coefficient of linear expansion, $\alpha = 15\times10^{-6}\ /\ °C$
[– 240 MN/m² at the smaller end]
2. If the bar in Problem 4 in Exercise 24 were prevented from moving axially by two rigid walls and subjected to a temperature rise of 10°C, what would be the maximum stress in the bar? Assume the 0.2 MN load is not acting, and $\alpha = 15\times10^{-6}\ /\ °C$.
[–480 MN/m² at the smaller end]

4.13 Compound bars

Compound bars are of much importance in a number of different branches of engineering, including reinforced concrete pillars, bimetallic bars, etc., and in this section the solution of such problems usually involves two important considerations, namely

(a) compatibility (or consideration of displacements)
(b) equilibrium.

NB It is necessary to introduce compatibility in this section as compound bars are, in general, *statically indeterminate*. To demonstrate the method of solution, the next two examples will be considered.

Problem 8. A solid bar of cross-sectional area A_1, elastic modulus E_1 and coefficient of linear expansion α_1 is surrounded co-axially by a hollow tube of cross-sectional area A_2, elastic modulus E_2 and coefficient of linear expansion α_2, as shown in Figure 4.30. If the two bars are secured firmly to each other, so that no slipping takes place with temperature change, determine expressions for the thermal stresses due to a temperature rise *T*. Both bars have an initial length *l*.

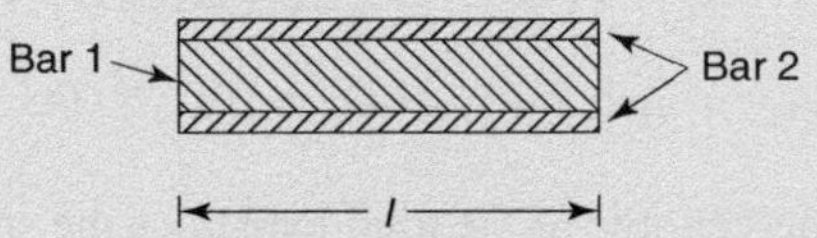

Figure 4.30 Compound bar

Compatibility considerations

There are two unknowns; therefore two simultaneous equations will be required. The first equation can be obtained by considering the *compatibility* (i.e. 'deflections') of the bars, with the aid of Figure 4.31

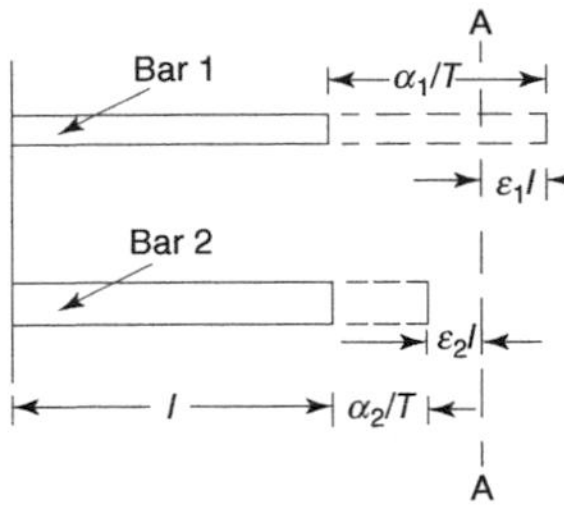

Figure 4.31 'Deflections' of compound bar

Free expansion of bar 1 = $\alpha_1 l\ T$
Free expansion of bar 2 = $\alpha_2 l\ T$

In practice, however, the final resting position of the compound bar will be *somewhere* between these two positions (i.e. at the position A-A). To achieve this, it will be necessary for bar 2 to be pulled out by a distance $\varepsilon_2 l$ and for bar 1 to be pushed in by a distance, $\varepsilon_1 l$, where

ε_1 = compressive strain in 1
and ε_2 = tensile strain in 2

From consideration of compatibility ('deflections') in Figure 4.30

$$\alpha_1 lT - \varepsilon_1 l = \alpha_2 lT + \varepsilon_2 l$$

or
$$\alpha_1 lT - \alpha_2 lT - \varepsilon_2 l = \varepsilon_1 l$$

or
$$\alpha_1 T - \alpha_2 T - \varepsilon_2 = \varepsilon_1$$

i.e. compressive strain in bar 1, $\varepsilon_1 = (\alpha_1 - \alpha_2)T - \varepsilon_2$

and since $\dfrac{\text{stress}}{\text{strain}} = E$,

stress, $\sigma_1 = \varepsilon_1 E_1 = (\alpha_1 - \alpha_2)E_1 T - \varepsilon_2 E_1$

However, tensile strain $\varepsilon_2 = \dfrac{\sigma_2}{E_2}$

Hence stress, $$\sigma_1 = (\alpha_1 - \alpha_2)E_1 T - \sigma_2 \frac{E_1}{E_2} \qquad (4.23)$$

Equilibrium considerations

To obtain the second simultaneous equation, it will be necessary to consider *equilibrium*.

Let F_1 = compressive force in bar 1
and F_2 = tensile force in bar 2

Now $F_1 = F_2$

i.e. $\sigma_1 A_1 = \sigma_2 A_2$

from which $$\sigma_1 = \sigma_2 \frac{A_2}{A_1} \quad (4.24)$$

Equating (4.23) and (4.24) gives

$$\sigma_2 \frac{A_2}{A_1} = (\alpha_1 - \alpha_2)E_1 T - \sigma_2 \frac{E_1}{E_2}$$

Therefore $$\sigma_2 \frac{A_2}{A_1} + \sigma_2 \frac{E_1}{E_2} = (\alpha_1 - \alpha_2)E_1 T$$

i.e. $$\sigma_2 \left(\frac{A_2}{A_1} + \frac{E_1}{E_2} \right) = (\alpha_1 - \alpha_2)E_1 T$$

and $$\sigma_2 = \frac{(\alpha_1 - \alpha_2)E_1 T}{\left(\frac{A_2}{A_1} + \frac{E_1}{E_2} \right)} = \frac{(\alpha_1 - \alpha_2)E_1 T}{\frac{A_2 E_2 + A_1 E_1}{A_1 E_2}}$$

Multiplying numerator and denominator by $A_1 E_2$ gives

i.e. $$\boldsymbol{\sigma_2 = \frac{(\alpha_1 - \alpha_2)E_1 E_2 A_1 T}{(A_1 E_1 + A_2 E_2)}} \text{ **(tensile)**} \quad (4.25)$$

Similarly

$$\boldsymbol{\sigma_1 = \frac{(\alpha_1 - \alpha_2)E_1 E_2 A_2 T}{(A_1 E_1 + A_2 E_2)}} \text{ **(compressive)**} \quad (4.26)$$

Equations (4.25) and (4.26) are used in Problem 11.

Problem 9. If the solid bar of Problem 8 did not undergo a temperature change, but instead was subjected to a tensile axial force P, as shown in Figure 4.32, determine expressions for stresses σ_2 and σ_2

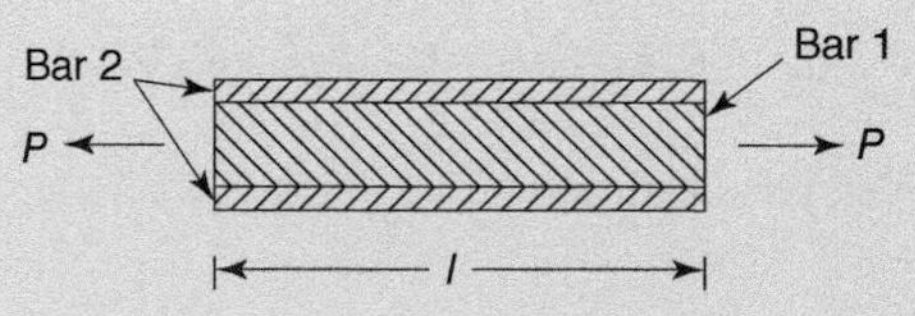

Figure 4.32 Compound bar under axial tension

There are two unknowns; therefore two simultaneous equations will be required. The first of these simultaneous equations can be obtained by considering *compatibility*, i.e.

deflection of bar 1 = deflection of bar 2

i.e. $$\varepsilon_1 l = \varepsilon_2 l$$

and $$\frac{\sigma_1}{E_1} = \frac{\sigma_2}{E_2}$$

Therefore $$\sigma_1 = \sigma_2 \frac{E_1}{E_2} \quad (4.27)$$

The second simultaneous equation can be obtained by considering *equilibrium*.

Let F_1 = tensile force in bar 1

F_2 = tensile force in bar 2

Now $P = F_1 + F_2$

i.e. $$P = \sigma_1 A_1 + \sigma_2 A_2 \quad (4.28)$$

Substituting (4.27) into (4.28) gives

$$P = \left(\sigma_2 \frac{E_1}{E_2} \right) A_1 + \sigma_2 A_2$$

i.e. $$P = \sigma_2 \left(\frac{E_1 A_1}{E_2} + A_2 \right) = \sigma_2 \left(\frac{E_1 A_1 + E_2 A_2}{E_2} \right)$$

from which $$\boldsymbol{\sigma_2 = \frac{P E_2}{(A_1 E_1 + A_2 E_2)}} \quad (4.29)$$

From equation (4.27)

$$\sigma_1 = \sigma_2 \frac{E_1}{E_2} = \frac{P E_2}{(A_1 E_1 + A_2 E_2)} \frac{E_1}{E_2}$$

i.e. $$\boldsymbol{\sigma_1 = \frac{P E_1}{(A_1 E_1 + A_2 E_2)}} \quad (4.30)$$

NB If P is a compressive force, then both σ_1 and σ_2 will be compressive stresses.

Equations (4.29) and (4.30) are used in the following worked problem.

Problem 10. A concrete pillar, which is reinforced with steel rods, supports a compressive axial load of 1 MN. Determine σ_1 and σ_2 given the following:

Steel: $A_1 = 3 \times 10^{-3}\ \text{m}^2$, $E_1 = 2 \times 10^{11}\ \text{N/m}^2$

Concrete: $A_2 = 0.1\ \text{m}^2$, $E_2 = 2 \times 10^{10}\ \text{N/m}^2$

What percentage of the total load does the steel reinforcement take?

From equation (4.30)

$$\sigma_1 = \frac{PE_1}{(A_1E_1 + A_2E_2)} = \frac{-1\times10^6\times2\times10^{11}}{(3\times10^{-3}\times2\times10^{11}+0.1\times2\times10^{10})}$$

$$= \frac{-2\times10^{17}}{(6\times10^8+2\times10^9)} = \frac{-2\times10^{17}}{2.6\times10^9}$$

i.e. $\sigma_1 = -76.92\times10^6\,\text{N/m}^2 = \mathbf{-76.92\,MN/m^2}$ (4.31)

From equation (4.29)

$$\sigma_2 = \frac{PE_2}{(A_1E_1 + A_2E_2)} = \frac{-1\times10^6\times2\times10^{10}}{(3\times10^{-3}\times2\times10^{11}+0.1\times2\times10^{10})}$$

$$= \frac{-2\times10^{16}}{(6\times10^8+2\times10^9)} = \frac{-2\times10^{16}}{2.6\times10^9}$$

i.e. $\sigma_2 = -7.69\times10^6\,\text{N/m}^2 = \mathbf{-7.69\,MN/m^2}$ (4.32)

Since stress $= \frac{\text{force}}{\text{area}}$, force = stress × area

Hence force on the steel, $F_1 = \sigma_1 \times A_1$

$$= -76.92\times10^6\,\text{N/m}^2\times3\times10^{-3}\,\text{m}^2$$

i.e. $F_1 = 2.308\times10^5\,\text{N}$

Therefore, **the percentage total load taken by the steel reinforcement** $= \frac{2.308\times10^5}{1\times10^6}\times100\%$

$$= \mathbf{23.08\%}$$

Problem 11. If the pillar of Problem 10 were subjected to a temperature rise of 30°C, what would be the values of σ_1 and σ_2? Assume that $\alpha_1 = 15\times10^{-6}\,/°\text{C}$ for steel, and $\alpha_2 = 12\times10^{-6}\,/°\text{C}$ for concrete.

As α_1 is larger than α_2, the effect of a temperature rise will cause the 'thermal stresses' in 1 to be compressive and those in 2 to be tensile.

From equation (4.26)

$$\boldsymbol{\alpha_1}\,\textbf{(thermal)} = \frac{(\alpha_1-\alpha_2)E_1E_2A_2T}{(A_1E_1+A_2E_2)}$$

$$= -\frac{(15\times10^{-6}-12\times10^{-6})\times2\times10^{11}\times2\times10^{10}\times0.1\times30}{2.6\times10^9}$$

$$= -\frac{3.6\times10^{16}}{2.6\times10^9} = -13.85\times10^6\,\text{N/m}^2$$

$$= \mathbf{-13.85\,MN/m^2} \quad (4.33)$$

From equation (4.25)

$$\boldsymbol{\alpha_2}\,\textbf{(thermal)} = \frac{(\alpha_1-\alpha_2)E_1E_2A_1T}{(A_1E_1+A_2E_2)}$$

$$= \frac{(15\times10^{-6}-12\times10^{-6})\times2\times10^{11}\times2\times10^{10}\times3\times10^{-3}\times30}{2.6\times10^9}$$

$$= \frac{1.08\times10^{15}}{2.6\times10^9} = 0.42\times10^6\,\text{N/m}^2 = \mathbf{0.42\,MN/m^2} \quad (4.34)$$

From equations (4.31) to (4.34),

$$\alpha_1 = -76.92 - 13.85 = \mathbf{-90.77\,MN/m^2}$$

and $\quad \alpha_2 = -7.69 + 0.42 = \mathbf{-7.27\,MN/m^2}$

Problem 12. A rigid horizontal bar is supported by three rods, where the outer rods are made from aluminium alloy and the middle rod from steel, as shown in Figure 4.33. If the temperature of all three rods is raised by 50°C, what will be the thermal stresses in the rods?

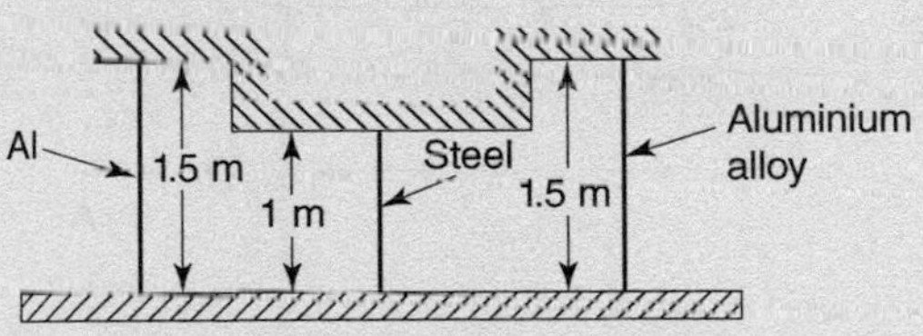

Figure 4.33 Compound bar

The following may be assumed:

$A_a = 3\times10^{-3}\,\text{m}^2$ = sectional area of one aluminium rod

$E_a = 7\times10^{10}\,\text{N/m}^2$ = elastic modulus of aluminium

$\alpha_a = 25\times10^{-6}\,/°\text{C}$ = coefficient of linear expansion of aluminium

$A_s = 2\times10^{-3}\,\text{m}^2$ = sectional area of steel rod

$E_s = 2\times10^{11}\,\text{N/m}^2$ = elastic modulus of steel

$\alpha_s = 15\times10^{-6}\,/°\text{C}$ = coefficient of linear expansion of steel

Free expansion of aluminium

$$= \alpha_a \times l \times T = 25 \times 10^{-6} \times 1.5 \times 50$$

$$= \mathbf{1.875 \times 10^{-3}\ m}$$

Free expansion of steel

$$= \alpha_s \times l \times T = 15 \times 10^{-6} \times 1 \times 50$$

$$= \mathbf{7.5 \times 10^{-4}\ m}$$

That is, as the free expansion of the aluminium is greater than that of steel, the aluminium will be in compression and the steel in tension, owing to a temperature rise.

Let ε_a = compressive strain in aluminium

and ε_s = tensile strain in steel

Now there are two unknowns; therefore two simultaneous equations will be required. The first of these can be obtained by considering *compatibility,* with the aid of Figure 4.34.

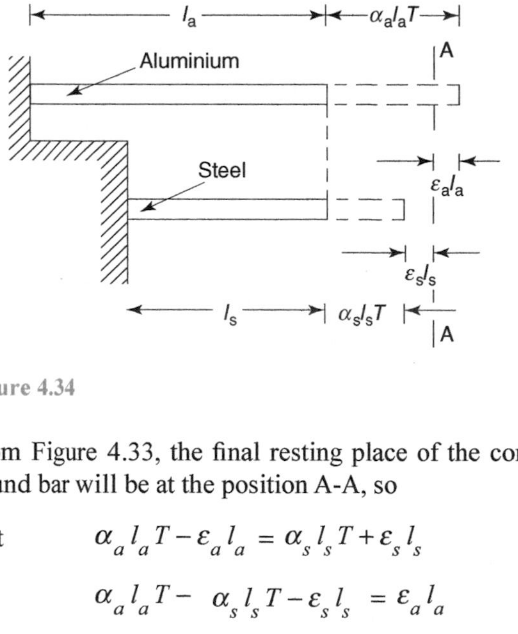

Figure 4.34

From Figure 4.33, the final resting place of the compound bar will be at the position A-A, so

that $\quad \alpha_a l_a T - \varepsilon_a l_a = \alpha_s l_s T + \varepsilon_s l_s$

or $\quad \alpha_a l_a T - \alpha_s l_s T - \varepsilon_s l_s = \varepsilon_a l_a$

and $\quad \varepsilon_a l_a = (\alpha_a l_a - \alpha_s l_s) T - \varepsilon_s l_s$

from which $\varepsilon_a = (\alpha_a l_a - \alpha_s l_s)\dfrac{T}{l_a} - \varepsilon_s \dfrac{l_s}{l_a}$ (i)

Since $\dfrac{\text{stress}(\sigma)}{\text{strain}(\varepsilon)} = E$, then $\sigma = E\varepsilon$

Hence $\quad \sigma_a = E_a \times \varepsilon_a$ and $\varepsilon_s = \dfrac{\sigma_s}{E_s}$

Using equation (i) $\sigma_a = E_a \left\{ \dfrac{(\alpha_a l_a - \alpha_s l_s) T - \dfrac{\sigma_s}{E_s} l_s}{l_a} \right\}$

$$= (7 \times 10^{10}) \left\{ \frac{(25 \times 10^{-6} \times 1.5 - 15 \times 10^{-6} \times 1)(50) - \dfrac{\sigma_s}{2 \times 10^{11}}(1)}{1.5} \right\}$$

$$= (7 \times 10^{10}) \left\{ \frac{(1.125 \times 10^{-3}) - \dfrac{\sigma_s}{2 \times 10^{11}}(1)}{1.5} \right\}$$

i.e. $\sigma_a = 5.25 \times 10^7 - 0.233\,\sigma_s$ (4.35)

The second equation can be obtained from *equilibrium* considerations, where tensile force in steel (F_s) = compressive force in aluminium (F_a)

i.e. $\quad F_s = F_a$

i.e. $\quad \sigma_s A_s = \sigma_a A_a$

and $\quad \sigma_a = \sigma_s \dfrac{A_s}{A_a} = \sigma_s \dfrac{2 \times 10^{-3}}{3 \times 10^{-3}}$

i.e. $\quad \sigma_a = 0.333\sigma_s$ (4.36)

Equating (4.35) and (4.36) gives

$$5.25 \times 10^7 - 0.233\,\sigma_s = 0.333\sigma_s$$

Therefore $\quad 5.25 \times 10^7 = 0.333\sigma_s + 0.233\,\sigma_s$

i.e. $\quad 5.25 \times 10^7 = 0.566\,\sigma_s$

Hence, stress in the steel $\sigma_s = \dfrac{5.25 \times 10^7}{0.566}$

$$= 92.76 \times 10^6\ \mathrm{N/m^2}$$

i.e. $\quad \boldsymbol{\sigma_s = 92.76\ \mathrm{MN/m^2}}$ **(tensile)**

and from equation (4.36) $\sigma_a = 0.333\sigma_s$

$$= 0.333(95.76 \times 10^6)$$

i.e. $\quad \boldsymbol{\sigma_a = 30.89\ \mathrm{MN/m^2}}$ **(compressive)**

Problem 13. An electrical cable consists of a copper core surrounded co-axially by a steel sheath, so that the whole acts as a compound bar. If this cable hangs vertically down a mineshaft, prove that the maximum stresses in the copper and the steel are given by:

σ_c = maximum stress in copper

$$= \frac{E_c(\rho_c A_c + \rho_s A_s)gl}{(A_c E_c + A_s E_s)}$$

and σ_s = maximum stress in steel

$$= \frac{E_s(\rho_c A_c + \rho_s A_s)gl}{(A_c E_c + A_s E_s)}$$

where ρ_c = density of copper, ρ_s = density of steel, E_c = elastic modulus of copper, E_s = elastic modulus of steel, g = acceleration due to gravity, A_c = cross-sectional area of copper, A_s = cross-sectional area of steel and l = length of cable

Consider compatibility

Let δ_c = maximum deflection of the copper core

and δ_s = maximum deflection of the steel sheath

Now $\delta_c = \delta_s$

or $l\varepsilon_c = l\varepsilon_s$

where ε_c = maximum strain in copper

and ε_s = maximum strain in steel

Since $\frac{\sigma}{\varepsilon} = E$, from which, $\frac{\sigma}{E} = \varepsilon$

Hence $\frac{\widehat{\sigma_c}}{E_c} = \frac{\widehat{\sigma_s}}{E_s}$

or $$\widehat{\sigma_c} = \widehat{\sigma_s}\frac{E_c}{E_s} \quad (4.37)$$

Consider equilibrium

Weight of cable $= (\rho_c A_c + \rho_s A_s) \times g \times l$

Resolving vertically gives

$$(\rho_c A_c + \rho_s A_s) \times g \times l = \widehat{\sigma_c} \times A_c + \widehat{\sigma_s} \times A_s \quad (4.38)$$

Substituting equation (4.37) into (4.38)

$$(\rho_c A_c + \rho_s A_s) \times g \times l = \widehat{\sigma_s}\frac{E_c}{E_s} \times A_c + \widehat{\sigma_s} \times A_s$$

$$\widehat{\sigma_s}\left(\frac{A_c E_c}{E_s} + A_s\right) = (\rho_c A_c + \rho_s A_s) \times g \times l$$

or $$\widehat{\sigma_s}\left(\frac{A_c E_c + A_s E_s}{E_s}\right) = (\rho_c A_c + \rho_s A_s) \times g \times l$$

from which maximum stress in the steel

$$\widehat{\sigma_s} = \frac{\boldsymbol{E_s(\rho_c A_c + \rho_s A_s)gl}}{\boldsymbol{(A_c E_c + A_s E_s)}} \quad (4.39)$$

Substituting equation (4.39) into (4.37) gives

$$\widehat{\sigma_c} = \widehat{\sigma_s}\frac{E_c}{E_s} = \frac{E_s(\rho_c A_c + \rho_s A_s)gl}{(A_c E_c + A_s E_s)} \times \frac{E_c}{E_s}$$

from which maximum stress in the copper

$$\widehat{\sigma_c} = \frac{\boldsymbol{E_c(\rho_c A_c + \rho_s A_s)gl}}{\boldsymbol{(A_c E_c + A_s E_s)}} \quad (4.40)$$

Problem 14. A compound bar consists of a steel bolt surrounded co-axially by an aluminium-alloy tube, as shown in Figure 4.35. Assuming that the nut on the end of the steel bolt is initially 'just' hand-tight, determine the strains in the bolt and the tube if the nut is rotated clockwise by an angle, θ, relative to the other end.

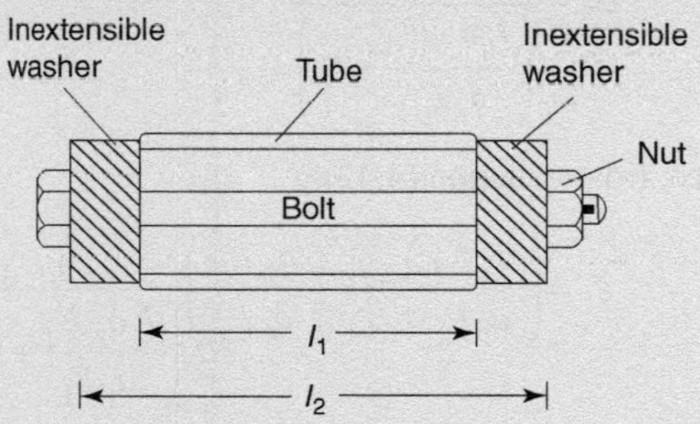

Figure 4.35 Bolt/tube compound bar

Let ε_1 = strain in the aluminium-alloy tube $= \frac{\delta_1}{l_1}$

ε_2 = strain in the steel bolt $= \frac{\delta_2}{l_2}$

a_1 = sectional area of aluminium alloy

a_2 = sectional area of steel

l_1 = length of aluminium-alloy tube

l_2 = length of steel bolt

δ = $\theta \times$ pitch of thread/360°

θ = rotation in degrees

E_1 = elastic modulus for aluminium alloy

E_2 = elastic modulus for steel

Now, when the nut is turned clockwise by an angle θ, the aluminium-alloy tube will decrease its length by δ_1, and the steel bolt will increase its length by δ_2, where

$$\delta = \delta_1 + \delta_2$$

Since $\varepsilon = \dfrac{\delta}{l}$

$$\delta = \varepsilon_1 l_1 + \varepsilon_2 l_2 \qquad (4.41)$$

From equilibrium considerations

$$\sigma_1 a_1 = \sigma_2 a_2$$

and $E_1 \varepsilon_1 a_1 = E_2 \varepsilon_2 a_2$ since $\dfrac{\sigma}{\varepsilon} = E$ and $\sigma = E\varepsilon$

from which
$$\varepsilon_1 = \left(\frac{a_2 E_2}{a_1 E_1}\right)\varepsilon_2 \qquad (4.42)$$

Substituting equation (4.42) into (4.41) gives

$$\delta = \left(\frac{a_2 E_2}{a_1 E_1}\right)\varepsilon_2 l_1 + \varepsilon_2 l_2 = \varepsilon_2\left(\frac{l_1 a_2 E_2}{a_1 E_1} + l_2\right)$$

Hence, **the tensile strain in the steel bolt,**

$$\varepsilon_2 = \frac{\delta}{\left(\dfrac{l_1 a_2 E_2}{a_1 E_1} + l_2\right)} \qquad (4.43)$$

Hence, from equation (4.42),

$$\varepsilon_1 = \left(\frac{a_2 E_2}{a_1 E_1}\right)\varepsilon_2 = \left(\frac{a_2 E_2}{a_1 E_1}\right)\frac{\delta}{\left(\dfrac{l_1 a_2 E_2}{a_1 E_1} + l_2\right)}$$

$$= \left(\frac{a_2 E_2}{a_1 E_1}\right)\frac{\delta}{\left(\dfrac{l_1 a_2 E_2 + l_2 a_1 E_1}{a_1 E_1}\right)}$$

$$= \frac{a_2 E_2 \delta}{\left(l_1 a_2 E_2 + l_2 a_1 E_1\right)}$$

$$= \frac{a_2 E_2 \delta}{a_2 E_2\left(l_1 + \dfrac{l_2 a_1 E_1}{a_2 E_2}\right)}$$

Hence, **the compressive strain in the aluminium-alloy tube,**

$$\varepsilon_1 = \frac{\delta}{\left(l_1 + \dfrac{l_2 a_1 E_1}{a_2 E_2}\right)} \qquad (4.44)$$

Now try the following Practice Exercise

Practice Exercise 26. Compound bars

1. An electrical cable consists of a copper core surrounded co-axially by a steel sheath, so that the two can be assumed to act as a compound bar. If the cable hangs down a vertical mine-shaft, determine the maximum permissible length of the cable, assuming the following apply:

   $A_c = 1\times10^{-4}\,\text{m}^2$ = sectional area of copper, $E_c = 1\times10^{11}\,\text{N/m}^2$ = elastic modulus of copper, ρ_s = 8960 kg/m³ = density of copper, Maximum permissible stress in copper = 30 MN/m², $A_s = 0.2\times10^{-4}\,\text{m}^2$ = sectional area of steel, $E_s = 2\times10^{11}\,\text{N/m}^2$ = elastic modulus of steel, ρ_s = 7860 kg/m³ = density of steel, maximum permissible stress in steel = 100 MN/m², g = 9.81 m/s²

   [406.5 m]

2. How much will the cable of Question 1 stretch, owing to self-weight?

   [61 mm]

3. If a weight of 100 kN were lowered into the sea, via a steel cable of cross-sectional area $8\times10^{-4}\,\text{m}^2$, what would be the maximum permissible depth that the weight could be lowered if the following apply?

   Density of steel = 7860 kg/m³, density of sea water = 1020 kg/m³, maximum permissible stress in steel = 200 MN/m², g = 9.81 m/s². Any buoyancy acting on the weight itself may be neglected.

   [1118 m]

4. A weightless rigid horizontal beam is supported by two vertical wires, as shown in Figure 4.36. If the following apply, determine the position from the left that a weight W can be suspended, so that the bar will remain horizontal when the wires stretch.

   Left wire: cross-sectional area = $2A$, elastic modulus = E, length = $2l$

   Right wire: cross-sectional area = A, elastic modulus = $3E$, length = l

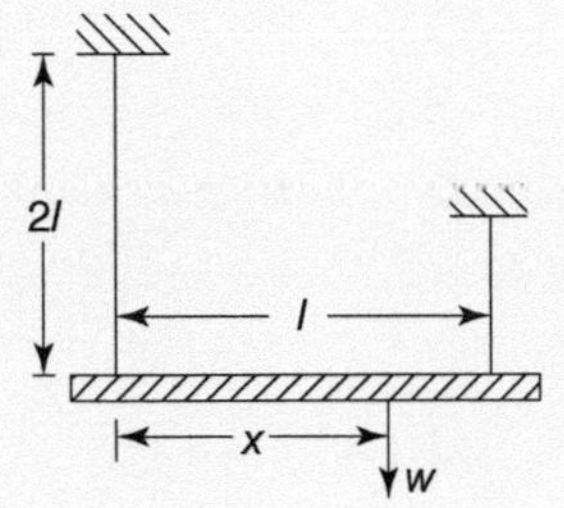

Figure 4.36

[0.75 l]

5. An electrical cable consists of a copper core surrounded co-axially by a steel sheath, so that the two can be assumed to act as a compound bar. If the cable hangs down a vertical mineshaft, determine the maximum permissible length of the cable, assuming the following apply:

   $A_c = 2\times10^{-4}$ m^2 = sectional area of copper, $E_c = 1\times10^{11}$ N/m^2 = elastic modulus of copper, $\rho_c = 8960$ kg/m^3 = density of copper, maximum permissible stress in copper = 30 MN/m^2, $A_s = 0.1\times10^{-4}$ m^2 = sectional area of steel, $E_s = 2\times10^{11}$ N/m^2 = elastic modulus of steel, $\rho_s = 7860$ kg/m^3 = density of steel, maximum permissible stress in steel = 100 MN/m^2, $g = 9.81$ m/s^2

   [359.7 m]

6. How much will the cable of Question 5 stretch, owing to self-weight? [53.9 mm]

7. If a weight of 95 kN were lowered into the sea via a steel cable of cross-sectional area 8×10^{-4} m^2, what would be the maximum permissible depth that the weight could be lowered if the following apply?

   Density of steel = 7860 kg/m^3, density of sea water = 1020 kg/m^3, maximum permissible stress in steel = 200 MN/m^2, $g = 9.81$ m/s^2. Any buoyancy acting on the weight itself may be neglected.

   [1211 m]

8. A weightless rigid horizontal beam is supported by two vertical wires, as shown in Figure 4.36. If the following apply, determine the position from the left that a weight W can suspended, so that the bar will remain horizontal when the wires stretch.

   Left wire: cross-sectional area = $3A$, elastic modulus = E, length – $2l$

   Right wire: cross-sectional area = $1.5A$, elastic modulus = $3E$, length = l

   [$x = 0.75\ l$]

4.14 Failure by fatigue

Under repeated cyclic loading, many structures are known to fail, despite the fact that the maximum calculated stress in the structure is well below the elastic limit. The reason for this mode of failure is that most structures have microscopic cracks in them, together with other stress concentrations. Under repeated cyclic loading, the maximum stresses in these cracks exceed the elastic limit and this causes the cracks to grow. The larger the crack becomes, the more rapidly it grows, until catastrophic failure takes place.

Materials such as mild steel and titanium will not fail through fatigue if the maximum stress in these materials does not exceed certain values. These values are called the **endurance limits**. Materials such as aluminium alloy and most other non-ferrous materials do not exhibit an endurance limit, as shown in Figure 4.37. For more on fatigue, see Chapter 23.

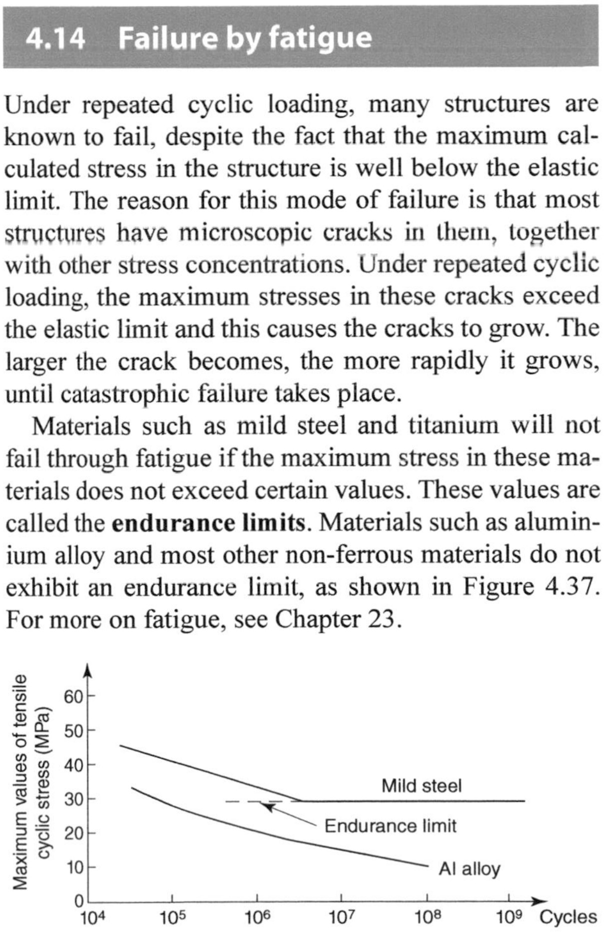

Figure 4.37 Fatigue failure

4.15 Failure due to creep

Under high temperatures, most materials lose some of their resistance to deformation and will yield at stresses lower than the so-called yield stress. Additionally, if the same load is continuously applied to the material, when it is subjected to this high temperature, the material will continue to deform with time, until failure takes place. This mode of failure is known as **creep**. Many plastic materials creep at room temperature, including polymethylmethacrylate (PMMA or 'perspex'). The rate of growth of the structure, when failing due to creep, is usually in three stages, namely the primary stage, the secondary stage and the tertiary stage, as shown in Figure 4.38.

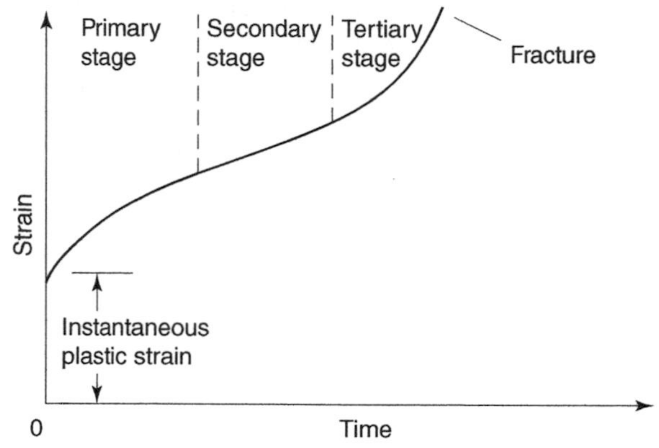

Figure 4.38 Creep failure

It can be seen from Figure 4.38 that the rate of strain during the primary stage is quite rapid, but slows when the secondary stage is reached. During the secondary stage, the rate of strain is quite linear, but during the tertiary and final stage of creep behaviour, the strain rate increases rapidly, until catastrophic failure eventually takes place. Depending on the value of the temperatures some structures can take days, or even months, to fail due to creep. For more on creep, see Chapter 23.

For fully worked solutions to each of the problems in Exercises 24 to 26 in this chapter, go to the website: www.routledge.com/cw/bird

Chapter 5

Geometrical properties of symmetrical sections

Why it is important to understand: **Geometrical properties of symmetrical sections**

The geometrical properties of sections are of great importance in a number of different branches of engineering, including stress analysis. For example, if a beam is subjected to bending action, its bending stiffness will depend not only on the material properties of the beam, but also on the geometrical properties of its cross-section. Typical beam cross-sections for symmetrical sections include circular, rectangular, triangular, I and T sections.

The strength of a structure is not only dependent on the material properties of the structure but also on its geometrical properties. For example, the strength of a wire in tension is also dependent on the size of its cross-section. Likewise the strength of a beam in bending is also dependent on its ability to withstand bending and torsion, besides its size. Since the invention of steel, the strength of a beam can be increased by the shape of its cross-section, in addition to the size of its cross-section. This has been achieved by the introduction of flanges, etc, in its cross-section, which could not be achieved with wooden structures in the past. Also, the ability of the water plane of a ship to withstand movement due to rolling and trimming, together with immersion, is also dependent on the size and shape of its water plane (See *Mechanical Engineering Principles* 4th Edition by Bird and Ross, Routledge, Chapter 25).

At the end of this chapter you should be able to:

- define a centroid, centroidal axes, centre of mass and central axes
- calculate the second moment of area of a section
- understand and use the parallel axis theorem
- understand and use the perpendicular axis theorem
- calculate the second moment of area using numerical integration – using the Naval architects' method and Ross's method
- use a computer program to determine centroids and second moments of area

Mathematical references

In order to understand the theory involved in this chapter on **geometrical properties of symmetrical sections,** knowledge of the following mathematical topics is required: ***integration, quadratic equations, trigonometric identities, double angles and Simpson's rule.*** If help/revision is needed in these areas, see page 49 for some textbook references.

5.1 Introduction

Some typical cross-sections for symmetrical sections are shown in Figure 5.1, where it is evident that providing the material properties of the section are the same, the bending resistances of the sections are dependent on their geometrical properties.

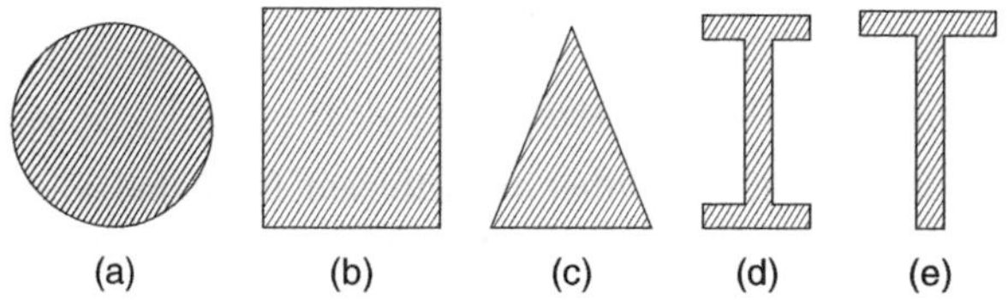

Figure 5.1 Some symmetrical cross-sections of beams: (a) circular section; (b) rectangular section; (c) triangular section; (d) 'I' section; (e) 'T' section

After many years of experience, structural engineers have found that if beams are made from steel or aluminium alloy, the cross-sections of Figure 5.1(d) and (e) usually provide a better strength to weight ratio than do the sections of Figures 5.1(a) to (c).
The section of Figure 5.1(d) is known as a ***rolled steel joist*** **(RSJ)** and that of Figure 5.1(e) is known as a ***tee section***.

5.2 Centroid

The centroid is the centre of the *moment of area* of a plane figure; or if the plane figure is of uniform thickness, this is the same position as the centre of gravity. This position is very important in elastic stress analysis. At the centroid, the following equations apply

$$\int y\,dA = 0 \text{ and } \int x\,dA = 0 \qquad (5.1)$$

where x is the horizontal axis, and y is the vertical axis, mutually perpendicular to x, as shown in Figure 5.2.

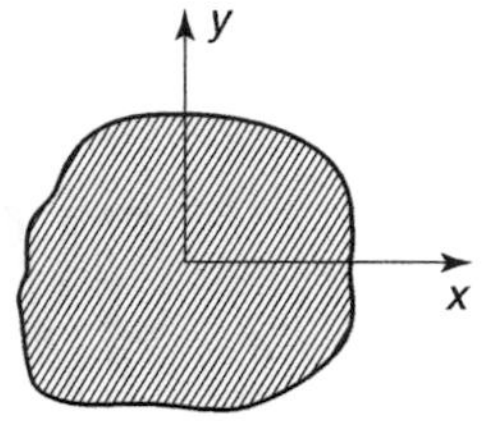

Figure 5.2 Plane figure

Centroidal axes

These are lines that pass through the centroid.

Centre of area

For a plane figure, the centre of area is obtained from the following considerations

Area above the horizontal central axis = area below the horizontal central axis (5.2)

Area to the left of the vertical central axis = area to the right of the vertical central axis (5.3)

Central axes

These are lines that pass through the centre of area.

5.3 Second moment of area

The second moment of area of the section of Figure 5.3 about XX is

$$I_{XX} = \int y^2\,dA \qquad (5.4)$$

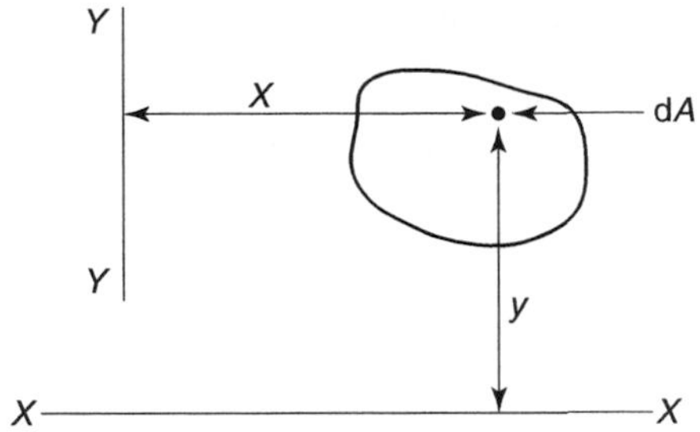

Figure 5.3

The second moment of area of the section of Figure 5.3 about YY is

$$I_{YY} = \int x^2\, dA \tag{5.5}$$

where $\int dA$ = area of section

5.4 Polar second moment of area

The polar second moment of area of the circular section of Figure 5.4, about its centre O is given by

$$J = \int_0^R r^2\, dA \tag{5.6}$$

where $dA = 2\pi r\, dr$

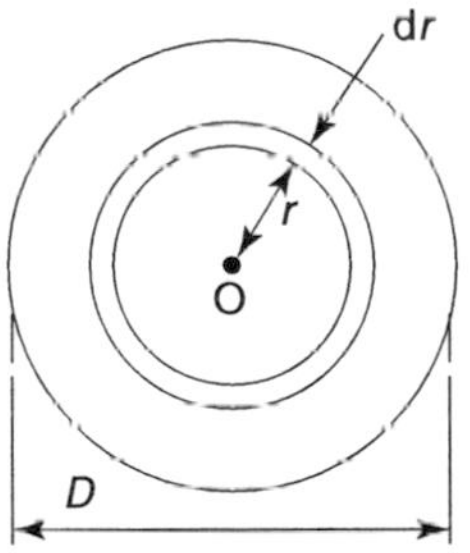

Figure 5.4 Circular section

J is of importance in the torsion of circular sections.

5.5 Parallel axis theorem

Consider the plane section of Figure 5.5, which has a second moment of area about its centroid equal to I_{xx} and suppose that it is required to determine the second moment of area about XX, where XX is parallel to xx, and that the perpendicular distance between the two axes is h.

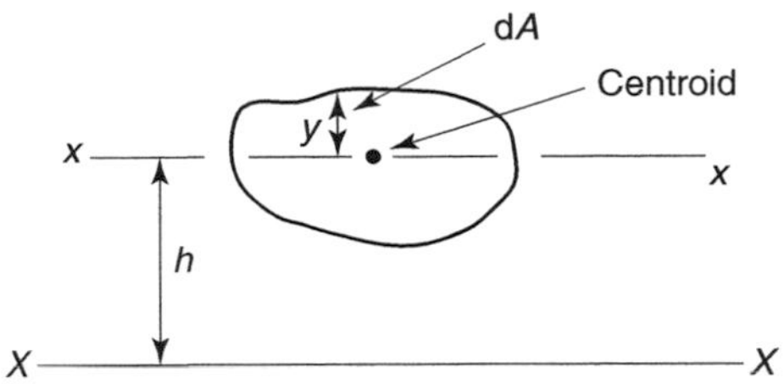

Figure 5.5 Parallel axes

Now $\quad I_{xx} = \int y^2\, dA$

and $\quad I_{XX} = \int (y+h)^2\, dA$

i.e. $\quad I_{XX} = \int (y^2 + 2hy + h^2)\, dA$

or $\quad I_{XX} = \int y^2\, dA + \int 2hy\, dA + \int h^2\, dA \quad (5.7)$

but as xx is at the centroid $\int 2hy\, dA = 0$

Therefore

$$\boldsymbol{I_{XX} = I_{xx} + h^2 \int dA} \tag{5.8}$$

Equation (5.8) is known as the ***parallel axis theorem*** and it is important in determining second moments of area for 'built-up' sections, such as RSJs, tees, etc.

5.6 Perpendicular axis theorem

From Figure 5.6, it can be seen that

$$I_{xx} = \int y^2\, dA \text{ and } I_{yy} = \int x^2\, dA$$

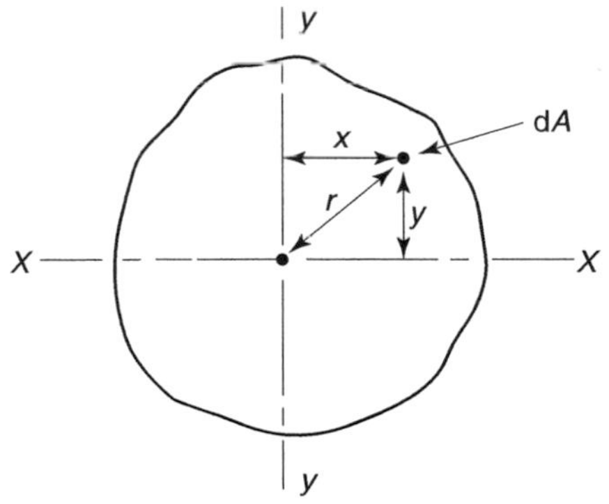

Figure 5.6

Now from equation (5.6) $I = \int_0^R r^2\, dA$

but as $x^2 + y^2 = r^2$ (from Pythagoras)

$$I = I_{xx} + I_{yy} \tag{5.9}$$

Equation (5.9) is known as the ***perpendicular axes theorem***, which states that the sum of the second moments of area of two mutually perpendicular axes is equal to the polar second moment of area about the point where these two axes cross.

To demonstrate the theories described in this section, the following worked problems will be considered.

Problem 1. Determine the positions of the centroidal and central axes for the isosceles triangle of Figure 5.7. Hence, or otherwise, determine the second moments of area about the centroid and the base *XX*. Verify the parallel axes theorem by induction.

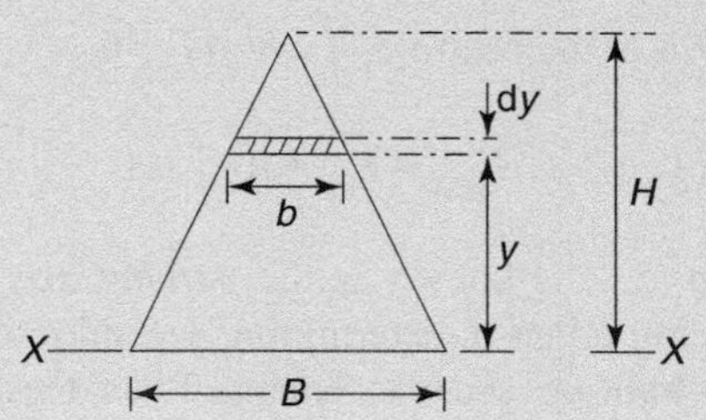

Figure 5.7 Isosceles triangle

To find the area, $\int dA = \int b\,dy$

The small triangle of base b and perpendicular height $(H - y)$, is similar to the whole triangle of base B and perpendicular height H.

Hence $\frac{H-y}{b} = \frac{H}{B}$ from which $B(H - y) = bH$

and $$b = \frac{B}{H}(H-y) = B - \frac{By}{H}$$

i.e. $$b = B\left(1 - \frac{y}{H}\right)$$

Therefore $$\int dA = \int b\,dy = \int_0^H B\left(1-\frac{y}{H}\right)dy$$

$$= B\left[y - \frac{y^2}{2H}\right]_0^H = B\left[\left(H - \frac{H^2}{2H}\right) - \left(0 - \frac{0^2}{2H}\right)\right]$$

$$= B\left[H - \frac{H}{2}\right]$$

i.e. $$\textbf{cross-sectional area} = \frac{\boldsymbol{BH}}{\mathbf{2}} \tag{5.10}$$

To find the centroidal axis

Let $\bar{y}$ be the distance of the centroid from *XX*. Therefore, the first moment of area about *XX* is

$$A\bar{y} = \int_0^H y\,b\,dy = \int_0^H y\,B\left(1-\frac{y}{H}\right)dy$$

$$= B\int_0^H \left(y - \frac{y^2}{H}\right)dy$$

$$= B\left[\frac{y^2}{2} - \frac{y^3}{3H}\right]_0^H = B\left[\left(\frac{H^2}{2} - \frac{H^3}{3H}\right) - (0)\right]$$

$$= B\left[\left(\frac{H^2}{2} - \frac{H^2}{3}\right)\right]$$

i.e. $$A\bar{y} = \frac{BH^2}{6}$$

from which $$\bar{y} = \frac{\frac{BH^2}{6}}{\frac{BH}{2}} = \frac{2BH^2}{6BH} = \frac{H}{3} \tag{5.11}$$

i.e. the distance of the centroid above *XX* is $\frac{H}{3}$

To find the central axis

Let $\bar{Y}$ be the distance of the central axis above *XX*. Consider the isosceles triangle of Figure 5.8 and equate areas above and below the central axis.

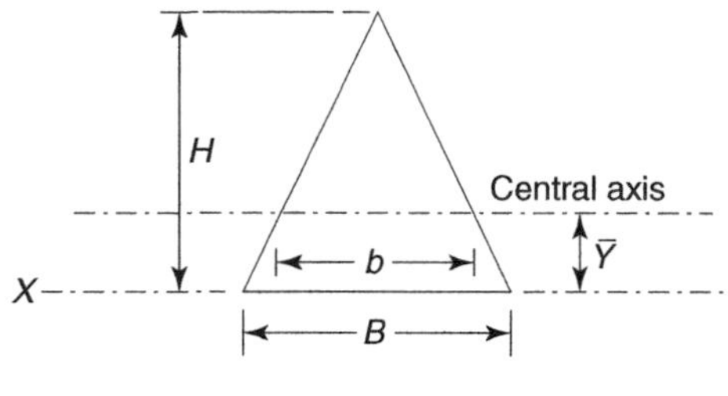

Figure 5.8 Central axis

Area above the central axis = area below the central axis

i.e. $\frac{b\left(H-\bar{Y}\right)}{2} = \frac{\left(B+b\right)\bar{Y}}{2}$ from the area of a triangle and a trapezium

i.e. $b\left(H-\bar{Y}\right) = \left(B+b\right)\bar{Y}$

but from earlier $b = B\left(1-\frac{\bar{Y}}{H}\right)$

Therefore $$B\left(1-\frac{\bar{Y}}{H}\right)\left(H-\bar{Y}\right) = \left(B + B\left(1-\frac{\bar{Y}}{H}\right)\right)\bar{Y}$$

i.e. $B\left(H-\bar{Y}-\bar{Y}+\frac{(\bar{Y})^2}{H}\right)=\bar{Y}\left(B+B-\frac{B\bar{Y}}{H}\right)$

and $\quad BH-2B\bar{Y}+\frac{B(\bar{Y})^2}{H}=2B\bar{Y}-\frac{B(\bar{Y})^2}{H}$

i.e. $\quad BH-4B\bar{Y}+\frac{2B(\bar{Y})^2}{H}=0$

Multiplying throughout by H and dividing throughout by B gives

$$H^2-4H\bar{Y}+2(\bar{Y})^2=0$$

or $\quad 2(\bar{Y})^2-4H\bar{Y}+H^2=0$

Using the quadratic formula

$$\bar{Y}=\frac{--4H\pm\sqrt{\left[(-4H)^2-4(2)(H^2)\right]}}{2(2)}$$

$$=\frac{4H\pm\sqrt{\left[16H^2-8H^2\right]}}{4}=\frac{4H\pm\sqrt{8H^2}}{4}$$

$$=\frac{4H\pm\sqrt{8}\,H}{4}=\frac{(4\pm\sqrt{8})}{4}H$$

i.e. $\bar{y}=0.293\,H$ (5.12)

or 1.707 H (which is ignored since $\bar{y}$ cannot be greater than H)

From equation (5.11) $\bar{y}=\frac{H}{3}=0.333\,H$

Hence, the central axis is $\left(\frac{0.333-0.293}{0.333}\right)\times 100\%$

= 12.01% below the centroidal axis in this case.

In stress analysis, the position of the central axis is only of importance in plasticity; so for the remainder of the present chapter, we will restrict our interest to the centroidal axis, which is of much interest in elastic theory.

To find I_{xx}

Let I_{xx} be the second moment of area about the centroidal axis, as in Figure 5.9.

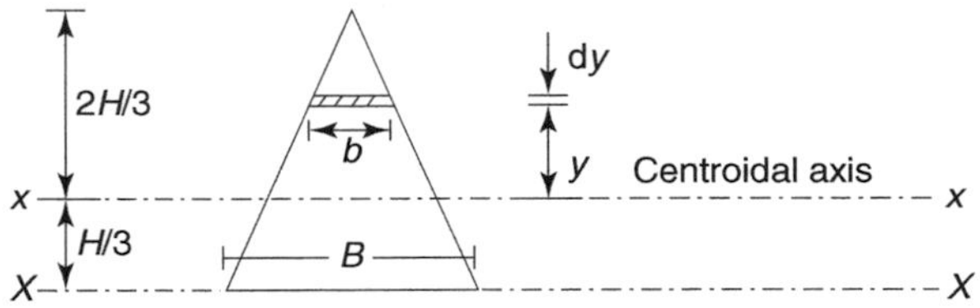

Figure 5.9

By similar triangles $\frac{b}{B}=\frac{\frac{2H}{3}-y}{H}$

from which $\quad b=\frac{B\left(\frac{2H}{3}-y\right)}{H}=\frac{\frac{2BH}{3}}{H}-\frac{By}{H}$

i.e. $\quad b=\frac{2B}{3}-\frac{By}{H}=B\left(\frac{2}{3}-\frac{y}{H}\right)$

and $\quad I_{xx}=\int_{-H/3}^{2H/3}y^2b\,dy$

$$=B\int_{H/3}^{2H/3}y^2\left(\frac{2}{3}-\frac{y}{H}\right)dy=B\int_{-H/3}^{2H/3}\left(\frac{2y^2}{3}-\frac{y^3}{H}\right)dy$$

$$=B\left[\frac{2y^3}{9}-\frac{y^4}{4H}\right]_{-H/3}^{2H/3}$$

$$=B\left[\left(\frac{2(2H/3)^3}{9}-\frac{(2H/3)^4}{4H}\right)-\left(\frac{2(-H/3)^3}{9}-\frac{(-H/3)^4}{4H}\right)\right]$$

$$=B\left[\left(\frac{16H^3}{9\times 27}-\frac{16H^3}{4\times 81}\right)-\left(\frac{-2H^3}{9\times 27}-\frac{H^3}{4\times 81}\right)\right]$$

$$=BH^3\left[\left(\frac{16}{243}-\frac{16}{324}\right)-\left(\frac{-2}{243}-\frac{1}{324}\right)\right]$$

i.e. $\boldsymbol{I_{xx}=\frac{BH^3}{36}}$ (5.13)

Similarly, from Figure 5.10

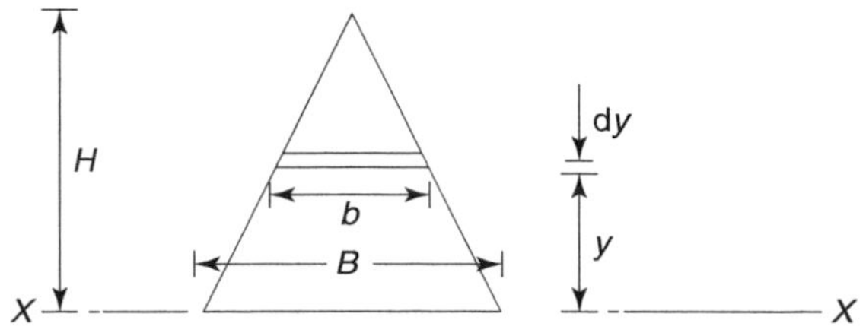

Figure 5.10

$$I_{XX} = \int_0^H y^2 b\,dy$$

By similar triangles $\frac{b}{B} = \frac{H-y}{H} = 1 - \frac{y}{H}$

from which $b = B\left(1 - \frac{y}{H}\right)$

Hence $I_{XX} = \int_0^H y^2\left(B\left(1-\frac{y}{H}\right)\right)dy$

$$= B\int_0^H y^2\left(1-\frac{y}{H}\right)dy = B\int_0^H\left(y^2 - \frac{y^3}{H}\right)dy$$

$$= B\left[\frac{y^3}{3} - \frac{y^4}{4H}\right]_0^H = B\left[\left(\frac{H^3}{3} - \frac{H^4}{4H}\right) - (0)\right]$$

$$= B\left[\left(\frac{H^3}{3} - \frac{H^3}{4}\right)\right]$$

i.e. $\boldsymbol{I_{XX} = \frac{BH^3}{12}}$ (5.14)

Check on parallel axis theorem

From equation (5.8) $I_{XX} = I_{xx} + h^2 \int dA$

i.e. $I_{XX} = \frac{BH^3}{36} + \left(\frac{H}{3}\right)^2\left(\frac{BH}{2}\right)$

i.e. $I_{XX} = \frac{BH^3}{36} + \frac{BH^3}{18}$

i.e. $\boldsymbol{I_{XX} = \frac{BH^3}{12}}$ as in equation (5.14)

Problem 2. Determine the area and second moments of area about the major and minor axes for the elliptical section of Figure 5.11.

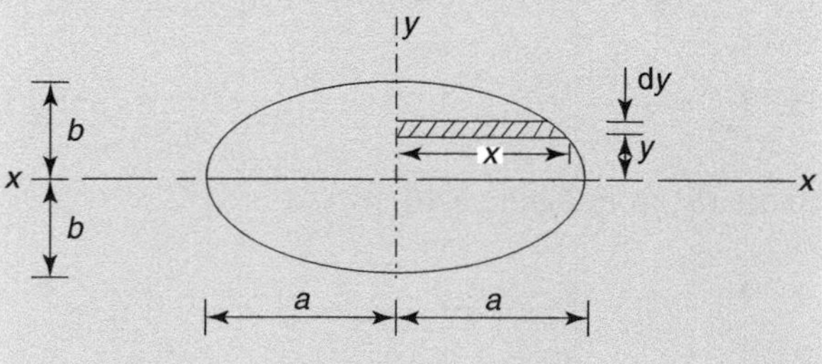

Figure 5.11 Elliptical section

The equation of an ellipse is $\frac{x^2}{a^2} + \frac{y^2}{b^2} = 1$

from which $\frac{y^2}{b^2} = 1 - \frac{x^2}{a^2}$

and $y^2 = \left(1 - x^2/a^2\right)b^2$

Let $x = a\cos\phi$ (5.15)

then $y^2 = \left(1 - \cos^2\phi\right)b^2$

Since $\cos^2\phi + \sin^2\phi = 1$ (from Chapter 1), then $\cos^2\phi = 1 - \sin^2\phi$

Hence $y^2 = \left(\sin^2\phi\right)b^2$

from which $y = b\sin\phi$ (5.16)

and $\frac{dy}{d\phi} = b\cos\phi$ or $dy = b\cos\phi\,d\phi$ (5.17)

If A = the area of the elliptical figure

then $A = 4\int x\,dy = 4\int_0^{\pi/2}(a\cos\phi)(b\cos\phi\,d\phi)$

$$= 4ab\int_0^{\pi/2}\cos^2\phi\,d\phi$$

From trigonometric identities $\cos 2\phi = 2\cos^2\phi - 1$

from which $1 + \cos 2\phi = 2\cos^2\phi$

and $\cos^2\phi = \frac{1+\cos 2\phi}{2}$

Therefore area

$$A = 4ab\int_0^{\pi/2}\cos^2\phi\,d\phi = 4ab\int_0^{\pi/2}\left(\frac{1+\cos 2\phi}{2}\right)d\phi$$

$$= 2ab\left[\phi + \frac{\sin 2\phi}{2}\right]_0^{\pi/2}$$

$$= 2ab\left[\left(\frac{\pi}{2} + \frac{\sin 2(\pi/2)}{2}\right) - \left(0 + \frac{\sin 0}{2}\right)\right]$$

$$= 2ab\left(\frac{\pi}{2}\right)$$

i.e. $\boldsymbol{A = \pi ab}$ (5.18)

From Figure 5.11, it can be seen that

$$I_{xx} = 4\int y^2\,dA \text{ from equation (5.4)}$$

$$= 4\int y^2 x\,dy$$

where $y = b\sin\phi$ and $x = a\cos\phi$ and $dy = b\cos\phi\,d\varphi$ from above

i.e. $I_{xx} = 4\int (b\sin\phi)^2(a\cos\phi)(b\cos\phi\,d\phi)$

i.e. $= 4ab^3\int \sin^2\phi\cos^2\phi\,d\phi$

From trigonometric identities, $\cos 2\phi = 2\cos^2\phi - 1$ from which, $\cos^2\phi = \dfrac{1+\cos 2\phi}{2}$ as above, and $\cos 2\phi = 1 - 2\sin^2\phi$ from which, $\sin^2\phi = \dfrac{1-\cos 2\phi}{2}$

Hence $$I_{xx} = 4ab^3 \int \sin^2\phi \cos^2\phi \, d\phi$$

$$= 4ab^3 \int \left(\frac{1-\cos 2\phi}{2}\right)\left(\frac{1+\cos 2\phi}{2}\right) d\phi$$

$$= ab^3 \int (1-\cos 2\phi)(1+\cos 2\phi)\, d\phi$$

$$= ab^3 \int (1-\cos^2 2\phi)\, d\phi$$

If $\cos^2\phi = \dfrac{1+\cos 2\phi}{2}$ then $\cos^2 2\phi = \dfrac{1+\cos 4\phi}{2}$

Hence $$I_{xx} = ab^3 \int_0^{\pi/2} \left(1 - \left(\frac{1+\cos 4\phi}{2}\right)\right) d\phi$$

$$= ab^3 \int_0^{\pi/2} \left(\frac{1}{2} - \frac{\cos 4\phi}{2}\right) d\phi$$

$$= ab^3 \left[\left(\frac{\phi}{2} - \frac{\sin 4\phi}{8}\right)\right]_0^{\pi/2}$$

$$= ab^3 \left[\left(\frac{(\pi/2)}{2} - \frac{\sin 4(\pi/2)}{8}\right) - \left(\frac{0}{2} - \frac{\sin 0}{8}\right)\right]$$

$$= ab^3 \left[\frac{\pi}{4}\right]$$

i.e. $$\boldsymbol{I_{xx} = \frac{\pi a b^3}{4}} \qquad (5.19)$$

Similarly it can be proven that $\boldsymbol{I_{yy} = \dfrac{\pi a^3 b}{4}}$ (5.20)

For a **circle of radius *R*** (or diameter D) $R = a = b$

Therefore, $$\boldsymbol{I_{xx} = I_{yy} = \frac{\pi R^4}{4} = \frac{\pi D^4}{64}} \qquad (5.21)$$

since $R = \dfrac{D}{2}$

and **polar second moment of area,**

$$\mathbf{J} = \frac{\pi R^4}{2} = \frac{\pi\left(\frac{D}{2}\right)^4}{2} = \boldsymbol{\frac{\pi D^4}{32}} \qquad (5.22)$$

Problem 3. Determine the second moments of area about axes xx and XX for the rectangle in Figure 5.12, and verify the parallel axis theorem by induction.

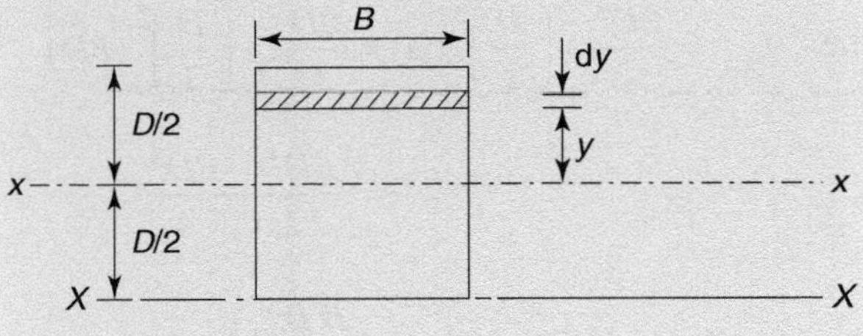

Figure 5.12 Rectangle

The second moment of area about xx is given by

$$I_{xx} = \int_{-D/2}^{D/2} y^2 B \left[\frac{y^3}{3}\right]_{-D/2}^{D/2}$$

$$= B\left[\left(\frac{(D/2)^3}{3}\right) - \left(\frac{(-D/2)^3}{3}\right)\right]$$

$$= B\left[\frac{D^3}{24} - -\frac{D^3}{24}\right]$$

i.e. $$\boldsymbol{I_{xx} = \frac{B D^3}{12}} \qquad (5.23)$$

The second moment of area about XX is given by

$$I_{XX} = \int_{-D/2}^{D/2} \left(y + \frac{D}{2}\right)^2 B\, dy$$

$$= B \int_{D/2}^{D/2} \left(y^2 + Dy + \frac{D^2}{4}\right) dy$$

$$= B\left[\frac{y^3}{3} + \frac{Dy^2}{2} + \frac{D^2 y}{4}\right]_{-D/2}^{D/2}$$

$$= B\left[\left(\frac{(D/2)^3}{3} + \frac{D(D/2)^2}{2} + \frac{D^2(D/2)}{4}\right) - \left(\frac{(-D/2)^3}{3} + \frac{D(-D/2)^2}{2} + \frac{D^2(-D/2)}{4}\right)\right]$$

$$= B\left[\left(\frac{D^3}{24} + \frac{D^3}{8} + \frac{D^3}{8}\right) - \left(-\frac{D^3}{24} + \frac{D^3}{8} - \frac{D^3}{8}\right)\right]$$

$$= BD^3\left[\frac{1}{24} + \frac{1}{8} + \frac{1}{8} + \frac{1}{24} - \frac{1}{8} + \frac{1}{8}\right]$$

i.e. $$\boldsymbol{I_{XX} = \frac{B D^3}{3}} \qquad (5.24)$$

Check on parallel axis theorem

From equation (5.8) $I_{XX} = I_{xx} + h^2 \int dA$

i.e. $I_{XX} = \frac{BD^3}{12} + \left(\frac{D}{2}\right)^2 (A) = \frac{BD^3}{12} + \left(\frac{D}{2}\right)^2 (BD)$

$$= \frac{BD^3}{12} + \frac{BD^3}{4}$$

i.e. $$\boldsymbol{I_{XX} = \frac{BD^3}{3}}$$

For the parallelogram of Figure 5.13, it can be proven that

$$\boldsymbol{I_{XX} = \frac{BD^3}{3}}$$

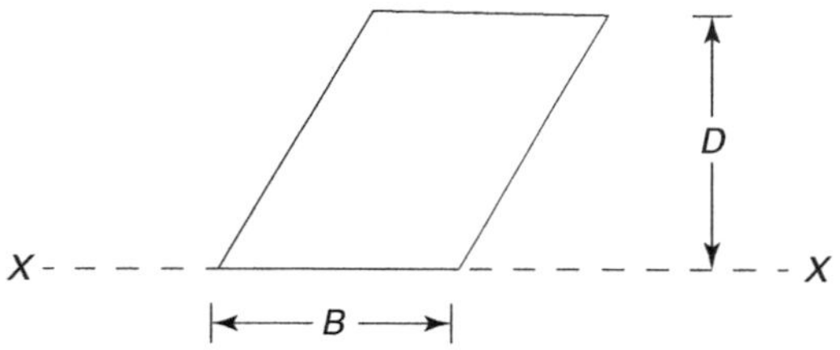

Figure 5.13 Parallelogram

Problem 4. Determine the polar second moment of area for the circular section in Figure 5.14.

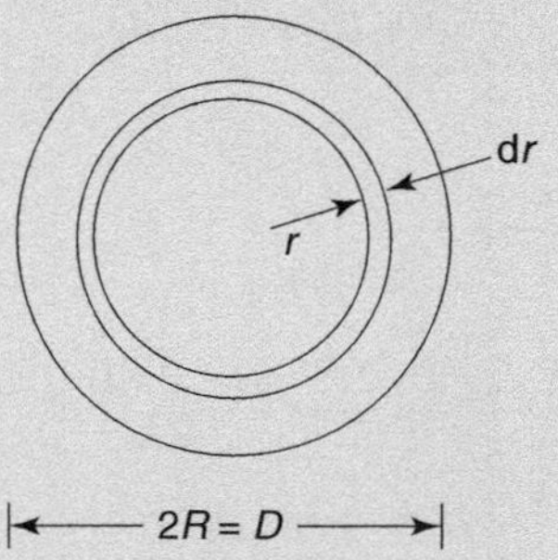

Figure 5.14 Circle

From equation (5.6), the polar second moment of area

$$J = \int_0^R r^2\, dA = \int_0^R r^2 (2\pi r\, dr)$$

$$= 2\pi \int_0^R r^3\, dr = 2\pi \left[\frac{r^4}{4}\right]_0^R$$

$$= \frac{\pi R^4}{2} = \frac{\pi \left(\frac{D}{2}\right)^4}{2} = \frac{\pi D^4}{32}$$

i.e. $$\boldsymbol{J = \frac{\pi R^4}{2} = \frac{\pi D^4}{32}} \qquad (5.25)$$

Problem 5. Determine the position of the centroidal axis xx and the second moment of area about this axis for the tee-bar shown in Figure 5.15.

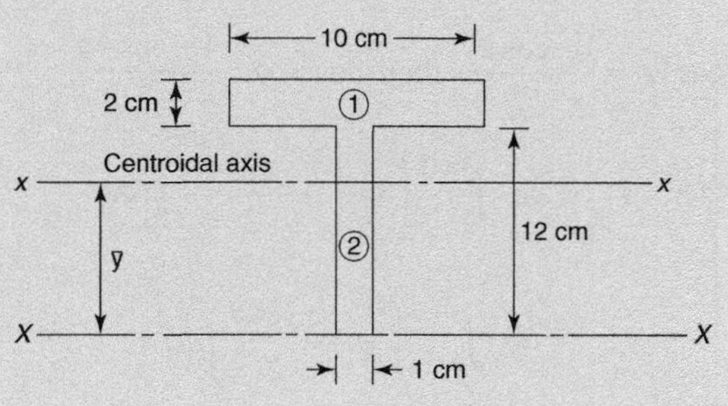

Figure 5.15 Tee-bar

Table 5.1 will be used to determine the geometrical properties of the tee-bar, where the rows of the table refer to the two rectangular elements (1) and (2).

a = area of an individual rectangular element

y = distance of local centroid of an individual rectangular element from XX

i = second moment of area of an individual rectangular element about its local centroid and parallel to XX.

ay = the product $a \times y$

$a y^2$ = the product $a \times y \times y$

$\sum$ = summation of the appropriate column

To determine the position of the centroid $\bar{y}$

$$\bar{y} \sum a = \sum a y \text{ from which, } \bar{y} = \frac{\sum ay}{\sum a}$$

and from Table 5.1 $$\bar{y} = \frac{\sum ay}{\sum a} = \frac{3.32 \times 10^{-4}}{3.2 \times 10^{-3}} \qquad (5.26)$$

i.e. $$\boldsymbol{\bar{y} = 0.104\ \text{m}}$$

Table 5.1 Geometrical calculations for tee-bar

Section	a (m^2)	y (m)	ay (m^3)	ay^2 (m^4)	i (m^4)
(1)	2×10^{-3}	0.13	2.6×10^{-4}	3.38×10^{-5}	$\frac{0.1\times0.02^3}{12}=6.67\times10^{-8}$
(2)	1.2×10^{-3}	0.06	7.2×10^{-5}	4.32×10^{-6}	$\frac{0.01\times0.12^3}{12}=1.44\times10^{-6}$
Σ	3.2×10^{-3}		3.32×10^{-4}	3.812×10^{-5}	1.51×10^{-6}

From the parallel axis theorem

$$I_{XX}=\sum ay^2+\sum i \qquad (5.27)$$

$$=3.812\times10^{-5}+1.51\times10^{-6}$$

i.e. $\mathbf{I_{XX}=3.963\times10^{-5}\,m^4}$

Also, from the parallel axis theorem

$$I_{xx}=I_{XX}-\left(\bar{y}\right)^2\sum a \qquad (5.28)$$

$$=3.963\times10^{-5}-(0.104)^2\times3.2\times10^{-3}$$

i.e. $\mathbf{I_{xx}=5.02\times10^{-6}\,m^4}$

Problem 6. Determine the position of the centroidal axis of the section in Figure 5.16 and also the second moment of area about this axis.

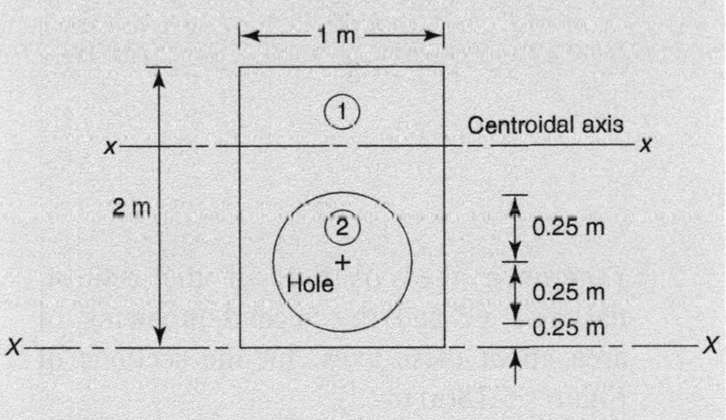

Figure 5.16

The determination of the geometrical properties of the section in Figure 5.16 will be aided with the calculations of Table 5.2, where the symbols are defined as in Problem 5.

From Table 5.2 and from equations (5.26) to (5.28)

$$\bar{y}=\frac{\sum ay}{\sum a}=\frac{1.9018}{1.8037}=\mathbf{1.054\ m}$$

Also $I_{XX}=\sum ay^2+\sum i$

$$=1.9509+0.6636=\mathbf{2.615\ m^4}$$

Finally, $I_{xx}=I_{XX}-\left(\bar{y}\right)^2\sum a$

$$=2.615-(1.054)^2\times1.8037=\mathbf{0.611\ m^4}$$

A summary of some standard results for the second moment of areas of regular sections is shown in Table 5.3 on page 120.

Table 5.2 Geometrical calculations

Section	a (m^2)	y (m)	ay (m^3)	ay^2 (m^4)	i (m^4)
(1)	2	1	2	2	$\frac{1\times2^3}{12}=0.66667$
(2)	–0.1963	0.5	–0.0982	–0.0491	$-\frac{\pi\times\left(5\times10^{-1}\right)^4}{64}=-3.07\times10^{-3}$
Σ	1.8037		1.9018	1.9509	0.6636

Table 5.3 Summary of standard results of the second moments of areas of regular sections

Shape	Position of axis	Second moment of area, I
Rectangle length D breadth B	(1) Coinciding with B	$\frac{BD^3}{3}$
	(2) Coinciding with D	$\frac{DB^3}{3}$
	(3) Through centroid, parallel to B	$\frac{BD^3}{12}$
	(4) Through centroid, parallel to D	$\frac{DB^3}{12}$
Triangle Perpendicular height H base B	(1) Coinciding with B	$\frac{BH^3}{12}$
	(2) Through centroid, parallel to base	$\frac{BH^3}{36}$
	(3) Through vertex, parallel to base	$\frac{BH^3}{4}$
Circle radius R diameter D	(1) Through centre perpendicular to plane (i.e. polar axis)	$\frac{\pi R^4}{2}$ or $\frac{\pi D^4}{32}$
	(2) Coinciding with diameter	$\frac{\pi R^4}{4}$ or $\frac{\pi D^4}{64}$
	(3) About a tangent	$\frac{5\pi R^4}{4}$ or $\frac{5\pi D^4}{64}$
Semicircle radius R	Coinciding with diameter	$\frac{\pi R^4}{8}$

Now try the following Practice Exercise

Practice Exercise 27. Second moment of area

1. Determine the second moments of area about the centroid of the squares shown in Figures 5.17(a) and (b).

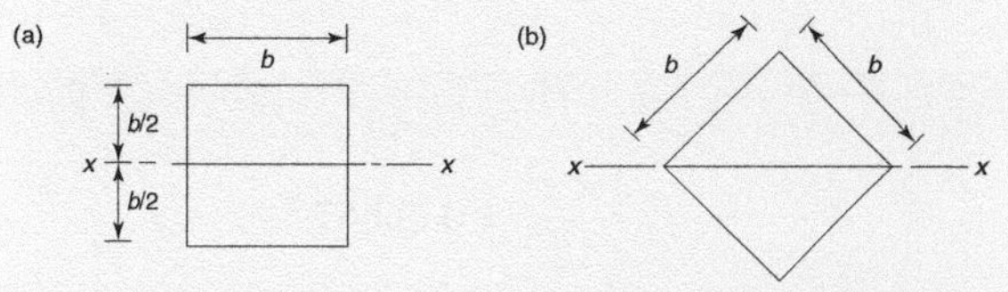

Figure 5.17

[(a) $b^4/12$ (b) $b^4/12$]

2. Determine the positions of the centroidal axis xx and the second moments of area about these axes, for the sections of Figures 5.18(a) to (d).

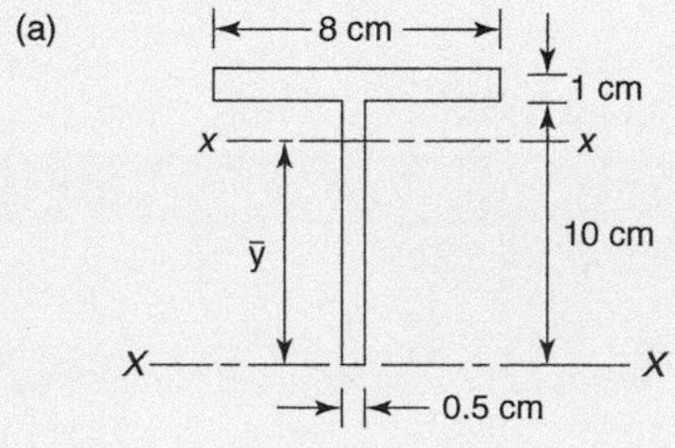

Figure 5.18

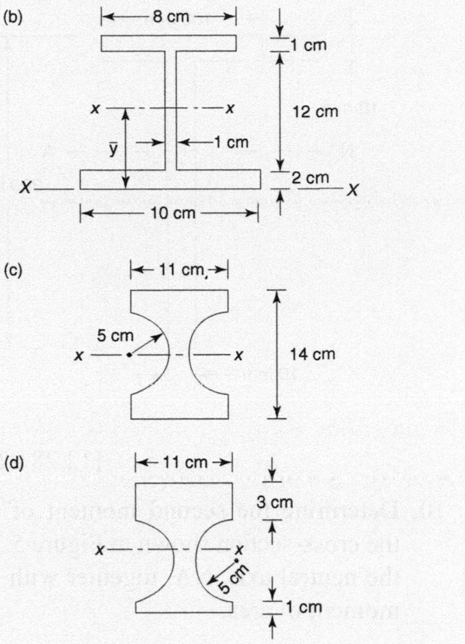

Figure 5.18 Continued

[(a) 8.385 cm, $1.353\times10^{-6}m^4$ (b) 5.80 cm, $1.276\times10^{-5}m^4$ (c) $2.024\times10^{-5}m^4$ (d) 8.04 cm, $1.865\times10^{-5}m^4$]

3. Determine the second moment of area of the section shown in Figure 5.19 about an axis passing through the centroid and parallel to the *xx* axis.

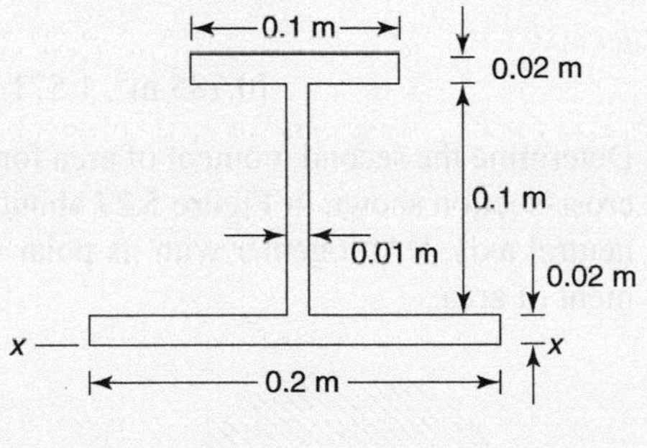

Figure 5.19

What would be the percentage reduction in second moment of area if the bottom flange was identical to the top flange?

[$2.054\times10^{-5}m^4$, 25.2%]

4. Determine the second moment of area for the cross-section shown in Figure 5.20 about the neutral axis, NA.

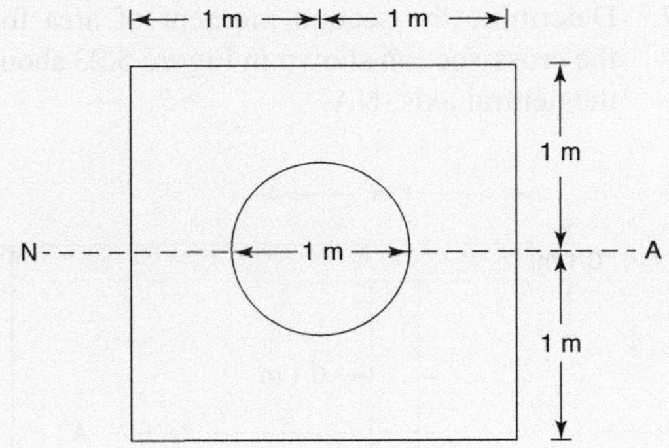

Figure 5.20

[1.284 m^4]

5. Determine the second moment of area for the cross-section shown in Figure 5.21 about the neutral axis, NA.

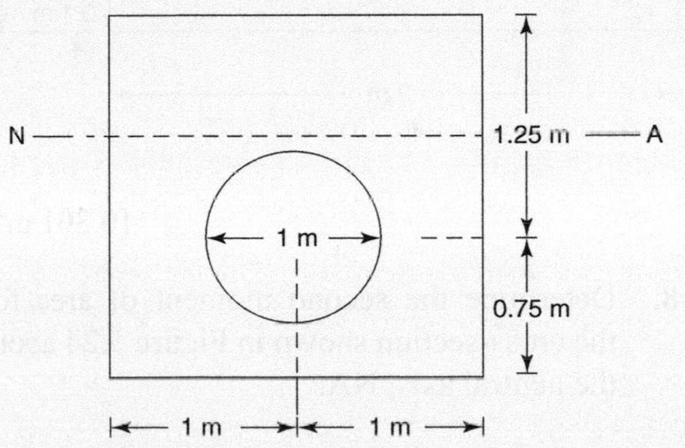

Figure 5.21

[1.220 m^4]

6. Determine the second moment of area for the cross-section shown in Figure 5.22 about the neutral axis, NA.

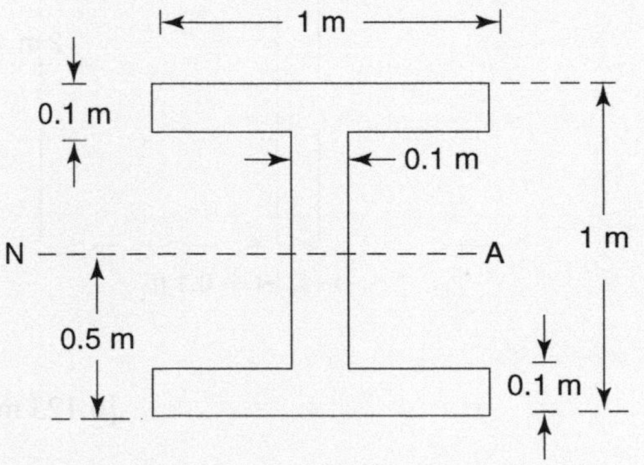

Figure 5.22

[0.0449 m^4]

7. Determine the second moment of area for the cross-section shown in Figure 5.23 about the neutral axis, NA.

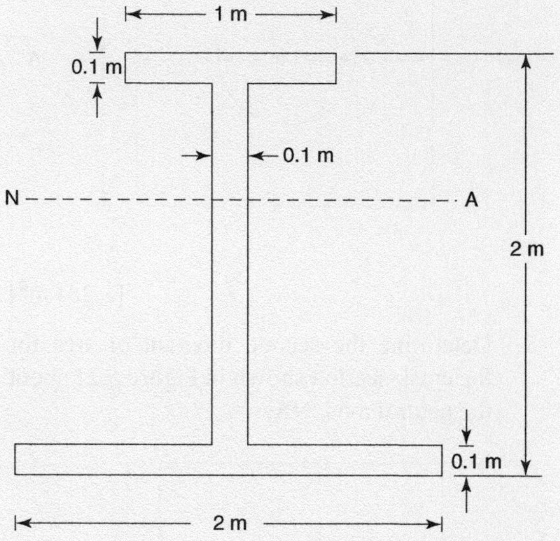

Figure 5.23

[0.301 m^4]

8. Determine the second moment of area for the cross-section shown in Figure 5.24 about the neutral axis, NA.

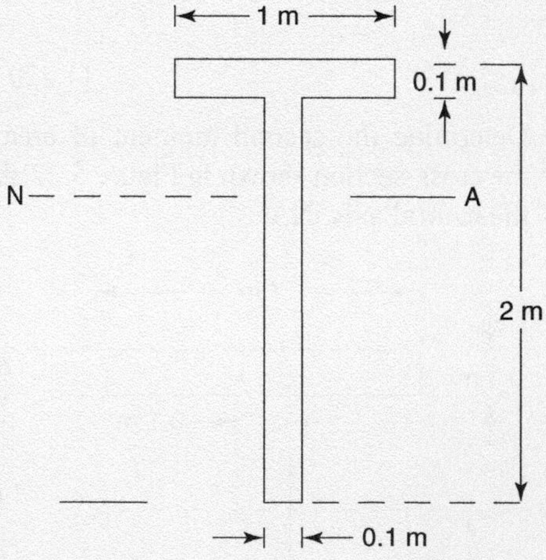

Figure 5.24

[0.123 m^4]

9. Determine the second moment of area for the cross-section shown in Figure 5.25 about the neutral axis, NA.

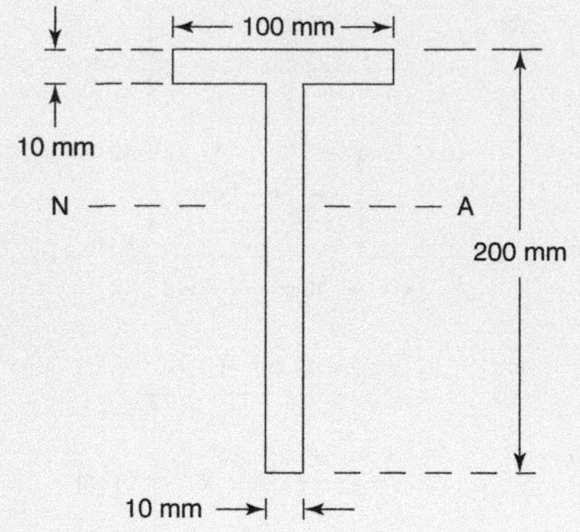

Figure 5.25

[12.28×10^{-6} m^4]

10. Determine the second moment of area for the cross-section shown in Figure 5.26 about the neutral axis, NA, together with its polar moment of area.

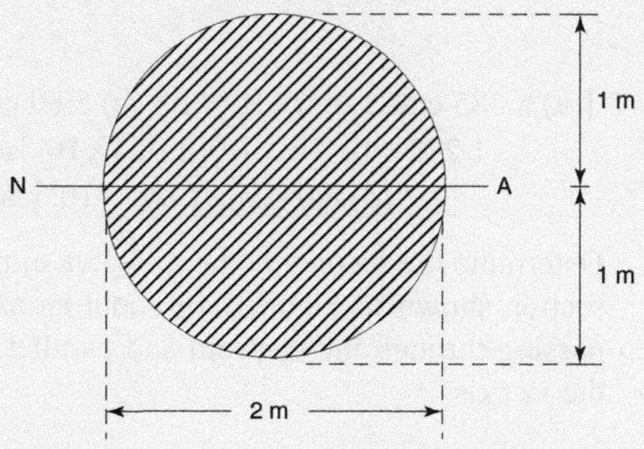

Figure 5.26

[0.785 m^4, 1.571 m^4]

11. Determine the second moment of area for the cross-section shown in Figure 5.27 about the neutral axis, NA, together with its polar moment of area.

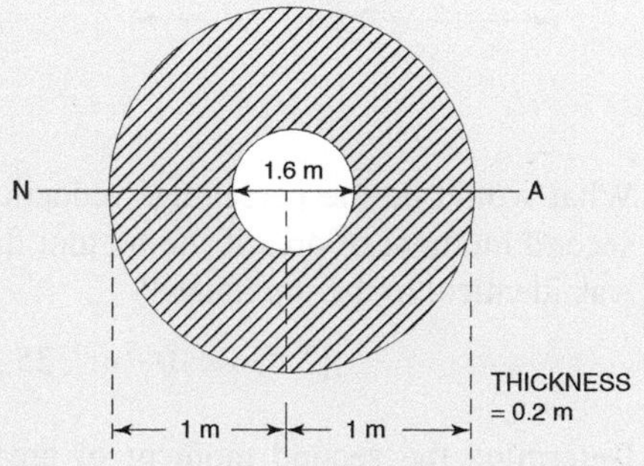

Figure 5.27

[0.464 m^4, 0.927 m^4]

5.7 Calculation of I through numerical integration

If I is required for an arbitrarily shaped section, such as that shown in Figure 5.28, the calculation for I can be carried out through numerical integration.

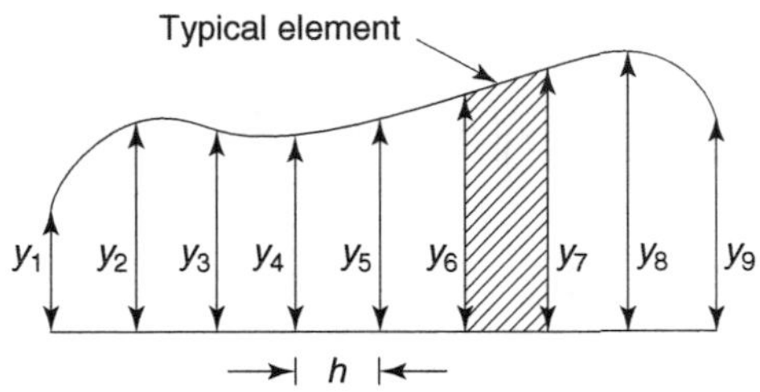

Figure 5.28 Arbitrarily shaped section

If the numerical integration is based on a Simpson's rule* approach, then the section must be divided into an equal number of elements, where the number of elements must be even.

Proof of Simpson's rule

Simpson's rule is based on employing a parabola to describe the function over any three stations, as shown in Figure 5.29. The equation is

$$y = a + bx + cx^2 \tag{5.29}$$

where a, b and c are arbitrary constants.

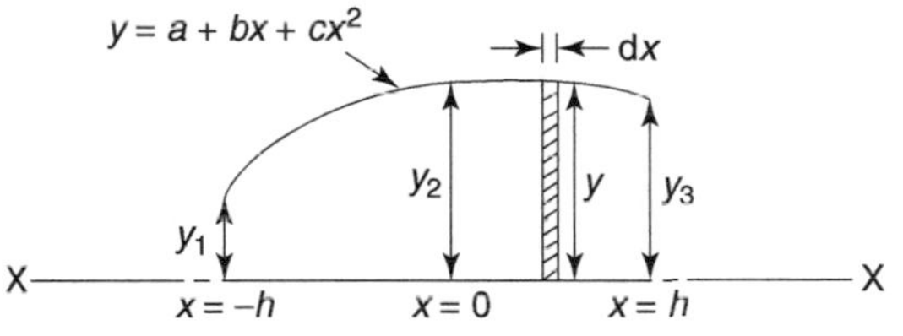

Figure 5.29 Parabolic variation

To obtain the three unknown constants, it will be necessary to obtain three simultaneous equations by putting in the *boundary values* for y in equation (5.29), as follows

At $x = 0$ $\quad y = y_2 = a$

Therefore $\quad a = y_2 \qquad (5.30)$

At $x = -h$ $\quad y = y_1$

i.e. $\quad y_1 = a + b(-h) + c(-h)^2$

*Thomas Simpson (20 August 1710 – 14 May 1761) was the British mathematician who invented Simpson's rule to approximate definite integrals. To find out more about Simpson go to www.routledge.com/cw/bird

i.e. $\quad y_1 = y_2 - bh + ch^2$

and $\quad y_1 - y_2 = -bh + ch^2 \qquad (5.31)$

At $x = h$ $\quad y = y_3$

i.e. $\quad y_3 = a + b(h) + c(h)^2$

i.e. $\quad y_3 = y_2 + bh + ch^2$

and $\quad y_3 - y_2 = bh + ch^2 \qquad (5.32)$

Adding (5.31) and (5.32) gives

$$y_1 - 2y_2 + y_3 = 2\,ch^2$$

from which $\quad c = \dfrac{(y_1 - 2y_2 + y_3)}{2h^2} \qquad (5.33)$

Substituting equation (5.33) into (5.32) gives

$$y_3 - y_2 = bh + \frac{(y_1 - 2y_2 + y_3)}{2h^2}h^2$$

i.e. $\quad y_3 - y_2 = bh + \dfrac{1}{2}(y_1 - 2y_2 + y_3)$

i.e. $\quad bh = y_3 - y_2 - \dfrac{1}{2}(y_1 - 2y_2 + y_3)$

i.e. $\quad bh = y_3 - y_2 - \dfrac{1}{2}y_1 + y_2 - \dfrac{1}{2}y_3$

i.e. $\quad bh = \dfrac{1}{2}y_3 - \dfrac{1}{2}y_1$

from which $\quad b = \dfrac{(-y_1 + y_3)}{2h} \qquad (5.34)$

Area of section $\quad A = \int_{-h}^{h} y\,dx$

$$= \int_{-h}^{h}(a + bx + cx^2)\,dx$$

$$= \left[ax + \frac{bx^2}{2} + \frac{cx^3}{3}\right]_{-h}^{h}$$

$$= \left[\left(ah + \frac{bh^2}{2} + \frac{ch^3}{3}\right) - \left(a(-h) + \frac{b(-h)^2}{2} + \frac{c(-h)^3}{3}\right)\right]$$

$$= \left[ah + \frac{bh^2}{2} + \frac{ch^3}{3} + ah - \frac{bh^2}{2} + \frac{ch^3}{3}\right]$$

$$= \left[2ah + \frac{2ch^3}{3}\right]$$

i.e. $\quad A = 2\left(ah + \dfrac{ch^3}{3}\right) \qquad (5.35)$

Substituting equations (5.30) and (5.33) into (5.35) gives:

$$A = 2\left(y_2 h + \frac{\frac{(y_1 - 2y_2 + y_3)}{2h^2} h^3}{3} \right)$$

$$= 2\left(y_2 h + \frac{(y_1 - 2y_2 + y_3)h}{6} \right)$$

$$= 2h\left(y_2 + \frac{1}{6}(y_1 - 2y_2 + y_3) \right)$$

$$= h\left(2y_2 + \frac{1}{3}y_1 - \frac{4}{6}y_2 + \frac{1}{3}y_3 \right)$$

$$= h\left(\frac{1}{3}y_1 + \frac{4}{3}y_2 + \frac{1}{3}y_3 \right)$$

i.e. $$\boldsymbol{A = \frac{h}{3}(y_1 + 4y_2 + y_3)} \quad (5.36)$$

Equation (5.36) is known as ***Simpson's rule*** for calculating areas.

The Naval architects' method of numerically calculating I_{XX} for a ship's water plane

The Naval architects' method of calculating I_{XX}, which is based on Simpson's rule, is given, without proof, by equation (5.37). This expression is reasonable for gentle curves with relatively small values of *h:*

$$\boldsymbol{I_{XX} = \frac{h}{9}(y_1^3 + 4y_2^3 + y_3^3)} \quad (5.37)$$

Ross's alternative method of numerically calculating I_{XX} for a ship's water plane

Strictly speaking equation (5.37) is incorrect, because for a rectangle of height y and width dx

$$I_{XX} = \int_{-h}^{h} \frac{y^3}{3}\, dx$$

where y is a function of x.

Hence, by substitution, equation (5.29) gives

$$I_{XX} = \frac{1}{3}\int_{-h}^{h} (a + bx + cx^2)^3\, dx$$

which is very different to equation (5.37).

$$(a + bx + cx^2)^2 = (a + bx + cx^2)(a + bx + cx^2)$$

$$= a^2 + abx + acx^2 + abx + b^2x^2 + bcx^3 + acx^2 + bcx^3 + c^2x^4$$

$$= a^2 + 2abx + 2acx^2 + b^2x^2 + 2bcx^3 + c^2x^4$$

and $(a + bx + cx^2)^3$

$$= (a + bx + cx^2)\left(a^2 + 2abx + 2acx^2 + b^2x^2 + 2bcx^3 + c^2x^4 \right)$$

$$= a^3 + 2a^2bx + 2a^2cx^2 + ab^2x^2 + 2abcx^3 + ac^2x^4$$
$$+ a^2bx + 2ab^2x^2 + 2abcx^3 + b^3x^3 + 2b^2cx^4 + bc^2x^5$$
$$+ a^2cx^2 + 2abcx^3 + 2ac^2x^4 + b^2cx^4 + 2bc^2x^5 + c^3x^6$$

$$= a^3 + 3a^2bx + 3a^2cx^2 + 3ab^2x^2 + 6abcx^3 + 3ac^2x^4 + b^3x^3 + 3b^2cx^4 + 3bc^2x^5 + c^3x^6$$

Hence $I_{XX} = \frac{1}{3}\int_{-h}^{h} (a + bx + cx^2)^3\, dx$

$$= \frac{1}{3}\int_{-h}^{h} a^3 + 3a^2bx + 3a^2cx^2 + 3ab^2x^2 + 6abcx^3 + 3ac^2x^4 + b^3x^3 + 3b^2cx^4 + 3bc^2x^5 + c^3x^6\, dx$$

$$= \frac{1}{3}\left[\left(a^3x + 3a^2b\left(\frac{x^2}{2}\right) + 3a^2c\left(\frac{x^3}{3}\right) + 3ab^2\left(\frac{x^3}{3}\right) + 6abc\left(\frac{x^4}{4}\right) + 3ac^2\left(\frac{x^5}{5}\right) + b^3\left(\frac{x^4}{4}\right) + 3b^2c\left(\frac{x^5}{5}\right) + 3bc^2\left(\frac{x^6}{6}\right) + c^3\left(\frac{x^7}{7}\right) \right) \right]_{-h}^{h}$$

$$= \frac{1}{3}\left[\begin{array}{l} \left(a^3h + \left(\frac{3a^2bh^2}{2}\right) + \left(\frac{3a^2ch^3}{3}\right) + \left(\frac{3ab^2h^3}{3}\right) + \left(\frac{6abch^4}{4}\right) + \left(\frac{3ac^2h^5}{5}\right) + \left(\frac{b^3h^4}{4}\right) + \left(\frac{3b^2ch^5}{5}\right) + \left(\frac{3bc^2h^6}{6}\right) + \left(\frac{c^3h^7}{7}\right) \right) \\ - \left(-a^3h + \left(\frac{3a^2bh^2}{2}\right) - \left(\frac{3a^2ch^3}{3}\right) - \left(\frac{3ab^2h^3}{3}\right) + \left(\frac{6abch^4}{4}\right) - \left(\frac{3ac^2h^5}{5}\right) + \left(\frac{b^3h^4}{4}\right) - \left(\frac{3b^2ch^5}{5}\right) + \left(\frac{3bc^2h^6}{6}\right) - \left(\frac{c^3h^7}{7}\right) \right) \end{array} \right]$$

$$= \frac{1}{3}\left[\left(\begin{array}{l} 2a^3h + 2a^2ch^3 + 2ab^2h^3 + \frac{6}{5}ac^2h^5 \\ \qquad + \frac{6}{5}b^2ch^5 + \frac{2}{7}c^3h^7 \end{array}\right)\right]$$

$$= \frac{2h}{3}\left[\left(\begin{array}{l} a^3 + a^2ch^2 + ab^2h^2 + \frac{3}{5}ac^2h^4 \\ \qquad + \frac{3}{5}b^2ch^4 + \frac{1}{7}c^3h^6 \end{array}\right)\right]$$

From equation (5.30) $a = y_2$

From equation (5.34) $b = \dfrac{(-y_1 + y_3)}{2h}$

and $b^2 = \dfrac{(-y_1 + y_3)^2}{4h^2} = \dfrac{y_1^2 - 2y_1y_3 + y_3^2}{4h^2}$

From equation (5.33) $c = \dfrac{(y_1 - 2y_2 + y_3)}{2h^2}$

and $c^2 = \dfrac{(y_1 - 2y_2 + y_3)^2}{4h^4}$

$$= \frac{y_1^2 - 4y_1y_2 + 2y_1y_3 + 4y_2^2 - 4y_2y_3 + y_3^2}{4h^4}$$

and $c^3 = \left(\dfrac{(y_1 - 2y_2 + y_3)}{2h^2}\right) \times \left(\dfrac{y_1^2 - 4y_1y_2 + 2y_1y_3 + 4y_2^2 - 4y_2y_3 + y_3^2}{4h^4}\right)$

$$= \frac{\begin{array}{r} y_1^3 - 4y_1^2y_2 + 2y_1^2y_3 + 4y_1y_2^2 - 4y_1y_2y_3 + y_1y_3^2 \\ -2y_1^2y_2 + 8y_1y_2^2 - 4y_1y_2y_3 - 8y_2^3 + 8y_2^2y_3 \\ -2y_2y_3^2 + y_1^2y_3 - 4y_1y_2y_3 + 2y_1y_3^2 + 4y_2^2y_3 \\ -4y_2y_3^2 + y_3^3 \end{array}}{8h^6}$$

$$= \frac{\begin{array}{r} y_1^3 - 6y_1^2y_2 + 3y_1^2y_3 + 12y_1y_2^2 - 12y_1y_2y_3 \\ +3y_1y_3^2 - 6y_2y_3^2 + 12y_2^2y_3 - 8y_2^3 + y_3^3 \end{array}}{8h^6}$$

Hence

$$I_{XX} = \frac{2h}{3}\left[\left(\begin{array}{l} y_2^3 + y_2^2\dfrac{(y_1 - 2y_2 + y_3)}{2h^2}h^2 + y_2\left(\dfrac{(-y_1 + y_3)}{2h}\right)^2 h^2 \\ +\dfrac{3}{5}y_2\left(\dfrac{(y_1 - 2y_2 + y_3)}{2h^2}\right)^2 h^4 + \dfrac{3}{5}\left(\dfrac{(-y_1 + y_3)}{2h}\right)^2 \\ \left(\dfrac{(y_1 - 2y_2 + y_3)}{2h^2}\right)h^4 + \dfrac{1}{7}\left(\dfrac{(y_1 - 2y_2 + y_3)}{2h^2}\right)^3 h^6 \end{array}\right)\right]$$

---

i.e. $$I_{XX} = \frac{2h}{3}\left[\left(\begin{array}{l} y_2^3 + \frac{1}{2}y_1y_2^2 - y_2^3 + \frac{1}{2}y_2^2y_3 + y_2\left(\dfrac{y_1^2 - 2y_1y_3 + y_3^2}{4h^2}\right)h^2 \\ \qquad + \frac{3}{5}y_2\left(\dfrac{y_1^2 - 4y_1y_2 + 2y_1y_3 + 4y_2^2 - 4y_2y_3 + y_3^2}{4h^4}\right)h^4 \\ \qquad + \frac{3}{5}\left(\dfrac{y_1^2 - 2y_1y_3 + y_3^2}{4h^2}\right)\left(\dfrac{(y_1 - 2y_2 + y_3)}{2h^2}\right)h^4 \\ + \frac{1}{7}\left(\dfrac{y_1^3 - 6y_1^2y_2 + 3y_1^2y_3 + 12y_1y_2^2 - 12y_1y_2y_3 + 3y_1y_3^2 - 6y_2y_3^2 + 12y_2^2y_3 - 8y_2^3 + y_3^3}{8h^6}\right)h^6 \end{array}\right)\right]$$

$$
=\frac{2h}{3}\left[\left(\begin{aligned}
&y_2^3+\frac{1}{2}y_1y_2^2-y_2^3+\frac{1}{2}y_2^2y_3+y_2\left(\frac{y_1^2-2y_1y_3+y_3^2}{4h^2}\right)h^2\\
&+\frac{3}{5}y_2\left(\frac{y_1^2-4y_1y_2+2y_1y_3+4y_2^2-4y_2y_3+y_3^2}{4h^4}\right)h^4\\
&+\frac{3}{5}\left(\frac{y_1^3-2y_1^2y_2+y_1^2y_3-2y_1^2y_3+4y_1y_2y_3-2y_1y_3^2+y_1y_3^2-2y_2y_3^2+y_3^3}{8h^4}\right)h^4\\
&+\frac{1}{7}\left(\frac{y_1^3-6y_1^2y_2+3y_1^2y_3+12y_1y_2^2-12y_1y_2y_3+3y_1y_3^2-6y_2y_3^2+12y_2^2y_3-8y_2^3+y_3^3}{8h^6}\right)h^6
\end{aligned}\right)\right]
$$

$$
=\frac{2h}{3}\left[\left(\begin{aligned}
&y_2^3+\frac{1}{2}y_1y_2^2-y_2^3+\frac{1}{2}y_2^2y_3+\frac{1}{4}y_1^2y_2-\frac{1}{2}y_1y_2y_3+\frac{1}{4}y_2y_3^2\\
&+\frac{3}{5}\left(\frac{1}{4}y_1^2y_2-y_1y_2^2+\frac{1}{2}y_1y_2y_3+y_2^3-y_2^2y_3+\frac{1}{4}y_2y_3^2\right)\\
&+\frac{3}{5}\left(\frac{1}{8}y_1^3-\frac{1}{4}y_1^2y_2+\frac{1}{8}y_1^2y_3-\frac{1}{4}y_1^2y_3+\frac{1}{2}y_1y_2y_3-\frac{1}{4}y_1y_3^2+\frac{1}{8}y_1y_3^2-\frac{1}{4}y_2y_3^2+\frac{1}{8}y_3^3\right)\\
&+\frac{1}{7}\left(\frac{1}{8}y_1^3-\frac{3}{4}y_1^2y_2+\frac{3}{8}y_1^2y_3+\frac{3}{2}y_1y_2^2-\frac{3}{2}y_1y_2y_3+\frac{3}{8}y_1y_3^2-\frac{3}{4}y_2y_3^2+\frac{3}{2}y_2^2y_3-y_2^3+\frac{1}{8}y_3^3\right)
\end{aligned}\right)\right]
$$

$$
=\frac{2h}{3}\left[\left(\begin{aligned}
&y_2^3+\frac{1}{2}y_1y_2^2-y_2^3+\frac{1}{2}y_2^2y_3+\frac{1}{4}y_1^2y_2-\frac{1}{2}y_1y_2y_3+\frac{1}{4}y_2y_3^2\\
&+\frac{3}{20}y_1^2y_2-\frac{3}{5}y_1y_2^2+\frac{3}{10}y_1y_2y_3+\frac{3}{5}y_2^3-\frac{3}{5}y_2^2y_3+\frac{3}{20}y_2y_3^2\\
&+\frac{3}{40}y_1^3-\frac{3}{20}y_1^2y_2+\frac{3}{40}y_1^2y_3-\frac{3}{20}y_1^2y_3+\frac{3}{10}y_1y_2y_3-\frac{3}{20}y_1y_3^2+\frac{3}{40}y_1y_3^2-\frac{3}{20}y_2y_3^2+\frac{3}{40}y_3^3\\
&+\frac{1}{56}y_1^3-\frac{3}{28}y_1^2y_2+\frac{3}{56}y_1^2y_3+\frac{3}{14}y_1y_2^2-\frac{3}{14}y_1y_2y_3+\frac{3}{56}y_1y_3^2-\frac{3}{28}y_2y_3^2+\frac{3}{14}y_2^2y_3\\
&-\frac{1}{7}y_2^3+\frac{1}{56}y_3^3
\end{aligned}\right)\right]
$$

$$
=\frac{2h}{3}\left[\begin{aligned}
&\frac{16}{35}y_2^3+\frac{4}{35}y_1y_2^2+\frac{4}{35}y_2^2y_3+\frac{1}{7}y_1^2y_2-\frac{4}{35}y_1y_2y_3+\frac{1}{7}y_2y_3^2+\frac{13}{140}y_1^3-\frac{3}{140}y_1^2y_3\\
&-\frac{3}{140}y_1y_3^2+\frac{13}{140}y_3^3
\end{aligned}\right]
$$

Combining similar terms gives

$$
I_{XX}=\frac{2h}{3}\left\{\begin{aligned}
&\frac{13}{140}\left(y_1^3+y_3^3\right)+\frac{16}{35}y_2^3+\frac{1}{7}\left(y_1^2y_2+y_2y_3^2\right)-\frac{3}{140}\left(y_1^2y_3+y_1y_3^2\right)\\
&+\frac{4}{35}\left(y_1y_2^2+y_2^2y_3\right)-\frac{4}{35}\left(y_1y_2y_3\right)
\end{aligned}\right\} \tag{5.38}
$$

Summarising, the second moment of area

by **the Naval architects' method** is given by $\boldsymbol{I_{XX}=\frac{h}{9}\left(y_1^3+4y_2^3+y_3^3\right)}$ (5.37) by **Ross's method** is given by

$$I_{XX}=\frac{2h}{3}\left\{\begin{array}{l}\frac{13}{140}\left(y_1^3+y_3^3\right)+\frac{16}{35}y_2^3+\frac{1}{7}\left(y_1^2y_2+y_2y_3^2\right)-\frac{3}{140}\left(y_1^2y_3+y_1y_3^2\right)\\ \qquad+\frac{4}{35}\left(y_1y_2^2+y_2^2y_3\right)-\frac{4}{35}\left(y_1y_2y_3\right)\end{array}\right\} \quad (5.38)$$

Problem 7. Calculate I_{XX} for the section in Figure 5.30 by equations (5.37) and (5.38), and then compare the two results.

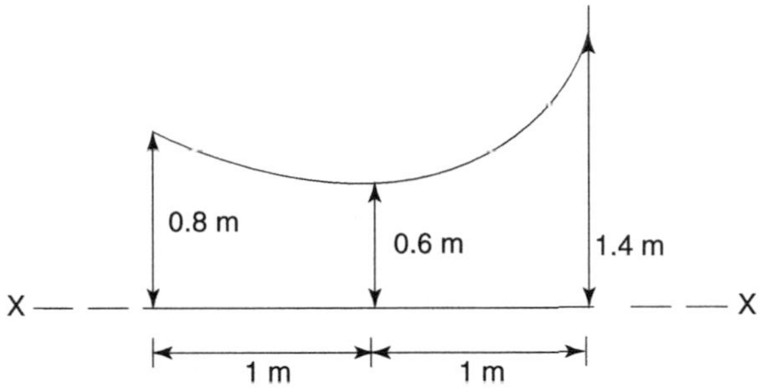

Figure 5.30

From Simpson's rule, i.e. the Naval architects' method, using equation (5.37),

$$I_{XX}=\frac{h}{9}\left(y_1^3+4y_2^3+y_3^3\right)=\frac{1}{9}\left(0.8^3+4\times0.6^3+1.4^3\right)$$

$$=\mathbf{0.4578\ m^4}$$

From the Ross's numerical method, equation (5.38),

$$I_{XX}=\frac{2(1)}{3}\left\{\begin{array}{l}\frac{13}{140}\left(0.8^3+1.4^3\right)+\frac{16}{35}0.6^3\\ +\frac{1}{7}\left(0.8^2(0.6)+(0.6)1.4^2\right)\\ -\frac{3}{140}\left(0.8^2(1.4)+(0.8)1.4^2\right)\\ +\frac{4}{35}\left((0.8)0.6^2+0.6^2(1.4)\right)\\ -\frac{4}{35}(0.8)(0.6)(1.4)\end{array}\right\}$$

$$=\frac{2}{3}\left\{\begin{array}{l}\frac{13}{140}\times3.256+\frac{16}{35}\times0.216+\frac{1}{7}\times1.56\\ -\frac{3}{140}\times2.464+\frac{4}{35}\times0.792-\frac{4}{35}\times0.672\end{array}\right\}$$

i.e. $\mathbf{I_{XX}=0.390\ m^4}$

Percentage difference between the two results

$$=\frac{0.4578-0.390}{0.390}\times100\% = \mathbf{17.38\%}$$

NB The water planes of ships are likely to have less error than the above, because they are not as curved as in Figure 5.30.

For comparisons between Ross's method, the Naval architects' method and the exact solution, the following worked problems will be considered.

Problem 8. Determine the second moment of area of the section enclosed by $y=2x^2$ and the x-axis between $x=0$ m and $x=2$ m by (a) integration, (b) the Naval architects' method and (c) by Ross's method

The area is shown in Figure 5.31 where x varies from 0 m to 2 m.

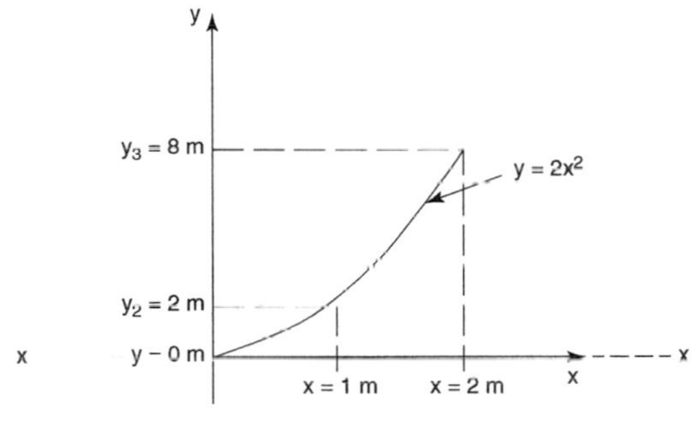

Figure 5.31

At $x=0$, $y_1=0$ m

At $x=1$ m, $y_2=2(1)^2=2$ m

At $x=2$ m, $y_3=2(2)^2=8$ m

(a) **By integration,** $I_{XX}=\int y\,dx\frac{y^2}{3}=\int\left(\frac{y^3}{3}\right)dx$

and from Figure 5.31

$$I_{XX}=\int_0^2\left(\frac{\left(2x^2\right)^3}{3}\right)dx=\frac{8}{3}\int_0^2\left(x^6\right)dx$$

$$= \frac{8}{3}\left[\frac{x^7}{7}\right]_0^2 = \frac{8}{3}\left[\frac{2^7}{7} - \frac{0^7}{7}\right] = \mathbf{48.76\ m^4}$$

Integration gives an exact value.

(b) **By the Naval architects' method**

From equation (5.37) $I_{XX} = \frac{h}{9}\left(y_1^3 + 4y_2^3 + y_3^3\right)$

where from Figure 5.31, $h = 1$ m, $y_1 = 0$, $y_2 = 2$ m and $y_3 = 8$ m

Hence $\quad I_{XX} = \frac{1}{9}\left(0^3 + 4(2)^3 + 8^3\right)$

$$= \frac{1}{9}\left(0^3 + 32 + 512\right) = \mathbf{60.44\ m^4}$$

(c) **By Ross's method,** from equation (5.38)

$$I_{XX} = \frac{2h}{3}\left\{\begin{array}{r} \frac{13}{140}\left(y_1^3 + y_3^3\right) + \frac{16}{35}y_2^3 + \frac{1}{7}\left(y_1^2 y_2 + y_2 y_3^2\right) \\ -\frac{3}{140}\left(y_1^2 y_3 + y_1 y_3^2\right) + \frac{4}{35}\left(y_1 y_2^2 + y_2^2 y_3\right) \\ -\frac{4}{35}\left(y_1 y_2 y_3\right) \end{array}\right\}$$

$$= \frac{2(1)}{3}\left\{\begin{array}{r} \frac{13}{140}\left(0^3 + 8^3\right) + \frac{16}{35}2^3 + \frac{1}{7}\left(0^2(2) + (2)(8)^2\right) \\ -\frac{3}{140}\left(0^2(8) + (0)(8)^2\right) + \frac{4}{35}\left((0)(2)^2 + (2)^2(8)\right) \\ -\frac{4}{35}\left((0)(2)(8)\right) \end{array}\right\}$$

$$= \frac{2}{3}\left\{\begin{array}{r} \frac{13}{140}(512) + \frac{16}{35}(8) + \frac{1}{7}(0 + 128) - \frac{3}{140}(0 + 0) \\ + \frac{4}{35}(0 + 32) - \frac{4}{35}(0) \end{array}\right\}$$

$$= \frac{2}{3}\{47.543 + 3.657 + 18.286 - 0 + 3.657 - 0\}$$

$= \mathbf{48.76\ m^4}$

From above it is seen that **Ross's method is exact** (i.e. the same as by integration), but the **Naval architects' method** is $\left(\frac{60.44 - 48.76}{48.76}\right) \times 100\% =$ **23.95% larger** than the exact solution.

Problem 9. Determine the second moment of area of the shape enclosed by $y = 2x^3$ and the x-axis between $x = 0$ m and $x = 2$ m by (a) integration, (b) the Naval architects' method and (c) by Ross's method

The area is shown in Figure 5.32 where x varies from 0 m to 2 m.

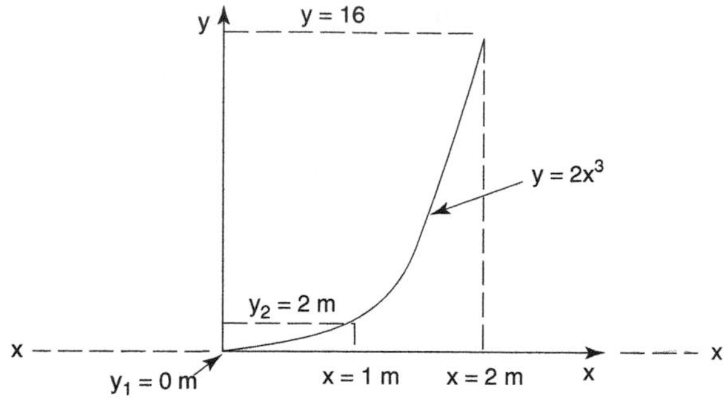

Figure 5.32

At $x = 0$, $y_1 = 0$ m

At $x = 1$ m, $y_2 = 2(1)^3 = 2$ m

At $x = 2$ m, $y_3 = 2(2)^3 = 16$ m

(a) **By integration,** $I_{XX} = \int y\,dx\frac{y^2}{3} = \int \frac{y^3}{3}\,dx$

and from Figure 5.32

$$I_{XX} = \int_0^2 \left(\frac{\left(2x^3\right)^3}{3}\right) dx = \frac{8}{3}\int_0^2 \left(x^9\right)$$

$$= \frac{8}{3}\left[\frac{x^{10}}{10}\right]_0^2 = \frac{8}{3}\left[\frac{2^{10}}{10} - \frac{0^{10}}{10}\right] = \mathbf{273.1\ m^4}$$

Integration gives an exact value.

(b) **By the Naval architects' method,**

From equation (5.37), $I_{XX} = \frac{h}{9}\left(y_1^3 + 4y_2^3 + y_3^3\right)$

where from Figure 5.32, $h = 1$ m, $y_1 = 0$, $y_3 = 2$ m and $y_3 = 16$ m

Hence

$$I_{XX} = \frac{1}{9}\left(0^3 + 4(2)^3 + 16^3\right)$$

$$= \frac{1}{9}\left(0^3 + 32 + 4096\right) = \mathbf{458.7\ m^4}$$

(c) **By Ross's method,** from equation (5.38)

$$I_{XX} = \frac{2h}{3}\left\{\begin{array}{l} \frac{13}{140}\left(y_1^3 + y_3^3\right) + \frac{16}{35}y_2^3 + \frac{1}{7}\left(y_1^2 y_2 + y_2 y_3^2\right) \\ -\frac{3}{140}\left(y_1^2 y_3 + y_1 y_3^2\right) + \frac{4}{35}\left(y_1 y_2^2 + y_2^2 y_3\right) \\ -\frac{4}{35}\left(y_1 y_2 y_3\right) \end{array}\right\}$$

$$= \frac{2(1)}{3}\left\{\begin{array}{l} \frac{13}{140}\left(0^3 + 16^3\right) + \frac{16}{35}2^3 + \frac{1}{7}\left(0^2(2) + (2)(16)^2\right) \\ -\frac{3}{140}\left(0^2(16) + (0)(16)^2\right) \\ +\frac{4}{35}\left((0)(2)^2 + (2)^2(16)\right) \\ -\frac{4}{35}\left((0)(2)(16)\right) \end{array}\right\}$$

$$= \frac{2}{3}\left\{\begin{array}{l} \frac{13}{140}(4096) + \frac{16}{35}(8) + \frac{1}{7}(0 + 512) - \frac{3}{140}(0 + 0) \\ +\frac{4}{35}(0 + 64) - \frac{4}{35}(0) \end{array}\right\}$$

$$= \frac{2}{3}\{380.343 + 3.657 + 73.143 - 0 + 7.314 - 0\}$$

$= \mathbf{309.6\ m^4}$

From above it is seen that **Ross's method** is $\left(\frac{309.6 - 273.1}{273.1}\right) \times 100\%$ = **13% larger than the exact method**, but the **Naval architects' method** is $\left(\frac{458.7 - 273.1}{273.1}\right) \times 100\%$ = **68% larger** than the exact solution.

Now try the following Practice Exercise

Practice Exercise 28. Calculation of I using numerical integration

1. Determine the second moment of area about XX of the section enclosed by $y = 2x + x^2$ and the x-axis between $x = 0$ m and $x = 2$ m as shown in Figure 5.33 by (a) the Naval architects' method and (b) by Ross's method. Compare the results with the exact solution obtained by integration.

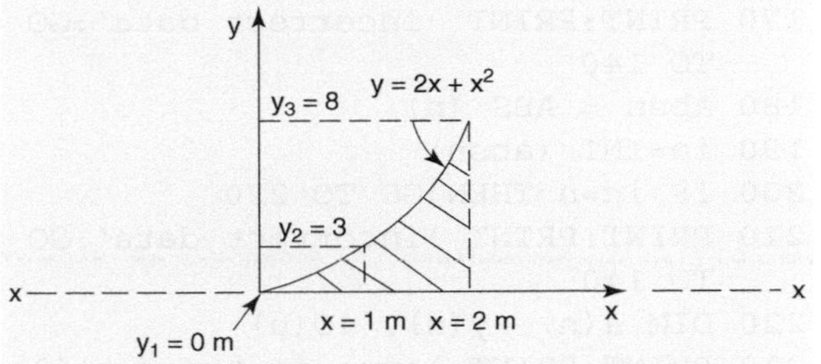

Figure 5.33

[(a) 68.89 m^4 (b) 63.70 m^4; by integration 63.70 m^4. Hence error in Naval architects' formula = 8.15% and Ross's method is exact]

2. Calculate the second moment of area, about the base XX (where $y = 0$), of the area enclosed by $y = e^x$ and the x-axis between the limits of $x = 0$ and $x = 2$ m, by the 'exact' integration method. Hence, or otherwise, obtain the approximate numerical values of this equation using (a) the Naval architects' method and (b) Ross's method.

[44.714 m^4 (a) 53.752 m^4 (b) 49.39 m^4]

5.8 Computer program for calculating $\bar{y}$ and I_{XX}

Listing 5.1 gives a computer program, in BASIC, for calculating $\bar{y}$ and I_{XX} for symmetrical sections, Listings 5.2 and 5.3 give the outputs for this program for questions 2(b) and 2(d) respectively of Practice Exercise 27 on page 120.

It should be noted from Listings 5.2 and 5.3 that the units being used must not be mixed (i.e. do not use a mixture of centimetres with metres or millimetres, etc.).

Listing 5.1 Computer program for calculating $\bar{y}$ and I_{XX}

```
100 CLS
110 REMark program for second moments
    of area for symmetrical sections
120 PRINT:PRINT 'program for second
    moments of area for symmetrical
    sections'
130 PRINT:PRINT 'copyright of
    Dr.C.T.F.ROSS'
140 PRINT:PRINT 'type in the number
    of sections'
150 INPUT n
```

```
160 IF n > 0 THEN GO TO 180
170 PRINT:PRINT 'incorrect data':GO
    TO 140
180 absn = ABS (n)
190 in=INT (absn)
200 IF in=n THEN GO TO 220
210 PRINT:PRINT 'incorrect data':GO
    TO 140
220 DIM a(n), y(n), iO(n)
230 PRINT:PRINT 'type in the details
    of each element':PRINT
240 area = 0:moment = 0:second = 0:
    i local = 0
250 FOR i = 1 TO n
260 PRINT 'elemental area
    (";i;"I=";:INPUT a(i)
270 PRINT 'element centroid from XX
    _ i.e. y (";i;"I=";:INPUT y(i)
280 PRINT 'elemental local 2nd moment
    of area iO (";i;")=";:INPUT iO (i)
290 area=area+a(i)
300 moment=moment+a(i)•y(i)
310 second=second+a(i)·y(i)*y(i)
320 ilocal=ilocal+iO(il
330 NEXT i
340 ybar=moment/area
350 ixx=ilocal+second
360 ina=ixx-ybar²* area
380 PRINT"number of elements=";n
390 FOR I=l to n
400 PRINT"element area
    a('';i;''j='';a(i)
410 PRINT"element centroid y
    (";i;")=";y(i)
420 PRINT"elemental local 2nd moment
    of area=";iO(i)
430 NEXT
440 PRINT
450 PRINT
460 PRINT"sectional area=";area
470 PRINT"sectional centroid (ybar)
    from XX=";ybar
480 PRINT"2nd moment of area of
    section about its centroid=";ina
500 STOP
```

Listing 5.2 Computer output for problem 2(b)

```
Number of elements=3
element area a(i) =8
element centroid y(i)=14.5
elemental local 2nd moment of
    area=.666667
element area a(2)=12
element centroid y(2)=8
elemental local 2nd moment of
    area=144
element area a(3)=20
element centroid y(3)=1
elemental local 2nd moment of
    area=6.6667
sectional area=40
section centroid (ybar) from XX=5.8
2nd moment of area of section about
    its centroid=1275.733
```

Listing 5.3 Computer output for problem 2 (d)

```
number of elements=2
element area a(i) =154
element centroid y(i)=7
elemental local 2nd moment of
    area=2515.3
element area a(2)=-78.54
element centroid y(2)=6
elemental local 2nd moment of area
    =-490.9
sectional area=75.46
section centroid (ybar) from
    XX=8.040816
2nd moment of area of section about
    its centroid=1864.1144
```

Program input

```
INPUT n - number of elements
FOR i = 1 TO n
INPUT a (i) - area of element i
INPUT y (i) - distance of centroid
    of element i from XX
INPUT iO (i) - local second moment
    of area of element i
NEXT i
```

Program output

area = $\sum A$

ybar = $\bar{y}$

ina = I_{XX}

5.9 Use of EXCEL spreadsheet in calculating geometrical properties of beams

Table 5.4 shows a printout of work carried out on one of the problems in this chapter, namely, the tee-bar of Figure 5.15, in calculating certain geometrical properties of built-up cross-sections of beams, using the Microsoft computer program, namely the EXCEL Spreadsheet, which comes with Microsoft Office. EXCEL is a spreadsheet, which not only allows tables to be made, but can also carry out mathematical functions such as calculating various items in different cells, together with multiplying quantities in the corresponding cells, in certain rows, with others in the same row of various other columns. Remember, a row is horizontal and a column is vertical. Thus, EXCEL is particularly useful for automatically mimicking the tabular hand calculations carried in this chapter. (For advice on how to use EXCEL, see, for example Reference 37, page 498).

An EXCEL spreadsheet is a calculator as well as a table. The appropriate formulae are initially inserted in the top row of the table. For example, in the top row of column 4, the formula $A = b*d$ is inserted or the first column, which contains values of 'b' is multiplied by the second column, which contains values of 'd', to give the result, namely 'A' in column 4. Column 4, which now contains 'A', is then multiplied into Column 3, which contains values of 'y', to give Ay; which is inserted into column 5, and now contains the product Ay. Column 5, which now contains Ay, is then multiplied into Column 3, which contains 'y', to give Column 6, which now contains the formula $Ay*y$; in the cell in the top row, of the 6th column. Finally, the formula for Column 7 is $b*d$^3/12; which is inserted in the top row. The appropriate columns are 'summed', very easily and simply, and the required formulae for determining 'ybar' and 'Ina' are inserted below. Thus, once the spreadsheet is prepared, fresh data in the form of 'b', 'd' and 'y' are put in the appropriate cells, below the top row, and the new information is calculated at the 'flick' of a button, and repeated for further new data, as and when required.

NB The dimension 'b' of an individual rectangular element **must be parallel to the axis of which the second moment of area is required**, and the dimension 'd' should be perpendicular to this axis.

Table 5.4 Second moment of area for rectangular sections

Section	1 Input b (mm)	2 Input d (mm)	3 Input y (mm)	4 Output A (mm^2)	5 Output Ay (mm^3)	6 Output Ay^2 (mm^4)	7 Output I (mm^4)
1	100	20	130	2000	2.60E + 05	3.38E + 07	6.667E + 04
2	10	120	60	1200	7.20E + 04	4.32E + 06	1.440E + 06
3	0	0	0	0	0.00E + 00	0.00E + 00	0.000E + 00
Summation				3200	332000	3.812E + 07	1506666.667
ybar =	1.038E + 02 mm						
Ioo =	3.963E + 07 mm^4		(i.e. I abt base of beam) = I_{XX}				
Iuu =	5.182E + 06 mm^4		(i.e. I abt NA of whole beam) = I_{NA}				

For fully worked solutions to each of the problems in Exercises 27 and 28 in this chapter, go to the website:
www.routledge.com/cw/bird

Chapter 6

Bending stresses in beams

Why it is important to understand: **Bending stresses in beams**

If a beam of symmetrical cross-section is subjected to a bending moment, then stresses due to bending action will occur. In pure or simple bending, the beam will bend into an arc of a circle. Due to couples, the upper layers of the beam will be in tension, because their lengths have been increased, and the lower layers of the beam will be in compression, because their lengths have been decreased. Somewhere in between these two layers lies a layer whose length has not changed, so that its stress due to bending is zero. This layer is called the neutral layer and its intersection with the beam's cross-section is called the neutral axis. The bending of beams is of much importance in designing buildings, bridges, ships, etc., and ensuring their safety under these loads.

In the case of the longitudinal strength of ships, due to sea waves, the maximum bending moments that occur along its length are due to either a big wave acting in the middle length of the ship, causing the vessel to hog, or alternatively, due to two big waves acting together at both the fore and aft ends of the ship at the same time, causing the vessel to sag at amidships. The transverse strength of ships is dependent on the buoyant forces acting upwards on the ship, and the self-weight of cargo, etc., acting downwards. In the case of bridges, bending stresses are caused by a number of different forces, including wind, snow, and the weight of automobiles and trains, together with the self-weight of the bridge. In the case of buildings, bending stresses can be caused by wind, snow and the weight of the contents of the building, together with its self-weight.

At the end of this chapter you should be able to:

- define neutral layer
- define the neutral axis of a beam's cross-section
- prove that $\frac{\sigma}{y} = \frac{M}{I} = \frac{E}{R}$
- calculate the stresses in a beam due to bending
- define anticlastic curvature
- analyse composite beams
- consider compatibility and horizontal and rotational equilibrium
- define a flitched beam
- understand composite ship structures

Mathematical references

In order to understand the theory involved in this chapter on **bending stresses in beams**, knowledge of the following mathematical topics is required: ***algebraic manipulation/transposition, integration and quadratic equations***. If help/revision is needed in these areas, see page 49 for some textbook references.

6.1 Introduction

It is evident that if a symmetrical-section beam is subjected to the bending action shown in Figure 6.1, known as sagging, the fibres at the top of the beam will decrease their lengths, whilst the fibres at the bottom of the beam will increase their lengths. The effect of this will be to cause compressive direct stresses to occur in the fibres at the top of the beam and tensile direct stresses to occur in the fibres at the bottom of the beam, these stresses being parallel to the axis of the beam.

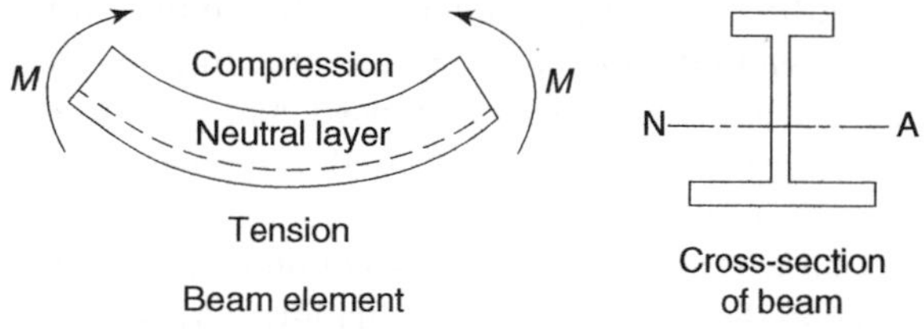

Figure 6.1 Beam element in bending

It is evident also from Figure 6.1 that, as the top layers of the beam decrease their lengths and the bottom fibres increase their lengths, somewhere between the two there will be a layer that will be in neither compression nor tension. This layer is called the ***neutral layer***, and its intersection with the beam's cross-section is called the ***neutral axis* (NA)**. The stress at the neutral axis is zero.

Assumptions made

The assumptions made in the theory of bending in this chapter are as follows:

(a) Deflections are small.
(b) The radius of curvature of the deformed beam is large compared with its other dimensions.
(c) The beam is initially straight.
(d) The cross-section of the beam is symmetrical.
(e) The effects of shear are negligible.
(f) Transverse sections of the beam, which are plane and normal before bending, remain plane and normal during bending.
(g) Elastic theory is obeyed, and the elastic modulus of the beam is the same in tension as it is in compression.
(h) The beam material is **homogeneous** and **isotropic.**

6.2 Proof of $\frac{\sigma}{y} = \frac{M}{I} = \frac{E}{R}$

In the equation

$$\frac{\sigma}{y} = \frac{M}{I} = \frac{E}{R} \tag{6.1}$$

σ = stress (due to the bending moment M) occurring at a distance y from the neutral axis

M = bending moment

I = second moment of area of the cross-section about its neutral axis (centroidal axis)

E = elastic modulus

R = radius of curvature of the neutral layer of the beam, when M is applied

Equation (6.1) is the fundamental expression that is used in the bending theory of beams, and it will be proven with the aid of Figure 6.2, which shows the deformed shape of an initially straight beam under the action of a sagging bending moment M. Initially, all layers of the beam element will be the same length as the neutral layer, AB, so that

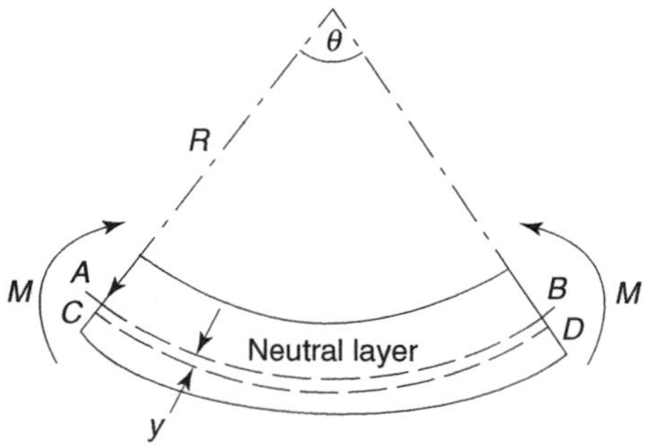

Figure 6.2 Beam element in bending

Initial length of $CD = AB = R\theta$ (i.e. the arc length of a circle = radius × angle subtended)

Final length of $CD = (R + y)\theta$

where CD is the length of the beam element at a distance y from the neutral layer.

Tensile strain of $CD = \dfrac{(CD - AB)}{AB}$

$$= \frac{(R+y)\theta - R\theta}{R\theta} = \frac{R\theta + y\theta - R\theta}{R\theta} = \frac{y\theta}{R\theta} = \frac{y}{R}$$

$\dfrac{\text{stress}}{\text{strain}}$ = Young's modulus,* E, from which,

stress = strain × E

Hence stress in the layer of $CD = \sigma = \dfrac{y}{R} \times E$

Rearranging gives $$\frac{\sigma}{y} = \frac{E}{R} \qquad (6.2)$$

* Thomas Young (13 June 1773 – 10 May 1829) was an English polymath. He is perhaps best known for his work on Egyptian hieroglyphics and the Rosetta Stone, but Young also made notable scientific contributions to the fields of vision, light, solid mechanics, energy, physiology, language and musical harmony. *Young's modulus* relates the stress in a body to its associated strain. To find out more about Young go to www.routledge.com/cw/bird

Equation (6.2) shows that the bending stress σ varies linearly with y and it will act on the section, as shown in Figure 6.3, where NA is the position of the neutral axis. Equation (6.2) also shows that the *largest stress* in magnitude occurs in the fibre which is the *furthest distance from the neutral axis.*

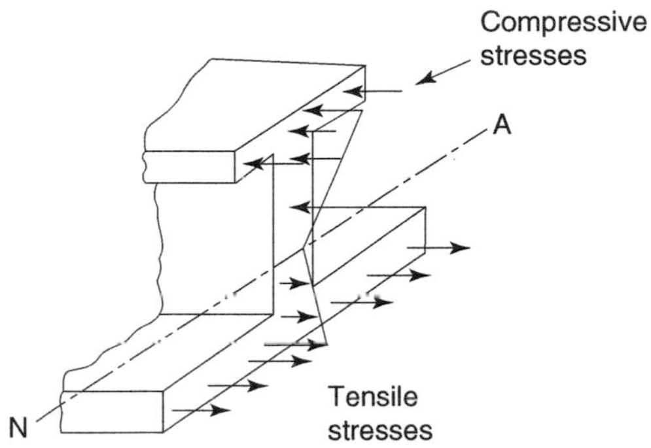

Figure 6.3 Bending stress distribution

It is evident also, from equilibrium considerations, that the longitudinal tensile force caused by the tensile stresses due to bending must be equal and opposite to the longitudinal compressive force caused by the compressive stresses due to bending, so that

$$\int \sigma \, dA = 0$$

where dA is the area of an element of the cross-section at a distance y from the neutral axis, but

$$\sigma = \frac{Ey}{R}$$

Therefore, at the neutral axis, $\displaystyle\int \frac{Ey}{R} dA = 0$

i.e. $\displaystyle\frac{E}{R}\int y \, dA = 0$

i.e. $$\int y \, dA = 0$$

or *the neutral axis* is at the *centroid of the beam's cross-section.*

Now it can be seen from Figure 6.3 that the bending stresses cause a couple which, from equilibrium considerations, must be equal and opposite to the externally applied moment M at the appropriate section, i.e.

$$M = \int \sigma \, y \, dA$$

From above $\sigma = \dfrac{Ey}{R}$ hence, $M = \displaystyle\int \frac{Ey}{R} y \, dA$

i.e. $$M = \frac{E}{R}\int y^2 \, dA$$

but $\int y^2\, dA = I$, the second moment of area about its neutral axis

Therefore $M = \frac{E}{R}\int y^2\, dA = \frac{E}{R} I$

from which $$\frac{M}{I} = \frac{E}{R} \qquad (6.3)$$

From equations (6.2) and (6.3)

$$\frac{\sigma}{y} = \frac{M}{I} = \frac{E}{R} \qquad (6.4)$$

6.3 Sectional modulus (Z)

From equation (6.2), it can be seen that the maximum stress due to bending occurs in the fibre which is the greatest distance from the neutral axis.

Let $\bar{y}$ = distance of the fibre in the cross-section of the beam which is the furthest distance from NA

Then, sectional modulus $Z = \frac{I}{\bar{y}}$ (6.5)

Therefore, from equation (6.4), the maximum bending stress is

$$\hat{\sigma} = \frac{M\bar{y}}{I} = \frac{M}{\left(\frac{I}{\bar{y}}\right)}$$

i.e. $$\sigma = \frac{M}{Z} \qquad (6.6)$$

Problem 1. A solid circular-section steel bar of diameter 2 cm and length 1 m is subjected to a pure bending moment of magnitude M. If the maximum permissible stress in the bar is 100 MN/m^2, determine the maximum permissible value of M. If the lateral deflection at the mid-point of this beam, relative to its two ends, is 6.25 mm, what will be the elastic modulus of the beam?

From Table 5.3, page 120, the second moment of area of a circle coinciding with its diameter

$$I = \frac{\pi D^4}{64} = \frac{\pi\left(2\times10^{-2}\right)^4}{64} = 7.854\times10^{-9}\ \text{m}^4$$

The maximum stress will occur at the fibre in the cross-section which is the furthest distance from the neutral axis, i.e.

$$\bar{y} = 1\ \text{cm} = 1\times10^{-2}\ \text{m}$$

and from equation (6.5),

$$Z = \frac{I}{\bar{y}} = \frac{7.854\times10^{-9}}{1\times10^{-2}} = 7.854\times10^{-7}\ \text{m}^3$$

From equation (6.6), $\hat{\sigma} = \frac{M}{Z}$, from which $M = \hat{\sigma} Z$

i.e. maximum permissible bending moment,

$\boldsymbol{M} = 100\times10^6\times7.854\times10^{-7} = \mathbf{78.54\ N\ m}$

Now under pure bending, the beam will bend into a perfect arc of a circle, as shown in Figure 6.4.

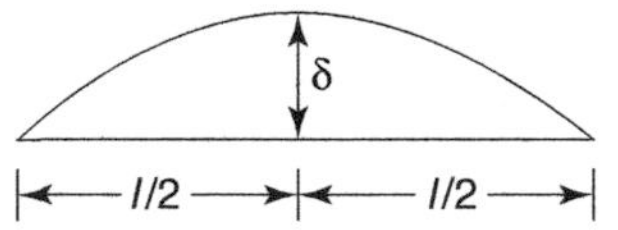

Figure 6.4

In the figure l = length of beam

δ = central deflection

Now from the properties of a circle

$$\delta(2R-\delta) = \frac{l}{2}\times\frac{l}{2}$$

or $$2R\delta - \delta^2 = \frac{l^2}{4}$$

However, as deflections are small, δ^2 is small compared with $2R\delta$. Therefore

$$2R\delta = \frac{l^2}{4} \text{ from which, } \delta = \frac{l^2}{8R}$$

or $$R = \frac{l^2}{8\delta} = \frac{l^2}{8\times6.25\times10^{-3}}$$

i.e. $\mathbf{R = 20\ m}$

From equation (6.4)

$$E = \frac{MR}{I} = \frac{78.54\times20}{7.854\times10^{-9}}$$

i.e. elastic modulus of the beam,

$$\boldsymbol{E = 2\times10^{11}\ N/m^2}$$

Problem 2. A beam of length 2 m and with the cross-section shown in Figure 5.15 (on page118) is simply supported at its ends and carries a uniformly distributed load w, spread over its entire length, as shown in Figure 6.5. Determine a suitable value for w, given that the maximum permissible tensile stress is 100 MN/m^2 and the maximum permissible compressive stress is 30 MN/m^2

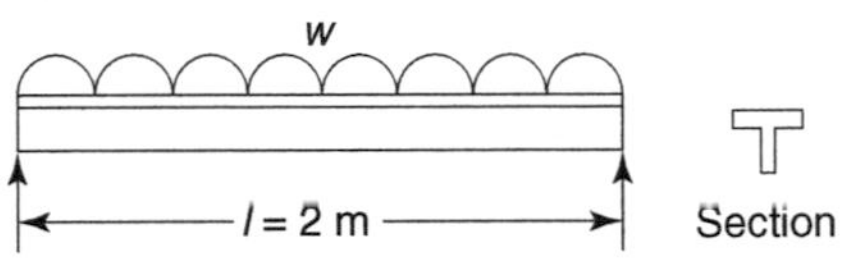

Figure 6.5

From Figure 6.5, the maximum bending moment is at mid-span, and is given by

$$\widehat{M} = \frac{wl}{2} \times \frac{l}{2} - \frac{wl}{2} \times \frac{l}{4} = \frac{wl^2}{4} - \frac{wl^2}{8}$$

i.e. $$\widehat{M} = \frac{wl^2}{8} \text{ at mid-span} \qquad (6.7)$$

The bottom of the beam will be in tension, and the top will be in compression.
Now from equation (5.26), page 118, the distance of the neutral axis is 0.104 m from the bottom, as shown in Figure 6.6.

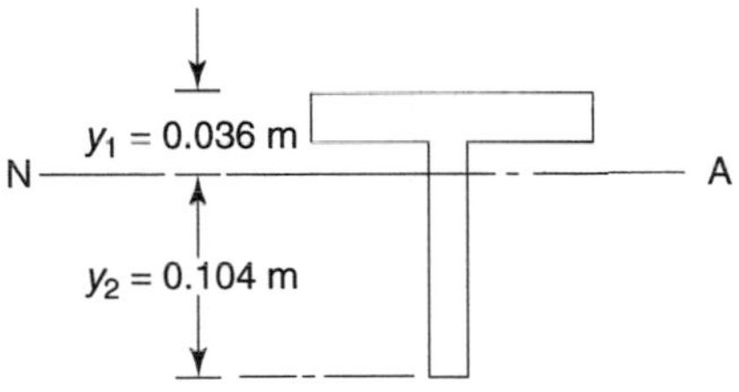

Figure 6.6 Beam cross-section

In the figure:

y_1 is used to determine the maximum compressive stress
y_2 is used to determine the maximum tensile stress

To determine the design criterion

If the tensile stress of 100 MN/m^2 is used in conjunction with y_2 then the maximum permissible compressive stress which is at the top is

$$\frac{0.036}{0.104} \times 100 = 34.6 \text{ MN/m}^2$$

That is, the compressive stress in the top of the flange will be exceeded if the tensile stress of 100 MN/m^2 is adopted as the design criterion; hence, the design criterion is the 30 MN/ m^2 *in the top flange.* If the top flange is under a compressive stress of 30 MN/m^2 then the tensile stress at the bottom of the web is 30 × 0.104/0.036 = 86.67 MN/m^2 in the bottom flange. Therefore, from equation (6.1)

$$\frac{\sigma}{y} = \frac{\widehat{M}}{I} \text{ and } \widehat{M} = \frac{\sigma I}{y} = \frac{30 \times 10^6 \times 5.02 \times 10^{-6}}{0.036}$$

from Chapter 5, Problem 5, page 118

i.e. $$\widehat{M} = \mathbf{4183.3 \text{ N m}} \qquad (6.8)$$

Equating (6.7) and (6.8) gives:

$$\widehat{M} = \frac{wl^2}{8} \text{ from which,}$$

$$w = \frac{8\widehat{M}}{l^2} = \frac{8 \times 4183.3}{2^2} = 8366.6 \text{ N/m}$$

i.e. $$\mathbf{w = 8.37 \text{ kN/m}}$$

Now try the following Practice Exercise

Practice Exercise 29. Bending stresses in beams

1. A concrete beam of uniform square cross-section, as shown in Figure 6.7, is to be lifted by its ends, so that it may be regarded as being equivalent to a horizontal beam, simply supported at its ends and subjected to a uniformly distributed load due to its self-weight. Determine the maximum permissible length of this beam, given the following:

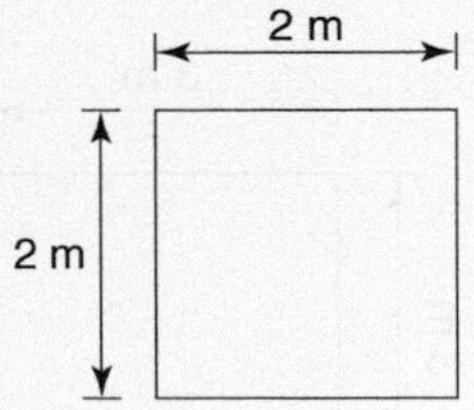

Figure 6.7 Cross-section of concrete beam

Density of concrete = 2400 kg/m^3

Maximum permissible tensile stress in the concrete = 1 MN/m^2
g = 9.81 m/s^2 [10.64 m]

2. If the concrete beam of question 1 had a hole in the bottom of its cross-section, as shown in Figure 6.8, what would be the maximum permissible length of the beam? [10.55 m]

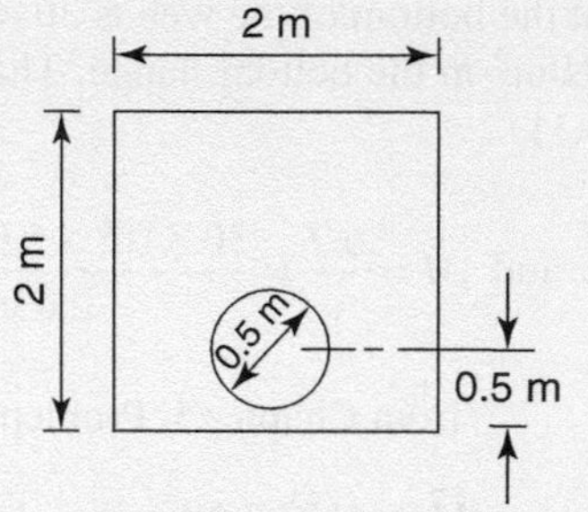

Figure 6.8 Cross-section with hole

3. What would be the maximum permissible length of the beam of questions 1 and 2, if the hole were at the top? [10.83 m]

4. A horizontal beam, of length 4 m, is simply supported at its ends and subjected to a vertically applied concentrated load of 10 kN at mid-span. Assuming that the width of the beam is constant and equal to 0.03 m, and neglecting the self-weight of the beam, determine an equation for the depth of the beam, so that the beam will be of uniform strength. The maximum permissible stress in the beam is 100 MN/m^2 [$d = 0.1x^{1/2}$ from 0 to 2 m]

5. A concrete beam of uniform square cross-section, as shown in Figure 6.9, is to be lifted by its ends, so that it may be regarded as being equivalent to a horizontal beam, simply supported at its ends and subjected to a uniformly distributed load due to its self-weight. Determine the maximum permissible length of this beam, given the following:

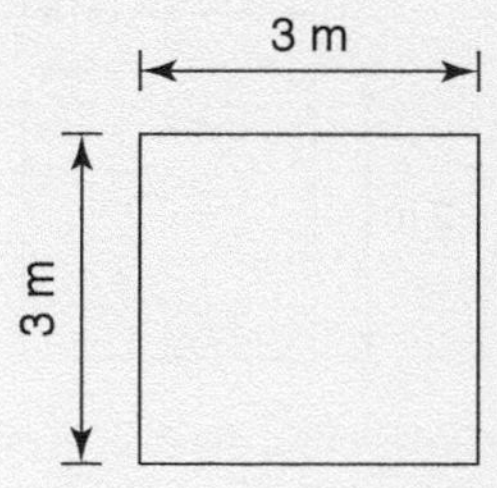

Figure 6.9 Cross-section of concrete beam

Density of concrete = 2400 kg/m^3
$g = 9.81$ m/s^2

Maximum permissible tensile stress in the concrete = 1 MN/m^2

[13.03 m]

6. If the concrete beam of problem 5 has a hole in the bottom of its cross-section, as shown in Figure 6.10, what would be the maximum permissible length of the beam? [8.74 m]

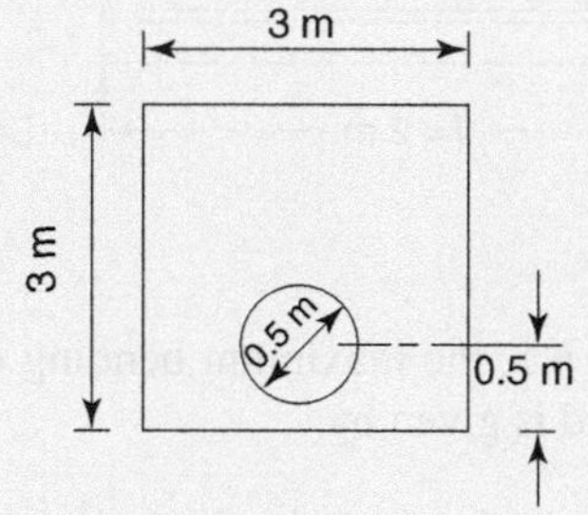

Figure 6.10 Cross-section with hole

7. What would be the maximum permissible length of the beam in problem 5, if the hole were at the top? [8.875 m]

8. The horizontal beam, of length 5 m, is simply supported at its ends and subjected to a vertically applied concentrated load of 10 kN at mid-span. Assuming that the width of the beam is constant and equal to 0.04 m, and neglecting the self-weight of the beam, determine an equation for the depth of the beam, so that the beam will be of uniform strength. The maximum permissible stress in the beam is 100 MN/m^2 [$d = 0.087x^{0.5}$ from 0 to 2.5 m]

6.4 Anticlastic curvature

If a beam of rectangular section is subjected to a pure bending moment, as shown in Figure 6.11(a), its cross-section changes shape, as shown in Figure 6.11(b). This curvature of the cross-section is known as **anticlastic curvature** and it is due to the Poisson effect.
If the radius of curvature of the beam is R, then the radius of curvature of the neutral axis of the beam due to anticlastic curvature is R/v, where v is Poisson's ratio.

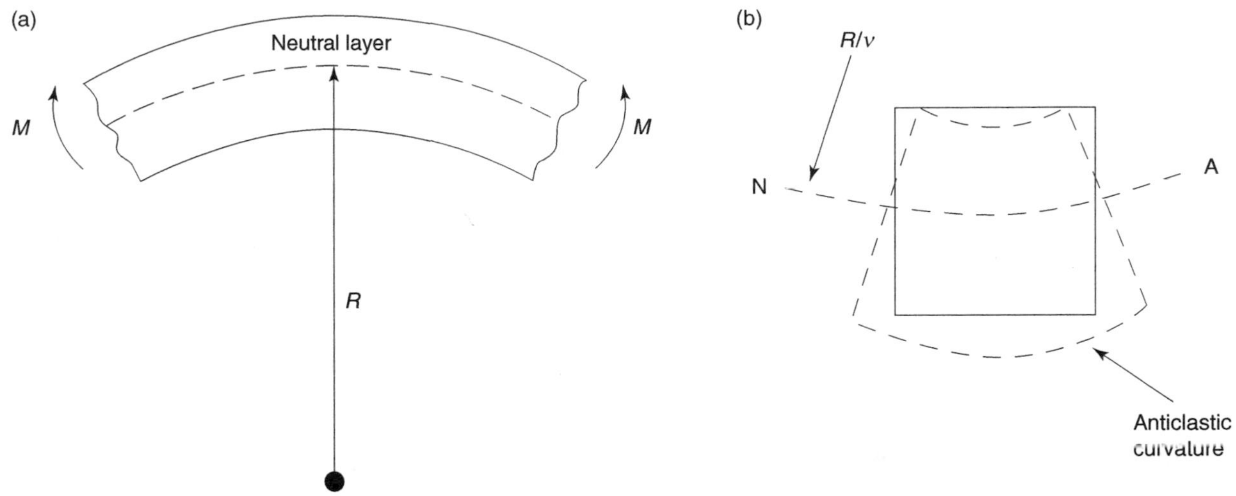

Figure 6.11 Anticlastic curvature: (a) beam under pure bending; (b) cross-section of beam

6.5 Composite beams

Composite beams occur in a number of different branches of engineering, and appear in the form of reinforced concrete beams, flitched beams, ship structures, glass-reinforced plastics, etc.

In the case of *reinforced concrete beams,* it is normal practice to reinforce the concrete with steel rods on the section of the beam where tensile stresses occur, leaving the unreinforced section of the beam to withstand compressive stresses, as shown in Figure 6.12. The reason for this practice is that, whereas concrete is strong in compression, it is weak in tension, but because the elastic modulus of steel is about 15 times greater than concrete, the steel will absorb the vast majority of the load on the tensile side of the beam. Furthermore, the alkaline content of the concrete reacts with the rust on the steel, causing the rust to form a tight protective coating around the steel reinforcement, thereby preventing further rusting. Thus, steel and concrete form a mutually compatible pair of materials which, when used together, actually improve each other's performance.

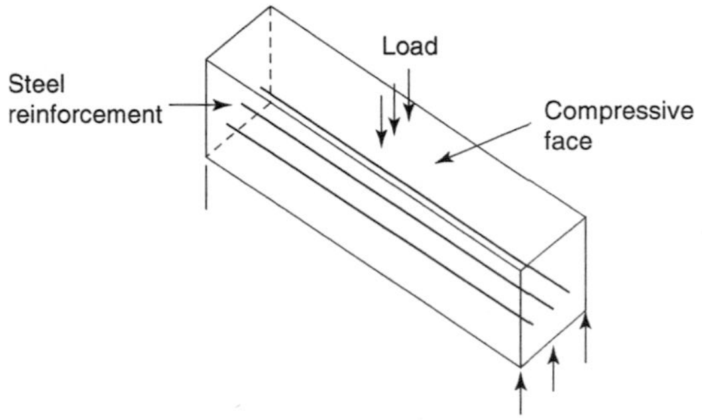

Figure 6.12 Reinforced concrete beam

The method of analysing reinforced concrete beams of the type shown in Figure 6.12 is to assume that all the tensile load is taken by the steel reinforcement and that all the compressive load is taken by that part of the concrete above the neutral axis, so that the stress distribution will be as shown in Figure 6.13. In this figure,

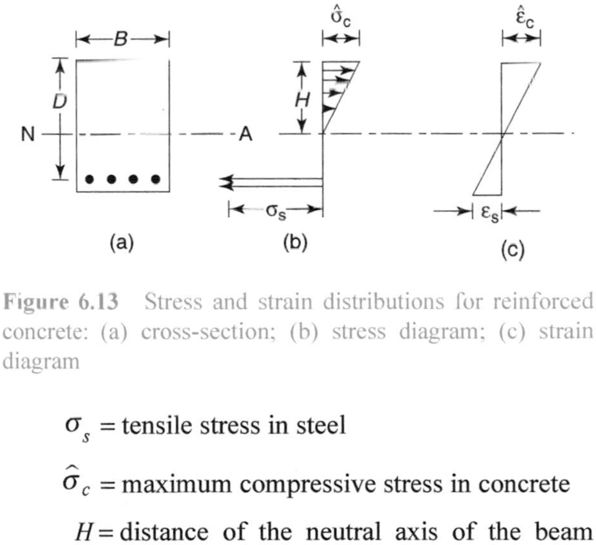

Figure 6.13 Stress and strain distributions for reinforced concrete: (a) cross-section; (b) stress diagram; (c) strain diagram

σ_s = tensile stress in steel

$\hat{\sigma}_c$ = maximum compressive stress in concrete

H = distance of the neutral axis of the beam from its top face

B = breadth of beam

D = distance between the steel reinforcement and the top of the beam

Let A_s = cross-sectional area of steel reinforcement

E_s = Young's modulus for steel

E_c = Young's modulus for concrete

$$m = \frac{E_s}{E_c} = \text{modular ratio}$$

ε_s = tensile strain in steel

$\hat{\varepsilon}_c$ = maximum strain in concrete

Now there are three unknowns in this problem, namely σ_s, $\hat{\sigma}_c$ and H, and as only two equations can be obtained from statical considerations, the problem is statically indeterminate, i.e. to obtain the third equation it will be necessary to consider compatibility, which, in this case, consists of strains.

Compatibility considerations

Consider similar triangles in the strain diagrams of Figure 6.13(c):

$$\frac{\hat{\varepsilon}_c}{H} = \frac{\varepsilon_s}{(D-H)}$$

or $$\frac{\hat{\sigma}_c}{E_c H} = \frac{\sigma_s}{E_s(D-H)}$$ since $\frac{\sigma}{\varepsilon} = E$ from which

$$\varepsilon = \frac{\sigma}{E}$$

or $$\hat{\sigma}_c = \frac{E_c}{E_s}\frac{H\sigma_s}{(D-H)}$$

and since $\frac{E_s}{E_c} = m$, $$\hat{\sigma}_c = \frac{H\sigma_s}{m(D-H)} \tag{6.9}$$

Considering 'horizontal equilibrium'

Tensile force in steel = compressive force in concrete

$$\sigma_s A_s = \left(\frac{\hat{\sigma}_c}{2}\right) BH$$

or $$\hat{\sigma}_c = \frac{2\sigma_s A_s}{BH} \tag{6.10}$$

Considering 'rotational equilibrium'

Externally applied moment M = moment of resistance of the section

i.e. $$M = \sigma_s A_s \times (D-H) + \left(\frac{\hat{\sigma}_c}{2}\right) \times BH \times \frac{2H}{3}$$

i.e. $$\boldsymbol{M = \sigma_s A_s (D-H) + \hat{\sigma}_c \frac{BH^2}{3}} \tag{6.11}$$

Equating (6.9) and (6.10) gives:

$$\frac{H\sigma_s}{m(D-H)} = \frac{2\sigma_s A_s}{BH}$$

from which $$BH^2 = 2m(D-H)A_s$$

i.e. $$BH^2 = 2mDA_s - 2mHA_s$$

and $$H^2 = 2mD\frac{A_s}{B} - 2mH\frac{A_s}{B}$$

Therefore $$H^2 + 2mH\frac{A_s}{B} - 2mD\frac{A_s}{B} = 0$$

Using the quadratic formula

$$H = \frac{-2mA_s/B \pm \sqrt{(2mA_s/B)^2 - 4(1)(-2mDA_s/B)}}{2(1)}$$

$$= \frac{-2mA_s/B \pm \sqrt{(2mA_s/B)^2 + (8mDA_s/B)}}{2}$$

i.e. $$H = \sqrt{(mA_s/B)^2 + 2mDA_s/B} - mA_s/B \tag{6.12}$$

Substituting equation (6.10) into (6.11) gives

$$M = \sigma_s A_s (D-H) + \frac{2\sigma_s A_s}{BH}\frac{BH^2}{3}$$

i.e. $$M = \sigma_s\left[A_s(D-H) + \frac{2A_s}{BH}\frac{BH^2}{3}\right]$$

$$= \sigma_s A_s\left[(D-H) + \frac{2H}{3}\right]$$

from which $$\sigma_s = \frac{M}{A_s[(D-H) + 2H/3]}$$

$$= \frac{M}{A_s[(D-H/3)]} \tag{6.13}$$

and from equation (6.10) $$\hat{\sigma}_c = \frac{2\sigma_s A_s}{BH}$$

$$= \frac{2\left(\frac{M}{A_s[(D-H/3)]}\right)A_s}{BH} = \frac{2\left(\frac{M}{[(D-H/3)]}\right)}{BH}$$

i.e. $$\hat{\sigma}_c = \frac{2M}{BH(D-H/3)} \tag{6.14}$$

Problem 3. A reinforced concrete beam, of rectangular section, is subjected to a bending moment, such that the steel reinforcement is in tension. Determine the maximum permissible value of this bending moment, given the following:

$D = 0.4$ m $\quad B = 0.3$ m $\quad m = 15$

Maximum permissible compressive stress in concrete = 10 MN/m^2

Maximum permissible tensile stress in steel = 150 MN/m^2

Diameter of each steel reinforcing rod = 2 cm

n = number of rods = 8

Cross-sectional area of steel reinforcement

$$= A_s = 8 \times \frac{\pi d^2}{4} = 8 \times \frac{\pi\left(2\times10^{-2}\right)^2}{4} = 2.513\times10^{-3}\ \text{m}^2$$

From equation (6.12)

$$H = \sqrt{\left(mA_s / B\right)^2 + 2mDA_s / B} - mA_s / B$$

$$\text{where } m = \frac{E_s}{E_c} = \frac{150\times10^6}{10\times10^6} = 15$$

$$= \sqrt{\left[15(2.513\times10^{-3})/(0.3)\right]^2 + 2(15)(0.4) - (2.513\times10^{-3})/(0.3)} - (15)(2.513\times10^{-3})/(0.3)$$

$$= \sqrt{0.0158 + 0.1005} - 0.12565$$

$$= 0.34103 - 0.12565$$

i.e. $\boldsymbol{H = 0.215}$ **m**

From equation (6.13)

$$\sigma_s = \frac{M}{A_s\left[(D - H/3)\right]} \text{ and}$$

$$M = \sigma_s A_s\left[(D - H/3)\right]$$

i.e. $M = \left(150\times10^6\right)\left(2.513\times10^{-3}\right)\left[0.4 - 0.215/3\right]$

i.e. **bending moment in steel rods,**

$$\boldsymbol{M} = 123.8\times10^3\,\text{kN m} = \mathbf{0.124\ MN\ m}$$

From equation (6.14)

$$\hat{\sigma}_c = \frac{2M}{BH(D - H/3)} \text{ from which,}$$

$$M = \frac{\hat{\sigma}_c BH(D - H/3)}{2}$$

i.e. $M = \dfrac{\left(10\times10^6\right)(0.3)(0.215)(0.4 - 0.215/3)}{2}$

i.e. **bending moment in concrete,**

$$\boldsymbol{M} = 105.9\times10^3\ \text{kN m} = \mathbf{0.106\ MN\ m}$$

i.e. the maximum stress in the concrete is the design criterion.

Therefore, **maximum permissible bending moment = 0.106 MN m**

NB The overall dimensions of the beam's cross-section should allow for the steel reinforcement to be covered by at least 5 cm of concrete.

Now try the following Practice Exercise

Practice Exercise 30. Bending stresses in composite beams

1. The cross-section of a reinforced concrete beam is as shown in Figure 6.14. Determine the maximum bending moment that this beam can sustain, assuming that the steel reinforcement is on the tensile side and that the following apply:

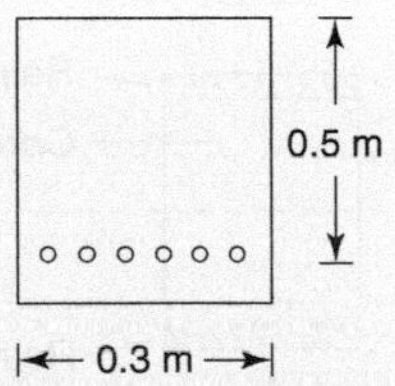

Figure 6.14

Maximum permissible compressive stress in concrete = 10 MN/m^2

Maximum permissible tensile stress in steel = 200 MN/m^2

Modular ratio = 15

Diameter of a steel rod = 2 cm

n = number of steel rods = 6 [0.145 MN m]

2. The cross-section of a reinforced concrete beam is as shown in Figure 6.15. Determine the maximum bending moment that this beam can sustain, assuming that the steel reinforcement is on the tensile side and that the following apply:

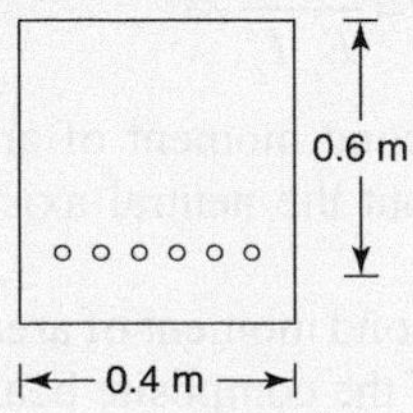

Figure 6.15

Maximum permissible compressive stress in concrete = 10 MN/m^2

Maximum permissible tensile stress in steel = 200 MN/m^2

Modular ratio = 14

Diameter of a steel rod = 2 cm

n = number of steel rods = 6 [0.198 MN m]

6.6 Flitched beams

A flitched beam is a common type of composite beam, where the reinforcements are relatively thin compared with the depth of the beam, and are usually attached to its outer surfaces, as shown in Figure 6.16. Typical materials used for flitched beams include a wooden core combined with external steel reinforcement and various types of plastic reinforcement combined with a synthetic porous core of low density.

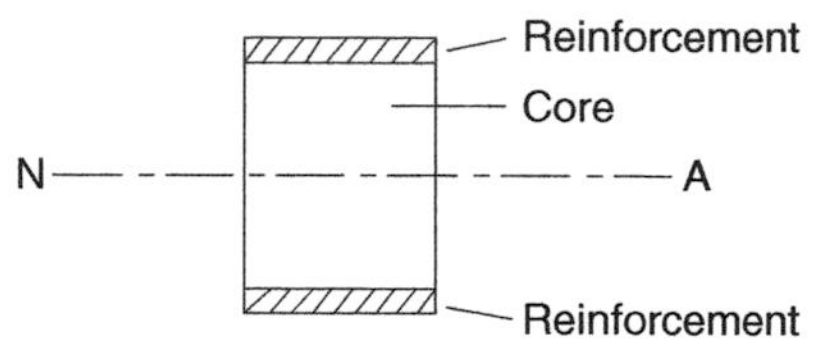

Figure 6.16 Cross-section of beam with horizontal reinforcement

Let M = applied moment at the section

M_r = moment of resistance of external reinforcement

M_c = moment of resistance of core

so that $$M = M_r + M_c \tag{6.15}$$

The main assumption made is that the *radius of curvature, R, is the same for the core as it is for the reinforcement,* i.e.

$$R = \frac{E_c I_c}{M_c} = \frac{E_r I_r}{M_r} \quad \text{from equation (6.1)}$$

from which $$M_r = \frac{E_r I_r}{E_c I_c} M_c \tag{6.16}$$

where I_r = second moment of area of the external reinforcement about the neutral axis of the composite beam

I_c = second moment of area of the core about the neutral axis of the composite beam

E_r = Young's modulus for the external reinforcement

E_c = Young's modulus for the core

Substituting equation (6.16) into (6.15) gives

$$M = \frac{E_r I_r}{E_c I_c} M_c + M_c = M_c\left(\frac{E_r I_r}{E_c I_c} + 1\right)$$

$$= M_c\left(\frac{E_r I_r + E_c I_c}{E_c I_c}\right)$$

i.e. $$M = \frac{M_c}{E_c I_c}\left(E_r I_r + E_c I_c\right)$$

from which $$\boldsymbol{M_c = \frac{E_c I_c M}{E_r I_r + E_c I_c}} \tag{6.17}$$

and from equation (6.16)

$$M_r = \frac{E_r I_r}{E_c I_c} M_c = \frac{E_r I_r}{E_c I_c}\left(\frac{E_c I_c M}{E_r I_r + E_c I_c}\right)$$

i.e. $$\boldsymbol{M_r = \frac{E_r I_r M}{E_r I_r + E_c I_c}} \tag{6.18}$$

Now $\sigma_r = \frac{M_r}{Z_r}$ and $\sigma_c = \frac{M_c}{Z_c}$ from equation (6.6)

where σ_r = maximum stress in the external reinforcement

σ_c = maximum stress in the core

Z_r = sectional modulus of the external reinforcement about NA

Z_c = sectional modulus of the core

Hence $$\sigma_r = \frac{M_r}{Z_r} = \frac{\frac{E_r I_r M}{E_r I_r + E_c I_c}}{Z_r} = \frac{\frac{E_r I_r M}{E_r I_r + E_c I_c}}{\frac{I_r}{y_r}}$$

from equation (6.5)

i.e. $$\sigma_r = \frac{E_r y_r M}{\left(E_r I_r + E_c I_c\right)} \tag{6.19}$$

Similarly $$\sigma_c = \frac{E_c y_c M}{\left(E_r I_r + E_c I_c\right)} \tag{6.20}$$

where y_r = distance of the outermost fibre of the external reinforcement from the neutral axis of the composite beam

y_c = distance of the outermost fibre of the core from the neutral axis of the composite beam

Problem 4. (a) A wooden beam of rectangular section is of depth 10 cm and width 5 cm. Determine the moment of resistance of this section given the following:

Young's modulus for wood = 1.4×10^{10} N/m^2

Maximum permissible stress in wood = 20 MN/m^2

(b) What percentage increase will there be in the moment of resistance of the beam section if it is reinforced by a 5 mm thick galvanised steel plate attached to both the top and the bottom surfaces of the beam, one plate at the bottom and the other plate on the top? Assume the following:

Young's modulus for steel = 2×10^{11} N/m^2

Maximum permissible stress in steel =150 MN/m^2

(a) From equation (6.4), $\frac{\sigma}{y} = \frac{M}{I}$ and $M = \frac{\sigma I}{y}$

I_w = second moment of area of wood about its neutral axis

$$= \frac{BD^3}{12} = \frac{(5 \times 10^{-2})(10 \times 10^{-2})^3}{12}$$

i.e. $I_w = 4.167 \times 10^{-6}$ m^4

Centroid $\bar{y} = \frac{1}{2}(10 \times 10^{-2}) = 5 \times 10^{-2}$ m

Therefore, moment of resistance of wood

$$M_w = \frac{\hat{\sigma}_w I_w}{\bar{y}} = \frac{(20 \times 10^6)(4.167 \times 10^{-6})}{5 \times 10^{-2}}$$

i.e. $\mathbf{M_w = 1666.7}$ **N m**

(b) Let R_w = radius of curvature of wood

R_s = radius of curvature of steel

σ_s = maximum stress in steel

σ_w = maximum stress in wood

E_s = elastic modulus of steel

E_w = elastic modulus of wood

y_s = distance of steel from NA

y_w = distance of outermost fibre of wooden core from NA

Now $R_s = R_w$

From equation (6.1), $\frac{E_s y_s}{\sigma_s} = \frac{E_w y_w}{\sigma_w}$

or $\frac{\sigma_s}{E_s y_s} = \frac{\sigma_w}{E_w y_w}$

from which $\sigma_s = \sigma_w \left(\frac{E_s y_s}{E_w y_w} \right)$ where y_s is the distance of the middle of each galvanised plate to the neutral axis of the wooden beam = 5.25 cm

$= 5.25 \times 10^{-2}$ m

i.e. $\sigma_s = \sigma_w \left(\frac{2 \times 10^{11} \times 5.25 \times 10^{-2}}{1.4 \times 10^{10} \times 5 \times 10^{-2}} \right)$

i.e. $\boldsymbol{\sigma_s = 15\sigma_w}$

From the design criterion of the question, the maximum permissible stress in steel,

σ_s = 150 MN/m^2

Hence $150 = 15\sigma_w$ from which,

$\sigma_w = \frac{150}{15} = 10$ MN/m^2

and $M_w = 1666.7 \times \frac{\sigma_w}{\hat{\sigma}_w}$, where σ_w is the design criterion stress in wood

Hence $M_w = 1666.7 \times \frac{10}{20} = 833.4$ N m

The second moment of area of the steel plate

$$I_s = 5 \times 10^{-2} \times 5 \times 10^{-3} \times (5.25 \times 10^{-2})^2 \times 2$$

$= 1.3781 \times 10^{-6}$ m^4

and $M_s = \frac{\hat{\sigma}_s I_s}{y_s} = \frac{150 \times 10^6 \times 1.3781 \times 10^{-6}}{5.25 \times 10^{-2}}$

= **3937.5 N m**

$M = M_s + M_w = 3937.5 + 833.4 =$ **4770.9 N m**

i.e. the percentage increase in moment of resistance of the flitched beam over the wooden beam is:

$$\left(\frac{4770.9-1666.7}{1666.7}\right)\times 100\% = 186.2\%$$

NB In this case, the chosen thickness for the steel plate was too small, as the stress in the wood was well below its permissible value.

Another type of horizontal flitched beam is shown in Figure 6.17, where the reinforcement is vertical.
In this case, the composite beam can be regarded as an equivalent wooden beam, as shown in Figure 6.18(b), or as an equivalent steel beam, as shown in Figure 6.18(c).

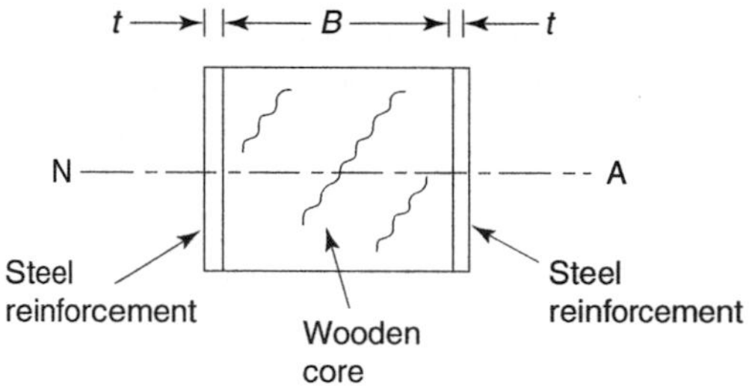

Figure 6.17 Cross-section of a horizontal beam with vertical reinforcement

If I_s = total second moment of area of the two steel beam reinforcements about NA
I_w = second moment of area of the wooden core about NA
M_s = total bending moment of resistance of the two steel reinforcements
M_w = bending moment of resistance of the wooden core

then from equation (6.1) $\frac{\sigma}{y}=\frac{M}{I}=\frac{E}{R}$

and $$M_s = \frac{E_s I_s}{R} \tag{6.21}$$

and $$M_w = \frac{E_w I_w}{R} \tag{6.22}$$

The total bending moment of resistance of the composite beam is

$$M = M_s + M_w$$

Therefore $$M = \frac{E_s I_s}{R} + \frac{E_w I_w}{R}$$

i.e. $$M = \frac{1}{R}\left(E_s I_s + E_w I_w\right) \tag{6.23}$$

where E_s = Young's modulus for steel reinforcement
E_w = Young's modulus for wooden core

Let E_I be the bending stiffness of the composite beam. Then

$$E_I = E_s I_s + E_w I_w \tag{6.24}$$

From equation (6.23)

$$\frac{1}{R} = \frac{M}{\left(E_s I_s + E_w I_w\right)} \tag{6.25}$$

and from equation (6.21)

$$\frac{1}{R} = \frac{M_s}{E_s I_s} \tag{6.26}$$

Equating (6.25) and (6.26) gives

$$\frac{M}{\left(E_s I_s + E_w I_w\right)} = \frac{M_s}{E_s I_s}$$

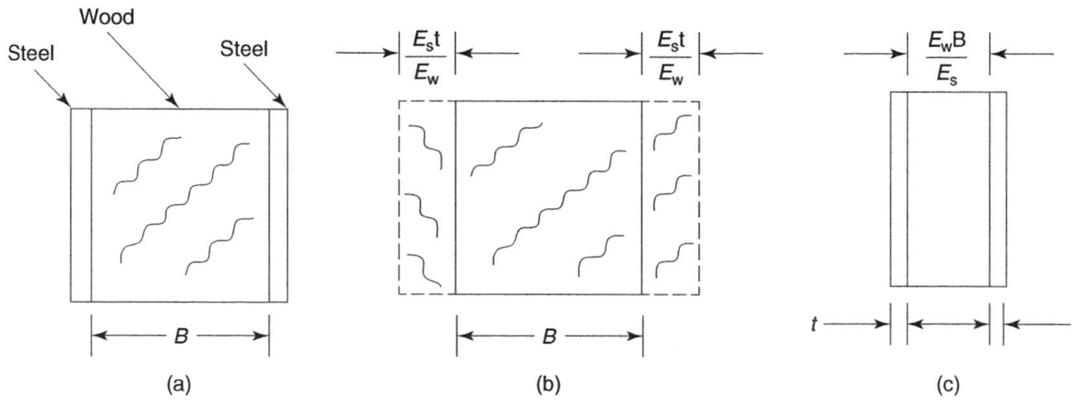

Figure 6.18 Cross-sections of horizontal beam: (a) actual cross-section; (b) equivalent wooden cross-section; (c) equivalent steel cross-section

from which $$M_s = \frac{E_s I_s M}{(E_s I_s + E_w I_w)}$$

i.e. $$M_s = \frac{M}{\left(\frac{E_s I_s + E_w I_w}{E_s I_s}\right)} = \frac{M}{\left(1 + \frac{E_w I_w}{E_s I_s}\right)} \quad (6.27)$$

From equation (6.1)

$$\frac{\sigma_s}{y} = \frac{M_s}{I_s} \text{ from which } \sigma_s = \frac{M_s y}{I_s} \quad (6.28)$$

Substituting equation (6.27) into (6.28) gives

$$\sigma_s = \frac{\frac{M y}{\left(1 + \frac{E_w I_w}{E_s I_s}\right)}}{I_s} = \frac{M y}{I_s\left(1 + \frac{E_w I_w}{E_s I_s}\right)}$$

i.e. $$\sigma_s = \frac{M y}{\left(I_s + \frac{E_w I_w}{E_s}\right)} \quad (6.29)$$

Now the bending strain on any fibre that is a distance y from NA is

$$\varepsilon = \frac{\sigma_s}{E_s} \quad (6.30)$$

i.e. $$\varepsilon = \frac{\frac{M y}{\left(I_s + \frac{E_w I_w}{E_s}\right)}}{E_s} = \frac{M y}{E_s\left(I_s + \frac{E_w I_w}{E_s}\right)}$$

$$= \frac{M y}{(E_s I_s + E_w I_w)} \quad (6.31)$$

but the strain will be the same in both the steel reinforcement and the wooden core at distance y from NA, so that $\sigma_w = E_w \varepsilon$

Therefore

$$\sigma_w = (E_w)\frac{M y}{(E_s I_s + E_w I_w)}$$

$$= \frac{M y}{\left(\frac{E_s I_s + E_w I_w}{E_w}\right)} = \frac{M y}{\left(\frac{E_s I_s}{E_w} + I_w\right)} \quad (6.32)$$

From equations (6.29) and (6.32), it can be seen that the denominators are equivalent to an equivalent steel second moment of area for the first equation, and to an equivalent wooden second moment of area for the second equation, as shown in Figures 6.18(c) and (b), respectively.

Now try the following Practice Exercise

Practice Exercise 31. Bending stresses in composite beams

1. (a) A wooden beam of rectangular section is 12 cm depth and 6 cm width. Determine the moment of resistance of this section, given the following: $E_{wood} = 1.4 \times 10^{10}$ N/m^2 and $\hat{\sigma}_{wood} = 20$ MPa.

   (b) What percentage increase will there be in the moment of resistance of this beam section if it is reinforced by a 6 mm thick galvanised steel plate attached to the top and bottom surfaces of the beam, one plate at the bottom and the other plate at the top. Assume the following:

   $E_{steel} = 2 \times 10^{11}$ N/m^2 and $\hat{\sigma}_{steel} = 150$ MPa

   [2880 N m, 136.3%]

6.7 Composite ship structures

Composite ship structures appear in the form of a steel hull, together with an aluminium-alloy superstructure. The reason for this combination is that steel is a suitable material for the main hull of a ship because of its ductility and good welding properties, but in order to keep the centre of gravity of a ship as low as possible, for the purposes of ship stability, it is convenient to use a material with a lower density than steel for the superstructure. In general, it is not suitable to use aluminium for the main hull, because aluminium has poor corrosion resistance to salt water.

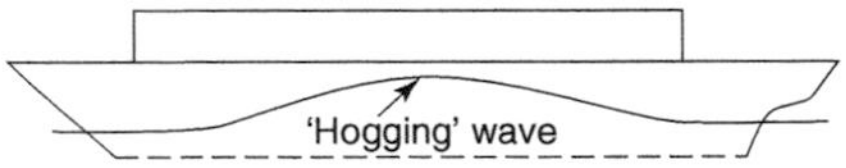

Figure 6.19 Longitudinal bending of a ship

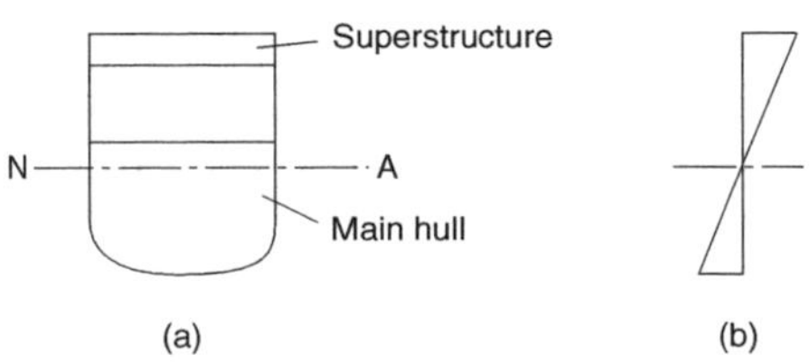

Figure 6.20 Cross-section of a ship: (a) transverse section; (b) strain distribution

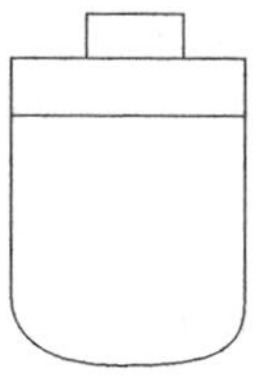

Figure 6.21 Equivalent section

Under longitudinal bending moments, caused by the self-weight of the ship and buoyant forces due to waves, as shown in Figure 6.19, the longitudinal strength of a ship can be based on beam theory, where the cross-section of the 'equivalent beam' is in fact the cross-section of the ship, as shown in Figure 6.20. The strain distribution across the transverse section of the ship is as shown in Figure 6.20(b), but as stress = $E \times$ strain, the equivalent moment of resistance of the aluminium alloy will be equivalent to E_a / E_s of a steel section of the same size. Thus, to calculate the position of the neutral axis (NA) and the second moment of area, the aluminium-alloy superstructure can be assumed to be equivalent to the form shown in Figure 6.21, where

E_a = Young's modulus for aluminium alloy

and E_s = Young's modulus for steel

Problem 5. A box-like cross-section consists of two parts, namely a steel bottom and an aluminium-alloy top, as shown in Figure 6.22. If the plate thickness of both the aluminium alloy and the steel is 1 cm, determine the maximum stress in both materials when the section is subjected to a bending moment of 100 MN m which causes it to bend about a horizontal plane (NA).

Assume E_a = Young's modulus for aluminium alloy

$$= 6.67 \times 10^{10} \mathrm{N/m^2}$$

and E_s = Young's modulus for steel

$$= 2 \times 10^{11} \mathrm{N/m^2}$$

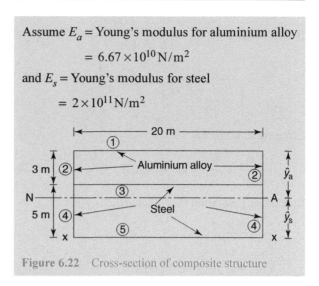

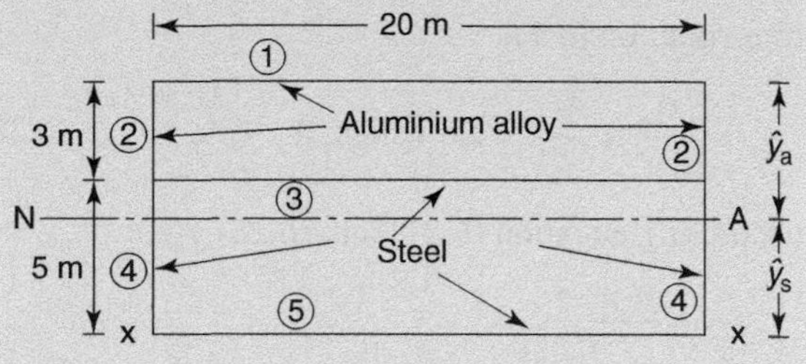

Figure 6.22 Cross-section of composite structure

To determine the position of the neutral axis and the second moment of area of the equivalent steel section, use will be made of Table 6.1.

$$\hat{y}_s = \frac{\sum Ay}{\sum A} = \frac{1.9134}{0.5867}$$

i.e. $\hat{y}_s$ **= 3.261 m**

Therefore, from Figure 6.22 $\hat{y}_a = 8 - 3.261 =$ **4.739 m**

Second moment of area about axis XX is given by

$$I_{xx} = \sum ay^2 + \sum i = 10.737 + 0.223$$

from the parallel axis theorem

i.e. I_{xx} **= 10.96 m^4**

Table 6.1

Section	A (m)	y (m)	Ay (m^3)	Ay^2 (m^4)	i (m^4)
(1)	0.0667	8	0.5334	4.267	$20 \times (1 \times 10^{-2})^3 / 36 = 0$
(2)	0.02	6.5	0.13	0.845	$1 \times 10^{-2} \times 3^3 \times 2 / 36 = 0.015$
(3)	0.2	5	1.0	5.0	$20 \times (1 \times 10^{-2})^3 / 12 = 0$
(4)	0.1	2.5	0.25	0.625	$1 \times 10^{-2} \times 5^3 \times 2 / 12 = 0.208$
(5)	0.2	0	0	0	$(1 \times 10^{-2})^3 \times 20 / 3 = 0$
Σ	0.5867	–	1.9134	10.737	0.223

Second moment of area about NA is given by

$$I_{NA} = I_{XX} - \left(\hat{y}_s\right)^2 \sum A = 10.96 - (3.261)^2 \times 0.5867$$

i.e. $\boldsymbol{I_{NA} = 4.721\ \mathrm{m}^4}$

The maximum stress in the aluminium alloy

$$\hat{\sigma}_a = \frac{M\hat{y}_a}{I_{NA}} \times \left(\frac{E_a}{E_s}\right)$$

$$= \frac{\left(100 \times 10^6\right)(4.739)}{4.721} \times \left(\frac{6.67 \times 10^{10}}{2 \times 10^{11}}\right)$$

i.e. $\boldsymbol{\hat{\sigma}_a = 33.48\ \mathrm{MN/m}^2}$

The maximum stress in the steel

$$\hat{\sigma}_s = \frac{M\hat{y}_s}{I_{NA}} = \frac{\left(100 \times 10^6\right)(3.261)}{4.721}$$

i.e. $\boldsymbol{\hat{\sigma}_s = 69.07\ \mathrm{MN/m}^2}$

It can be seen from the above calculations that, despite the fact that the aluminium-alloy deck is further away from NA than is the steel bottom, the stress in the aluminium alloy is less than in the steel, because its elasticity is three times greater than that of the steel.

6.8 Composite structures

The use of composites is of much interest in structures varying from car bodies to boat hulls and from chairs to ship superstructures. Composites appear in many and various forms from glass-reinforced plastics (GRP) and carbon-fibre-reinforced plastics (CFRP) to metal matrix composites (MMC), etc. Analysis of such structures is given in Chapter 19, and for further study the reader should consult References 2 and 3, on page 497. This chapter has shown that the sensible use of composites can improve the structural efficiency of many types of structure in various engineering applications.

6.9 Combined bending and direct stress

The case of combined bending and direct stress occurs in a number of engineering situations, including the eccentric loading of short columns, as shown in Figure 6.23. By placing the equal and opposite forces on the centre line of the strut, as shown in Figure 6.23(b), the loading condition of Figure 6.23(a) is unaltered. However, it can be seen that the column of Figure 6.23(b) is in fact subjected to a centrally applied force P and a couple $P\Delta$, as shown in Figure 6.23(c). Furthermore, from Figure 6.23(c), it can be seen that owing to the centrally applied direct load P, the whole of the strut will be subjected to a direct compressive stress σ_d, but owing to the couple $P\Delta$, the side AB will be in tension and the side CD will be in compression.

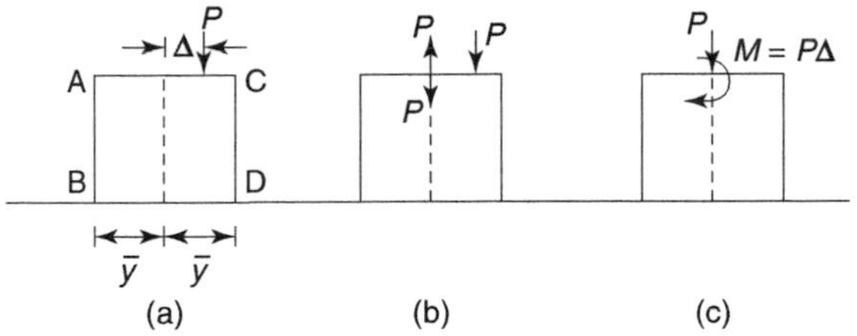

Figure 6.23 Eccentrically loaded short column

Thus, the effect of M will be to cause the stress to be increased in magnitude on face CD, and to be decreased in magnitude on face AB, i.e.

$$\text{stress on face AB} = \sigma_{AB} = -\frac{P}{A} + \frac{M\hat{y}}{I} \qquad (6.33)$$

$$\text{and} \quad \text{stress on face CD} = \sigma_{CD} = -\frac{P}{A} + \frac{M\hat{y}}{I} \qquad (6.34)$$

where P is the compressive load and A is the cross-sectional area of the short strut.

It is evident, therefore, that in general the stress due to the combined effects of a bending moment M and a tensile load P will be given by

$$\sigma = \sigma_d \pm \sigma_b$$

where σ_d = direct stress (tensile is positive)

and σ_b = bending stress

For a practical video demonstration of combined bending and torsion of circular section shafts go to www.routledge.com/cw/bird and click on the menu for 'Mechanics of Solids 3rd Edition'

Eccentrically loaded concrete columns

As concrete is weak in tension, it is desirable to determine how eccentric a load can be so that no part of a short column is in tension.

The next two examples will be used to demonstrate the calculations usually associated with eccentrically loaded short columns.

Problem 6. Determine the position in which an eccentrically applied vertical compressive load can be placed, so that no tension occurs in a short vertical column of square cross-section.

By applying a compressive force P at the point shown in Figure 6.24 it can be seen that the face AB is likely to develop tensile stresses due to the bending action about the YY axis.

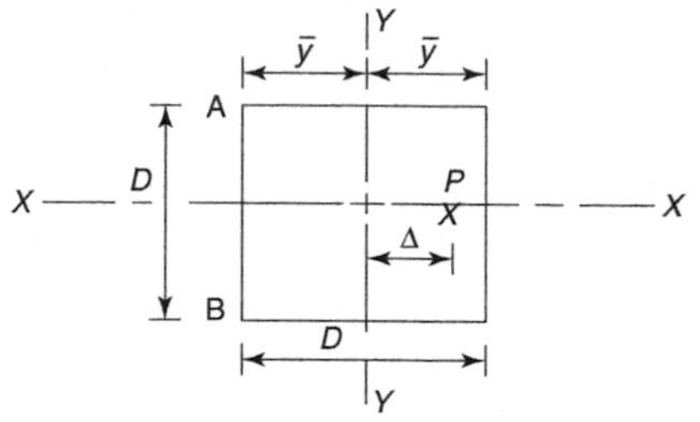

Figure 6.24 Cross-section of concrete column

To satisfy the requirements of the example, let the stress on the face AB equal zero, so that

$$0 = -\frac{P}{A} + \frac{P\Delta\hat{y}}{I} \quad \text{as there is no tensile stress}$$

i.e. $$0 = -\frac{1}{A} + \frac{\Delta\hat{y}}{I} = -\frac{1}{D^2} + \frac{\Delta\left(\frac{D}{2}\right)}{\frac{D \times D^3}{12}} = -\frac{1}{D^2} + \frac{6\Delta}{D^3}$$

from which $$\frac{1}{D^2} = \frac{6\Delta}{D^3} \quad \text{or} \quad \frac{D^3}{D^2} = 6\Delta$$

and $$\mathbf{\Delta} = \frac{\mathbf{D}}{\mathbf{6}} \tag{6.35}$$

From equation (6.35), it can be seen that for no tension to occur in the short column, owing to an eccentrically applied compressive load, the eccentrically applied load must be applied within the mid-third area of the centre of the square section. For this reason, this rule is known as the ***mid-third rule***.

Problem 7. Determine the position in which an eccentrically applied vertical compressive load can be placed, so that no tension occurs in a short vertical column of circular cross-section.

By applying a compressive force P to the point shown in Figure 6.25, it can be seen that tension is most likely to occur at the point C, so that to satisfy the requirements of the example, the stress at C equals zero.

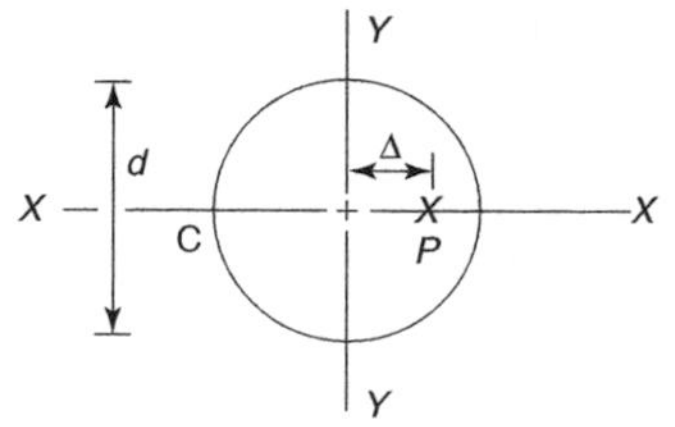

Figure 6.25 Cross-section of concrete column

As there is no tension

$$\sigma = 0 = -\frac{P}{A} + \frac{P\Delta\hat{y}}{I} = -\frac{P}{\left(\frac{\pi d^2}{4}\right)} + \frac{P\Delta\left(\frac{d}{2}\right)}{\left(\frac{\pi d^4}{64}\right)}$$

$$= -\frac{4P}{\pi d^2} + \frac{32P\Delta}{\pi d^3}$$

i.e. $$\frac{4P}{\pi d^2} = \frac{32P\Delta}{\pi d^3} \quad \text{and} \quad 4 = \frac{32\Delta}{d}$$

i.e. $$\mathbf{\Delta} = \frac{\mathbf{d}}{\mathbf{8}} \tag{6.36}$$

Equation (6.36) shows that for no tensile stress to occur in a short concrete column of circular cross-section, the eccentricity of the load must not exceed $d/8$. For this reason, this is known as the ***mid-quarter rule***.

Problem 8. Determine the maximum tensile and compressive stresses in the clamp of Figure 6.26

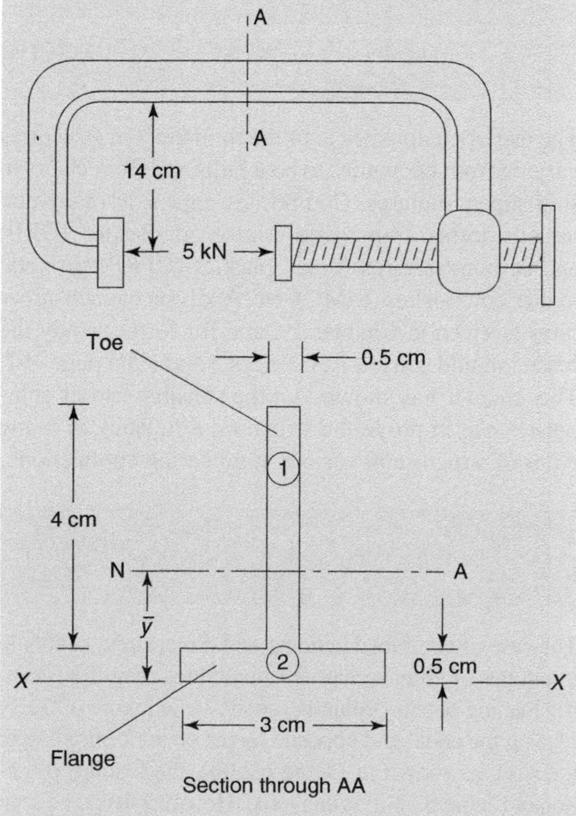

Figure 6.26 Clamp under loading

Table 6.2

Section	A (m)	y (m)	Ay (m³)	Ay^2 (m⁴)	i_0 (m⁴)
(1)	2×10^{-4}	2.5×10^{-2}	5×10^{-6}	1.25×10^{-7}	2.667×10^{-8}
(2)	1.5×10^{-4}	0.25×10^{-2}	3.75×10^{-7}	9.375×10^{-10}	3.125×10^{-8}
Σ	3.5×10^{-4}	-	5.375×10^{-6}	1.259×10^{-7}	2.7×10^{-8}

The relevant geometrical properties of the tee-bar are calculated with the aid of Table 6.2.

$$\bar{y} = \frac{\sum Ay}{\sum A} = \frac{5.375\times10^{-6}}{3.5\times10^{-4}}$$

i.e. $\bar{y}$ **= 0.0154 m**

Second moment of area about axis XX is given by

$$I_{xx} = \sum Ay^2 + \sum i_o = 1.259\times10^{-7} + 2.7\times10^{-8}$$

from the parallel axis theorem

i.e. $\mathbf{I_{xx} = 1.529\times10^{-7}\ m^4}$

Second moment of area about NA is given by

$$I_{NA} = I_{XX} - (\bar{y})^2 \sum A - 1.529\times10^{-7} - (0.0154)^2\times3.5\times10^{-4}$$

i.e. $\mathbf{I_{NA} = 6.989\times10^{-8}\ m^4}$

Let $\widehat{M}$ = the maximum bending moment on the 'top' member of the clamp (or at AA)

Hence, from Figure 6.24

$$\widehat{M} = 5\text{ kN}\times\left(14\times10^{-2}\text{m} + \bar{y}\right)$$

$$= 5\text{ kN}\times\left(14\times10^{-2} + 0.0154\right)$$

i.e. $\widehat{M}$ **= 0.777 kN m**

The stress in the 'top' (toe) of the tee

$$\sigma_T = -\frac{\widehat{M}\times\hat{y}}{I} + \frac{5\text{ kN}}{A}\text{ where}$$

$$\hat{y} = 4.5\times10^{-2} - \bar{y} = 0.045 - 0.0154 = 0.0296\text{m}$$

i.e.
$$\sigma_T = -\frac{0.777\text{ kN m}\times0.0296\text{m}}{6.989\times10^{-8}\text{m}^4} + \frac{5\text{ kN}}{A}$$

$$= -329.077\frac{\text{kN}}{\text{m}^2} + \frac{5\text{ kN}}{3.5\times10^{-4}\text{m}^2}$$

$$= -399077 + 14286 = -314791\text{ kN/m}^2$$

i.e. σ_T **= – 314.8 MN/m²**

The stress in the 'bottom' (flange) of the tee

$$\sigma_B = +\frac{\widehat{M}\times\bar{y}}{I_{\text{NA}}} + \frac{5\text{ kN}}{A}$$

$$\sigma_B = \left(\frac{0.777\times0.0154}{6.989\times10^{-8}} + \frac{5}{3.5\times10^{-4}}\right)\text{kN/m}^2$$

$$= 171209 + 14286$$

$$= 185495\text{ kN/m}^2$$

i.e. σ_B **= 185.5 MN/m²**

NB By placing the flange of the tee-bar on the inner part of the clamp, the maximum stresses have been reduced. This has been achieved by 'lowering' the neutral axis towards the tensile face, where the bending and direct stresses are additive. Movement of the neutral axis towards the tensile face also had the effect of lowering the maximum bending moment, as the lever arm that the load was acting on was decreased.

Now try the following Practice Exercise

Practice Exercise 32. Bending stresses in composite structures

1. A short steel column, of circular cross-section, has an external diameter of 0.4 m and a wall thickness of 0.1 m and carries a compressive but axially applied eccentric load. Two linear strain gauges, which are mounted longitudinally at opposite sides on the external surface of the column but in the plane of the load, record strains of 50×10^{-6} and -200×10^{-6}. Determine the magnitude and eccentricity of the axial load. Assume that $E = 2 \times 10^{11} \text{N/m}^2$.

   [– 1.414 MN, 0.104 m]

2. Determine the maximum tensile and compressive stresses in the clamp of Figure 6.27.

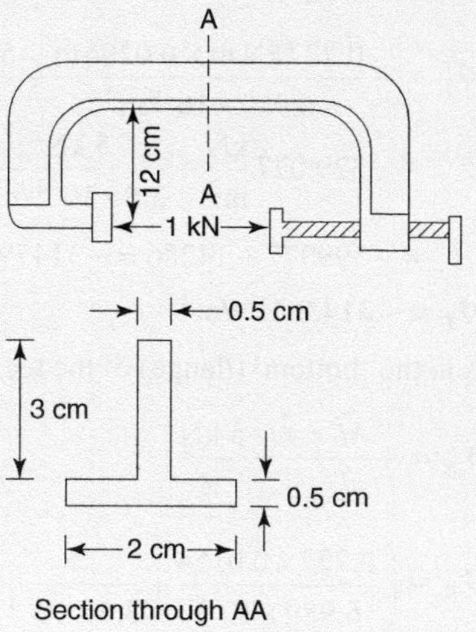

Figure 6.27

   [61.96 MN/m^2, – 94.1 MN/m^2]

3. A short steel column, of circular cross-section, has an external diameter of 0.5 m and a wall thickness of 0.1 m and carries a compressive but axially applied eccentric load. Two linear strain gauges, which are mounted longitudinally at opposite sides on the external surface of the column but in the plane of the load, record strains of 60×10^{-6} and -300×10^{-6}. Determine the magnitude and eccentricity of the axial load. Assume that $E = 2 \times 10^{11} \text{N/m}^2$.

   [–3.016 MN, 0.469 m]

4. Determine the maximum tensile and compressive stresses in the clamp of Figure 6.28.

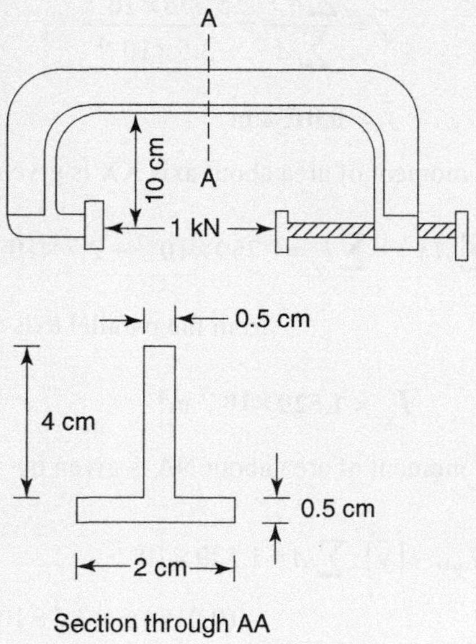

Figure 6.28

   [37.24 MPa, – 50 MPa]

For fully worked solutions to each of the problems in Exercises 29 to 32 in this chapter, go to the website:

www.routledge.com/cw/bird

Chapter 7

Beam deflections due to bending

Why it is important to understand: **Beam deflections due to bending**

When designing beams to withstand lateral loads, the designer must consider both bending stresses and also lateral deflections of the beam. Very often if the material of construction of the beam is steel and the length of the beam is small, the lateral deflections are usually relatively small and not of much importance. If, however, the beam's length is large then lateral deflections of the beam become important, partly due to the fact that very often the deflection of the beam is proportional to the cube of its length. Moreover, if the Young's modulus of the beam is relatively small, such as if the beam were made of wood, then the design criterion of the beam may become its bending deflections, rather than its bending stresses. One example of such a case is if a plank of wood were attached to a brick wall with the aid of metal brackets to make a bookcase, the design criterion is usually the bending deflections of the wooden plank, rather than its bending stresses.

At the end of this chapter you should be able to:

- appreciate that beam deflections are due to bending and shear
- understand and use the repeated integration method to determine maximum beam deflection
- understand and use Macaulay's method to determine maximum beam deflection
- perform calculations on statically determinate and indeterminate beams
- understand and use the moment-area method
- derive and use slope-deflection equations

Mathematical references

In order to understand the theory involved in this chapter on **beam deflections due to bending**, knowledge of the following mathematical topics is required: ***algebraic manipulation/transposition, integration and integration by parts***. If help/revision is needed in these areas, see page 49 for some textbook references.

7.1 Introduction

Beam deflections are usually due to bending and shear, but only those due to the former will be considered in this chapter.

The radius of curvature R of a beam in terms of its deflection y, at a distance x along the length of the beam, is given by

$$\frac{1}{R} = \frac{\frac{d^2y}{dx^2}}{\left[1+\left(\frac{dy}{dx}\right)^2\right]^{3/2}} \qquad (7.1)$$

However, if the deflections are small, as is the usual requirement in structural design, then $(dy/dx)^2$ is negligible compared with unity, so that equation (7.1) can be approximated to

$$\frac{1}{R} = \frac{d^2y}{dx^2} \qquad (7.2)$$

However, from equation (6.1) on page 134

$$\frac{M}{I} = \frac{E}{R} \text{ from which } M = \frac{EI}{R} = EI\left(\frac{1}{R}\right)$$

Therefore, bending moment $M = EI\dfrac{d^2y}{dx^2}$ (7.3)

Equation (7.3) is a very important expression for the bending of beams, and there are a number of different ways of solving it, but only three methods will be considered here, namely repeated integration and Macaulay's method, the moment-area method and use of the slope-deflection equations.

7.2 Repeated integration method

This is a **boundary value method** which depends on integrating equation (7.3) twice and then substituting boundary conditions to determine the arbitrary constants, together with other unknowns. It will be demonstrated through detailed solutions of the following worked problems.

Problem 1. Determine an expression for the maximum deflection of a cantilever of uniform section, under a concentrated load at its free end, as shown in Figure 7.1.

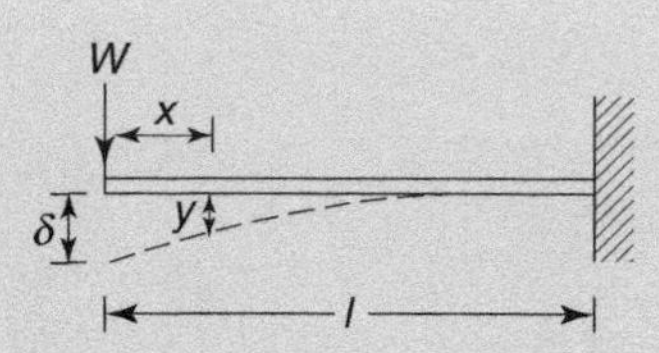

Figure 7. 1 Cantilever with end load

$$EI\frac{d^2y}{dx^2} = M = -Wx$$

Integrating both sides of the equation with respect to x gives

$$\int EI\frac{d^2y}{dx^2}dx = \int(-Wx)dx$$

i.e. $$EI\frac{dy}{dx} = -W\frac{x^2}{2} + A \qquad (7.4)$$

Integrating again gives

$$\int EI\frac{dy}{dx}dx = \int\left(-W\frac{x^2}{2} + A\right)dx$$

i.e. $$EIy = -W\frac{x^3}{6} + Ax + B \qquad (7.5)$$

There are two unknowns, namely the arbitrary constants A and B; therefore, two boundary conditions will be required, as follows

at $x = l$, $\dfrac{dy}{dx} = 0$ (i.e. the slope at the built-in end is zero at the right end)

Hence, from equation (7.4) $EI(0) = -W\dfrac{l^2}{2} + A$

i.e. $$0 = -W\frac{l^2}{2} + A$$

or $$\boldsymbol{A = W\frac{l^2}{2}} \qquad (7.6)$$

The other boundary condition is that

at $x = l$, $y = 0$ (i.e. the deflection y is zero at the built-in end, i.e. at $x = l$)

Hence from equation (7.5)

$$EI(0) = -W\frac{l^3}{6} + Al + B$$

From equation (7.6) $A = W\frac{l^2}{2}$

hence, $0 = -W\frac{l^3}{6} + \left(W\frac{l^2}{2}\right)l + B$

i.e. $0 = -W\frac{l^3}{6} + W\frac{l^3}{2} + B$

and $B = W\frac{l^3}{6} - W\frac{l^3}{2}$

i.e. $\boldsymbol{B = -W\frac{l^3}{3}}$ (7.7)

Substituting equations (7.6) and (7.7) into equation (7.5), the deflection y *at* any distance x along the length of the beam is given by

$$EIy = -W\frac{x^3}{6} + Ax + B$$

$$= -W\frac{x^3}{6} + \left(W\frac{l^2}{2}\right)x - W\frac{l^3}{3}$$

i.e. $EIy = -W\left[\frac{x^3}{6} - \frac{l^2}{2}x + \frac{l^3}{3}\right]$

and $\boldsymbol{y = -\frac{W}{EI}\left[\frac{x^3}{6} - \frac{l^2 x}{2} + \frac{l^3}{3}\right]}$ (7.8)

By inspection, the **maximum deflection** δ occurs at $x = 0$ at the free end on the left

i.e. $\boldsymbol{\delta} = -\frac{W}{EI}\left[\frac{l^3}{3}\right] = \boldsymbol{-\frac{Wl^3}{3EI}}$ (7.9)

The negative sign denotes that the deflection is downward.

Problem 2. Determine an expression for the maximum deflection of a cantilever under a uniformly distributed load w, as shown in Figure 7.2.

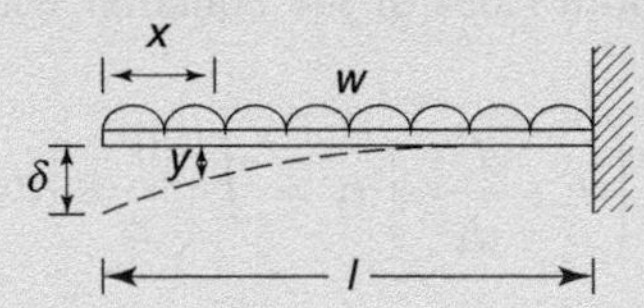

Figure 7.2 Cantilever with a UDL

At any distance x along the length of the beam, the bending moment M is given by

$$M = -w \times x \times \frac{x}{2} = -\frac{wx^2}{2}$$

From equation (7.3) $M = EI\frac{d^2y}{dx^2}$

hence $EI\frac{d^2y}{dx^2} = -\frac{wx^2}{2}$ (7.10)

Integrating both sides of the equation with respect to x gives

$$\int EI\frac{d^2y}{dx^2}dx = \int\left(-\frac{wx^2}{2}\right)dx$$

i.e. $EI\frac{dy}{dx} = -w\frac{x^3}{6} + A$

Integrating again gives

$$\int EI\frac{dy}{dx}dx = \int\left(-w\frac{x^3}{6} + A\right)dx$$

i.e. $EIy = -w\frac{x^4}{24} + Ax + B$ (7.11)

There are two unknowns, namely A and B, and therefore two boundary conditions will be required, as follows

at $x = l$, $\frac{dy}{dx} = 0$ (i.e. the slope at the built-in end is zero)

Hence, from above $EI\frac{dy}{dx} = -w\frac{x^3}{6} + A$

i.e. $EI(0) = -w\frac{l^3}{6} + A$

from which $A = \frac{wl^3}{6}$ (7.12)

The other boundary condition is that

at $x = l$, $y = 0$ (i.e. the deflection y is zero at the built-in end)

Hence from equation (7.11)

$$EI(0) = -w\frac{l^4}{24} + \left(w\frac{l^3}{6}\right)l + B$$

i.e. $B = w\frac{l^4}{24} - w\frac{l^4}{6} = -\frac{wl^4}{8}$ (7.13)

Substituting equations (7.12) and (7.13) into equation (7.11), the following expression is obtained for the deflection y of the beam at any distance x from its free end

$$EIy = -w\frac{x^4}{24} + Ax + B = -w\frac{x^4}{24} + \left(\frac{wl^3}{6}\right)x - \frac{wl^4}{8}$$

i.e. $$EIy = -w\left(\frac{x^4}{24} - \left(\frac{l^3}{6}\right)x + \frac{l^4}{8}\right)$$

and $$\boldsymbol{y = -\frac{w}{EI}\left(\frac{x^4}{24} - \frac{l^4 x}{6} + \frac{l^4}{8}\right)} \quad (7.14)$$

By inspection, the **maximum deflection** δ occurs at $x = 0$, i.e. at the free end, where

$$\delta = -\frac{w}{EI}\left(\frac{(0)^4}{24} - \frac{l^3(0)}{6} + \frac{l^4}{8}\right)$$

i.e. $$\boldsymbol{\delta = -\frac{wl^4}{8EI}} \quad (7.15)$$

The negative sign denotes that the deflection is downward.

Alternative method for determining δ

From Section 3.11, page 81,

$$\frac{d^2M}{dx^2} = w$$

However, from equation (7.3) $M = EI\dfrac{d^2y}{dx^2}$

Then $$\frac{dM}{dx} = EI\frac{d^3y}{dx^3} \text{ and } \frac{d^2M}{dx^2} = EI\frac{d^4y}{dx^4}$$

Therefore $$EI\frac{d^4y}{dx^4} = w \quad (7.16)$$

In this case, *w, is downward;* therefore equation (7.16) becomes

$$EI\frac{d^4y}{dx^4} = -w$$

Integrating both sides of the equation with respect to x gives

$$\int EI\frac{d^4y}{dx^4}dx = \int(-w)dx$$

i.e. $$EI\frac{d^3y}{dx^3} = F = -wx + A$$ [where F = the shearing force]

Integrating both sides of the equation with respect to x gives

$$\int EI\frac{d^3y}{dx^3}dx = \int(-wx + A)dx$$

i.e. $$EI\frac{d^2y}{dx^2} = M = -w\frac{x^2}{2} + Ax + B$$

At $x = 0$, $F = 0$ at the free end, so that $0 = -w(0) + A$ i.e. $A = 0$

At $x = 0$, $M = 0$ at the free end, so that

$$0 = -w\frac{(0)^2}{2} + A(0) + B \quad \text{i.e.} \quad B = 0$$

Hence $$EI\frac{d^2y}{dx^2} = -\frac{wx^2}{2} \quad (7.17)$$

which is identical to equation (7.10).

Equation (7.16) is particularly useful for beams with distributed loads of complex form, such as those met in determining the longitudinal strengths of ships, owing to the combined effects of self-weight and buoyant forces caused by waves.

Problem 3. Determine an expression for the maximum deflection of a uniform-section beam, simply supported at its ends and subjected to a centrally placed concentrated load W, as shown in Figure 7.3.

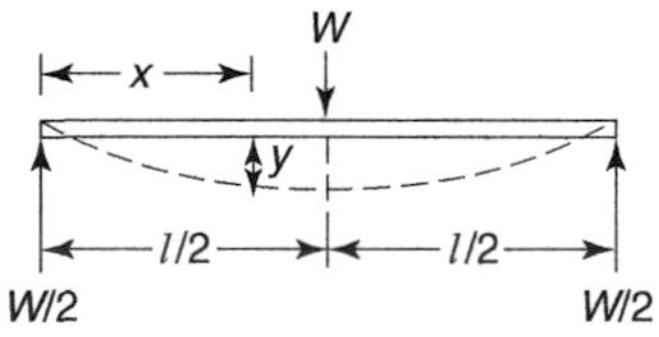

Figure 7.3

In this case, there is a discontinuity for bending moment distribution at mid-span, hence equation (7.3), together with its boundary conditions, can only be applied between $x = 0$ and $x = \dfrac{l}{2}$

i.e. $$EI\frac{d^2y}{dx^2} = M = \frac{W}{2}x$$

Integrating both sides of the equation with respect to x gives

$$\int\left(EI\frac{d^2y}{dx^2}\right)dx = \int\left(\frac{W}{2}x\right)dx$$

i.e. $$EI\frac{dy}{dx} = \frac{Wx^2}{4} + A$$

Integrating both sides of the equation with respect to x gives

$$\int\left(EI\frac{dy}{dx}\right)dx = \int\left(\frac{Wx^2}{4}+A\right)dx$$

i.e. $$EIy = \frac{Wx^3}{12} + Ax + B \qquad (7.18)$$

At $x = 0$, i.e. the left end, $y = 0$; therefore, $\boldsymbol{B = 0}$

At $x = \frac{l}{2}$, i.e. at mid-span, $\frac{dy}{dx} = 0$

Therefore $$EI(0) = \frac{W\left(\frac{l}{2}\right)^2}{4} + A$$

i.e. $$0 = \frac{Wl^2}{16} + A$$

and $$\boldsymbol{A = -\frac{Wl^2}{16}}$$

Hence, from equation (7.18)

$$EIy = \frac{Wx^3}{12} + \left(-\frac{Wl^2}{16}\right)x + 0$$

from which $$y = \frac{W}{EI}\left(\frac{x^3}{12} - \frac{l^2x}{16}\right)$$

By inspection, **the maximum deflection δ** occurs at $x = \frac{l}{2}$, i.e. at mid-length, where

$$\delta = \frac{W}{EI}\left(\frac{\left(\frac{l}{2}\right)^3}{12} - \frac{l^2\left(\frac{l}{2}\right)}{16}\right) = \frac{W}{EI}\left(\frac{l^3}{96} - \frac{l^3}{32}\right)$$

i.e. $$\boldsymbol{\delta = -\frac{Wl^3}{48EI}} \qquad (7.19)$$

Problem 4. Determine an expression for the maximum deflection of a uniform-section beam, simply supported at its ends and subjected to a uniformly distributed load w, as shown in Figure 7.4.

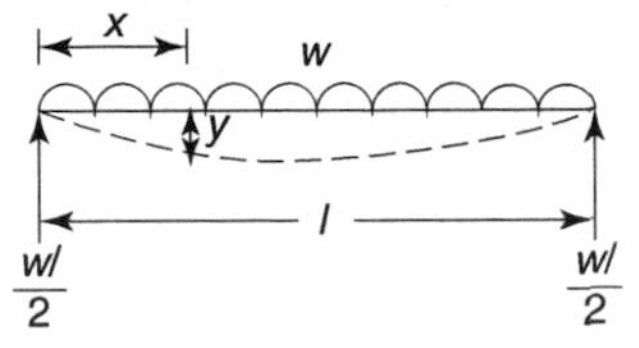

Figure 7.4

At distance x from the left end, $M = \frac{wlx}{2} - \frac{wx^2}{2}$

Thus, from equation (7.3)

$$EI = \frac{d^2y}{dx^2} = \frac{wl}{2}x - \frac{w}{2}x^2 \qquad (7.20)$$

Integrating both sides of the equation with respect to x gives

$$\int\left(EI\frac{d^2y}{dx^2}\right)dx = \int\left(\frac{wl}{2}x - \frac{w}{2}x^2\right)dx$$

i.e. $$EI\frac{dy}{dx} = \frac{wl}{2}\left(\frac{x^2}{2}\right) - \frac{w}{2}\left(\frac{x^3}{3}\right) + A$$

i.e. $$EI\frac{dy}{dx} = \frac{wlx^2}{4} - \frac{wx^3}{6} + A$$

Integrating both sides of the equation with respect to x gives

$$\int\left(EI\frac{dy}{dx}\right)dx = \int\left(\frac{wlx^2}{4} - \frac{wx^3}{6} + A\right)dx$$

i.e. $$EIy = \frac{wl}{4}\left(\frac{x^3}{3}\right) - \frac{w}{6}\left(\frac{x^4}{4}\right) + Ax + B$$

i.e. $$EIy = \frac{wlx^3}{12} - \frac{wx^4}{24} + Ax + B$$

At $x = 0$, i.e. at the left end, $y = 0$, therefore, $\boldsymbol{B = 0}$

At $x = l$, i.e. at the right end, $y = 0$

Therefore $$EI(0) = \frac{wl^4}{12} - \frac{wl^4}{24} + Al + 0$$

i.e. $$Al = \frac{wl^4}{24} - \frac{wl^4}{12} = -\frac{wl^4}{24}$$

and $$\boldsymbol{A = -\frac{wl^3}{24}}$$

Hence, from above $$EIy = \frac{wlx^3}{12} - \frac{wx^4}{24} + Ax + B$$

i.e. $$EIy = \frac{wlx^3}{12} - \frac{wx^4}{24} + \left(-\frac{wl^3}{24}\right)x + 0$$

and $$\boldsymbol{y = \frac{w}{EI}\left(\frac{lx^3}{12} - \frac{x^4}{24} - \frac{xl^3}{24}\right)} \qquad (7.21)$$

By inspection, the maximum deflection, δ occurs at $x = \frac{l}{2}$, i.e. at mid-length, where

$$\delta=\frac{w}{EI}\left(\frac{l\left(\frac{l}{2}\right)^3}{12}-\frac{\left(\frac{l}{2}\right)^4}{24}-\frac{\left(\frac{l}{2}\right)l^3}{24}\right)$$

$$=\frac{w}{EI}\left(\frac{l^4}{96}-\frac{l^4}{24(16)}-\frac{l^4}{48}\right)$$

i.e. $\boldsymbol{\delta=-\frac{5wl^4}{384EI}}$

An alternative method of solving Problem 4 above

From equation (7.16), $EI\frac{d^4y}{dx^4}=w$

On integrating once, $EI\frac{d^3y}{dx^3}=F=-wx+A$

At $x=0$, i.e. the left end, $F=+\frac{wl}{2}$ positive – according to the definition in Chapter 3,

Therefore, since $EI\frac{d^3y}{dx^3}=F=-wx+A$

then $\frac{wl}{2}=-wx+A$

and $\frac{wl}{2}+wx=A$

i.e. $\frac{wl}{2}+0=A$ and $\boldsymbol{A=\frac{wl}{2}}$

Hence $EI\frac{d^3y}{dx^3}=-wx+\frac{wl}{2}$

Integrating gives

$$EI\frac{d^2y}{dx^2}=M=-w\left(\frac{x^2}{2}\right)+\frac{wl}{2}x+B$$

At $x=0$, i.e. at the left end, $M=0$; therefore, $\boldsymbol{B=0}$

Therefore $$EI\frac{d^2y}{dx^2}=-\frac{wx^2}{2}+\frac{wlx}{2} \quad (7.22)$$

which is identical to equation (7.20).

Now try the following Practice Exercise

Practice Exercise 33. Beam deflections due to bending using the repeated integration method

1. Obtain an expression for the deflection y at any distance x from the left end of the uniform section beam of Figure 7.5.

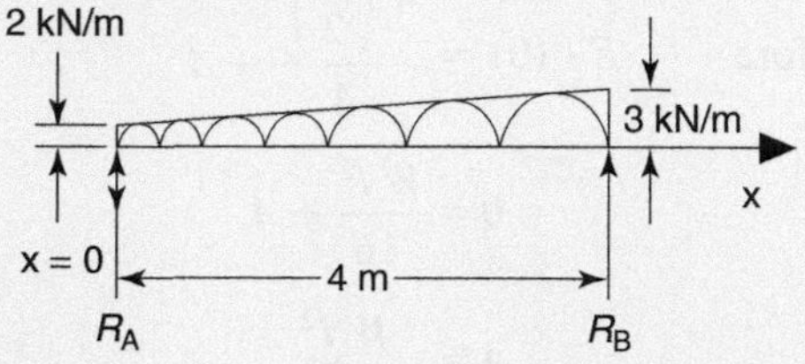

Figure 7.5

$$\left[y=\frac{1}{EI}(0.778x^2-0.0833x^4 -2.086\times10^{-3}x^5-6.583x)\right]$$

2. Determine the value of the reactions and end fixing moments for the uniform-section beam of Figure 7.6. Hence, or otherwise, obtain an expression for the deflection y at any distance x from the left end of the beam by the double integration method.

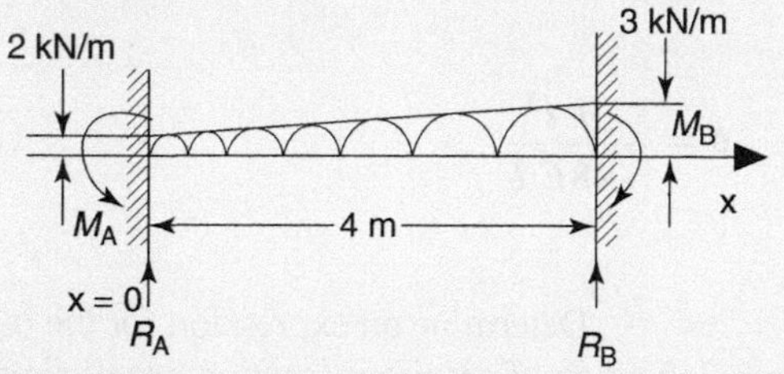

Figure 7.6

$$\left[R_A=4.62\text{ kN}, M_A=3.24\text{ kN m}, R_B=5.38\text{ kN}, M_B=3.43\text{ kN m}; y=\frac{1}{EI}\{0.77x^3-1.62x^2-0.0833x^4-2.086\times10^{-3}x^5\}\right]$$

3. Obtain an expression for the deflection y at any distance x from the left end of the uniform section beam of Figure 7.7.

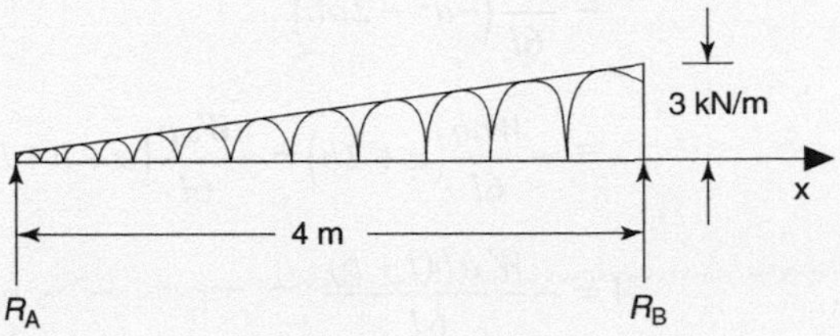

Figure 7.7

$$\left[y = \frac{1}{EI}\left(0.333x^3 - 0.00643x^5 - 3.7325x\right)\right]$$

4. Determine the value of the reactions and end fixing moments for the uniform-section beam of Figure 7.8. Hence, or otherwise, obtain an expression for the deflection y at any distance x from the left end of the beam.

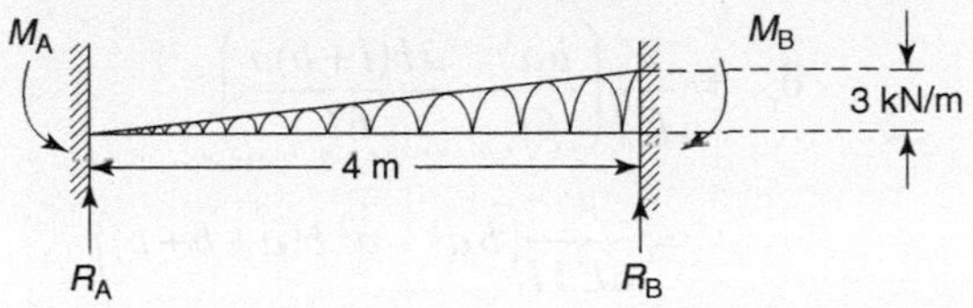

Figure 7.8

$$\left[R_A = 1.8 \text{ kN}, M_A = 1.6 \text{ kN m},\right.$$

$$R_B = 4.2 \text{ kN}, M_B = -2.4 \text{ kN m};$$

$$\left. y = \frac{1}{EI}\left(0.3x^3 - 0.8x^2 - 0.00625x^5\right)\right]$$

7.3 Macaulay's method

This method will be given without proof, as a number of proofs already exist in numerous texts (see References 3 and 4 on page 497). In the case of Problem 3 above, it can be seen that the equation for M only applied between $x = 0$ and $x = l/2$, and that both boundary conditions had to be applied within these limits. Furthermore, because of symmetry, the boundary condition $dy/dx = 0$ at $x = l/2$ also applied.

However, if the beam were not symmetrically loaded, it would not be possible to obtain the second boundary condition, namely $dy/dx = 0$ at $x = l/2$.

To demonstrate how to overcome this problem, the following examples will be considered, which are based on Macaulay's method.†

Problem 5. Determine an expression for the deflection under the load W for the uniform-section beam of Figure 7.9.

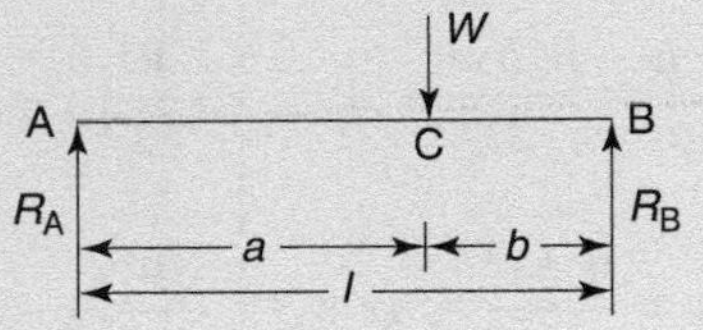

Figure 7.9

First, it will be necessary to determine the value of R_A which can be obtained by taking moments about the point B:

$$R_A \times l = W \times b$$

from which $$\boldsymbol{R_A = \frac{W \times b}{l}}$$

Macaulay's method is to use separate bending moment equations for each section of the beam, but to integrate the equations via the '**Macaulay brackets**', so that the constants of integration apply to all sections of the beam. It must, however, be emphasised that if the term within the *Macaulay bracket is negative,* then *that part of the expression does not apply* for boundary conditions, etc.

For the present problem, the bending moment between the points A and C will be different to that between the points C and B; hence, it will be necessary to separate the two expressions by the dashed line as shown below

$$x = a$$

$$EI\frac{d^2y}{dx^2} = R_A x \quad \vdots \quad - W[x - a]$$

where x varies from 0 at the left end to $x = l$ at the right end

Now, $R_A = \dfrac{Wb}{l}$ from above

†William Herrick Macaulay (16 November, 1853 – 28 November, 1936) was a British mathematician, known for his work in engineering, and proposed a mechanical technique for structural analysis. To find out more about Macaulay go to www.routledge.com/cw/bird

$$x = a$$

hence $$EI\frac{d^2y}{dx^2} = \frac{Wbx}{l} \quad \vdots \quad -W[x-a] \qquad (7.23)$$

$$EI\frac{dy}{dx} = \frac{Wbx^2}{2l} + A \quad \vdots \quad -\frac{W}{2}\left[x-a\right]^2 \qquad (7.24)$$

$$EIy = \frac{Wbx^3}{6l} + Ax + B \quad \vdots \quad -\frac{W}{6}\left[x-a\right]^3 \qquad (7.25)$$

The brackets [ ] which appear in equations (7.23) to (7.25) are known as Macaulay brackets, and their integration must be carried out in the manner shown in equations (7.24) and (7.25), i.e. from $x = 0$ to $x = a$, and from $x = a$ to $x = l$, so that the arbitrary constants A and B apply to both sides of the beam.

Now in setting the boundary conditions and in obtaining values for dy/dx and y, *if the terms within the Macaulay brackets become negative, then they do not apply.*

Boundary conditions

The first boundary condition is as follows: at $x = 0$, $y = 0$ which, when applied to equation (7.25), reveals that

$$\boldsymbol{B = 0}$$

NB The expression $\left[x-a\right]^3$ does not apply when the above boundary condition is substituted into equation (7.25), because the term within the Macaulay brackets, [ ], is negative and the term $\left[x-a\right]^3$ applies when x varies from 0 to l.

The second boundary condition is as follows: at $x = l$, $y = 0$

Therefore $$0 = \frac{Wbl^2}{6} + Al - \frac{Wb^3}{6} \quad \text{where } b = (l - a)$$

from which $$Al = \frac{Wb^3}{6} - \frac{Wbl^2}{6}$$

and $$A = \frac{Wb}{l}\left(\frac{b^2}{6} - \frac{l^2}{6}\right) = \frac{Wb}{6l}\left(b^2 - l^2\right)$$

$$= \frac{Wb}{6l}\left(b^2 - (a+b)^2\right)$$

i.e. $$A = \frac{Wb}{6l}\left(b^2 - (a^2 + 2ab + b^2)\right)$$

$$= \frac{Wb}{6l}\left(b^2 - a^2 - 2ab - b^2\right)$$

$$= \frac{Wb}{6l}\left(-a^2 - 2ab\right)$$

$$= -\frac{Wab}{6l}\left(a + 2b\right) = -\frac{Wab}{6l}\left(a + b + b\right)$$

i.e. $$A = -\frac{Wab(l+b)}{6l}$$

i.e. the deflection y at a distance x along the length of the beam is given by equation (7.26), providing the term within the Macaulay brackets, [ ], does not become negative

$$x = a$$

$$EIy = \frac{Wbx^3}{6l} - \frac{Wab(l+b)x}{6l} \quad \vdots \quad -\frac{W}{6}\left[x-a\right]^3 \qquad (7.26)$$

The deflection under the load δ_C, where $x = a$, is given by

$$\delta_C = \frac{W}{EI}\left(\frac{ba^3}{6l} - \frac{ab(l+b)a}{6l}\right)$$

$$= \frac{W}{6EIl}\left(ba^3 - a^2b(a+b+b)\right)$$

$$= \frac{W}{6EIl}\left(ba^3 - a^3b - a^2b^2 - a^2b^2\right)$$

i.e. $$\boldsymbol{\delta_C = -\frac{Wa^2b^2}{3EIl}} \qquad (7.27)$$

If W is at the mid-length of the beam, then $a = b = l/2$ in equation (7.27)

thus $$\boldsymbol{\delta_C = -\frac{Wl^3}{48EI}}$$ (as obtained in Problem 3 on page 154)

Problem 6. Determine an expression for the deflection distribution for the simply supported beam of Figure 7.10. Hence, or otherwise, obtain the position and value of the maximum deflection, given the following:

$$E = 2\times10^{11}\ \text{N/m}^2 \text{ and } I = 2\times10^{-8}\ \text{m}^4$$

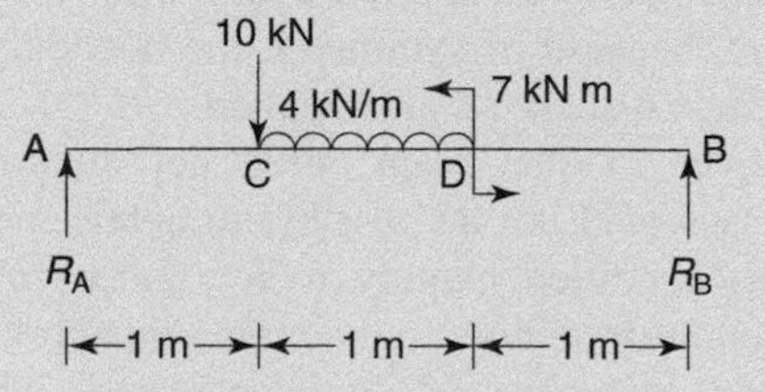

Figure 7.10

First, it is necessary to determine R_A, which can be obtained by taking moments about the point B:

$$R_A \times 3 = 10 \times 2 + 4 \times 1 \times 1.5 + 7$$

i.e. $$R_A \times 3 = 20 + 6 + 7 = 33$$

and $$\boldsymbol{R_A} = \frac{33}{3} = \mathbf{11\ kN}$$

In applying Macaulay's method to this beam, and remembering that *the negative terms inside the Macaulay brackets must be ignored*, it is necessary to make the distributed load of Figure 7.10 equivalent to that of Figure 7.11, which is essentially the same as that of Figure 7.10 (where the 4 kN/m load has to be cancelled out between $x = 2$ m and $x = 3$ m; thus it has to be applied *upwards* as well as *downwards*).

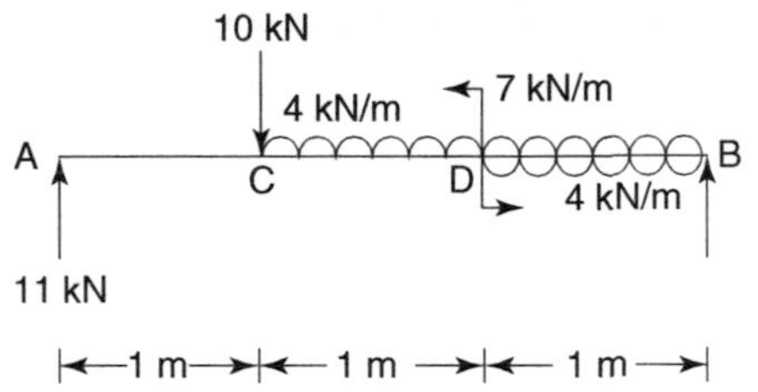

Figure 7.11

As the bending moment expression is different for sections AC, CD and DB, it will be necessary to apply Macaulay's method to each of these sections, as follows (where x varies from 0 at the left end, to $x = 3$ m at the right end)

$x = 1$ m | $x = 2$ m

$$EI\frac{d^2y}{dx^2} = 11x \;\Big|\; -10[x-1] \;\Big|\; +\frac{4}{2}[x-2]^2$$
$$\Big|\; -\frac{4}{2}[x-1]^2 \;\Big|\; -7[x-2]^0 \qquad (7.28)$$

$$EI\frac{dy}{dx} = \frac{11x^2}{2} + A \;\Big|\; -\frac{10}{2}[x-1]^2 \;\Big|\; +\frac{2}{3}[x-2]^3$$
$$\Big|\; -\frac{2}{3}[x-1]^3 \;\Big|\; -7[x-2] \qquad (7.29)$$

$$EIy = \frac{11x^3}{6} + Ax + B \;\Big|\; -\frac{5}{3}[x-1]^3 \;\Big|\; +\frac{1}{6}[x-2]^4$$
$$\Big|\; -\frac{1}{6}[x-1]^4 \;\Big|\; -\frac{7}{2}[x-2]^2 \qquad (7.30)$$

NB The Macaulay bracket for the couple must be written as in equation (7.28), so that integration can be carried out as in equations (7.29) and (7.30).

Boundary conditions

A suitable boundary condition is as follows

At $x = 0$, $y = 0$; therefore $\boldsymbol{B} = \mathbf{0}$

NB As the terms in the Macaulay brackets in the second and third columns are negative, they must be ignored when applying the above boundary condition to equation (7.30).

Another suitable boundary condition is as follows
At $x = 3$, $y = 0$, hence

$$0 = \frac{11 \times 3^3}{6} + 3A - \frac{5}{3} \times 2^3 - \frac{1}{6} \times 2^4 + \frac{1}{6} \times 1^4 - \frac{7}{2} \times 1^2$$

i.e. $$-3A = 49.5 - 13.333 - 2.667 + 0.1667 - 3.5 = 30.1667$$

from which $$A = \frac{30.1667}{-3} = \mathbf{-10.056}$$

Substituting the above boundary conditions into equation (7.30), the following is obtained for the deflection y at any point x along the length of the beam

$x = 1$ m | $x = 2$ m

$$EIy = \frac{11x^3}{6} - 10.056x \;\Big|\; -\frac{5}{3}[x-1]^3 - \frac{1}{6}[x-1]^4 \;\Big|\; +\frac{1}{6}[x-2]^4 - \frac{7}{2}[x-2]^2$$

The maximum deflection may occur in the span CD, where the condition $dy/dx = 0$ must be satisfied, i.e.

$$EI\frac{dy}{dx} = \frac{11x^2}{2} - 10.056 - 5[x-1]^2 - \frac{2}{3}[x-2]^3$$

or $$0 = 5.5x^2 - 10.056 - 5(x^2 - 2x + 1) - \frac{2}{3}(x^3 - 3x^2 + 3x - 1)$$

i.e. $$-0.667x^3 + 2.5x^2 + 8x - 14.389 = 0$$

which has three real roots, as follows

$x_1 = -2.913$ m $\quad x_2 = 5.250$ m $\quad x_3 = 1.411$ m by calculator or smartphone.

It is evident that the root of interest is $x_3 = \mathbf{1.411\ m}$, as this is the only one that applies within the span CD,

i.e. δ = maximum deflection

$$= \frac{1}{2\times10^{11}\times2\times10^{-8}}\left(\frac{11\times1.411^3}{6} - 10.05\times1.411 - \frac{5}{3}(0.411)^3 - \frac{0.411^4}{6}\right)$$

i.e. $\boldsymbol{\delta = -2.288\ m}$

Now try the following Practice Exercise

Practice Exercise 34. Beam deflections due to bending using Macaulay's method

1. Determine the position (from the left end) and the value of the maximum deflection for the uniform-section beam of Figure 7.12.

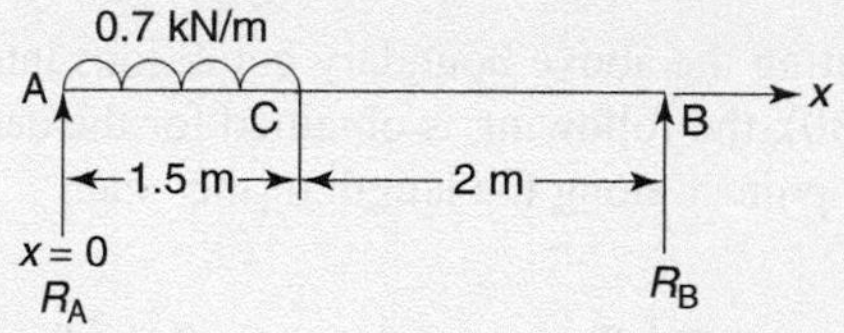

Figure 7.12

$$\left[x = 1.54\ \text{m},\ \delta = -\frac{0.543}{EI}\right]$$

2. Determine the deflections at the points C and D for the uniform-section beam of Figure 7.13, given that $EI = 4300$ kN m^2.

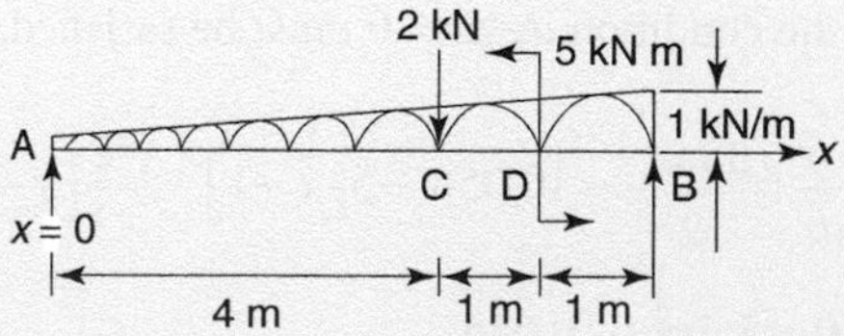

Figure 7.13

$$\left[\delta_C = -5.60\times10^{-3}\ \text{m},\ \delta_D = -3.32\times10^{-3}\ \text{m}\right]$$

3. Determine the position and value of the maximum deflection of the simply supported beam shown in Figure 7.14, given that $E\ I$ = 100 kN m^2.

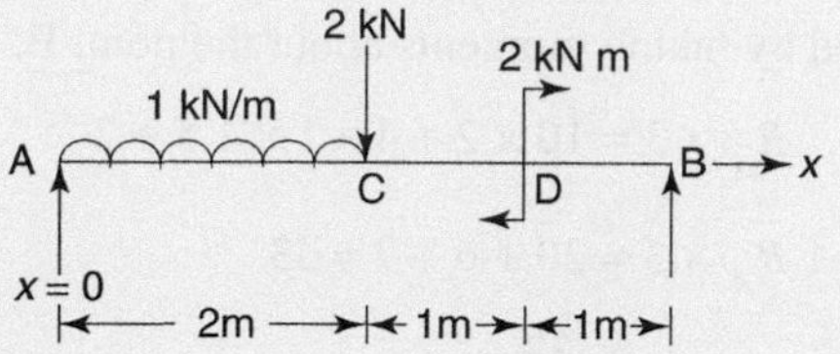

Figure 7.14

$[x = 1.87\ \text{m}, \delta = -0.0285\ \text{m}]$

4. The beam CAB is simply supported at the points A and B and is subjected to a concentrated load W at the point C, together with a uniformly distributed load w between the points A and B, as shown in Figure 7.15. Determine the relationship between W and wl, so that no deflection will occur at the point C.

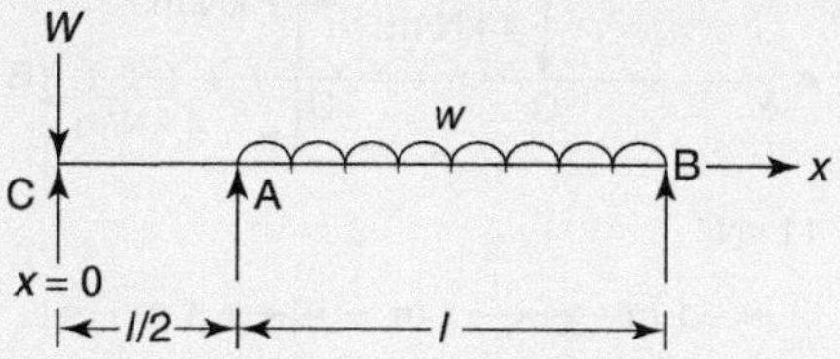

Figure 7.15

$[w = 6W/l]$

5. Determine the position (from the left end) and the value of the maximum deflection for the uniform-section beam of Figure 7.16.

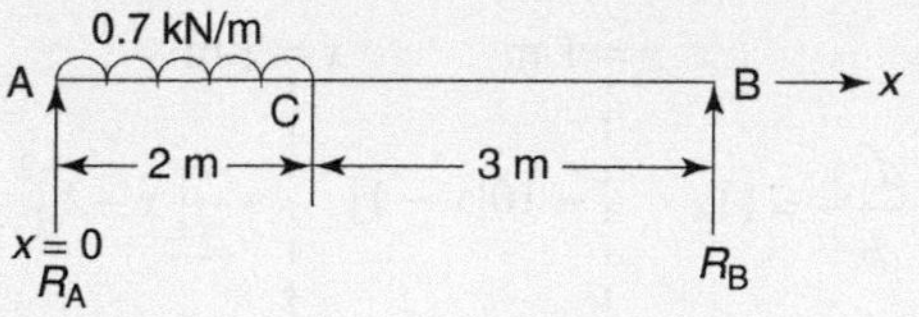

Figure 7.16

$$\left[x = 2.231\ \text{m},\ \delta = -\frac{1.98}{EI}\right]$$

6. Determine the deflections at the points C and D for the uniform-section beam shown in Figure 7.17, given that $EI = 4300$ kN m^2.

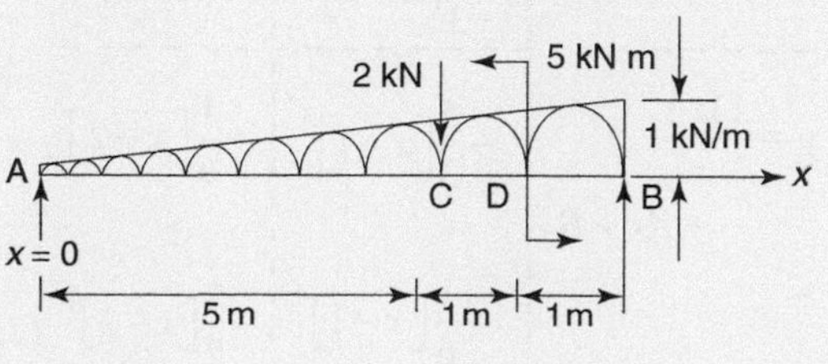

Figure 7.17

$[\delta_C = -8.48 \times 10^{-3}\,\text{m},\ \delta_D = -5.03 \times 10^{-3}\,\text{m}]$

7. Determine the maximum deflection for the uniform-section beam of Figure 7.18, given that $EI = 4300$ kN m^2.

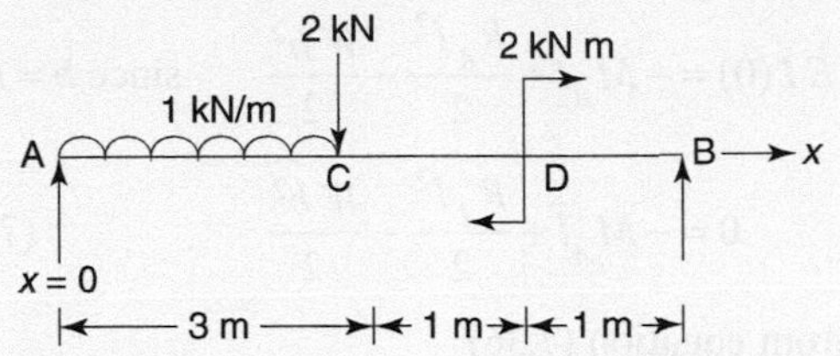

Figure 7.18

[0.00394 m]

8. The beam CAB is simply supported at the points A and B and is subjected to a concentrated load W at the point C, together with a uniformly distributed load w between the points A and B, as shown in Figure 7.19. Determine the relationship between W and wl, so that no deflection will occur at the point B.

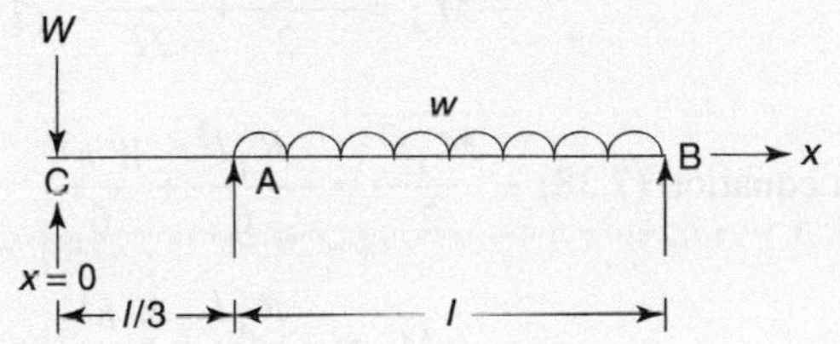

Figure 7.19

$[w = 3.582W/l]$

7.4 Statically indeterminate beams

So far, the beams that have been analysed were statically determinate; that is their reactions and bending moments were determined solely from statical considerations. For statically indeterminate beams, their analysis is more difficult, as their reactions and bending moments cannot be obtained from statical considerations alone. To demonstrate the method of analysing statically indeterminate beams, the following two simple cases will be considered.

Problem 7. Determine the end fixing moments, M_F, and the maximum deflection for the encastré beam of Figure 7.20, where x varies from 0 at the left end to $x = l$ at the right end.

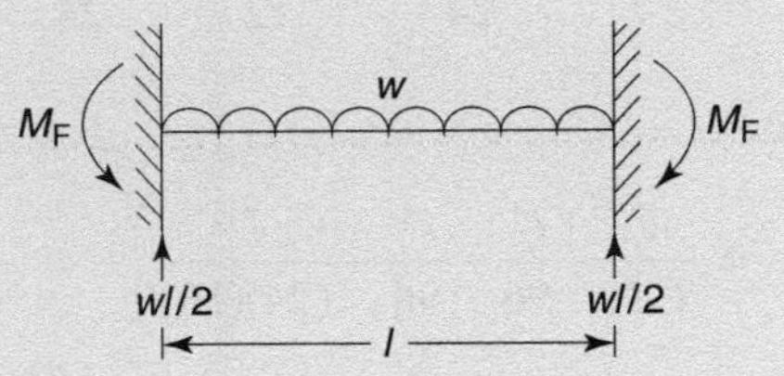

Figure 7.20

$$EI\frac{d^2y}{dx^2} = -M_F + \frac{wl}{2}x - \frac{w}{2}x^2$$

Integrating gives

$$EI\frac{dy}{dx} = -M_F x + \frac{wlx^2}{4} - \frac{wx^3}{6} + A \qquad (7.31)$$

At $x = 0$, $\frac{dy}{dx} = 0$; therefore $\boldsymbol{A = 0}$

At $x = l$, $\frac{dy}{dx} = 0$; therefore $0 = -M_F l + \frac{wl^3}{4} - \frac{wl^3}{6}$

from which $$M_F l = \frac{wl^3}{4} - \frac{wl^3}{6} = \frac{wl^3}{12}$$

and $$\boldsymbol{M_F = \frac{wl^2}{12}} \qquad (7.32)$$

i.e. the end fixing moment, $M_F = \frac{wl^2}{12}$

On integrating equation (7.31),

$$EIy = -\frac{M_F x^2}{2} + \frac{wlx^3}{12} - \frac{wx^4}{24} + B$$

At $x = 0, y = 0$; therefore $\boldsymbol{B = 0}$

i.e. $$y = \frac{1}{EI}\left(-\frac{\left(\frac{wl^2}{12}\right)x^2}{2} + \frac{wlx^3}{12} - \frac{wx^4}{24}\right)$$

i.e. $$y = \frac{w}{EI}\left(-\frac{l^2x^2}{24} + \frac{lx^3}{12} - \frac{x^4}{24}\right)$$

By inspection, the maximum deflection δ occurs at $x = l/2$, where

$$\delta = \frac{w}{EI}\left(-\frac{l^2\left(\frac{l}{2}\right)^2}{24}+\frac{l\left(\frac{l}{2}\right)^3}{12}-\frac{\left(\frac{l}{2}\right)^4}{24}\right)$$

$$= \frac{w}{EI}\left(-\frac{l^4}{96}+\frac{l^4}{96}-\frac{l^4}{(24)(16)}\right)$$

i.e. $$\boldsymbol{\delta = -\frac{wl^4}{384EI}} \qquad (7.33)$$

Equation (7.33) shows that the central deflection of an encastré beam is only one-fifth of that of the simply supported case, which in the latter is: $\delta = -\dfrac{5wl^4}{384EI}$ from Problem 4 on page 155/156.

Problem 8. Determine the end fixing moments and reactions and the deflection under the load for the encastré beam of Figure 7.21.

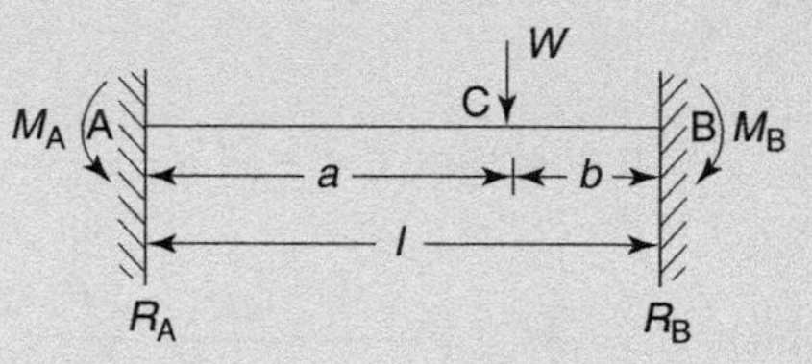

Figure 7.21

$$x = a$$

$$EI\frac{d^2y}{dx^2} = -M_A + R_A x \quad \Big| \quad - W[x-a] \qquad (7.34)$$

$$EI\frac{dy}{dx} = -M_A x + \frac{R_A x^2}{2} + A \quad \Big| \quad -\frac{W}{2}[x-a]^2 \qquad (7.35)$$

$$EIy = -\frac{M_A x^2}{2} + \frac{R_A x^3}{6} + Ax + B \quad \Big| \quad -\frac{W}{6}[x-a]^3 \qquad (7.36)$$

To determine A, B, R_A and M_A, it will be necessary to apply four boundary conditions to equations (7.35) and (7.36), as follows

At $x = 0$, $y = 0$, therefore from equation (7.36), $\boldsymbol{B = 0}$

At $x = 0$, $\dfrac{dy}{dx} = 0$, therefore from equation (7.35), $\boldsymbol{A = 0}$

At $x = l$, $\dfrac{dy}{dx} = 0$ and $y = 0$,

therefore from equation (7.35)

$$EI(0) = -M_A l + \frac{R_A l^2}{2} - \frac{W}{2}[l-a]^2$$

i.e. $$EI(0) = -M_A l + \frac{R_A l^2}{2} - \frac{Wb^2}{2} \qquad \text{since } b = l - a$$

i.e. $$0 = -M_A l + \frac{R_A l^2}{2} - \frac{Wb^2}{2} \qquad (7.37)$$

and from equation (7.36)

$$EI(0) = -\frac{M_A l^2}{2} + \frac{R_A l^3}{6} - \frac{W}{6}[l-a]^3$$

i.e. $$0 = -\frac{M_A l^2}{2} + \frac{R_A l^3}{6} - \frac{Wb^3}{6} \qquad (7.38)$$

From equation (7.37) $$-M_A l = -\frac{R_A l^2}{2} + \frac{Wb^2}{2}$$

and $$-M_A = -\frac{R_A l}{2} + \frac{Wb^2}{2l} \qquad (7.39)$$

From equation (7.38) $$-\frac{M_A l^2}{2} = -\frac{R_A l^3}{6} + \frac{Wb^3}{6}$$

and $$-M_A = -\frac{R_A l}{3} + \frac{Wb^3}{3l^2} \qquad (7.40)$$

Equating equations (7.39) and (7.40) gives

$$-\frac{R_A l}{2} + \frac{Wb^2}{2l} = -\frac{R_A l}{3} + \frac{Wb^3}{3l^2}$$

$$+\frac{Wb^2}{2l} - \frac{Wb^3}{3l^2} = -\frac{R_A l}{3} + \frac{R_A l}{2}$$

and $$\frac{3Wb^2 l - 2Wb^3}{6l^2} = \frac{R_A l}{6}$$

from which $$R_A = \frac{6\left(3Wb^2 l - 2Wb^3\right)}{6l^3} = \frac{3Wb^2 l - 2Wb^3}{l^3} = \frac{Wb^2(3l-2b)}{l^3}$$

However, $b = l - a$, thus $R_A = \dfrac{Wb^2\left(3l - 2(l - a)\right)}{l^3}$

$$= \frac{Wb^2\left(3l - 2l + 2a\right)}{l^3}$$

i.e. $$\boldsymbol{R_A = \frac{Wb^2\left(l + 2a\right)}{l^3}} \quad (7.41)$$

and from equation (7.39)

$$M_A = \frac{R_A l}{2} - \frac{W b^2}{2l} = \frac{\left(\dfrac{Wb^2\left(l+2a\right)}{l^3}\right)l}{2} - \frac{W b^2}{2l}$$

i.e. $$M_A = \frac{\left(Wb^2\left(l+2a\right)\right)}{2l^2} - \frac{W b^2}{2l}$$

$$= \frac{\left(Wb^2 l + 2Wab^2\right)}{2l^2} - \frac{W b^2}{2l}$$

$$= \frac{Wb^2 l}{2l^2} + \frac{2Wab^2}{2l^2} - \frac{W b^2}{2l}$$

$$= \frac{Wb^2}{2l} + \frac{Wab^2}{l^2} - \frac{W b^2}{2l}$$

i.e. $$\boldsymbol{M_A = \frac{Wab^2}{l^2}} \quad (7.42)$$

Expressions for R_B and M_B can now be obtained from statical considerations, as follows.

Resolving vertically

$$R_A + R_B = W$$

Therefore $R_B = W - R_A = W - \dfrac{Wb^2\left(l+2a\right)}{l^3}$ from equation (7.41)

$$= \frac{W l^3 - Wb^2\left(l+2a\right)}{l^3}$$

$$= \frac{W\left[\left(a+b\right)^2 l - b^2\left(l+2a\right)\right]}{l^3}$$

$$= \frac{W}{l^3}\left(l a^2 + l b^2 + 2abl - b^2 l - 2ab^2\right)$$

$$= \frac{W}{l^3}\left(l a^2 + 2abl - 2ab^2\right)$$

$$= \frac{W}{l^3}\left[\left(a+b\right)a^2 + 2ab\left(a+b\right) - 2ab^2\right]$$

$$= \frac{W}{l^3}\left[a^3 + a^2 b + 2a^2 b + 2ab^2 - 2ab^2\right]$$

$$= \frac{W}{l^3}\left(a^3 + 3a^2 b\right) = \frac{W a^2}{l^3}\left(a + 3b\right)$$

$$= \frac{W a^2}{l^3}\left(l - b + 3b\right)$$

i.e. $$\boldsymbol{R_B = \frac{Wa^2(l+2b)}{l^3}} \quad (7.43)$$

and $$\boldsymbol{M_b = \frac{Wa^2 b}{l^2}} \quad (7.44)$$

Substituting equations (7.41) and (7.42) into equation (7.36), with a value of $x = a$

$$E I y = -\frac{\left(\dfrac{Wab^2}{l^2}\right)a^2}{2} + \frac{\left(\dfrac{Wb^2\left(l+2a\right)}{l^3}\right)a^3}{6}$$

From equation (7.36), with $A = B = 0$

$$E I y = -\frac{M_A x^2}{2} + \frac{R_A x^3}{6} \quad \Big|_{x=a} \quad -\frac{W}{6}\left[x - a\right]^3$$

and at $x = a$, $y_C = \delta_C$

Hence $$\delta_C = \frac{1}{EI}\left(-\frac{M_A a^2}{2} + \frac{R_A a^3}{6}\right)$$

$$= \frac{1}{EI}\left(-\frac{\left(\dfrac{Wab^2}{l^2}\right)a^2}{2} + \frac{\left(\dfrac{Wb^2\left(l+2a\right)}{l^3}\right)a^3}{6}\right)$$

from equations (7.42) and (7.43)

$$= \frac{W}{EI}\left(-\frac{ab^2 a^2}{2l^2} + \frac{b^2\left(l+2a\right)a^3}{6l^3}\right)$$

$$= \frac{W}{E I l^3}\left(-\frac{a^3 b^2 l}{2} + \frac{b^2\left(l+2a\right)a^3}{6}\right)$$

$$= \frac{W}{E I l^3}\left(-\frac{a^3 b^2 l}{2} + \frac{b^2 l a^3}{6} + \frac{2b^2 a^4}{6}\right)$$

$$= \frac{W}{E I l^3}\left(-\frac{a^3 b^2 l}{3} + \frac{a^4 b^2}{3}\right)$$

$$= \frac{W}{3EIl^3}\left(-a^3b^2l + (l-b)a^3b^2\right) \text{ since } a = l - b$$

$$= \frac{W}{3EIl^3}\left(-a^3b^2l + a^3b^2l - a^3b^3\right)$$

i.e. $$\delta_C = -\frac{Wa^3b^3}{3EIl^3} \quad (7.45)$$

If the beam of Figure 7.21 were loaded symmetrically, so that $a = b = l/2$, then

$$M_A = M_B = \frac{Wl}{8} \quad (7.46)$$

and $$\delta_C = -\frac{Wl^3}{192EI} \quad (7.47)$$

From equation (7.47), it can be seen that the central deflection for a centrally loaded beam, with encastré ends, is one-quarter of the value for a similar beam with simply supported ends.

Now try the following Practice Exercise

Practice Exercise 35. Statically indeterminate beams

1. Determine the position (from the left end) and the maximum deflection for the uniform-section beam of Figure 7.22 together with the reactions and end fixing forces by the double integration method.

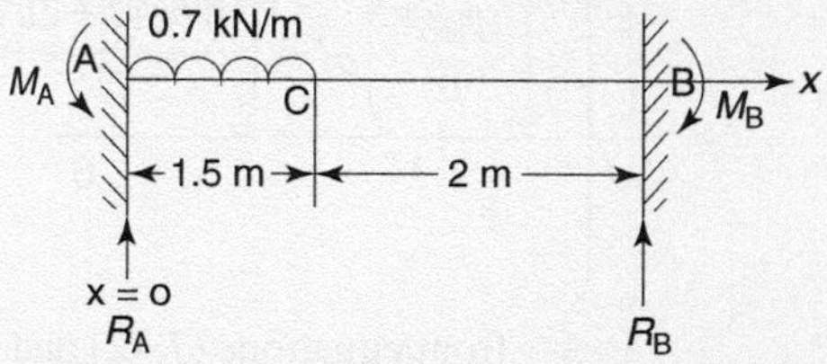

Figure 7.22

$\left[R_A = 0.897 \text{ kN}, R_B = 0.153 \text{ kN}, M_A = 0.407 \text{ kN m}, M_A = 0.155 \text{ kN m}; x = 1.47 \text{ m}, \delta = -\frac{0.101}{EI}\right]$

2. Determine the end fixing moments and reactions for the encastré beam shown in Figure 7.23. The beam may be assumed to be of uniform section.

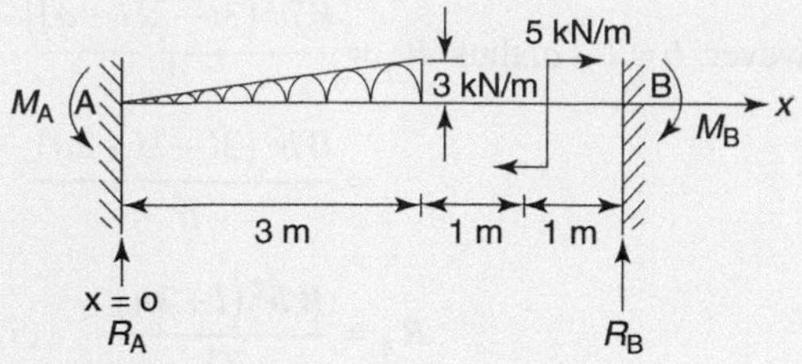

Figure 7.23

$\left[R_A = 1.888 \text{ kN}, R_B = 2.612 \text{ kN}, M_A = 1.445 \text{ kN m}, M_B = 0.505 \text{ kN m}\right]$

3. Determine the end fixing moments and reactions for the uniform-section encastré beam of Figure 7.24.

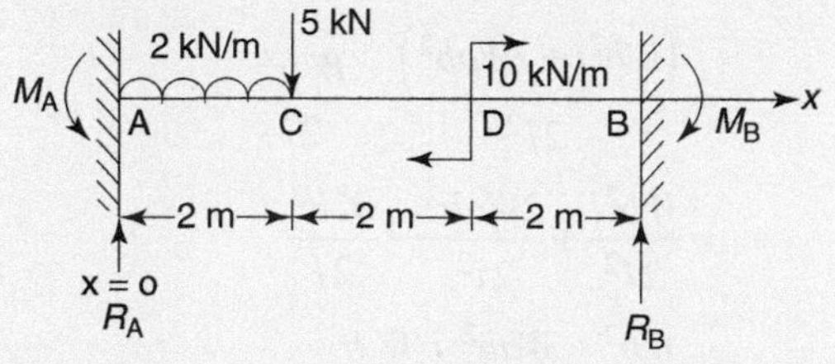

Figure 7.24

$\left[R_A = 5.11 \text{ kN}, R_B = 3.89 \text{ kN}, M_A = 3.556 \text{ kN m}, M_B = 2.896 \text{ kN m}\right]$

4. Determine the end reactions (from the left end) and the maximum deflection for the uniform-section beam of Figure 7.25 together with the reactions and end fixing forces.

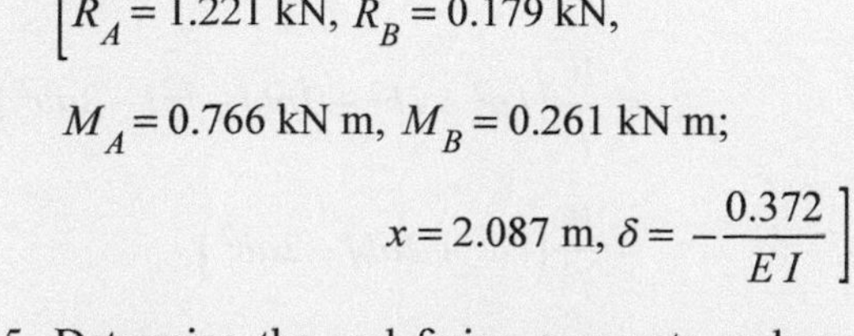

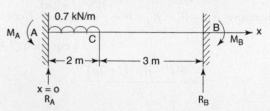

Figure 7.25

$\left[R_A = 1.221 \text{ kN}, R_B = 0.179 \text{ kN}, M_A = 0.766 \text{ kN m}, M_B = 0.261 \text{ kN m}; x = 2.087 \text{ m}, \delta = -\frac{0.372}{EI}\right]$

5. Determine the end fixing moments and reactions for the encastré beam shown in

Figure 7.26. The beam may be assumed to be of uniform section.

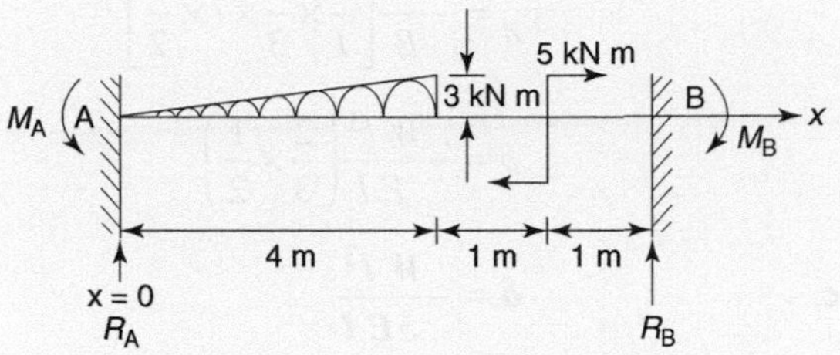

Figure 7.26

$\left[R_A = 2.728 \text{ kN}, R_B = 3.272 \text{ kN}, M_A = 3.017 \text{ kN m}, M_B = 1.649 \text{ kN m}\right]$

6. Determine the end fixing moments and reactions for the uniform-section encastré beam of Figure 7.27.

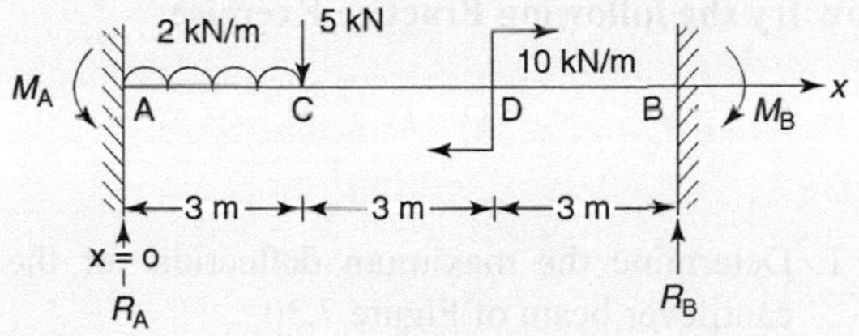

Figure 7.27

$\left[R_A = 7.375 \text{ kN}, R_B = 3.625 \text{ kN}, M_A = 7.875 \text{ kN m}, M_B = 6.50 \text{ kN m}\right]$

7.5 Moment-area method

This method is particularly useful if there is step variation with the sectional properties of the beam.

Now $$EI\frac{d^2y}{dx^2} = M$$

or $$\frac{d^2y}{dx^2} = \frac{M}{EI} \tag{7.48}$$

Consider the deformed beam of Figure 7.28, and apply equation (7.48) between the points A and B:

$$\int_{x_A}^{x_B}\left(\frac{d^2y}{dx^2}\right)dx = \int_{x_B}^{x_B}\left(\frac{M}{EI}\right)dx$$

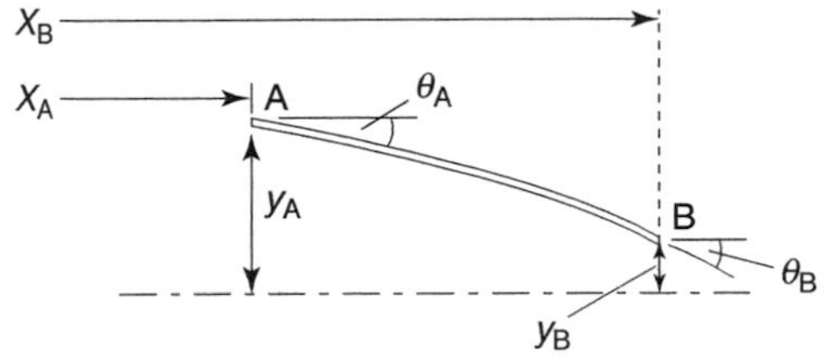

Figure 7.28 Deformed beam element

i.e. $$\left[\frac{dy}{dx}\right]_{x_A}^{x_B} = \int_{x_B}^{x_B}\left(\frac{M}{EI}\right)dx$$

i.e. $$\theta_B - \theta_A = \int_{x_B}^{x_B}\left(\frac{M}{EI}\right)dx \tag{7.49}$$

= area of (M/EI) between x_A and x_B if EI is constant

$$\theta_B - \theta_A = \frac{1}{EI} \times \text{area of bending moment diagram between } x_A \text{ and } x_B$$

Furthermore, it can be seen that if both sides of equation (7.48) are multiplied by x, then

$$x\frac{d^2y}{dx^2} = \frac{M}{EI}x$$

or $$\int_{x_A}^{x_B}\left(x\frac{d^2y}{dx^2}\right)dx = \int_{x_A}^{x_B}\left(\frac{M}{EI}\right)x\,dx$$

[To integrate the left-hand integral requires a technique called 'integration by parts'.

This states $$\int u\frac{dv}{dx}dx = uv - \int v\frac{du}{dx}dx$$

Thus, letting $u = x$, from which $\frac{du}{dx} = 1$

and letting $\frac{dv}{dx} = \frac{d^2y}{dx^2}$, from which

$$v = \int \frac{d^2y}{dx^2}dx = \frac{dy}{dx}$$

Hence $$\int\left(x\frac{d^2y}{dx^2}\right)dx = (x)\left(\frac{dy}{dx}\right) - \int\left(\frac{dy}{dx}\right)(1)\,dx$$

$$= x\frac{dy}{dx} - y\,]$$

Thus, if $$\int_{x_A}^{x_B}\left(x\frac{d^2y}{dx^2}\right)dx = \int_{x_A}^{x_B}\left(\frac{M}{EI}\right)x\,dx$$

Then $\left[x\frac{dy}{dx}-y\right]_{x_A}^{x_B} = \int_{x_A}^{x_B}\left(\frac{M}{EI}\right)x\,dx$ (7.50)

Suitable use of equations (7.49) and (7.50) can be used to solve many problems, and if EI is constant, then:

$\left[x\frac{dy}{dx}-y\right]_{x_A}^{x_B} = \frac{1}{EI}$ × moment of area of the bending moment diagram about the point A

Problem 9. Determine the end deflection for the cantilever loaded with a point load at its free end, as shown in Figure 7.29.

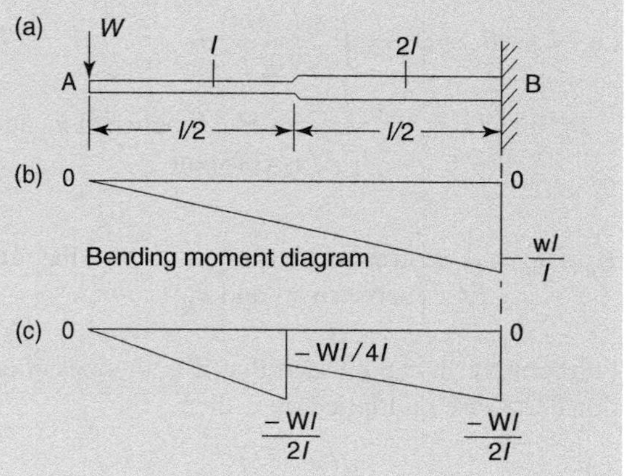

Figure 7.29 *M/I* diagram

Take the point B to be at the built-in end and the point A to be at the free end.
From equation (7.50)

$\left[x\frac{dy}{dx}-y\right]_{0}^{l} = \frac{1}{EI}$ × moment of area of the *M/I* diagram about A

i.e. $\left[\left(l\theta_B - y_B\right)-\left(0\times\theta_A - y_A\right)\right]=$

$$-\frac{1}{E}\left[\frac{Wl}{2I}\times\frac{2}{3}\times\frac{l}{2}\times\frac{l}{4}+\frac{Wl}{4I}\times\frac{3l}{4}\times\frac{l}{2}\right.$$

$$\left.+\frac{Wl}{4I}\times\frac{l}{4}\times\left(\frac{l}{2}+\frac{2}{3}\times\frac{l}{2}\right)\right]$$

However $\theta_B = y_B = 0$

Hence $$y_A = -\frac{Wl^3}{EI}\left[\frac{1}{24}+\frac{3}{32}+\frac{5}{96}\right]$$

i.e. $$\boldsymbol{\delta = y_A = -\frac{3Wl^3}{16EI}}$$

If I were constant throughout the length of the cantilever, then from Figure 7.29(b)

$$y_A = -\frac{W}{E}\left[\frac{l}{I}\times\frac{2}{3}\times l\times\frac{l}{2}\right]$$

$$\delta = -\frac{Wl^3}{EI}\left(\frac{2}{3}\times\frac{1}{2}\right)$$

i.e. $$\boldsymbol{\delta = -\frac{Wl^3}{3EI}}$$

To solve beam problems, other than statically determinate ones or the simpler cases of statically indeterminate ones, is difficult and cumbersome, and for such cases it is better to use computer methods (see References 1, 5 and 6 on page 497). Some of these computer methods are based on the *slope-deflection equations,* which are derived in the next section.

Now try the following Practice Exercise

Practice Exercise 36. Moment-area method

1. Determine the maximum deflection for the cantilever beam of Figure 7.30.

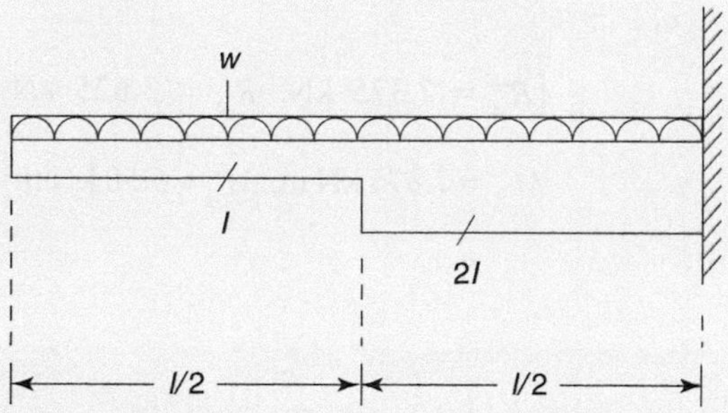

Figure 7.30

$$-\frac{3wl^3}{2EI}$$

7.6 Slope-deflection equations

Slope-deflection equations are a boundary value problem and are dependent on the displacement boundary conditions of Figure 7.31, where

Y_1 = vertical force at node 1 Y_2 = vertical force at node 2

M_1 = clockwise moment at node 1 M_2 = clockwise moment at node 2

Y_1 = vertical deflection at node 1 Y_2 = vertical deflection at node 2

θ_1 = rotation at node 1 (clockwise) θ_2 = rotation at node 2 (clockwise)

NB A node is defined as a point.

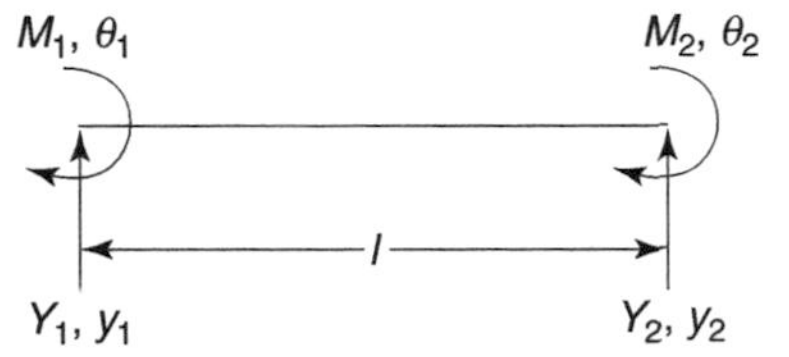

Figure 7.31

Resolving vertically

$$Y_1 = -Y_2 \qquad (7.51)$$

Taking moments about the right end

$$Y_1 = -\frac{(M_1 + M_2)}{l} \qquad (7.52)$$

and $$Y_2 = \frac{(M_1 + M_2)}{l}$$

From equation (7.3) $$EI\frac{d^2y}{dx^2} = Y_1 x + M_1$$

$$= -\frac{(M_1 + M_2)}{l}x + M_1$$

Integrating both sides gives

$$EI\frac{dy}{dx} = -\frac{(M_1 + M_2)x^2}{2l} + M_1 x + A$$

At $x = 0$, $\frac{dy}{dx} = -\theta_1$ therefore $\boldsymbol{A = -EI\theta_1}$ (7.53)

Hence $$EI\frac{dy}{dx} = -\frac{(M_1 + M_2)x^2}{2l} + M_1 x - EI\theta_1$$

Integrating again gives

$$EIy = -\frac{(M_1 + M_2)x^3}{6l} + \frac{M_1 x^2}{2} - EI\theta_1 x + B$$

At $x = 0$, $y = y_1$ therefore $\boldsymbol{B = EIy_1}$ (7.54)

At $x = l$, $\frac{dy}{dx} = -\theta_2$ therefore

$$-EI\theta_2 = -\frac{(M_1 + M_2)l}{2} + M_1 l - EI\theta_1 \qquad (7.55)$$

At $x = l$, $y = y_2$ therefore

$$EIy_2 = -\frac{(M_1 + M_2)l^2}{6} + \frac{M_1 l^2}{2} - EI\theta_1 l + EIy_1 \qquad (7.56)$$

From equations (7.51) to (7.56), the *slope-deflection equations* are obtained as follows

$$\left.\begin{aligned}
M_1 &= \frac{4EI\theta_1}{l} + \frac{2EI\theta_2}{l} - \frac{6EI}{l^2}(y_1 - y_2) \\
Y_1 &= -\frac{6EI\theta_1}{l^2} - \frac{6EI\theta_2}{l^2} + \frac{12EI}{l^3}(y_1 - y_2) \\
M_2 &= \frac{2EI\theta_1}{l} + \frac{4EI\theta_2}{l} - \frac{6EI}{l^2}(y_1 - y_2) \\
Y_2 &= \frac{6EI\theta_1}{l^2} + \frac{6EI\theta_2}{l^2} - \frac{12EI}{l^3}(y_1 - y_2)
\end{aligned}\right\} \qquad (7.57)$$

Equations (7.57) lend themselves to satisfactory computer analysis and form the basis of the finite element method, which is discussed in Chapter 21 and is also described in detail in a number of other texts (see References 5 to 7 on page 497).

For fully worked solutions to each of the problems in Exercises 33 to 36 in this chapter, go to the website:
www.routledge.com/cw/bird

Chapter 8

Torsion

Why it is important to understand: **Torsion**

The torsion of shafts appears in a number of different branches of engineering, including propeller shafts for ships and aircraft, shafts driving the blades of a helicopter, shafts driving the rear wheels of an automobile and shafts driving food mixers, washing machines, tumble dryers, dishwashers and so on. If the shaft is overstressed due to a torque, so that the maximum shear stress in the shaft exceeds the yield shear stress of the shaft's material, the shaft can fracture. This is an undesirable phenomenon and normally it should be designed out; hence the need for the theory contained in this chapter. The twisting of shafts is particularly important for designing the propulsion systems for ships, automobiles and helicopters, and also for food mixers etc. Torsion is also important when a rolled steel joint is connected to another rolled steel joint which is at right angles to it, and out-of-plane. In this case if one beam bends, it can transmit a torque to the other beam, but this phenomenon is beyond the scope of this text.

At the end of this chapter you should be able to:

- define torque
- prove that $\frac{\tau}{r} = \frac{T}{J} = \frac{G\theta}{l}$ and perform practical calculations
- understand flanged couplings and perform practical calculations on them
- understand keyed couplings and perform practical calculations on them
- understand compound shafts and perform practical calculations on them
- understand close-coiled helical springs
- understand torsion of thin-walled non-circular sections and perform practical calculations on them
- understand torsion of thin-walled rectangular sections and perform practical calculations on them
- understand torsion of thin-walled open sections and perform practical calculations on them
- appreciate elastic-plastic torsion of circular-section shafts

Mathematical references

In order to understand the theory involved in this chapter on **torsion**, knowledge of the following mathematical topics is required: ***algebraic manipulation/transposition, integration and the Newton-Raphson method***. If help/revision is needed in these areas, see page 49 for some textbook references.

8.1 Introduction

In engineering, it is often required to transmit power via a circular-section shaft, and some typical examples of this are given below:

(a) propulsion of a ship or a boat by a screw propeller, via a shaft;
(b) transmission of power to the rear wheels of an automobile, via a shaft;
(c) transmission of power from an electric motor to various types of machinery, via a shaft.

8.2 Torque (T)

A torque is defined as a twisting moment that acts on the shaft in an axial direction, as shown in Figure 8.1, where T is according to the right-hand screw rule. (The right-hand screw rule defines the direction of torque by pointing the right hand in the required direction and turning the right hand in a clockwise direction.) This torque causes the end B to rotate by an angle θ, relative to A, where

θ = angle of twist

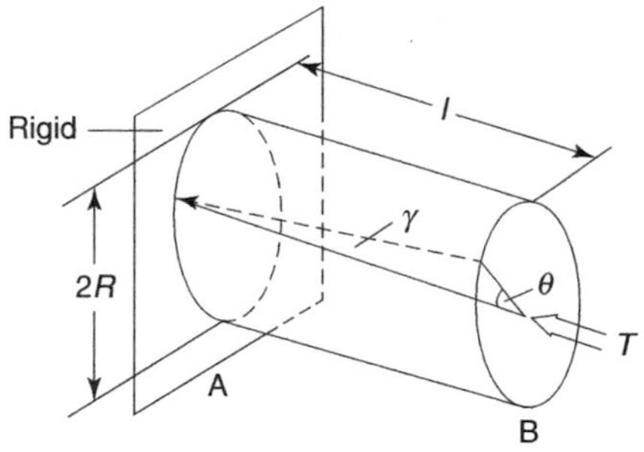

Figure 8.1 Shaft under torque

Assuming that no other forces act on the shaft, the effect of T will be to cause the shaft to be subjected to a system of shearing stresses, as shown in Figure 8.2. The system of shearing stresses acting on an element of the shaft of Figure 8.2 is known as *pure shear*, as these shearing stresses are unaccompanied by direct stresses.

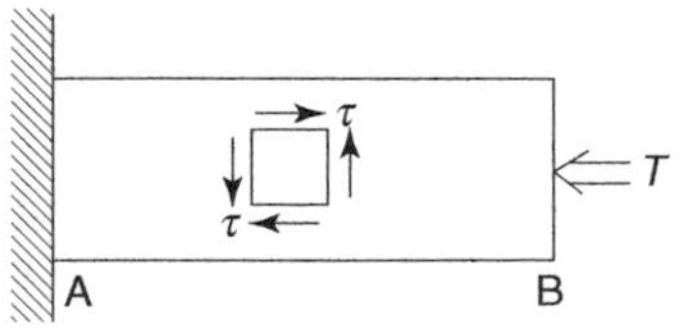

Figure 8.2 Shaft in pure shear

Later, it will be proven that, providing the angle of twist is small and the limit of proportionality is not exceeded, the shearing stresses τ will have a maximum value on the external surface of the shaft and their magnitude will decrease linearly to zero at the centre of the shaft.

8.3 Assumptions made in circular shaft theory

The following assumptions are made in the theory of circular section shafts:

(a) the shaft is of circular cross-section;
(b) the cross-section of the shaft is uniform throughout its length;
(c) the shaft is straight;
(d) the material is homogeneous, isotropic and obeys Hooke's law;
(e) rotations are small and the limit of proportionality is not exceeded;
(f) plane cross-sections remain plane during twisting;
(g) radial lines across the shaft's cross-section remain radial during twisting.

8.4 Proof of $\frac{\tau}{r} = \frac{T}{J} = \frac{G\theta}{l}$

The following relationships, which are used in the torsional theory of circular-section shafts, will now be proven

$$\frac{\tau}{r} = \frac{T}{J} = \frac{G\theta}{l}$$

where τ = shearing stress at any radius r

T = applied torque

J = polar second moment of area

G = rigidity or shear modulus

θ = angle of twist over a length l

From Figure 8.1, it can be seen that

γ = shear strain

and that $\gamma\, l = R\theta$ provided θ is small,

or $\left(\frac{\tau}{G}\right) l = R\theta$ from equation (4.9), page 96.

Therefore $$\frac{\tau}{R} = \frac{G\theta}{l}$$

If the radial lines across the section remain radial on twisting, then it follows that the shearing stress is proportional to any radius r, so that

$$\frac{\boldsymbol{\tau}}{\boldsymbol{r}}=\frac{\boldsymbol{G\theta}}{\boldsymbol{l}} \tag{8.1}$$

Similarly $$\frac{\tau}{r}=G\frac{d\theta}{dx} \tag{8.2}$$

where $\frac{d\theta}{dx}$ is the change of the angle of twist over a length dx.

Consider a cylindrical shell element of radius r and thickness dr, as shown by the shaded area of Figure 8.3. The shearing stresses acting on the cross-section of this cylindrical shell are shown in Figure 8.4, where they can be seen to act tangentially to the cross-section of the cylindrical shell.

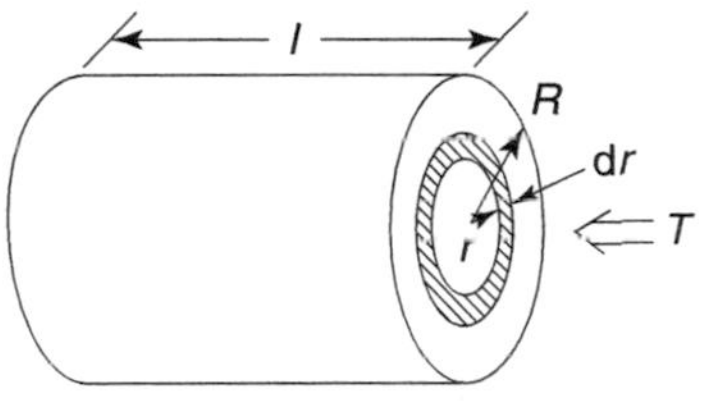

Figure 8.3

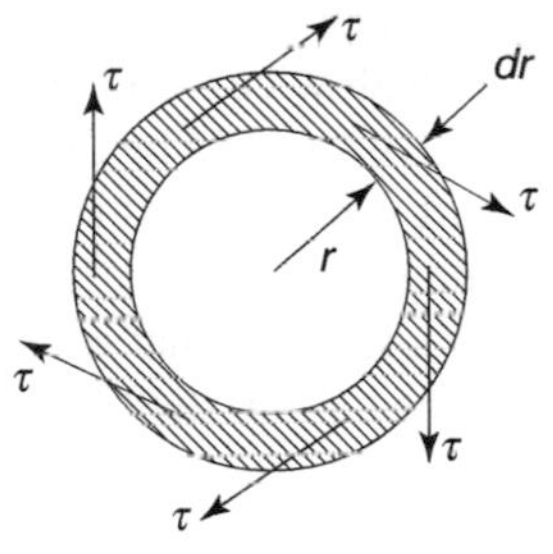

Figure 8.4 Shearing stresses acting on annular element

From Figure 8.4, it can be seen that these shearing stresses cause an elemental torque, δT, where

$$\delta T=\tau\times 2\pi r\,dr\times r=2\pi\,\tau\,r^2\,dr$$

but the total torque T is the sum of all the elemental torques acting on the section

i.e. $$T=\sum\delta T=\int 2\pi\,\tau\,r^2\,dr$$

i.e. $$T=2\pi\int r^2\,\tau\,dr \tag{8.3}$$

Substituting τ from equation (8.1) into equation (8.3) gives

$$T=2\pi\int_0^R r^2\left(\frac{G\theta r}{l}\right)dr=2\pi\int_0^R\left(\frac{G\theta}{l}\right)r^3\,dr$$

i.e. $$T=2\pi\left(\frac{G\theta}{l}\right)\left[\frac{r^4}{4}\right]_0^R=2\pi\left(\frac{G\theta}{l}\right)\left(\frac{R^4}{4}-0\right)$$

$$=\frac{G\theta}{l}\left(\frac{\pi R^4}{2}\right)$$

However, $\frac{\pi R^4}{2}=J$ the polar second moment of area of a circular section, from Table 5.3, page 120.

Therefore $$T=\frac{G\theta}{l}(J)$$

i.e. $$\frac{T}{J}=\frac{G\theta}{l} \tag{8.4}$$

From equations (8.1) and (8.4)

$$\frac{\boldsymbol{\tau}}{\boldsymbol{r}}=\frac{\boldsymbol{T}}{\boldsymbol{J}}=\frac{\boldsymbol{G\theta}}{\boldsymbol{l}} \tag{8.5}$$

From equation (8.5), it can be seen that τ is linearly proportional to the radius r, so that the shear stress at the centre is zero and it reaches a maximum value on the outermost surface of the shaft, as shown in Figure 8.5.

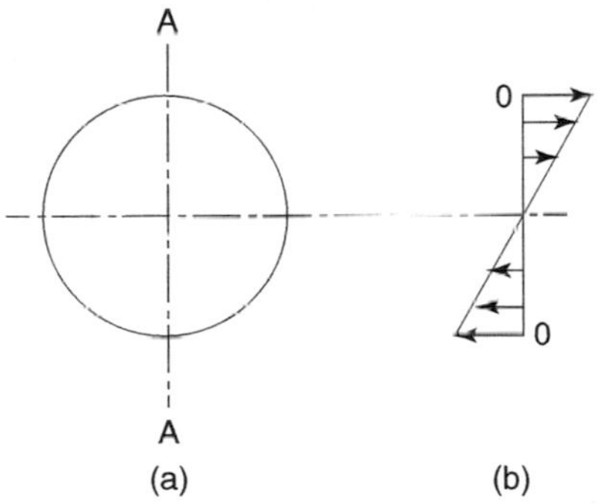

Figure 8.5 Distribution of τ: (a) section; (b) shear stress distribution at AA

Hollow circular sections

Equation (8.5) is also applicable to circular-section tubes of uniform thickness, except that the polar second moment of area of an annulus is given by

$$J=\frac{\pi\left(R_2^4-R_1^4\right)}{2}$$

where R_2 = external radius of tube, and R_1 = internal radius of tube.

Using the above value of J, equation (8.5) can be applied to a circular-section tube, between the radii R_1 and R_2, and the shear stress distribution will have the form shown in Figure 8.6.

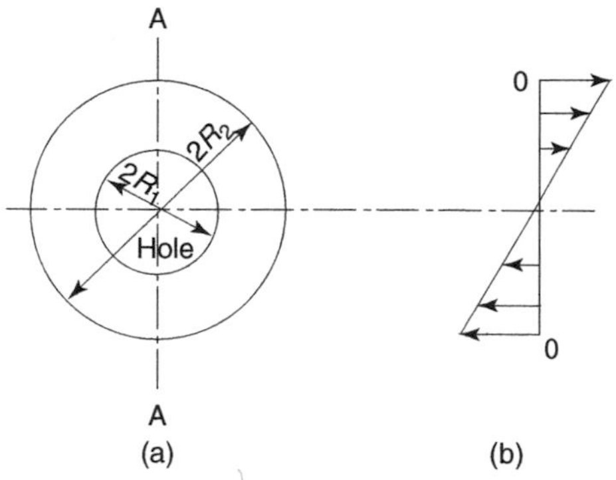

Figure 8.6 Distribution of τ across a hollow circular section: (a) section; (b) distribution of τ across AA

The **torsional stiffness k of a shaft** is given by

$$k = \frac{GJ}{l} \tag{8.6}$$

To demonstrate the use of equation (8.5), several worked problems will be considered.

Problem 1. An internal combustion engine transmits 40 horsepower (hp) at 200 rev/min (rpm) to the rear wheels of an automobile. Neglecting transmission losses, determine the minimum diameter of a solid circular-section shaft, if the maximum permissible shear stress in the shaft is 60 MN/m^2. Hence, or otherwise, determine the angle of twist over a length of 2 m, given that 1 hp = 745.7 W and $G = 7.7 \times 10^{10}$ N/m^2

40 hp = 40 hp × 745.7 W/hp = 29.83 kW

Now, from *Mechanical Engineering Science*, 4th Edition, Chapter 11,

$$\text{power, } P = T\omega \tag{8.7}$$

where P is in units of watts
T = torque (N m)
and ω = speed of rotation (rad/s)

In this case $\omega = 200\frac{\text{rev}}{\text{min}} \times 2\pi\frac{\text{rad}}{\text{rev}} \times \frac{1\,\text{min}}{60\,\text{s}}$

$= \mathbf{20.94\ rad/s}$

Hence, from equation, (8.7) $29.83 \times 10^3 = T \times 20.94$

from which **torque,** $T = \frac{29.83 \times 10^3}{20.94} = 1425$ N m

$$= \mathbf{1.425\ kN\ m} \tag{8.8}$$

Now the maximum permissible shearing stress is 60 MN/m^2, which occurs at the external radius R.

Hence, from $\frac{\tau}{r} = \frac{T}{J}$, $\frac{60 \times 10^6}{R} = \frac{1.425 \times 10^3}{\left(\frac{\pi R^4}{2}\right)}$

i.e. $\frac{60 \times 10^6}{1} = \frac{2 \times 1.425 \times 10^3}{\pi R^3}$

i.e. $R^3 = \frac{2 \times 1.425 \times 10^3}{\pi \times 60 \times 10^6}$

and $\boldsymbol{R} = \sqrt[3]{\left(\frac{\mathbf{2 \times 1.425 \times 10^3}}{\boldsymbol{\pi} \times \mathbf{60 \times 10^6}}\right)} = \mathbf{0.0247\ m}$

Hence, **the shaft diameter** = 2 × R = 2 × 0.0247

= 0.0495 m = **4.95 cm**

From $\frac{\tau}{r} = \frac{G\theta}{l}$

angle of twist, $\boldsymbol{\theta} = \frac{\tau l}{GR} = \frac{60 \times 10^6 \times 2}{7.7 \times 10^{10} \times 0.0247}$

$= \mathbf{0.0631\ rad} = 0.0631 \times \frac{180^\circ}{\pi} = \mathbf{3.62^\circ}$

Problem 2. If the shaft of Problem 1 above were in the form of a circular-section tube, where the external diameter had twice the magnitude of the internal diameter, what would be the weight saving if this hollow shaft were adopted in place of the solid one, assuming that both shafts had the same maximum shearing stress? What is the resulting angle of twist of the hollow shaft?

Let R be the external radius of the hollow shaft. Then polar second moment of area,

$$J = \frac{\pi\left(R^4 - (R/2)^4\right)}{2} = \frac{\pi\left(R^4 - \frac{R^4}{16}\right)}{2} = \frac{\pi\left(\frac{15}{16}R^4\right)}{2}$$

i.e. $\boldsymbol{J} = \mathbf{1.473}\boldsymbol{R}^{\mathbf{4}}$

From $\frac{\tau}{r} = \frac{T}{J}$ $\quad \frac{60 \times 10^6}{R} = \frac{1.425 \times 10^3}{1.473R^4}$

from which $\frac{1.473R^4}{R} = \frac{1.425 \times 10^3}{60 \times 10^6}$

and $R^3 = \frac{1.425 \times 10^3}{60 \times 10^6 \times 1.473}$

i.e. $R = \sqrt[3]{\left(\dfrac{1.425 \times 10^3}{60 \times 10^6 \times 1.473}\right)} = 0.02526 \text{ m} = 2.526 \text{ cm}$

Hence, external diameter of hollow shaft

$= 2 \times 2.526 = \mathbf{5.05\ cm}$

internal diameter of hollow shaft $= 0.5 \times 5.05 = \mathbf{2.53\ cm}$

weight of solid shaft $= \rho \times g \times (\pi \times 0.0247^2 \times l)$

$R = 0.0247$ m from Problem 1

$= 0.0247^2 \rho g \pi l$

weight of hollow shaft

$= \rho \times g \times \left(\pi \times \left(R^2 - (R/2)^2\right) \times l\right)$

$R = 0.02526$ m from above

$= \frac{3}{4} R^2 \rho g \pi l = \frac{3}{4}(0.02526)^2 \rho g \pi l$

Percentage weight saving

$$= \frac{0.0247^2 \rho g \pi l - \frac{3}{4}(0.02526)^2 \rho g \pi l}{0.0247^2 \rho g \pi l} \times 100$$

$$= \frac{0.0247^2 - \frac{3}{4}(0.02526)^2}{0.0247^2} \times 100 = \mathbf{21.6\%}$$

From $\dfrac{\tau}{r} = \dfrac{G\theta}{l}$ $\quad \theta = \dfrac{\tau \times l}{r \times G} = \dfrac{60 \times 10^6 \times 2}{0.0253 \times 7.7 \times 10^{10}}$

i.e. **angle of twist, θ = 0.0616 rad** $= 0.0616 \times \dfrac{180°}{\pi}$

$= \mathbf{3.53°}$

For a practical video demonstration of torsion of circular section shafts go to www.routledge.com/cw/bird and click on the menu for 'Mechanics of Solids 3rd Edition'

8.5 Flanged couplings

In some engineering situations, it is necessary to join together two shafts made from dissimilar materials. In such cases, it is usually undesirable to weld together the two shafts, and one method of overcoming this problem is by the use of a flanged coupling, as shown in Figure 8.7. In this case the torque is transmitted from one shaft to the other by the action of shearing forces δF acting on the bolts, as shown in Figure 8.8.

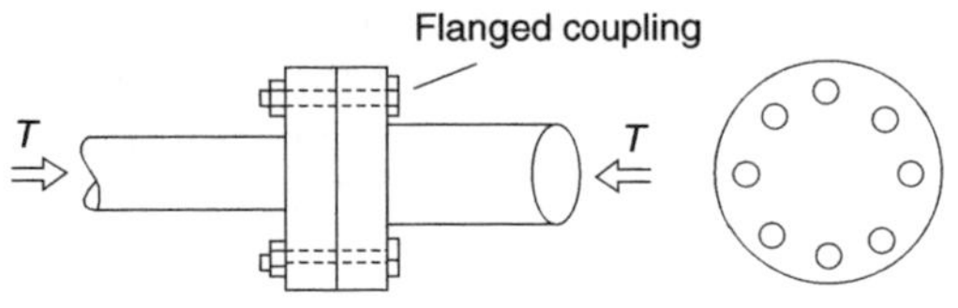

Figure 8.7 Flanged coupling

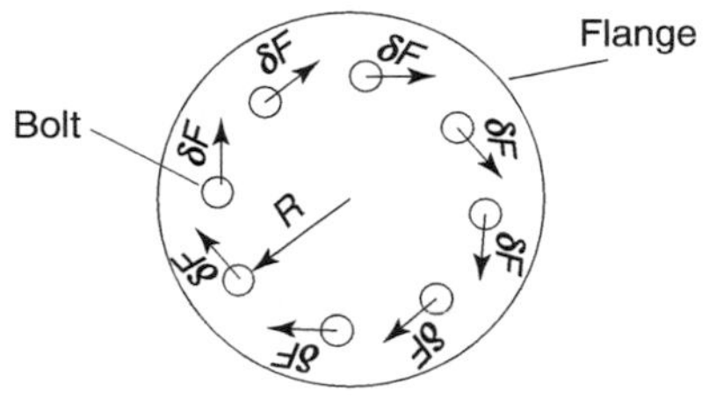

Figure 8.8 Shearing forces on bolts

Thus, if there are n bolts

$$\text{applied torque, } T = n \times \delta F \times R \tag{8.9}$$

where

$$\delta F = \tau_b \times \frac{\pi d^2}{4} \tag{8.10}$$

R = pitch circle radius of bolt

τ_b = shearing stress in bolt

and d = diameter of bolt

In Figure 8.8, it is assumed that all the bolts carry an equal shearing force δF; this may not be the case in practice, owing to manufacturing imperfections.

Problem 3. A torque of 10 kN m is to be transmitted from a hollow phosphor bronze shaft, whose external diameter is twice its internal diameter, to a solid mild-steel shaft, through a flanged coupling with 12 bolts made from a high tensile steel. Determine the dimensions of the shafts, the pitch circle diameter (PCD), and the diameters of the bolts, given that

maximum permissible shear stress in phosphor bronze = 20 MN/m^2

maximum permissible shear stress in mild steel = 30 MN/m^2

maximum permissible stress in high tensile steel = 60 MN/m^2

Consider the phosphor bronze shaft

Let R_p be the external radius of the phosphor bronze shaft.

Then, polar second moment of area

$$J = \frac{\pi\left(R_p^{\,4} - (R_p/2)^4\right)}{2} = \frac{\pi\left(R^4 - \frac{R^4}{16}\right)}{2} = \frac{\pi\left(\frac{15}{16}R^4\right)}{2}$$

i.e. $\quad \mathbf{J = 1.473 R_p^4}$

From $\frac{\tau}{r} = \frac{T}{J}$ $\quad \frac{20 \times 10^6}{R_p} = \frac{10 \times 10^3}{1.473 R_p^4}$

from which $\quad \frac{1.473 R_p^4}{R_p} = \frac{10 \times 10^3}{20 \times 10^6}$

and $\quad R_p^3 = \frac{10 \times 10^3}{20 \times 10^6 \times 1.473}$

i.e. $\quad \boldsymbol{R_p} = \sqrt[3]{\frac{10 \times 10^3}{20 \times 10^6 \times 1.473}} = \mathbf{0.0698\ m}$

and *external diameter of phosphor bronze shaft*

$= 2 \times 0.0698 = 0.1396$ m $= 13.96$ cm, *say,* 14 cm

internal diameter of phosphor bronze shaft

$= 0.0698$ m $= 6.98$ cm, *say,* 7 cm

Consider the steel shaft

Let R_s be the external radius of the steel shaft. Then, polar second moment of area

$$J = \frac{\pi R_s^4}{2} = 1.571 R_s^4$$

From $\frac{\tau}{r} = \frac{T}{J}$ $\quad \frac{30 \times 10^6}{R_s} = \frac{10 \times 10^3}{1.571 R_s^4}$

from which $\quad \frac{1.571 R_s^4}{R_s} = \frac{10 \times 10^3}{30 \times 10^6}$

and $\quad R_s^3 = \frac{10 \times 10^3}{30 \times 10^6 \times 1.571}$

i.e. $\quad \boldsymbol{R_s} = \sqrt[3]{\frac{10 \times 10^3}{30 \times 10^6 \times 1.571}} = \mathbf{0.05964\ m}$

and *diameter of steel shaft* $= 2 \times 0.05964$

$= 0.1193$ m $= 11.9$ cm, *say* 12 cm

Bolts on flanged coupling

As the external diameter of the hollow shaft is 14 cm, it will be necessary to assume that the PCD of the bolts on the flanged coupling is larger than this, so that the bolts can be accommodated.

Assume that $\quad D = \text{PCD}$

$= 20$ cm (to allow for fitting)

$n =$ number of bolts $= 12$

From equation (8.9), torque $T = n \times \delta F \times R$

where $\quad R = D/2 = 20/2 = 10$ cm $= 0.1$ m

from which $\quad \delta F = \frac{T}{nR} = \frac{10\,\text{kN m}}{12 \times 0.1\,\text{m}} = \mathbf{8.33\ kN}$

From equation (8.10) $\quad \delta F = \tau_b \times \frac{\pi d^2}{4}$

from which $\quad d^2 = \frac{4\delta F}{\pi \times \tau_b} = \frac{4 \times 8.33 \times 10^3}{\pi \times 60 \times 10^6}$

and $\quad d = \sqrt{\left(\frac{4 \times 8.33 \times 10^3}{\pi \times 60 \times 10^6}\right)} = 0.0133$ m

i.e. $\quad$ **diameter of bolt, d = 1.33 cm, say, 1.5 cm**

Now try the following Practice Exercise

Practice Exercise 37. Torsion of shafts

1. A circular-section steel shaft consists of three elements, two solid and one hollow, as shown in Figure 8.9, and it is subjected to a torque of 3 kN m.

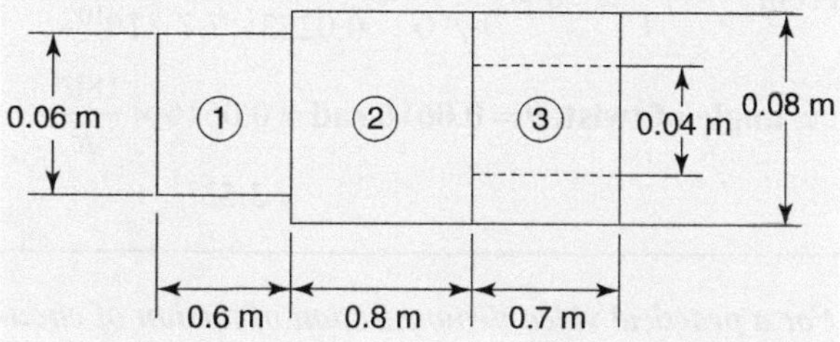

Figure 8.9

Determine (a) the angle of twist of one end relative to the other, and (b) the maximum shearing stresses in each element. It may be assumed that $G = 7.7 \times 10^{10}$ N/m^2

[(a) 1.91° (b) 70.75 MN/m^2, 29.84 MN/m^2, 31.83 MN/m^2]

2. A circular-section steel shaft consists of three elements, two solid and one hollow, as shown in Figure 8.9, and it is subjected to a torque of 4 kN m.
Determine (a) the angle of twist of one end relative to the other, and (b) the maximum shearing stresses in each element. It may be assumed that $G = 7.7 \times 10^{10}$ N/m^2

[(a) 2.55° (b) 94.34 MN/m^2, 39.79 MN/m^2, 42.44 MN/m^2]

3. If the shaft of question 1 were constructed from three separate materials, namely steel in element (1), aluminium alloy in element (2) and manganese bronze in element (3), determine (a) the total angle of twist, and (b) the maximum shearing stresses in each element. Assume that $G_1 = 7.7 \times 10^{10}$ N/m^2, $G_2 = 2.5 \times 10^{10}$ N/m^2 and $G_3 = 3.9 \times 10^{10}$ N/m^2

   [(a) 3.24° (b) 70.75 MN/m^2, 29.84 MN/m^2, 31.83 MN/m^2]

4. If the shaft of question 2 were constructed from three separate materials, namely steel in element (1), aluminium alloy in element (2) and manganese bronze in element (3), determine (a) the total angle of twist, and (b) the maximum shearing stresses in each element. Assume that $G_1 = 7.7 \times 10^{10}$ N/m^2, $G_2 = 2.5 \times 10^{10}$ N/m^2 and $G_3 = 3.9 \times 10^{10}$ N/m^2

   [(a) 4.314° (b) 94.34 MN/m^2, 39.79 MN/m^2, 42.44 MN/m^2]

5. Determine the output torque of an electric motor which supplies 5 kW at 25 rev/s.
   (a) If this torque is transmitted through a tube of external diameter 20 mm, determine the internal diameter if the maximum permissible shearing stress in the shaft is 35 MN/m^2.
   (b) If this shaft is to be connected to another one, via a flanged coupling, determine a suitable bolt diameter if four bolts are used on a pitch circle diameter of 30 mm, and the maximum permissible shearing stresses in the bolts equal 45 MN/m^2.

   [31.83 N m; (a) 16.11 mm, say 16 mm (b) 3.87 mm, say 4 mm]

6. Determine the output torque of an electric motor which supplies 8 kW at 25 rev/s.
   (a) If this torque is transmitted through a tube of external diameter 22 mm, determine the internal diameter if the maximum permissible shearing stress in the shaft is 35 MN/m^2.
   (b) If this shaft is to be connected to another one, via a flanged coupling, determine a suitable bolt diameter if four bolts are used on a pitch circle diameter of 32 mm, and the maximum permissible shearing stresses in the bolts equal 45 MN/m^2.

   [50.93 N m; (a) 16.34 mm, say 16.5 mm (b) 4.745 mm, say 5.0 mm]

8.6 Keyed couplings

Another method of transmitting power through shafts of dissimilar materials is through the use of keyed couplings, as shown in Figure 8.10. In such cases, the key is the male portion of the coupling, and the keyway, which is the female portion of the coupling, is on the shaft. It is evident that the manufacture of the 'key' and the 'keyway' has to be precise, so that the former fits snugly into the latter. Precise analysis of keyed couplings is difficult and beyond the scope of this book.

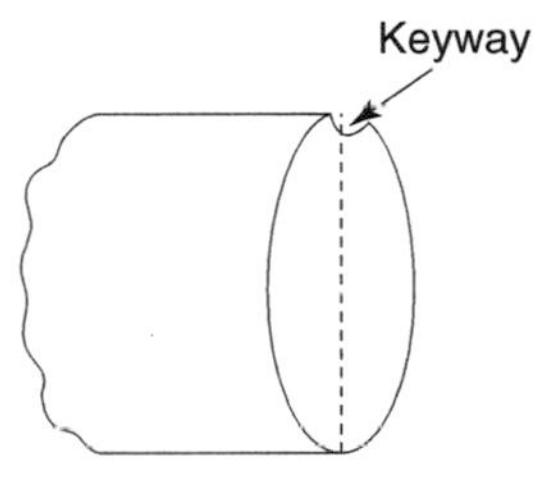

Figure 8.10 Keyway on a shaft

Problem 4. Determine the torque diagram for the shaft AB, which is subjected to an intermediate torque of magnitude T at the point C, as shown in Figure 8.11. The shaft may be assumed to be firmly fixed at its ends. From Newton's third law of motion, it is evident that the intermediate torque T will be resisted by the end torques T_1 and T_2, acting in an opposite direction to T, where the 'direction' of the torques are shown according to the right-hand screw rule (RHS rule). The RHS rule is defined by: *point the right hand in the direction of the double-tailed arrow and rotating the hand in a clockwise direction.*

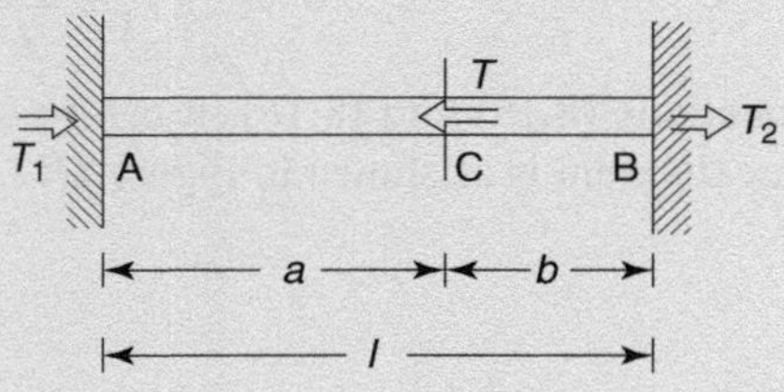

Figure 8.11 Shaft with intermediate torque

$$T_1 + T_2 = T \qquad (8.11)$$

Let θ_C be the rotation at the point C.

Then, from $\frac{T}{J} = \frac{G\theta}{l}$ $\qquad T_1 = \frac{GJ\theta_C}{a} \qquad (8.12)$

and $\qquad T_2 = \frac{GJ\theta_C}{b} \qquad (8.13)$

Dividing (8.12) by (8.13) gives

$$\frac{T_1}{T_2} = \frac{\frac{GJ\theta_C}{a}}{\frac{GJ\theta_C}{b}} = \frac{GJ\theta_C}{a} \times \frac{b}{GJ\theta_C}$$

i.e. $\qquad \frac{T_1}{T_2} = \frac{b}{a}$

or $\qquad T_1 = \frac{bT_2}{a} \qquad (8.14)$

Substituting equation (8.14) into equation (8.11) gives

$$\frac{bT_2}{a} + T_2 = T$$

i.e. $\qquad T_2\left(\frac{b}{a} + 1\right) = T$

or $\qquad T_2\left(\frac{b+a}{a}\right) = T$

i.e. $\qquad T_2\left(\frac{l}{a}\right) = T$

from which $\qquad \boldsymbol{T_2 = \frac{Ta}{l}} \qquad (8.15)$

and from equation (8.14) $\qquad T_1 = \frac{bT_2}{a} = \frac{b\left(\frac{Ta}{l}\right)}{a}$

$$= \frac{bTa}{la}$$

i.e. $\qquad \boldsymbol{T_1 = \frac{Tb}{l}} \qquad (8.16)$

From equations (8.15) and (8.16), it can be seen that the torque diagram is as shown in Figure 8.12.

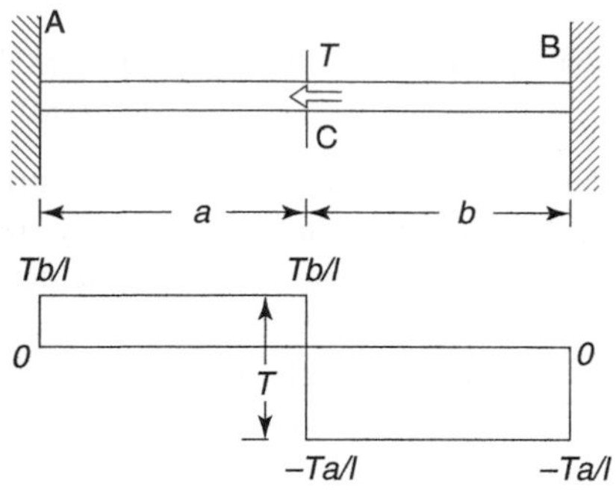

Figure 8.12 Torque diagram

If the shaft of Figure 8.12 were made from two different sections, joined together at C, so that

J_1 = polar second moment of area of shaft AC

J_2 = polar second moment CB of area

R_1 = external radius of shaft AC

R_2 = external radius CB of shaft

τ_1 = maximum shear stress in shaft AC

τ_2 = maximum shear stress in shaft CB

then

$$\frac{\tau_1}{R_1} = \frac{T_1}{J_1} = \frac{G\theta_C}{a} \qquad (8.17)$$

and

$$\frac{\tau_2}{R_2} = \frac{T_2}{J_2} = \frac{G\theta_C}{b} \qquad (8.18)$$

From equations (8.17) and (8.18), the two shafts can be designed.

8.7 Compound shafts

Composite shafts are made up from several smaller shafts with different material properties. The use of different materials for the manufacture of shafts is required when a shaft has to pass through different fluids, some of which are hostile to certain materials.

Composite shafts are usually either in series, as shown in Figure 8.13, or in parallel as shown in Figure 8.14.

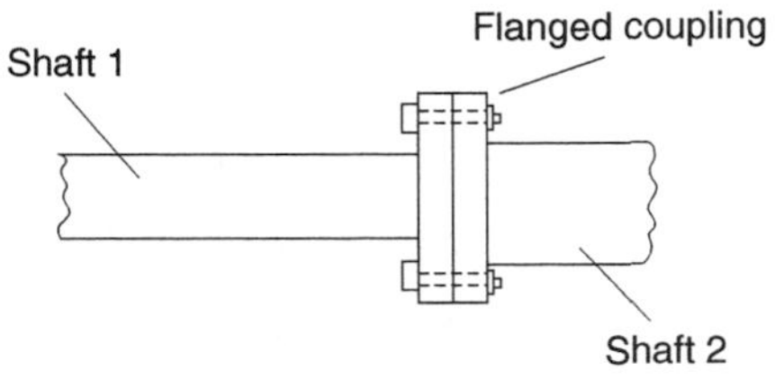

Figure 8.13 Composite shaft in series

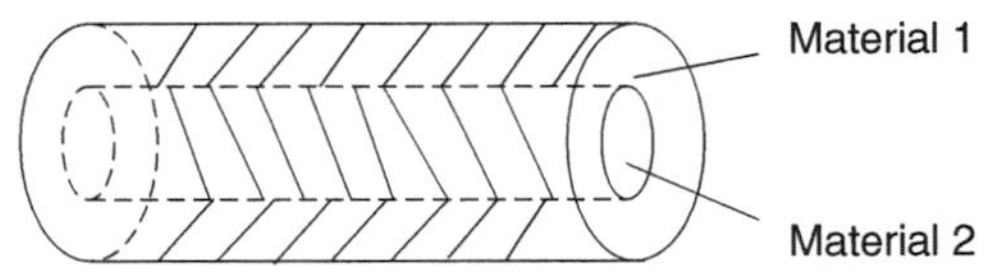

Figure 8.14 Composite shafts in parallel

Problem 5. A compound shaft ACB is fixed at the points A and B and subjected to a torque of 5 kN m at the point C, as shown in Figure 8.15. If both parts of the shaft are of solid circular sections, determine the angle of twist at the point C and the maximum shearing stresses in each section, assuming the following apply:

Shaft 1: $G_1 = 2.6 \times 10^{10}\,\text{N/m}^2$ $R_1 = 6$ cm $a = 1$ m

Shaft 2: $G_2 = 7.8 \times 10^{10}\,\text{N/m}^2$ $R_2 = 4$ cm $b = 1.5$ m

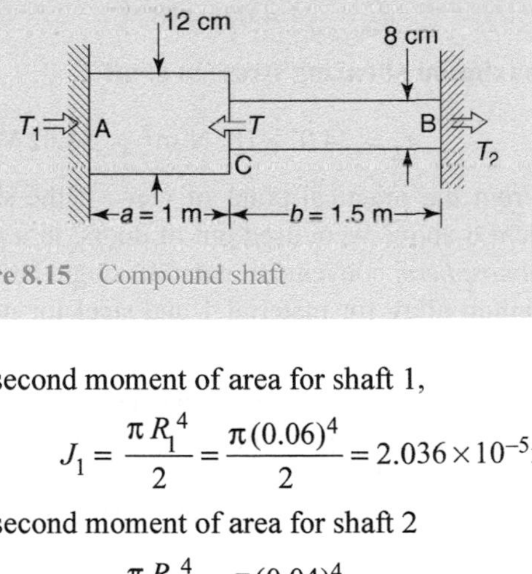

Figure 8.15 Compound shaft

Polar second moment of area for shaft 1,

$$J_1 = \frac{\pi R_1^4}{2} = \frac{\pi (0.06)^4}{2} = 2.036 \times 10^{-5}\text{m}^4$$

Polar second moment of area for shaft 2

$$J_2 = \frac{\pi R_2^4}{2} = \frac{\pi (0.04)^4}{2} = 4.021 \times 10^{-6}\text{m}^4$$

Let θ_C be the angle of twist at the point C.

Then, from $\frac{T}{J} = \frac{G\theta}{l}$

$$\theta_C = \frac{T_1 a}{G_1 J_1} = \frac{T_1 \times 1}{2.6 \times 10^{10} \times 2.036 \times 10^{-5}}$$

i.e. $\boldsymbol{\theta_C = 1.889 \times 10^{-6} T_1}$ (8.19)

Similarly $\theta_C = \frac{T_2 b}{G_2 J_2} = \frac{T_2 \times 1.5}{7.8 \times 10^{10} \times 4.021 \times 10^{-6}}$

i.e. $\boldsymbol{\theta_C = 4.783 \times 10^{-6} T_2}$ (8.20)

Equating equations (8.19) and (8.20) gives

$$1.889 \times 10^{-6} T_1 = 4.783 \times 10^{-6} T_2$$

from which $T_1 = \frac{4.783 \times 10^{-6} T_2}{1.889 \times 10^{-6}}$

i.e. $\boldsymbol{T_1 = 2.532\, T_2}$ (8.21)

Now $T = T_1 + T_2$

i.e. $5\text{ kN m} = 2.532 T_2 + T_2$ from equation (8.21)

i.e. $5\text{ kN m} = T_2\,(1 + 2.532)$

from which $\boldsymbol{T_2} = \frac{5}{3.532} = \mathbf{1.416\ kN\ m}$

and $\boldsymbol{T_1} = 2.532 T_2 = 2.532 \times 1.416 = \mathbf{3.585\ kN\ m}$

From equation (8.20)

angle of twist at point C, $\theta_C = 4.783 \times 10^{-6} T_2$

$$= 4.783 \times 10^{-6} \times 1.416 \times 10^3 = \mathbf{6.773 \times 10^{-3}\ rad}$$

$$= 6.773 \times 10^{-3} \times \frac{180°}{\pi} = \mathbf{0.388°}$$

From $\frac{\tau}{r} = \frac{T}{J}$ $\tau_1 = \frac{T_1 R_1}{J_1} = \frac{3.585 \times 10^3 \times 0.06}{2.036 \times 10^{-5}}$

i.e. **maximum shearing stress in shaft 1,**

$$\tau_1 = 10.56 \times 10^6\,\text{N/m}^2 = \mathbf{10.56\,MN/m^2}$$

and $\tau_2 = \frac{T_2 R_2}{J_2} = \frac{1.416 \times 10^3 \times 0.04}{4.021 \times 10^{-6}}$

i.e. **maximum shearing stress in shaft 2,**

$$\tau_2 = 14.09 \times 10^6\,\text{N/m}^2 = \mathbf{14.09\,MN/m^2}$$

Problem 6. The compound shaft of Figure 8.16 is subjected to a torque of 5 kN m. Determine the angle of twist and the maximum shearing stresses in materials 1 and 2, given the following values for the moduli of rigidity: material 1: $G_1 = 2.5 \times 10^{10}\,\text{N/m}^2$, material 2: $G_1 = 7.8 \times 10^{10}\,\text{N/m}^2$

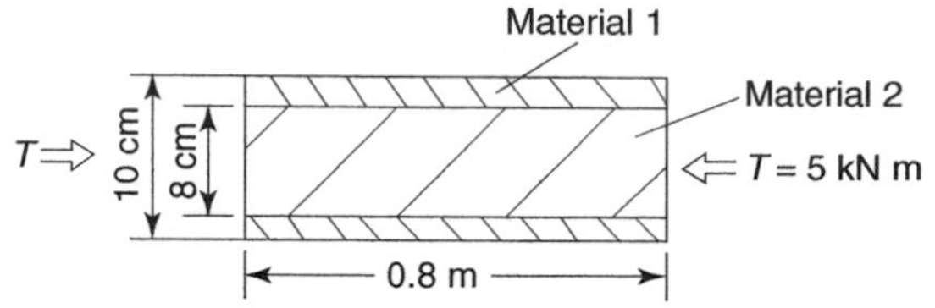

Figure 8.16 Compound shaft

Assumptions made

(a) No slipping takes place at the common interface.
(b) Radial lines remain straight on twisting.

From (b), angle of twist for material 1 = angle of twist for material 2

i.e. $\theta_1 = \theta_2$

Let $R_2 = \dfrac{10}{2} = 5$ cm = 0.05 m,

and let $R_1 = \dfrac{8}{2} = 4$ cm = 0.04 m

Polar second moment of area for material 1,

$$J_1 = \frac{\pi\left(R_2{}^4 - R_1{}^4\right)}{2} = \frac{\pi\left(0.05^4 - 0.04^4\right)}{2}$$

$$= \mathbf{5.796 \times 10^{-6}\,m^4}$$

Polar second moment of area for material 2,

$$J_2 = \frac{\pi R_2{}^4}{2} = \frac{\pi(0.04)^4}{2} = \mathbf{4.021 \times 10^{-6}\,m^4}$$

NB Although the thickness of material 1 is only 1 cm, $J_1 > J_2$ which shows that a hollow shaft has a better strength to weight ratio than a solid one.

From $\dfrac{T}{J} = \dfrac{G\theta}{l}$, $\theta = \dfrac{T_1 l}{G_1 J_1} = \dfrac{T_1 \times 0.8}{2.5 \times 10^{10} \times 5.796 \times 10^{-6}}$

i.e. $\boldsymbol{\theta = 5.521 \times 10^{-6} T_1}$ (8.22)

However, $\theta = \dfrac{T_2 l}{G_2 J_2} = \dfrac{T_2 \times 0.8}{7.8 \times 10^{10} \times 4.021 \times 10^{-6}}$

i.e. $\boldsymbol{\theta = 2.551 \times 10^{-6} T_2}$ (8.23)

Equating equations (8.22) and (8.23) gives

$$5.521 \times 10^{-6} T_1 = 2.551 \times 10^{-6} T_2$$

from which $T_1 = \dfrac{2.551 \times 10^{-6} T_2}{5.521 \times 10^{-6}}$

i.e. $\boldsymbol{T_1 = 0.462 T_2}$

Now $T = T_1 + T_2$

i.e. $5\text{ kN m} = 0.462 T_2 + T_2$

i.e. $5\text{ kN m} = T_2(1 + 0.462)$

from which $\boldsymbol{T_2} = \dfrac{5}{1.462} = \mathbf{3.420\ kN\ m}$

and $\boldsymbol{T_1} = 0.462 T_2 = 0.462 \times 3.420 = \mathbf{1.580\ kN\ m}$

From equation (8.23),

angle of twist at point C, $\boldsymbol{\theta_C} = 2.551 \times 10^{-6} T_2$

$$= 2.551 \times 10^{-6} \times 3.420 \times 10^3 = \mathbf{8.724 \times 10^{-3}\,rad}$$

$$= 8.724 \times 10^{-3} \times \frac{180^\circ}{\pi} = \mathbf{0.50^\circ}$$

From $\dfrac{\tau}{r} = \dfrac{T}{J}$ $\quad \tau_1 = \dfrac{T_1 R_1}{J_1} = \dfrac{1.580 \times 10^3 \times 0.05}{5.796 \times 10^{-6}}$

i.e. **maximum shearing stress in material 1,**

$$\boldsymbol{\tau_1} = 13.63 \times 10^6\,\text{N/m}^2 = \mathbf{13.63\,MN/m^2}$$

and $\tau_2 = \dfrac{T_2 R_2}{J_2} = \dfrac{3.420 \times 10^3 \times 0.04}{4.021 \times 10^{-6}}$

i.e. **maximum shearing stress in shaft 2,**

$$\boldsymbol{\tau_2} = 34.02 \times 10^6\,\text{N/m}^2 = \mathbf{34.02\,MN/m^2}$$

NB From the practical point of view, if the shaft of Problem 6 above were used out of doors, in a *normal UK atmosphere,* convenient materials might have been aluminium alloy for material 1 and steel for material 2. The reasons for such a choice would be as follows:

(a) both materials are relatively inexpensive;
(b) mild steel is, in general, stronger and stiffer than aluminium alloy;
(c) aluminium alloy does not rust; hence its use as a sheath over a steel core.

To *manufacture* the compound shaft of Figure 8.16, it is necessary to make the external diameter of the steel shaft slightly larger than the internal diameter of the aluminium-alloy shaft and to 'join' the two together by either heating the aluminium-alloy shaft and/or cooling the steel shaft.

Now try the following Practice Exercise

Practice Exercise 38. Compound shafts

1. A compound shaft consists of two equal length hollow shafts joined together in series and subjected to a torque *T*, as shown in Figure 8.17. If the shaft on the left is made from steel, and the shaft on the right of the figure is made from aluminium alloy, determine the maximum permissible value of *T*, given the following:

For steel: $G = 7.7 \times 10^{10}$ N/m^2, maximum permissible shear stress = 140 MN/m^2

For aluminium alloy: $G = 2.6 \times 10^{10}$ N/m^2, maximum permissible shear stress = 90 MN/m^2

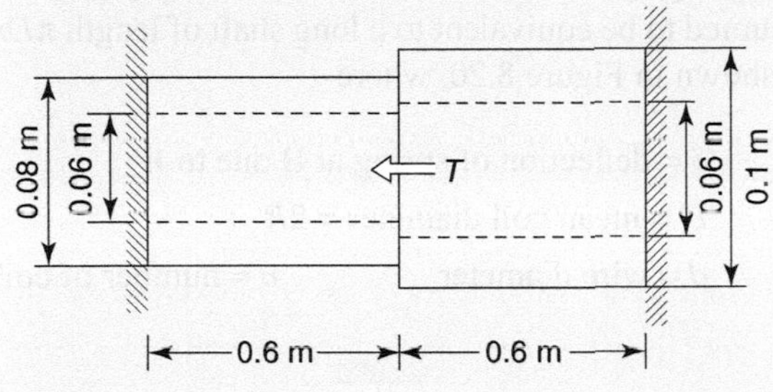

Figure 8.17

[9.62 kN m]

2. A compound shaft consists of a solid aluminium-alloy cylinder of length 1 m and diameter 0.1 m, connected in series to a steel tube of the same length and external diameter, and of thickness 0.02 m. The shaft is fixed at its ends and is subjected to an intermediate torque of 9 kN m at the joint. Determine the angle of twist and the maximum shear stress in the two halves. Assume that $G_{(steel)} = 7.7 \times 10^{10}$ N/m^2 and $G_{(Al\ alloy)} = 2.6 \times 10^{10}$ N/m^2

$\left[0.565°, \tau_{(Al\ alloy)} = 12.81 \text{ MPa}, \tau_{(steel)} = 37.94 \text{ MPa}\right]$

3. A compound shaft consists of two elements of equal length, joined together in series. If one element of the shaft is constructed from gunmetal tube and the other from solid steel, where the external diameter of the steel shaft and the internal diameter of the gunmetal shaft equal 50 mm, determine the external diameter of the gunmetal shaft if the two shafts are to have the some torsional stiffness. Determine also the maximum permissible torque that can be applied to the shaft, given the following:

For gunmetal: $G = 3 \times 10^{10}$ N/m^2, maximum permissible shear stress = 45 MN/m^2

For steel: $G = 7.5 \times 10^{10}$ N/m^2, maximum permissible shear stress = 90 MN/m^2

[68.4 mm, 2209 N m]

4. A compound shaft consists of a solid steel core, which is surrounded co-axially by an aluminium bronze sheath. Determine suitable values for the diameters of the shafts if the steel core is to carry two-thirds of a total torque of 500 N m and if the limiting stresses are not exceeded.

For aluminium alloy: $G = 3.8 \times 10^{10}$ N/m^2, maximum permissible shear stress = 18 MN/m^2

For steel: $G = 7.7 \times 10^{10}$ N/m^2, maximum permissible shear stress = 36 MN/m^2

[38.6 mm, 38.1 mm, where the stress in the bronze shaft is the design criterion]

5. A compound shaft consists of two equal length hollow shafts joined together in series and subjected to a torque *T*, as shown in Figure 8.18. If the shaft on the left is made from steel, and the shaft on the right of the figure is made from aluminium alloy, determine the maximum permissible value of *T*, given the following:

For steel: $G = 7.7 \times 10^{10}$ N/m^2, maximum permissible shear stress = 140 MN/m^2

For aluminium alloy: $G = 2.6 \times 10^{10}$ N/m^2, maximum permissible shear stress = 90 MN/m^2

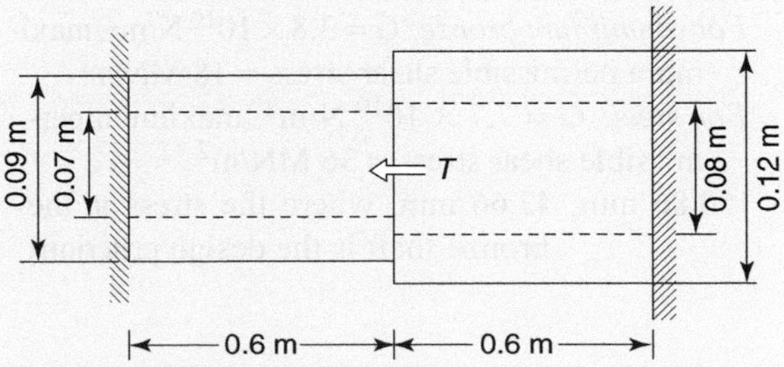

Figure 8.18

[2.56 kN m]

6. A compound shaft consists of a solid aluminium-alloy cylinder of length 1 m and diameter 0.1 m, connected in series to a steel tube of the same length and external diameter, and of thickness 0.02 m. The shaft is fixed at its ends and is subjected to an intermediate torque of 10 kN m at the joint. Determine the angle of twist and the maximum shear stress in the two halves. Assume that $G_{(\text{steel})} = 7.7 \times 10^{10}$ N/m^2 and $G_{(\text{Al alloy})} = 2.6 \times 10^{10}$ N/m^2

   [0.627°, $\tau_{(\text{Al alloy})} = 14.23$ MN/m^2, $\tau_{(\text{steel})} = 42.16$ MN/m^2]

7. A compound shaft consists of two elements of equal length, joined together in series. If one element of the shaft is constructed from gunmetal tube and the other from solid steel, where the external diameter of the steel shaft and the internal diameter of the gunmetal shaft equal 60 mm, determine the external diameter of the gunmetal shaft if the two shafts are to have the same torsional stiffness. Determine also the maximum permissible torque that can be applied to the shaft, given the following:

   For gunmetal: $G = 3 \times 10^{10}$ N/m^2, maximum permissible shear stress = 45 MN/m^2

   For steel: $G = 7.5 \times 10^{10}$ N/m^2, maximum permissible shear stress = 90 MN/m^2

   [82 mm, 6877 N m]

8. A compound shaft consists of a solid steel core, which is surrounded co-axially by an aluminium bronze sheath. Determine suitable values for the diameters of the shafts if the steel core is to carry two thirds of a total torque of 700 N m and if the limiting stresses are not exceeded.

   For aluminium bronze: $G = 3.8 \times 10^{10}$ N/m^2, maximum permissible shear stress = 18 MN/m^2

   For steel: $G = 7.7 \times 10^{10}$ N/m^2, maximum permissible shear stress = 36 MN/m^2

   [50.86 mm, 42.66 mm, where the stress in the bronze shaft is the design criterion]

8.8 Tapered shafts

There are a number of texts that treat the analysis of tapered shafts by using a simple extension of equation (8.5), but this method of analysis is incorrect, as the longitudinal shearing stresses produced on torsion are not parallel to the axis of the shaft.

For an analysis of tapered shafts, see Reference 8 on page 497.

8.9 Close-coiled helical springs

Equation (8.5) can be used for the stress analysis of close-coiled helical springs, providing the angle of helix and the deflections are small.

For the close-coiled helical spring of Figure 8.19, most of the coils will be under torsion, where the torque T equals $WD/2$, so that the maximum deflection at B, caused by W, would be due to the combined effect of all the rotations of all the circular sections of the spring. Thus, the close-coiled helical spring can be assumed to be equivalent to a long shaft of length πDn, as shown in Figure 8.20, where

δ = deflection of spring at B due to W

D = mean coil diameter = $2R$

d = wire diameter $\quad n$ = number of coils

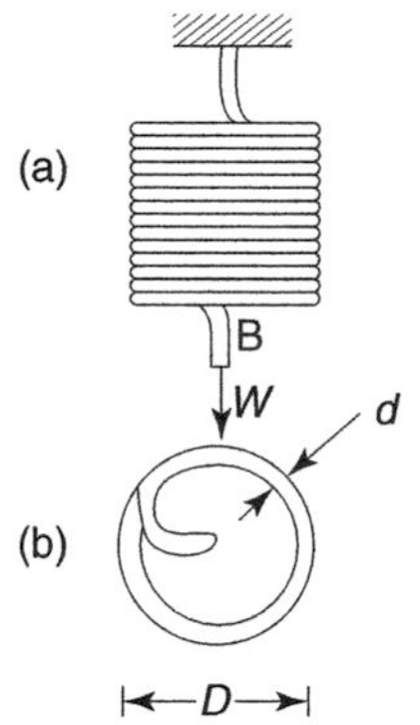

Figure 8.19 Close-coiled helical spring: (a) front elevation; (b) plan (looking upwards)

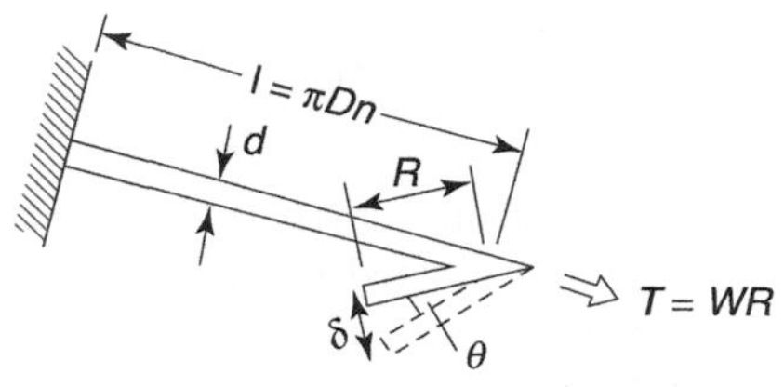

Figure 8.20

From Figure 8.19, $\theta = \dfrac{T}{3} \times \dfrac{l}{G} = \dfrac{T}{J} \times \dfrac{\pi D n}{G}$

or
$$\theta = \frac{W \times R \times \pi \times 2Rn}{\dfrac{\pi d^4}{32} \times G} = \frac{64WR^2 n}{G D^4}$$

Hence, deflection of spring at B due to W

$$\delta = R\theta = \frac{\mathbf{64}\boldsymbol{WR^3n}}{\boldsymbol{GD^4}}$$

If τ is the maximum shearing stress in the spring, then

$$\tau = \frac{T \times r}{J} = \frac{T \times \left(\dfrac{d}{2}\right)}{\pi \times \left(\dfrac{d^4}{32}\right)} = \frac{16T}{\pi d^3} = \frac{16WR}{\pi d^3}$$

which occurs throughout the coils on the external surface of the spring. For the torsion of non-circular sections, see References 8 to 11 on page 497.

8.10 Torsion of thin-walled non-circular sections

The torsion of non-circular sections involves the solution of Poisson's equation (see References 8 to 11 on page 497), which is beyond the scope of the present book. However, for the torsion of thin-walled open non-circular sections, a much simpler theory is available, which is now presented. These thin-walled open sections appear in the form of angle bars, tees and channels.

Consider a thin-walled non-circular-section tube to be subjected to the torques shown in Figure 8.21.

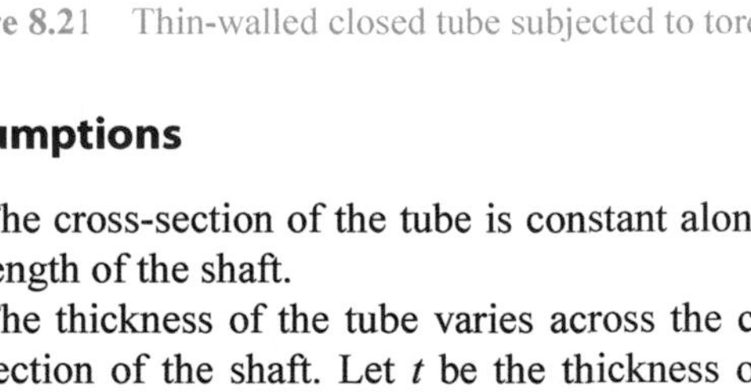

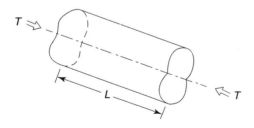

Figure 8.21 Thin-walled closed tube subjected to torques

Assumptions

(a) The cross-section of the tube is constant along the length of the shaft.
(b) The thickness of the tube varies across the cross-section of the shaft. Let t be the thickness of the tube at any point in the cross-section.

Under a torque T, the shearing stress at any point in the cross-section can be assumed to be τ. Now the shearing force/unit circumferential length at any point in the cross-section equals

$$q = \tau \times t = \text{shear flow}$$

Consider the equilibrium of q on the shell ABCD, as shown in Figure 8.22(a).

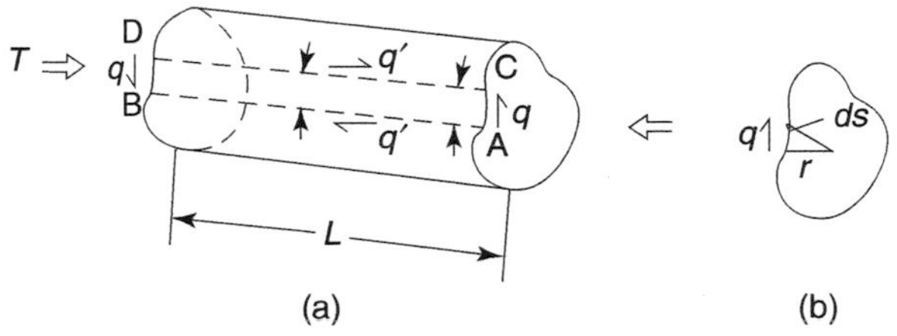

Figure 8.22 Shear flow in a closed tube: (a) shear flow in tube; (b) cross-section of tube

From this figure it can be seen from equilibrium considerations that q is the same on faces AC and BD, and that q' is the same on faces AB and CD.

Taking moments about A

clockwise moments = counter-clockwise moments

i.e. $$q' \times L \times b = q \times b \times L$$

Therefore, $$q' = q$$

That is, the shear flow q is constant at any point on the section of the tube.

To determine the relationship between *T* and *q*

From Figure 8.22(b), it can be seen that the torsional resistance of the tube is

$$T = \oint q\, ds\, r$$ [note that $\oint$ means an integral taken over a closed path]

but $$q = \tau \times t = \text{a constant}$$

Therefore $$T = \tau\, t \oint r\, ds$$

However $$\oint r\, ds = 2A$$

where A is the enclosed area of the tube.

Therefore $$T = 2A\tau t$$

or
$$\tau = \frac{T}{2At} \quad (8.24)$$

From equation (8.24), it can be seen that the shearing stress in a closed non-circular section is maximum where the thickness is the smallest.

To obtain the relationship between T and θ

From equation (11.23), page 247, the strain energy per unit volume due to τ is

$$U=\frac{\tau^2}{2G}\times\text{volume}$$

$$=\oint\frac{\tau^2}{2G}\times L\times t\times ds$$

$$=\oint\frac{q^2}{2G}\times L\frac{ds}{t}\quad\text{since } q=\tau\times t \text{ from which,}$$

$$\tau=\frac{q}{t}\ \text{and}\ \tau^2=\frac{q^2}{t^2}$$

but $\dfrac{q^2L}{2G}=\text{constant}$

Therefore, $$U=\frac{q^2L}{2G}\oint\frac{ds}{t}\qquad(8.25)$$

From equation (8.24) $\tau=\dfrac{T}{2At}$ and since $q=\tau\times \text{t}$

then $$q=\left(\frac{T}{2At}\right)\times t=\frac{T}{2A}\qquad(8.26)$$

Substituting equation (8.26) into (8.25),

$$U=\frac{\left(\frac{T}{2A}\right)^2L}{2G}\oint\frac{ds}{t}=\frac{T^2L}{8A^2G}\oint\frac{ds}{t}\qquad(8.27)$$

If θ is the rotation of the cross-section over a length L then

$$U=\frac{1}{2}T\theta\qquad(8.28)$$

Equating equation (8.27) to (8.28) gives

$$\frac{T^2L}{8A^2G}\oint\frac{ds}{t}=\frac{1}{2}T\theta$$

from which, $$\theta=\frac{2T^2L}{8TA^2G}\oint\frac{ds}{t}$$

i.e. $$\theta=\frac{TL}{4A^2G}\oint\frac{ds}{t}\qquad(8.29)$$

If t is constant over the whole circumference and S is the length of the perimeter of the cross-section, then

$$\theta=\frac{TLS}{4A^2Gt}\qquad(8.30)$$

Since $\dfrac{T}{J}=\dfrac{G\theta}{L}$ then $J=\dfrac{TL}{G\theta}$

i.e. $$J=\frac{TL}{G\left(\frac{TLS}{4A^2Gt}\right)}\quad\text{from equation (8.30)}$$

i.e. $$J=\frac{TL\left(4A^2Gt\right)}{G\left(TLS\right)}$$

i.e. $$J=\frac{4A^2t}{S}=\text{torsional constant}\qquad(8.31)$$

It can be seen from equation (8.31) that the second polar moment of area cannot be used for the torsion of non-circular sections. For a thin-walled open section, such as an angle bar, a tee, or a channel, $J=\dfrac{b_i t_i^3}{3}$, where 'i' varies from 1 to n, and n is the number of legs (2 for an angle bar, 3 for a channel section). Note $b_i/t_i\ge 5$.

8.11 Torsion of thin-walled rectangular sections

Consider the thin-walled rectangular cross-section of Figure 8.23, where $b/t\ge 5$.

Assume that the shearing stress due to torsion varies as shown to the right of Figure 8.23; this will be proven to be correct a little later in this section. Analysis of the rectangular cross-section can be achieved if the theory for thin-walled closed tubes is applied to the thin tube, shown shaded in Figure 8.23.

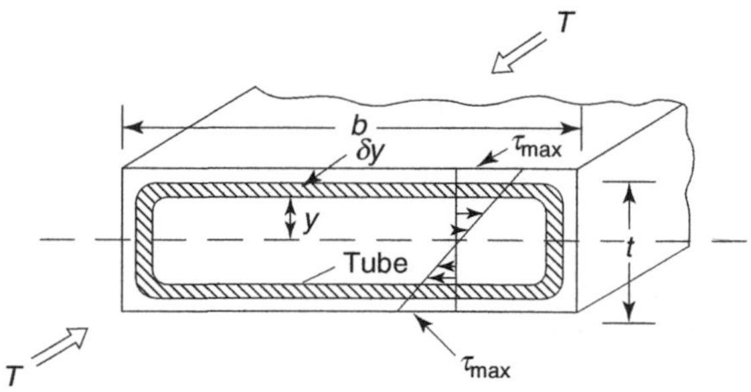

Figure 8.23 Rectangular cross-section

Now from equation (8.24) $\tau=\dfrac{T}{2At}$

which, when applied to the thin tube of Figure 8.21, gives

$$\tau = \frac{\delta T}{2 \times b \times 2y \times \delta y}$$

or $$\frac{dT}{dy} = 4by\tau \qquad (8.32)$$

From equation (8.30), $\theta = \dfrac{TLS}{4A^2Gt}$

which, when applied to our thin-walled tube, gives

$$\theta = \frac{\delta T\, L(2b+4y)}{4 \times 4b^2y^2G\,\delta y}$$

but $2b \gg 4y$, so $\theta = \dfrac{\delta T}{\delta y}\dfrac{L(2b)}{4 \times 4b^2y^2G} = \dfrac{\delta T}{\delta y}\dfrac{L}{8by^2G}$

or $$\frac{\delta T}{\delta y} = \frac{8by^2G\theta}{L} \qquad (8.33)$$

Equating equations (8.32) and (8.33),

$$4by\tau = \frac{8by^2G\theta}{L}$$

from which, $$\tau = \frac{8by^2G\theta}{4byL} = \frac{2yG\theta}{L} \qquad (8.34)$$

i.e. τ varies linearly with y and is a maximum when

$y = \dfrac{t}{2}$

Therefore, $$\tau_{max} = \frac{2\left(\frac{t}{2}\right)G\theta}{L} = \frac{tG\theta}{L}$$

or $$\tau_{max} = Gt\left(\frac{\theta}{L}\right) \qquad (8.35)$$

To obtain the relationship between T and θ

From equation (8.32), $\dfrac{dT}{dy} = 4by\tau$

from which, $$dT = 4by\tau\, dy \qquad (8.36)$$

Substituting equation (8.34) into (8.36),

$$dT = 4by\left(\frac{2yG\theta}{L}\right)dy$$

$$= 8bG\left(\frac{\theta}{L}\right)y^2dy$$

and $$T = 8bG\left(\frac{\theta}{L}\right)\int_{-t/2}^{+t/2} y^2dy$$

$$= 8bG\left(\frac{\theta}{L}\right)\left[\frac{y^3}{3}\right]_{-t/2}^{+t/2} = 8bG\left(\frac{\theta}{L}\right)\left[\frac{\left(\frac{t}{2}\right)^3}{3} - \frac{\left(-\frac{t}{2}\right)^3}{3}\right]$$

$$= 8bG\left(\frac{\theta}{L}\right)\left[\frac{t^3}{24} - \frac{-t^3}{24}\right]$$

$$= 8bG\left(\frac{\theta}{L}\right)\left[\frac{t^3}{12}\right]$$

i.e. $$T = 2G\left(\frac{bt^3}{3}\right)\left(\frac{\theta}{L}\right) \qquad (8.37)$$

Now from equation (8.4), $\dfrac{T}{J} = \dfrac{G\theta}{l}$

from which, $T = GJ\left(\dfrac{\theta}{L}\right)$

Therefore $$J = \text{torsional constant} = \frac{2bt^3}{3} \qquad (8.38)$$

which does not equal the polar second moment of area! For very thin-walled rectangles, where $b_i / t_i \geq 5$, assumed $J = \dfrac{bt^3}{3}$

8.12 Torsion of thin-walled open sections

The theory of Section 8.11 can be used for built-up open sections, such as angle bars, channel and tee sections, as shown in Table 8.1.

Table 8.1

Section	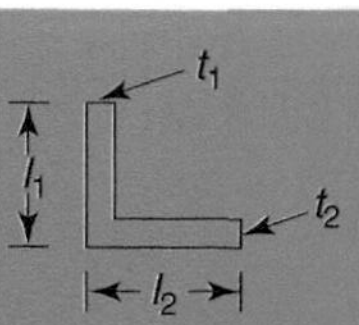 Figure 8.24	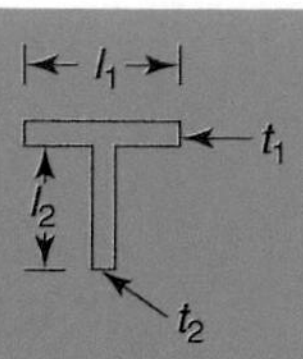 Figure 8.25	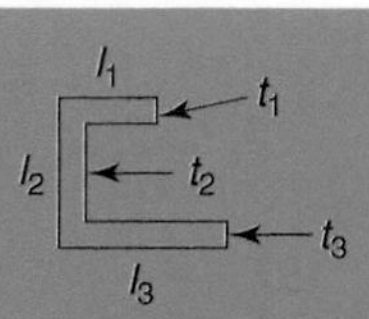 Figure 8.26
J	$\frac{l_1 t_1^3}{3} + \frac{l_2 t_2^3}{3}$	$\frac{l_1 t_1^3}{3} + \frac{l_2 t_2^3}{3}$	$\frac{l_1 t_1^3}{3} + \frac{l_2 t_2^3}{3} + \frac{l_3 t_3^3}{3}$

For a built-up section

$$\tau_{max} = G t_{max} \left(\frac{\theta}{L} \right) \tag{8.39}$$

where $\left(\frac{\theta}{L} \right)$ is the same for each rectangular element of the cross-section.

Additionally, $T = GJ\left(\frac{\theta}{L} \right)$ from which $\frac{T}{J} = G\left(\frac{\theta}{L} \right)$

$$\tag{8.40}$$

where J is calculated as described in Table 8.1.

From equation (8.39), $\frac{\tau_{max}}{t_{max}} = G\left(\frac{\theta}{L} \right)$

Hence, from equations (8.39) and (8.40), $\frac{\tau_{max}}{t_{max}} = \frac{T}{J}$

from which, $$\tau_{max} = \frac{T t_{max}}{J} \tag{8.41}$$

8.13 Elastic-plastic torsion of circular-section shafts

Equation (8.5) is based on elastic theory, but in practice a circular-section shaft can withstand a much larger torque than that predicted by this theory, because for most materials the shaft can become fully plastic before failure.

In this section the theory will be based on the material behaving as an ideally elastic-plastic material, as shown by Figure 8.27, where τ_{yp} is the shear yield stress.

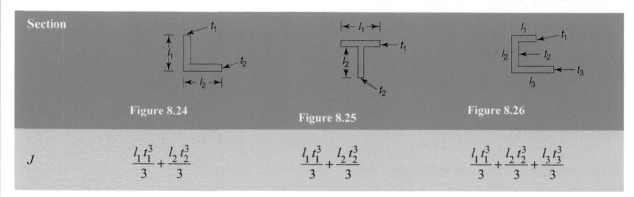

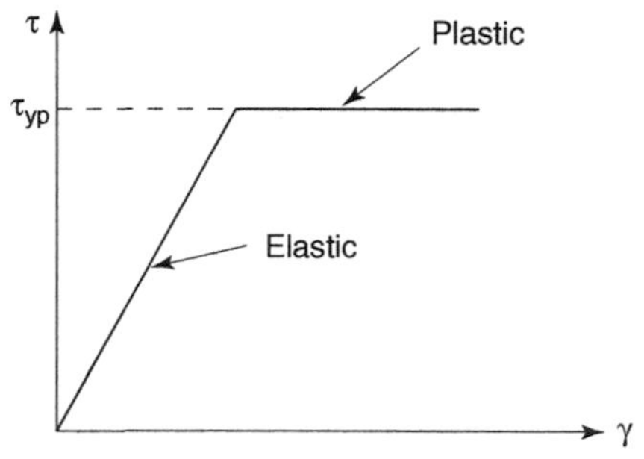

Figure 8.27 Shear stress-shear strain relationship

Problem 7. A solid steel shaft of diameter 0.1 m and length 0.7 m is subjected to a torque of 0.05 MN m, causing the shaft to suffer elastic-plastic deformation.

Determine (a) the depth of plastic penetration, (b) the angle of twist on application of the torque, (c) the residual angle of twist on release of the torque, (d) the full plastic torsional resistance of a similar shaft (T_p), and (e) the ratio of T_p to torque at first yield (T_{yp}).

The shaft may be assumed to have the following material properties

τ_{yp} = shear yield stress = 200 MN/m^2 and

G = rigidity modulus = 7.7×10^{10} N / m^2

(a) As the shaft section is partially elastic and partially plastic, the shear stress distribution will be as shown in Figure 8.28, where it can be seen that there is an elastic core.

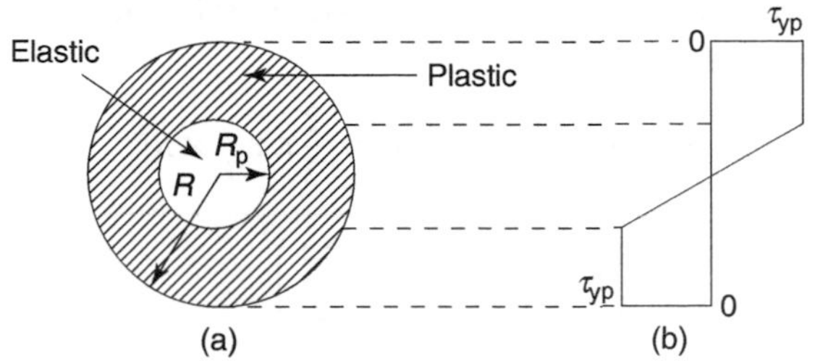

Figure 8.28 Circular-section shaft: (a) section; (b) shear stress distribution

Let R = radius of shaft

R_p = outer radius of elastic core

T = applied torque = $T_e + T_p$

T_e = torsional resistance of elastic core

T_p = torsional resistance of plastic portion of section (shaded in Figure 8.28(a))

J_e = polar second moment of area of elastic core

τ_{yp} = shear stress at yield

To calculate angle of twist θ

First, it will be necessary to calculate T_e and T_p
From Reference 1 on page 497, the torsional equation for circular-section shafts can be applied to the elastic core,

i.e. $$\frac{\tau_{yp}}{R_p} = \frac{T_e}{J_e} = \frac{G\theta}{l} \tag{8.42}$$

where G = rigidity or shear modulus

θ = angle of twist over the length l of the shaft

and $$J_e = \frac{\pi R_p^{\ 4}}{2}$$

Therefore, $$T_e = \frac{\tau_{yp} \times J_e}{R_p} = \frac{200\times10^6 \times \left(\frac{\pi R_p^{\ 4}}{2}\right)}{R_p}$$

i.e. $$T_e = 100\times10^6 \times \pi R_p^{\ 3} = 314.16\times10^6 R_p \tag{8.43}$$

To calculate T_p consider an annular element of radius r and thickness dr in the plastic zone, where the shear stress is of constant value τ_{yp}, as shown in Figure 8.29.

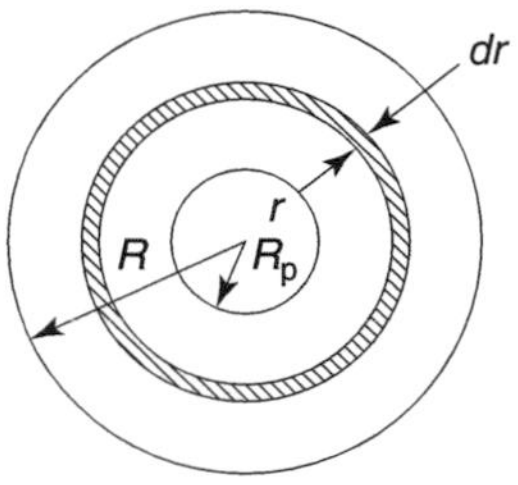

Figure 8.29 Plastic section of shaft

From Figure 8.26 $$T_p = \int_{R_p}^{R} \tau_{yp} \times (2\pi r \times dr) \times r$$

$$= 2\pi\tau_{yp}\int_{R_p}^{R} r^2 dr = 2\pi\tau_{yp}\left[\frac{r^3}{3}\right]_{R_p}^{R}$$

$$= \frac{2}{3}\pi\tau_{yp}\left(R^3 - R_p^{\ 3}\right) \tag{8.44}$$

i.e. $$T_p = \frac{2}{3}\pi \times 200\times10^6\left(\left(\frac{0.1}{2}\right)^3 - R_p^{\ 3}\right)$$

i.e. $$T_p = 418.9\times10^6\left(1.25\times10^{-4} - R_p^3\right)$$

Now $T = T_e + T_p$

i.e. $$0.05\times10^6 = 314.16\times10^6 R_p^3 + 418.9\times10^6\left(1.25\times10^{-4} - R_p^3\right)$$

from equation (8.43) and (8.44)

i.e. $$0.05\times10^6 = 314.16\times10^6 R_p^3 + 52.4\times10^3 - 418.9\times10^6 R_p^3$$

i.e. $$418.9\times10^6 R_p^3 - 314.16\times10^6 R_p^3 = 52.4\times10^3 - 0.05\times10^6$$

i.e. $$104.74\times10^6 R_p^3 = 2400$$

from which, $$R_p^3 = \frac{2400}{104.74\times10^6}$$

and $$R_p = \sqrt[3]{\frac{2400}{104.74\times10^6}}$$

i.e. $$\mathbf{R_p = 0.028\ m} \tag{8.45}$$

Therefore, ***depth of plastic penetration***
= 0.05 – 0.028 = **0.022 m**

(b) To calculate θ_1 the angle of twist due to T

From equation (8.42) $\dfrac{\tau_{yp}}{R_p} = \dfrac{G\theta_1}{l}$

from which $\theta_1 = \dfrac{\tau_{yp} \times l}{R_p \times G} = \dfrac{200 \times 10^6 \times 0.7}{0.028 \times 7.7 \times 10^{10}}$

i.e. $\boldsymbol{\theta_1 = 0.065}$ **rad** $= 0.065 \times \dfrac{180°}{\pi} = \mathbf{3.72°}$

(c) To calculate the residual angle of twist (θ_R)

The T-θ relationship on loading and on unloading is shown in Figure 8.30 where on removal of the applied torque, the shaft behaves elastically.

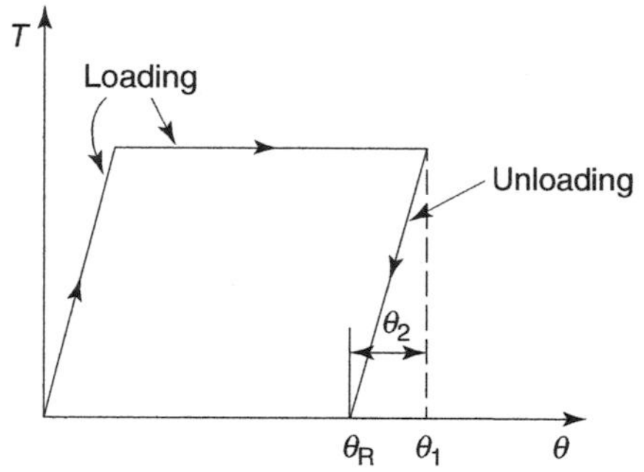

Figure 8.30 T–θ relationship for shaft

From Figure 8.27, it can be seen that θ_2 can be calculated by applying the full torque T to the whole section, and assuming elastic behaviour on unloading

i.e. $\theta_2 = \dfrac{Tl}{GJ} = \dfrac{0.05 \times 10^6 \times 0.7}{7.7 \times 10^{10} \times J}$

However, $J = \dfrac{\pi R^4}{2} = \dfrac{\pi \times \left(\dfrac{0.1}{2}\right)^4}{2} = 9.817 \times 10^{-6}\ \text{m}^4$

Hence, $\theta_2 = \dfrac{0.05 \times 10^6 \times 0.7}{7.7 \times 10^{10} \times 9.817 \times 10^{-6}}$

$= \mathbf{0.046}$ **rad** or $\mathbf{2.65°}$

Therefore, **the residual angle of twist,**

$\theta_R = \theta_1 - \theta_2 = 3.72° - 2.65° = \mathbf{1.07°}$

(d) To determine fully plastic torsional resistance (T_p)

From equation (8.44), $T_p = \dfrac{2}{3}\pi\tau_{yp} \times R^3$

$= \dfrac{2}{3}\pi \times 200 \times 10^6 \times \left(\dfrac{0.1}{2}\right)^3$

i.e. $= 418.9 \times 10^6 \times 1.25 \times 10^{-4}$

i.e. $\boldsymbol{T_p} = 52.4 \times 10^3\ \text{N m} = 52.4\ \text{kN m}$

$= \mathbf{0.0524\ MN\ m}$

(e) To determine the ratio T_p / T_{yp}

T_{yp} = maximum torsional resistance up to first yield

From equation (8.42), $\dfrac{\tau_{yp}}{R} = \dfrac{T_{yp}}{J}$

from which, $T_{yp} = \dfrac{\tau_{yp} \times J}{R} = \dfrac{200 \times 10^6 \times \left(\dfrac{\pi(0.1)^4}{32}\right)}{0.05 \times 10^6}$

i.e. $\boldsymbol{T_{yp}} = \mathbf{0.0393\ MN\ m}$

Therefore, the ratio $\dfrac{\boldsymbol{T_p}}{\boldsymbol{T_{yp}}} = \dfrac{0.0524 \times 10^6}{0.0393 \times 10^6} = \mathbf{1.333}$

Problem 8. A compound shaft of length 0.7 m consists of an aluminium-alloy core, of diameter 0.07 m, surrounded co-axially by a steel tube of external diameter 0.1 m. Determine (a) the torque that can be applied without causing yield, (b) the resulting angle of twist on applying a torque of 36000 N m, (c) the residual angle of twist remaining on removal of this torque, (d) the fully plastic torsional resistance of this shaft, and (e) the ratio of the torque obtained from (d) to that obtained from (c). The following may be assumed to apply:

Steel: τ_{yps} = shear yield stress in steel = 200 MN/m^2,

G_s = rigidity modulus = $7.7 \times 10^{10}\ \text{N/m}^2$

Aluminium alloy: τ_{ypa} = shear yield stress in aluminium alloy = 100 MN/m^2, G_a = rigidity modulus = $2.6 \times 10^{10}\ \text{N/m}^2$

The assumption can be made that radial lines remain straight on application or removal of the torque, i.e. θ, the angle of twist for the steel and the aluminium alloy, is constant.

(a) To determine T_{yp}

Prior to determining T_{yp} it will be necessary to determine whether the aluminium alloy or the steel will yield first.

Considering the angle of twist in the steel:

From equation (8.42) $\frac{\tau_{yp_s}}{R} = \frac{G\theta}{l}$

from which $\theta = \frac{\tau_{yp_s} \times l}{R \times G} = \frac{200 \times 10^6 \times 0.7}{0.05 \times 7.7 \times 10^{10}}$

i.e. $\boldsymbol{\theta = 0.0364}$ **rad** (8.46)

Considering the angle of twist in the aluminium alloy:

$$\theta = \frac{\tau_{ypa} \times l}{r \times G} = \frac{100 \times 10^6 \times 0.7}{0.035 \times 2.6 \times 10^{10}}$$

i.e. $\boldsymbol{\theta = 0.0769}$ **rad** (8.47)

That is, if the aluminium alloy were allowed to reach yield, then the steel would become plastic; therefore, the yield stress in the steel is the design criterion.

i.e. **'design'** $\boldsymbol{\theta = 0.0364}$ **rad** (8.48)

Let J_s = polar second moment of area of the steel tube

J_a = polar second moment of the aluminium alloy shaft

T_s = elastic torque in steel due to the application of T_{yp}

T_a = elastic torque in the aluminium alloy due to the application of T_{yp}

$$J_s = \frac{\pi\left(0.1^4 - 0.07^4\right)}{32} = \mathbf{7.46 \times 10^{-6}\,m^4}$$

$$J_a = \frac{\pi \times 0.07^4}{32} = \mathbf{2.357 \times 10^{-6}\,m^4}$$

From equation (8.42), $\frac{T_a}{J_a} = \frac{G_a\theta}{l}$

from which,

$$T_a = \frac{G_a\theta \times J_a}{l} = \frac{2.6 \times 10^{10} \times 0.0364 \times 2.357 \times 10^{-6}}{0.7}$$

i.e. $\boldsymbol{T_a = 3187}$ **N m**

Similarly,

$$T_s = \frac{G_s\theta \times J_s}{l} = \frac{7.7 \times 10^{10} \times 0.0364 \times 7.46 \times 10^{-6}}{0.7}$$

i.e. $\boldsymbol{T_s = 29870}$ **N m**

Now $T_{yp} = T_a + T_s = 3187 + 29870$

i.e. $\boldsymbol{T_{yp} = 33057}$ **N m**

(b) To determine the angle of twist θ due to a torque of 36000 N m

On application of the torque, both the steel and the aluminium alloy may be assumed to go plastic, as shown in Figure 8.31.

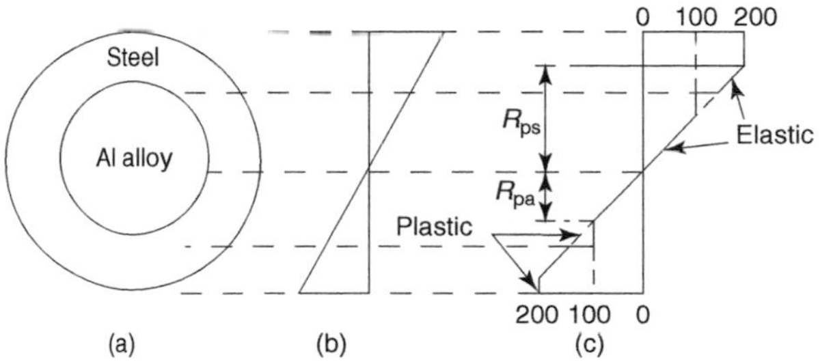

Figure 8.31 Shear stress and shear strain distributions in compound shaft: (a) section; (b) shear strain distribution; (c) shear stress distribution

From equation (8.44)

$T_{p's}$ = torque contribution from the part of the steel that becomes plastic

$$= \frac{2}{3}\pi\tau_{yps}\left(0.05^3 - R_{ps}^{\ 3}\right)$$

$$= \frac{2}{3}\pi\left(200 \times 10^6\right)\left(0.05^3 - R_{ps}^{\ 3}\right)$$

i.e. $\boldsymbol{T_{p's} = 418.9 \times 10^6 \times \left(1.25 \times 10^{-4} - R_{ps}^{\ 3}\right)}$ (8.49)

Also from equation (8.44)

$\boldsymbol{T_{p'a}}$ = torque contribution from the part of the aluminium alloy that becomes plastic

$$= \frac{2}{3}\pi\tau_{ypa}\left(0.035^3 - R_{pa}^{\ 3}\right)$$

$$= \frac{2}{3}\pi\left(100 \times 10^6\right)\left(0.05^3 - R_{ps}^{\ 3}\right)$$

i.e. $\boldsymbol{T_{p'a} = 209.4 \times 10^6 \times \left(4.288 \times 10^{-5} - R_{pa}^{\ 3}\right)}$ (8.50)

The elastic components of torque in the steel (T_{es}) and in the aluminium alloy (T_{ea}) can be calculated from elementary elastic theory, as follows.

From equation (8.42), $\frac{\tau_{yps}}{R_{ps}} = \frac{T_{es}}{J_{es}}$

hence $T_{es} = \dfrac{\tau_{yps} \times J_{es}}{R_{ps}} = \dfrac{\tau_{yps}}{R_{ps}} \times \dfrac{\pi\left(R_{ps}^{\ 4} - 0.035^4\right)}{2}$ (8.51)

$$= \frac{200 \times 10^6}{R_{ps}} \times \frac{\pi\left(R_{ps}^{\ 4} - 1.50 \times 10^{-6}\right)}{2}$$

i.e. $T_{es} = 314.2 \times 10^6 \times \dfrac{\left(R_{ps}^{\ 4} - 1.50 \times 10^{-6}\right)}{R_{ps}}$ (8.52)

Also $T_{ea} = \dfrac{\tau_{ypa} \times J_{ea}}{R_{pa}} = \dfrac{\tau_{ypa}}{R_{pa}} \times \dfrac{\pi R_{pa}^{\ 4}}{2}$

$$= \frac{100 \times 10^6}{R_{pa}} \times \frac{\pi R_{pa}^{\ 4}}{2}$$

i.e. $T_{ea} = 157.1 \times 10^6 \times R_{pa}^{\ 3}$ (8.53)

Now, the total torque is $T = T_{p's} + T_{p'a} + T_{es} + T_{ea}$

Hence, $36000 = 418.9 \times 10^6 \times \left(1.25 \times 10^{-4} - R_{ps}^{\ 3}\right)$

$$+ 209.4 \times 10^6 \times \left(4.288 \times 10^{-5} - R_{pa}^{\ 3}\right)$$

$$+ 314.2 \times 10^6 \times \frac{\left(R_{ps}^{\ 4} - 1.50 \times 10^{-6}\right)}{R_{ps}}$$

$$+ 157.1 \times 10^6 \times R_{pa}^{\ 3}$$

i.e. $\dfrac{36000}{1 \times 10^6} = 0.0524 - 418.9 R_{ps}^{\ 3}$

$$+ 8.979 \times 10^{-3} - 209.4 R_{pa}^{\ 3}$$

$$+ 314.2 R_{ps}^{\ 3} - \frac{4.713 \times 10^{-4}}{R_{ps}} + 157.1 R_{pa}^{\ 3}$$

i.e. $0.036 = 0.0614 - 104.7 R_{ps}^{\ 3} - 52.3 R_{pa}^{\ 3}$

$$- \frac{4.713 \times 10^{-4}}{R_{ps}}$$

i.e. $0.0254 = 104.7 R_{ps}^{\ 3} + 52.3 R_{pa}^{\ 3}$

$$+ \frac{4.713 \times 10^{-4}}{R_{ps}} \quad (8.54)$$

This problem is statically indeterminate; hence, it will be necessary to consider compatibility

i.e. from $\dfrac{\tau}{R} = \dfrac{G\theta}{l}$, $\theta = \text{constant} = \dfrac{\tau \times l}{G \times R}$

Therefore, $\theta = \dfrac{\tau_{yps}}{R_{ps}} \times \dfrac{l}{G_s} = \dfrac{\tau_{ypa}}{R_{pa}} \times \dfrac{l}{G_a}$

i.e. $\dfrac{200 \times 10^{-6}}{R_{ps} \times 7.7 \times 10^{10}} = \dfrac{100 \times 10^{-6}}{R_{pa} \times 2.6 \times 10^{10}}$

Therefore, $R_{ps} = \dfrac{200 \times 10^{-6} \times R_{pa} \times 2.6 \times 10^{10}}{100 \times 10^6 \times 7.7 \times 10^{10}}$

i.e. $\boldsymbol{R_{ps} = 0.675 R_{pa}}$ (8.55)

However, R_{ps} cannot be less than R_{pa}; therefore the aluminium alloy must be completely elastic, so that:

$\boldsymbol{R_{pa} = 0.035}$ **m and** $\boldsymbol{T_{p'a} = 0}$ (8.56)

Furthermore, it can no longer be assumed that the maximum shear stress in the aluminium alloy will reach τ_{ypa} so that it will be necessary to determine a new expression for T_{ea}

T_{ea} can be obtained in terms of R_{ps} by considering the compatibility condition that

$$\theta = \text{constant} = \frac{\tau \times l}{G \times R}$$

Therefore, $\dfrac{\tau_{yps}}{R_{ps}} \times \dfrac{l}{G_s} = \dfrac{\tau_a}{0.035} \times \dfrac{l}{G_a}$

where τ_a is the maximum shear stress in the aluminium alloy.

Therefore, $\tau_a = \dfrac{0.035 \times \tau_{yps} \times G_a}{G_s \times R_{ps}}$

i.e. $\tau_a = \dfrac{0.035 \times 200 \times 10^6 \times 2.6 \times 10^{10}}{7.7 \times 10^{10} \times R_{ps}}$

i.e. $\boldsymbol{\tau_a = \dfrac{2.364 \times 10^6}{R_{ps}}}$ (8.57)

and since $\dfrac{\tau_a}{R_{pa}} = \dfrac{T_{ea}}{J_{ea}}$

then $T_{ea} = \dfrac{\tau_a \times J_a}{0.035} = \dfrac{\tau_a}{0.035} \times \dfrac{\pi\left(0.035^4\right)}{2}$

i.e. $$T_{ea} = \frac{\left(\dfrac{2.364\times10^6}{R_{ps}}\right)}{0.035}\times\frac{\pi\left(0.035^4\right)}{2}$$

i.e. $$\boldsymbol{T_{ea} = \frac{159.2}{R_{ps}}} \qquad (8.58)$$

Hence, from equations (8.49), (8.51) and (8.58),

$$T = T_{p's} + T_{es} + T_{ea}$$

i.e. $$T = 418.9\times10^6\times\left(1.25\times10^{-4} - R_{ps}^{\ 3}\right)$$

$$+\frac{\tau_{yps}}{R_{ps}}\times\frac{\pi\left(R_{ps}^{\ 4} - 0.035^4\right)}{2} + \frac{159.2}{R_{ps}}$$

Hence, $$36000 = 418.9\times10^6\times\left(1.25\times10^{-4} - R_{ps}^{\ 3}\right)$$

$$+\frac{200\times10^6}{R_{ps}}\times\frac{\pi\left(R_{ps}^{\ 4} - 0.035^4\right)}{2} + \frac{159.2}{R_{ps}}$$

i.e. $$\frac{36000}{1\times10^6} = 0.0524 - 418.9R_{ps}^{\ 3}$$

$$+\,314.2\times10^6 R_{pa}^{\ 3} - \frac{471.4\times10^{-6}}{R_{ps}} + \frac{159.2\times10^{\ 6}}{R_{ps}}$$

Therefore, $$0 = 0.0164 - 104.7R_{ps}^{\ 3} - \frac{3.122\times10^{-4}}{R_{ps}}$$

i.e. $104.7R_{ps}^{\ 4} \times 0.0164R_{ps} + 3.122\times10^{-4} = 0$ (8.59)

Solving equation (8.59), the four roots of the equation were found (by the Newton-Raphson method) to be as follows

$$R_{ps} = 0.02\text{ m, }0.0448\text{ m and }(-0.0325 \pm 0.0475\text{j})$$

The root of interest is $\boldsymbol{R_{ps} = 0.0448}$ **m**

Hence, from equation (8.58) $T_{ea} = \dfrac{159.2}{R_{ps}} = \dfrac{159.2}{0.0448}$

i.e. $\boldsymbol{T_{ea}} = 3.554\times10^3\,\text{N m} = \mathbf{3.553\ kN\ m}$

The angle of twist,

$$\theta = \frac{Tl}{GJ} = \frac{3.553\times10^3\times0.7}{2.6\times10^{10}\times\left(\dfrac{\pi\times0.035^4}{2}\right)} = 0.0406\text{ rad}$$

i.e. $$\theta = 0.0406\times\frac{180^\circ}{\pi} = \mathbf{2.326^\circ}$$

(c) To determine the residual angle of twist θ_R

On release of the torque of 36000 N m, the shaft is assumed to behave elastically. Thus (see Figure 8.32):

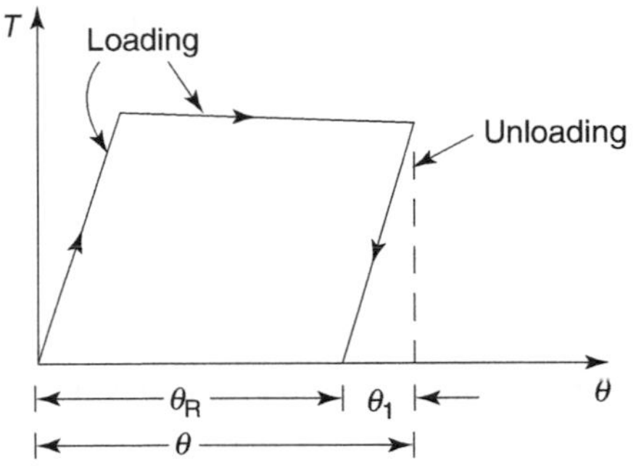

Figure 8.32 T–θ relationship for the compound shaft

$$\theta_1 = \frac{Tl}{GJ} = \frac{T_s l}{G_s J_s} = \frac{T_a l}{G_a J_a}$$

Therefore,

$$T_s = \frac{T_a G_s J_s}{G_a J_a} = \frac{T_a\times7.7\times10^{10}\times7.46\times10^{-6}}{2.6\times10^{10}\times2.357\times10^{-6}}$$

i.e. $\boldsymbol{T_s = 9.373T_a}$

and since $T = T_s + T_a$ then $T = 9.373T_a + T_a$

i.e. $36000 = 10.373T_a$

from which, $\boldsymbol{T_a} = \dfrac{36000}{10.373} = \mathbf{3470\ N\ m}$

The angle of twist,

$$\theta_1 = \frac{T_a l}{G_a J_a} = \frac{3470\times0.7}{2.6\times10^{10}\times2.357\times10^{-6}} = \mathbf{0.0396\ rad}$$

i.e. $$\theta = 0.0396\times\frac{180^\circ}{\pi} = \mathbf{2.269^\circ}$$

and **the residual angle of twist,** θ_R is given by

$$\boldsymbol{\theta_R} = \theta - \theta_1 = 2.326^\circ - 2.269^\circ = \mathbf{0.057^\circ}$$

(d) To determine the fully plastic torsional resistance of the compound shaft (T_p)

The steel tube will first become fully plastic and then it will rotate, until the aluminium alloy becomes fully plastic.

From equation (8.44), $$T_{pa} = \frac{2}{3}\pi\times\tau_{ypa}\left(\frac{0.07}{2}\right)^3$$

$$= \frac{2}{3}\pi \times 100 \times 10^6 \times \left(\frac{0.07}{2}\right)^3$$

i.e. $\quad T_{pa} = \mathbf{8980\ N\ m}$

Similarly, $T_{ps} = \frac{2}{3}\pi \times \tau_{yps}\left(0.05^3 - 0.035^3\right)$

i.e. $\quad = \frac{2}{3}\pi \times 200 \times 10^6 \times \left(0.05^3 - 0.035^3\right)$

i.e. $\quad T_{ps} = \mathbf{34400\ N\ m}$

Therefore, the total fully plastic moment of resistance is

$$T_p = 8980 + 34400 = \mathbf{43380\ N\ m}$$

(e) Torque ratio $= \frac{T_p}{T_{yp}} = \frac{\mathbf{43380}}{\mathbf{33057}} = \mathbf{1.312}$

Now try the following Practice Exercise

Practice Exercise 39. Elastic-plastic torsion of circular-section shafts

1. A circular-section shaft of diameter 0.2 m and length 1 m is subjected to a torque that causes an angle of twist of 3.5°. Determine this torque and the residual angle of twist on removal of this torque. Assume that $G = 7.7 \times 10^{10}\ \text{N/m}^2$ and $\tau_{yp} = 180\ \text{MN/m}^2$

   [0.372 MN m, 2.161°]

2. Determine the maximum possible torque that the cross-section of Figure 8.33 can withstand, given that yield shear stress = 170 MPa.

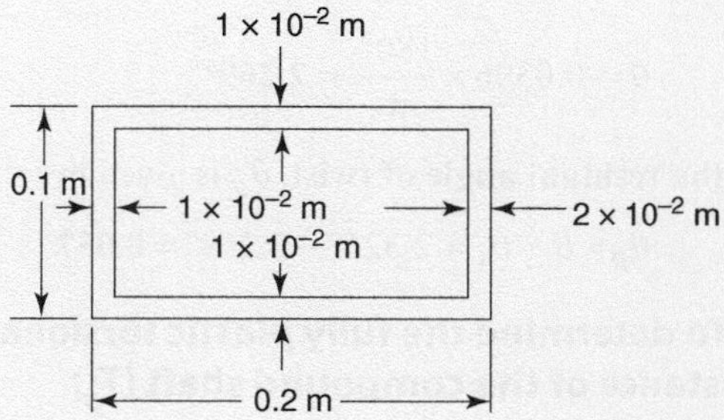

Figure 8.33 Closed tube

   What would be the angle of twist per unit length if $G = 7.7 \times 10^{10}$ Pa?

   [68000 N m, 1.739°/m]

3. Determine the maximum possible torque that the cross-section of Figure 8.34 can withstand given that yield shear stress = 170 MPa.

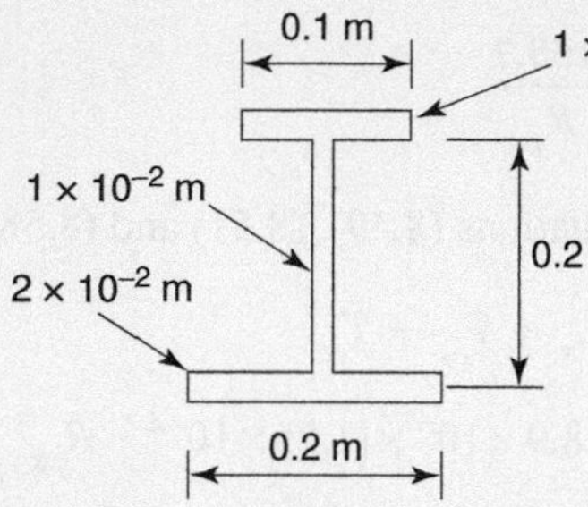

Figure 8.34 Open section

   What would be the angle of twist per unit length if $G = 7.7 \times 10^{10}$ Pa?

   [5383 N m, 6.325°/m]

4. A circular-section shaft of diameter 0.25 m and length 1 m is subjected to a torque that causes an angle of twist of 3.5°. Determine this torque and the residual angle of twist on removal of this torque. Assume that $G = 7.7 \times 10^{10}\ \text{N/m}^2$ and $\tau_{yp} = 180\ \text{MN/m}^2$

   [0.731 MN m, 1.58°]

5. Determine the maximum possible torque that the cross-section of Figure 8.35 can withstand, given that yield shear stress = 170 MPa.

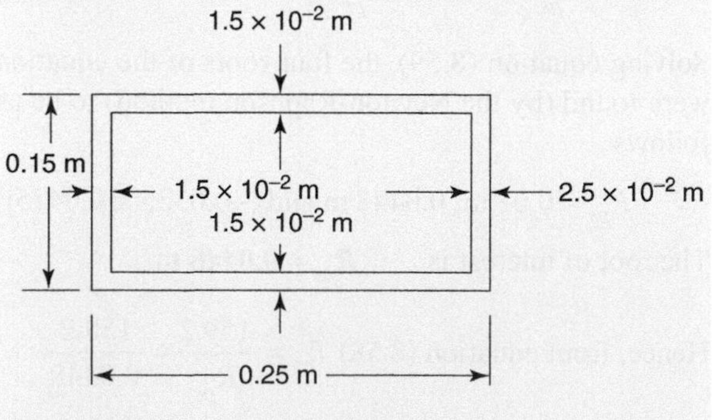

Figure 8.35 Closed tube

   What would be the angle of twist per unit length if $G = 7.7 \times 10^{10}$ Pa?

   [191.25 kN m, 3.92°/m]

For fully worked solutions to each of the problems in Exercises 37 to 39 in this chapter, go to the website:

www.routledge.com/cw/bird

Multiple-Choice Questions Test 3

This test covers the material in Chapters 3 to 8. All questions have only one correct answer (answers on page 496). All answers are rounded to 3 significant figures.

1. Choose the correct free body diagram from Figure M3.2 for the cantilever shown in Figure M3.1.

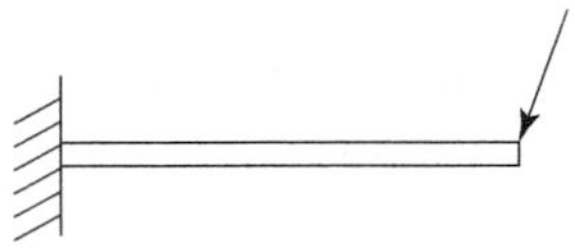

Figure M3.1 Cantilever

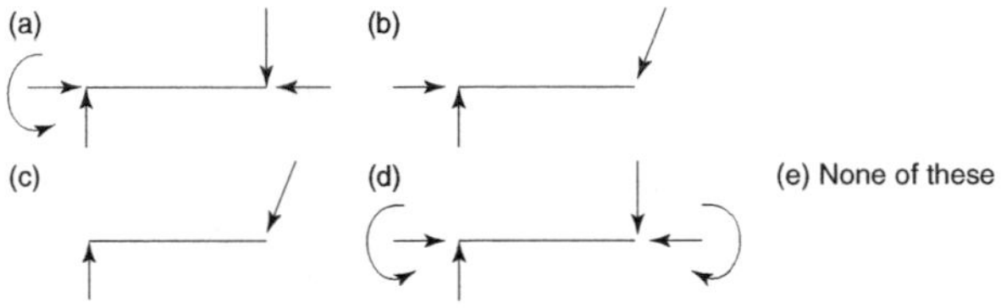

Figure M3.2

2. Choose the correct free body diagram from Figure M3.4 for the simply supported beam shown in Figure M3.3.

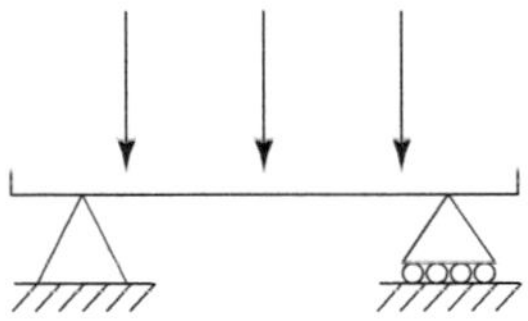

Figure M3.3 Simply supported

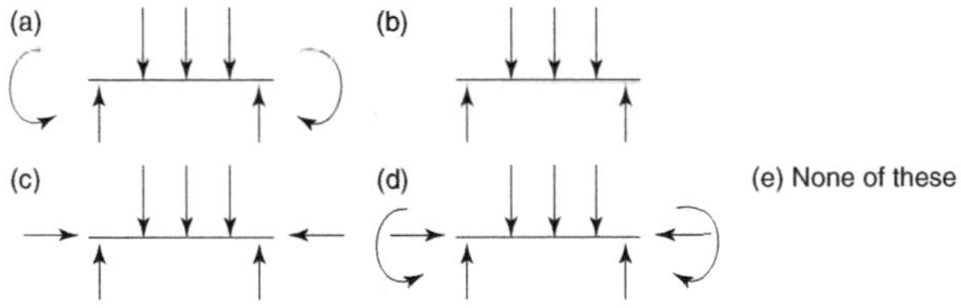

Figure M3.4

3. Choose the correct free body diagram from Figure M3.6 for the propped cantilever shown in Figure M3.5.

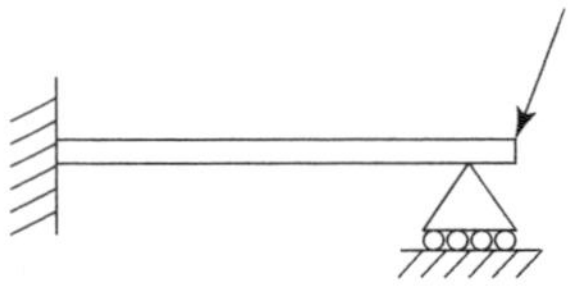

Figure M3.5 Propped Cantilever

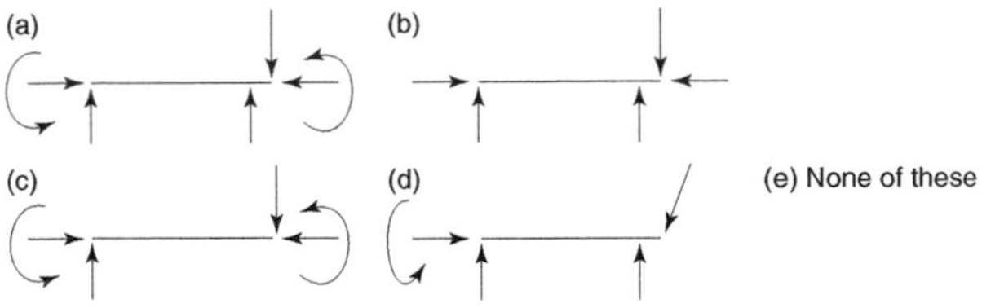

Figure M3.6

4. The reaction in the x-direction at joint A in Figure M3.7 is:
   (a) – 5.77 kN (b) 5.77 kN
   (c) 5 kN (d) 0 kN
   (e) None of these

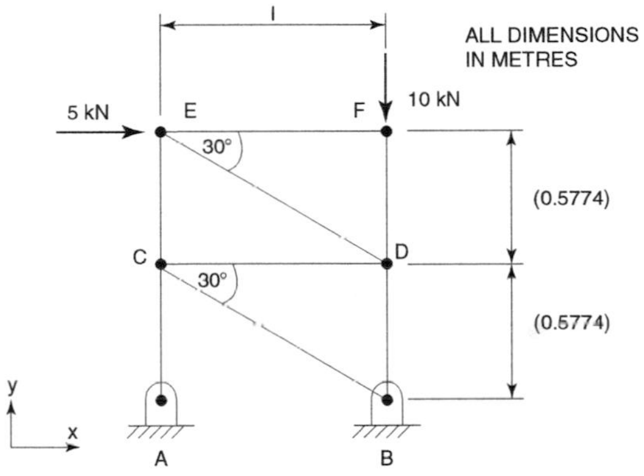

Figure M3.7

5. The reaction in the y-direction at joint A in Figure M3.7 is:
   (a) – 5.77 kN (b) 5.77 kN
   (c) 5 kN (d) 0 kN
   (e) None of these

6. The reaction in the x-direction at joint B in Figure M3.7 is:
   (a) – 5.774 kN (b) 5.774 kN
   (c) – 5 kN (d) 0 kN
   (e) None of these

7. The reaction in the y-direction at joint B in Figure M3.7 is:
   (a) 15.8 kN (b) – 5.78 kN
   (c) – 5 kN (d) 0 kN
   (e) None of these

8. TWO methods/techniques for analysing loads within a pin-jointed framework are:
   (a) Pins & Points
   (b) Joints & Points
   (c) Joints & Sections
   (d) Nodes & Elements
   (e) None of these

9. What is the load in member EF in Figure M3.7?
   (a) – 5.77 kN (b) 5.77 kN
   (c) 5 kN (d) 0 kN
   (e) None of these

10. What is the load in member ED in Figure M3.7?
   (a) – 5.77 kN (b) 5.77 kN
   (c) 5 kN (d) 0 kN
   (e) None of these

11. Forces of 7 N and 8 N as well as a 20 N m couple act on the rigid lever as shown in Figure M3.8. The resulting moment about point O has a magnitude of:
   (a) 180 N mm (b) 260 N mm
   (c) 1220 N mm (d) 350 N mm
   (e) None of these

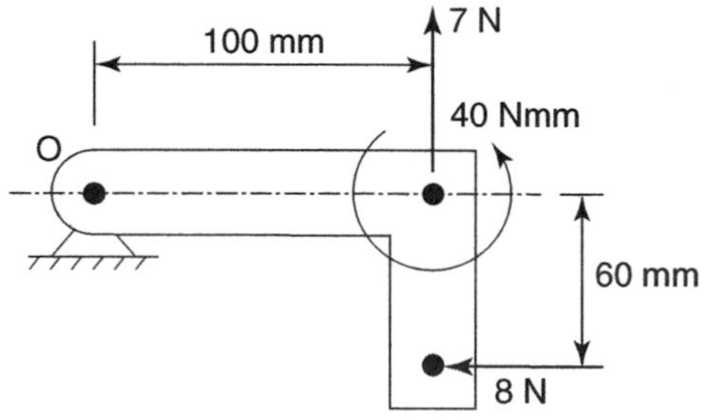

Figure M3.8

12. Forces of 700 N and 500 N act at point A as shown in Figure M3.9. Choose the diagram in Figure M3.10 which represents the resultant force at point A

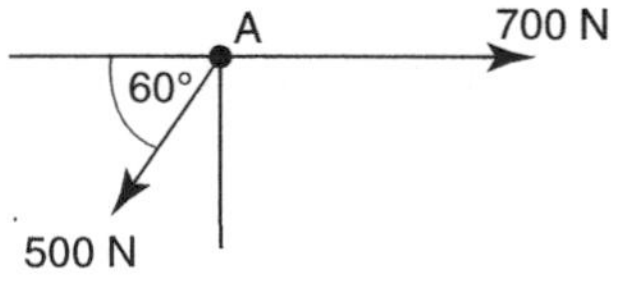

Figure M3.9

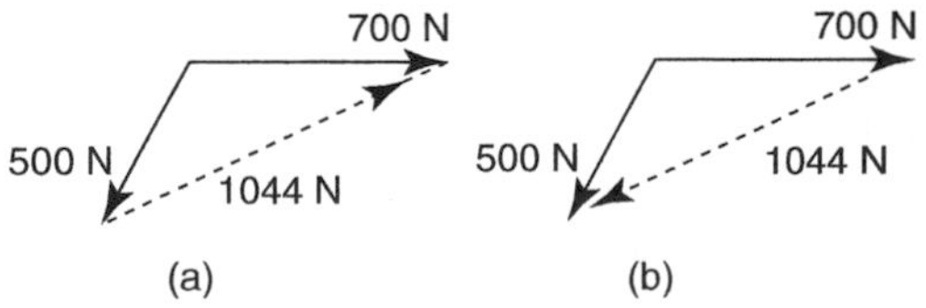

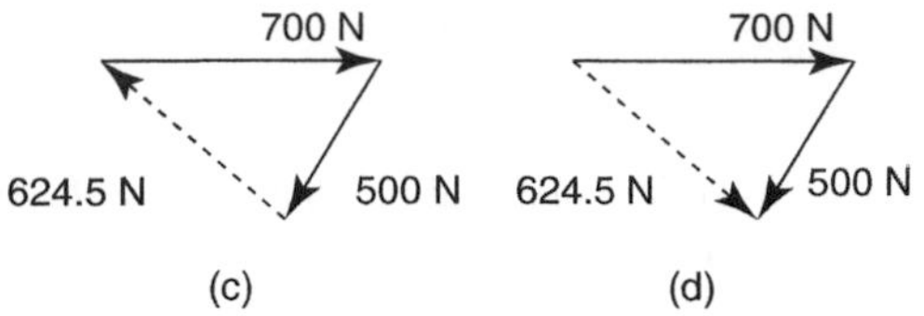

Figure M3.10

13. The *x* and *y* components of the 300 N force shown in Figure M3.11 are:

	x-component	*y*-component
(a)	246	172
(b)	172	246
(c)	–246	172
(d)	–172	246
(e)	None of these	

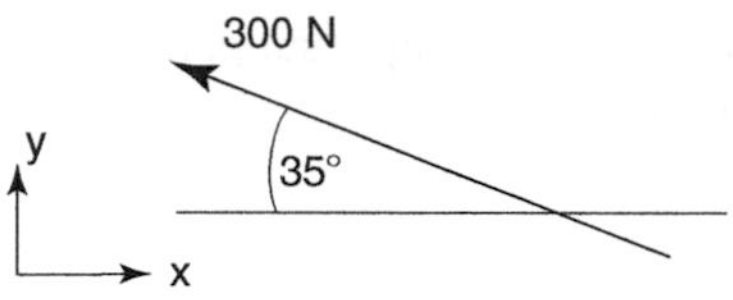

Figure M3.11

14. A rigid body is pivoted about point O and is acted on by two external forces as shown in Figure M3.12. The value of the moment M that must be applied at point O to maintain rotational equilibrium about O is:

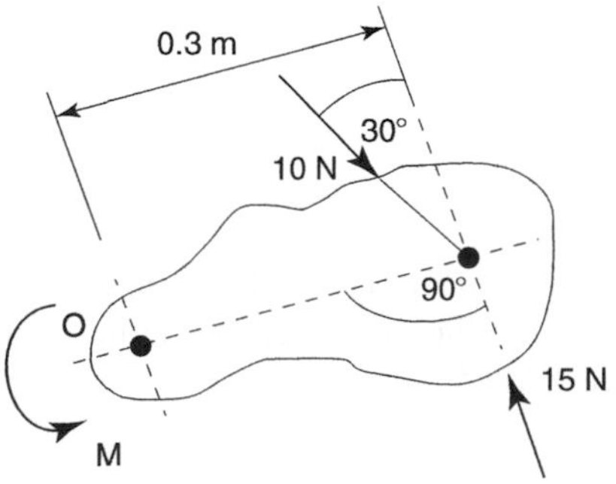

Figure M3.12

   (a) 1.50 N m (b) 1.90 N m
   (c) 3.00 N m (d) 4.50 N m
   (e) None of these

15. The resultant of the forces shown in Figure M3.13 is:

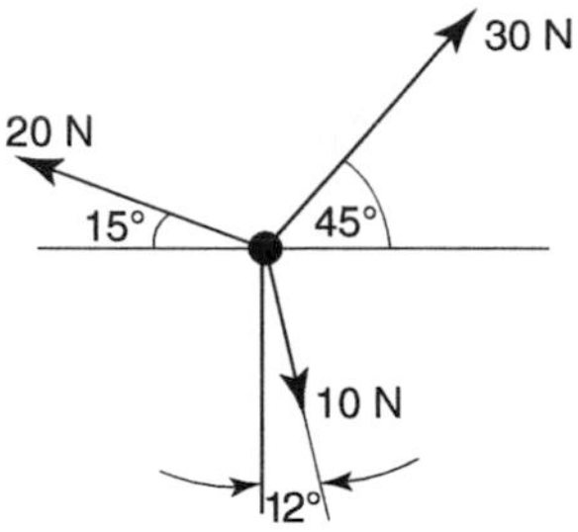

Figure M3.13

(a) 55.9 N (b) 17.1 N
(c) 6.25 N (d) 45.5 N
(e) None of these

16. Choose the correct free body diagram from Figure M3.15 for the portion of the beam shown to the left of the section XX in Figure M3.14.

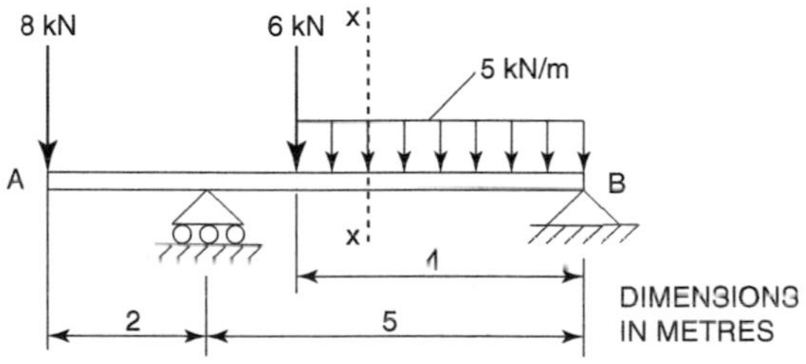

Figure M3.14

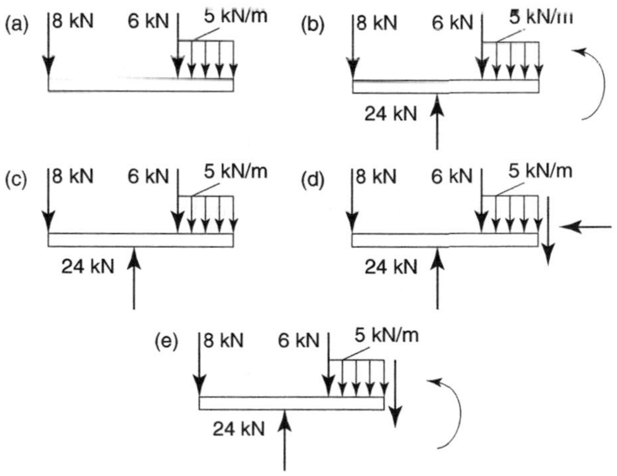

Figure M3.15

17. The free body diagram for part of a rigid beam is as shown in Figure M3.16. For the beam to be in equilibrium the magnitude of the moment M must be:

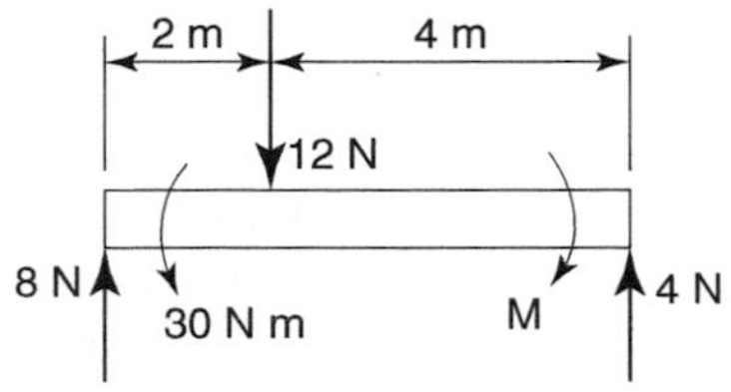

Figure M3.16

(a) 30.0 N m (b) 6.0 N m
(c) 6.6 N m (d) 15.5 N m
(e) None of these

18. A 50 mm gauge length is marked on a tensile test specimen which has a working diameter of 5 mm. A force of 8 kN is applied to the specimen and, with that load still applied, the gauge length is found to be 50.20 mm.
The modulus of the material is:
(a) 2.04 GPa (b) 102 GPa
(c) 200 GPa (d) 85.5 GPa
(e) None of these

19. A particular alloy has a limit of proportionality of 60 MPa, a yield strength of 130 MPa and ultimate tensile strength of 215 MPa.

An axial tensile force of 40 kN is applied to a 25 mm diameter rod of that material. It can be expected that:

(a) The rod will break in tension.
(b) The rod will return to exactly its original length when the load is removed.
(c) The rod will not return to its original length when the load is removed. It will be slightly stretched.
(d) The rod will fail in shear.
(e) There is not enough information to conclude any of the above.

20. A force of 14 kN is only just sufficient to punch a rectangular hole in an aluminium-alloy sheet. The rectangular hole is 10 mm long by 6 mm wide, and the aluminium-alloy sheet is 2 mm thick. The ultimate shear stress of the aluminium alloy is:
(a) 219 MPa (b) 233 MPa
(c) 275 MPa (d) 200 GPa
(e) 255 MPa

21. A rod has a diameter of 8 mm. It is brazed into a 5 mm thick plate to form a bracket, as shown in

Figure M3.17. The brazing process results in a thin layer of braze being formed around the outside of the rod and inside of the hole in the plate. The braze joins to both and creates the joint. The bracket can be considered rigid compared to the rod and the brazed joint. If the material of the rod has an ultimate tensile strength of 150 MPa and the braze has an ultimate shear strength of 85 MPa. When the load F is steadily increased with the bracket held fixed, failure will occur when:

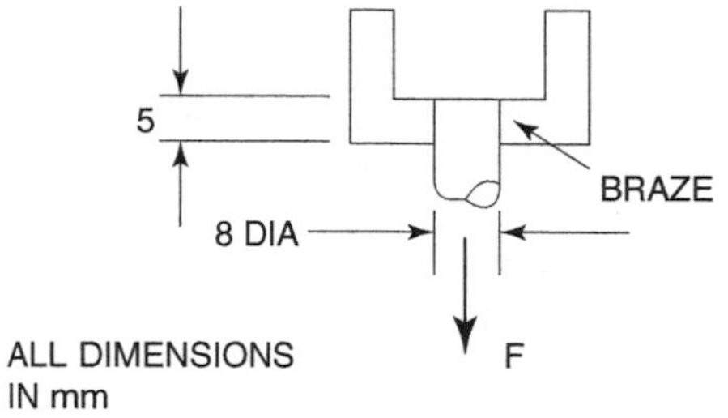

Figure M3.17

(a) The rod breaks when load F = 7.54 kN
(b) The rod breaks when load F = 4.27 kN
(c) The braze shears when load F = 18.8 kN
(d) The braze shears when load F = 7.54 kN
(e) The braze shears when load F = 4.27 kN

22. A 400 mm long stainless steel bar has a diameter of 25 mm. It is welded to rigid end supports, fixing it at each end. Its temperature is then raised by 60°C. The thermal expansion coefficient (α) and the Young's modulus (E) for stainless steel are 17.3×10^{-6} /°C and 185 GPa respectively.
The supports at the ends of the bar can be considered to be rigid and unaffected by the change in temperature, so that the length of the bar remains constant. The thermal stress will be:
(a) 92.4 MPa (b) 0 MPa
(c) − 192 MPa (d) − 92.4 MPa
(e) 192 MPa

23. A pin-jointed framework is shown in Figure M3.18. The load in member FG is:
(a) 30.0 kN (b) 38.2 kN
(c) 20.0 kN (d) 45.8 kN
(e) 57.3 kN

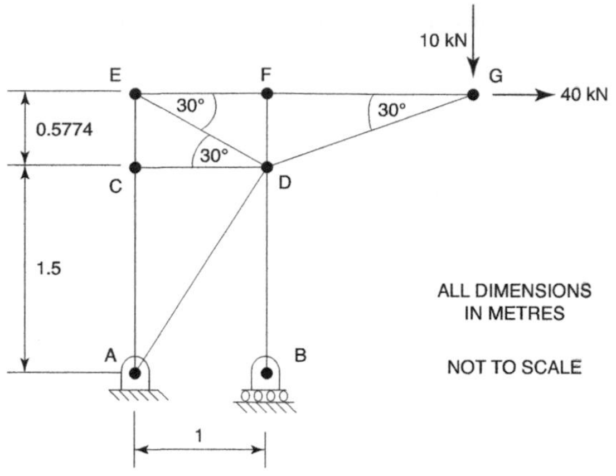

Figure M3.18

24. Correct, to 3 significant figures, the x and y components of the 120 N force shown in Figure M3.19 are:

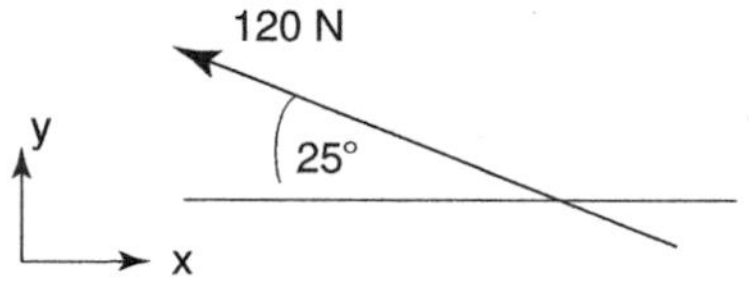

Figure M3.19

	x-component (N)	y-component (N)
(a)	− 109	− 50.7
(b)	50.7	109
(c)	− 50.7	109
(d)	− 109	50.7
(e)	109	50.7

25. The free body diagram for part of a rigid beam AB is as shown in Figure M3.20. For the beam to be in equilibrium the magnitude of the moment M must be:

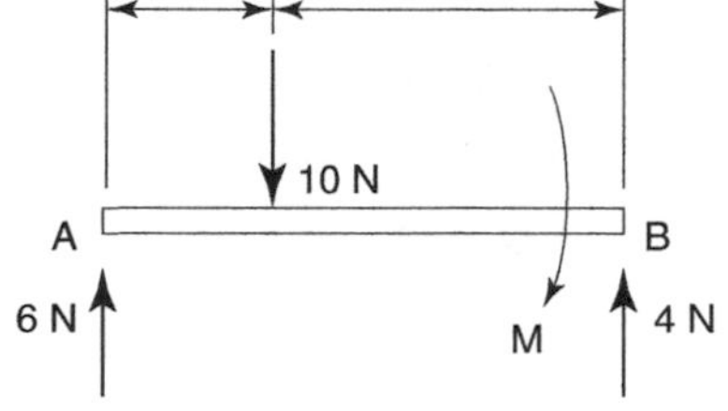

Figure M3.20

(a) 10.0 N m (b) 6.0 N m
(c) 4.0 N m (d) 8.0 N m
(e) − 6 N m

Revision Test 3 Specimen examination questions for Chapters 3 to 8

This Revision Test covers the material in Chapters 3 to 8. *The marks for each question are shown in brackets at the end of each question.*

1. Calculate and plot the bending moment and shearing force diagrams for the simply supported beam shown in Figure RT3.1. (22)

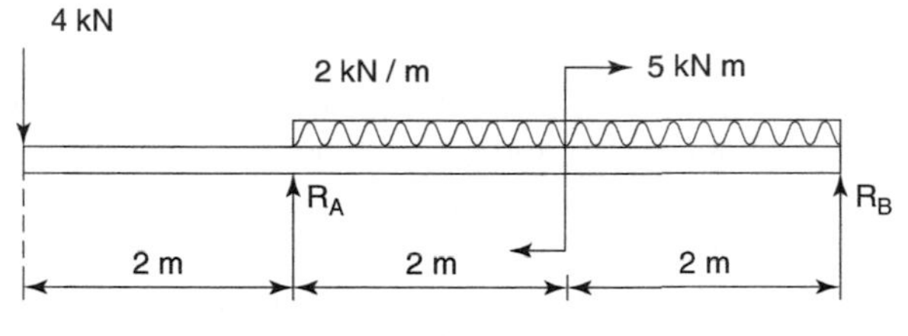

Figure RT3.1

2. Calculate and plot the bending moment and shearing force diagrams for the beam shown in Figure RT3.2. (15)

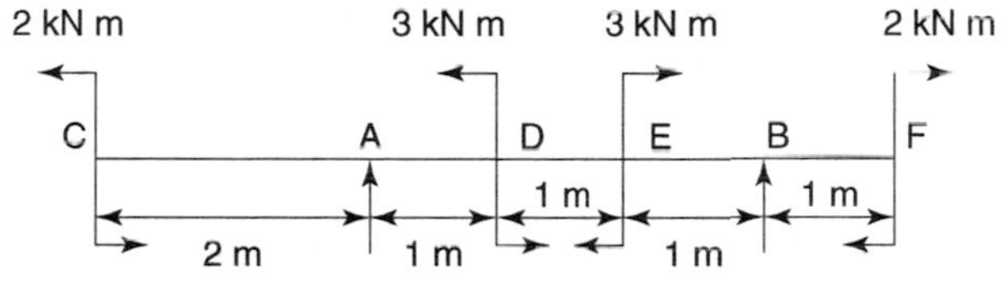

Figure RT3.2

3. Determine the forces in the plane pin-jointed truss of Figure RT3.3, due to the applied loads at joint A. (20)

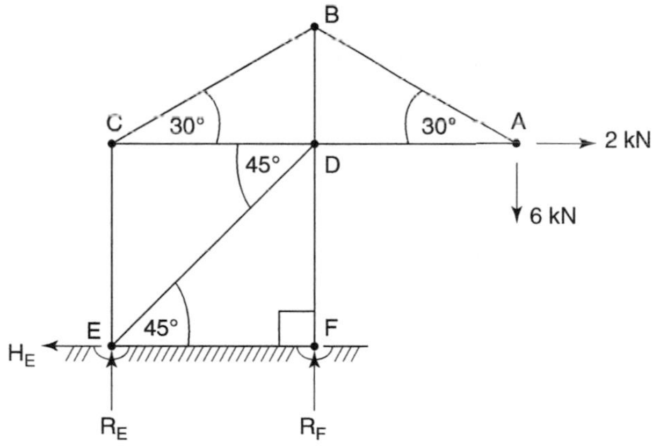

Figure RT3.3

4. Calculate the maximum stress in the tapered bar of circular cross-section shown in Figure RT3.4. If the bar is made of steel, with a Young's modulus of 1.9×10^{11} N/m^2, determine the extension of the bar due to this load. (11)

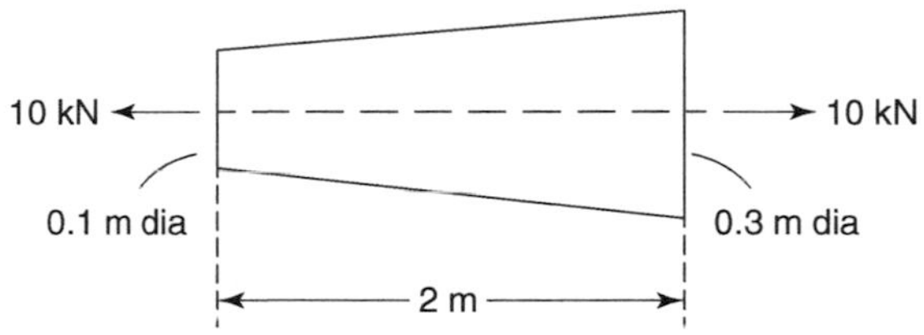

Figure RT3.4

5. An electrical cable consists of a copper core, tightly surrounded by a steel sheath cover, so that the cable acts as a compound bar. If the cable hangs vertically down a mine shaft of depth 0.5 km, what will be the maximum stresses in the two materials, given that:
   External diameter of copper core = 5 cm,
   External diameter of steel sheath = 8 cm
   For copper: Young's modulus, $E = 1 \times 10^{11}$ Pa, density, $\rho = 8960$ kg/m^3
   For steel: Young's modulus, $E = 1.9 \times 10^{11}$ Pa, density, $\rho = 7860$ kg/m^3
   $g = 9.81$ m/s^2 (12)

6. Determine the second moment of area about the neutral axis (NA) of the cross-section shown in Figure RT3.5. (12)

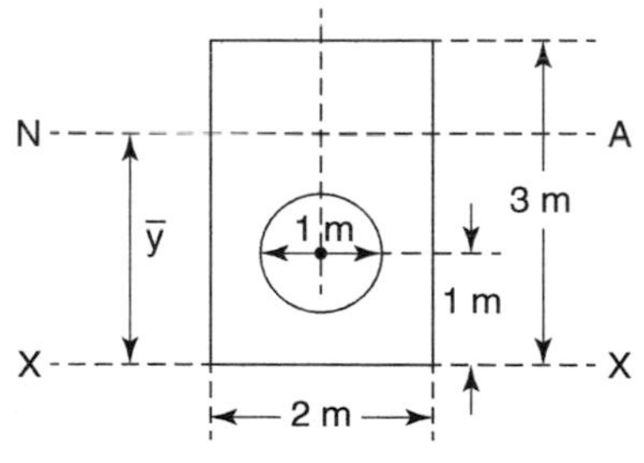

Figure RT3.5

7. Determine the second moment of area about the neutral axis (NA) of the '*I*' beam shown in Figure RT3.6. (16)

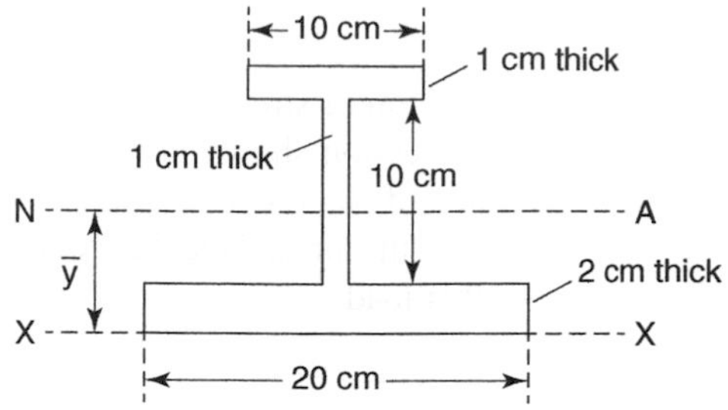

Figure RT3.6

8. Determine the second moment of area about the neutral axis (NA) of the cross-section shown in Figure RT3.7. (12)

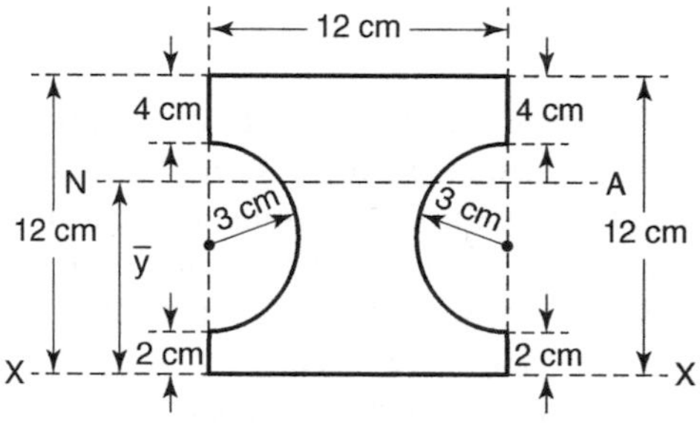

Figure RT3.7

9. Determine the deflections at the points 'C' and 'D' for the constant section laterally loaded beam of Figure RT3.8. Assume that: $E = 2 \times 10^{11}$ Pa and $I = 5 \times 10^{-8}\ \text{m}^4$ (22)

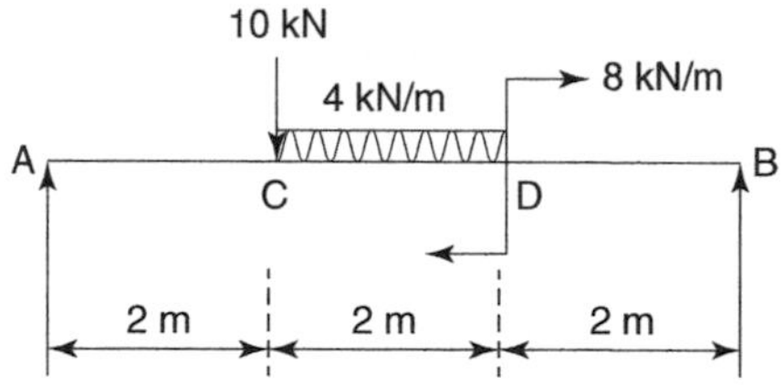

Figure RT3.8

10. Obtain an expression for the lateral deflection '*y*' at any distance '*x*' from the left end of the uniform section beam of Figure RT3.9, in terms of '*E*' and '*I*', the Young's modulus and the second moment of area of the section respectively. Hence, or otherwise, obtain an expression for the beam's maximum deflection. (28)

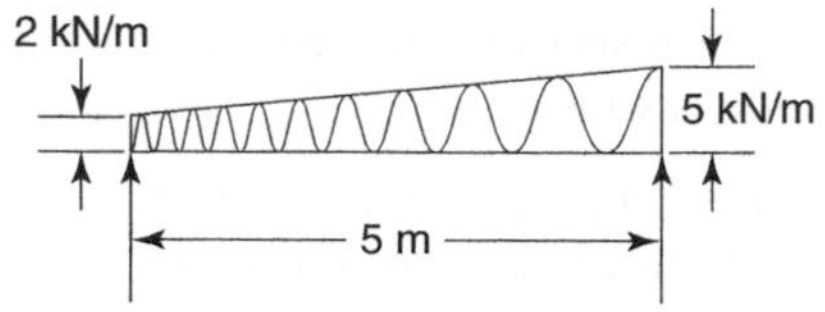

Figure RT3.9

11. Determine the output torque of an electric motor which supplies 800 kW at 20 rev/s.
    (a) If this torque is transmitted through a circular-section tube of external diameter 25 cm, determine the internal diameter if the maximum permissible shear stress in the shaft is 40 MPa.
    (b) If the shaft is to be connected to another similar shaft, via a flanged coupling, determine a suitable bolt diameter, if 6 bolts are used on a pitch circle diameter of 30 mm, where the maximum permissible shearing stresses for the bolt's material is 50 MPa. (12)

12. A compound shaft consists of two hollow shafts, joined together in series, and subjected to an intermediate torque *T*, as shown in Figure RT3.10. If the shaft on the left is made of steel and the shaft on the right is made of aluminium alloy, determine the maximum permissible value of *T*, given that:

    For steel: Modulus of rigidity, $G = 7.8 \times 10^{10}$ Pa
    Permissible shear stress, $\tau = 120$ MPa
    For aluminium alloy: Modulus of rigidity, $G = 2.8 \times 10^{10}$ Pa
    Permissible shear stress, $\tau = 80$ MPa (18)

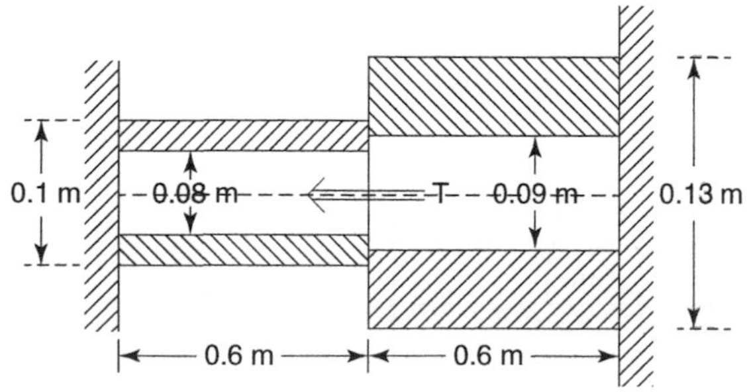

Figure RT3.10

For lecturers/instructors/teachers, fully worked solutions to each of the problems in Revision Test 3, together with a full marking scheme, are available at the website:
www.routledge.com/cw/bird

Multiple-Choice Questions Test 4

This test covers the material in Chapters 3 to 8. All questions have only one correct answer (answers on page 496). All answers are rounded to 3 significant figures.

1. The distance from the base to the neutral axis of a beam with the cross-section shown in Figure M4.1 is:

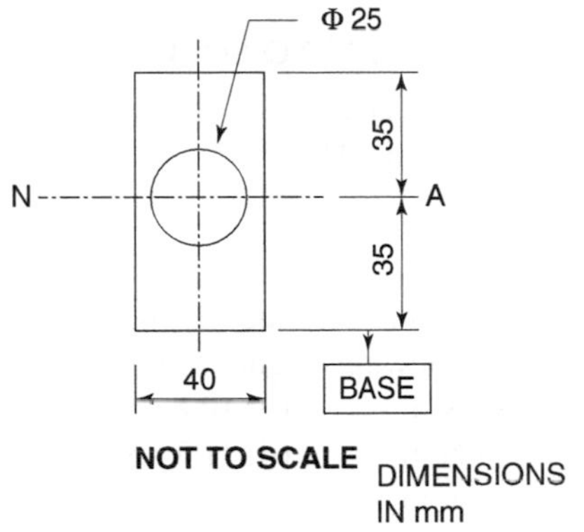

Figure M4.1

(a) 30 mm (b) 40 mm (c) 35 mm (d) 45 mm
(e) None of these

2. The second moment of area about its neutral axis, of a beam with the cross-section shown in Figure M4.1 is:
(a) 1.32×10^6 mm^4 (b) 1.45×10^6 mm^4
(c) 1.26×10^6 mm^4 (d) 1.12×10^6 mm^4
(e) None of these

3. The distance from the base to the neutral axis of a beam with the cross-section shown in Figure M4.2 is:

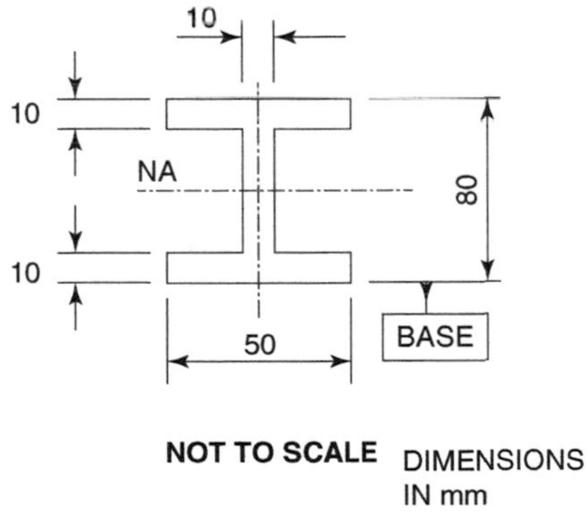

Figure M4.2

(a) 30 mm (b) 40 mm (c) 35 mm (d) 45 mm
(e) None of these

4. The second moment of area about its neutral axis, of a beam with the cross-section shown in Figure M4.2 is:
(a) 1.17×10^6 mm^4 (b) 1.29×10^6 mm^4
(c) 1.35×10^6 mm^4 (d) 1.41×10^6 mm^4
(e) None of these

5. The distance from the base to the neutral axis of the beam shown in Figure M4.3 is:
(a) 79.4 mm (b) 77.1 mm
(c) 70.6 mm (d) 66.5 mm
(e) None of these

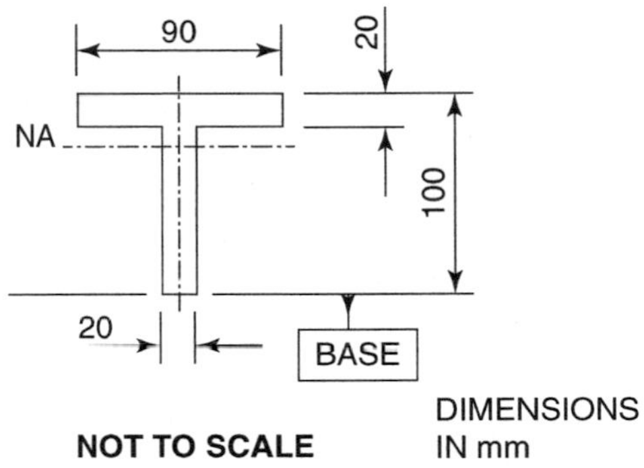

Figure M4.3

6. The location of the neutral axis (NA) of a tee beam is given in Figure M4.4. The second moment of area of this beam about its neutral axis is:

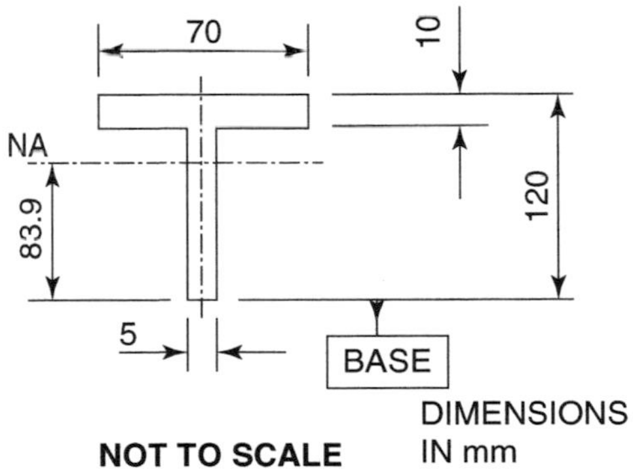

Figure M4.4

(a) 1.67×10^6 mm^4 (b) 1.22×10^6 mm^4
(c) 1.05×10^6 mm^4 (d) 1.48×10^6 mm^4
(e) None of these

7. The maximum shear force on the light beam AB shown in Figure M4.5 is:
(a) 6.75 kN (b) 3.0 kN
(c) 10.0 kN (d) 3.67 kN
(e) None of these

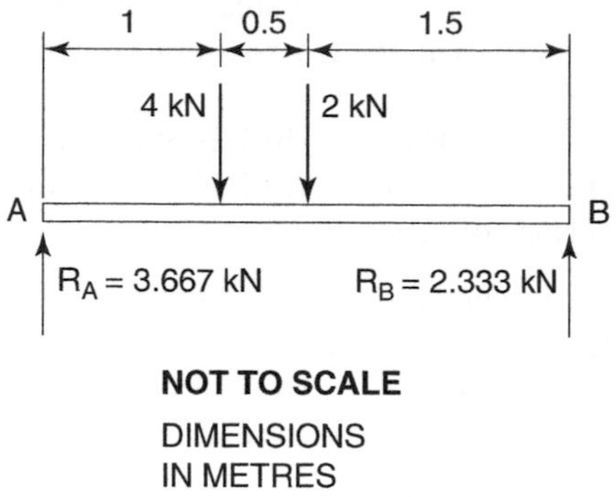

Figure M4.5

8. The maximum bending moment on the light beam AB shown in Figure M4.5 is:
(a) 3.67 kN m (b) 2.33 kN m
(c) 5.3 kN m (d) 2.67 kN m
(e) None of these

9. The maximum shear force on the light beam AB shown in Figure M4.6 is:
(a) 10.8 N (b) 14.2 N
(c) 28.3 N (d) 17.8 N
(e) None of these

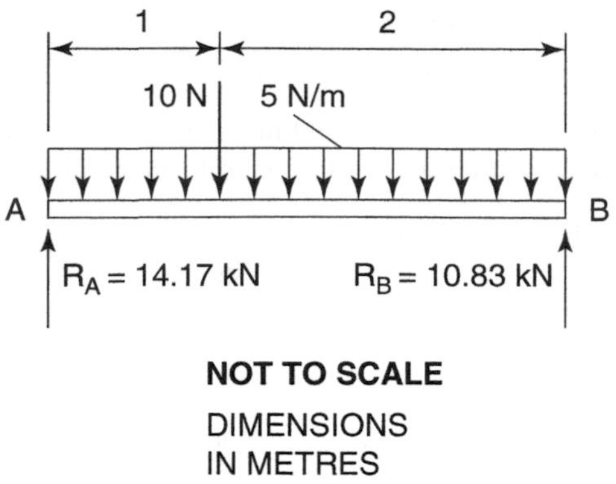

Figure M4.6

10. The maximum bending moment on the light beam AB shown in Figure M4.6 is:
(a) 29.3 kN m (b) 10.8 kN m
(c) 14.2 kN m (d) 11.7 kN m
(e) None of these

11. A drive shaft is made from a steel with a shear modulus of 76 GPa. The shaft is hollow and its external diameter is stepped as shown in Figure M4.7. When the shaft is transmitting a torque of 285 N m the maximum shear stress in the shaft will be:
(a) 13.3 MPa (b) 58.4 MPa
(c) 33.2 MPa (d) 22.7 MPa
(e) 26.6 MPa

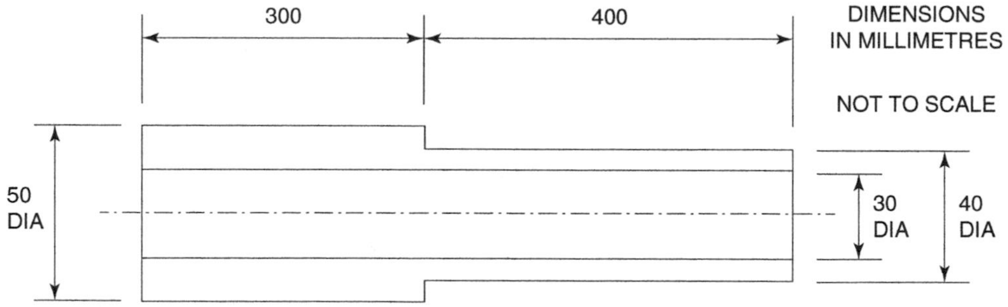

Figure M4.7

12. Figure M4.8 shows the cross-section of a beam, together with the location of its horizontal neutral axis relative to its base (0-0 shown in Figures M4.8).

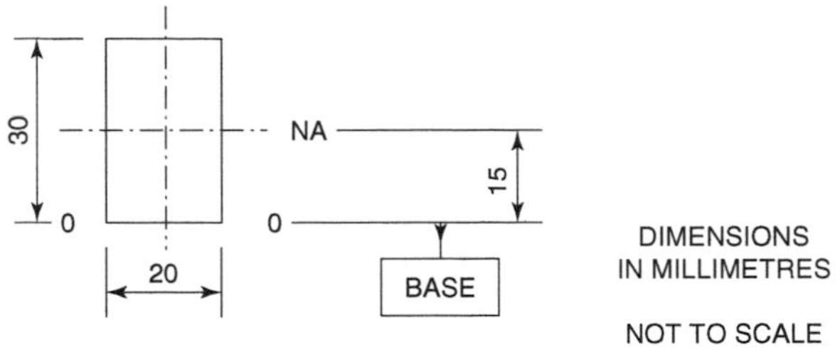

Figure M4.8

The second moment of area of this rectangular beam about its base, I_{00} is:
(a) 54000 mm^4 (b) 160000 mm^4
(c) 45600 mm^4 (d) 45000 mm^4
(e) 180000 mm^4

13. A light beam AD carries a point load at B and a pure moment at point C, as shown in Figure M4.9. The beam is simply supported and the reactions at A and at D are 30 kN and 10 kN respectively (both forces acting upwards). The shear force acting on the beam at point C (i.e. when $x = 2$ m) is:

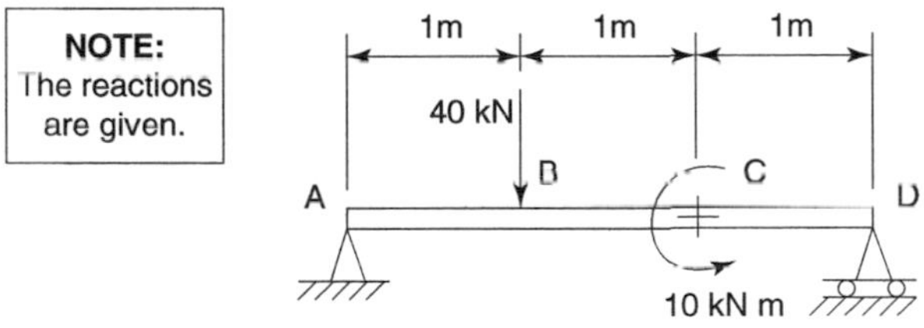

Figure M4.9

(a) −20 kN (b) 30 kN
(c) 20 kN (d) −10 kN
(e) −40 kN

14. The free body diagram of a light beam AB subject to a load and a pure moment is shown in Figure M4.10. Using the reactions given, the bending moment at mid-span of this beam is:
(a) 7 kN m (b) 16 kN m
(c) 21 kN m (d) 12 kN m
(e) 9 kN m

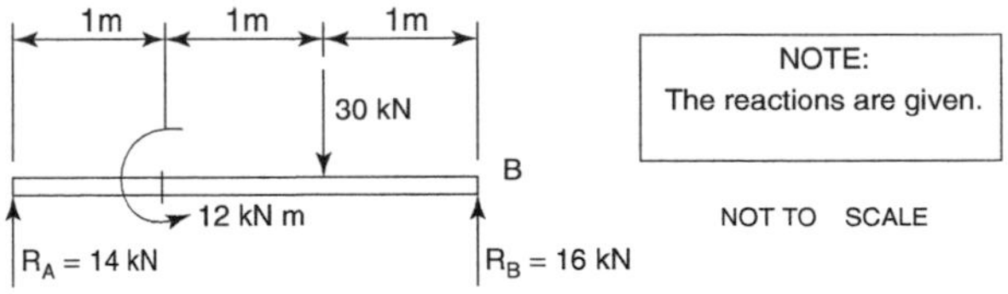

Figure M4.10

15. The free body diagram of a light beam AB subject to a uniformly distributed load is shown in Figure M4.11 and the reactions at A and B are given. The value of x, measured from A, where the maximum bending moment occurs will be:
(a) 2.00 (b) 1.33 m (c) 4.00 m (d) 2.20 m
(e) 1.00 m

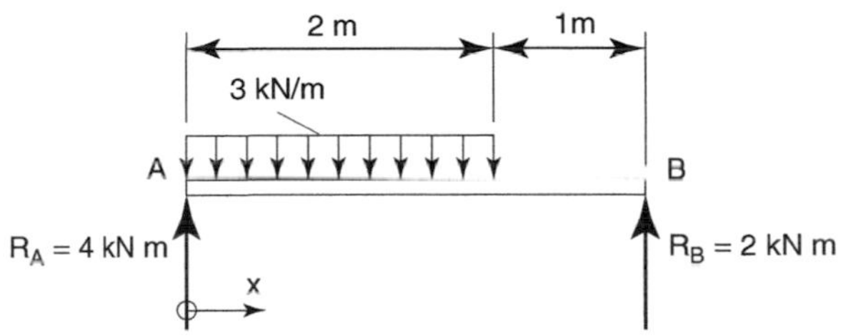

Figure M4.11

12. Figure M4.8 shows the cross-section of a beam, together with the location of its horizontal neutral axis relative to its base (0-0) shown in Figures M4.8.

[illegible]

(a) 1880 mm[illegible] (b) [illegible]

(c) 1780 mm[illegible] (d) [illegible]

(e) [illegible]00 mm[illegible]

13. [illegible]

14. The beam [illegible] to a load and [illegible] moment [illegible] M4.10. Using the [illegible]

(a) [illegible] kN m (b) [illegible]

(c) 1.21 kN m (d) 1.32 kN m

(e) 1.25 m

[illegible]

15. The [illegible] diagram [illegible]

Chapter 9

Complex stress and strain

Why it is important to understand: **Complex stress and strain**

Two- and three-dimensional systems of stress and strain are often met in the field of structural analysis. In the case of two-dimensional systems of stress and strain, such cases are often met on the decks and bulkheads of ships, together with those on rudders and propeller blades. Similar cases are met with aircraft and submarine structures, on the fuselages and wings of aircraft, and the pressure hulls and bulkheads of submarines. In these cases, even though strain gauges can be used to measure a two-dimensional system of stress or strain, very often the strain gauges cannot readily determine the maximum or principal stresses and strains, because they lie on angles which may be at a different angle to the measured strains. Two-dimensional and principal stresses also occur on boilers, condensers, nuclear reactors silos and so on.

At the end of this chapter you should be able to:

- understand principal stresses and strains, together with the maximum shear strain
- calculate principal stresses and strains, and maximum shear strains
- solve problems involving combined bending and torsion of circular section shafts
- understand pure shear, plane stress and plane strain
- understand the use and applications of strain gauge rosettes
- calculate principal stresses and principal strains from strain gauge readings
- understand the basic equations involving materials made from composites

Mathematical references

In order to understand the theory involved in this chapter on **complex stress and strain**, knowledge of the following mathematical topics is required: ***trigonometry, trigonometric identities, double angles, standard differentiation, partial differentiation and matrices.*** If help/revision is needed in these areas, see page 49 for some textbook references.

9.1 Introduction

A typical system of two-dimensional complex stresses, acting on an infinitesimally small rectangular lamina, is shown in Figure 9.1, where the direct stresses σ_x and σ_y are accompanied by a set of shearing stresses τ_{xy} acting in the *x-y* plane. These stresses are known as *co-ordinate stresses*, and Figure 9.1 shows the positive signs for the direct stresses σ_x and σ_y

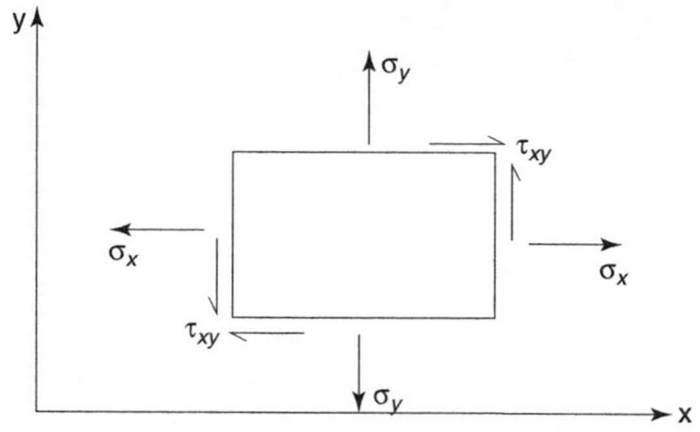

Figure 9.1 Complex stress system

Positive shearing stresses are said to act clockwise, as shown by the horizontal faces of Figure 9.1, and negative shearing stresses are said to act counter-clockwise, as shown by the vertical faces of Figure 9.1.

The reason for choosing a lamina of rectangular shape is to simplify mathematical computation. That is, σ_x has no component in the *y* direction and σ_y has no component in the *x* direction, and also because τ_{xy} is complementary and equal.

In Figure 9.1, the existence of the direct stresses is self-evident, but the reader may not as readily accept the system of shearing stresses. These, however, can be explained with the aid of Section 4.3 on page 91, where it can be seen that shearing stresses are complementary and equal; thus, for completeness, it is necessary to assume the shear stress system of Figure 9.1.

In practice, however, it will be useful to have relationships for the direct and shearing stresses at any angle θ, in terms of the **co-ordinate stresses.**

Consider the stress system acting on the sub-element abc of Figure 9.2, where ac is at an angle θ to the *y* axis and σ_θ is at an angle θ (anticlockwise) to the *x* axis.

Let t = thickness of lamina

σ_θ = direct stress acting on the plane ac

and τ_θ = shear stress acting on the plane ac

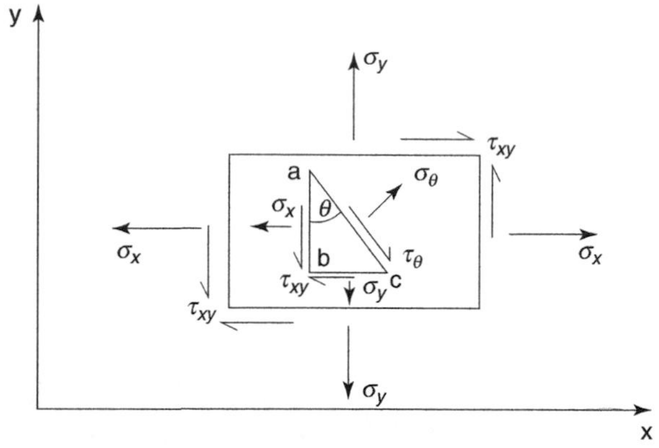

Figure 9.2 Stress system on sub-element abc

9.2 To obtain σ_θ in terms of the co-ordinate stresses

Resolving perpendicular to ac,

$$\sigma_\theta \times ac \times t = \sigma_x \times ab \times t \times \cos\theta + \sigma_y \times bc \times t \times \sin\theta + \tau_{xy} \times ab \times t \times \sin\theta + \tau_{xy} \times bc \times t \times \cos\theta$$

Therefore,

$$\sigma_\theta = \sigma_x \times \frac{ab}{ac}\cos\theta + \sigma_y \times \frac{bc}{ac}\sin\theta + \tau_{xy} \times \frac{ab}{ac}\sin\theta + \tau_{xy} \times \frac{bc}{ac}\cos\theta$$

From Figure 9.2, $\frac{ab}{ac} = \cos\theta$ and $\frac{bc}{ac} = \sin\theta$

Hence $\sigma_\theta = \sigma_x \times \cos\theta\cos\theta + \sigma_y \times \sin\theta\sin\theta + \tau_{xy} \times \cos\theta\sin\theta + \tau_{xy} \times \sin\theta\cos\theta$

i.e. $\sigma_\theta = \sigma_x \cos^2\theta + \sigma_y \sin^2\theta + 2\tau_{xy} \sin\theta\cos\theta$

From double angles in Chapter 2, $\cos 2\theta = 2\cos^2\theta - 1$

from which, $\cos^2\theta = \frac{1}{2}(1 + \cos 2\theta)$

or $\cos 2\theta = 1 - 2\sin^2\theta$ from which,

$\sin^2\theta = \frac{1}{2}(1 - \cos 2\theta)$

Hence $$\sigma_\theta = \frac{\sigma_x}{2}(1 + \cos 2\theta) + \frac{\sigma_y}{2}(1 - \cos 2\theta) + 2\tau_{xy}\sin\theta\cos\theta$$

Also from Chapter 2, $\sin 2\theta = 2\sin\theta\cos\theta$

Thus, $$\boldsymbol{\sigma_\theta = \frac{1}{2}(\sigma_x + \sigma_y) + \frac{1}{2}(\sigma_x - \sigma_y)\cos 2\theta + \tau_{xy}\sin 2\theta} \qquad (9.1)$$

To determine τ_θ in terms of the co-ordinate stresses

Resolving parallel to ac,

$$\tau_\theta \times ac = \sigma_x \times ab \times \sin\theta - \tau_{xy} \times ab \times \cos\theta - \sigma_y \times bc \times \cos\theta + \tau_{xy} \times bc \times \sin\theta$$

NB The effects of t can be ignored, as t appears on both sides of the equation and therefore cancels out.

Therefore
$$\tau_\theta = \sigma_x \times \frac{ab}{ac}\sin\theta - \tau_{xy} \times \frac{ab}{ac}\cos\theta - \sigma_y \times \frac{bc}{ac}\cos\theta + \tau_{xy} \times \frac{bc}{ac}\sin\theta$$

$$= \sigma_x \cos\theta\sin\theta - \tau_{xy}\cos\theta\cos\theta - \sigma_y\sin\theta\cos\theta + \tau_{xy}\sin\theta\sin\theta$$

$$= \sigma_x \cos\theta\sin\theta - \sigma_y\sin\theta\cos\theta - \tau_{xy}\cos^2\theta + \tau_{xy}\sin^2\theta$$

$$= \frac{\sigma_x}{2}\sin 2\theta - \frac{\sigma_y}{2}\sin 2\theta - \tau_{xy}\left(\frac{1}{2}(1+\cos 2\theta)\right) + \tau_{xy}\left(\frac{1}{2}(1-\cos 2\theta)\right)$$

$$= \frac{\sigma_x}{2}\sin 2\theta - \frac{\sigma_y}{2}\sin 2\theta - \tau_{xy}\cos 2\theta$$

i.e.
$$\boldsymbol{\tau_\theta = \frac{1}{2}\left(\sigma_x - \sigma_y\right)\sin 2\theta - \tau_{xy}\cos 2\theta} \qquad (9.2)$$

One of the main reasons for obtaining equations (9.1) and (9.2) is to determine the magnitudes and directions of the maximum values of σ_θ and τ_θ

9.3 Principal stresses (σ_1 and σ_2)

The maximum and minimum values of σ_θ occur when

$$\frac{d\sigma_\theta}{d\theta} = 0$$

Hence, from equation (9.1),

$$\frac{d\sigma_\theta}{d\theta} = -\left(\sigma_x - \sigma_y\right)\sin 2\theta + 2\tau_{xy}\cos 2\theta = 0$$

from which
$$2\tau_{xy}\cos 2\theta = \left(\sigma_x - \sigma_y\right)\sin 2\theta$$

and
$$\frac{2\tau_{xy}}{\left(\sigma_x - \sigma_y\right)} = \frac{\sin 2\theta}{\cos 2\theta}$$

i.e.
$$\tan 2\theta = \frac{2\tau_{xy}}{\left(\sigma_x - \sigma_y\right)} \qquad (9.3)$$

since $\frac{\sin ax}{\cos ax} = \tan ax$

To determine the maximum and minimum values of σ_θ (i.e. $\hat{\sigma}_\theta$)

If equation (9.3) is represented by the mathematical triangle of Figure 9.3 then

$$\sin 2\theta = \pm\frac{2\tau_{xy}}{\sqrt{\left(\sigma_x - \sigma_y\right)^2 + 4\tau_{xy}^2}} \qquad (9.4)$$

and
$$\cos 2\theta = \pm\frac{\left(\sigma_x - \sigma_y\right)}{\sqrt{\left(\sigma_x - \sigma_y\right)^2 + 4\tau_{xy}^2}} \qquad (9.5)$$

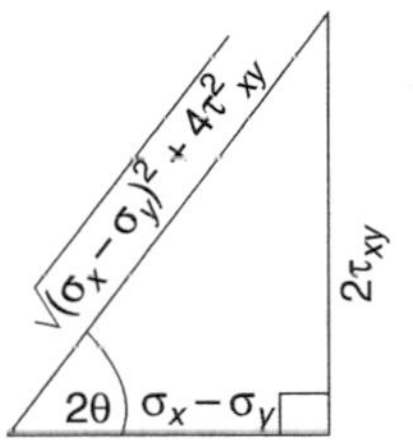

Figure 9.3 Mathematical triangle

Substituting equations (9.4) and (9.5) into equation (9.1) gives:

$$\hat{\sigma}_\theta = \frac{1}{2}\left(\sigma_x + \sigma_y\right) \pm \frac{1}{2}\left(\sigma_x - \sigma_y\right)\left(\frac{\left(\sigma_x - \sigma_y\right)}{\sqrt{\left(\sigma_x - \sigma_y\right)^2 + 4\tau_{xy}^2}}\right) \pm \tau_{xy}\left(\frac{2\tau_{xy}}{\sqrt{\left(\sigma_x - \sigma_y\right)^2 + 4\tau_{xy}^2}}\right)$$

$$= \frac{1}{2}\left(\sigma_x + \sigma_y\right) \pm \frac{1}{2}\left(\frac{\left(\sigma_x - \sigma_y\right)\left(\sigma_x - \sigma_y\right)}{\sqrt{\left(\sigma_x - \sigma_y\right)^2 + 4\tau_{xy}^2}}\right)$$

(*cont.*)

$$\pm\left(\frac{2\tau_{xy}^{\,2}}{\sqrt{\left(\sigma_x-\sigma_y\right)^2+4\tau_{xy}^{\,2}}}\right)$$

or $$\hat{\sigma}_\theta=\frac{1}{2}\left(\sigma_x+\sigma_y\right)\pm\left(\frac{\frac{1}{2}\left(\sigma_x-\sigma_y\right)^2+2\tau_{xy}^{\,2}}{\sqrt{\left(\sigma_x-\sigma_y\right)^2+4\tau_{xy}^{\,2}}}\right)$$

$$=\frac{1}{2}\left(\sigma_x+\sigma_y\right)\pm\frac{1}{2}\left(\frac{\left(\sigma_x-\sigma_y\right)^2+4\tau_{xy}^{\,2}}{\sqrt{\left(\sigma_x-\sigma_y\right)^2+4\tau_{xy}^{\,2}}}\right)$$

i.e. $$\hat{\sigma}_\theta=\frac{1}{2}\left(\sigma_x+\sigma_y\right)\pm\frac{1}{2}\sqrt{\left(\sigma_x-\sigma_y\right)^2+4\tau_{xy}^{\,2}}$$

i.e. $\hat{\sigma}_\theta$ has two values: a maximum value σ_1 and a minimum value σ_2, where

σ_1 **= maximum principal stress**

$$=\frac{1}{2}\left(\sigma_x+\sigma_y\right)+\frac{1}{2}\sqrt{\left(\sigma_x-\sigma_y\right)^2+4\tau_{xy}^2} \quad (9.6)$$

and σ_2 **= minimum principal stress**

$$=\frac{1}{2}\left(\sigma_x+\sigma_y\right)-\frac{1}{2}\sqrt{\left(\sigma_x-\sigma_y\right)^2+4\tau_{xy}^2} \quad (9.7)$$

NB Even though σ_2 is the minimum principal stress, if it is negative it can be larger in magnitude than σ_1

Maximum shear stress ($\hat{\tau}$)

$\hat{\tau}$ occurs when $\dfrac{d\tau_\theta}{d\theta}=0$

i.e. from equation (9.2),

if $$\tau_\theta=\frac{1}{2}\left(\sigma_x-\sigma_y\right)\sin 2\theta-\tau_{xy}\cos 2\theta$$

then $$\frac{d\tau_\theta}{d\theta}=\frac{1}{2}\left(\sigma_x-\sigma_y\right)\left(2\cos 2\theta\right)+\tau_{xy}\left(2\sin 2\theta\right)$$

$$=0 \text{ for maximum shear stress}$$

Hence $$\left(\sigma_x-\sigma_y\right)\left(\cos 2\theta\right)=-2\tau_{xy}\sin 2\theta$$

and $$\frac{\sin 2\theta}{\cos 2\theta}=-\frac{\left(\sigma_x-\sigma_y\right)}{2\tau_{xy}}$$

from which $$\tan 2\theta=\frac{\left(\sigma_y-\sigma_x\right)}{2\tau_{xy}} \quad (9.8)$$

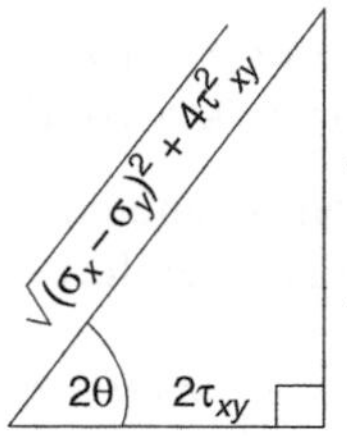

Figure 9.4 Mathematical triangle

Equation (9.8) can be represented by the mathematical triangle of Figure 9.4, where it can be seen that:

$$\cos 2\theta=\pm\frac{2\tau_{xy}}{\sqrt{\left(\sigma_y-\sigma_x\right)^2+4\tau_{xy}^{\,2}}} \quad (9.9)$$

and $$\sin 2\theta=\pm\frac{\left(\sigma_y-\sigma_x\right)}{\sqrt{\left(\sigma_y-\sigma_x\right)^2+4\tau_{xy}^{\,2}}} \quad (9.10)$$

Substituting equations (9.9) and (9.10) into equation (9.2) gives:

$$\hat{\tau}=\pm\frac{1}{2}\left(\sigma_x-\sigma_y\right)\left(\frac{\left(\sigma_y-\sigma_x\right)}{\sqrt{\left(\sigma_y-\sigma_x\right)^2+4\tau_{xy}^{\,2}}}\right)$$

$$\pm\tau_{xy}\left(\frac{2\tau_{xy}}{\sqrt{\left(\sigma_y-\sigma_x\right)^2+4\tau_{xy}^{\,2}}}\right)$$

$$=\pm\left(\frac{\frac{1}{2}\left(\sigma_x-\sigma_y\right)\left(\sigma_y-\sigma_x\right)}{\sqrt{\left(\sigma_y-\sigma_x\right)^2+4\tau_{xy}^{\,2}}}\right)$$

$$\pm\left(\frac{2\tau_{xy}^{\,2}}{\sqrt{\left(\sigma_y-\sigma_x\right)^2+4\tau_{xy}^{\,2}}}\right)$$

$$=\pm\left(\frac{\frac{1}{2}\left(\sigma_x-\sigma_y\right)\left(\sigma_y-\sigma_x\right)+2\tau_{xy}^{\,2}}{\sqrt{\left(\sigma_y-\sigma_x\right)^2+4\tau_{xy}^{\,2}}}\right)$$

$$= \pm\frac{1}{2}\left(\frac{-(\sigma_x-\sigma_y)^2+4\tau_{xy}^2}{\sqrt{(\sigma_y-\sigma_x)^2+4\tau_{xy}^2}}\right)$$

$$= \pm\frac{1}{2}\left(\frac{(\sigma_y-\sigma_x)^2+4\tau_{xy}^2}{\sqrt{(\sigma_y-\sigma_x)^2+4\tau_{xy}^2}}\right)$$

i.e. $$\hat{\tau} = \pm\frac{1}{2}\sqrt{(\sigma_y-\sigma_x)^2+4\tau_{xy}^2}$$

$$= \pm\sqrt{\frac{1}{4}(\sigma_y-\sigma_x)^2+\tau_{xy}^2} \quad (9.11)$$

If Figure 9.4 is compared with Figure 9.3, it can be seen that $\hat{\tau}$ occurs on planes which are at 45° to the planes of the principal stresses. Furthermore, it can be seen that if equation (9.3) is substituted into equation (9.2), then $\tau_\theta = 0$ on the principal planes. That is, *there are no shearing stresses on a principal plane,* and the maximum shearing stresses occur on planes at 45° to the principal planes.

It can also be seen from equations (9.6), (9.7) and (9.11) that

$$\hat{\tau} = \frac{(\sigma_1-\sigma_2)}{2} \quad (9.12)$$

9.4 Mohr's stress circle

Equations (9.1) and (9.2) can be represented in terms of principal stresses, by substituting

$$\sigma_x = \sigma_1$$

and $$\sigma_y = \sigma_2$$

on a principal plane, and

$$\tau_{xy} = 0$$

so that equation (9.1) now becomes:

$$\sigma_\theta = \frac{1}{2}(\sigma_1+\sigma_2)+\frac{1}{2}(\sigma_1-\sigma_2)\cos 2\theta \quad (9.13)$$

and equation (9.2) becomes:

$$\tau_\theta = \frac{1}{2}(\sigma_1-\sigma_2)\sin 2\theta \quad (9.14)$$

A careful study of equations (9.13) and (9.14) reveals that they can be represented by a circle of radius $(\sigma_1 - \sigma_2)/2$, if σ_θ is the horizontal axis and τ_θ is the vertical axis, as shown in Figure 9.5.

[In the circle shown, by Pythagoras,

$$r^2 = \left(\frac{\sigma_1-\sigma_2}{2}\sin 2\theta\right)^2 + \left(\frac{\sigma_1-\sigma_2}{2}\cos 2\theta\right)^2$$

i.e. $$r^2 = \left(\frac{\sigma_1-\sigma_2}{2}\right)^2 \sin^2 2\theta + \left(\frac{\sigma_1-\sigma_2}{2}\right)^2 \cos^2 2\theta$$

from which, $$r^2 = \left(\frac{\sigma_1-\sigma_2}{2}\right)^2 \left(\sin^2 2\theta + \cos^2 2\theta\right)$$

i.e. $$r^2 = \left(\frac{\sigma_1-\sigma_2}{2}\right)^2 \quad \text{since } \sin^2 ax + \cos^2 ax = 1$$

from which radius, $r = \dfrac{\sigma_1-\sigma_2}{2}$]

The centre of the circle is at a distance $(\sigma_1 + \sigma_2)/2$ from the origin as shown in Figure 9.5.

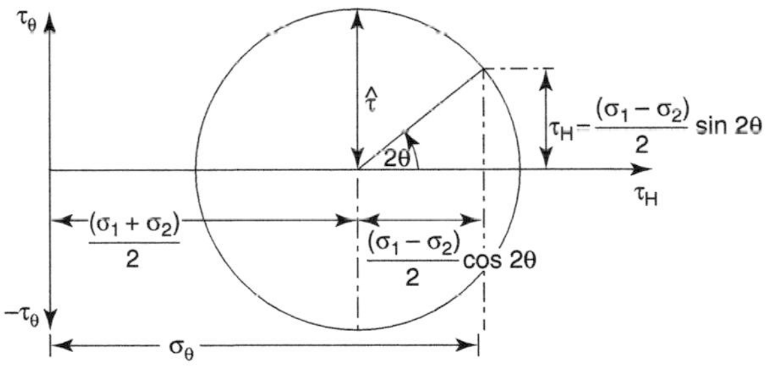

Figure 9.5 Mohr's stress circle

From Figure 9.5, it can be seen that the maximum principal stress is on the right of the circle, and the minimum principal stress is on the left of the circle, but it must be pointed out that the magnitude of the minimum principal stress can be larger than the magnitude of the maximum principal stress, if the former is compressive, as shown in Figure 9.6.

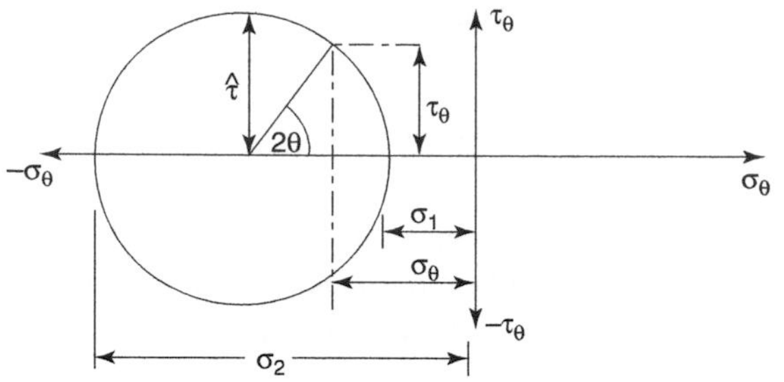

Figure 9.6 Mohr's stress circle for compressive principal stresses

Problem 1. The state of stress in a point of material is shown in Figure 9.7. Determine the direction and magnitudes of the principal stresses.

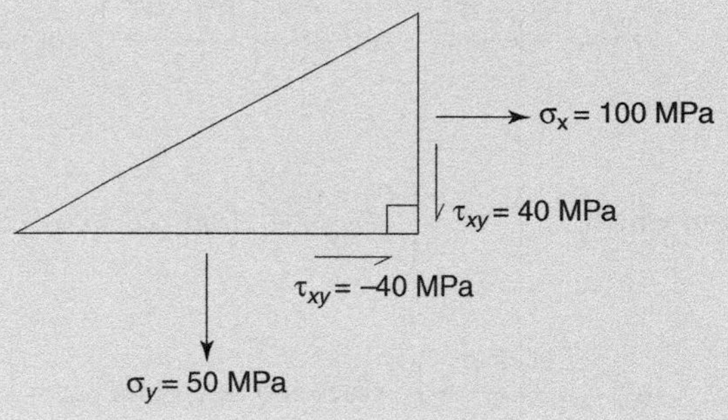

Figure 9.7 Stress at a point

Now from Figure 9.7, it can be seen that the shear stresses associated with σ_x are positive because they act clockwise and that the shear stresses associated with σ_y are negative, because they act counter-clockwise.

The principal stresses can be calculated with the aid of Mohr's stress circle in Figure 9.8, where

$$OE = \sigma_x \quad OD = \sigma_y \quad \text{and} \quad AE = \tau_{xy}$$

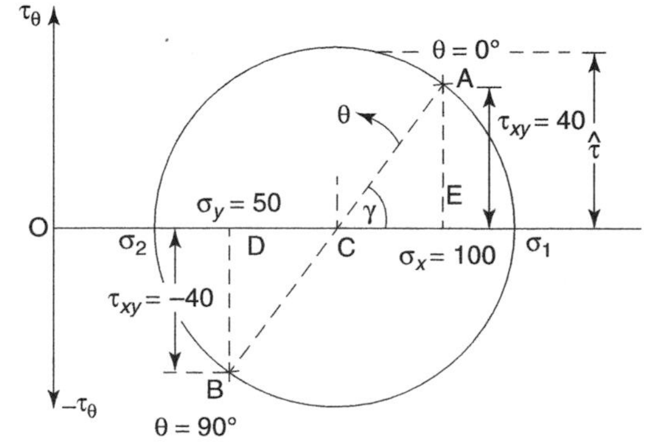

Figure 9.8

The point A on Mohr's stress circle* is obtained from the values of σ_x and τ_{xy} as shown in Figure 9.8. Similarly, the point B on Mohr's stress circle can be obtained from the values of σ_y and $-\tau_{xy}$ as shown in Figure 9.8; point B lies at 180° to point A.

From Figure 9.8 $\quad OC = (OE + OD)/2 = 150/2 = 75$

$$AC^2 = CE^2 + AE^2$$

where $\quad CE = OE - OC = 25$

Therefore $\quad AC^2 = 25^2 + 40^2$

from which $\quad AC = \sqrt{25^2 + 40^2} = 47.17$ MPa

It is evident that $\quad \sigma_1 = OC + AC = 75 + 47.17$

i.e. $\quad \boldsymbol{\sigma_1 = 122.17}$ **MPa**

*Christian Otto Mohr (8 October 1835 – 2 October 1918) was a German civil engineer. In 1882, he famously developed the graphical method for analysing stress known as Mohr's circle and used it to propose an early theory of strength based on shear stress. To find out more about Mohr go to www.routledge.com/cw/bird

Similarly $\sigma_2 = OC - AC = 75 - 47.17$

i.e. $\boldsymbol{\sigma_2 = 27.83}$ **MPa**

Also from Figure 9.8 $\gamma = \tan^{-1}$ (AE/CE)

i.e. $\gamma = \tan^{-1}(40/25) = \tan^{-1}(1.6) = 58^\circ$

Therefore $\theta = -\frac{\gamma}{2} = -\frac{58}{2} = \mathbf{-29^\circ}$

from σ_x to σ_1 in Figure 9.9.

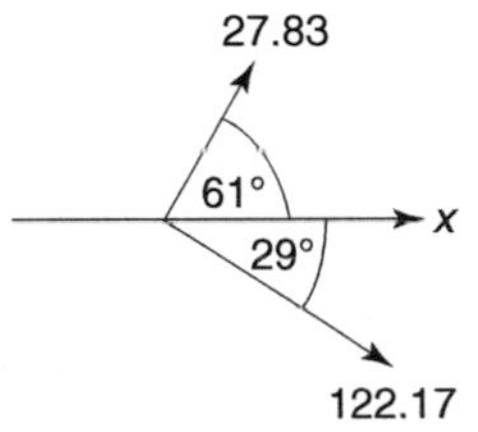

Figure 9.9

Problem 2. The state of stress in a point of material is as shown in Figure 9.10. Determine the direction and magnitudes of the principal stresses.

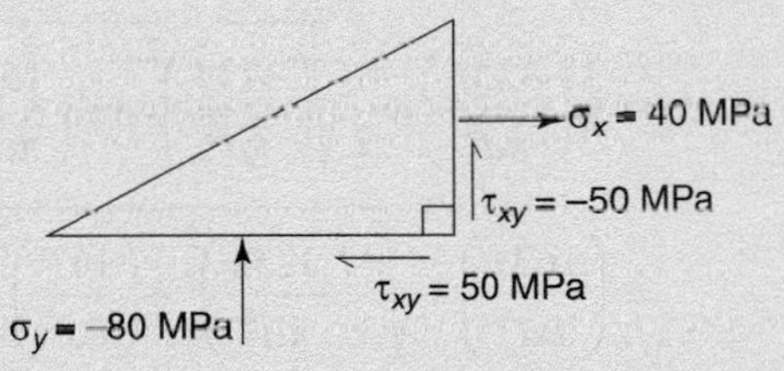

Figure 9.10 Stress acting at a point

From Figure 9.10, it can be seen that as σ_x is tensile, it is positive, and also that as σ_y is compressive, it is negative. Additionally the shearing stress that is acting on the vertical face is causing a counter-clockwise couple, so that it is negative, and that the shearing stress acting on the horizontal face is causing a clockwise couple, so that it is positive.

Thus, point A on Mohr's stress circle is obtained by plotting a positive σ_x and a negative τ_{xy}, as shown in Figure 9.11. Similarly, point B on Mohr's stress circle, which lies at 180° to point A, is obtained by plotting a negative σ_y with a positive τ_{xy}, as shown in Figure 9.11, where OE = σ_x OD = σ_y and AE = 50.

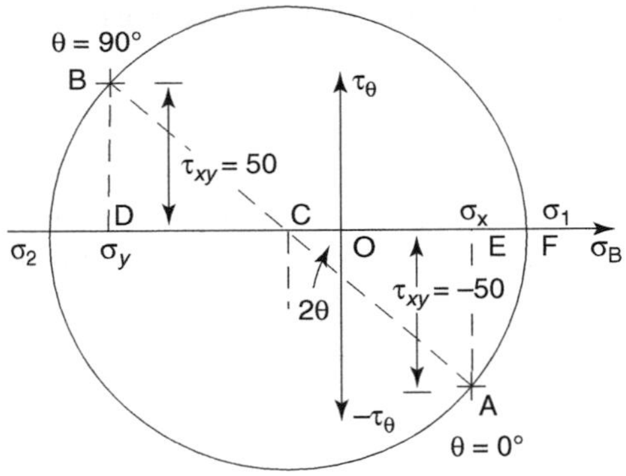

Figure 9.11

From Figure 9.11 $OC = (OE + OD)/2$

i.e. $OC = (40 - 80)/2 = -20$ MPa

$CE = -OC + OE = 60$ MPa

$AC^2 = CE^2 + AE^2$

Therefore $AC^2 = 60^2 + 50^2$

from which $AC = \sqrt{60^2 + 50^2} = 78.10$ MPa

Now $\sigma_1 = CF + OC$

but $CF = AC$

Therefore $\sigma_1 = 78.1 - 20 = \mathbf{58.1}$ **MPa**

Similarly, from Figure 9.11

$\sigma_2 = -78.1 - 20 = \mathbf{-98.1}$ **MPa**

Also, $2\theta = \tan^{-1}(50/60) = 39.81^\circ$

from which $\theta = 19.90^\circ$ from σ_x to σ_1 in Figure 9.12.

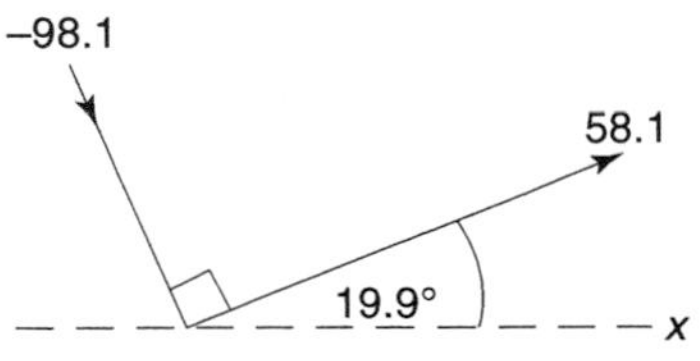

Figure 9.12

The maximum shear stress $\hat{\tau} = \frac{(\sigma_1 - \sigma_2)}{2} = \frac{58.1 - -98.1}{2}$

i.e. $\hat{\tau} = \mathbf{78.1}$ **MPa**

Now try the following Practice Exercise

Practice Exercise 40. Mohr's stress circle

1. In a two-dimensional system of stress, $\sigma_x = 100$ MPa, $\sigma_y = 50$ MPa and $\tau_{xy} = 40$ MPa. Determine the principal stresses in this stress system, together with their directions, and the maximum shearing stresses.

   [122.17 MPa, 27.83 MPa, $\theta = 29°$, 47.17 MPa]

2. In a two-dimensional system of stress, $\sigma_x = 60$ MPa, $\sigma_y = -120$ MPa and $\tau_{xy} = 50$ MPa. Determine the principal stresses in this stress system, together with their directions, and the maximum shearing stresses.

   [72.96 MPa, – 132.96 MPa, $\theta = 14.53°$, 102.96 MPa]

9.5 Combined bending and torsion

Problems in this category frequently occur in circular-section shafts, particularly those of ships. In such cases, the combination of shearing stresses due to torsion and direct stresses due to bending causes a complex system of stress.

Consider a circular-section shaft subjected to a combined bending moment M and a torque T, as shown in Figure 9.13.

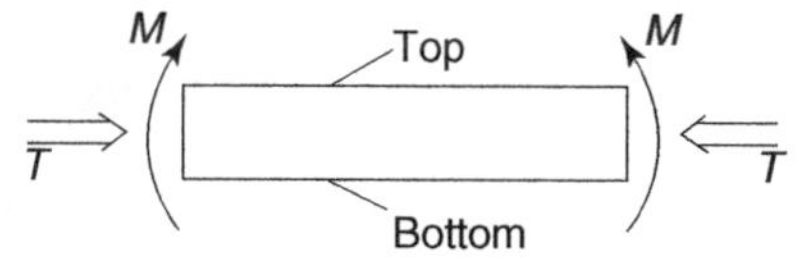

Figure 9.13 Shaft under combined bending and torsion

The largest bending stresses due to M will be at both the top and the bottom of the shaft, and the combined effects of bending stress and shear stress at those positions will be as shown in Figure 9.14.

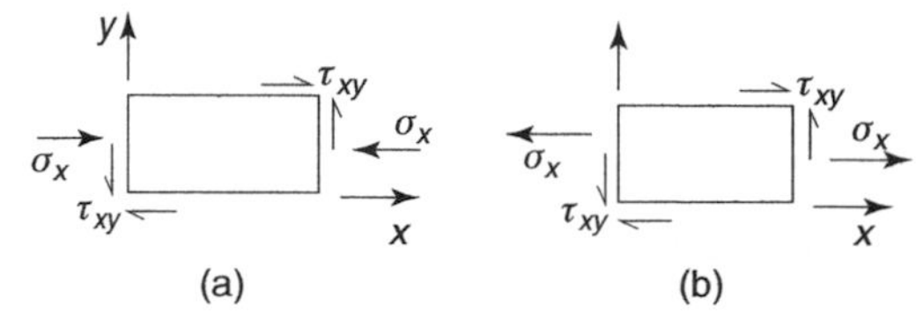

Figure 9.14 Complex stress system due to M and T: (a) top of shaft (looking down): (b) bottom of shaft (looking up)

Now σ_x is entirely due to M, and τ_{xy} is entirely due to T.

From equation (6.1), page 134, $\dfrac{\sigma}{y} = \dfrac{M}{I}$

from which, $\sigma = \dfrac{M}{I}y$

Hence $\sigma_x = \dfrac{M}{\left(\dfrac{\pi d^4}{64}\right)}\left(\dfrac{d}{2}\right) = M\left(\dfrac{64}{\pi d^4}\right)\dfrac{d}{2}$

from Table 5.3, page 120

i.e. $$\sigma_x = \frac{32M}{\pi d^3} \quad (9.15)$$

From equation (8.5), page 171, $\dfrac{\tau}{r} = \dfrac{T}{J}$

from which, $\tau = \dfrac{T}{J}r$

Hence $\tau_{xy} = \dfrac{T}{\left(\dfrac{\pi d^4}{32}\right)}\dfrac{d}{2}$ from Table 5.3, page 120

i.e. $$\tau_{xy} = \frac{16T}{\pi d^3} \quad (9.16)$$

From equilibrium considerations $\sigma_y = 0$

Substituting equations (9.15) and (9.16) into equations (9.6) and (9.7) gives:

$$\sigma_1 = \frac{1}{2}\left(\frac{32M}{\pi d^3}\right) + \frac{1}{2}\sqrt{\left(\frac{32M}{\pi d^3}\right)^2 + 4\left(\frac{16T}{\pi d^3}\right)^2}$$

$$= \left(\frac{16M}{\pi d^3}\right) + \sqrt{\frac{1}{4}\left(\frac{32M}{\pi d^3}\right)^2 + \left(\frac{16T}{\pi d^3}\right)^2}$$

$$= \frac{16M}{\pi d^3} + \sqrt{\left(\frac{32M}{2\pi d^3}\right)^2 + \left(\frac{16T}{\pi d^3}\right)^2}$$

$$= \frac{16M}{\pi d^3} + \sqrt{\left(\frac{16M}{\pi d^3}\right)^2 + \left(\frac{16T}{\pi d^3}\right)^2}$$

$$= \frac{16M}{\pi d^3} + \sqrt{\left(\frac{16}{\pi d^3}\right)^2 \left(M^2 + T^2\right)}$$

$$= \frac{16M}{\pi d^3} + \frac{16}{\pi d^3}\sqrt{\left(M^2 + T^2\right)}$$

i.e. $$\boldsymbol{\sigma_1 = \frac{16}{\pi d^3}\left(M + \sqrt{\left(M^2 + T^2\right)}\right)} \quad (9.17)$$

Similarly $$\boldsymbol{\sigma_2 = \frac{16}{\pi d^3}\left(M - \sqrt{(M^2+T^2)}\right)}$$

From equation (9.11) $$\hat{\tau} = \pm\sqrt{\frac{1}{4}\left(\sigma_y - \sigma_x\right)^2 + \tau_{xy}^2}$$

Since $\sigma_x = \frac{32M}{\pi d^3}$, $\tau_{xy} = \frac{16T}{\pi d^3}$ and $\sigma_y = 0$,

$$\hat{\tau} = \pm\sqrt{\frac{1}{4}\left(-\frac{32M}{\pi d^3}\right)^2 + \left(\frac{16T}{\pi d^3}\right)^2}$$

$$= \pm\sqrt{\left(-\frac{32M}{2\pi d^3}\right)^2 + \left(\frac{16T}{\pi d^3}\right)^2}$$

$$= \pm\sqrt{\left(-\frac{16M}{\pi d^3}\right)^2 + \left(\frac{16T}{\pi d^3}\right)^2}$$

$$= \pm\sqrt{\left(\frac{16M}{\pi d^3}\right)^2 + \left(\frac{16T}{\pi d^3}\right)^2} \quad \text{since } (-x)^2 = (x)^2$$

i.e. $$\boldsymbol{\hat{\tau} = \pm\frac{16}{\pi d^3}\sqrt{M^2+T^2}} \qquad (9.18)$$

If the equivalent bending moment

$$M_e = \frac{1}{2}\left(M \pm \sqrt{M^2+T^2}\right) \qquad (9.19)$$

then equation (9.17) may be written as:

$$\sigma_1, \sigma_2 = \frac{16}{\pi d^3}\left(M \pm \sqrt{(M^2+T^2)}\right) = \boldsymbol{\frac{32M_e}{\pi d^3}} \qquad (9.20)$$

If the equivalent torque, $T_e = \sqrt{M^2+T^2}$ (9.21)

then equation (9.18) may be written as:

$$\hat{\tau} = \pm\frac{16}{\pi d^3}\sqrt{M^2+T^2} = \pm\boldsymbol{\frac{16T_e}{\pi d^3}} \qquad (9.22)$$

Problem 3. A ship's propeller shaft is of a solid circular cross-section, of diameter 0.25 m. If the shaft is subjected to an axial thrust of 1 MN, together with a bending moment of 0.02 MN m and a torque of 0.05 MN m, determine the magnitude of the largest direct stress.

Second moment of area,

$$I = \frac{\pi d^4}{64} = \frac{\pi(0.25)^4}{64} = 1.917\times10^{-4}\ \text{m}^4$$

Polar second moment of area,

$$J = \frac{\pi(0.25)^4}{32} = 3.835\times10^{-4}\ \text{m}^4$$

Cross-sectional area,

$$A = \frac{\pi d^2}{4} = \frac{\pi(0.25)^2}{4} = 0.0491\ \text{m}^2$$

The value of σ_x due to bending, which is of interest, is the negative value, because the axial thrust causes a compressive stress, and if these two are added together, the value of the largest σ_x will be given by:

σ_x = compressive axial stress + compressive bending stress

$$\sigma_x = -\frac{1\times10^6}{0.0491} - \frac{0.02\times10^6}{1.917\times10^{-4}}\times0.125$$

as one component of the bending stresses is in compression

$$= -20.37\times10^6 - 13.04\times10^6$$

i.e. $$\boldsymbol{\sigma_x = -33.41\ \text{MN/m}^2} \qquad (9.23)$$

$$\boldsymbol{\sigma_y = 0} \qquad (9.24)$$

$$\tau_{xy} = \frac{0.05\times10^6}{3.835\times10^{-4}}\times0.125$$

$$= \boldsymbol{16.30\ \text{MN/m}^2} \qquad (9.25)$$

Substituting equations (9.23) to (9.25) into equations (9.6) and (9.7) gives:

$$\sigma_1 = \frac{1}{2}\left(\sigma_x + \sigma_y\right) + \frac{1}{2}\sqrt{\left(\sigma_x - \sigma_y\right)^2 + 4\tau_{xy}^{\ 2}}$$

$$= \frac{1}{2}(-33.41) + \frac{1}{2}\sqrt{(-33.41)^2 + 4(16.30)^2}$$

$$= \boldsymbol{6.63\ \text{MN/m}^2}$$

and $$\sigma_2 = \frac{1}{2}(-33.41) - \frac{1}{2}\sqrt{(-33.41)^2 + 4(16.30)^2}$$

$$= \boldsymbol{-40.04\ \text{MN/m}^2}$$

i.e. **the largest magnitude of direct stress is compressive and equal to −40.06 MN/m²**

Problem 4. The stresses σ_x, τ_{xy} and σ_1 are shown on the triangular element abc of Figure 9.15. Determine the principal stresses and also σ_y and α.

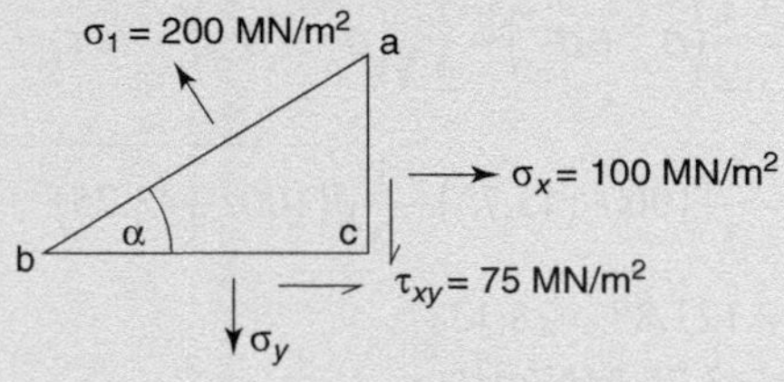

Figure 9.15

As there is no shear stress on the plane ab, σ_1 is a principal stress and as σ_1 is greater than σ_x, σ_1 must be the maximum principal stress:

i.e $\quad$ $\boldsymbol{\sigma_1 = 200\ \mathrm{MN/m^2}}$

Resolving horizontally gives:

$$200 \times \mathrm{ab}\sin\alpha = 75 \times \mathrm{bc} + 100 \times \mathrm{ac}$$

i.e. $$200 = 75 \times \frac{\mathrm{bc}}{\mathrm{ab}\sin\alpha} + 100 \times \frac{\mathrm{ac}}{\mathrm{ab}\sin\alpha}$$

$$= 75 \times \frac{\cos\alpha}{\sin\alpha} + 100 \times \frac{\sin\alpha}{\sin\alpha}$$

Thus $$200 = 75\cot\alpha + 100$$

i.e. $$200 - 100 = 75\cot\alpha$$

and $$\cot\alpha = \frac{100}{75} \text{ from which, } \tan\alpha = \frac{75}{100} = 0.75$$

and $$\boldsymbol{\alpha} = \tan^{-1} 0.75 = \mathbf{36.87^\circ}$$

Resolving vertically gives:

$$200 \times \mathrm{ab}\cos\alpha = 75 \times \mathrm{ac} + \sigma_y \times \mathrm{bc}$$

$$\sigma_y \times \mathrm{bc} = 200 \times \mathrm{ab}\cos\alpha - 75 \times \mathrm{ac}$$

from which $$\sigma_y = 200 \times \frac{\mathrm{ab}\cos\alpha}{\mathrm{bc}} - 75 \times \frac{\mathrm{ac}}{\mathrm{bc}}$$

$$= 200 \times \frac{\cos\alpha}{\cos\alpha} - 75\tan\alpha$$

i.e. $$\sigma_y = 200 - 75\tan\alpha$$

i.e. $$\sigma_y = 200 - \tan 36.87^\circ = \mathbf{143.75\ MN/m^2}$$

From equation (9.7)

$$\sigma_2 = \frac{1}{2}\left(\sigma_x + \sigma_y\right) - \frac{1}{2}\sqrt{\left(\sigma_x - \sigma_y\right)^2 + 4\tau_{xy}^{\ 2}}$$

$$= \frac{1}{2}(100 + 143.75) - \frac{1}{2}\sqrt{(100 - 143.75)^2 + 4(75)^2}$$

$$= 121.88 - 78.13$$

i.e. $\boldsymbol{\sigma_2 = 43.75\ \mathrm{MN/m^2}}$

Now try the following Practice Exercise

Practice Exercise 41. Combined bending and torsion

1. At a point in a two-dimensional stress system, the known stresses are as shown in Figure 9.16. Determine the principal stresses σ_y , σ and $\hat{\tau}$.

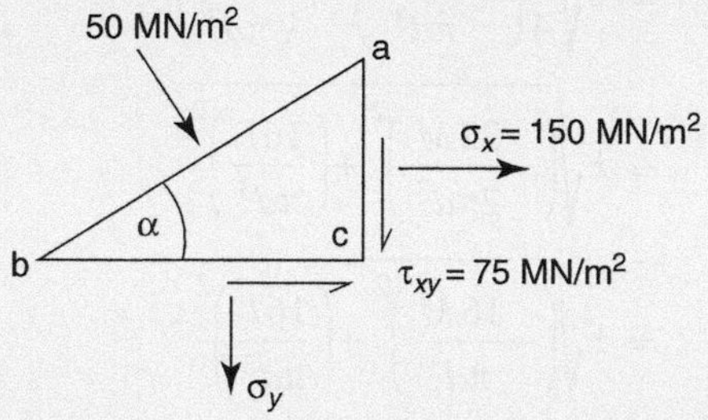

Figure 9.16

[−20.56°, σ_1 = 178.1 MN/m², σ_2 = −50 MN/m², σ_y = −21.88 MN/m²]

2. At a certain point A in a piece of material, the magnitudes of the direct stresses are −10 MN/m², 30 MN/m² and 40 MN/m², as shown in Figure 9.17. Determine the magnitude and direction of the principal stresses and the maximum shear stress.

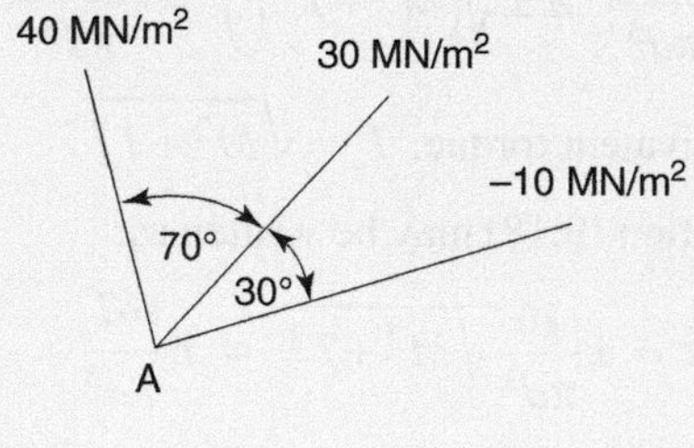

Figure 9.17

[−21.35°, σ_1 = 62.68 MN/m², σ_2 = −21.10 MN/m², $\hat{\tau}$ = 41.89 MN/m²]

3. A circular shaft of diameter 40 cm is subjected to a combined bending moment, *M*, of 3 MN m, together with a torsional moment, *T*, of 1 MN m. Determine the maximum stresses due to the combined effects of '*M*' and '*T*'.

[σ_1 = 490.4 MN/m², σ_2 = −12.89 MN/m², $\hat{\tau}$ = 251.6 MN/m²]

4. If for the shaft in Problem 3, $M = 4$ MN m and $T = 2$ MN m, what are the maximum stresses due to these effects?

   $[\sigma_1 = 674.2\ \text{MN/m}^2,\ \sigma_2 = -37.56\ \text{MN/m}^2,$
   $\hat{\tau} = 355.9\ \text{MN/m}^2]$

9.6 Two-dimensional strain systems

In a number of practical situations, particularly with experimental strain analysis, it is more convenient to make calculations with equations involving strains. Hence, for such cases, it will be necessary to obtain the expressions for strains, rather similar to those obtained for stresses in Section 9.3, page 203.

Consider an infinitesimal rectangular elemental lamina of material OABC, in the *x-y* plane, which is subjected to an in-plane stress system that causes the lamina to strain, as shown by the deformed quadrilateral OA'B'C' in Figure 9.18. Let the co-ordinate strains ε_x, ε_y and γ_{xy} be defined as follows:

ε_x = direct strain in the *x* direction

ε_y = direct strain in the *y* direction

and γ_{xy} = shear strain in the *x-y* plane

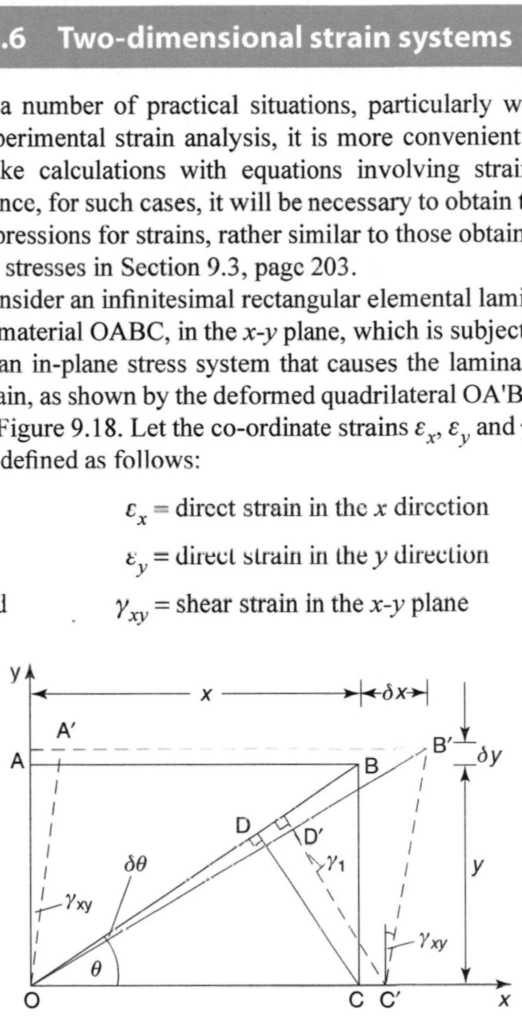

Figure 9.18 Strained quadrilateral

To obtain ε_θ in terms of the co-ordinate strains

Let ε_θ = direct strain at angle θ

and $OB = r$

so that $BC = y = r \sin\theta$

and $OC = x = r\cos\theta$

Hence, $\delta y = r\sin\theta\,\varepsilon_y$ (i)

and $\delta x = r\cos\theta\,\varepsilon_x + r\sin\theta\,\gamma_{xy}$ (ii)

Consider the movement of *B* parallel to *OB*

Now $(OB)^2 = x^2 + y^2 = r^2$

and $(OB')^2 = (x+\delta x)^2 + (y+\delta y)^2 = (r+\delta r)^2$

i.e. $x^2 + 2x\delta x + (\delta x)^2 + y^2 + 2y\delta y + (\delta y)^2 = r^2 + 2r\delta r + (\delta r)^2$

or $x^2 + y^2 + 2x\delta x + 2y\delta y = r^2 + 2r\delta r$

since $(\delta x)^2$, $(\delta y)^2$ and $(\delta r)^2$ are very small

but as $x^2 + y^2 = r^2$

then $2x\delta x + 2y\delta y = 2r\delta r$

and $\delta r = \frac{x}{r}\delta x + \frac{y}{r}\delta y$

or $\delta r = \delta x\cos\theta + \delta y \sin\theta$ from Figure 9.18

Since the strains are small

$$\varepsilon_\theta = \frac{\delta r}{r} = \frac{\delta x}{r}\cos\theta + \frac{\delta y}{r}\sin\theta$$

Substituting for δx and δy from (i) and (ii) above

$$\varepsilon_\theta = \frac{r\cos\theta\,\varepsilon_x + r\sin\theta\,\gamma_{xy}}{r}\cos\theta + \frac{r\sin\theta\,\varepsilon_y}{r}\sin\theta$$

$$= \varepsilon_x\cos^2\theta + \gamma_{xy}\sin\theta\cos\theta + \varepsilon_y\sin^2\theta$$

$$= \varepsilon_x\left(\frac{1+\cos 2\theta}{2}\right) + \gamma_{xy}\left(\frac{1}{2}\sin 2\theta\right) + \varepsilon_y\left(\frac{1-\cos 2\theta}{2}\right)$$

from double angle formulae (see page 6)

$$= \frac{1}{2}\varepsilon_x + \frac{1}{2}\varepsilon_x\cos 2\theta + \frac{1}{2}\gamma_{xy}\sin 2\theta + \frac{1}{2}\varepsilon_y - \frac{1}{2}\varepsilon_y\cos 2\theta$$

i.e. $$\boldsymbol{\varepsilon_\theta = \frac{1}{2}(\varepsilon_x + \varepsilon_y) + \frac{1}{2}(\varepsilon_x - \varepsilon_y)\cos 2\theta + \frac{1}{2}\gamma_{xy}\sin 2\theta} \quad (9.26)$$

This is similar in form to the equation for stress at any angle θ.

To determine the shearing strain γ_ε in terms of co-ordinate strains

To evaluate shearing strain at an angle θ, we note that D is displaced to D' (see Figure 9.18).

Now, as $BB' = \varepsilon_\theta r$

and $OD = OC \cos\theta = r\cos^2\theta$

then $DD' = \varepsilon_\theta OD = \varepsilon_\theta r\cos^2\theta$

During straining, the line CD rotates anticlockwise through a small angle γ_1, where, from Figure 9.18,

$$\gamma_1 = \left(\frac{CC'\cos\theta - DD'}{CD}\right) = \left(\frac{\varepsilon_x\cos^2\theta - \varepsilon_\theta\cos^2\theta}{\cos\theta\sin\theta}\right)$$

Dividing the denominator into the numerator,

$$\gamma_1 = \left(\varepsilon_x - \varepsilon_\theta\right)\cot\theta \qquad \text{(iii)}$$

At the same time, OB rotates in a clockwise direction through a small angle, $\delta\theta$, as shown in Figure 9.19.

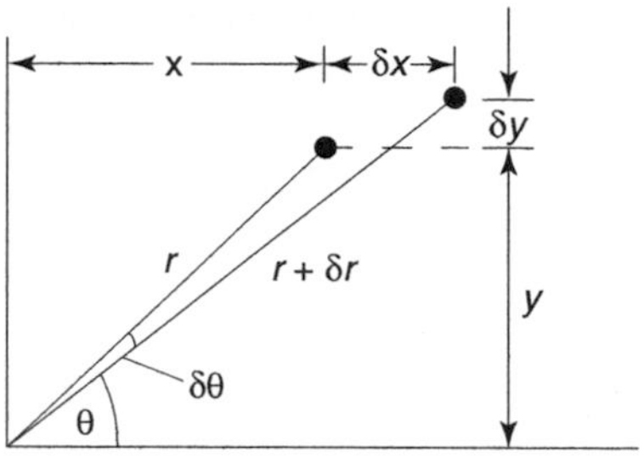

Figure 9.19

$$\theta = \tan^{-1}\left(\frac{x}{y}\right)$$

and

$$\delta\theta = -\left(\frac{\partial\theta}{\partial x}\delta x + \frac{\partial\theta}{\partial y}\delta y\right)$$

[NB $\frac{\partial\theta}{\partial x}$ means the partial derivative of θ with respect to x, with y remaining constant. Partial differentiation, indicated by the 'curly' dee, together with differentiation of inverse trigonometric functions is explained in the references given on pages 49 and 50.]

Now

$$\frac{\partial\theta}{\partial x} = \frac{-\frac{y}{x^2}}{1+\left(\frac{y}{x}\right)^2} = \frac{-\frac{y}{x^2}}{\frac{x^2+y^2}{x^2}} = \left(-\frac{y}{x^2}\right)\left(\frac{x^2}{x^2+y^2}\right)$$

$$= \frac{-y}{x^2+y^2} = \frac{-y}{r^2} = \frac{-y/r}{r} = \frac{-\sin\theta}{r}$$

and

$$\frac{\partial\theta}{\partial y} = \frac{\frac{1}{x}}{1+\left(\frac{y}{x}\right)^2} = \frac{\frac{1}{x}}{\frac{x^2+y^2}{x^2}} = \left(\frac{1}{x}\right)\left(\frac{x^2}{x^2+y^2}\right)$$

$$= \frac{x}{x^2+y^2} = \frac{x}{r^2} = \frac{-x/r}{r} = \frac{\cos\theta}{r}$$

Therefore

$$\delta\theta = -\left(\frac{-\sin\theta}{r}\delta x + \frac{\cos\theta}{r}\delta y\right)$$

$$= \frac{\delta x\sin\theta - \delta y\cos\theta}{r}$$

$$= \frac{\left(r\cos\theta\,\varepsilon_x + r\sin\theta\,\gamma_{xy}\right)\sin\theta - \left(r\sin\theta\,\varepsilon_y\right)\cos\theta}{r}$$

from equations (i) and (ii) above

$$= \left(\varepsilon_x\cos\theta + \gamma_{xy}\sin\theta\right)\sin\theta - \left(\varepsilon_y\sin\theta\right)\cos\theta \qquad \text{(iv)}$$

and the shear strain at any angle θ is

$$\gamma_\theta = \gamma_1 + \delta\theta = \left(\varepsilon_x - \varepsilon_\theta\right)\cot\theta$$

$$+ \left(\varepsilon_x\cos\theta + \gamma_{xy}\sin\theta\right)\sin\theta - \left(\varepsilon_y\sin\theta\right)\cos\theta$$

from equations (iii) and (iv).

Substituting for ε_θ from equation (7.26) gives:

$$\gamma_\theta = \left(\varepsilon_x - \left[\begin{array}{l}\frac{1}{2}\left(\varepsilon_x + \varepsilon_y\right) + \frac{1}{2}\left(\varepsilon_x - \varepsilon_y\right)\cos 2\theta \\ +\frac{1}{2}\gamma_{xy}\sin 2\theta\end{array}\right]\right)\cot\theta$$

$$+\left(\varepsilon_x\cos\theta + \gamma_{xy}\sin\theta\right)\sin\theta - \left(\varepsilon_y\sin\theta\right)\cos\theta$$

$$= \varepsilon_x\cot\theta - \frac{1}{2}\left(\varepsilon_x + \varepsilon_y\right)\cot\theta - \frac{1}{2}\left(\varepsilon_x - \varepsilon_y\right)\cot\theta\cos 2\theta$$

$$-\frac{1}{2}\gamma_{xy}\cot\theta\sin 2\theta + \varepsilon_x\sin\theta\cos\theta + \gamma_{xy}\sin^2\theta$$

$$-\varepsilon_y\sin\theta\cos\theta$$

$$= \varepsilon_x\cot\theta - \frac{1}{2}\varepsilon_x\cot\theta - \frac{1}{2}\varepsilon_y\cot\theta$$

$$-\frac{1}{2}\left(\varepsilon_x - \varepsilon_y\right)\cot\theta\cos 2\theta - \frac{1}{2}\gamma_{xy}\frac{\cos\theta}{\sin\theta}2\sin\theta\cos\theta$$

$$+\varepsilon_x\sin\theta\cos\theta + \gamma_{xy}\sin^2\theta - \varepsilon_y\sin\theta\cos\theta$$

$$= \frac{1}{2}\left(\varepsilon_x - \varepsilon_y\right)\cot\theta - \frac{1}{2}\left(\varepsilon_x - \varepsilon_y\right)\cot\theta\cos 2\theta$$

$$-\gamma_{xy}\cos^2\theta + \left(\varepsilon_x - \varepsilon_y\right)\sin\theta\cos\theta + \gamma_{xy}\sin^2\theta$$

$$= \frac{1}{2}\left(\varepsilon_x - \varepsilon_y\right)\cot\theta - \frac{1}{2}\left(\varepsilon_x - \varepsilon_y\right)\cot\theta\cos 2\theta$$

$$-\gamma_{xy}\left(\cos^2\theta - \sin^2\theta\right) + \left(\varepsilon_x - \varepsilon_y\right)\sin\theta\cos\theta$$

$$= \frac{1}{2}(\varepsilon_x - \varepsilon_y)\cot\theta - \frac{1}{2}(\varepsilon_x - \varepsilon_y)\cot\theta\cos 2\theta$$

$$-\gamma_{xy}\cos 2\theta + \frac{1}{2}(\varepsilon_x - \varepsilon_y)\sin 2\theta$$

$$= \frac{1}{2}(\varepsilon_x - \varepsilon_y)\left[\cot\theta - \cot\theta\cos 2\theta + \sin 2\theta\right] - \gamma_{xy}\cos 2\theta$$

$$= \frac{1}{2}(\varepsilon_x - \varepsilon_y)\left[\frac{\cos\theta}{\sin\theta} - \frac{\cos\theta}{\sin\theta}\left(1 - 2\sin^2\theta\right) + 2\sin\theta\cos\theta\right]$$

$$-\gamma_{xy}\cos 2\theta$$

$$= \frac{1}{2}(\varepsilon_x - \varepsilon_y)\left[\frac{\cos\theta}{\sin\theta} - \frac{\cos\theta}{\sin\theta} + 2\sin\theta\cos\theta + 2\sin\theta\cos\theta\right]$$

$$-\gamma_{xy}\cos 2\theta$$

$$= \frac{1}{2}(\varepsilon_x - \varepsilon_y)\left[4\sin\theta\cos\theta\right] - \gamma_{xy}\cos 2\theta$$

$$= \frac{1}{2}(\varepsilon_x - \varepsilon_y)2\sin 2\theta - \gamma_{xy}\cos 2\theta$$

i.e. $$\gamma_\theta = (\varepsilon_x - \varepsilon_y)\sin 2\theta - \gamma_{xy}\cos 2\theta \qquad (9.27)$$

Equation (9.27) is often written in the form:

$$\frac{\gamma_\theta}{2} = \frac{1}{2}(\varepsilon_x - \varepsilon_y)\sin 2\theta - \frac{1}{2}\gamma_{xy}\cos 2\theta \qquad (9.28)$$

9.7 Principal strains (ε_1 and ε_2)

Principal strains are the maximum and minimum values of direct strain, and they are obtained by satisfying the equation

$$\frac{d\varepsilon_\theta}{d\theta} = 0$$

Hence, from equation (9.26)

since $$\varepsilon_\theta = \frac{1}{2}(\varepsilon_x + \varepsilon_y) + \frac{1}{2}(\varepsilon_x - \varepsilon_y)\cos 2\theta$$

$$+\frac{1}{2}\gamma_{xy}\sin 2\theta$$

then $$\frac{d\varepsilon_\theta}{d\theta} = 0 + \frac{1}{2}(\varepsilon_x - \varepsilon_y)(-2\sin 2\theta)$$

$$+\frac{1}{2}\gamma_{xy}(2\cos 2\theta)$$

Hence $$0 = -(\varepsilon_x - \varepsilon_y)\sin 2\theta + \gamma_{xy}\cos 2\theta$$

from which $$(\varepsilon_x - \varepsilon_y)\sin 2\theta = \gamma_{xy}\cos 2\theta$$

Rearranging gives $$\frac{\sin 2\theta}{\cos 2\theta} = \frac{\gamma_{xy}}{(\varepsilon_x - \varepsilon_y)}$$

i.e. $$\tan 2\theta = \frac{\gamma_{xy}}{(\varepsilon_x - \varepsilon_y)} \qquad (9.29)$$

Equation (9.29) can be represented by the mathematical triangle of Figure 9.20. From this figure:

$$\sin 2\theta = \pm\frac{\gamma_{xy}}{\sqrt{(\varepsilon_x - \varepsilon_y)^2 + \gamma_{xy}^2}} \qquad (9.30)$$

and $$\cos 2\theta = \pm\frac{(\varepsilon_x - \varepsilon_y)}{\sqrt{(\varepsilon_x - \varepsilon_y)^2 + \gamma_{xy}^2}} \qquad (9.31)$$

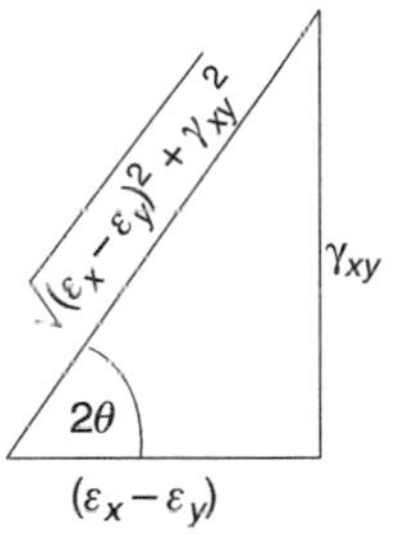

Figure 9.20 Mathematical triangle

Substituting equations (9.30) and (9.31) into equation (9.28) gives:

$$\frac{\gamma_\theta}{2} = \frac{1}{2}(\varepsilon_x - \varepsilon_y)\left(\frac{\gamma_{xy}}{\sqrt{(\varepsilon_x - \varepsilon_y)^2 + \gamma_{xy}^2}}\right)$$

$$-\frac{1}{2}\gamma_{xy}\left(\frac{(\varepsilon_x - \varepsilon_y)}{\sqrt{(\varepsilon_x - \varepsilon_y)^2 + \gamma_{xy}^2}}\right)$$

from which $$\gamma_\theta = 0$$

i.e. the shear strain on a principal plane is zero, as is the case for shear stress.

Furthermore, as $\tau = G\gamma$, it follows that the *planes for principal strains are the same for principal stresses.*

To obtain the expressions for the principal strains in terms of the co-ordinate strains

The values of the principal strains can be obtained by substituting equations (9.30) and (9.31) into equation (9.26):

$$\varepsilon_\theta = \frac{1}{2}\left(\varepsilon_x + \varepsilon_y\right) + \frac{1}{2}\left(\varepsilon_x - \varepsilon_y\right)\left(\pm\frac{\left(\varepsilon_x - \varepsilon_y\right)}{\sqrt{\left(\varepsilon_x - \varepsilon_y\right)^2 + \gamma_{xy}^2}}\right)$$

$$+\frac{1}{2}\gamma_{xy}\left(\frac{\gamma_{xy}}{\sqrt{\left(\varepsilon_x - \varepsilon_y\right)^2 + \gamma_{xy}^2}}\right)$$

$$= \frac{1}{2}\left(\varepsilon_x + \varepsilon_y\right) \pm \frac{1}{2}\left(\frac{\left(\varepsilon_x - \varepsilon_y\right)^2}{\sqrt{\left(\varepsilon_x - \varepsilon_y\right)^2 + \gamma_{xy}^2}}\right)$$

$$\pm\frac{1}{2}\left(\frac{\gamma_{xy}^2}{\sqrt{\left(\varepsilon_x - \varepsilon_y\right)^2 + \gamma_{xy}^2}}\right)$$

$$= \frac{1}{2}\left(\varepsilon_x + \varepsilon_y\right) \pm \frac{1}{2}\left(\frac{\left(\varepsilon_x - \varepsilon_y\right)^2 + \gamma_{xy}^2}{\sqrt{\left(\varepsilon_x - \varepsilon_y\right)^2 + \gamma_{xy}^2}}\right)$$

$$= \frac{1}{2}\left(\varepsilon_x + \varepsilon_y\right) \pm \frac{1}{2}\sqrt{\left(\varepsilon_x - \varepsilon_y\right)^2 + \gamma_{xy}^2}$$

i.e. $$\boldsymbol{\varepsilon_1 = \frac{1}{2}\left(\varepsilon_x + \varepsilon_y\right) + \frac{1}{2}\sqrt{\left(\varepsilon_x - \varepsilon_y\right)^2 + \gamma_{xy}^2}} \qquad (9.32)$$

and $$\boldsymbol{\varepsilon_2 = \frac{1}{2}\left(\varepsilon_x + \varepsilon_y\right) - \frac{1}{2}\sqrt{\left(\varepsilon_x - \varepsilon_y\right)^2 + \gamma_{xy}^2}}$$

To determine the value and direction of the maximum shear strain ($\hat{\gamma}$)

The direction of $\hat{\gamma}$ can be obtained by satisfying the condition

$$\frac{d\gamma_\theta}{d\theta} = 0$$

Hence, from equation (9.28)

since $$\frac{\gamma_\theta}{2} = \frac{1}{2}\left(\varepsilon_x - \varepsilon_y\right)\sin 2\theta - \frac{1}{2}\gamma_{xy}\cos 2\theta$$

then $$\frac{d\gamma_\theta}{d\theta} = \frac{1}{2}\left(\varepsilon_x - \varepsilon_y\right)\left(2\cos 2\theta\right) - \frac{1}{2}\gamma_{xy}\left(-2\sin 2\theta\right)$$

i.e. $$0 = \left(\varepsilon_x - \varepsilon_y\right)\cos 2\theta + \gamma_{xy}\sin 2\theta$$

Rearranging gives $$\gamma_{xy}\sin 2\theta = -\left(\varepsilon_x - \varepsilon_y\right)\cos 2\theta$$

and $$\frac{\sin 2\theta}{\cos 2\theta} = -\frac{\left(\varepsilon_x - \varepsilon_y\right)}{\gamma_{xy}}$$

i.e. $$\tan 2\theta = -\frac{\left(\varepsilon_x - \varepsilon_y\right)}{\gamma_{xy}} \qquad (9.33)$$

Equation (9.33) can be represented by the mathematical triangle of Figure 9.21. From this figure:

$$\sin 2\theta = \pm\frac{\left(\varepsilon_y - \varepsilon_x\right)}{\sqrt{\left(\varepsilon_x - \varepsilon_y\right)^2 + \gamma_{xy}^2}} \qquad (9.34)$$

and $$\cos 2\theta = \pm\frac{\gamma_{xy}}{\sqrt{\left(\varepsilon_x - \varepsilon_y\right)^2 + \gamma_{xy}^2}} \qquad (9.35)$$

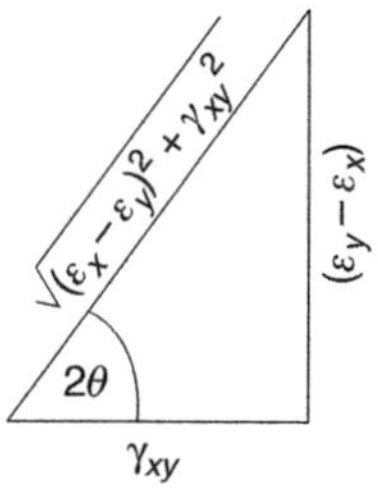

Figure 9.21 Mathematical triangle

Substituting equations (9.34) and (9.35) into equation (9.28) gives:

$$\frac{\gamma_\theta}{2} = \frac{1}{2}\left(\varepsilon_x - \varepsilon_y\right)\left(\pm\frac{\left(\varepsilon_y - \varepsilon_x\right)}{\sqrt{\left(\varepsilon_x - \varepsilon_y\right)^2 + \gamma_{xy}^2}}\right)$$

$$-\frac{1}{2}\gamma_{xy}\left(\pm\frac{\gamma_{xy}}{\sqrt{\left(\varepsilon_x - \varepsilon_y\right)^2 + \gamma_{xy}^2}}\right)$$

$$= \pm\frac{1}{2}\left(\frac{\left(\varepsilon_y - \varepsilon_x\right)^2}{\sqrt{\left(\varepsilon_x - \varepsilon_y\right)^2 + \gamma_{xy}^2}}\right) \pm \frac{1}{2}\left(\frac{\gamma_{xy}^2}{\sqrt{\left(\varepsilon_x - \varepsilon_y\right)^2 + \gamma_{xy}^2}}\right)$$

$$= \pm\frac{1}{2}\left(\frac{\left(\varepsilon_y - \varepsilon_x\right)^2 + \gamma_{xy}^2}{\sqrt{\left(\varepsilon_x - \varepsilon_y\right)^2 + \gamma_{xy}^2}}\right)$$

i.e.

$$\hat{\gamma} = \sqrt{\left(\varepsilon_x - \varepsilon_y\right)^2 + \gamma_{xy}^2} \qquad (9.36)$$

which, when compared with equation (9.32), reveals that, from equation (9.36), similar to 2D-stress,

$$\hat{\gamma} = \pm\left(\varepsilon_x - \varepsilon_y\right) \qquad (9.37)$$

If Figure 9.21 is compared with Figure 9.20, it can be seen that the maximum shear strain occurs at 45° to the principal planes.

9.8 Mohr's circle of strain

On a principal plane,

$$\varepsilon_x = \varepsilon_1$$

$$\varepsilon_y = \varepsilon_2$$

and

$$\gamma_{xy} = 0$$

which, when substituted into equations (9.26) and (9.28), yield the following expressions:

$$\varepsilon_\theta = \frac{1}{2}\left(\varepsilon_1 + \varepsilon_2\right) + \frac{1}{2}\left(\varepsilon_1 - \varepsilon_2\right)\cos 2\theta \qquad (9.38)$$

and

$$\frac{\gamma_\theta}{2} = \frac{1}{2}\left(\varepsilon_1 - \varepsilon_2\right)\sin 2\theta \qquad (9.39)$$

Equations (9.38) and (9.39) can be represented by a circle if ε_θ is taken as the horizontal axis and $\gamma_\theta/2$ as the vertical axis, as shown in Figure 9.22.

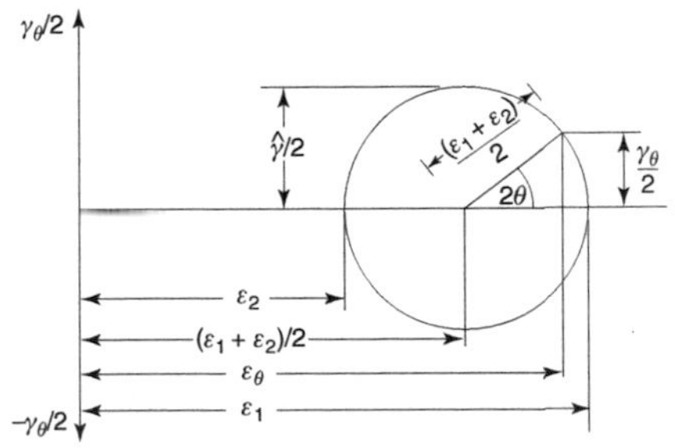

Figure 9.22 Mohr's circle of strain

Plane stress is a two-dimensional system of stress and a three-dimensional system of strain, where the stresses σ_x and σ_y act in the plane of the plate, as shown in Figure 9.23.

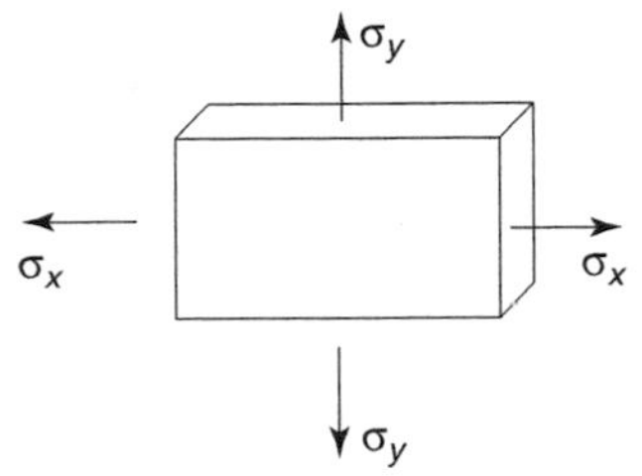

Figure 9.23 Plane stress

In addition to causing strains in the x and y directions, these stresses will cause an out-of-plane strain due to the Poisson effect.

9.9 Stress-strain relationships for plane stress

In stress-strain relationships for plane stress, the following is assumed:

E_x = Young's modulus in the x direction

E_y = Young's modulus in the y direction

ν_x = Poisson's ratio in the x direction due to σ_y

ν_y = Poisson's ratio in the y direction due to σ_x

and

$$\nu_x E_y = \nu_y E_x \qquad (9.40)$$

From Figure 9.22, it can be seen that the strain in the x direction due to σ_x is σ_x/E, and the strain in the x direction due to σ_y is $-\nu\sigma_y/E$, so that ε_x, the strain in the x direction due to the combined stresses, is given by:

$$\varepsilon_x = \frac{\left(\sigma_x - \nu\sigma_y\right)}{E} \qquad (9.41)$$

Similarly, ε_y, the strain in the y direction due to the combined stress, is given by

$$\varepsilon_y = \frac{\left(\sigma_y - \nu\sigma_x\right)}{E} \qquad (9.42)$$

The stress-strain relationships of equations (9.41) and (9.42) can be put in the alternative form of equations (9.43) and (9.44):

$$\sigma_x = \frac{E}{\left(1 - \nu^2\right)}\left(\varepsilon_x + \nu\varepsilon_y\right) \qquad (9.43)$$

$$\sigma_y = \frac{E}{\left(1 - \nu^2\right)}\left(\varepsilon_y + \nu\varepsilon_x\right) \qquad (9.44)$$

For an orthotropic material (see Chapter 19), equations (9.43) and (9.44) can be put in the form

$$\sigma_x = \frac{1}{\left(1 - \nu_x \nu_y\right)}\left(E_x \varepsilon_x + \nu_x E_y \varepsilon_y\right) \qquad (9.45)$$

$$\sigma_y = \frac{1}{\left(1 - \nu_x \nu_y\right)}\left(E_y \varepsilon_y + \nu_y E_x \varepsilon_x\right) \qquad (9.46)$$

A typical case for plane stress is that of a thin plate under in-plane forces.

9.10 Stress-strain relationships for plane strain

Plane strain is a three-dimensional system of stress and a two-dimensional system of strain, as shown in Figure 9.24, where the out-of-plane stress σ_x, is related to the in-plane stresses σ_x and σ_y, and Poisson's ratio, so that the out-of-plane strain is zero.

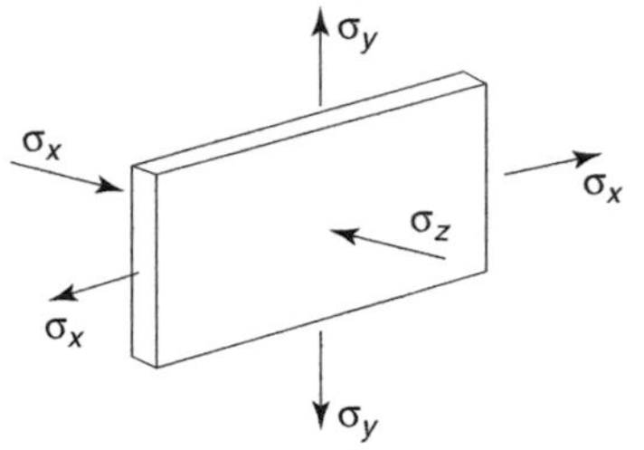

Figure 9.24 Plane strain

From Figure 9.24

$$\varepsilon_x = \frac{\left(\sigma_x - \nu\sigma_y - \nu\sigma_z\right)}{E} \qquad (9.47)$$

$$\varepsilon_y = \frac{\left(\sigma_y - \nu\sigma_x - \nu\sigma_z\right)}{E} \qquad (9.48)$$

$$\varepsilon_z = 0 = \frac{\left(\sigma_z - \nu\sigma_x - \nu\sigma_y\right)}{E} \qquad (9.49)$$

From equation (9.49) $\sigma_z = \nu\left(\sigma_x + \sigma_y\right)$ (9.50)

Substituting equation (9.50) into equations (9.47) and (9.48), these last two equations can be put in the form of equations (9.51) and (9.52), which is usually found to be more convenient:

$$\sigma_x = \frac{E}{(1+\nu)(1-2\nu)}\left[(1-\nu)\varepsilon_x + \nu\varepsilon_y\right] \qquad (9.51)$$

$$\sigma_y = \frac{E}{(1+\nu)(1-2\nu)}\left[(1-\nu)\varepsilon_y + \nu\varepsilon_x\right] \qquad (9.52)$$

For both plane stress and plane strain,

$$\tau_{xy} = G\gamma_{xy} \qquad (9.53)$$

A typical case for plane strain occurs at the mid-length of the cross-section of a gravity dam.

9.11 Pure shear

A system of pure shear is shown by Figure 9.25, where the shear stresses τ are not accompanied by direct stresses on the same planes. As these shear stresses are maximum shear stresses at the point, the principal stresses will lie at 45° to the direction of these shear stresses, as shown by Figure 9.26, where σ_1 = maximum principal stress, and σ_2 = minimum principal stress

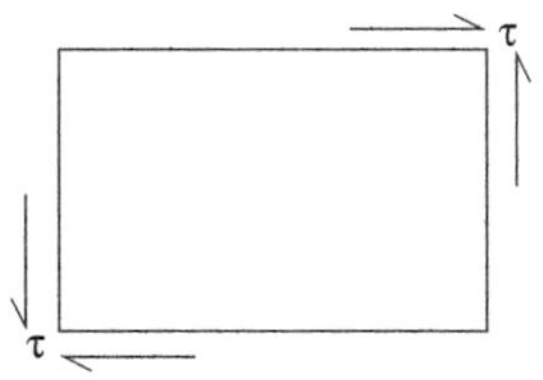

Figure 9.25 Pure shear

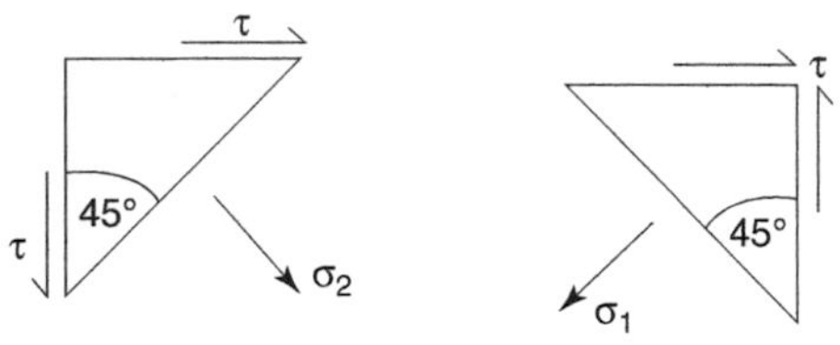

Figure 9.26

By resolution it can be seen that

$$\sigma_1 = \tau \text{ and } \sigma_2 = -\tau \qquad (9.54)$$

i.e. the system of shear stresses of Figure 9.25 is equivalent to the system of Figure 9.27. Thus, if shear strain is required to be measured, the strain gauges have to be placed at 45° to the directions of shear stresses, as shown in Figure 9.28. These strains will measure the principal strains, ε_1 and ε_2, and such strain gauges are known as shear pairs (see Chapter 17).

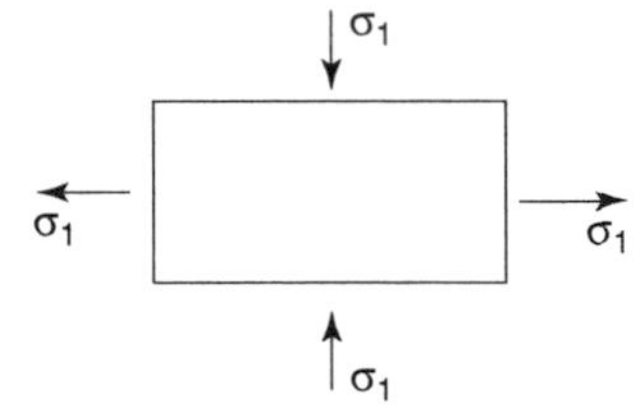

Figure 9.27 Pure shear

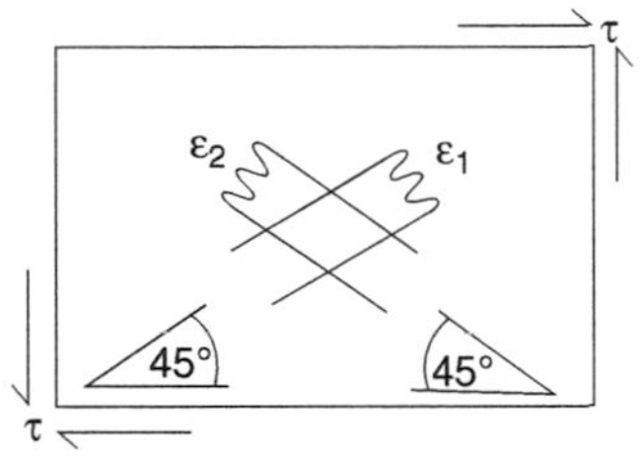

Figure 9.28 Pure shear

From equation (9.37),

$$\gamma = \varepsilon_1 - \varepsilon_2 \tag{9.55}$$

and

$$\tau = G\gamma \tag{9.56}$$

NB If strain gauges were placed in the direction of x or y, then the gauges would simply change their shapes, as shown in Figure 9.29 and would not measure strain.

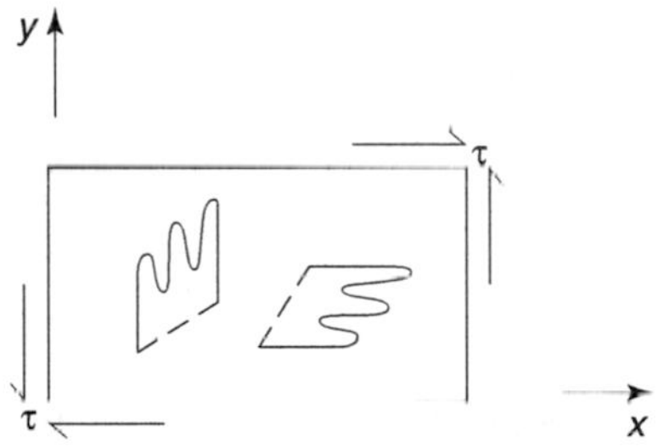

Figure 9.29 Incorrect method of shear strain measurement

Relationships between elastic constants

From equations (9.55) and (9.56),

$$\tau = G(\varepsilon_1 - \varepsilon_2) \tag{9.57}$$

but

$$\sigma_1 = \tau \tag{9.58}$$

From Chapter 4, equation (4.15),

$$G = \frac{E}{2(1+v)}$$

Therefore

$$\varepsilon_1 = \frac{\sigma_1(1+v)}{E} \tag{9.59}$$

and

$$\varepsilon_2 = \frac{-\sigma_1(1+v)}{E} \tag{9.60}$$

Substituting equations (9.58) to (9.60) into (9.57) gives:

$$\tau = G(\varepsilon_1 - \varepsilon_2)$$

hence

$$\sigma_1 = \left(\frac{\sigma_1(1+v)}{E} - \frac{-\sigma_1(1+v)}{E}\right)$$

from which

$$\sigma_1 = \frac{G\sigma_1}{E}((1+v) - -(1+v))$$

$$= \frac{G\sigma_1}{E}(2(1+v))$$

and

$$G = \frac{E}{2(1+v)} = \text{modulus of rigidity}$$

From Section 4.8, page 98,

$$\text{volumetric strain} = \varepsilon_v = \varepsilon_x + \varepsilon_y + \varepsilon_z$$

But for an elemental cube under a hydrostatic stress σ,

$$\varepsilon_x = \varepsilon_y = \varepsilon_z = \frac{\sigma(1-2v)}{E}$$

Therefore

$$\varepsilon_v = \frac{3\sigma(1-2v)}{E} \tag{9.61}$$

Now

$$\frac{\text{volumetric stress } (\sigma)}{\text{volumetric strain } (\varepsilon_v)} = \text{bulk modulus } (K)$$

[NB $K = \dfrac{E}{3(1-2v)}$ from Chapter 4]

Therefore

$$\varepsilon_v = \frac{\sigma}{K} \tag{9.62}$$

Equating (9.61) and (9.62)

$$\frac{3\sigma(1-2v)}{E} = \frac{\sigma}{K}$$

from which

$$K = \frac{E}{3(1-2v)} \tag{9.63}$$

Problem 5. A solid circular-section rotating shaft, of diameter 0.2 m, is subjected to combined bending, torsion and axial load, where the maximum direct stresses due to bending occur on the top and bottom surfaces of the shaft. If a shear pair is attached to the shaft, then the strains recorded from this pair are as follows:

$$\left.\begin{aligned}\varepsilon_1^T &= 200\times10^{-6}\\ \varepsilon_2^T &= 80\times10^{-6}\end{aligned}\right\}\text{ when the shear pair is at the top}$$

$$\left.\begin{aligned}\varepsilon_1^B &= 100\times10^{-6}\\ \varepsilon_2^B &= -20\times10^{-6}\end{aligned}\right\}\text{ when the shear pair is at the bottom}$$

$$E = 2\times10^{11}\text{N/m}^2 \quad \text{and } v = 0.3$$

Using the above information, determine the applied bending moment, torque and axial load.

Let $\varepsilon_1{}^T{}_T$ and $\varepsilon_2{}^T{}_T$ = strain in gauges 1 and 2 due to T at the top

$\varepsilon_1{}^B{}_T$ and $\varepsilon_2{}^B{}_T$ = strain in gauges 1 and 2 due to T at the bottom

$\varepsilon_1{}^T{}_M$ and $\varepsilon_2{}^T{}_M$ = strain in gauges 1 and 2 due to M at the top

$\varepsilon_1{}^B{}_M$ and $\varepsilon_2{}^B{}_M$ = strain in gauges 1 and 2 due to M at the bottom

$\varepsilon_1{}^T{}_D$ and $\varepsilon_2{}^T{}_D$ = strain in gauges 1 and 2 due to the axial load at the top

$\varepsilon_1{}^B{}_D$ and $\varepsilon_2{}^B{}_D$ = strain in gauges 1 and 2 due to the axial load at the bottom

By inspection it can be deduced that

$$\varepsilon_1{}^T{}_T = -\varepsilon_2{}^T{}_T \qquad \varepsilon_1{}^B{}_T = -\varepsilon_2{}^B{}_T$$

$$\varepsilon_1{}^T{}_T = \varepsilon_1{}^B{}_T \qquad \varepsilon_2{}^T{}_T = \varepsilon_2{}^B{}_T$$

$$\varepsilon_1{}^T{}_M = \varepsilon_2{}^T{}_M \qquad \varepsilon_1{}^B{}_M = \varepsilon_2{}^B{}_M$$

$$\varepsilon_1{}^T{}_M = -\varepsilon_1{}^B{}_M \qquad \varepsilon_2{}^T{}_M = -\varepsilon_2{}^B{}_M$$

$$\varepsilon_1{}^T{}_D = \varepsilon_2{}^T{}_D = \varepsilon_1{}^B{}_D = \varepsilon_2{}^B{}_D$$

Hence, *at the top*

$$\varepsilon_1^T = \varepsilon_1{}^T{}_T + \varepsilon_1{}^T{}_M + \varepsilon_1{}^T{}_D \qquad (9.64)$$

and

$$\varepsilon_2^T = -\varepsilon_1{}^T{}_T + \varepsilon_1{}^T{}_M + \varepsilon_1{}^T{}_D \qquad (9.65)$$

and *at the bottom*,

$$\varepsilon_1^B = \varepsilon_1{}^T{}_T - \varepsilon_1{}^T{}_M + \varepsilon_1{}^T{}_D \qquad (9.66)$$

$$\varepsilon_2^B = -\varepsilon_1{}^T{}_T - \varepsilon_1{}^T{}_M + \varepsilon_1{}^T{}_D \qquad (9.67)$$

To obtain *T*

Taking equation (9.65) from equation (9.64), or equation (9.67) from equation (9.66) gives:

$$\gamma = \varepsilon_1^T - \varepsilon_2^T = 200\times10^{-6} - 80\times10^{-6}$$

i.e. $$\gamma = \mathbf{120\times10^{-6}}$$

Now, modulus of rigidity $G = \dfrac{E}{2(1+v)} = \dfrac{2\times10^{11}}{2(1+0.3)}$

i.e. $$G = \mathbf{7.69\times10^{10}\,N/m^2}$$

Therefore, from equation (9.57)

$$\tau = G\gamma = 7.69\times10^{10}\times120\times10^{-6}$$

i.e. $$\tau = 9.228\times10^6 = \mathbf{9.228\,MN/m^2}$$

Now, the polar second moment of area

$$J = \frac{\pi d^4}{32} = \frac{\pi\times0.2^4}{32} = 1.571\times10^{-4}\,\text{m}^4$$

and since $\dfrac{\tau}{r} = \dfrac{T}{J}$

then $$T = \frac{\tau J}{r} = \frac{9.228\times10^6\times1.571\times10^{-4}}{\frac{0.2}{2}}$$

i.e. **torque, T** = 14497 N m = **14.497 kN m**

To obtain *M*

Taking equation (9.66) from equation (9.64), or equation (9.67) from equation (9.65) gives:

$$2\,\varepsilon_1{}^T{}_M = \varepsilon_1^T - \varepsilon_1^B = \varepsilon_2^T - \varepsilon_2^B$$

or $$\varepsilon_1{}^T{}_M = \frac{1}{2}\left(100\times10^{-6}\right)$$

i.e. $$\boldsymbol{\varepsilon_1{}^T{}_M = 50\times10^6}$$

Let σ_b' be the stress due to bending, acting along the strain gauges, as shown in Figure 9.30.

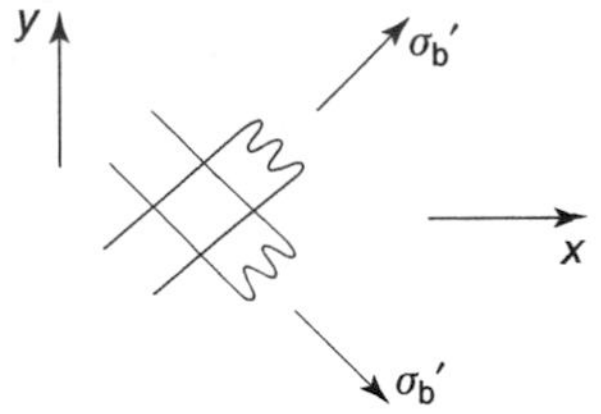

Figure 9.30

Consider the equilibrium of the triangle in Figure 9.31:

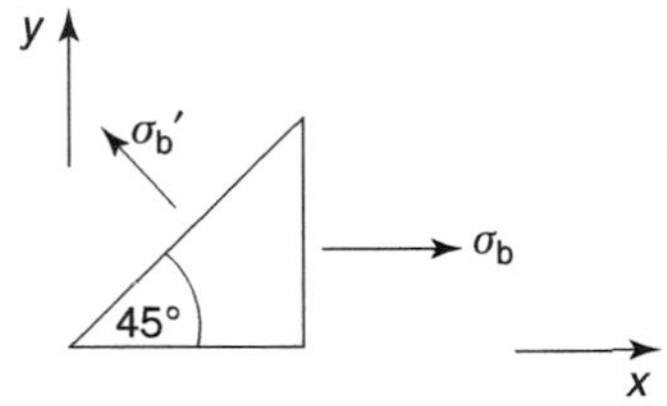

Figure 9.31

If σ_b is the bending stress in the x direction due to M,

$$\sigma_b' \times \sqrt{2} = \frac{\sigma_b}{2}$$

from which $$\boldsymbol{\sigma_b' = \frac{\sigma_b}{2}} \qquad (9.68)$$

From equation (9.40), $\varepsilon_x = \dfrac{(\sigma_x - \nu\sigma_y)}{E}$

i.e. $$\varepsilon_1{}^M{}_M = \frac{1}{E}(\sigma_b' - \nu\sigma_b')$$

i.e. $$\varepsilon_1{}^M{}_M = \frac{\sigma_b'}{E}(1-\nu) = \frac{\sigma_b(1-\nu)}{2E} = 50 \times 10^{-6} \qquad (9.69)$$

Therefore,

$$\sigma_b = \frac{50 \times 10^{-6} \times 2E}{(1-\nu)} = \pm\frac{50 \times 10^{-6} \times 2 \times 2 \times 10^{11}}{(1-0.3)}$$

i.e. $$\boldsymbol{\sigma_b = +28.57\ \text{MN/m}^2}$$

Now $\dfrac{\sigma}{y} = \dfrac{M}{I}$ from which, $M = \dfrac{\sigma \times I}{y}$ where

$$I = \frac{\pi \times 0.2^4}{64} = 7.854 \times 10^{-5}$$

i.e. $$M = \frac{28.57 \times 10^6 \times 7.854 \times 10^{-5}}{\frac{0.2}{2}}$$

i.e. $$\boldsymbol{M = 22.44\ \text{kN m}}$$

To obtain direct load

$$\varepsilon_1{}^T + \varepsilon_2{}^T = 2\varepsilon_1{}^T{}_M + 2\varepsilon_1{}^T{}_D \qquad (9.70)$$

and $$\varepsilon_1{}^B + \varepsilon_2{}^B = -2\varepsilon_1{}^T{}_M + 2\varepsilon_1{}^T{}_D \qquad (9.71)$$

Adding equations (9.70) and (9.71) gives:

$$\varepsilon_1{}^T + \varepsilon_2{}^T + \varepsilon_1{}^B + \varepsilon_2{}^B = 4\varepsilon_1{}^T{}_D$$

Therefore $$\varepsilon_1{}^T{}_D = \frac{\varepsilon_1{}^T + \varepsilon_2{}^T + \varepsilon_1{}^B + \varepsilon_2{}^B}{4}$$

$$= \frac{200 \times 10^{-6} + 80 \times 10^{-6} + 100 \times 10^{-6} - 20 \times 10^{-6}}{4}$$

i.e. $$\boldsymbol{\varepsilon_1{}^T{}_D} = \frac{360 \times 10^{-6}}{4} = \boldsymbol{90 \times 10^{-6}} \qquad (9.72)$$

In a manner similar to that adopted for the derivation of equations (9.68) and (9.69),

$$\varepsilon_1{}^T{}_D = \frac{1}{2E}(\sigma_d - \nu\sigma_d) \qquad (9.73)$$

where σ_d is the stress due to the axial load.
From equations (9.72) and (9.73),

$$\varepsilon_1{}^T{}_D = \frac{\sigma_d}{2E}(1-\nu)$$

from which $$\boldsymbol{\sigma_d} = \frac{2E\varepsilon_1{}^T{}_D}{(1-\nu)} = \frac{2 \times 2 \times 10^{11} \times 90 \times 10^6}{(1-0.3)}$$

$$= \mathbf{51.43\ MN/m^2}$$

Therefore, since stress $= \dfrac{\text{force}}{\text{area}}$

then force = stress × cross-sectional area,

$$\textbf{axial load} = 51.43 \times 10^6 \times \frac{\pi \times 0.2^2}{4}$$

$$= \mathbf{1.62\ MN}\ \text{(tensile)}$$

Now try the following Practice Exercise

Practice Exercise 42. Pure shear

1. In a two-dimensional system of stress, $\sigma_x = 100$ MPa, $\sigma_y = 50$ MPa and $\tau_{xy} = 40$ MPa. Determine the magnitudes of the principal strains, and the magnitude of the maximum shear strain, given that: $E = 2 \times 10^{11}$ N/m² and $\nu = 0.3$.

   [5.691×10^{-4}, 5.260×10^{-5}, 5.165×10^{-4}]

2. In a two-dimensional system of stress, $\sigma_x = 60$ MPa, $\sigma_y = -120$ MPa and $\tau_{xy} = 50$ MPa. If $E = 2 \times 10^{11}$ N/m^2 and $v = 0.33$, what is the magnitude of the principal strains, and the magnitude of the maximum shear strain?

[5.642×10^{-4}, -7.742×10^{-4}, 1.338×10^{-3}]

9.12 Strain rosettes

In practice, strain systems are usually very complicated, and when it is required to determine experimentally the stresses at various points in a plate, at least three strain gauges have to be used at each point of interest. The reason for requiring at least three strain gauges is that there are three unknowns at each point, namely the two principal strains and their direction. Thus, by inputting the values of the three strains, together with their 'positions', into equation (9.26), three simultaneous equations will result, the solution of which will yield the two principal strains, ε_1 and ε_2, and their 'direction' θ.

To illustrate the method of determining ε_1, ε_2 and θ, consider the strains of Figure 9.32.

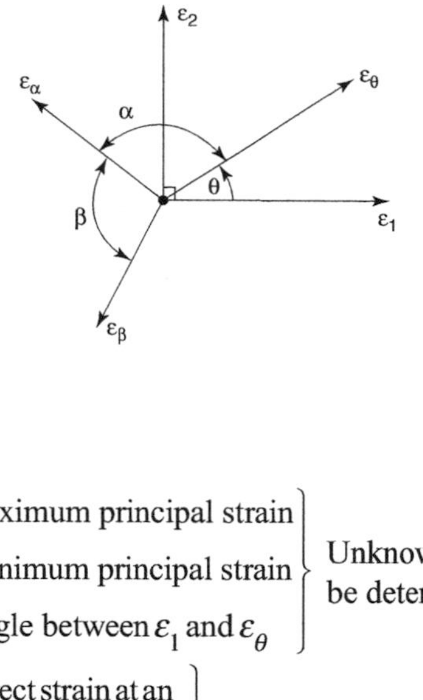

Figure 9.32

Let

$$\left.\begin{aligned}\varepsilon_1 &= \text{maximum principal strain}\\ \varepsilon_2 &= \text{minimum principal strain}\\ \theta &= \text{angle between } \varepsilon_1 \text{ and } \varepsilon_\theta\end{aligned}\right\}\text{Unknowns to be determined}$$

$$\left.\begin{aligned}\varepsilon_\theta &= \text{direct strain at an angle } \theta \text{ from } \varepsilon_1\\ \varepsilon_\alpha &= \text{direct strain at an angle } \alpha \text{ from } \varepsilon_\theta\\ \varepsilon_\beta &= \text{direct strain at an angle } \beta \text{ from } \varepsilon_\alpha\end{aligned}\right\} \quad (9.74)$$

These are experimentally measured strains

Substituting each measured value of equation (9.74), in turn, into equation (9.26), together with its 'direction' the following three simultaneous equations are obtained:

$$\varepsilon_\theta = \frac{1}{2}(\varepsilon_1 + \varepsilon_2) + \frac{1}{2}(\varepsilon_1 - \varepsilon_2)\cos 2\theta \quad (9.75)$$

$$\varepsilon_\alpha = \frac{1}{2}(\varepsilon_1 + \varepsilon_2) + \frac{1}{2}(\varepsilon_1 - \varepsilon_2)\cos[2(\theta + \alpha)] \quad (9.76)$$

$$\varepsilon_\beta = \frac{1}{2}(\varepsilon_1 + \varepsilon_2) + \frac{1}{2}(\varepsilon_1 - \varepsilon_2)\cos[2(\theta + \alpha + \beta)] \quad (9.77)$$

From equations (9.75) to (9.77), it can be seen that the only unknowns are ε_1, ε_2 and θ, which can be readily determined.

For mathematical convenience, manufacturers of strain gauges normally supply strain rosettes, where the angles α and β have values of 45° or 60° or 120°, because the cosine and sine of these angles are readily known. (The strain gauge technique is described in greater detail in Chapter 17.)

If, however, the experimentalist has difficulty in attaching a standard rosette to an awkward zone, then he or she can attach three linear strain gauges at this point, and choose suitable values for α and β. In such cases, it will be necessary to use equations (9.75) to (9.77) to determine the required unknowns.

The 45° rectangular rosette (see Figure 9.33)

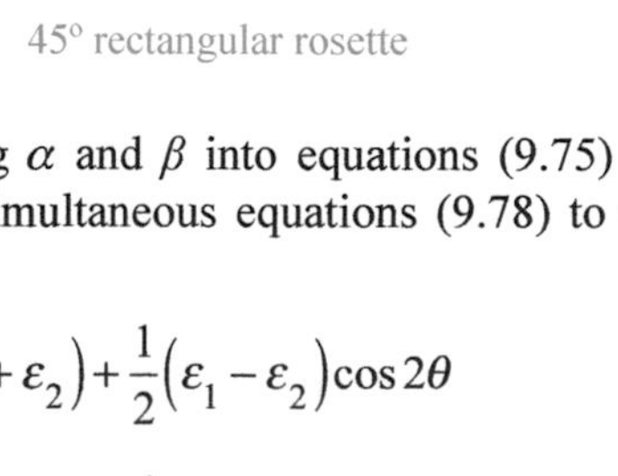

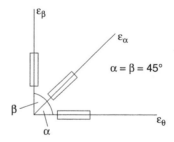

Figure 9.33 45° rectangular rosette

Substituting α and β into equations (9.75) to (9.77), the three simultaneous equations (9.78) to (9.80) are obtained:

$$\varepsilon_\theta = \frac{1}{2}(\varepsilon_1 + \varepsilon_2) + \frac{1}{2}(\varepsilon_1 - \varepsilon_2)\cos 2\theta \quad (9.78)$$

$$\varepsilon_\alpha = \frac{1}{2}(\varepsilon_1 + \varepsilon_2) - \frac{1}{2}(\varepsilon_1 - \varepsilon_2)\sin 2\theta \quad (9.79)$$

$$\varepsilon_\beta = \frac{1}{2}(\varepsilon_1 + \varepsilon_2) - \frac{1}{2}(\varepsilon_1 - \varepsilon_2)\cos 2\theta \quad (9.80)$$

Adding equations (9.78) and (9.80) gives:

$$\varepsilon_\theta + \varepsilon_\beta = \frac{1}{2}(\varepsilon_1 + \varepsilon_2) + \frac{1}{2}(\varepsilon_1 - \varepsilon_2)\cos 2\theta$$

$$+\frac{1}{2}(\varepsilon_1 + \varepsilon_2) - \frac{1}{2}(\varepsilon_1 - \varepsilon_2)\cos 2\theta$$

i.e. $$\varepsilon_\theta + \varepsilon_\beta = \varepsilon_1 + \varepsilon_2 \qquad (9.81)$$

Equation (9.81) is known as the *first invariant of strain*, which states that the sum of mutually perpendicular direct strains at a point is constant.
From equations (9.79) and (9.80)

$$-\frac{1}{2}(\varepsilon_1 - \varepsilon_2)\sin 2\theta = \varepsilon_\alpha - \frac{1}{2}\varepsilon_1 - \frac{1}{2}\varepsilon_2 \qquad (9.82)$$

$$-\frac{1}{2}(\varepsilon_1 - \varepsilon_2)\cos 2\theta = \varepsilon_\beta - \frac{1}{2}\varepsilon_1 - \frac{1}{2}\varepsilon_2 \qquad (9.83)$$

Substituting equation (9.81) into (9.82) and (9.83), and then dividing equation (9.82) by equation (9.83) gives:

$$\tan 2\theta = \frac{\varepsilon_\alpha - \frac{1}{2}\varepsilon_\theta - \frac{1}{2}\varepsilon_\beta}{\varepsilon_\beta - \frac{1}{2}\varepsilon_\theta - \frac{1}{2}\varepsilon_\beta}$$

i.e. $$\tan 2\theta = \frac{2\varepsilon_\alpha - \varepsilon_\theta - \varepsilon_\beta}{2\varepsilon_\beta - \varepsilon_\theta - \varepsilon_\beta} = \frac{\varepsilon_\theta - 2\varepsilon_\alpha + \varepsilon_\beta}{\varepsilon_\theta - 2\varepsilon_\beta + \varepsilon_\beta}$$

i.e. $$\tan 2\theta = \frac{(\varepsilon_\theta - 2\varepsilon_\alpha + \varepsilon_\beta)}{(\varepsilon_\theta - \varepsilon_\beta)} \qquad (9.84)$$

To determine ε_1 and ε_2

Taking equation (9.80) from equation (9.78) gives:

$$\varepsilon_\theta - \varepsilon_\beta = \frac{1}{2}(\varepsilon_1 + \varepsilon_2) + \frac{1}{2}(\varepsilon_1 - \varepsilon_2)\cos 2\theta$$

$$-\left[\frac{1}{2}(\varepsilon_1 + \varepsilon_2) - \frac{1}{2}(\varepsilon_1 - \varepsilon_2)\cos 2\theta\right]$$

i.e. $$\varepsilon_\theta - \varepsilon_\beta = (\varepsilon_1 - \varepsilon_2)\cos 2\theta \qquad (9.85)$$

Hence, from equation (9.85)

$$\frac{\varepsilon_\theta - \varepsilon_\beta}{\cos 2\theta} = \varepsilon_1 - \varepsilon_2 \qquad (a)$$

From equation (9.81)

$$\varepsilon_\theta + \varepsilon_\beta = \varepsilon_1 + \varepsilon_2 \qquad (b)$$

Adding equations (a) and (b) gives:

$$\varepsilon_\theta + \varepsilon_\beta + \frac{\varepsilon_\theta - \varepsilon_\beta}{\cos 2\theta} = 2\varepsilon_1$$

from which $$\varepsilon_1 = \frac{\varepsilon_\theta + \varepsilon_\beta}{2} + \frac{\varepsilon_\theta - \varepsilon_\beta}{2\cos 2\theta} \qquad (9.86)$$

Equation (b) – equation (a) gives:

$$\varepsilon_2 = \frac{\varepsilon_\theta + \varepsilon_\beta}{2} - \frac{\varepsilon_\theta - \varepsilon_\beta}{2\cos 2\theta} \qquad (9.87)$$

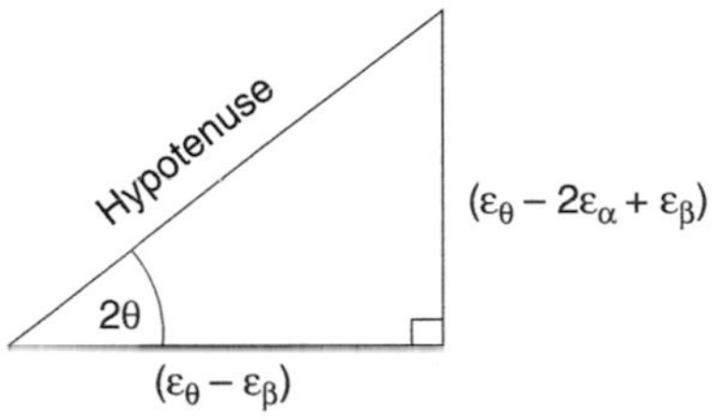

Figure 9.34 Mathematical triangle

Equation (9.84) can be represented by the mathematical triangle of Figure 9.34 where the hypotenuse

$$= \sqrt{(\varepsilon_\theta - 2\varepsilon_\alpha + \varepsilon_\beta)^2 + (\varepsilon_\theta - \varepsilon_\beta)^2}$$

$$= \sqrt{(\varepsilon_\theta - 2\varepsilon_\alpha + \varepsilon_\beta)(\varepsilon_\theta - 2\varepsilon_\alpha + \varepsilon_\beta) + (\varepsilon_\theta - \varepsilon_\beta)(\varepsilon_\theta - \varepsilon_\beta)}$$

$$= \sqrt{\left(\begin{aligned}&\varepsilon_\theta^2 - 2\varepsilon_\theta\varepsilon_\alpha + \varepsilon_\theta\varepsilon_\beta - 2\varepsilon_\theta\varepsilon_\alpha + 4\varepsilon_\alpha^2 - 2\varepsilon_\alpha\varepsilon_\beta\\ &+\varepsilon_\theta\varepsilon_\beta - 2\varepsilon_\alpha\varepsilon_\beta + \varepsilon_\beta^2 + \varepsilon_\theta^2 - \varepsilon_\theta\varepsilon_\beta - \varepsilon_\theta\varepsilon_\beta + \varepsilon_\beta^2\end{aligned}\right)}$$

$$= \sqrt{(2\varepsilon_\theta^2 - 4\varepsilon_\theta\varepsilon_\alpha + 4\varepsilon_\alpha^2 - 4\varepsilon_\alpha\varepsilon_\beta + 2\varepsilon_\beta^2)}$$

$$= \sqrt{(2\varepsilon_\theta^2 - 4\varepsilon_\theta\varepsilon_\alpha + 2\varepsilon_\alpha^2) + (2\varepsilon_\beta^2 - 4\varepsilon_\alpha\varepsilon_\beta + 2\varepsilon_\alpha^2)}$$

$$= \sqrt{2(\varepsilon_\theta - \varepsilon_\alpha)^2 + 2(\varepsilon_\beta - \varepsilon_\alpha)^2}$$

$$= \sqrt{2}\sqrt{(\varepsilon_\theta - \varepsilon_\alpha)^2 + (\varepsilon_\beta - \varepsilon_\alpha)^2} \qquad (9.88)$$

Then $$\cos 2\theta = \frac{\varepsilon_\theta - \varepsilon_\beta}{\sqrt{2}\sqrt{(\varepsilon_\theta - \varepsilon_\alpha)^2 + (\varepsilon_\beta - \varepsilon_\alpha)^2}} \qquad (9.89)$$

and $$\sin 2\theta = \frac{\varepsilon_\theta - 2\varepsilon_\alpha + \varepsilon_\beta}{\sqrt{2}\sqrt{\left(\varepsilon_\theta - \varepsilon_\alpha\right)^2 + \left(\varepsilon_\beta - \varepsilon_\alpha\right)^2}} \quad (9.90)$$

Substituting equation (9.89) into equation (9.86) gives:

$$\varepsilon_1 = \frac{\varepsilon_\theta + \varepsilon_\beta}{2} + \frac{\varepsilon_\theta - \varepsilon_\beta}{2\left[\dfrac{\varepsilon_\beta - \varepsilon_\alpha}{\sqrt{2}\sqrt{\left(\varepsilon_\theta - \varepsilon_\alpha\right)^2 + \left(\varepsilon_\beta - \varepsilon_\alpha\right)^2}}\right]}$$

i.e. $$\varepsilon_1 = \frac{1}{2}\left(\varepsilon_\theta + \varepsilon_\beta\right) + \frac{\sqrt{2}}{2}\sqrt{\left(\varepsilon_\theta - \varepsilon_\alpha\right)^2 + \left(\varepsilon_\beta - \varepsilon_\alpha\right)^2} \quad (9.91)$$

Similarly, substituting equation (9.89) into equation (9.87) gives:

$$\varepsilon_2 = \frac{1}{2}\left(\varepsilon_\theta + \varepsilon_\beta\right) - \frac{\sqrt{2}}{2}\sqrt{\left(\varepsilon_\theta - \varepsilon_\alpha\right)^2 + \left(\varepsilon_\beta - \varepsilon_\alpha\right)^2} \quad (9.91)'$$

The 120° equiangular rosette (see Figure 9.35)

For greater precision, 120° equiangular rosettes are preferred to 45° rectangular rosettes.

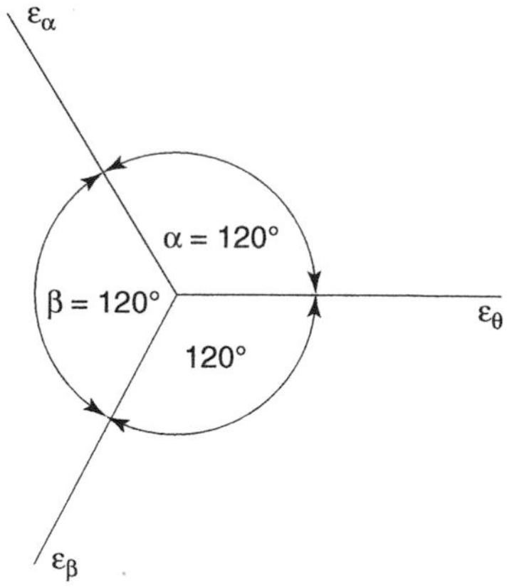

Figure 9.35 120° equiangular rosette

Substituting α and β into equations (9.75) to (9.77), the simultaneous equations of (9.92) to (9.94) are obtained:

$$\varepsilon_\theta = \frac{1}{2}\left(\varepsilon_1 + \varepsilon_2\right) + \frac{1}{2}\left(\varepsilon_1 - \varepsilon_2\right)\cos 2\theta \quad (9.92)$$

$$\varepsilon_\alpha = \frac{1}{2}\left(\varepsilon_1 + \varepsilon_2\right) + \frac{1}{2}\left(\varepsilon_1 - \varepsilon_2\right)\cos[2\left(\theta + 120°\right)] \quad (9.93)$$

$$\varepsilon_\beta = \frac{1}{2}\left(\varepsilon_1 + \varepsilon_2\right) + \frac{1}{2}\left(\varepsilon_1 - \varepsilon_2\right)\cos[2\left(\theta + 240°\right)] \quad (9.94)$$

Equations (9.93) and (9.94) can be rewritten (using compound angle formulae) in the forms:

$$\varepsilon_\alpha = \frac{1}{2}\left(\varepsilon_1 + \varepsilon_2\right) + \frac{1}{2}\left(\varepsilon_1 - \varepsilon_2\right)\left(\cos 2\theta \cos 240° - \sin 2\theta \sin 240°\right)$$

which is equivalent to

$$\varepsilon_\alpha = \frac{1}{2}\left(\varepsilon_1 + \varepsilon_2\right) + \frac{1}{2}\left(\varepsilon_1 - \varepsilon_2\right)\left(-\cos 2\theta \cos 60° + \sin 2\theta \sin 60°\right) \quad (9.95)$$

since cos 240° ≡ –cos 60° and sin 240° ≡ –sin 60°

and $$\varepsilon_\beta = \frac{1}{2}\left(\varepsilon_1 + \varepsilon_2\right) + \frac{1}{2}\left(\varepsilon_1 - \varepsilon_2\right)\left(\cos 2\theta \cos 480° - \sin 2\theta \sin 480°\right)$$

which is equivalent to

$$\varepsilon_\beta = \frac{1}{2}\left(\varepsilon_1 + \varepsilon_2\right) + \frac{1}{2}\left(\varepsilon_1 - \varepsilon_2\right)\left(-\cos 2\theta \cos 60° - \sin 2\theta \sin 60°\right) \quad (9.96)$$

since cos 480° ≡ –cos 60° and sin 480° ≡ +sin 60°

Adding together equations (9.92), (9.95) and (9.96) gives:

$$\varepsilon_\theta + \varepsilon_\alpha + \varepsilon_\beta = \frac{1}{2}\left(\varepsilon_1 + \varepsilon_2\right) + \frac{1}{2}\left(\varepsilon_1 - \varepsilon_2\right)\cos 2\theta$$

$$+ \frac{1}{2}\left(\varepsilon_1 + \varepsilon_2\right) + \frac{1}{2}\left(\varepsilon_1 - \varepsilon_2\right)\left(-\cos 2\theta \cos 60° + \sin 2\theta \sin 60°\right)$$

$$+ \frac{1}{2}\left(\varepsilon_1 + \varepsilon_2\right) + \frac{1}{2}\left(\varepsilon_1 - \varepsilon_2\right)\left(-\cos 2\theta \cos 60° - \sin 2\theta \sin 60°\right)$$

i.e $$\varepsilon_\theta + \varepsilon_\alpha + \varepsilon_\beta = \frac{3}{2}\left(\varepsilon_1 + \varepsilon_2\right) + \frac{1}{2}\left(\varepsilon_1 - \varepsilon_2\right)\begin{bmatrix}\cos 2\theta - \cos 2\theta \cos 60° + \sin 2\theta \sin 60° \\ - \cos 2\theta \cos 60° - \sin 2\theta \sin 60°\end{bmatrix}$$

i.e. $$\varepsilon_\theta + \varepsilon_\alpha + \varepsilon_\beta = \frac{3}{2}\left(\varepsilon_1 + \varepsilon_2\right) + \frac{1}{2}\left(\varepsilon_1 - \varepsilon_2\right)\left[\cos 2\theta - 2\cos 2\theta \cos 60°\right]$$

i.e. $$\varepsilon_\theta + \varepsilon_\alpha + \varepsilon_\beta = \frac{3}{2}\left(\varepsilon_1 + \varepsilon_2\right)$$

or $$\varepsilon_1 + \varepsilon_2 = \frac{2}{3}\left(\varepsilon_\theta + \varepsilon_\alpha + \varepsilon_\beta\right) \qquad (9.97)$$

Taking equation (9.96) from (9.95) gives:

$$\begin{aligned}\varepsilon_\alpha - \varepsilon_\beta &= \frac{1}{2}\left(\varepsilon_1 - \varepsilon_2\right)\left(-\cos 2\theta \cos 60° + \sin 2\theta \sin 60°\right)\\ &\quad - \frac{1}{2}\left(\varepsilon_1 - \varepsilon_2\right)\left(-\cos 2\theta \cos 60° - \sin 2\theta \sin 60°\right)\end{aligned}$$

i.e. $\varepsilon_\alpha - \varepsilon_\beta$

$$= \frac{1}{2}\left(\varepsilon_1 - \varepsilon_2\right)\begin{pmatrix}-\cos 2\theta \cos 60° + \sin 2\theta \sin 60°\\ +\cos 2\theta \cos 60° + \sin 2\theta \sin 60°\end{pmatrix}$$

i.e. $$\varepsilon_\alpha - \varepsilon_\beta = \frac{1}{2}\left(\varepsilon_1 - \varepsilon_2\right)\left(2\sin 2\theta \sin 60°\right)$$

i.e. $$\varepsilon_\alpha - \varepsilon_\beta = \left(\varepsilon_1 - \varepsilon_2\right)\sin 2\theta \sin 60° \qquad (9.98)$$

Taking equation (9.96) from (9.92) gives:

$$\begin{aligned}\varepsilon_\theta - \varepsilon_\beta &= \frac{1}{2}\left(\varepsilon_1 - \varepsilon_2\right)\cos 2\theta\\ &\quad - \frac{1}{2}\left(\varepsilon_1 - \varepsilon_2\right)\left(-\cos 2\theta \cos 60° - \sin 2\theta \sin 60°\right)\end{aligned}$$

$$= \frac{1}{2}\left(\varepsilon_1 - \varepsilon_2\right)\left[\cos 2\theta + \cos 2\theta \cos 60° + \sin 2\theta \sin 60°\right]$$

i.e. $$\varepsilon_\theta - \varepsilon_\beta = \frac{1}{2}\left(\varepsilon_1 - \varepsilon_2\right)\left[\frac{3}{2}\cos 2\theta + \sin 2\theta \sin 60°\right] \qquad (9.99)$$

Dividing equation (9.99) by (9.98) gives:

$$\frac{\varepsilon_\theta - \varepsilon_\beta}{\varepsilon_\alpha - \varepsilon_\beta} = \frac{\frac{1}{2}\left(\varepsilon_1 - \varepsilon_2\right)\left[\frac{3}{2}\cos 2\theta + \sin 2\theta \sin 60°\right]}{\left(\varepsilon_1 - \varepsilon_2\right)\sin 2\theta \sin 60°}$$

$$= \frac{\frac{1}{2}\left[\frac{3}{2}\cos 2\theta + \sin 2\theta \sin 60°\right]}{\sin 2\theta \sin 60°}$$

$$= \frac{1}{2}\left[\frac{\frac{3}{2}\cos 2\theta}{\sin 2\theta \sin 60°} + \frac{\sin 2\theta \sin 60°}{\sin 2\theta \sin 60°}\right]$$

$$= \frac{1}{2}\left[\frac{3}{2}\frac{\cot 2\theta}{\sin 60°} + 1\right]$$

i.e. $$\frac{2\left(\varepsilon_\theta - \varepsilon_\beta\right)}{\varepsilon_\alpha - \varepsilon_\beta} = \left[\frac{3}{2}\frac{\cot 2\theta}{\frac{\sqrt{3}}{2}} + 1\right] = \sqrt{3}\cot 2\theta + 1$$

Hence

$$\begin{aligned}\sqrt{3}\cot 2\theta &= \frac{2\left(\varepsilon_\theta - \varepsilon_\beta\right)}{\varepsilon_\alpha - \varepsilon_\beta} - 1 = \frac{2\left(\varepsilon_\theta - \varepsilon_\beta\right)}{\varepsilon_\alpha - \varepsilon_\beta} - \frac{\varepsilon_\alpha - \varepsilon_\beta}{\varepsilon_\alpha - \varepsilon_\beta}\\ &= \frac{2\varepsilon_\theta - 2\varepsilon_\beta - \varepsilon_\alpha + \varepsilon_\beta}{\varepsilon_\alpha - \varepsilon_\beta} = \frac{\left(2\varepsilon_\theta - \varepsilon_\beta - \varepsilon_\alpha\right)}{\left(\varepsilon_\alpha - \varepsilon_\beta\right)}\end{aligned}$$

i.e. $$\frac{\sqrt{3}}{\tan 2\theta} = \frac{\left(2\varepsilon_\theta - \varepsilon_\beta - \varepsilon_\alpha\right)}{\left(\varepsilon_\alpha - \varepsilon_\beta\right)}$$

from which, $$\mathbf{\tan 2\theta} = \frac{\sqrt{3}\left(\varepsilon_\alpha - \varepsilon_\beta\right)}{\left(2\varepsilon_\theta - \varepsilon_\beta - \varepsilon_\alpha\right)} \qquad (9.100)$$

To determine ε_1 and ε_2

Equation (9.100) can be represented by the mathematical triangle of Figure 9.36, where the hypotenuse

$$\sqrt{2}\sqrt{\left[\left(2\varepsilon_\theta - \varepsilon_\alpha - \varepsilon_\beta\right)^2 + \left(\sqrt{3}\left(\varepsilon_\alpha - \varepsilon_\beta\right)\right)^2\right]}$$

may be shown to be equivalent to

$$\sqrt{2}\sqrt{\left(\varepsilon_\theta - \varepsilon_\alpha\right)^2 + \left(\varepsilon_\alpha - \varepsilon_\beta\right)^2 + \left(\varepsilon_\theta - \varepsilon_\beta\right)^2}$$

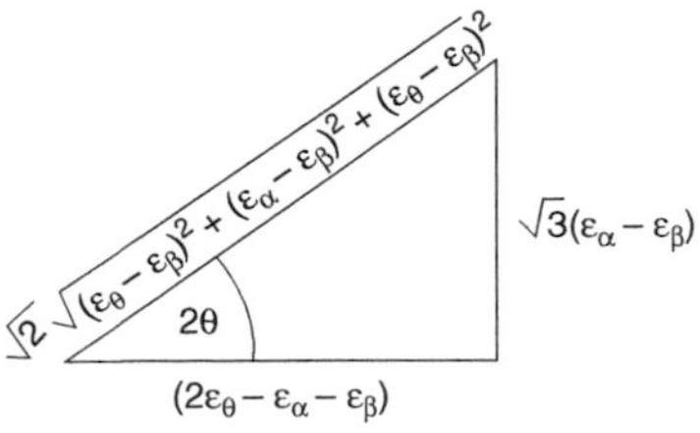

Figure 9.36 Mathematical triangle

From this figure

$$\cos 2\theta = \frac{2\varepsilon_\theta - \varepsilon_\beta - \varepsilon_\alpha}{\sqrt{2}\sqrt{\left(\varepsilon_\theta - \varepsilon_\alpha\right)^2 + \left(\varepsilon_\alpha - \varepsilon_\beta\right)^2 + \left(\varepsilon_\theta - \varepsilon_\beta\right)^2}} \tag{9.101}$$

and

$$\sin 2\theta = \frac{\sqrt{3}\left(\varepsilon_\alpha - \varepsilon_\beta\right)}{\sqrt{2}\sqrt{\left(\varepsilon_\theta - \varepsilon_\alpha\right)^2 + \left(\varepsilon_\alpha - \varepsilon_\beta\right)^2 + \left(\varepsilon_\theta - \varepsilon_\beta\right)^2}} \tag{9.102}$$

Substituting equation (9.102) into (9.98) gives:

$$\varepsilon_\alpha - \varepsilon_\beta = \left(\varepsilon_1 - \varepsilon_2\right)\left(\frac{\sqrt{3}\left(\varepsilon_\alpha - \varepsilon_\beta\right)}{\sqrt{2}\sqrt{\left(\varepsilon_\theta - \varepsilon_\alpha\right)^2 + \left(\varepsilon_\alpha - \varepsilon_\beta\right)^2 + \left(\varepsilon_\theta - \varepsilon_\beta\right)^2}}\right)\sin 60^\circ$$

from which

$$\left(\varepsilon_1 - \varepsilon_2\right) = \frac{\sqrt{2}\sqrt{\left(\varepsilon_\theta - \varepsilon_\alpha\right)^2 + \left(\varepsilon_\alpha - \varepsilon_\beta\right)^2 + \left(\varepsilon_\theta - \varepsilon_\beta\right)^2}}{\sqrt{3}\sin 60^\circ}$$

i.e.

$$\left(\varepsilon_1 - \varepsilon_2\right) = \frac{2\sqrt{2}}{3}\sqrt{\left(\varepsilon_\theta - \varepsilon_\alpha\right)^2 + \left(\varepsilon_\alpha - \varepsilon_\beta\right)^2 + \left(\varepsilon_\theta - \varepsilon_\beta\right)^2} \tag{9.103}$$

Adding equation (9.97) to (9.103) gives:

$$2\varepsilon_1 = \frac{2}{3}\left(\varepsilon_\theta + \varepsilon_\alpha + \varepsilon_\beta\right) + \frac{2\sqrt{2}}{3}\sqrt{\left(\varepsilon_\theta - \varepsilon_\alpha\right)^2 + \left(\varepsilon_\alpha - \varepsilon_\beta\right)^2 + \left(\varepsilon_\theta - \varepsilon_\beta\right)^2}$$

and

$$\varepsilon_1 = \frac{1}{3}\left(\varepsilon_\theta + \varepsilon_\alpha + \varepsilon_\beta\right) + \frac{\sqrt{2}}{3}\sqrt{\left(\varepsilon_\theta - \varepsilon_\alpha\right)^2 + \left(\varepsilon_\alpha - \varepsilon_\beta\right)^2 + \left(\varepsilon_\theta - \varepsilon_\beta\right)^2} \tag{9.104}$$

Similarly, equation (9.97) – equation (9.103) gives:

$$\varepsilon_2 = \frac{1}{3}\left(\varepsilon_\theta + \varepsilon_\alpha + \varepsilon_\beta\right) - \frac{\sqrt{2}}{3}\sqrt{\left(\varepsilon_\theta - \varepsilon_\alpha\right)^2 + \left(\varepsilon_\alpha - \varepsilon_\beta\right)^2 + \left(\varepsilon_\theta - \varepsilon_\beta\right)^2} \tag{9.105}$$

Other types of strain gauge rosette

These include the 60° delta of Figure 9.37, the four-gauge 45° fan of Figure 9.38 and the four-gauge T-delta of Figure 9.39.

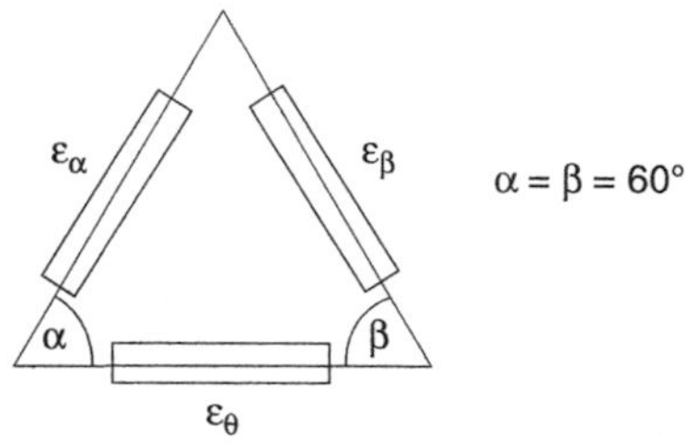

Figure 9.37 60° delta rosette

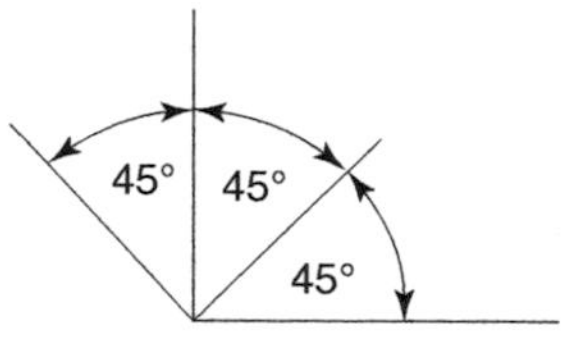

Figure 9.38 Four-gauge 45° fan rosette

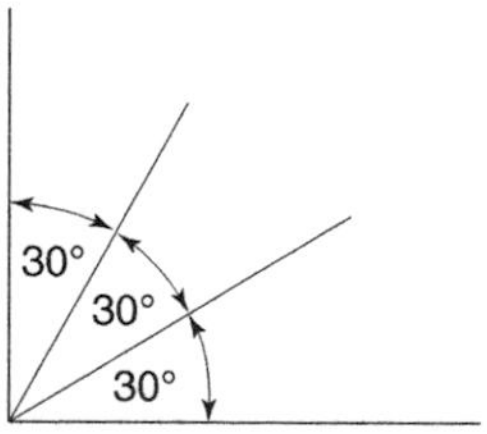

Figure 9.39 T-delta rosette

The advantages of using four-gauge rosettes are that they can lead to greater precision, particularly if one of the strains is zero.

Problem 6. A 120° equiangular rosette records the following values of strain at a point:

$\varepsilon_\theta = 300\times10^{-6}$ $\quad \varepsilon_\alpha = -100\times10^{-6}$

$\varepsilon_\beta = 150\times10^{-6}$

Determine the directions and magnitudes of the values of principal stresses, together with the maximum shear stress at this point, for plane stress and plane strain conditions. Assume that modulus of rigidity, $E = 2\times10^{11}\,\mathrm{N/m^2}$ and Poisson's ratio, $v = 0.3$.

To determine θ

From equation (9.100)

$$\tan 2\theta = \frac{\sqrt{3}\left(\varepsilon_\alpha - \varepsilon_\beta\right)}{\left(2\varepsilon_\theta - \varepsilon_\beta - \varepsilon_\alpha\right)}$$

$$= \frac{\sqrt{3}\left(-100\times10^{-6} - 150\times10^{-6}\right)}{\left(2\times300\times10^{-6} - 150\times10^{-6} - 100\times10^{-6}\right)}$$

$$= \frac{\sqrt{3}(-250)}{550} = -0.7873$$

and $\quad 2\theta = \tan^{-1}(-0.7873) = -38.21°$

from which $\quad \boldsymbol{\theta = -19.11°}$

For the directions of σ_1 and σ_2, see Figure 9.40.

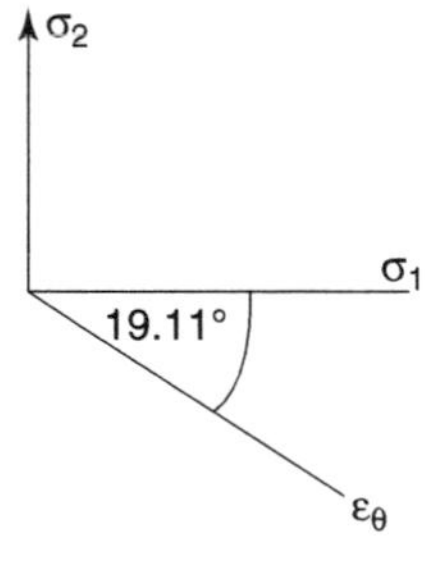

Figure 9.40 Directions of σ_1 and σ_2

From equation (9.104),

$$\varepsilon_1 = \frac{1}{3}\left(\varepsilon_\theta + \varepsilon_\alpha + \varepsilon_\beta\right) + \frac{\sqrt{2}}{3}\sqrt{\left(\varepsilon_\theta - \varepsilon_\alpha\right)^2 + \left(\varepsilon_\alpha - \varepsilon_\beta\right)^2 + \left(\varepsilon_\theta - \varepsilon_\beta\right)^2}$$

$$= \frac{1}{3}(300 - 100 + 150)\times10^6 + \frac{\sqrt{2}}{3}\sqrt{\left((300-100)\times10^6\right)^2 + \left((-100-150)\times10^{-6}\right)^2 + \left((300-150)\times10^{-6}\right)^2}$$

$$= \frac{1}{3}(350)\times10^{-6} + \frac{\sqrt{2}}{3}\sqrt{160\times10^{-9} + 62.5\times10^{-9} + 22.5\times10^{-9}}$$

$$= 116.67\times10^{-6} + 233.33\times10^{-6} = \mathbf{350\times10^6}$$

Similarly, from equation (9.105),

$$\varepsilon_2 = 116.67\times10^{-6} - 233.33\times10^{-6} = \mathbf{-116.67\times10^6}$$

To determine σ_1, σ_2 and $\hat{\tau}$ for plane stress

From equations (9.42) and (9.43)

$$\sigma_1 = \frac{E}{\left(1-v^2\right)}\left(\varepsilon_1 + v\varepsilon_2\right)$$

$$= \frac{2\times10^{11}}{\left(1-0.3^2\right)}\left(350\times10^{-6} + (0.3)\left(-116.67\times10^{-6}\right)\right)$$

$$= \mathbf{69.23\ MN/m^2}$$

and $$\sigma_2 = \frac{E}{\left(1-v^2\right)}\left(\varepsilon_2 + v\varepsilon_1\right)$$

$$= \frac{2\times10^{11}}{\left(1-0.3^2\right)}\left(-116.67\times10^{-6} + (0.3)\left(350\times10^{-6}\right)\right)$$

$$= \mathbf{-2.56\ MN/m^2}$$

Maximum shear stress,

$$\hat{\tau} = \frac{\sigma_1 - \sigma_2}{2} = \frac{69.23 - 2.56}{2} = \mathbf{33.34\ MN/m^2}$$

To determine σ_1, σ_2 and $\hat{\tau}$ for plane strain

From equations (9.51) and (9.52),

$$\sigma_1 = \frac{E}{(1+v)(1-2v)}\left[(1-v)\varepsilon_1 + v\varepsilon_2\right]$$

$$= \frac{2\times10^{11}}{(1+0.3)(1-2(0.3))}\begin{bmatrix}(1-0.3)(350\times10^{6})+\\(0.3)(-116.67\times10^{-6})\end{bmatrix}$$

$$= \mathbf{80.77\ MN/m^2}$$

$$\sigma_2 = \frac{E}{(1+\nu)(1-2\nu)}\left[(1-\nu)\varepsilon_2+\nu\varepsilon_1\right]$$

$$= \mathbf{8.97\ MN/m^2}$$

Maximum shear stress, $\hat{\tau} = \frac{\sigma_1-\sigma_2}{2} = \frac{80.77-8.97}{2}$

$$= \mathbf{35.90\ MN/m^2}$$

Now try the following Practice Exercise

Practice Exercise 43. Strain rosettes

1. Prove that the following relationships apply to the 60° delta rosette of Figure 9.37 (page 224):

$$\tan 2\theta = \frac{\sqrt{3}(\varepsilon_\alpha-\varepsilon_\beta)}{(2\varepsilon_\theta-\varepsilon_\beta-\varepsilon_\alpha)}$$

$$\varepsilon_1,\varepsilon_2 = \frac{1}{3}(\varepsilon_\theta+\varepsilon_\alpha+\varepsilon_\beta) \pm\frac{\sqrt{2}}{3}\sqrt{(\varepsilon_\theta-\varepsilon_\alpha)^2+(\varepsilon_\alpha-\varepsilon_\beta)^2+(\varepsilon_\theta-\varepsilon_\beta)^2}$$

2. The web of a rolled steel joist has three linear strain gauges attached to a point A and indicating strains as shown in Figure 9.41. Determine the magnitude and direction of the principal stresses, and the value of the maximum shear stress at this point, assuming the following to apply:

   Elastic modulus = $2\times10^{11}\text{N/m}^2$ and Poisson's ratio = 0.3

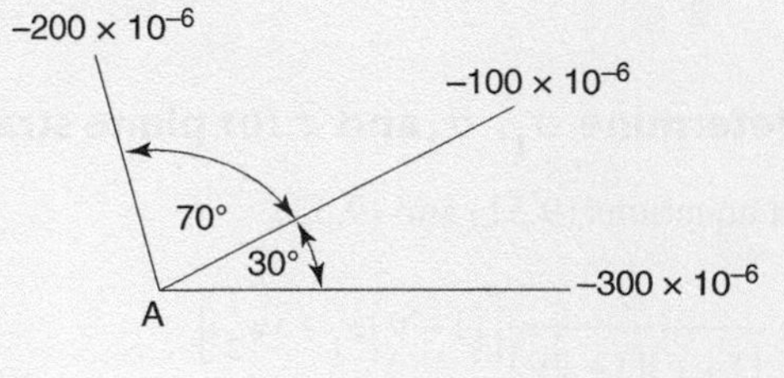

Figure 9.41

[σ_1 = 62.2 MN/m², σ_2 = −66.7 MN/m², − 21.36°, $\hat{\tau}$ = 64.44 MN/m²]

3. A solid stainless steel propeller shaft of a power boat is of diameter 2 cm. A 45° rectangular rosette is attached to the shaft, where the central gauge (Gauge No. 2) is parallel to the axis of the shaft. Assuming bending and thermal stresses are negligible, determine the thrust and torque that the shaft is subjected to, given that the recorded strains are as follows:

   $\varepsilon_1 = -300\times10^{-6}$

   $\varepsilon_2 = -142.9\times10^{-6}$ (Gauge No. 2)

   $\varepsilon_3 = 200\times10^{-6}$

   $E = 200\times10^{6}\text{N/m}^2$ $\nu = 0.3$

   [−8.98 kN, 60.40 MN m]

4. A solid circular-section steel shaft, of 0.03 m diameter, is simply supported at its ends and is subjected to a torque and a load that is radial to its axis. A small 60° strain gauge rosette is attached to the underside of the mid-span, as shown in Figure 9.42.

   Determine (a) the applied torque, (b) the length of the shaft, and (c) the direction and value of the maximum principal stress.

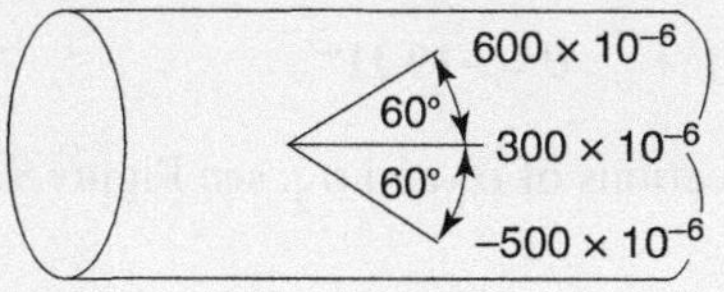

Figure 9.42

   Assume that $E = 2.1\times10^{11}\text{N/m}^2$, $\nu = 0.28$, $\rho = 7860\text{kg/m}^3$ and $g = 9.81\text{m/s}^2$

   [(a) 552.3 N m (b) 5.08 m (c) 37.65° anticlockwise from the middle gauge, 146.5 MN/m², −68.9 MN/m²]

5. A solid cylindrical aluminium-alloy shaft of 0.03 m diameter is subjected to a combined radial and axial load and a torque. A 120° strain gauge rosette is attached to the shaft and records the strains shown in Figure 9.43. Determine the values of the axial load and torque, given the following: $E = 1\times10^{11}\text{N/m}^2$ and $\nu = 0.32$

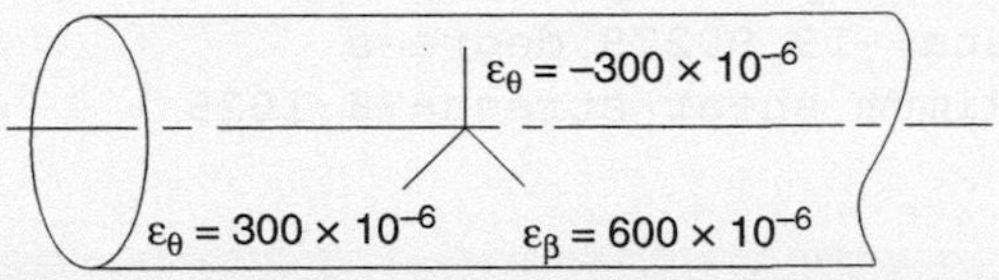

Figure 9.43

[15.40 kN, 208 N m]

9.13 Computer program for principal stresses and strains

Listing 9.1 gives a computer program in BASIC, for determining principal stresses and strains from co-ordinate stresses or co-ordinate strains.

The *input* for determining principal stresses from co-ordinate stresses is as follows:

Type in σ_x (direct stress in the x direction)

Type in σ_y (direct stress in the y direction)

Type in τ_{xy} (shear stress in the x-y plane)

Typical input and output values are given in Listings 9.2 and 9.3 for Problems 1 and 2 (on page 228).

The input for determining principal stresses and strains from co-ordinate strains is as follows:

Type in ε_x (direct strain in x direction)

Type in ε_y (direct strain in y direction)

Type in γ_{xy} (shear strain in the x-y plane)

Type in E (elastic modulus)

Type in v (Poisson's ratio)

NB When this part of the program is being used, the condition of *plane stress* is assumed.

Listing 9.1 Computer program for calculating σ_1, σ_2, etc, from either the co-ordinate stresses or the co-ordinate strains

```
100 REMark principal stresses &
    strains
110 CLS
120 PRINT:PRINT``principal stresses
    & strains''
130 PRINT:PRINT``copyright of
    C.T.F.Ross'':PRINT'
140 PRINT``if inputting CO-ORDINATE
    STRESSES type 1; else if
    inputting CO-ORDINATE STRAINS,
    TYPE 0'':PRINT
150 INPUT str
160 IF str=1 OR str=0 THEN GO TO 180
170 PRINT:PRINT``incorrect
    data'':PRINT:GO to 140
180 IF str=0 THEN GO TO 300
190 PRINT:PRINT``stress in
    x-direction='';:INPUT sigmax
200 PRINT``stress in
    y-direction='';:INPUT sigmay
210 PRINT``shear stress in x-y
    plane='';:INPUT tauxy
220 const=.5*SQRT((sigmax-sigmay)^2
    +4*tauxy*^2)
230 sigma1=.5*(sigmax+sigmay)+const
240 sigma2=.5*(sigmax+sigmay)-const
250 theta=.5*ATAN(2*tauxy/(sigmax-
    sigmay))
260 GO TO 600
300 PRINT:PRINT``strain in x
    direction='';:INPUT ex
310 PRINT``strain in y
    direction='';:INPUT ey
320 PRINT``shear in x-y
    plain='':INPUT gxy
330 const=.5*SQRT((ex-ey)*^2+gxy*^2)
340 e1=.5*(ex+ey)+const
350 e2=.5*(ex+ey)-const
360 theta=.5*ATAN(gxy/(ex-ey))
370 PRINT:PRINT``elastic
    modulus='';:INPUT e
380 PRINT``poisson's ratio='';:INPUT
    nu
600 PI=3.14159
610 IF str=1 THEN PRINT:PRINT:PRINT,
    ``principal stresses'':PRINT
620 IF STR=0 THEN PRINT:PRINT:PRINT,
    ``principal stresses & principal
    strains'':PRINT: GO to 800
630 PRINT``stress in
    x-direction='';sigmax
640 PRINT``stress in
    y-direction='';sigmay
650 PRINT``shear in x-y
    plane='';tauxy
660 PRINT:PRINT
670 PRINT``maximum principal
    stress='';sigma1
680 PRINT``minimum principal
    stress='';sigma2
690 PRINT``theta='';theta*180/PI;
    ``degrees''
```

```
695 PRINT``maximum shear stress=´´;(sigma1-sigma2) /2
700 GO TO 1000
800 PRINT``strain in x direction=´´;ex
810 PRINT``strain in y direction=´´;ey
820 PRINT``shear strain in x-y plane=´´;gxy
830 PRINT``elastic modulus=´´;e
840 PRINT``poisson's ratio=´´;nu
850 PRINT:PRINT
860 PRINT``maximum principal strain=´´;e1
870 PRINT``maximum principal strain=´´;e2
875 PRINT``theta=´´;theta*180/PI;``degrees´´
880 PRINT``maximum shear strain=´´;e1-e2
890 sigma1=e*(e1+nu*e2)/(1-nu*2)
900 sigma2=e*(e2+nu*e1)/(1-nu*2)
910 PRINT``maximum principal stress=´´;sigma1
920 PRINT``minimum principal stress=´´;sigma2
930 PRINT``maximum shear stress=´´;(sigma1-sigma2)/2
1000 PRINT:PRINT: PRINT:PRINT
1020 STOP
```

Listing 9.2 Computer output for worked Problem 1 (on p. 206)

```
principal stresses

stress in x-direction=100
stress in y-direction=50
shear in x-y plane=40

maximum principal stress=122.1699
minimum principal stress=27.83009
theta=28.99731 degrees
maximum shear stress=47.16991
```

Listing 9.3 Computer output for worked Problem 2 (on p. 207)

```
principal stresses
stress in x-direction=40
stress in y-direction=-80
shear in x-y plane=-50
maximum principal stress=58.1025
minimum principal stress=-98.1025
theta=-19.90279 degrees
maximum shear stress=78.1025
```

9.14 The constitutive laws for a lamina of a composite in global co-ordinates

To obtain the constitutive laws, in global co-ordinates, for a lamina of a composite, consider a lamina with orthogonal properties as shown in Figure 9.44.

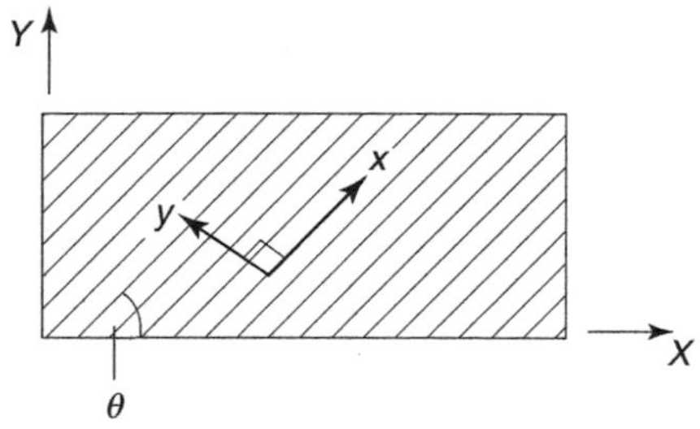

Figure 9.44 A lamina from a composite

In the figure, the local axes of the lamina are x and y and the global axes are X and Y, where the local x axis lies at an angle θ to the global X axis. The global X and Y axes are also known as reference axes.

From equations (4.19) and (4.20) the relationship between the local strains and stresses is given by:

$$\begin{pmatrix} \varepsilon_x \\ \varepsilon_y \\ \gamma_{xy} \end{pmatrix} = \begin{pmatrix} S_{11} & S_{12} & S_{13} \\ S_{21} & S_{22} & S_{23} \\ S_{31} & S_{32} & S_{33} \end{pmatrix} \begin{pmatrix} \sigma_x \\ \sigma_y \\ \tau_{xy} \end{pmatrix} \tag{9.106}$$

$$\left(\varepsilon_{xy}\right) = \left(S\right)\left(\sigma_{xy}\right) \tag{9.107}$$

where (S) is a matrix of compliance functions for the lamina, where for plane stress,

$$S_{11} = \frac{1}{E_x} \qquad S_{12} = S_{21} = \frac{-\nu_x}{E_x} = \frac{-\nu_y}{E_y}$$

$$S_{22} = \frac{1}{E_y} \qquad S_{33} = G_{xy} \qquad S_{13} = S_{31} = S_{23} = S_{32} = 0$$

and for plane strain,

$$S_{11} = \frac{\left(1 - \nu_{xz}\nu_{zx}\right)}{E_x} \qquad S_{12} = \frac{-\left(\nu_{xy} + \nu_{xz}\nu_{zy}\right)}{E_x}$$

$$S_{22} = \frac{\left(1 - \nu_{yz}\nu_{zy}\right)}{E_y}$$

$$S_{33} = \frac{1}{G_{xy}} \qquad S_{13} = S_{23} = 0$$

where

$$\nu_{yx} = \nu_y = \frac{\nu_{xy}E_y}{E_x} \qquad \nu_x = \nu_{xy}$$

$$\nu_{zx} = \frac{\nu_{xz}E_z}{E_x} \qquad \nu_{zy} = \frac{\nu_{yz}E_z}{E_y}$$

$$\left(\varepsilon_{xy}\right) = \begin{pmatrix} \varepsilon_x \\ \varepsilon_y \\ \gamma_{xy} \end{pmatrix} \qquad \left(\sigma_{xy}\right) = \begin{pmatrix} \sigma_x \\ \sigma_y \\ \sigma_{xy} \end{pmatrix}$$

Similarly, from equation (4.21), the relationship between the local stresses and strains is given by

$$\begin{pmatrix} \sigma_x \\ \sigma_y \\ \tau_{xy} \end{pmatrix} = \begin{pmatrix} k_{11} & k_{12} & k_{13} \\ k_{21} & k_{22} & k_{23} \\ k_{31} & k_{32} & k_{33} \end{pmatrix} \begin{pmatrix} \varepsilon_x \\ \varepsilon_y \\ \gamma_{xy} \end{pmatrix} \tag{9.108}$$

or

$$\left(\sigma_{xy}\right) = \left(k\right)\left(\varepsilon_{xy}\right) \tag{9.109}$$

where $(k) = \left(S^{-1}\right)$ is a material matrix for plane stress, and

$$k_{11} = \frac{E_x}{\left(1 - \nu_x\nu_y\right)} \qquad k_{22} = \frac{E_y}{\left(1 - \nu_x\nu_y\right)}$$

$$k_{12} = k_{21} = \frac{\nu_x E_y}{\left(1 - \nu_x\nu_y\right)} = \frac{\nu_y E_x}{\left(1 - \nu_x\nu_y\right)}$$

$$k_{33} = 0 \qquad k_{13} = k_{31} = k_{23} = k_{32} = 0$$

Now from Section 9.2, page 202, the relationships between local and global stresses are given by:

$$\sigma_x = \sigma_X \cos^2\theta + \sigma_Y \sin^2\theta + 2\tau_{XY} \sin\theta \cos\theta$$

$$\sigma_y = \sigma_{X+90^\circ} = \sigma_X \sin^2\theta + \sigma_Y \cos^2\theta + 2\tau_{XY} \sin\theta \cos\theta$$

$$\tau_{xy} = -\sigma_X \cos\theta \sin\theta + \sigma_Y \sin\theta \cos\theta + \tau_{XY}\left(\cos^2\theta - \sin^2\theta\right)$$

where σ_x, σ_y and τ_{xy} are local stresses and σ_X, σ_Y and τ_{XY} are global or reference stresses, or in matrix form:

$$\begin{pmatrix} \sigma_x \\ \sigma_y \\ \tau_{xy} \end{pmatrix} = \begin{pmatrix} C^2 & S^2 & 2SC \\ S^2 & C^2 & -2SC \\ -SC & SC & \left(C^2 - S^2\right) \end{pmatrix} \begin{pmatrix} \sigma_X \\ \sigma_Y \\ \tau_{XY} \end{pmatrix} \tag{9.110}$$

where $S = \sin\theta$ and $C = \cos\theta$

$$\left(\sigma_{xy}\right) = \left(DC\right)\left(\sigma_{XY}\right) \tag{9.111}$$

or

$$\left(\sigma_{XY}\right) = \left(DC\right)^{-1}\left(\sigma_{xy}\right) \tag{9.112}$$

where $\left(DC\right) = \begin{pmatrix} C^2 & S^2 & 2SC \\ S^2 & C^2 & -2SC \\ -SC & SC & \left(C^2 - S^2\right) \end{pmatrix}$ and

$$\left(DC\right)^{-1} = \begin{pmatrix} C^2 & S^2 & -2SC \\ S^2 & C^2 & 2SC \\ SC & -SC & \left(C^2 - S^2\right) \end{pmatrix} \tag{9.113}$$

Similarly $\left(\varepsilon_{xy}\right) = \left(DC_1\right)\left(\varepsilon_{XY}\right)$ and

$$\left(DC_1\right) = \begin{pmatrix} C^2 & S^2 & SC \\ S^2 & C^2 & -SC \\ -2SC & 2SC & \left(C^2 - S^2\right) \end{pmatrix} \tag{9.114}$$

Now, from equation (9.109)

$$\left(\sigma_{xy}\right) = \left(k\right)\left(\varepsilon_{xy}\right) \tag{9.115}$$

Substituting equation (9.114) into equation (9.115) gives:

$$\left(\sigma_{xy}\right) = \left(k\right)\left(DC_1\right)\left(\varepsilon_{XY}\right) \tag{9.116}$$

Substituting equation (9.116) into equation (9.112) gives:

$$\left(\sigma_{XY}\right)=\left(DC\right)^{-1}\left(k\right)\left(DC_1\right)\left(\varepsilon_{XY}\right) \qquad (9.117)$$

Equation (9.117) is the equivalent global stress-strain relationship for the lamina of Figure 9.44 on page 228, in terms of the global or reference areas X and Y. Equation (9.117) can be rewritten as

$$\left(\sigma_{XY}\right)=\left(k'\right)\left(\varepsilon_{XY}\right) \qquad (9.118)$$

where $\left(k'\right)=\left(DC\right)^{-1}\left(k\right)\left(DC_1\right)$ is a stiffness matrix

$$\left(k'\right)=\begin{pmatrix} k'_{11} & k'_{12} & k'_{13} \\ k'_{21} & k'_{22} & k'_{23} \\ k'_{31} & k'_{32} & k'_{33} \end{pmatrix} \qquad (9.119)$$

and

$$k'_{11}=\frac{1}{\gamma}\begin{bmatrix} E_x\cos^4\theta+E_y\sin^4\theta \\ +\left(2v_xE_y+4\gamma G\right)\cos^2\theta\sin^2\theta \end{bmatrix}$$

$$k'_{12}=k'_{21}=\frac{1}{\gamma}\begin{bmatrix} v_xE_y\left(\cos^4\theta+\sin^4\theta\right) \\ +\left(E_x+E_y-4\gamma G\right)\cos^2\theta\sin^2\theta \end{bmatrix}$$

$$k'_{13}=k'_{31}=\frac{1}{\gamma}\begin{bmatrix} \cos^3\theta\sin\theta\left(E_x-v_xE_y-2\gamma G\right) \\ -\cos\theta\sin^3\theta\left(E_y-v_xE_y-2\gamma G\right) \end{bmatrix}$$

$$k'_{22}=\frac{1}{\gamma}\begin{bmatrix} E_y\cos^4\theta+E_x\sin^4\theta \\ +\sin^2\theta\cos^2\theta\left(2v_xE_y+4\gamma G\right) \end{bmatrix}$$

$$k'_{23}=k'_{32}=\frac{1}{\gamma}\begin{bmatrix} \cos\theta\sin^3\theta\left(E_x-v_xE_y-2\gamma G\right) \\ -\cos^3\theta\sin\theta\left(E_y-v_xE_y-2\gamma G\right) \end{bmatrix}$$

$$k'_{33}=\frac{1}{\gamma}\begin{bmatrix} \sin^2\theta\cos^2\theta\left(E_x+E_y-2v_xE_y-2\gamma G\right) \\ +\gamma G\left(\cos^4\theta+\sin^4\theta\right) \end{bmatrix}$$

where $\gamma=\left(1-v_xv_y\right)$

Similarly, to obtain the global strains of the lamina of Figure 9.44 in terms of the global stresses, consider equation (9.107), i.e.

$$\left(\varepsilon_{xy}\right)=\left(S\right)\left(\sigma_{xy}\right) \qquad (9.120)$$

Substituting equation (9.111) into equation (9.120) gives:

$$\left(\varepsilon_{xy}\right)=\left(S\right)\left(DC\right)\left(\sigma_{XY}\right) \qquad (9.121)$$

Equating equations (9.114) and (9.121) gives:

$$\left(\varepsilon_{XY}\right)=\left(DC_1\right)^{-1}\left(S\right)\left(DC\right)\left(\sigma_{XY}\right)=\left(S'\right)\left(\varepsilon_{XY}\right) \qquad (9.122)$$

where $\left(S'\right)=\left(DC_1\right)^{-1}\left(S\right)\left(DC\right)$ is an overall compliance matrix

$$\left(S'\right)=\begin{pmatrix} S'_{11} & S'_{12} & S'_{13} \\ S'_{21} & S'_{22} & S'_{23} \\ S'_{31} & S'_{32} & S'_{33} \end{pmatrix}$$

and

$$S'_{11}=S_{11}\cos^4\theta+S_{22}\sin^4\theta+\left(2S_{12}+S_{33}\right)\cos^2\theta\sin^2\theta$$

$$S'_{12}=S'_{21}=\left(S_{11}+S_{22}-S_{33}\right)\cos^2\theta\sin^2\theta + S_{12}\cos^4\theta\sin^4\theta$$

$$S'_{13}=S'_{31}=\left(2S_{22}-2S_{12}-S_{33}\right)\cos^3\theta\sin\theta - \left(2S_{22}-2S_{12}-S_{33}\right)\sin^3\theta\cos\theta$$

$$S'_{22}=S_{11}\sin^4\theta+S_{22}\cos^4\theta+\left(2S_{12}+S_{33}\right)\cos^2\theta\sin^2\theta$$

$$S'_{23}=S'_{32}=\left(2S_{11}-2S_{12}-S_{33}\right)\cos\theta\sin^3\theta - \left(2S_{22}-2S_{12}-S_{33}\right)\sin\theta\cos^3\theta$$

$$S'_{33}=4\left(S_{11}-2S_{12}+S_{33}\right)\cos^3\theta\sin^2\theta + S_{33}\left(\cos^2\theta-\sin^2\theta\right)^2$$

Some numerical examples using the above equations appear in Chapter 19.

For fully worked solutions to each of the problems in Exercises 40 to 43 in this chapter, go to the website:

www.routledge.com/cw/bird

Chapter 10

Membrane theory for thin-walled circular cylinders and spheres

Why it is important to understand: **Membrane theory for thin-walled circular cylinders and spheres**

Thin-walled circular cylinders appear in a number of different branches of engineering, including ocean engineering, aeronautical engineering and civil engineering, and in food packaging, boiler technology and medicine. In ocean engineering, such structures include submarine pressure hulls, offshore drilling rigs and other submersibles. In aeronautical engineering, such structures appear as fuselages for aeroplanes and rockets. In civil engineering, such structures appear as silos, cooling towers, nuclear reactors and so on. In food packaging, such structures are used to contain a number of different foods, including liquids, fish, meat, vegetables and so on, and in medicine, such structures appear as needles, 'vacutainers', thermometers, rubber tubes and so on. In this text, only pressure vessels subjected to internal pressure will be considered, because pressure vessels imploding under external pressure are generally of a postgraduate nature and are covered elsewhere (*Pressure Vessels: External Pressure Technology*, C.T.F. Ross, Elsevier, UK).

At the end of this chapter you should be able to:

- define a thin-walled pressure vessel
- calculate the hoop stress in a thin-walled circular cylinder under uniform pressure
- calculate the longitudinal stress in a thin-walled circular cylinder under uniform pressure
- calculate the membrane stresses in a thin-walled spherical shell, under pressure
- understand volumetric stress and strain
- determine the additional compressible liquid that can be pumped into a thin-walled pressure vessel
- design a simple hemispherical dome/cylinder joint

Mathematical references
In order to understand the theory involved in this chapter on **membrane theory for thin-walled circular cylinders and spheres**, knowledge of the following mathematical topics is required: ***integration and double angles.*** If help/revision is needed in these areas, see page 49 for some textbook references.

10.1 Introduction

The use of thin-walled circular cylinders and spheres for containing gases or liquids under pressure is a popular industrial requirement. This is partly because of the nature of fluid pressure and partly because such loads can be most efficiently resisted by in-plane membrane stresses, acting in curved shells.

In general, thin-walled shells have a very small bending resistance in comparison with their ability to resist loads in membrane tension. Although thin-walled shells also have a relatively large capability to resist pressure loads in membrane compression, the possibility arises that in such cases, failure can take place owing to structural instability (buckling) at stresses which may be a small fraction of that to cause yield, as shown in Figure 10.1.

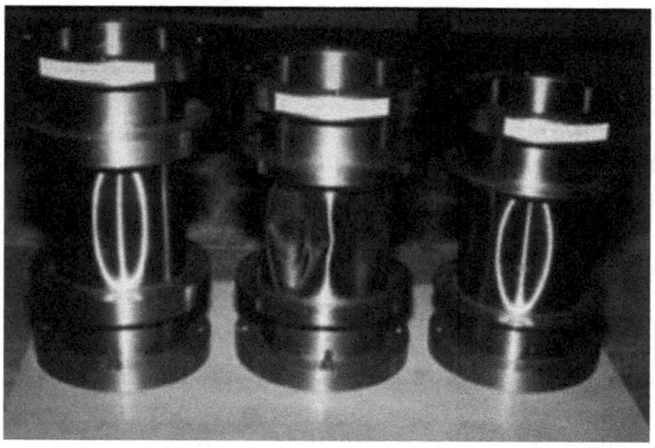

Figure 10.1 Buckled forms of thin-walled cylinders under uniform external pressure

Thin-walled circular cylinders and spheres appear in many forms, including submarine pressure hulls, the legs of offshore drilling rigs, containment vessels for nuclear reactors, boilers, condensers, storage tanks, gas holders, pipes, pumps and many other different types of pressure vessel.

Ideally, from a structural viewpoint, the perfect vessel to withstand uniform internal pressure is a thin-walled spherical shell, but such a shape may not necessarily be the most suitable from other considerations. For example, a submarine pressure hull, in the form of a spherical shell, is not a suitable shape for hydrodynamic purposes, nor for containing large quantities of equipment or large numbers of personnel, and in any case, docking a spherically shaped vessel may present problems.

Furthermore, pressure vessels of spherical shape may present difficulties in housing or storage, or in transport, particularly if the pressure vessel is being carried on the back of a lorry ('truck' in the USA).

Another consideration in deciding whether the pressure vessel should be cylindrical or spherical is from the point of view of its cost of manufacture. For example, although a spherical pressure vessel may be more structurally efficient than a similar cylindrical pressure vessel, the manufacture of the former may be considerably more difficult than the latter, so that additional labour costs of constructing a spherical pressure vessel may be much greater than any material savings that may be gained, especially as extruded cylindrical tubes can often be purchased 'off the shelf'.

Circular cylindrical shells are usually blocked off by domes, but can be blocked off by circular plates; however, if circular plates are used, their thickness is relatively large in comparison with the thickness of the shell dome ends.

For a practical video demonstration of the collapse of submarine pressure hulls go to www.routledge.com/cw/bird and click on the menu for 'Mechanics of Solids 3rd Edition'

10.2 Is it possible for humans to inhabit the moon?

In April 2012, Ross was emailed by someone, who asked him whether or not it was possible to build a submarine which can also be used as an aircraft. Ross's reply was, 'Yes, if you use carbon nanotubes/graphene for the construction of the pressure hull & the associated structures'. Carbon nanotubes/graphene have about 300:1 to about 600: 1 of the strength to weight ratio of high-tensile steel. This led Ross to think that it may be possible to colonise the Moon with a spherical pressure vessel filled with air under an internal pressure of 1 atmosphere. Ross thought that if you design a steel spherical pressure hull of (say) 50 miles diameter, to house humans, in a 1 atmosphere internal pressure vessel, on the Moon, then its wall thickness will be a massive (un-buildable) 100 ft – over 30 m (see his book on *Pressure Vessels: External Pressure Technology,* 2nd edn, Table 1.5 and Section 11.11.8)! If, however, you design the same specification spherical shell, of 50 miles diameter, using carbon nanotubes/graphene for the structure, its wall thickness need only be about 1 ft (or 30 cm)! Moreover, its mass will be about 1/600th to 1/300th of that of the comparable high-tensile steel vessel. If the diameter of the carbon nanotube/graphene pressure vessel is (say) 20 miles, its wall thickness need only be about 5 ins (or 12.5 cm) – the wall thickness of many present day submarines!!! **Thus humankind can inhabit the Moon and other planets!** We will have lift-off, albeit, when it happens, Ross will probably be 'off the planet'!

Some people think that Mars could be colonised, but the average distance of Mars from Planet Earth is about 140 million miles, while the Moon is only a distance of about 240,000 miles; thus it is better to attempt to colonise the Moon, before attempting to colonise Mars. Some people say we can grow lettuce on Mars, because its atmosphere is about 95% CO_2. However, the atmosphere of Mars is only at a pressure of about 1/100th of an atmosphere, so if you place water ice on Mars under pressure, it will vaporise directly into water vapour without turning into pure water, because Mars' atmosphere is in a vacuum state; so how are you going to get liquid water to irrigate the lettuce? Others say we can colonise Venus, but Venus is very hot – its surface is so hot, it can melt lead! Moreover, Venus' atmosphere is at a pressure of about 100 atmospheres, which is so high that it is equivalent to diving to a depth of about 1000 m in the oceans of Planet Earth! Venus has an atmosphere of about 96% CO_2 and is about 106 million miles from Earth. Nevertheless, to conquer the Moon and these planets, we will need a thin-walled spherical pressure vessel, made of composite – as explained in Chapter 19!

10.3 Circular cylindrical shells under uniform internal pressure

The two major methods of failure of a circular cylindrical shell under uniform pressure are as follows.

(1) Failure due to circumferential or hoop stress (σ_H)

If failure is due to the hoop stress, then fracture occurs along a longitudinal seam, as shown in Figure 10.2.

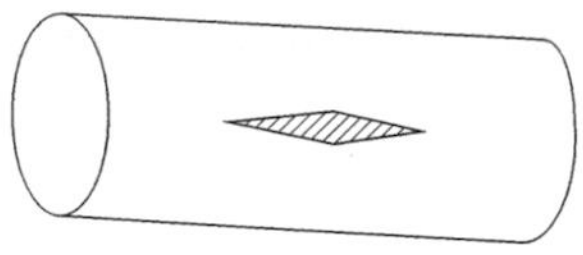

Figure 10.2 Fracture due to hoop stress

Consider the circular cylinder of Figure 10.3, which may be assumed to split in half, along two longitudinal seams, owing to the hoop stress, σ_H. To determine σ_H in terms of the applied pressure and the geometrical properties of the cylinder, consider the equilibrium of one half of the cylindrical shell, as shown in Figure 10.4.

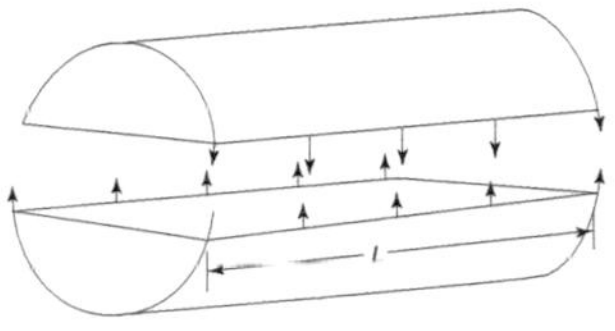

Figure 10.3 Failure due to hoop stress

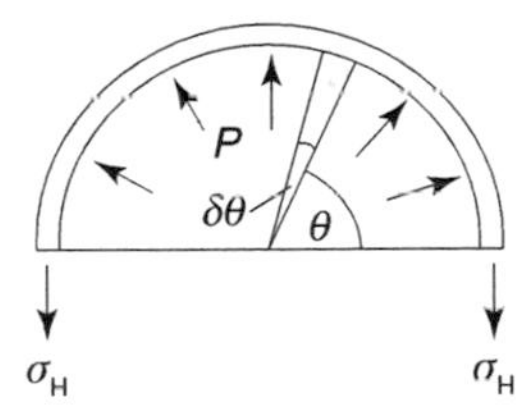

Figure 10.4 Equilibrium of circular cylinder

Given

σ_H = hoop stress, which under internal pressure is a maximum principal stress

P = internal pressure R = internal radius

t = wall thickness of cylinder L = length of cylinder

then resolving vertically in Figure 10.4 gives:

downward forces = upward component of force

i.e. $$\sigma_H \times t \times 2 \times L = \int_0^{\pi} P \times R \times d\theta \times \sin\theta \times L$$

i.e. $$\sigma_H \times t \times 2 = PR\left[-\cos\theta\right]_0^{\pi}$$

$$= PR\left[-\cos\pi - -\cos 0\right]$$

i.e. $$\sigma_H \times t \times 2 = PR[--1 --1] = PR[2] = 2PR$$

Therefore $$\sigma_H = \frac{2PR}{t \times 2} = \frac{PR}{t} \tag{10.1}$$

If η_L is the structural efficiency of a longitudinal joint of the cylinder ($\eta_L \leq 1$), then

$$\sigma_H = \frac{PR}{\eta_L t} \tag{10.2}$$

(2) Failure due to longitudinal stress (σ_L)

If failure is due to the longitudinal stress, then fracture will occur along a circumferential seam, as shown by Figure 10.5.

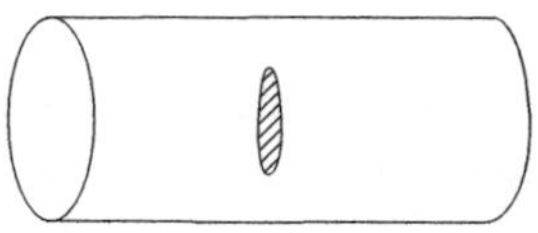

Figure 10.5 Fracture due to longitudinal stress

Consider the circular cylinder of Figure 10.6, which may be assumed to split in two along a circumferential seam, owing to the longitudinal stress, σ_L

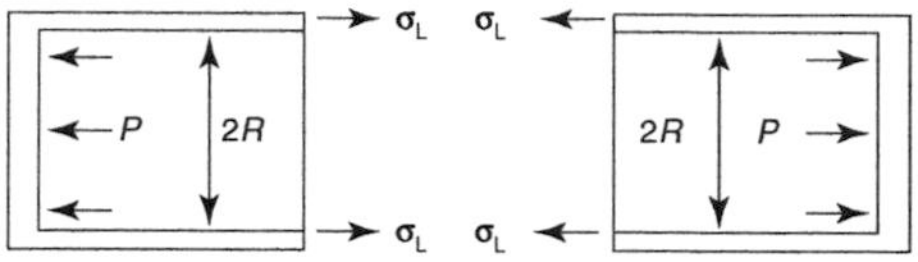

Figure 10.6 Failure along a circumferential seam

Resolving the left 'half' horizontally in Figure 10.6 gives:

forces to the left = forces to the right

i.e. $$P \times \pi R^2 = \sigma_L \times 2\pi R t$$

Therefore $$\sigma_L = \frac{P \times \pi R^2}{2\pi R t}$$

i.e. $$\sigma_L = \frac{PR}{2t} \qquad (10.3)$$

where σ_L is the longitudinal stress or the minimum principal stress, if the pressure is internal.
Comparing equations (10.2) and (10.3), it can be seen that the *hoop stress has twice the magnitude of the longitudinal stress.*
If η_c is the structural efficiency of a circumferential joint of the cylinder ($\eta_c \leq 1$), then

$$\sigma_L = \frac{PR}{2\eta_c t} \qquad (10.4)$$

For edge effects, see Section 10.4, page 235.

Problem 1. Two identical extruded tubes of internal radius 5 m and of wall thickness 2 cm are joined together along a circumferential seam to form the main body of a pressure vessel. Assuming that the ends of the vessel are blocked off by very thick inextensible plates, determine the maximum stress in the vessel when it is subjected to an internal pressure of 0.4 MPa.
It may be assumed that the joint efficiency equals 52%.

Now as the tubes are extruded, there is no longitudinal joint, so that

$$\eta_L = 1$$

From equation (10.2)

$$\sigma_H = \frac{PR}{\eta_L t} = \frac{0.4 \times 10^6 \times 5}{2 \times 10^{-2}}$$

i.e **hoop stress, $\sigma_H = 100$ MN/m²**

Now $$\eta_c = 52\% = 0.52$$

From equation (10.4)

$$\sigma_L = \frac{PR}{2\eta_c t} = \frac{0.4 \times 10^6 \times 5}{2 \times 0.52 \times 2 \times 10^{-2}}$$

i.e. **longitudinal stress, $\sigma_L = 96.2$ MN/m²**

and **maximum stress = $\sigma_H = 100$ MN/m²**

Problem 2. A circular cylindrical pressure vessel of internal diameter 10 m is blocked off at its ends by thick plates, and it is to be designed to sustain an internal pressure of 1 MPa.
Determine a suitable value for the wall thickness of the cylinder, assuming the following apply:

Longitudinal joint efficiency = 98%
Circumferential joint efficiency = 46%
Maximum permissible stress = 100 MN/m²

Consideration of hoop stress

From equation (10.2)

$$\sigma_H = \frac{PR}{\eta_L t}$$

from which $$t = \frac{PR}{\eta_L \sigma_H} = \frac{1 \times 10^6 \times \left(\frac{10}{2}\right)}{0.98 \times 100 \times 10^6}$$

i.e. **wall thickness, $t = 0.051$ m**

Consideration of longitudinal stress

From equation (10.4)

$$\sigma_L = \frac{PR}{2\eta_c t}$$

from which $$t = \frac{PR}{2\eta_c \sigma_L} = \frac{1 \times 10^6 \times 5}{2 \times 0.46 \times 100 \times 10^6}$$

i.e. **wall thickness, $t = 0.0544$ m**
and **design wall thickness = 5.44 cm,** the larger of the two design thicknesses

Now try the following Practice Exercise

Practice Exercise 44. Circular cylindrical shells under uniform internal pressure

1. A thin-walled circular cylinder of internal diameter 10 m is subjected to a maximum internal pressure of 50 bar. Determine its wall thickness, if $\eta_c = 0.8$ and $\eta_L = 0.4$. Assume that the maximum permissible stress is 300 MPa. (Note that 1 bar = 10^5 Pa)

   [208 mm]

2. Assuming that a submarine pressure hull of external diameter 12 m, and wall thickness 5 cm, can be designed using internal pressure theory, and neglecting failure due to buckling, determine its permissible diving depth for a safety factor of 2.

   Assume that the density of sea water = 1020 kg/m^3, $g = 9.81$ m/s^2, $\eta_L = 0.48$, $\eta_c = 0.5$ and the yield stress of the material of construction is 400 MPa.

   [159.9 m]

3. Design the wall thickness for an aircraft fuselage of internal diameter 6 m, subjected to an internal pressure of 0.5 bar. Assume that $\eta_L = 0.48$, $\eta_c = 0.5$ and $\sigma_{yp} = 200$ MPa. (Note that 1 bar = 10^5 Pa)

   [1.56 mm]

For a practical video demonstration of thin-walled circular cylinders under internal pressure go to www.routledge.com/cw/bird and click on the menu for 'Mechanics of Solids 3rd Edition'

10.4 Thin-walled spherical shells under uniform internal pressure

Under uniform internal pressure, a thin-walled spherical shell of constant thickness will have a constant membrane tensile stress, where all such stresses will be principal stresses.

Let σ be the membrane stress in the spherical shell. For such structures, fracture will occur along a diameter, as shown in Figure 10.7.

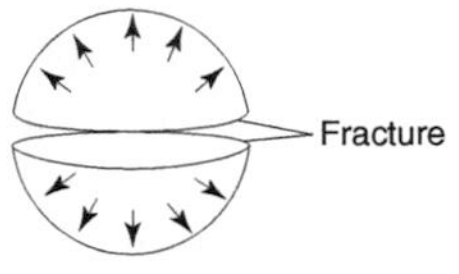

Figure 10.7 Fracture of spherical shell

To determine the membrane principal stress σ, in terms of the applied pressure P and the geometry of the spherical shell, consider the equilibrium of one half of the spherical shell, as shown in Figure 10.8.

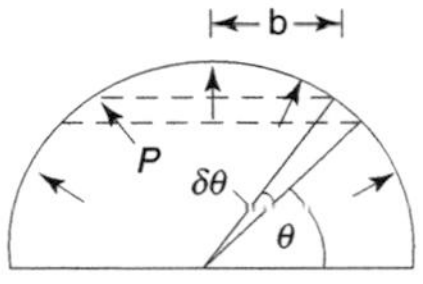

Figure 10.8 Equilibrium of hemispherical shell

Resolving vertically in the top half

upward forces = downward components of forces

i.e. $$\sigma \times 2\pi Rt = \int_0^{\pi/2} P \times 2\pi b \times R d\theta \times \sin\theta$$

From Figure 10.8, $\cos\theta = \frac{b}{R}$ from which, $b = R\cos\theta$

Hence

$$\sigma \times 2\pi Rt = \int_0^{\pi/2} P \times 2\pi\left(R\cos\theta\right) \times R d\theta \times \sin\theta$$

i.e. $$\sigma \times 2\pi Rt = 2\pi PR^2 \int_0^{\pi/2} \sin\theta\cos\theta\, d\theta$$

i.e. $$\sigma \times t = RP\int_0^{\pi/2} \frac{1}{2}\sin 2\theta\, d\theta$$

since $\sin 2\theta = 2\sin\theta\cos\theta$

i.e.

$$\sigma \times t = RP\left[-\frac{1}{4}\cos 2\theta\right]_0^{\pi/2} = -\frac{1}{4}RP\left[\cos 2\theta\right]_0^{\pi/2}$$

$$= -\frac{1}{4}RP\left[\cos\left(2 \times \frac{\pi}{2}\right) - \cos 0\right] = -\frac{1}{4}RP\left[-1-1\right]$$

Hence $$\sigma \times t = -\frac{1}{4}RP\left[-2\right] = \frac{RP}{2}$$

Therefore $$\sigma = \frac{PR}{2t} \quad (10.5)$$

From equations (10.1) and (10.5), it can be seen that the maximum principal stress in a spherical shell has half the value of the maximum principal stress in a circular cylinder of the same radius.

If η is the structural efficiency of a joint on a diameter of the spherical shell ($\eta \le 1$), then

$$\sigma = \frac{PR}{2\eta t} \quad (10.6)$$

Problem 3. A thin-walled spherical vessel, of internal diameter 10 m, is to be designed to withstand an internal pressure of 1 MN/m^2. Determine a suitable value of the wall thickness, assuming the following apply:
Joint efficiency = 75%
Maximum permissible stress = 100 MN/m^2

From equation (10.6)

$$\sigma = \frac{PR}{2\eta t}$$

from which **wall thickness,**

$$t = \frac{PR}{2\eta\sigma} = \frac{1\times 10^6 \times \left(\frac{10}{2}\right)}{2\times 0.75\times 100\times 10^6}$$

$$= 0.033 \text{ m} = \mathbf{3.3\ cm}$$

Problem 4. A submarine pressure hull of external diameter 10 m may be assumed to be composed of a long cylindrical shell, blocked off by two hemispherical shell domes. Neglecting buckling due to external pressure, and the effects of discontinuity at the intersections between the domes and the cylinder, determine suitable thicknesses for the cylindrical shell body and the hemispherical dome ends. The following may be assumed:
Maximum permissible stress = 200 MN/m^2
Diving depth of submarine = 250 m
Density of sea water = 1020 kg/m^3
Acceleration due to gravity, g = 9.81 m/s^2
Longitudinal joint efficiency = 90%
Circumferential joint efficiency = 70%

Cylindrical body

From equation (10.2)

$$\sigma_H = \frac{PR}{\eta_L t} \text{ from which, } t = \frac{PR}{\eta_L \sigma_H}$$

where pressure,

$$P = \rho \times g \times h = 1020\frac{\text{kg}}{\text{m}^3} \times 9.81\frac{\text{m}}{\text{s}^2} \times 250\text{m}$$

$$= 2.5 \times 10^6 \text{ Pa} = 2.5\text{MPa}$$

Hence, **thickness of cylindrical shell,**

$$t = \frac{PR}{\eta_L \sigma_H} = \frac{2.5\times 10^6 \times 5}{0.9\times 200\times 10^6} = 0.0694 \text{ m} = \mathbf{6.94cm}$$

Dome ends

From equation (10.6)

$$\sigma = \frac{PR}{2\eta t}$$

from which $$t = \frac{PR}{2\eta_c \sigma} = \frac{2.5\times 10^6 \times 5}{2\times 0.7\times 200\times 10^6}$$

i.e. **thickness of dome,** t = 0.0446 m = **4.46 cm**

Hence, the wall thickness of the cylindrical body is 6.94 cm and the wall thickness of the dome ends is 4.46 cm.

NB From the above calculations, it can be seen that the required wall thickness of a submarine pressure hull increases roughly in proportion to its diving depth. Thus, to ensure that the submarine has a sufficient reserve buoyancy for a given diameter, it is necessary to restrict its diving depth or to use a material of construction which has a better strength to weight ratio. It should also be noted that under external pressure, the cylindrical section of a submarine pressure hull can buckle at a pressure which may be a fraction of that to cause yield, as shown in Figure 10.1. In a similar manner, the dome ends can also buckle at a pressure which may be a small fraction of that to cause yield, as shown in Figures 10.9 and 10.10.

Figure 10.9 Axisymmetric buckling of an oblate dome under uniform external pressure

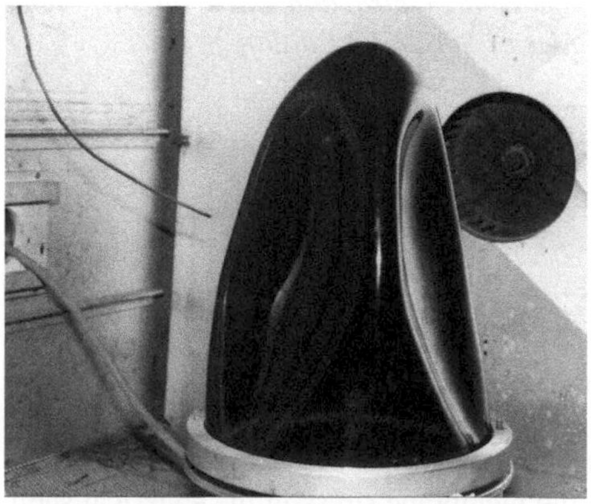

Figure 10.10 Lobar buckling of a hemispherical or prolate dome under uniform external pressure

Liquid required to raise the pressure inside a circular cylinder (based on small-deflection elastic theory)

In the case of a circular cylinder, assuming that it is just filled with liquid, the additional liquid that will be required to raise the internal pressure will be partly as a result of the swelling of the structure and partly as a result of the compression of the liquid itself, as shown by the following components:

(a) longitudinal extension of the cylinder
(b) radial extension of the cylinder
(c) compressibility of the liquid.

Consider a thin-walled circular cylinder, blocked off at its ends by thick inextensible plates, and just filled with the liquid.

The calculation for the additional liquid to raise the internal pressure to P will be composed of the following three components.

(a) Longitudinal extension of the circular cylinder

Let u = the longitudinal movement of one end of the cylinder, relative to the other, as shown in Figure 10.11.

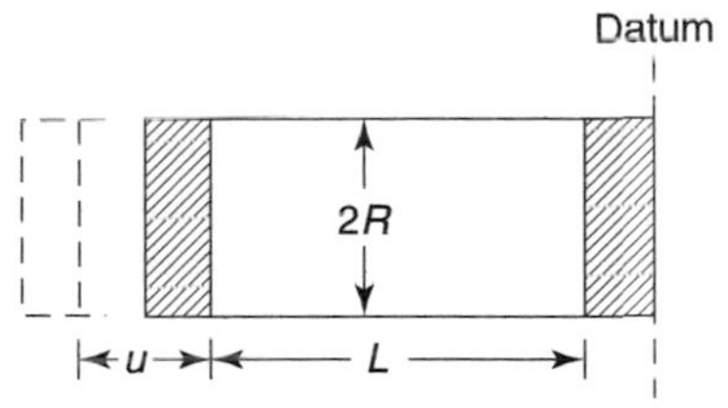

Figure 10.11 Longitudinal movement of the cylinder

and ε_L = longitudinal strain in cylinder $= \dfrac{u}{L}$ (10.7)

The change of volume of the cylinder due to u is given by:

$$\delta V_1 = \pi R^2 u$$

From equation (10.7), $u = \varepsilon_L L$

hence, $$\boldsymbol{\delta V_1 = \pi R^2 \varepsilon_L L} \qquad (10.8)$$

(b) Radial extension of cylinder

Let w = radial deflection of cylinder, as shown in Figure 10.12,

and $$\varepsilon_H = \text{hoop strain} = \frac{2\pi(R+w) - 2\pi R}{2\pi R}$$

$$= \frac{2\pi R + 2\pi w - 2\pi R}{2\pi R} = \frac{2\pi w}{2\pi R} = \frac{w}{R} \qquad (10.9)$$

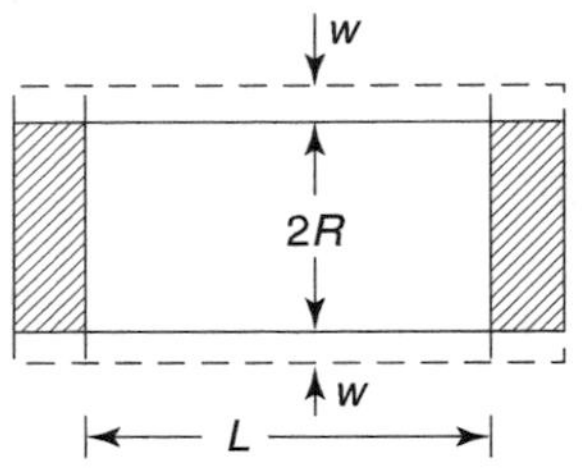

Figure 10.12 Radial deflection of the cylinder

The change of volume of the cylinder due to w is given by:

$$\delta V_2 = 2\pi RLw$$

From equation (10.9), $w = \varepsilon_H R$

hence $$\boldsymbol{\delta V_2 = 2\pi R^2 L \varepsilon_H} \qquad (10.10)$$

[Alternatively, from Section 4.8, the total volumetric strain of the circular cylindrical shell is given by:

$$\frac{\delta V}{V} = \left(\varepsilon_x + \varepsilon_y + \varepsilon_z\right)$$

or $$\frac{\delta V_s}{V} = \left(\varepsilon_L + \varepsilon_H + \varepsilon_H\right)$$

or $$\delta V_s = \pi R^2 L\left(\varepsilon_L + 2\varepsilon_H\right)$$

$$= \delta V_1 + \delta V_2 \text{ (as obtained above)}$$

which is the change in volume due to swelling of the shell.]

(c) Compressibility of liquid

The liquid will compress owing to the raising of its pressure by P. From Section 4.8,

$$\frac{\text{volumetric stress}}{\text{volumetric strain}} = K = \text{bulk modulus}$$

Let δV_3 be the change in volume of the liquid due to its compressibility. Now, P is the volumetric stress and the volumetric strain is $\dfrac{\delta V_3}{V}$ where V is the internal volume of the cylinder.

Therefore $$\delta V_3 = \frac{PV}{K} \qquad (10.11)$$

From equations (10.8), (10.10) and (10.11), the additional volume of liquid that is required to be pumped in to raise the internal pressure of the cylinder by P is

$$\delta V = \delta V_1 + \delta V_2 + \delta V_3$$

i.e. $$\boldsymbol{\delta V = \pi R^2 \varepsilon_L L + 2\pi R^2 L \varepsilon_H + \frac{PV}{K}} \qquad (10.12)$$

Problem 5. A thin-walled circular cylinder of internal diameter 10 m, wall thickness 2 cm and length 5 m is just filled with water. Determine the additional water that is required to be pumped into the vessel to raise its pressure by 0.5 MPa. Assume that: $E = 2 \times 10^{11}\text{N/m}^2$, $v = 0.3$ and $K = 2 \times 10^9\text{N/m}^2$

From equation (10.8)

$$\delta V_1 = \pi R^2 \varepsilon_L L$$

but $$\varepsilon_L = \frac{1}{E}(\sigma_L - v\sigma_H) = \frac{1}{E}\left(\frac{PR}{2t} - v\frac{PR}{t}\right)$$

from equations (10.1) and (10.3)

i.e.

$$\varepsilon_L = \frac{PR}{Et}\left(\frac{1}{2} - v\right) = \frac{0.5\times10^6 \times \left(\frac{10}{2}\right)}{2\times10^{11}\times2\times10^{-2}}\left(\frac{1}{2} - 0.3\right)$$

i.e. $$\varepsilon_L = 1.25\times10^{-4}$$

Therefore $$\delta V_1 = \pi R^2 \varepsilon_L L = \pi \times 5^2 \times 1.25\times10^{-4}\times 5$$

i.e. $$\boldsymbol{\delta V_1 = 0.049\,\text{m}^3} \qquad (10.13)$$

From equation (10.10)

$$\delta V_2 = 2\pi R^2 L \varepsilon_H$$

but $$\varepsilon_H = \frac{1}{E}(\sigma_H - v\sigma_L) = \frac{1}{E}\left(\frac{PR}{t} - v\frac{PR}{2t}\right)$$

i.e.

$$\varepsilon_H = \frac{PR}{Et}\left(1 - \frac{v}{2}\right) = \frac{0.5\times10^6 \times \left(\frac{10}{2}\right)}{2\times10^{11}\times2\times10^{-2}}\left(1 - \frac{0.3}{2}\right)$$

i.e. $$\varepsilon_H = 5.313\times10^{-4}$$

Therefore

$$\delta V_2 = 2\pi R^2 \varepsilon_H L = 2\pi \times 5^2 \times 5.313\times10^{-4}\times 5$$

i.e. $$\delta V_2 = 0.417\,\text{m}^3 \qquad (10.14)$$

From equation (10.11)

$$\delta V_3 = \frac{PV}{K} = \frac{0.5\times10^6 \times \pi R^2 L}{K} = \frac{0.5\times10^6\times\pi(5)^2\times5}{2\times10^9}$$

i.e. $$\boldsymbol{\delta V_3 = 0.098\,\text{m}^3} \qquad (10.15)$$

From equations (10.13) to (10.15), the additional volume of water required to be pumped into the vessel to raise its internal pressure by 0.5 MPa is:

$$\delta V = \delta V_1 + \delta V_2 + \delta V_3 = 0.049 + 0.417 + 0.098$$

i.e. $$\boldsymbol{\delta V = 0.564\,\text{m}^3} \qquad (10.16)$$

From equation (10.16), it can be seen that the bulk of the additional water required to raise the pressure was because of the radial extension of the cylinder.

Additional liquid required to raise the internal pressure of a thin-walled spherical shell (based on small-deflection elastic theory)

Assume that the spherical shell is just filled with liquid, and let w be the radial deflection of the sphere due to an internal pressure increase of P, as shown in Figure 10.13.

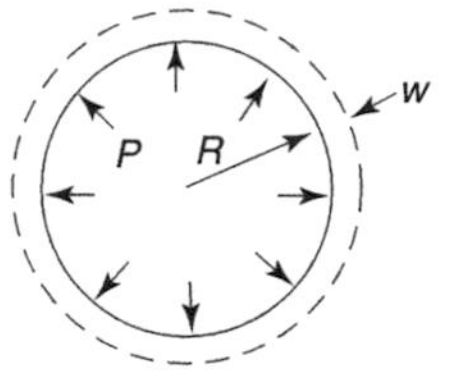

Figure 10.13 Deflected form of spherical shell

The additional liquid that is required to be pumped into the vessel will be because of the following:

(a) swelling of the structure due to the application of P
(b) compressibility of the liquid itself.

If ε is the membrane strain due to P, then

$$\varepsilon = \frac{2\pi(R+w) - 2\pi R}{2\pi R} = \frac{2\pi R + 2\pi w - 2\pi R}{2\pi R} = \frac{2\pi w}{2\pi R} = \frac{w}{R} \qquad (10.17)$$

but $\varepsilon = \frac{1}{E}(\sigma - \nu\sigma) = \frac{1}{E}\left(\frac{PR}{2t} - \nu\frac{PR}{2t}\right)$

$$= \frac{PR}{2Et}(1-\nu) \quad (10.18)$$

The change in volume due to swelling of the shell is:

$$\delta V_s = 4\pi R^2 w = 4\pi R^2(\varepsilon R) = 4\pi R^3\varepsilon \quad \text{from equation (10.17)}$$

$$= 4\pi R^3\left(\frac{PR}{2Et}(1-\nu)\right) \quad \text{from equation (10.18)}$$

i.e. $$\boldsymbol{\delta V_s = \frac{2\pi P R^4(1-\nu)}{Et}} \quad (10.19)$$

[Alternatively, from Section 4.8, the total volumetric strain due to swelling of the shell is:

$$\frac{\delta V}{V} = (\varepsilon_x + \varepsilon_y + \varepsilon_z)$$

or $$\frac{\delta V_s}{V} = (\varepsilon + \varepsilon + \varepsilon)$$

or $$\delta V_s = V(3\varepsilon) = \frac{4}{3}\pi R^3(3\varepsilon)$$

i.e. $$\delta V_s = 4\pi R^3\varepsilon \text{ (as obtained above)]}$$

The compressibility of the liquid can be calculated from

$$\delta V_L = \frac{PV}{K} \quad (10.20)$$

where δV_L = change in the volume of the liquid due to compression,
and V = internal volume of sphere = $\frac{4}{3}\pi R^3$

Problem 6. A thin-walled spherical shell, of diameter 10 m and a wall thickness 2 cm, is just filled with water. Determine the additional water that is required to be pumped into the vessel to raise its internal pressure by 0.5 MPa.

Assume that: $E = 2\times10^{11}\,\text{N/m}^2$, $\nu = 0.3$ and $K = 2\times10^9\,\text{N/m}^2$

From equation (10.19), the increase in volume of the spherical shell due to its swelling under pressure is:

$$\delta V_s = \frac{2\pi P R^4(1-\nu)}{Et} = \frac{2\pi\times0.5\times10^6\times\left(\frac{10}{2}\right)^4(1-0.3)}{2\times10^{11}\times2\times10^{-2}}$$

i.e. $$\boldsymbol{\delta V_s = 0.344\,\text{m}^3} \quad (10.21)$$

From equation (10.20), the additional volume of water to be pumped in, owing to the compressibility of the water, is:

$$\delta V_L = \frac{PV}{K} = \frac{0.5\times10^6\times\frac{4}{3}\pi R^3}{K} = \frac{0.5\times10^6\times\frac{4}{3}\pi\left(\frac{10}{2}\right)^3}{2\times10^9}$$

i.e. $$\boldsymbol{\delta V_L = 0.131\,\text{m}^3} \quad (10.22)$$

i.e. **the total additional quantity of water that is required to be pumped in to raise the pressure is:**

$$\delta V = \delta V_s + \delta V_L$$

$$= 0.344 + 0.131$$

i.e. $$\boldsymbol{\delta V = 0.475\,\text{m}^3}$$

From equations (10.21) and (10.22), it can be seen that for this spherical shell, the bulk of the additional liquid that was required to raise the pressure by 0.5 MPa was due to the swelling of the shell.

Now try the following Practice Exercise

Practice Exercice 45. Thin-walled spherical shells under uniform internal pressure

1. A boiler, which may be assumed to be composed of a thin-walled cylindrical shell body of internal diameter 4 m, is blocked off by two thin-walled hemispherical dome ends. Neglecting the effects of discontinuity at the intersection between the dome and cylinder, determine suitable thicknesses for the cylindrical shell body and the hemispherical dome ends. The following may be assumed:
   Maximum permissible stress = 100 MN/m^2
   Design pressure = 1 MPa
   Longitudinal joint efficiency = 75%
   Circumferential joint efficiency = 50%
   [Cylinder t = 2.67 cm; dome t = 2 cm]
2. If the vessel of Problem 1 is just filled with water, determine the additional water that is required to be pumped in, to raise the pressure by 1 MPa. The following may be assumed to apply:
   Length of cylindrical portion of vessel = 6 m, $E = 2\times10^{11}$ N/m^2, $\nu = 0.3$ and $K = 2\times10^9$ N/m^2
   [0.122 m^3]

3. A copper pipe of internal diameter 1.25 cm and wall thickness 0.16 cm is to transport water from a tank that is situated 30 m above it. Determine the maximum stress in the pipe, given the following: density of water = 1000 kg/m^3, $g = 9.81$ m/s^2

[1.148 MN/m^2]

4. What would be the change in diameter of the pipe of question 3 due to the applied head of water? Assume that for copper: $E = 1 \times 10^{11}$ N/m^2, $v = 0.33$ and $K = 2 \times 10^9$ N/m^2

[0.12 μm]

5. A thin-walled spherical pressure vessel of 1 m internal diameter is fed by a pipe of internal diameter 3 cm and wall thickness 0.16 cm. Assuming that the material of construction of the spherical pressure vessel has a yield stress of 0.7 of that of the pipe, determine the wall thickness of the spherical shell.

[3.81 cm]

6. A spherical pressure vessel of internal diameter 2 m is constructed by bolting together two hemispherical domes with flanges. Assuming that the number of bolts used to join the two hemispheres together is 12, determine the wall thickness of the dome and the diameter of the bolts, given the following:

Maximum applied pressure = 0.7 MPa
Permissible stress in spherical shell = 50 MPa
Permissible stress in bolts = 200 MPa

[t = 0.70 cm, $d = 3.42$ cm]

7. A thin-walled circular cylinder, blocked off by inextensible end plates, contains a liquid under zero gauge pressure. Show that the additional liquid that is required to be pumped into the vessel, to raise its internal gauge pressure by P, is the same under the following two conditions:

(a) when axial movement of the cylinder is completely free, and
(b) when the vessel is totally restrained from axial movement.

It may be assumed that Poisson's ratio (v) for the cylinder material is 0.25.

10.5 Bending stresses in circular cylinders under uniform pressure

The theory presented in this chapter neglects the effect of bending stresses at the edges of the cylinder, where the vessel may be firmly clamped, as shown in Figure 10.14.

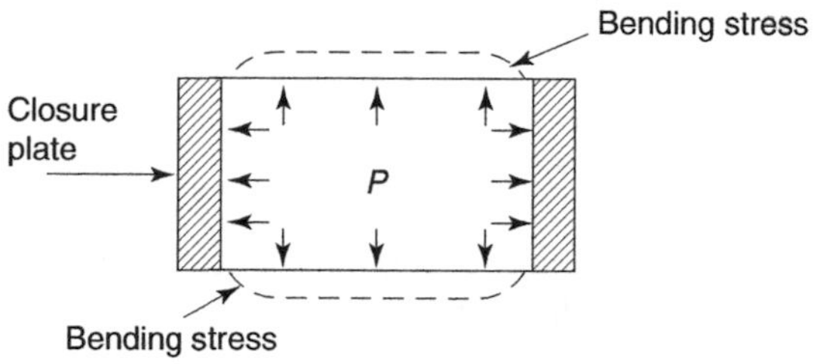

Figure 10.14 Deflected form of cylinder under internal pressure

Figure 10.15 shows the theoretical and experimental values (References 12 and 13, page 497) for the radial deflection of a thin-walled steel cylinder clamped at its ends. The theory was based on the solution of a fourth-order shell differential equation, which is beyond the scope of this book.

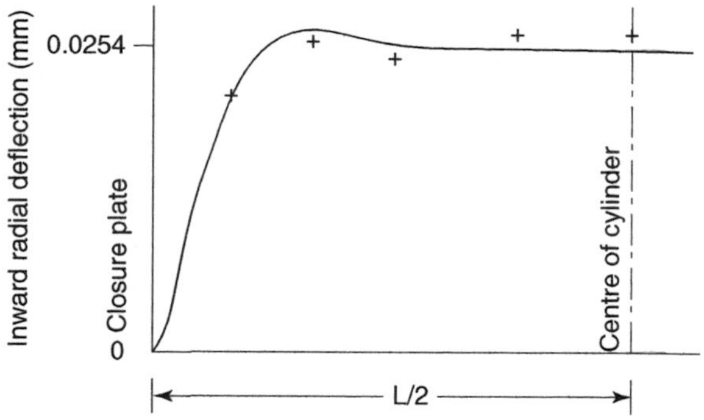

Figure 10.15 Theoretical and experimental values for the radial deflection of model No. 7 at 0.6895 MPa.

The vessel was firmly clamped at its ends, but free to move longitudinally, and it had the following properties:
Internal diameter = 26.04 cm
Wall thickness = 0.206 cm
Length of cylindrical shell = L = 25.4 cm
External pressure = 0.6895 MPa
$E = 1.93 \times 10^{11}$ N/m^2
$v = 0.3$ (assumed)
Initial out-of-roundness = 0.0102 cm

Plots of the theoretical and experimental stresses are shown in Figures 10.16 to 10.19 from one end of the vessel (closure plate) to its mid-span.

From Figures 10.15 to 10.19, it can be seen that the effects of bending are very localised.

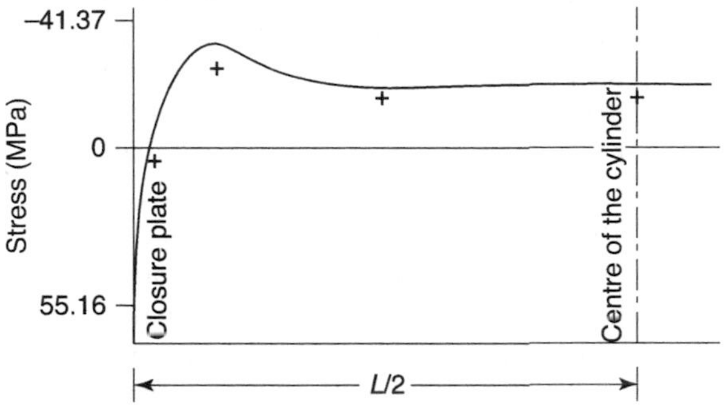

Figure 10.16 Longitudinal stress of the outermost fibre at 0.6895 MPa (model No. 7)

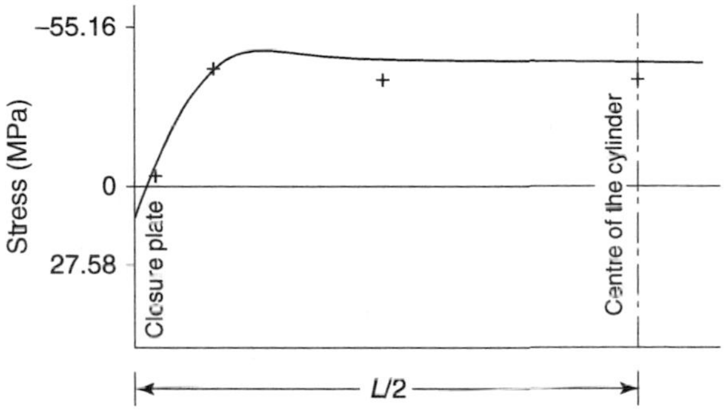

Figure 10.17 Circumferential stress of the outermost fibre at 0.6895 MPa (model No. 7)

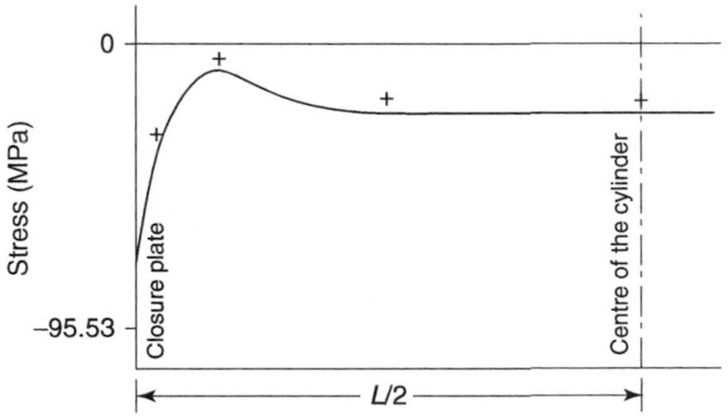

Figure 10.18 Longitudinal stress of the innermost fibre at 0.6895 MPa (model No. 7)

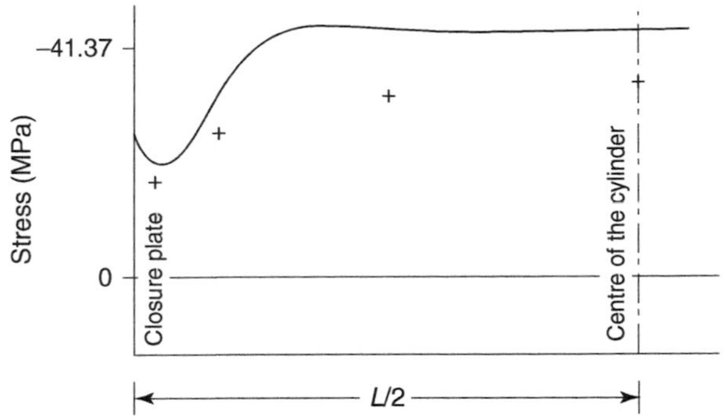

Figure 10.19 Circumferential stress of the innermost fibre at 0.6895 MPa (model No. 7)

10.6 Circular cylindrical shell with hemispherical ends

It is quite common to seal the ends of a circular cylindrical pressure vessel with hemispherical end caps as shown in Figure 10.20.

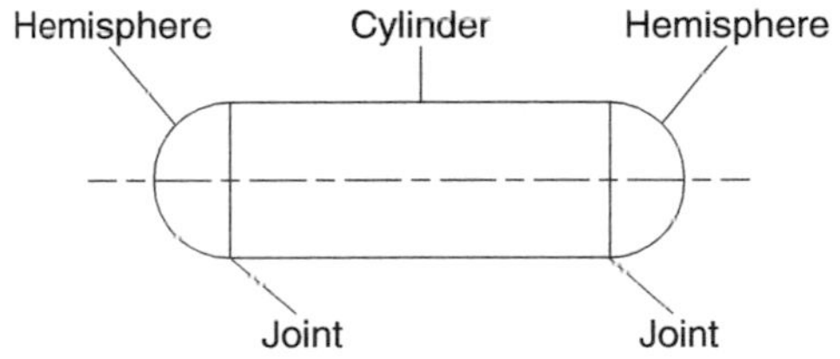

Figure 10.20

When such a vessel is subjected to uniform internal pressure, bending stresses can occur at the joints, because if both the circular cylindrical shell and the hemispherical end caps are made from the same material and of the same wall thickness, the circular cylinder will tend to have a larger radial deflection than the hemispherical shell; this will cause bending stresses at the joints. To overcome or reduce the bending stresses at the joints, the wall thickness of the hemispherical shell must be less than the wall thickness of the circular cylindrical shell. A theory for this will now be produced.

Let w_c = radial deflection of the circular cylinder at the joint due to an internal pressure P

w_s = radial deflection of the hemispherical shell at the joint due to an internal pressure P

R = radius of the circular cylinder

t_c = wall thickness of the circular cylinder

and t_s = wall thickness of the hemispherical shell

Let ε_H be the hoop strain in the cylinder, which from equation (10.9) is:

$$\varepsilon_H = \frac{w_c}{R} = \frac{1}{E}(\sigma_H - \nu\sigma_L)$$

$$= \frac{1}{E}\left(\frac{PR}{t_c} - \nu\frac{PR}{2t_c}\right)$$

from equations (10.1) and (10.3)

from which $\quad w_c = \frac{R}{E}\frac{PR}{t_c}\left(1 - \frac{\nu}{2}\right)$

i.e. $$w_c = \frac{PR^2}{Et_c}\left(1 - \frac{\nu}{2}\right) \qquad (10.23)$$

Let ε be the circumferential strain in the hemispherical shell equal to the meridional strain in the hemispherical shell, which from equation (10.17) is given by:

$$\varepsilon = \frac{w_s}{R} = \frac{1}{E}(\sigma - \nu\sigma)$$

$$= \frac{1}{E}\left(\frac{PR}{2t_s} - \nu\frac{PR}{2t_s}\right) = \frac{PR}{2t_sE}(1 - \nu)$$

from which $$w_s = \frac{PR^2}{2t_sE}(1 - \nu) \qquad (10.24)$$

For no bending to occur at the joint between the circular cylinder and the hemispherical shell, $w_c = w_s$, or equations (10.23) and (10.24) are equal:

$$\frac{PR^2}{Et_c}\left(1 - \frac{\nu}{2}\right) = \frac{PR^2}{2t_sE}(1 - \nu)$$

from which $$\frac{t_s}{t_c} = \frac{(1 - \nu)}{2\left(1 - \frac{\nu}{2}\right)}$$

If $\nu = 0.3$, then $$\frac{t_s}{t_c} = \frac{(1 - 0.3)}{2\left(1 - \frac{0.3}{2}\right)} = \frac{0.7}{1.7} = 0.412$$

i.e. $$\mathbf{t_s = 0.412\, t_c} \qquad (10.25)$$

Now try the following Practice Exercise

Practice Exercise 46. Circular cylindrical shell with hemispherical ends

1. A circular cylinder is to be blocked off by hemispherical ends and the whole is subjected to an internal pressure. Given that the internal diameter of the cylinder and hemisphere are 5 m, and the internal pressure is 2.5 bar, and if the joint efficiency is 100%, determine suitable thicknesses for the cylinder and hemispherical ends, assuming that the cylinder ends deflect the same as the hemispherical ends. Assume that the yield stress = 200 MPa and $\nu = 0.3$ (note that 1 bar $= 10^5$ Pa).

   [$t_{cyl} = 3.125$ mm, $t_s = 1.288$ mm]

And finally ...

In connection with pressure vessels, Carl Ross realised that in the USA alone, they consume about 600 million beverage cans per day and according to Omega Research Associates, Pittsburgh, Pennsylvania, USA, about 100 billion food cans are consumed per annum, worldwide. Thus, if the wall thicknesses of food cans could be decreased, large savings could be made in their manufacture. Also, it would be an environmentally friendly invention. Hence, Ross thoroughly analysed four standard food cans and found that he could make substantial savings on their wall thicknesses; in one case he was able to decrease the wall thickness of a food can by 57.1%, without it losing any of its strength! He tried to patent his invention through his employer in about 1993, but failed to do so. So he published a paper on this topic in the *Journal of Thin-Walled Structures* in 1996 (http://dl.dropbox.com/u/39907336/CorrugatedFoodCanRedesign_2.pdf).
Some ten years later, Ross has noted that the saving in the manufacture of food cans, worldwide, is about £800 million per annum.

For fully worked solutions to each of the problems in Exercises 44 to 46 in this chapter, go to the website:
www.routledge.com/cw/bird

Chapter 11

Energy methods

Why it is important to understand: **Energy methods**

Energy methods are very important in the branch of engineering science known as structural mechanics. Such methods can be used for calculating the deflections of thin curved beams and frameworks. The methods can be used for statically determinate structures and also statically indeterminate structures considering both the elastic and plastic analyses of these components. The biggest breakthrough with energy methods is their applications in the finite element method, where they are used to analyse large and complex shapes in both solid and fluid mechanics. A brief introduction to the finite element method is given in Chapters 20 and 21. (The finite element method is a computer-based method, based on energy, which allows us to analyse such complex shapes as ships to submarine structures, and tall tower blocks to long bridges, and pollution analysis to weather forecasting, and so on).

At the end of this chapter you should be able to:

- appreciate the method of minimum potential (Rayleigh-Ritz)
- state and appreciate the principle of virtual work
- state and appreciate the principle of complementary virtual work
- state and appreciate Castigliano's first and second theorems
- determine the strain energy stored in a rod under axial loading
- determine the strain energy stored in a beam subjected to couples at each end
- determine the strain energy due to a torque stored in a uniform circular-section shaft
- determine the strain energy due to a system of complementary shear stresses
- use Castigliano's theorems in calculations involving the deflection of thin curved beams
- understand and use the unit load method
- appreciate and calculate the effects of suddenly applied and impact loads
- define resilience
- understand plastic collapse of beams
- define plastic neutral axis, load factor and shape factor
- appreciate and calculate residual stresses in beams

Mathematical references

In order to understand the theory involved in this chapter on **energy methods**, knowledge of the following mathematical topics **is** required: ***differentiation, partial differentiation, integration, double angles and quadratic equations***. If help/revision is needed in these areas, see page 50 for some textbook references.

11.1 Introduction

Energy methods in structural mechanics are some of the most useful methods of theoretical analysis, as they lend themselves to computer solutions of complex structural problems (see References 5 and 14 on page 497), and they can also be extended to computer solutions of many other problems in engineering science (see References 15 to 22 and Chapters 20 and 21).

There are many energy theorems and principles (see Reference 14), but only the most popular methods will be considered in this chapter, and applications of some of these will be made to practical problems.

11.2 The method of minimum potential (Rayleigh-Ritz)*

This states that to satisfy the elasticity and equilibrium equations of an elastic body, the partial derivative of the total potential, with respect to the displacements, must be zero,

i.e. $$\frac{\partial \pi_p}{\partial u} = 0 \qquad (11.1)$$

where π_p = total potential, and u = displacement

11.3 The principle of virtual work

This states that if an elastic body under a system of external forces is given a small virtual displacement, then the increase in external virtual work done to the body is equal to the increase in internal virtual strain energy stored in the body.

***John William Strutt, 3rd Baron Rayleigh**, (12 November 1842 – 30 June 1919, pictured above) was an English physicist who, with William Ramsay, discovered argon, an achievement for which he earned the Nobel Prize for Physics in 1904. He also discovered the phenomenon now called Rayleigh scattering, which can be used to explain why the sky is blue, and predicted the existence of the surface waves now known as Rayleigh waves. To find out more about Rayleigh go to www.routledge.com/cw/bird

***Walther Ritz** (22 February 1878 – 7 July 1909) was a Swiss theoretical physicist. He is most famous for his work with Johannes Rydberg on the Rydberg-Ritz combination principle. Ritz is also known for the variational method named after him, the Ritz method. To find out more about Ritz go to www.routledge.com/cw/bird

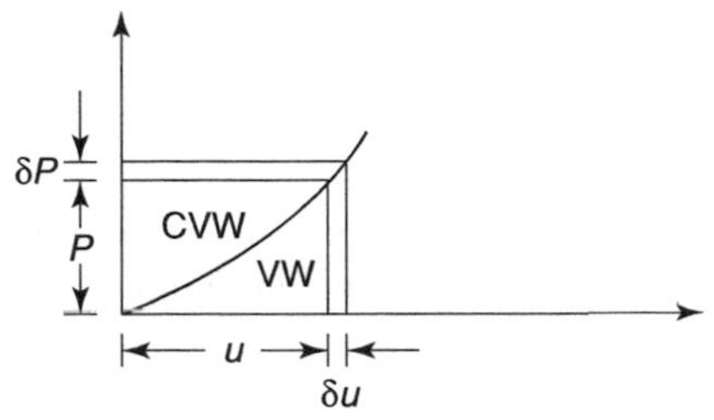

Figure 11.1 Force–displacement relationship

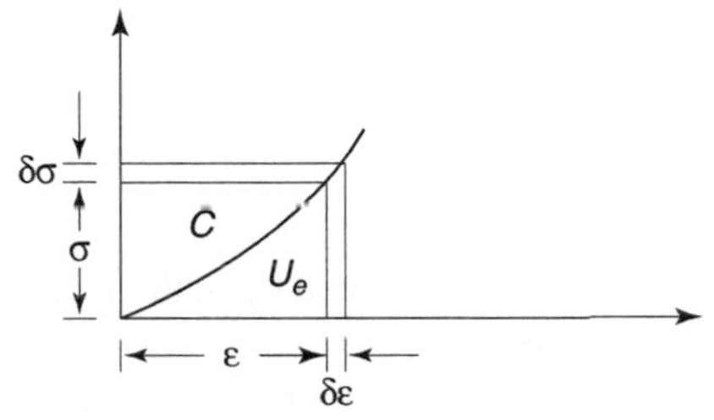

Figure 11.2 Stress–strain relationship

Consider Figure 11.1 and 11.2, where

δu = a small virtual displacement

$\delta\varepsilon$ = a small virtual strain, as a result of δu

P = external force

σ = stress due to P

From Figure 11.1, the increase in virtual work is:

$$d(VW) = \text{area of vertical trapezium} \quad (11.2)$$

and from Figure 11.2, the increase in virtual strain energy is:

$$d(U_e) = \text{area of vertical trapezium} \quad (11.3)$$

Equating equations (11.2) and (11.3) gives:

$$\boldsymbol{P \times \delta u = \sigma \times \delta\varepsilon \times \textbf{volume of the body}} \quad (11.4)$$

11.4 The principle of complementary virtual work

This theorem states that if an elastic body under a system of external forces is subjected to a small virtual force, then the increase in complementary virtual work is equal to the increase in complementary strain energy.

From Figure 11.1, CVW = complementary virtual work

δP = virtual force

u = displacement due to the external forces

and from Figure 11.2, C = complementary strain energy

$\delta\sigma$ = virtual stress due to δP

ε = strain due to the external forces

Hence, from Figure 11.1, the increase in complementary virtual work is

$$d(\text{CVW}) = \text{area of horizontal trapezium} = \delta P \times u \quad (11.5)$$

Similarly, from Figure 11.2, the increase in complementary strain energy is

$$d(\text{C}) = \text{area of horizontal trapezium} = \delta\sigma \times \varepsilon \times \text{volume of body} \quad (11.6)$$

Equating equations (11.5) and (11.6) gives:

$$\boldsymbol{\delta P \times u = \delta\sigma \times \varepsilon \times \textbf{volume of body}} \quad (11.7)$$

11.5 Castigliano's first theorem*

This is really an extension to the principle of complementary virtual work, and applies to bodies that behave in a *linear elastic* manner.

Now when a body is linear elastic

$$\delta(\text{C}) = \delta(U_e) = \delta P \times u$$

so that, in the limit

$$\frac{\partial U_e}{\partial P} = u \quad (11.8)$$

11.6 Castigliano's second theorem

This theorem is an extension of Castigliano's first theorem, and it is particularly suitable for analysing statically indeterminate frameworks.

Castigliano's second theorem simply states that for a framework with redundant members or forces,

$$\frac{\partial U_e}{\partial R} = \lambda \quad (11.9)$$

where U_e = the strain energy of the whole frame

λ = initial lack of fit of the members of the framework

and R = the force in any redundant member, to be determined

*Carlo Alberto Castigliano (9 November 1847 – 25 October 1884) was an Italian mathematician and physicist known for Castigliano's method for determining displacements in a linear-elastic system based on the partial derivatives of strain energy. To find out more about Castigliano go to www.routledge.com/cw/bird

If there is no initial lack of fit (i.e. the members of the framework have been made precisely), then

$$\frac{\partial U_e}{\partial R} = 0$$

For most *pin-jointed frames,* the loads are axial; hence the strain energy is given by:

$$U = \sum\left(\frac{\sigma_i^2}{2E}\right)\times \text{volume} = \sum\left(\frac{P_i}{A_i}\right)^2 \times \frac{A_i l_i}{2E_i}$$

or $$U = \sum_{i=0}^{n} \frac{P_i^2 l_i}{2A_i E_i}$$

$$\frac{\partial U}{\partial R} = \frac{\partial U}{\partial P_i}\frac{\partial P_i}{\partial R}$$

Therefore $$\frac{\partial U}{\partial R} = \sum_{i=0}^{n} \frac{P_i l_i}{A_i E_i}\frac{\partial P_i}{\partial R} = \text{initial lack of fit}$$

$$= 0 \text{ (for most undergraduate problems)} \qquad (11.10)$$

Thus, by applying equation (11.10) to each redundant member or 'force', in turn, the required number of simultaneous equations is obtained, and hence the unknown redundant 'forces' can be determined. Once the redundant 'forces' are known, the other 'forces' can be determined through statics.

11.7 Strain energy stored in a rod under axial loading

Consider the load-displacement relationship of a uniform-section rod, as shown in Figure 11.3.

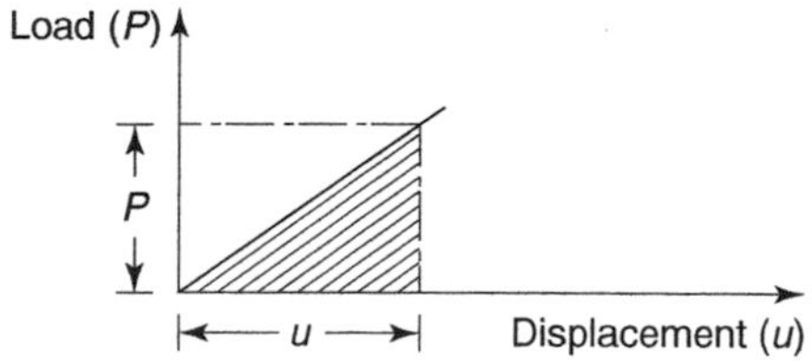

Figure 11.3 Load–displacement relationship for a rod

Now when the rod is displaced by u, the force required to achieve this displacement is P. Furthermore, as the average force during this load-displacement relationship is $P/2$, then:

work done = the shaded area of Figure 11.3

$$= \frac{P}{2}\times u \qquad (11.11)$$

However, this work done will be stored by the rod in strain energy U_e, so that

$$U_e = \frac{P}{2}\times u \qquad (11.12)$$

However $$P = \sigma \times A \qquad (11.13)$$

and $$u = \varepsilon \times l = \sigma \times \frac{l}{E} \qquad (11.14)$$

where σ = stress due to P,
ε = strain due to P
A = cross-sectional area of rod, and
l = length of rod

Substituting equations (11.13) and (11.14) into (11.12) gives:

$$U_e = \frac{P}{2}\times u = \frac{\sigma\times A}{2}\times\sigma\times\frac{l}{E}$$

i.e. $$U_e = \frac{\sigma^2}{2E}\times Al$$

i.e. $$\boldsymbol{U_e = \frac{\sigma^2}{2E}\times \textbf{volume of rod}} \qquad (11.15)$$

For an element of the rod of length dx,

$$\boldsymbol{d(U_e) = \frac{\sigma^2}{2E}\times A\times dx} \qquad (11.16)$$

where $d(U_e)$ is the strain energy in the rod element.

11.8 Strain energy stored in a beam subjected to couples of magnitude *M* at its ends

Consider an element of the beam of length dx and assume it is subjected to two end couples, each of magnitude M, as shown in Figure 11.4.

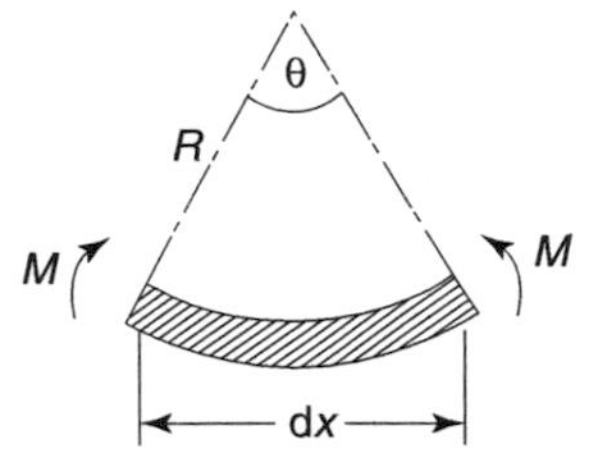

Figure 11.4 Element of beam in pure bending

The work done, WD, in bending the element dx of the beam of Figure 11.4 is:

$$\mathrm{WD} = \frac{1}{2}M\theta \quad \text{for variable moments}$$

(Note: WD = $M\theta$ if M is constant).

However $\quad \theta = \dfrac{dx}{R}$ (since the arc length of a circle, $dx = R\theta$)

Therefore $\quad \mathrm{WD} = \dfrac{1}{2}M \times \dfrac{dx}{R} \quad$ (11.17)

Now from equation (6.1) $\quad \dfrac{M}{I} = \dfrac{E}{R}$

from which $\quad \dfrac{1}{R} = \dfrac{M}{EI} \quad$ (11.18)

Substituting equation (11.18) into equation (11.17) gives:

$$\mathbf{WD} = \boldsymbol{d(U_b)} = \frac{\boldsymbol{M^2}}{\boldsymbol{2EI}} \times \boldsymbol{dx} \qquad (11.19)$$

where $d(U_b)$ is the bending strain energy in the beam element.

11.9 Strain energy due to a torque *T* stored in a uniform circular-section shaft

Let $\quad \theta$ = angle of twist of one end of the shaft relative to the other

and $\quad T$ = applied torque

Then $\quad \mathrm{WD} = U_T = \dfrac{1}{2}T \times \theta \quad$ (11.20)

However, from equation (8.5) $\quad \dfrac{T}{J} = \dfrac{G\theta}{l}$

hence, when $l = dx$ $\quad \theta = \dfrac{T\,dx}{GJ} \quad$ (11.21)

Therefore, substituting equation (11.21) into (11.20) gives:

$$U_T = \frac{1}{2}T \times \frac{T\,dx}{GJ}$$

i.e. $\quad \boldsymbol{U_T = \dfrac{T^2 \times dx}{2G \times J}} \quad$ (11.22)

where U_T is the strain energy due to torsion in an element dx.

Shear strain energy due to a system of complementary shear stresses τ

Consider the rectangular element of Figure 11.5, which is in a state of pure shear.

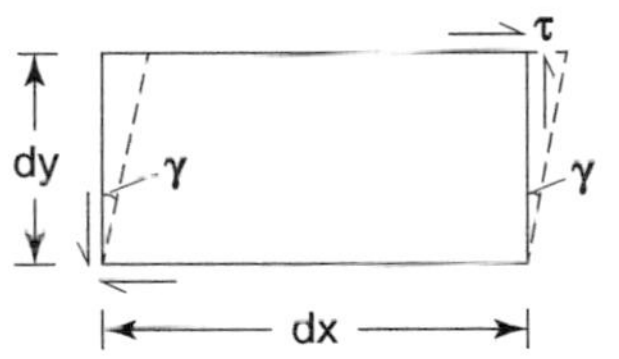

Figure 11.5 Rectangular element in pure shear

From Figure 11.5, it can be seen that the WD by the shear stresses τ, in changing the shape of the element, is

$$\mathrm{WD} = \text{shearing force} \times \frac{\text{displacement in Figure 11.5}}{2}$$

$$= U_s = \frac{\tau}{2} \times (t \times dx) \times (dy \times \gamma)$$

However, shear strain, $\gamma = \dfrac{\tau}{G}$

Therefore $\quad U_s = \dfrac{\tau}{2} \times (t \times dx) \times \left(dy \times \dfrac{\tau}{G}\right)$

i.e. $\quad U_s = \dfrac{\tau^2}{2G} \times (t \times dx \times dy)$

i.e. $\quad \boldsymbol{U_s = \dfrac{\tau^2}{2G} \times \text{volume}} \quad$ (11.23)

where U_s = shear strain energy, t = thickness of plate, and volume = $t \times dx \times dy$

Problem 1. A rod consists of three elements, each of length 0.5 m, joined firmly together. If this rod is subjected to an axial tensile force of 1 MN, determine the total extension of the rod, using strain energy principles and given that the cross-sectional areas of the elements are:

Section 1: $A_1 = 5\times10^{-3}\,m^2$

Section 2: $A_2 = 3\times10^{-3}\,m^2$

Section 3: $A_3 = 1\times10^{-2}\,m^2$

Elastic modulus = 1×10^{11} N/m²

Stress = $\frac{\text{force}}{\text{area}}$, hence

for section 1: $\sigma_1 = \frac{1\times10^6\,N}{5\times10^{-3}\,m^2} = \mathbf{200\ MN/m^2}$

for section 2: $\sigma_2 = \frac{1\times10^6\,N}{3\times10^{-3}\,m^2} = \mathbf{333.3\ MN/m^2}$

for section 3: $\sigma_3 = \frac{1\times10^6\,N}{1\times10^{-2}\,m^2} = \mathbf{100\ MN/m^2}$

Let δ be the total deflection of the rod, so that

$$WD = \frac{1}{2}\times1\times10^6\times\delta \qquad (11.24)$$

From equation (11.15)

$$U_e = \frac{\sigma_1^2}{2E}\times A_1\times l + \frac{\sigma_2^2}{2E}\times A_2\times l + \frac{\sigma_3^2}{2E}\times A_3\times l \qquad (11.25)$$

Equating equations (11.24) and (11.25) gives:

$$\frac{1}{2}\times1\times10^6\times\delta = \frac{\sigma_1^2}{2E}\times A_1\times l + \frac{\sigma_2^2}{2E}\times A_2\times l + \frac{\sigma_3^2}{2E}\times A_3\times l$$

i.e. $$\frac{1}{2}\times1\times10^6\times\delta = \frac{(200\times10^6)^2}{2(1\times10^{11})}\times5\times10^{-3}\times0.5 + \frac{(333.3\times10^6)^2}{2(1\times10^{11})}\times3\times10^{-3}\times0.5 + \frac{(100\times10^6)^2}{2(1\times10^{11})}\times1\times10^{-2}\times0.5$$

i.e. $0.5\times10^6\times\delta = 500 + 833.167 + 250$

from which $$\delta = \frac{500+833.167+250}{0.5\times10^6} = \frac{1583.167}{0.5\times10^6} = 3.166\times10^{-3}\,m$$

i.e. **total extension of the rod, δ = 3.17 mm**

Problem 2. Determine the maximum deflection of the end-loaded cantilever of Figure 11.6 using strain energy principles.

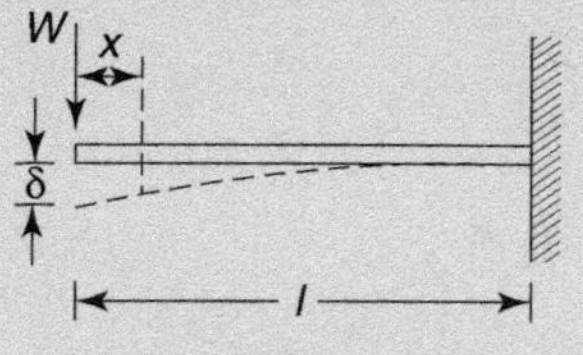

Figure 11.6 End-loaded cantilever

Let I = second moment of area of the beam section, about a horizontal axis

$$\text{WD by the load} = \frac{1}{2}W\times\delta \qquad (11.26)$$

U_b = bending strain energy in the beam

$$= \int_0^l \frac{M^2}{2EI}dx \quad \text{(from equation (11.19))}$$

However $M = -Wx$

Therefore $$U_b = \int_0^l \frac{M^2}{2EI}dx = \int_0^l \frac{(-Wx)^2}{2EI}dx = \frac{W^2}{2EI}\int_0^l x^2\,dx = \frac{W^2}{2EI}\left[\frac{x^3}{3}\right]_0^l = \frac{W^2l^3}{6EI} \qquad (11.27)$$

Equating equations (11.26) and (11.27) gives:

$$\frac{1}{2}W\times\delta = \frac{W^2l^3}{6EI}$$

from which $$\delta = \frac{W^2l^3}{6EI\left(\frac{1}{2}W\right)}$$

i.e. **maximum deflection of the cantilever, $\delta = \frac{Wl^3}{3EI}$**

(as required)

Problem 3. Determine the deflection at mid-span for the centrally loaded beam of Figure 11.7, which is simply supported at its ends.

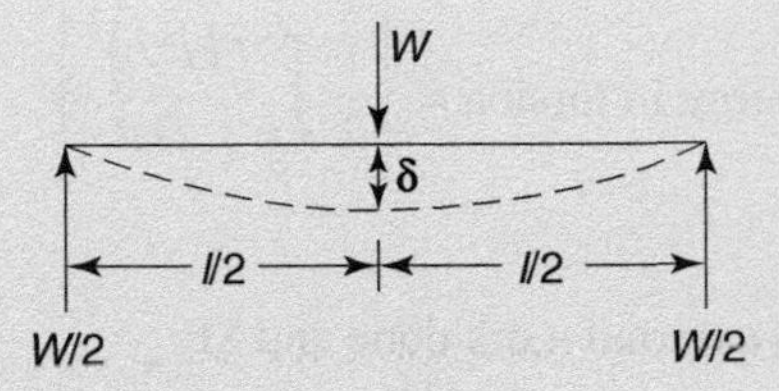

Figure 11.7 Centrally loaded beam

$$\text{WD by load} = \frac{1}{2} W \delta \qquad (11.28)$$

$$U_b = \int \frac{M^2}{2EI} dx$$

However $M = \dfrac{Wx}{2}$

Hence $$U_b = \int_0^{l/2} \frac{\left(\dfrac{Wx}{2}\right)^2}{2EI} dx \times 2$$

Note that this is done in two halves – because the bending moment equation is Wx from 0 to $l/2$, and $\dfrac{W}{2}(x) - W\left(x - \dfrac{l}{2}\right)$ from $l/2$ to l, i.e. the distribution of M is in two symmetrical halves.

Therefore $$U_b = \frac{1}{2EI}\int_0^{l/2}\left(\frac{Wx}{2}\right)^2 dx \times 2 \qquad (11.29)$$

$$= \frac{W^2}{4EI}\int_0^{l/2} x^2\, dx$$

$$= \frac{W^2}{4EI}\left[\frac{x^3}{3}\right]_0^{l/2} = \frac{W^2}{4EI}\left[\frac{(l/2)^3}{3} - 0\right]$$

i.e. $$\boldsymbol{U_b = \frac{W^2 l^3}{96EI}} \qquad (11.30)$$

Equating equations (11.28) and (11.30) gives:

$$\frac{1}{2} W \delta = \frac{W^2 l^3}{96EI}$$

from which, **mid-span deflection,** $\boldsymbol{\delta = \dfrac{W l^3}{48EI}}$

NB It should be noted that in equation (11.29), the upper limit of the integral was $l/2$. It was necessary to integrate in this manner because the expression for the bending moment, namely $M = Wx/2$, applied only between $x = 0$ and $x = l/2$.

Problem 4. Force P acts horizontally at point C on the light framework shown in Figure 11.8. Find the maximum horizontal deflection of point C, including the effects of tension, in terms of area A, second moment of area I and Young's modulus E. Ignore the self-weight of the framework.

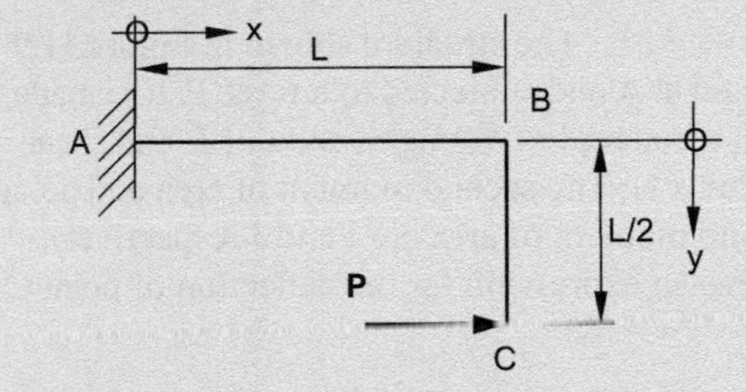

Figure 11.8

Bending moment in AB, $M_{AB} = P(L/2)$

Tension in AB, $N_{AB} = P$

Bending moment in CB, $M_{BC} = \text{Py}$

External word done $= \dfrac{1}{2} P\delta$

$$\text{Strain energy} = SE = \frac{1}{2}\int_0^L \frac{M_{AB}^{\ 2}}{EI} dx + \frac{1}{2}\int_0^L \frac{N_{AB}^{\ 2}}{AE} dx + \frac{1}{2}\int_0^{L/2} \frac{M_{BC}^{\ 2}}{EI} dy$$

Hence, $$\frac{1}{2}P\delta = \frac{1}{2}\int_0^L \frac{(PL/2)^2}{EI} dx + \frac{1}{2}\int_0^L \frac{(P)^2}{AE} dx + \frac{1}{2}\int_0^{L/2} \frac{(Py)^2}{EI} dy$$

and $$\delta = \frac{1}{P}\int_0^L \frac{(PL/2)^2}{EI} dx + \frac{1}{P}\int_0^L \frac{(P)^2}{AE} dx + \frac{1}{P}\int_0^{L/2} \frac{(Py)^2}{EI} dy$$

i.e. $$\delta = \frac{P}{EI}\left[\frac{L^2 x}{4}\right]_0^L + \frac{P}{AE}\left[x\right]_0^L + \frac{P}{EI}\left[\frac{y^3}{3}\right]_0^{L/2}$$

$$= \frac{P}{4EI}\left[L^3\right] + \frac{P}{AE}\left[L\right] + \frac{P}{3EI}\left[\frac{L^3}{8}\right]$$

$$=\frac{PL^3}{4EI}+\frac{PL}{AE}+\frac{PL^3}{24EI}=\frac{6PL^3}{24EI}+\frac{PL}{AE}+\frac{PL^3}{24EI}$$

i.e. the maximum horizontal deflection of point C,

$$\delta=\frac{\mathbf{7PL^3}}{\mathbf{24EI}}+\frac{\mathbf{PL}}{\mathbf{AE}}$$

Problem 5. The structure shown in Figure 11.9 is fixed at A and subjected to a force P. It is made from a material of Young's modulus E and shear modulus G. The second moment of area and polar second moment of area are I and J respectively. Derive an expression for the deflection of point C.

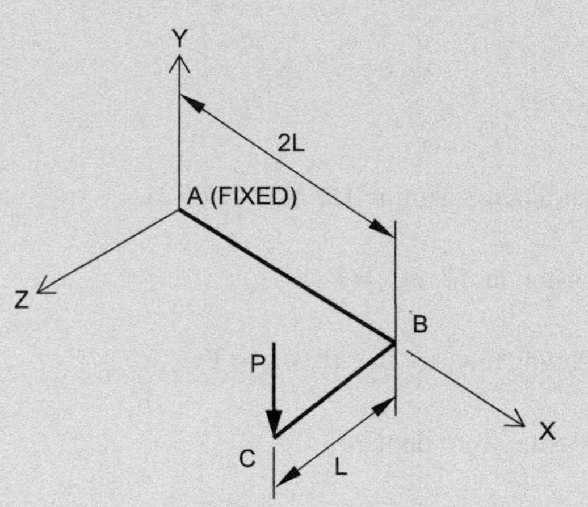

Figure 11.9

Energy is stored as bending energy in *AB* and *BC* and there is also torsional energy in *AB*.

Bending moment in $AB = Px$

Bending moment in $BC = Px$

Torsional moment in $AB = PL$

Strain energy in bending $=\frac{1}{2}\int\frac{M^2}{EI}dx$

Strain energy in torsion $=\frac{1}{2}\int\frac{T^2}{GJ}dx$

Strain energy in bending $=\left\{\frac{1}{2}\int_0^{2L}\frac{(Px)^2}{EI}dx\right\}_{in\,AB}+\left\{\frac{1}{2}\int_0^{L}\frac{(Px)^2}{EI}dx\right\}_{in\,BC}$

Strain energy in torsion $=\left\{\frac{1}{2}\int_0^{2L}\frac{(PL)^2}{GJ}\right\}_{in\,AB}$

Equating external work done and SE:

$$\frac{1}{2}P\delta=\left\{\frac{1}{2}\int_0^{2L}\frac{(Px)^2}{EI}dx\right\}_{in\,AB}+\left\{\frac{1}{2}\int_0^{L}\frac{(Px)^2}{EI}dx\right\}_{in\,BC}+\left\{\frac{1}{2}\int_0^{2L}\frac{(PL)^2}{GJ}\right\}_{in\,AB}$$

and $$\delta=\left\{\frac{1}{P}\int_0^{2L}\frac{(Px)^2}{EI}dx\right\}_{in\,AB}+\left\{\frac{1}{P}\int_0^{L}\frac{(Px)^2}{EI}dx\right\}_{in\,BC}+\left\{\frac{1}{P}\int_0^{2L}\frac{(PL)^2}{GJ}\right\}_{in\,AB}$$

i.e. $$\delta=\left\{\int_0^{2L}\frac{P(x)^2}{EI}dx\right\}_{in\,AB}+\left\{\int_0^{L}\frac{P(x)^2}{EI}dx\right\}_{in\,BC}+\left\{\int_0^{2L}\frac{P(L)^2}{GJ}dx\right\}_{in\,AB}$$

and $$\delta=P\left\{\left[\frac{x^3}{3EI}\right]_0^{2L}+\left[\frac{x^3}{3EI}\right]_0^{L}+\left[\frac{L^2x}{GJ}\right]_0^{2L}\right\}$$

Hence, **the deflection of point C,**

$$\boldsymbol{\delta=P\left\{\left[\frac{8L^3}{3EI}\right]+\left[\frac{L^3}{3EI}\right]+\left[\frac{2L^3}{GJ}\right]\right\}}$$

Now try the following Practice Exercise

Practice Exercise 47. Strain energy and work done

1. Determine the maximum deflection for the cantilever beam shown in Figure 11.10.

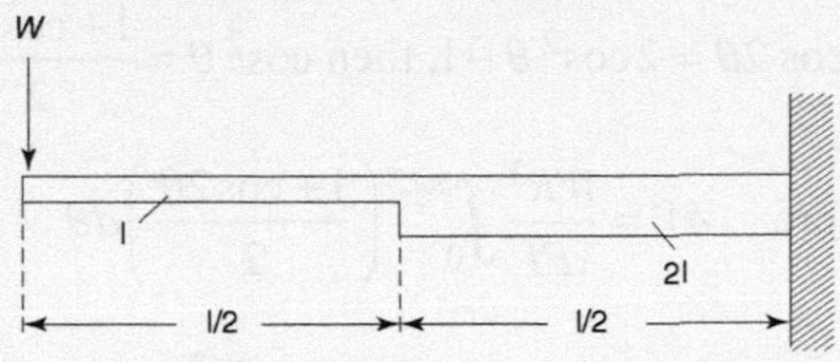

Figure 11.10 Stepped cantilever beam

$$\left[\frac{3Wl^3}{16EI}\right]$$

2. Determine the maximum deflection for the cantilever beam shown in Figure 11.11.

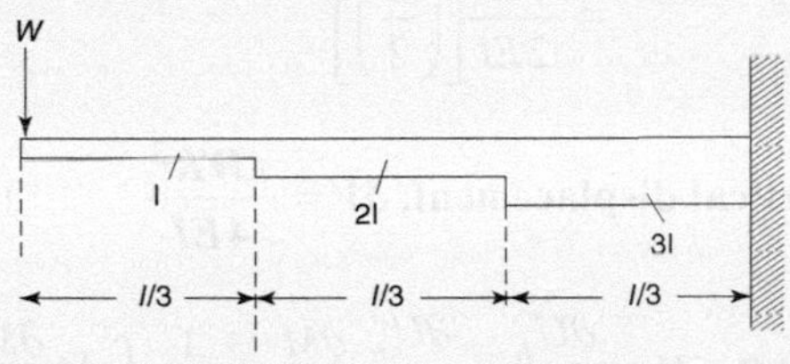

Figure 11.11 Stepped cantilever beam

$$\left[\frac{65Wl^3}{486EI}\right]$$

3. The structure shown in Figure 11.12 is fixed at A and subjected to a force P at D. It is made from a material of Young's modulus E and shear modulus G. The second moment of area and polar second moment of area are I and J respectively. Derive an expression for the deflection of point D.

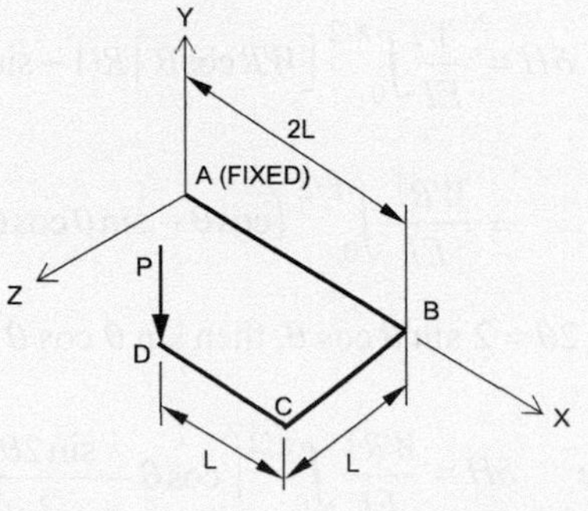

Figure 11.12

$$\left[\delta = \frac{4PL^3}{3EI} + \frac{3PL^3}{GJ}\right]$$

4. A plane structure made in the form of a quadrant of a circle lies in a horizontal plane as shown in Figure 11.13. It is subjected to a force P of 1000N at the free end, point A; the other end, B, is fixed. The structure is made from 25 mm diameter bar. The Young's modulus for the material is 200 GPa and the shear modulus is 76.9 GPa. If the radius of the quadrant, R is 300 mm, determine the vertical displacement of the free end A (i.e. in the direction of force P).

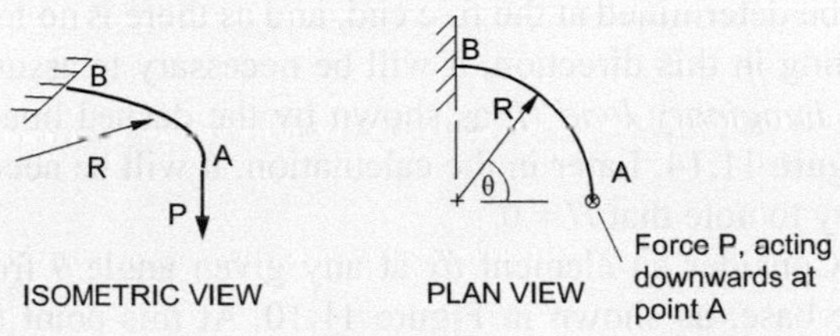

Figure 11.13

[δ = 8.79 mm]

11.10 Deflection of thin curved beams

In this theory, it will be assumed that the effects of shear strain energy and axial strain energy are negligible, and that all the work done by external loads on these beams is in fact absorbed by them in the form of bending strain energy.

Castigliano's theorems will be used in this section, and the method will be demonstrated by applying it to a number of examples.

Castigliano's first theorem requires a *load* to *act in the direction of the required displacement.* When the value of such a displacement is to be determined, at a point where no load points in the same direction, it will be necessary to assume an imaginary load to act in the direction of the required displacement, and, later, to set this imaginary load to zero.

Problem 6. A thin curved beam is in the form of a quadrant of a circle, as shown in Figure 11.14. One end of this quadrant is firmly fixed to a solid base, whilst the other end is subjected to a point load *W*, acting vertically downwards. Determine expressions for the vertical and horizontal displacements at the free end.

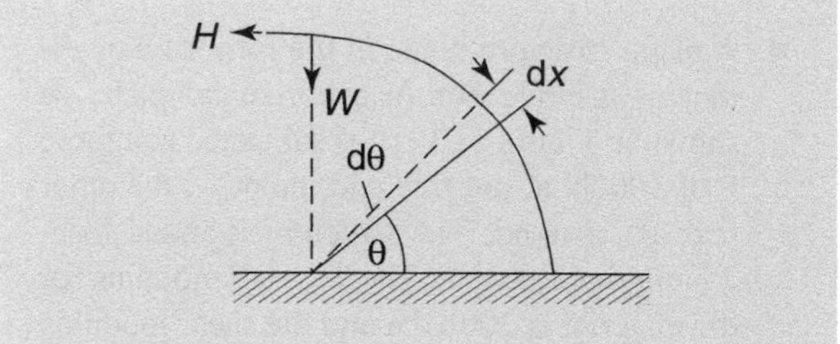

Figure 11.14 Thin curved beam quadrant

As the value of the horizontal displacement is required to be determined at the free end, and as there is no load acting in this direction, it will be necessary to assume *an imaginary load H,* as shown by the dashed line in Figure 11.14. Later in the calculation, it will be necessary to note that $H = 0$.

Consider an element dx at any given angle θ from the base, as shown in Figure 11.10. At this point the bending moment is:

$$M = -WR\cos\theta - HR(1 - \sin\theta)$$

NB The *sign convention* for bending moment is not important when using Castigliano's first theorem, providing that moments increasing curvature are of opposite sign to moments decreasing curvature.

Let δV = downward vertical displacement at the free end

δH = horizontal deflection (to the left) at the free end

From equation (11.19), the bending strain energy of the element dx is

$$d(U_b) = \frac{M^2}{2EI} \times dx$$

or

$$U_b = \int \frac{M^2}{2EI} dx$$

From Castigliano's first theorem,

$$\delta V = \frac{\partial U_b}{\partial W} = \frac{\partial U_b}{\partial M}\frac{\partial M}{\partial W}$$

$$= \frac{1}{EI}\int M \frac{\partial M}{\partial W} dx \qquad (11.31)$$

Therefore,

$$\delta V = \frac{1}{EI}\int_0^{\pi/2} \left[-WR\cos\theta - HR(1 - \sin\theta)\right]\left[-R\cos\theta\right] R\, d\theta$$

$$= \frac{1}{EI}\int_0^{\pi/2} \left[WR\cos\theta + HR(1 - \sin\theta)\right] R\cos\theta\, R\, d\theta$$

However, $H = 0$

Hence $$\delta V = \frac{1}{EI}\int_0^{\pi/2} \left[W\,R\cos\theta\right] R\cos\theta\, R\, d\theta$$

$$= \frac{WR^3}{EI}\int_0^{\pi/2} \cos^2\theta\, d\theta$$

Since $\cos 2\theta = 2\cos^2\theta - 1$, then $\cos^2\theta = \dfrac{1+\cos 2\theta}{2}$

Therefore $$\delta V = \frac{WR^3}{EI}\int_0^{\pi/2} \left(\frac{1+\cos 2\theta}{2}\right) d\theta$$

$$= \frac{WR^3}{2EI}\left[\theta + \frac{\sin 2\theta}{2}\right]_0^{\pi/2}$$

$$= \frac{WR^3}{2EI}\left[\left(\frac{\pi}{2} + \frac{\sin \pi}{2}\right) - \left(0 + \frac{\sin 0}{2}\right)\right]$$

$$= \frac{WR^3}{2EI}\left[\left(\frac{\pi}{2}\right)\right]$$

i.e. **vertical displacement,** $\boldsymbol{\delta V = \dfrac{\pi W R^3}{4EI}}$ (11.32)

Similatrly $$\delta H = \frac{\partial U_b}{\partial H} = \frac{\partial U_b}{\partial M}\frac{\partial M}{\partial H} = \frac{1}{EI}\int M \frac{\partial M}{\partial H} dx$$

$$= \frac{1}{EI}\int_0^{\pi/2} \left[-WR\cos\theta - HR(1 - \sin\theta)\right]\left[-R(1 - \sin\theta)\right] R d\theta$$

$$= \frac{1}{EI}\int_0^{\pi/2} \left[WR\cos\theta + HR(1 - \sin\theta)\right] R(1 - \sin\theta)\, R d\theta$$

However, $H = 0$

Hence $$\delta H = \frac{1}{EI}\int_0^{\pi/2} \left[WR\cos\theta\right] R(1 - \sin\theta)\, R d\theta$$

$$= \frac{WR^3}{EI}\int_0^{\pi/2} \left(\cos\theta - \sin\theta\cos\theta\right) d\theta$$

Since $\sin 2\theta = 2\sin\theta\cos\theta$, then $\sin\theta\cos\theta = \dfrac{\sin 2\theta}{2}$

Therefore $$\delta H = \frac{WR^3}{EI}\int_0^{\pi/2} \left(\cos\theta - \frac{\sin 2\theta}{2}\right) d\theta$$

$$= \frac{WR^3}{EI}\left[\sin\theta + \frac{\cos 2\theta}{4}\right]_0^{\pi/2}$$

$$= \frac{WR^3}{EI}\left[\left(\sin\pi/2+\frac{\cos\pi}{4}\right) - \left(\sin 0+\frac{\cos 0}{4}\right)\right]$$

$$= \frac{WR^3}{EI}\left[\left(1-\frac{1}{4}\right)-\left(0+\frac{1}{4}\right)\right]$$

$$= \frac{WR^3}{EI}\left[\left(\frac{1}{2}\right)\right]$$

i.e. **horizontal displacement,** $\boldsymbol{\delta H = \frac{WR^3}{2EI}}$ (11.33)

Thus, it can be seen that, owing to W, there is a horizontal displacement in addition to a vertical displacement. The value of the resultant displacement can be obtained by considerations of Pythagoras' theorem, together with elementary trigonometry.

Problem 7. A thin curved beam, which is in the form of a semi-circle, is firmly fixed at its base and is subjected to point loads at its free end, as shown in Figure 11.15. Determine expressions for the vertical and horizontal displacements at its free end.

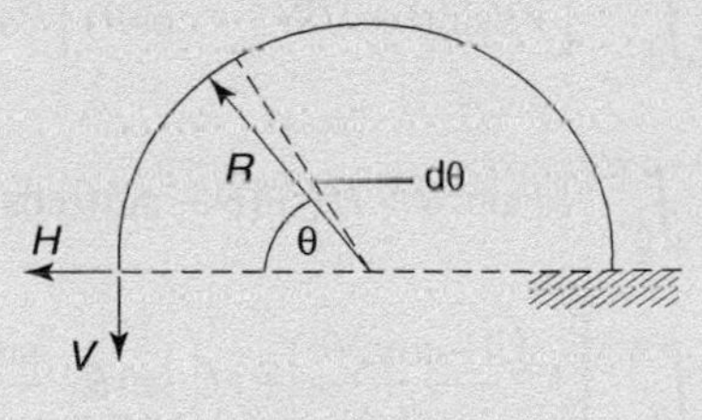

Figure 11.15 Thin curved beam (semi-circle)

Let δV = deflection in the direction of V

δH = deflection in the direction of H

At any angle θ, the bending moment acting on an element of the beam, from Figure 11.15, of length $Rd\theta$, is:

$$M = VR(1 - \cos\theta) - HR\sin\theta$$

Now $$\delta V = \frac{1}{EI}\int M\frac{\partial M}{\partial V}R\,d\theta$$

and $$\delta H = \frac{1}{EI}\int M\frac{\partial M}{\partial H}R\,d\theta$$

Therefore $$\delta V = \frac{1}{EI}\int_0^{\pi}\left\{\begin{matrix}[VR(1-\cos\theta) - HR\sin\theta] \\ R(1-\cos\theta)R\,d\theta\end{matrix}\right\}$$

$$= \frac{R^3}{EI}\int_0^{\pi}\begin{bmatrix}V(1-\cos\theta)^2 \\ -H\sin\theta(1-\cos\theta)\end{bmatrix}d\theta$$

$$= \frac{R^3}{EI}\int_0^{\pi}\begin{bmatrix}V(1-2\cos\theta+\cos^2\theta) \\ -H(\sin\theta-\sin\theta\cos\theta)\end{bmatrix}d\theta$$

$$= \frac{R^3}{EI}\int_0^{\pi}\begin{bmatrix}V\left(1-2\cos\theta+\dfrac{1+\cos 2\theta}{2}\right) \\ -H\left(\sin\theta-\dfrac{\sin 2\theta}{2}\right)\end{bmatrix}d\theta$$

$$= \frac{R^3}{EI}\begin{bmatrix}V\left(\theta-2\sin\theta+\dfrac{\theta}{2}+\dfrac{\sin 2\theta}{4}\right) \\ -H\left(-\cos\theta+\dfrac{\cos 2\theta}{4}\right)\end{bmatrix}_0^{\pi}$$

$$= \frac{R^3}{EI}\begin{bmatrix}\left\{\begin{matrix}V\left(\pi-2\sin\pi+\dfrac{\pi}{2}+\dfrac{\sin 2\pi}{4}\right) \\ -H\left(-\cos\pi+\dfrac{\cos 2\pi}{4}\right)\end{matrix}\right\} \\ -\left\{\begin{matrix}V\left(0-2\sin 0+0+\dfrac{\sin 0}{4}\right) \\ -H\left(-\cos 0+\dfrac{\cos 0}{4}\right)\end{matrix}\right\}\end{bmatrix}$$

$$= \frac{R^3}{EI}\begin{bmatrix}\left\{V\left(\pi-0+\dfrac{\pi}{2}+0\right)-H\left(--1+\dfrac{1}{4}\right)\right\} \\ -\left\{V(0)-H\left(-1+\dfrac{1}{4}\right)\right\}\end{bmatrix}$$

$$= \frac{R^3}{EI}\left[\left\{\frac{3\pi}{2}V - H\left(1\frac{1}{4}\right)\right\}-\left\{-H\left(-\frac{3}{4}\right)\right\}\right]$$

i.e. **vertical displacement,**

$$\boldsymbol{\delta V = \frac{R^3}{EI}\left(\frac{3\pi}{2}V - 2H\right)} \quad (11.34)$$

$$\delta H = \frac{1}{EI}\int M\frac{\partial M}{\partial H}R\,d\theta$$

$$= \frac{1}{EI}\int_0^{\pi}[VR(1-\cos\theta) - HR\sin\theta](-R\sin\theta)R\,d\theta$$

$$= \frac{R^3}{EI}\int_0^{\pi}[V(1-\cos\theta)(-\sin\theta) - H\sin\theta(-\sin\theta)]d\theta$$

$$= \frac{R^3}{EI}\int_0^{\pi}\left[V(-\sin\theta+\sin\theta\cos\theta)+H\sin^2\theta\right]d\theta$$

Since $\cos 2\theta = 1 - 2\sin^2\theta$, then $\sin^2\theta = \frac{1-\cos 2\theta}{2}$

Hence

$$\delta H = \frac{R^3}{EI}\int_0^{\pi}\left[V\left(-\sin\theta + \frac{\sin 2\theta}{2}\right) + H\left(\frac{1-\cos 2\theta}{2}\right)\right]d\theta$$

$$= \frac{R^3}{EI}\left[V\left(\cos\theta - \frac{\cos 2\theta}{4}\right) + H\left(\frac{\theta}{2} - \frac{\sin 2\theta}{4}\right)\right]_0^{\pi}$$

$$= \frac{R^3}{EI}\left[\left\{V\left(\cos\pi - \frac{\cos 2\pi}{4}\right) + H\left(\frac{\pi}{2} - \frac{\sin 2\pi}{4}\right)\right\} - \left\{V\left(\cos 0 - \frac{\cos 0}{4}\right) + H\left(0 - \frac{\sin 0}{4}\right)\right\}\right]$$

$$= \frac{R^3}{EI}\left[\left\{V\left(-1 - \frac{1}{4}\right) + H\left(\frac{\pi}{2}\right)\right\} - \left\{V\left(1 - \frac{1}{4}\right) + H(0)\right\}\right]$$

i.e. **horizontal displacement,**

$$\boldsymbol{\delta H = \frac{R^3}{EI}\left(-2V + \frac{\pi H}{2}\right)} \qquad (11.35)$$

Problem 8. Determine expressions for the deflections at the free end of the thin curved beam shown in Figure 11.16.

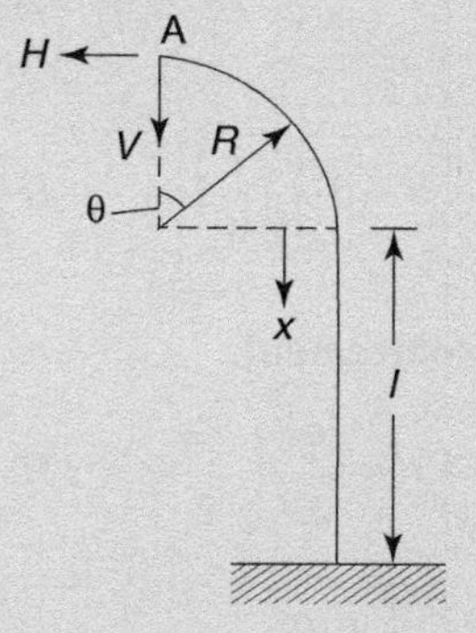

Figure 11.16 Thin beam (part curved and part straight)

For convenience, this beam will be considered in two sections, i.e. a curved section and a straight section.

Let δV = deflection at the free end in the direction of V

$= \delta V_c + \delta V_s$

δH = deflection at free end in the direction of H

$= \delta H_c + \delta H_s$

where δV_c = component of deflection at the free end in the direction of V due to the curved section

δH_c = component of deflection at the free end in the direction of H due to the curved section

δV_s = component of deflection at the free end in the direction of V due to the straight section

δH_s = component of deflection at the free end in the direction of H due to the straight section

As the problem is a linear elastic one, it is perfectly acceptable to add together the appropriate components of δV and δH.

Curved section

At any angle θ, the bending moment is:

$$M = VR\sin\theta + HR(1 - \cos\theta)$$

and $$\delta V_c = \frac{1}{EI}\int_0^{\pi/2} M\frac{\partial M}{\partial V}R\,d\theta$$

Therefore

$$\delta V_c = \frac{1}{EI}\int_0^{\pi/2}\left[VR\sin\theta + HR(1-\cos\theta)\right]R\sin\theta\,R\,d\theta$$

$$= \frac{R^3}{EI}\int_0^{\pi/2}\left[V\sin^2\theta + H(\sin\theta - \sin\theta\cos\theta)\right]d\theta$$

$$= \frac{R^3}{EI}\int_0^{\pi/2}\left[V\left(\frac{1-\cos 2\theta}{2}\right) + H\left(\sin\theta - \frac{\sin 2\theta}{2}\right)\right]d\theta$$

$$= \frac{R^3}{EI}\left[V\left(\frac{\theta}{2} - \frac{\sin 2\theta}{4}\right) + H\left(-\cos\theta + \frac{\cos 2\theta}{4}\right)\right]_0^{\pi/2}$$

$$= \frac{R^3}{EI}\left[\left\{V\left(\frac{\pi/2}{2} - \frac{\sin\pi}{4}\right) + H\left(-\cos\pi/2 + \frac{\cos\pi}{4}\right)\right\} - \left\{V\left(0 - \frac{\sin 0}{4}\right) + H\left(-\cos 0 + \frac{\cos 0}{4}\right)\right\}\right]$$

$$= \frac{R^3}{EI}\left[\left\{V\left(\frac{\pi}{4}\right) + H\left(-0 - \frac{1}{4}\right)\right\} - \left\{V(0) + H\left(-1 + \frac{1}{4}\right)\right\}\right]$$

i.e. $$\boldsymbol{\delta V_c = \frac{R^3}{EI}\left[\frac{\pi V}{4} + \frac{H}{2}\right]} \qquad (11.36)$$

Similarly $$\delta H_c = \frac{1}{EI}\int_0^{\pi/2} M\frac{\partial M}{\partial H}R\,d\theta$$

Therefore

$$\delta H_c = \frac{1}{EI}\int_0^{\pi/2}\left\{\begin{matrix}\left[VR\sin\theta + HR(1-\cos\theta)\right]\\ R(1-\cos\theta)R\end{matrix}\right\}d\theta$$

$$= \frac{R^3}{EI}\int_0^{\pi/2}\left[\begin{matrix}V\sin\theta(1-\cos\theta)\\ +H(1-\cos\theta)(1-\cos\theta)\end{matrix}\right]d\theta$$

$$= \frac{R^3}{EI}\int_0^{\pi/2}\left[\begin{matrix}V(\sin\theta - \sin\theta\cos\theta)\\ +H(1-2\cos\theta+\cos^2\theta)\end{matrix}\right]d\theta$$

$$= \frac{R^3}{EI}\int_0^{\pi/2}\left[\begin{matrix}V\left(\sin\theta - \frac{\sin 2\theta}{2}\right)\\ +H\left(1-2\cos\theta+\frac{1+\cos 2\theta}{2}\right)\end{matrix}\right]d\theta$$

$$= \frac{R^3}{EI}\left[\left\{\begin{matrix}V\left(-\cos\theta+\frac{\cos 2\theta}{4}\right)\\ +H\left(\theta-2\sin\theta+\frac{\theta}{2}+\frac{\sin 2\theta}{4}\right)\end{matrix}\right\}\right]_0^{\pi/2}$$

$$= \frac{R^3}{EI}\left[\begin{matrix}\left\{\begin{matrix}V\left(-\cos\pi/2+\frac{\cos\pi}{4}\right)\\ +H\left(\pi/2-2\sin\pi/2+\frac{\pi/2}{2}+\frac{\sin\pi}{4}\right)\end{matrix}\right\}\\ -\left\{\begin{matrix}V\left(-\cos 0+\frac{\cos 0}{4}\right)\\ +H\left(0-2\sin 0+0+\frac{\sin 0}{4}\right)\end{matrix}\right\}\end{matrix}\right]$$

$$= \frac{R^3}{EI}\left[\begin{matrix}\left\{V\left(-0-\frac{1}{4}\right)+H\left(\pi/2-2+\frac{\pi}{4}+0\right)\right\}\\ -\left\{V\left(-1+\frac{1}{4}\right)+H(0)\right\}\end{matrix}\right]$$

i.e. $$\boldsymbol{\delta H_c = \frac{R^3}{EI}\left[\frac{V}{2}+H\left(\frac{3\pi}{4}-2\right)\right]} \qquad (11.37)$$

Straight section

At any distance x, from Figure 11.16, the bending moment is:

$$M = VR + H(R+x)$$

Now $$\delta V_s = \frac{1}{EI}\int_0^l M\frac{\partial M}{\partial V}dx$$

$$= \frac{1}{EI}\int_0^l \left[VR + H(R+x)\right]R\,dx$$

$$= \frac{R}{EI}\int_0^l \left[VR+HR+Hx\right]dx$$

$$= \frac{R}{EI}\left[VRx+HRx+\frac{Hx^2}{2}\right]_0^l$$

$$= \frac{R}{EI}\left[\left\{VRl+HRl+\frac{Hl^2}{2}\right\}-\{0\}\right]$$

i.e. $$\boldsymbol{\delta V_s = \frac{R}{EI}\left[VRl+H\left(Rl+\frac{l^2}{2}\right)\right]} \qquad (11.38)$$

Similarly $$\delta H_s = \frac{1}{EI}\int_0^l M\frac{\partial M}{\partial H}dx$$

$$= \frac{1}{EI}\int_0^l\left[VR(R+x)+H(R+x)^2\right]dx$$

$$= \frac{1}{EI}\int_0^l\left[VR^2+VRx+H\left(R^2+2Rx+x^2\right)\right]dx$$

$$= \frac{1}{EI}\int_0^l\left[VR^2+VRx+HR^2+2HRx+Hx^2\right]dx$$

$$= \frac{1}{EI}\int_0^l\left[VR^2+VRx+HR^2+2HRx+Hx^2\right]dx$$

$$= \frac{1}{EI}\left[VR^2x+\frac{VRx^2}{2}+HR^2x+\frac{2HRx^2}{2}+\frac{Hx^3}{3}\right]_0^l$$

$$= \frac{1}{EI}\left[\begin{matrix}\left\{VR^2l+\frac{VRl^2}{2}+HR^2l+\frac{2HRl^2}{2}+\frac{Hl^3}{3}\right\}\\ -\{0\}\end{matrix}\right]$$

i.e.

$$\boldsymbol{\delta H_s = \frac{1}{EI}\left[VRl\left(R+\frac{l}{2}\right)+Hl\left(R^2+Rl+\frac{l^2}{3}\right)\right]} \qquad (11.39)$$

From equations (11.36) and (11.38)

$$\delta V = \delta V_c + \delta V_s = \frac{R^3}{EI}\left[\frac{\pi V}{4}+\frac{H}{2}\right]+\frac{R}{EI}\left[VRl+H\left(Rl+\frac{l^2}{2}\right)\right]$$

$$\text{i.e. } \boldsymbol{\delta V = \frac{1}{EI}\left[R^3\left(\frac{\pi V}{4}+\frac{H}{2}\right)+R\left\{VRl+Hl\left(R+\frac{l}{2}\right)\right\}\right]}$$

(11.40)

From equations (11.37) and (11.39),

$$\delta H = \delta H_c + \delta H_s = \frac{R^3}{EI}\left[\frac{V}{2} + H\left(\frac{3\pi}{4} - 2\right)\right]$$

$$+ \frac{1}{EI}\left[VRl\left(R + \frac{l}{2}\right) + Hl\left(R^2 + Rl + \frac{l^2}{3}\right)\right]$$

i.e.
$$\boldsymbol{\delta H} = \frac{\mathbf{1}}{\boldsymbol{EI}}\left[\begin{array}{l} \boldsymbol{R^3}\left\{\frac{\boldsymbol{V}}{\mathbf{2}} + \boldsymbol{H}\left(\frac{\mathbf{3\pi}}{\mathbf{4}} - \mathbf{2}\right)\right\} + \boldsymbol{VRl}\left(\boldsymbol{R} + \frac{\boldsymbol{l}}{\mathbf{2}}\right) \\ \qquad + \boldsymbol{Hl}\left(\boldsymbol{R^2} + \boldsymbol{Rl} + \frac{\boldsymbol{l^2}}{\mathbf{3}}\right) \end{array}\right] \qquad (11.41)$$

Problem 9. Determine the deflection at the free end of the thin curved beam shown in Figure 11.17, which has an out-of-plane concentrated load applied to its free end.

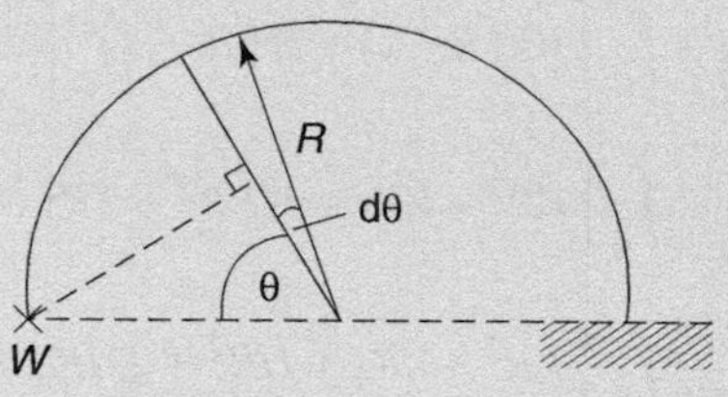

Figure 11.17 Thin curved beam under an out-of-plane load

In this case, the beam is subjected to both bending and torsion; hence, it will be necessary to consider bending strain energy in addition to torsional strain energy.

At any angle θ, in Figure 11.17, the element $Rd\theta$ of the beam is subjected to a bending moment M and a torque T, which are evaluated as follows:

$$M = WR\sin\theta$$

$$T = WR(1 - \cos\theta)$$

From equations (11.19) and (11.22), the total strain energy, U, is given by

$$U = \int \frac{M^2}{2EI}dx + \int \frac{T^2}{2GJ}dx$$

Let δW be the out-of-plane deflection under the load W, acting at the free end. Now

$$\delta W = \frac{1}{EI}\int M\frac{\partial M}{\partial W}R\,d\theta + \frac{1}{GJ}\int T\frac{\partial T}{\partial W}R\,d\theta$$

$$= \frac{R}{EI}\int_0^{\pi}(WR\sin\theta)R\sin\theta\,d\theta$$

$$+ \frac{R}{GJ}\int_0^{\pi}(WR(1-\cos\theta))R(1-\cos\theta)\,d\theta$$

$$= \frac{R^3}{EI}\int_0^{\pi}(W\sin^2\theta)\,d\theta + \frac{R^3}{GJ}\int_0^{\pi}(W(1-\cos\theta)^2)\,d\theta$$

$$= \frac{WR^3}{EI}\int_0^{\pi}\left(\frac{1-\cos 2\theta}{2}\right)d\theta$$

$$+ \frac{WR^3}{GJ}\int_0^{\pi}\left(1 - 2\cos\theta + \cos^2\theta\right)d\theta$$

$$= \frac{WR^3}{EI}\int_0^{\pi}\left(\frac{1-\cos 2\theta}{2}\right)d\theta$$

$$+ \frac{WR^3}{GJ}\int_0^{\pi}\left(1 - 2\cos\theta + \frac{1+\cos 2\theta}{2}\right)d\theta$$

$$= \frac{WR^3}{EI}\left[\frac{\theta}{2} - \frac{\sin 2\theta}{4}\right]_0^{\pi}$$

$$+ \frac{WR^3}{GJ}\left[\theta - 2\sin\theta + \frac{\theta}{2} + \frac{\sin 2\theta}{4}\right]_0^{\pi}$$

$$= \frac{WR^3}{EI}\left[\left\{\frac{\pi}{2} - \frac{\sin 2\pi}{4}\right\} - \left\{0 - \frac{\sin 0}{4}\right\}\right]$$

$$+ \frac{WR^3}{GJ}\left[\begin{array}{l}\left\{\pi - 2\sin\pi + \frac{\pi}{2} + \frac{\sin 2\pi}{4}\right\} \\ \qquad - \left\{0 - 2\sin 0 + 0 + \frac{\sin 0}{4}\right\}\end{array}\right]$$

$$= \frac{WR^3}{EI}\left[\left\{\frac{\pi}{2} - 0\right\} - \{0\}\right] + \frac{WR^3}{GJ}\left[\left\{\pi - 0 + \frac{\pi}{2} + 0\right\} - \{0\}\right]$$

$$= \frac{WR^3}{EI}\left[\frac{\pi}{2}\right] + \frac{WR^3}{GJ}\left[\frac{3\pi}{2}\right]$$

i.e.
$$\boldsymbol{\delta W} = \boldsymbol{WR^3}\left(\frac{\boldsymbol{\pi}}{\mathbf{2}\boldsymbol{EI}} + \frac{\mathbf{3}\boldsymbol{\pi}}{\mathbf{2}\boldsymbol{GJ}}\right) \qquad (11.42)$$

Problem 10. Determine expressions for the values of the bending moments at the points A and B of the thin ring shown in Figure 11.18, and also obtain expressions for the changes in diameter of the ring in the directions of the applied loads. Hence, or otherwise, sketch the bending moment diagram, around the circumference of the ring, when $H = 0$.

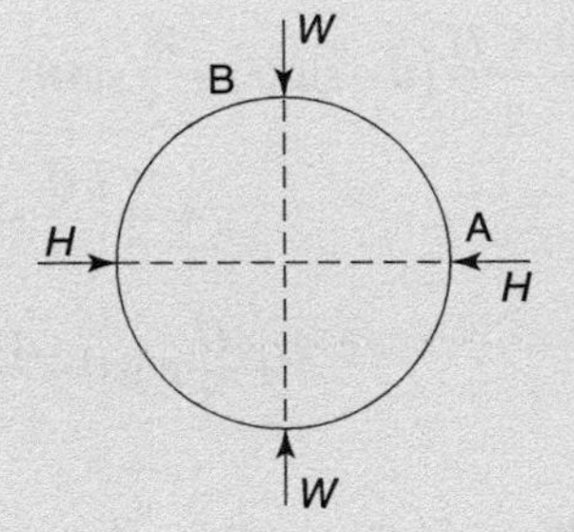

Figure 11.18 Thin ring, under diametral loads

Because of symmetry, it is only necessary to consider a quadrant of the ring, as shown in Figure 11.19.

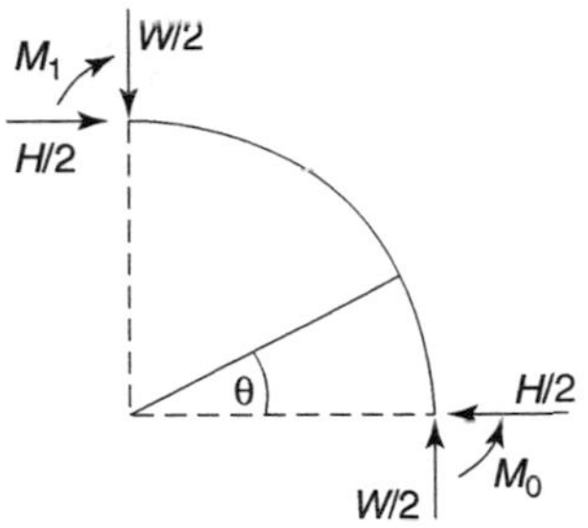

Figure 11.19 Quadrant of ring

Let M_0 = unknown bending moment at A, which can be assumed to be the redundancy

M_1 = unknown bending moment at B, which can be obtained from M_0 together with considerations of elementary statics

As this problem is statically indeterminate, it will be necessary to use Castigliano's second theorem to determine the redundancy M_0.

At angle θ, in Figure 11.19, the bending moment acting on an element $Rd\theta$ is:

$$M = M_0 + \frac{WR}{2}(1-\cos\theta) - \frac{HR}{2}\sin\theta \qquad (11.43)$$

To find M_0

From Castigliano's second theorem, $\frac{\partial U_b}{\partial M_0}$, the initial lack of fit, is zero (in this case, as the ring is assumed to be geometrically perfect)

i.e. $$\frac{\partial U_b}{\partial M_0} = \frac{\partial U_b}{\partial M}\frac{\partial M}{\partial M_0}$$

i.e. $$0 = \frac{\partial U_b}{\partial M}\frac{\partial M}{\partial M_0} = \frac{1}{EI}\int M\frac{\partial M}{\partial M_0}Rd\theta$$

Now, as there are four quadrants

$$0 = 4\times\frac{1}{EI}\int_0^{\pi/2}\begin{pmatrix} M_0 + \frac{WR}{2}(1-\cos\theta) \\ -\frac{HR}{2}\sin\theta \end{pmatrix}Rd\theta$$

i.e. $$0 = \frac{4R}{EI}\left[M_0\theta + \frac{WR}{2}(\theta - \sin\theta) + \frac{HR}{2}\cos\theta\right]_0^{\pi/2}$$

i.e. $$0\left(\frac{EI}{4R}\right) = \left[\begin{matrix} \left\{\begin{matrix} M_0\left(\frac{\pi}{2}\right) + \frac{WR}{2}\left(\frac{\pi}{2} - \sin\frac{\pi}{2}\right) \\ + \frac{HR}{2}\cos\frac{\pi}{2} \end{matrix}\right\} \\ -\left\{\begin{matrix} 0 + \frac{WR}{2}(0 - \sin 0) \\ + \frac{HR}{2}\cos 0 \end{matrix}\right\} \end{matrix}\right]$$

i.e. $$0 = \left[\left\{M_0\left(\frac{\pi}{2}\right) + \frac{WR}{2}\left(\frac{\pi}{2} - 1\right) + 0\right\} - \left\{\frac{HR}{2}\right\}\right]$$

i.e. $$0 = \left[M_0\left(\frac{\pi}{2}\right) + WR\frac{\pi}{4} - \frac{WR}{2} - \frac{HR}{2}\right]$$

from which $$M_0\left(\frac{\pi}{2}\right) = -WR\frac{\pi}{4} + \frac{WR}{2} + \frac{HR}{2}$$

and $$M_0 = \frac{2}{\pi}\left(-WR\frac{\pi}{4} + \frac{WR}{2} + \frac{HR}{2}\right)$$

i.e. $$M_0 = -\frac{WR}{2} + \frac{WR}{\pi} + \frac{HR}{\pi}$$

i.e. $$\boldsymbol{M_0 = (W+H)\frac{R}{\pi} - \frac{WR}{2}} \qquad (11.44)$$

An expression for M_1 can be determined from equation (11.43) by setting θ = 90° and by the use of equation (11.44)

i.e. $$M_1 = M_0 + \frac{WR}{2}\left(1 - \cos\frac{\pi}{2}\right) - \frac{HR}{2}\sin\frac{\pi}{2}$$

i.e. $$M_1 = M_0 + \frac{WR}{2}(1-0) - \frac{HR}{2}$$

i.e. $$M_1 = M_0 + \frac{WR}{2} - \frac{HR}{2}$$

$$= \left\{(W+H)\frac{R}{\pi} - \frac{WR}{2}\right\} + \frac{WR}{2} - \frac{HR}{2}$$

from equation (11.44)

$$=\frac{WR}{\pi}+\frac{HR}{\pi}-\frac{WR}{2}+\frac{WR}{2}-\frac{HR}{2}$$

i.e. $$\boldsymbol{M_1}=\frac{\boldsymbol{WR}}{\pi}+\boldsymbol{HR}\left(\frac{\mathbf{1}}{\pi}-\frac{\mathbf{1}}{\mathbf{2}}\right) \quad (11.45)$$

To find the *change in the diameter* of the ring in the direction of W

$$\partial V=\frac{1}{EI}\int M\frac{\partial M}{\partial W}dx$$

Now, as there are four quadrants

$$\partial V=4\times\frac{1}{EI}\int_0^{\pi/2}\begin{pmatrix}M_0+\frac{WR}{2}(1-\cos\theta)\\-\frac{HR}{2}\sin\theta\end{pmatrix}\frac{R}{2}(1-\cos\theta)Rd\theta$$

from equation (11.43)

$$=\frac{4}{EI}\int_0^{\pi/2}\begin{pmatrix}(W+H)\frac{R}{\pi}-\frac{WR}{2}\\+\frac{WR}{2}(1-\cos\theta)\\-\frac{HR}{2}\sin\theta\end{pmatrix}\frac{R}{2}(1-\cos\theta)Rd\theta$$

from equation (11.44)

$$=\frac{4R^3}{EI}\int_0^{\pi/2}\begin{pmatrix}\frac{W}{\pi}+\frac{H}{\pi}-\frac{W}{2}+\frac{W}{2}\\-\frac{W}{2}\cos\theta-\frac{H}{2}\sin\theta\end{pmatrix}\frac{1}{2}(1-\cos\theta)d\theta$$

$$=\frac{2R^3}{EI}\int_0^{\pi/2}\begin{pmatrix}\frac{W+H}{\pi}(1-\cos\theta)\\-\frac{W}{2}\cos\theta(1-\cos\theta)\\-\frac{H}{2}\sin\theta(1-\cos\theta)\end{pmatrix}d\theta$$

$$=\frac{2R^3}{EI}\int_0^{\pi/2}\begin{pmatrix}\frac{W+H}{\pi}(1-\cos\theta)-\frac{W}{2}\cos\theta\\+\frac{W}{2}\cos^2\theta-\frac{H}{2}\sin\theta\\+\frac{H}{2}\sin\theta\cos\theta\end{pmatrix}d\theta$$

$$=\frac{2R^3}{EI}\int_0^{\pi/2}\begin{pmatrix}\frac{W+H}{\pi}(1-\cos\theta)-\frac{W}{2}\cos\theta\\+\frac{W}{2}\left(\frac{1+\cos2\theta}{2}\right)-\frac{H}{2}\sin\theta\\+\frac{H}{2}\left(\frac{\sin2\theta}{2}\right)\end{pmatrix}d\theta$$

$$=\frac{2R^3}{EI}\left[\left(\begin{matrix}\frac{W+H}{\pi}(\theta-\sin\theta)-\frac{W}{2}\sin\theta\\+\frac{W}{2}\left(\frac{\theta}{2}+\frac{\sin2\theta}{4}\right)\\+\frac{H}{2}\cos\theta-\frac{H\cos2\theta}{8}\end{matrix}\right)\right]_0^{\pi/2}$$

$$=\frac{2R^3}{EI}\left[\begin{Bmatrix}\frac{W+H}{\pi}\left(\frac{\pi}{2}-\sin\frac{\pi}{2}\right)-\frac{W}{2}\sin\frac{\pi}{2}\\+\frac{W}{2}\left(\frac{\pi}{4}+\frac{\sin\pi}{4}\right)\\+\frac{H}{2}\cos\frac{\pi}{2}-\frac{H\cos\pi}{8}\end{Bmatrix}-\begin{Bmatrix}\frac{W+H}{\pi}(0-\sin0)-\frac{W}{2}\sin0\\+\frac{W}{2}\left(0+\frac{\sin0}{4}\right)\\+\frac{H}{2}\cos0-\frac{H\cos0}{8}\end{Bmatrix}\right]$$

$$=\frac{2R^3}{EI}\left[\begin{Bmatrix}\frac{W+H}{\pi}\left(\frac{\pi}{2}-1\right)\\-\frac{W}{2}+\frac{W}{2}\left(\frac{\pi}{4}+0\right)\\+\frac{H}{2}(0)+\frac{H}{8}\end{Bmatrix}-\left\{\frac{H}{2}-\frac{H}{8}\right\}\right]$$

$$=\frac{2R^3}{EI}\left[\begin{Bmatrix}\frac{W}{\pi}\left(\frac{\pi}{2}-1\right)+\frac{H}{\pi}\left(\frac{\pi}{2}-1\right)\\-\frac{W}{2}+\frac{W}{2}\left(\frac{\pi}{4}\right)+\frac{H}{8}\end{Bmatrix}-\left\{\frac{3H}{8}\right\}\right]$$

$$=\frac{2R^3}{EI}\left[\frac{W}{2}-\frac{W}{\pi}+\frac{H}{2}-\frac{H}{\pi}-\frac{W}{2}+\frac{W\pi}{8}+\frac{H}{8}-\frac{3H}{8}\right]$$

$$=\frac{2R^3}{EI}\left[-\frac{W}{\pi}+\frac{H}{2}-\frac{H}{\pi}+\frac{W\pi}{8}-\frac{H}{4}\right]$$

i.e. $$\partial V=\frac{\mathbf{2R^3}}{\boldsymbol{EI}}\left[\boldsymbol{W}\left(\frac{\pi}{\mathbf{8}}-\frac{\mathbf{1}}{\pi}\right)+\boldsymbol{H}\left(\frac{\mathbf{1}}{\mathbf{4}}-\frac{\mathbf{1}}{\pi}\right)\right] \quad (11.46)$$

By a similar process, the change in diameter of the ring in the direction of H can be found:

i.e. $$\partial\boldsymbol{H}=\frac{\boldsymbol{R^3}}{\boldsymbol{EI}}\left[\boldsymbol{H}\left(\frac{\pi}{\mathbf{8}}-\frac{\mathbf{1}}{\pi}\right)+\boldsymbol{W}\left(\frac{\mathbf{1}}{\mathbf{4}}-\frac{\mathbf{1}}{\pi}\right)\right] \quad (11.47)$$

To determine the bending moment diagram, when H = 0

From equation (11.43)

$$M = M_0 + \frac{WR}{2}(1 - \cos\theta) \text{ when } H = 0$$

$$= (W + H)\frac{R}{\pi} - \frac{WR}{2} + \frac{WR}{2}(1 - \cos\theta)$$

from equation (11.44)

$$= \frac{WR}{\pi} - \frac{WR}{2} + \frac{WR}{2}(1 - \cos\theta) \quad \text{when } H = 0$$

$$= \frac{WR}{\pi} - \frac{WR}{2} + \frac{WR}{2} - \frac{WR}{2}\cos\theta$$

$$= \frac{WR}{\pi} - \frac{WR}{2}\cos\theta$$

At $\theta = \frac{\pi}{2}$, $M = M_1$

hence $$M_1 = \frac{WR}{\pi} - \frac{WR}{2}\cos\frac{\pi}{2} = \frac{WR}{\pi} - \frac{WR}{2}(0)$$

i.e. $$\boldsymbol{M_1 = \frac{WR}{\pi}}$$

At $\theta = 0$, $M = M_0$

hence $$M_0 = \frac{WR}{\pi} - \frac{WR}{2}\cos 0$$

i.e. $$\boldsymbol{M_0 = \left(\frac{WR}{\pi} - \frac{WR}{2}\right)}$$

The bending moment diagram is shown in Figure 11.20.

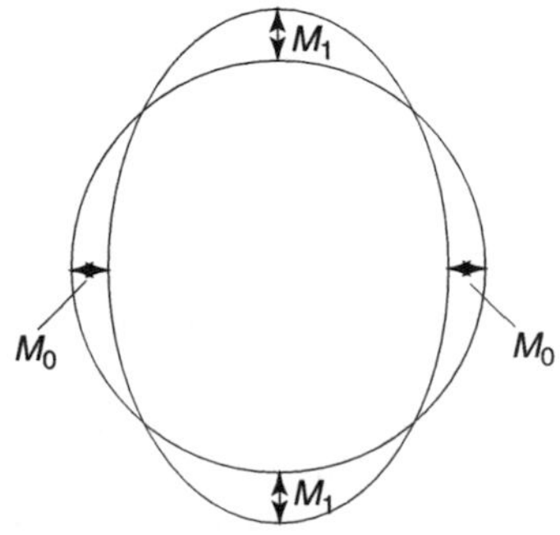

Figure 11.20 Bending moment distribution, when $H = 0$

Problem 11. Using Castigliano's method, find the maximum deflection of the cantilever loaded as shown in Figure 11.21 (ignoring self-weight). The beam has a second moment of area, I and a Young's modulus, E.

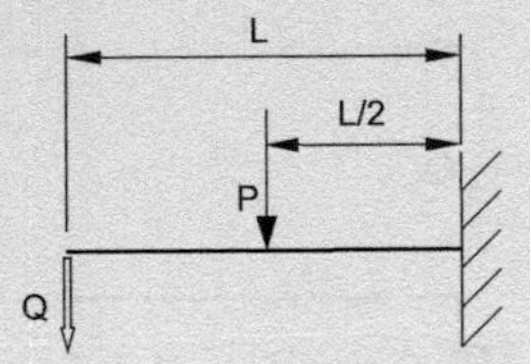

Figure 11.21

Apply a virtual load Q at the free end.

Q is applied at the point and in the direction of the desired deflection.

At the free end of cantilever, $\Delta = \int_0^L M \frac{\partial M}{\partial Q}\frac{dx}{EI}$

$0 < x < L/2$

$M = -Qx$

$\frac{\partial M}{\partial Q} = -x$

$L/2 < x < L$

$M = -Qx - P(x - L/2)$

$\frac{\partial M}{\partial Q} = -x$

Hence, maximum deflection, $$\delta = \int_0^{L/2} \frac{(-Qx)(-x)}{EI}dx + \int_{L/2}^{L} \frac{\{-Qx - P(x - L/2)\}(-x)}{EI}dx$$

However, $Q = 0$

Note: Put $Q = 0$ after carrying out the partial differentiation

Hence, $$\delta = 0 + \int_{L/2}^{L} \frac{\{P(x - L/2)\}(x)}{EI}dx$$

$$= \frac{P}{EI}\int_{L/2}^{L}\left(x^2 - Lx/2\right)dx$$

i.e $$\delta = \frac{P}{EI}\left[\frac{x^3}{3} - \frac{Lx^2}{4}\right]_{L/2}^{L}$$

$$= \frac{P}{EI}\left\{\left(\frac{L^3}{3} - \frac{L^3}{4}\right) - \left(\frac{L^3}{24} - \frac{L^3}{16}\right)\right\}$$

i.e. **the maximum deflection,** $\boldsymbol{\delta = \frac{5}{48}\frac{PL^3}{EI}}$

Problem 12. Use Castigliano's theorem to find the maximum deflection of a cantilever with a pure moment applied at the free end shown in Figure 11.22. Self-weight can be ignored. The beam has a second moment of area, I and a Young's modulus, E.

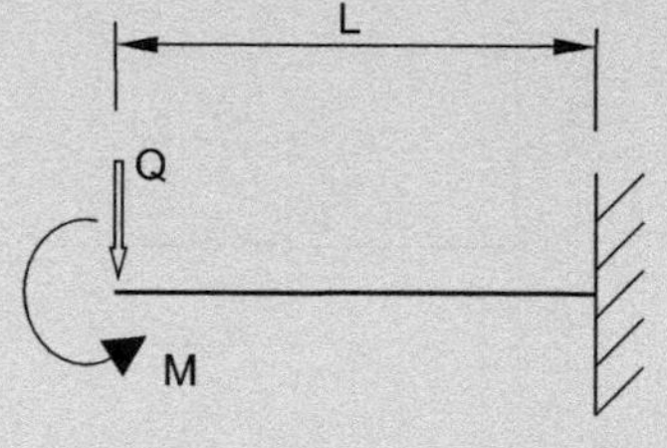

Figure 11.22

Apply a virtual load Q at the free end.

Q is applied at the point and in the direction of the desired deflection.

At the end of the cantilever, $\delta = \int_0^L M \frac{\partial M}{\partial Q} \frac{dx}{EI}$

$M = -M - Qx \qquad \frac{\partial M}{\partial Q} = -x$

This then gives a positive δ, i.e. in the direction of the virtual load.

Maximum deflection of the cantilever,

$$\delta = \int_0^L \frac{(-M - Qx)(-x)}{EI} dx$$

Note: Put Q = 0 after partially differentiating

Hence, $\delta = \int_0^L \frac{(-M)(-x)}{EI} dx = \frac{M}{EI}\left[\frac{x^2}{2}\right]_0^L$

i.e. **the maximum deflection of the cantilever,**

$$\delta = \frac{ML^2}{2EI}$$

Now try the following Practice Exercise

Practice Exercise 48. Deflection of thin curved beams

1. Determine expressions for the horizontal and vertical deflections of the free end of the thin curved beams shown in Figures 11.23(a) to (c).

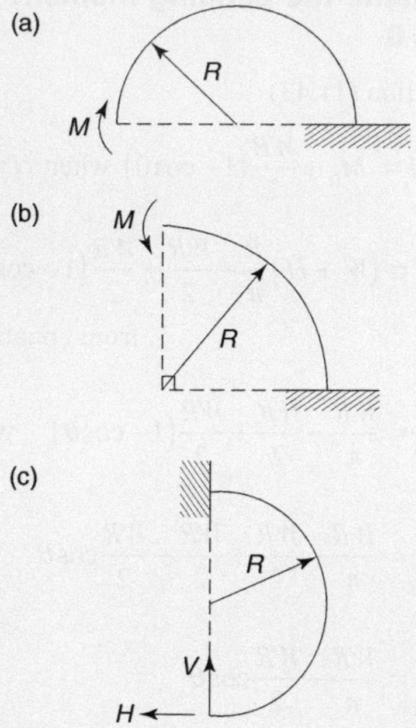

Figure 11.23

[(a) horizontal deflection $= \frac{2MR^2}{EI}$ to the left, vertical deflection $= -\frac{\pi MR^2}{EI}$ upwards; (b) horizontal deflection $= \frac{MR^2}{EI}\left(\frac{\pi}{2} - 1\right)$ to the left, vertical deflection $= \frac{MR^2}{EI}$ downwards; (c) horizontal deflection $= \delta H = \frac{R^3\left[(3\pi/2)H + 2V\right]}{EI}$ to the left, vertical deflection $= \delta V = \frac{R^3(2H + \pi V/2)}{EI}$ upwards.]

2. A thin curved beam consists of a length *AB*, which is of semi-circular form, and a length *BC*, which is a quadrant, as shown in Figure 11.24. Determine the vertical deflection at A due to a downward load *W*, applied at this point. The end C is firmly fixed, and the beam's cross-section may be assumed to be uniform.

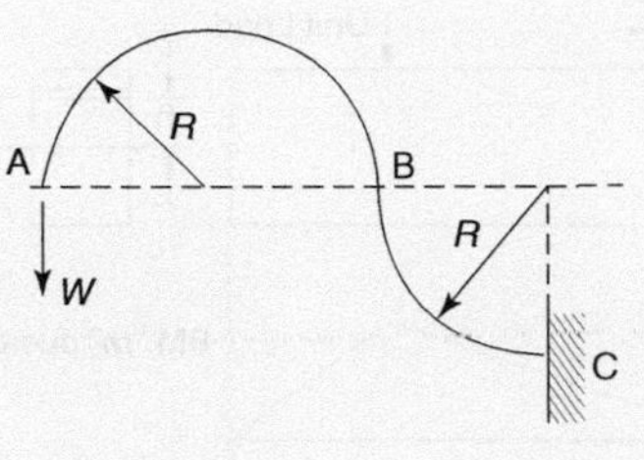

Figure 11.24

$$\left[\delta V = \frac{WR^3}{EI}\left(\frac{25\pi}{4} - 6\right)\right]$$

3. A clip, which is made from a length of wire of uniform section, is subjected to the load shown in Figure 11.25. Determine an expression for the distance by which the free ends of the clip subjected to this load.

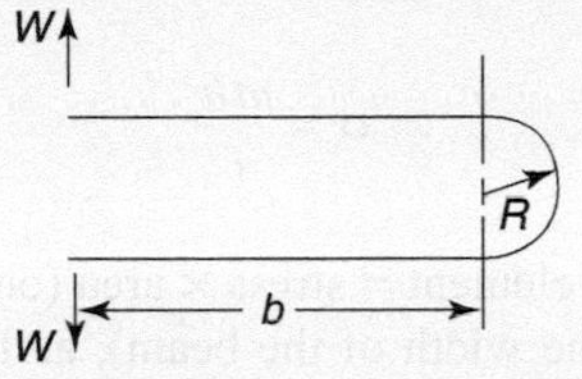

Figure 11.25

$$\left[\frac{2W}{EI}\left\{\frac{b^3}{3} + R\left(\frac{b^2\pi}{2} + \frac{R^2\pi}{4} + 2bR\right)\right\}\right]$$

4. Determine the vertical and horizontal displacements at the free end of the rigid-jointed frame shown in Figure 11.26.

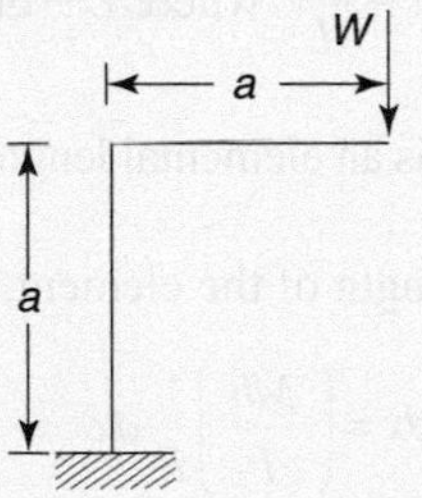

Figure 11.26

$$\left[\delta V = \frac{4Wa^3}{3EI} \text{ downwards, } \delta H = \frac{Wa^3}{2EI} \text{ to the right}\right]$$

5. Determine the horizontal deflection at the free end A of the rigid-jointed frame of Figure 11.27.

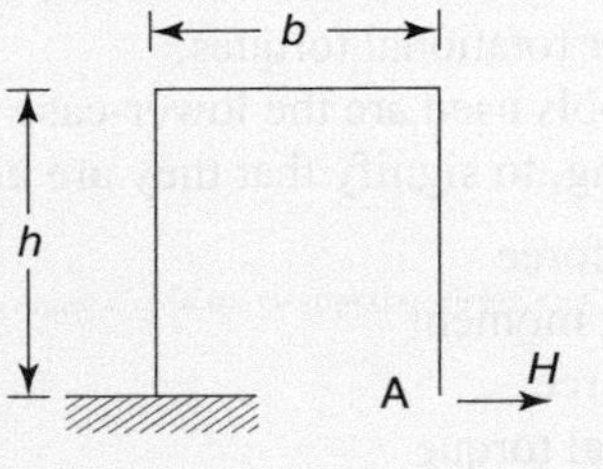

Figure 11.27

$$\left[\delta H = \frac{Hh^2}{EI}\left(\frac{2h}{3} + b\right)\right]$$

6. If the framework of Figure 11.27 were prevented from moving vertically at A, so that it was statically indeterminate to the first degree and $h = b$, what force would be required to prevent this movement?

[0.75 H, acting downwards]

7. Use Castigliano's theorem to find the central deflection of the simply supported beam shown in Figure 11.28. The beam carries a uniformly distributed load and has a second moment of area I and a Young's modulus E

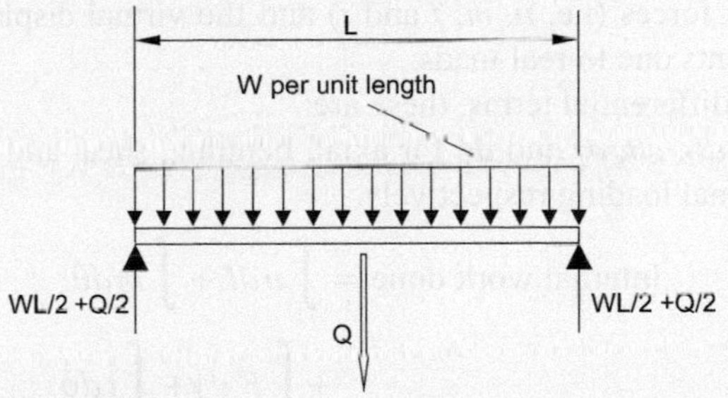

Figure 11.28

Hint: Put a virtual load Q at midspan and set Q = 0 after differentiating.

$$\left[\delta = \frac{5WL^4}{384EI}\right]$$

11.11 Unit load method

This is a powerful method for finding the deflections of structures. Two systems of loadings are considered to apply to the structure, the actual loads causing the displacements and the 'unit load'. The latter is chosen to act at the place where the deflection is required, and in the same direction. This gives rise to a system of internal loading, which can be forces, moments, shear forces and/or rotational torques.

The symbols used are the lower-case versions of the actual loading, to signify that they are unit loads:

n = normal force
m = bending moment
f = shear force
t = rotational torque

This gives a system of unit load reactions and internal loads that must be in equilibrium.

As with most strain energy the bending and torsion loading causes the greatest deflections. Axial loadings are sometimes of interest, but energy stored in shear is usually negligible.

Using the principle of virtual work, the system has an additional virtual displacement. The true displacements are chosen to be the virtual displacements of the actual loading. During this virtual displacement, the only work done on the structure is the unit load multiplied by the true displacement, δ.

$$\text{Work done} = 1\ \delta$$

where δ is the deflection at the point due to actual loads. The internal work is the product of the unit load internal forces (i.e. n, m, f and t) and the virtual displacements due to real loads.
In differential terms, these are:

dL, $d\theta$, $d\gamma$ and $d\phi$ for axial, bending, shear and torsional loading respectively.

$$\text{Internal work done} = \int n\,dL + \int m\,d\theta + \int F\,d\gamma + \int t\,d\phi$$

$$1\delta = \int n\,dL + \int m\,d\theta + \int F\,d\gamma + \int t\,d\phi$$

If the structure remains elastic, and follows Hook's law:

$$\delta = \int \frac{Nn}{AE}dx + \int \frac{Mm}{EI}dx + \int \frac{Ff}{GA}dx + \int \frac{Tt}{GJ}dx \tag{11.48}$$

where N, M, F and T are due to the main loading.

The sign convention must be consistent and then the displacement, δ, will be positive when in the same sense as the unit load.

Consider a beam with a bending moment (BM) due to a main load as well as a unit load, as shown in Figure 11.29. The object is to find the deflection at any point of an elastic structure.

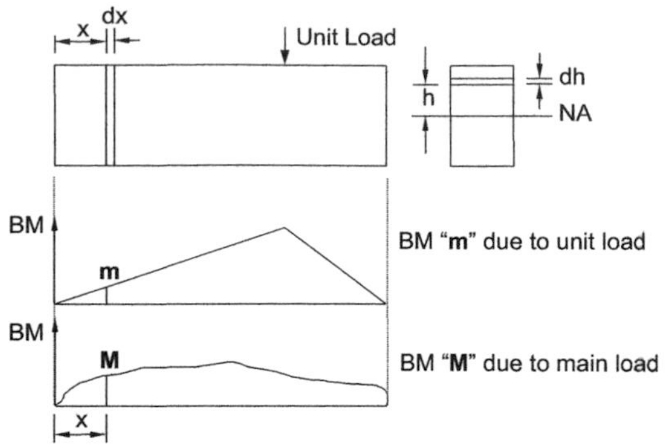

Figure 11.29

The unit load is applied at the point and in the direction in which the 'δ' (i.e. displacement) is required.

Let 'm' = BM on the element due to the unit load

The bending stress, σ, at distance h from the neutral axis (NA) is:

$$\sigma = \frac{mh}{I} \tag{11.49}$$

Force on the element = stress × area (on an elemental area across the width of the beam), and B = width of the beam

So, the force $= \left(\frac{mh}{I}\right)(b\,dh)$ due to the unit load

Regarding the unit load as being in operation, we now apply the main load system.

Let 'M' = BM on the element due to the main load.

Now the strain, $\varepsilon = \frac{\delta L}{L}$ where L = original length

When the length is an elemental length, dx, then $\varepsilon = \frac{\delta L}{dx}$

The change in length of the element,

$$\delta L = \varepsilon\,dx = \left(\frac{\sigma}{E}\right)dx = \left(\frac{Mh}{I}\right)\frac{1}{E}dx$$

from equation (11.49)

Work done on the element = force × distance

i.e. $$\text{WD} = \{stress \times area\}\{change\ of\ length\}$$

$$= \left\{\left(\frac{mh}{I}\right)(b\,dh)\right\}\left\{\left(\frac{Mh}{I}\right)\left(\frac{dx}{E}\right)\right\}$$

The work done on section dx is obtained by integrating w.r.t. dh

i.e. work done on element $dx = \frac{Mm}{EI^2}\int \left(b h^2 dh\right) dx$

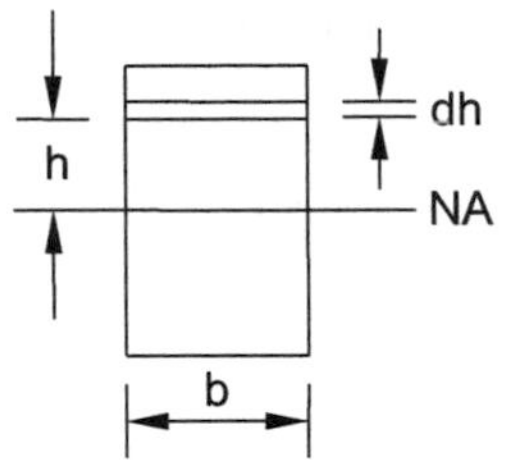

Figure 11.30

However, I is defined as $\int h^2\, dA - \int b h^2\, dx$

Work done on the section $= \frac{Mm}{EI} dx$

Work done on the whole beam $= \int_0^L \frac{Mm}{EI} dx$

This work done equals the external work done by the unit load, i.e. $WD =$ (unit load) $\times\, \delta$

i.e. $$\boldsymbol{\delta = \int_0^L \frac{Mm}{EI} dx} \qquad (11.50)$$

where M is the moment due to the load, and m is the moment due to the unit load.

Note that integration must be performed along the centreline of the beam.

Problem 13. Find the maximum deflection of the cantilever beam shown in Figure 11.31 using the unit load method. It is loaded by a force P acting at the free end, as shown. Ignore the self-weight of the beam. The beam has a second moment of area, I and a Young's modulus, E

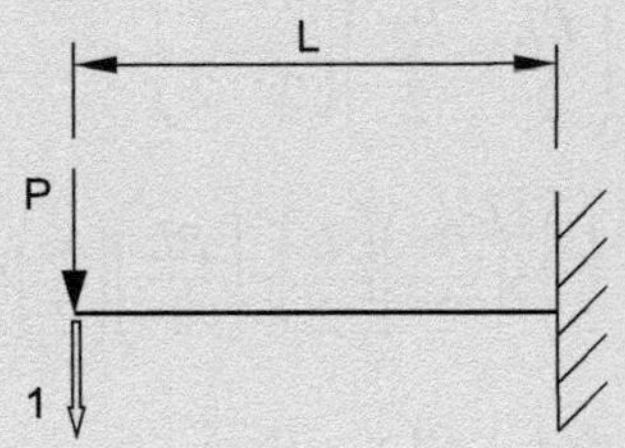

Figure 11.31

At the free end of the cantilever, $\delta = \int_0^L \frac{Mm}{EI} dx$

$M = -P\,x$ and $m = -1\,x$

Then $\delta = \int_0^L \frac{Mm}{EI} dx = \int_0^L \frac{(-Px)(-1x)}{EI} dx = \int_0^L \frac{Px^2}{EI} dx$

i.e. $\delta = \left[\frac{Px^3}{3EI}\right]_0^L$ i.e. **maximum deflection,** $\boldsymbol{\delta = \frac{PL^3}{3EI}}$

Problem 14. Using the unit load method, find the maximum deflection of the cantilever loaded as shown in Figure 11.32 (ignoring self-weight). The beam has a second moment of area, I and a Young's modulus, E.

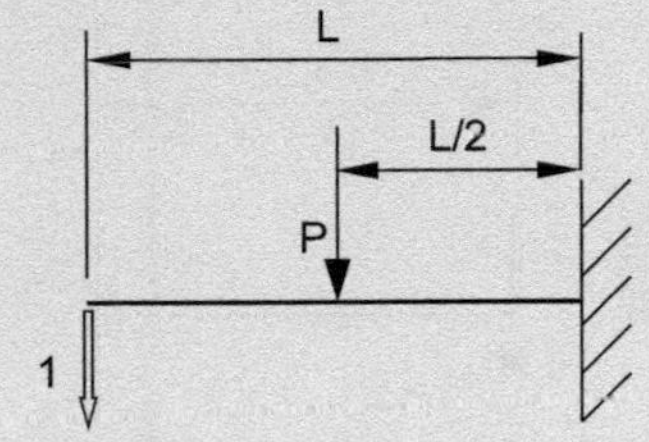

Figure 11.32

At the end of the cantilever, $\delta = \int_0^L \frac{Mm}{EI} dx$

$0 < x < L/2$ $\quad M = 0$ and $m = -1\,x$

$L/2 < x < L$ $\quad M = -P(x - L/2)$ and $m = -1\,x$

Then

$$\delta = \int_0^L \frac{Mm}{EI} dx$$

$$= \int_0^{L/2} \frac{(0)(-1x)}{EI} dx + \int_{L/2}^L \frac{\left[\left(-P\left(x - L/2\right)\right)\right]\left(-1x\right)}{EI} dx$$

i.e. $$\delta = 0 + \frac{P}{EI}\int_{L/2}^L \left(x^2 - Lx/2\right) dx$$

$$= \frac{P}{EI}\left[\frac{x^3}{3} - \frac{Lx^2}{4}\right]_{L/2}^L$$

$$= \frac{P}{EI}\left\{\left(\frac{L^3}{3} - \frac{L^3}{4}\right) - \left(\frac{L^3}{24} - \frac{L^3}{16}\right)\right\}$$

i.e. $$\delta = \frac{P}{EI}\left\{\left(\frac{4L^3}{12} - \frac{3L^3}{12}\right) - \left(\frac{2L^3}{48} - \frac{3L^3}{48}\right)\right\}$$

$$= \frac{P}{EI}\left\{\left(\frac{L^3}{12}\right) - \left(-\frac{L^3}{48}\right)\right\}$$

i.e **maximum deflection,** $\boldsymbol{\delta = \frac{5}{48}\frac{PL^3}{EI}}$

Problem 15. A light cantilever is loaded as shown in Figure 11.33. Use the unit load method to find the maximum deflection, δ. Take the second moment of area to be I and the Young's modulus to be E.

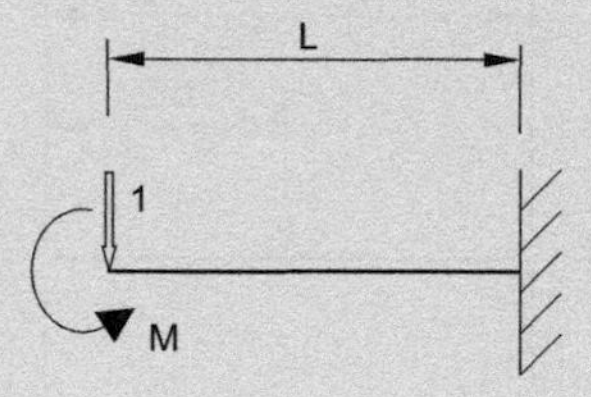

Figure 11.33

At the end of the cantilever, $\delta = \int_0^L \frac{Mm}{EI}dx$

$M = -M$ and $m = -1\,x$

Then $$\delta = \int_0^L \frac{Mm}{EI}dx = \int_0^L \frac{(-M)(-1x)}{EI}dx$$

$$= \int_0^L \frac{Mx}{EI}dx$$

i.e. $\delta = \frac{M}{EI}\left[\frac{x^2}{2}\right]_0^L$ i.e. $\boldsymbol{\delta = \frac{ML^2}{2EI}}$

Problem 16. Find the central deflection of a light, simply supported beam that carries a central load P, as shown in Figure 11.34. The second moment of area is I and the Young's modulus is E.

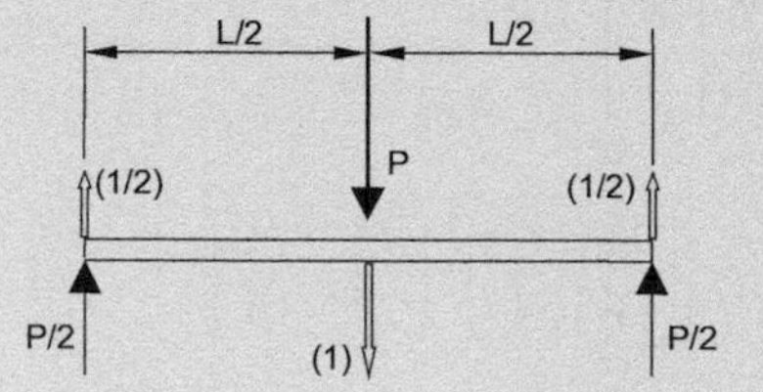

Figure 11.34

$0 < x < L/2$ $M = (P/2)(x)$ and $m = x/2$

The moment, m, is caused by the unit load.

Due to symmetry, half of the beam may be considered, and then the result is multiplied by 2.

$$\delta = 2\int_0^{L/2} \frac{Mm}{EI}dx = 2\int_0^{L/2} \frac{(P/2)(x)(x/2)}{EI}dx$$

$$= \frac{P}{2EI}\int_0^{L/2} x^2\,dx$$

$$= \frac{P}{2EI}\left[\frac{x^3}{3}\right]_0^{L/2} = \frac{P}{2EI}\left[\frac{L^3}{24}\right]$$

i.e. **central deflection,** $\boldsymbol{\delta = \frac{PL^3}{48\,EI}}$

If symmetry is not used, then the solution is much longer, as shown below.

$$\delta = \int_0^{L/2} \frac{Mm}{EI}dx + \int_{L/2}^{L} \frac{Mm}{EI}dx = I_1 + I_2$$

$0 < x < L/2$

$$I_1 = \int_0^{L/2} \left(\frac{Px}{2}\right)\left(\frac{x}{2}\right)\frac{dx}{EI} = \frac{P}{4EI}\left[\frac{x^3}{3}\right]_0^{L/2}$$

$$= \frac{P}{12EI}\left[\frac{L^3}{8}\right] = \frac{PL^3}{96EI}$$

$L/2 < x < L$

$$M = \frac{Px}{2} - P\left(x - \frac{L}{2}\right) = \frac{Px}{2} - Px + \frac{PL}{2} = \frac{PL}{2} - \frac{Px}{2}$$

and $$m = \frac{1x}{2} - 1\left(x - \frac{L}{2}\right) = \frac{x}{2} - x + \frac{L}{2} = \frac{L}{2} - \frac{x}{2}$$

Then $$Mm = \left(\frac{PL}{2} - \frac{Px}{2}\right)\left(\frac{L}{2} - \frac{x}{2}\right)$$

$$= \left\{\left(\frac{PL}{2}\right)\left(\frac{L}{2}\right) - \left(\frac{Px}{2}\right)\left(\frac{L}{2}\right) - \left(\frac{PL}{2}\right)\left(\frac{x}{2}\right) + \left(\frac{Px}{2}\right)\left(\frac{x}{2}\right)\right\}$$

$$= \left\{\frac{PL^2}{4} - \frac{Px\,L}{4} - \frac{PL\,x}{4} + \frac{Px^2}{4}\right\}$$

$$= \left\{\frac{PL^2}{4} - \frac{Px\,L}{2} + \frac{Px^2}{4}\right\}$$

i.e $$Mn = \frac{P}{4}\left\{L^2 - 2xL + x^2\right\}$$

Hence, $$I_2 = \int_{L/2}^{L} \frac{Mm}{EI}dx = \int_{L/2}^{L} \frac{P}{4}\left\{L^2 - 2xL + x^2\right\}\frac{dx}{EI}$$

$$= \frac{P}{4EI}\int_{L/2}^{L}\left(L^2 - 2xL + x^2\right)dx$$

$$= \frac{P}{4EI}\left[L^2x - 2\left(\frac{x^2}{2}\right)L + \frac{x^3}{3}\right]_{L/2}^{L}$$

$$= \frac{P}{4EI}\left[\left(L^3 - L^3 + \frac{L^3}{3}\right) - \left(\frac{L^3}{2} - \frac{L^3}{4} + \frac{L^3}{24}\right)\right]$$

$$= \frac{P}{4EI}\left[\frac{L^3}{3} - \frac{L^3}{2} + \frac{L^3}{4} - \frac{L^3}{24}\right]$$

$$= \frac{PL^3}{4EI}\left[\frac{1}{3} - \frac{1}{2} + \frac{1}{4} - \frac{1}{24}\right]$$

$$= \frac{PL^3}{4EI}\left[\frac{8-12+6-1}{24}\right] = \frac{PL^3}{4EI}\left(\frac{1}{24}\right)$$

i.e $$I_2 = \frac{PL^3}{96EI}$$

Then $$\delta = I_1 + I_2 = \frac{PL^3}{96EI} + \frac{PL^3}{96EI}$$

i.e $$\boldsymbol{\delta = \frac{PL^3}{48EI}}$$

Problem 17. Force P acts vertically at point A on the light framework shown in Figure 11.35. All sections have a second moment of area, I and Young's modulus E. Using the unit load method, derive an expression for maximum vertical deflection of point A.

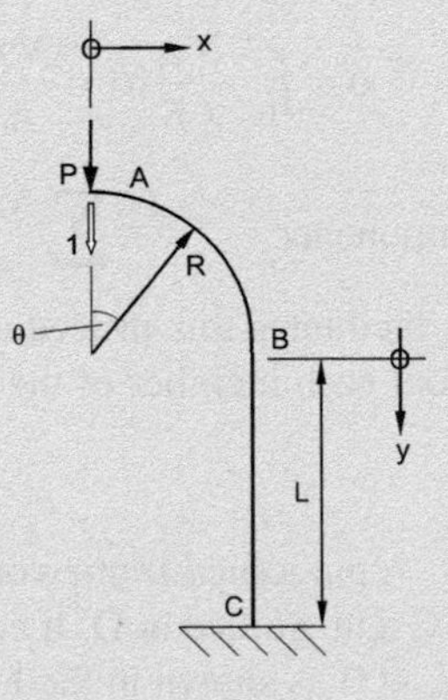

Figure 11.35

Vertical deflection,

$$\delta_{VERT} = \delta_{AB\ VERT} + \delta_{BC\ VERT}$$

For AB:

$M = -PR\sin\theta$

$m = (-1)R\sin\theta$

$dx = R\,d\theta$

$$\delta_{AB\,Vert} = \int_0^{\pi/2}\frac{Mm}{EI}dx = \int_0^{\pi/2}\frac{Mm}{EI}R\,d\theta$$

$$= \int_0^{\pi/2}\frac{(-PR\sin\theta)(-R\sin\theta)}{EI}R\,d\theta$$

$$= \int_0^{\pi/2}\frac{\left(PR^2\sin^2\theta\right)}{EI}R\,d\theta \qquad (11.51)$$

Now, $\cos 2\theta = 1 - 2\sin^2\theta$ from which,

$2\sin^2\theta = 1 - \cos 2\theta$ and $\sin^2\theta = \frac{1}{2} - \frac{\cos 2\theta}{2}$

Then $\int \sin^2\theta\,d\theta = \int\left(\frac{1}{2} - \frac{\cos 2\theta}{2}\right)d\theta = \frac{\theta}{2} - \frac{\sin 2\theta}{4}$

Hence, from equation (11.51),

$$= \delta_{AB\,vert} = \int_0^{\pi/2}\frac{\left(PR^2\sin^2\theta\right)}{EI}R\,d\theta$$

$$= \frac{PR^3}{EI}\left[\frac{\theta}{2} - \frac{\sin 2\theta}{4}\right]_0^{\pi/2}$$

$$= \frac{PR^3}{EI}\left[\frac{\pi}{4} - 0\right]$$

i.e. $$\delta_{AB\,vert} = \frac{\pi PR^3}{4EI}$$

For BC:

$M = -PR$

$m = (-1)R$

$$\delta_{BC\,Vert} = \int_0^{L}\frac{Mm}{EI}dy = \int_0^{L}\frac{(-PR)(-R)}{EI}dy$$

$$= \int_0^{L}\frac{PR^2}{EI}dy = \frac{PR^2}{EI}(y)_0^L$$

i.e. $\delta_{BC\,Vert} = \dfrac{PR^2 L}{EI}$

and **the maximum vertical deflection of point A,**

$$\boldsymbol{\delta_{vert} = \frac{\pi P R^3}{4EI} + \frac{PR^2 L}{EI}}$$

Problem 18. A thin curved beam is in the form of a quadrant of a circle, as shown in Figure 11.36. One end of this quadrant is firmly fixed to a solid base, whilst the other end is subjected to a point load *W*, acting vertically downwards. Determine expressions for the vertical and horizontal displacements at the free end by applying the unit load method

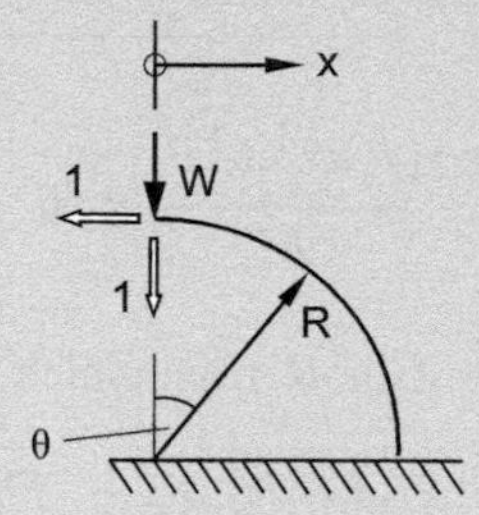

Figure 11.36 Thin curved beam quadrant

Note that this is the same as Problem 6 on pages 251/252, but using a different method of solution.
To find δ_{VERT}, apply a vertical unit load at the free end.

$$M = -W\,R\cos\theta \qquad (11.52)$$

$$m = (-1)\,R\cos\theta \qquad (11.53)$$

$dx = R\,d\theta$, so $\delta_{VERT} = \int_0^L \frac{Mm}{EI}dx = \int_0^{\pi/2} \frac{Mm}{EI} R d\theta$

$$= \int_0^{\pi/2} \frac{(-W\;R\sin\theta)(-R\sin\theta)}{EI} R d\theta$$

$$= \int_0^{\pi/2} \frac{(W\;R^3 \sin^2\theta)}{EI} d\theta$$

Using the standard integral for

$$\int \sin^2\theta \to \frac{\theta}{2} - \frac{\sin 2\theta}{4} + C$$

$$\delta_{AB\,VERT} = \frac{WR^3}{EI}\left[\frac{\theta}{2} - \frac{\sin 2\theta}{4}\right]_0^{\pi/2} = \frac{WR^3}{EI}\left[\frac{\pi}{4} - 0\right]$$

i.e. $\boldsymbol{\delta_{AB\,VERT} = \dfrac{\pi W R^3}{4EI}}$ which is identical to δV in Problem 6, page 252

To find δ_{HORIZ}, apply a horizontal unit load at free end.

$$M = -W\,R\cos\theta \qquad (11.54)$$

$$m = (-1)\,(R - R\sin\theta) \qquad (11.55)$$

$dx = R\,d\theta$, so $\delta_{HORIZ} = \int_0^L \frac{Mm}{EI}dx = \int_0^{\pi/2} \frac{Mm}{EI} R d\theta$

$$= \frac{1}{EI}\int_0^{\pi/2} (-W\;R\cos\theta)[-1(R - R\sin\theta)]Rd\theta$$

$$= \frac{WR^3}{EI}\int_0^{\pi/2} (\cos\theta - \cos\theta\,\sin\theta)d\theta$$

Now, $\sin 2\theta = 2\cos\theta\,\sin\theta$ and so $\cos\theta\,\sin\theta = \dfrac{\sin 2\theta}{2}$

$$\delta_{HORIZ} = \frac{WR^3}{EI}\int_0^{\pi/2}\left(\cos\theta - \frac{\sin 2\theta}{2}\right)d\theta$$

$$= \frac{WR^3}{EI}\left[\sin\theta - \frac{1}{2}\left(-\frac{1}{2}\cos 2\theta\right)\right]_0^{\pi/2}$$

$$= \frac{WR^3}{EI}\left[\sin\theta + \frac{1}{4}\cos 2\theta\right]_0^{\pi/2}$$

$$= \frac{WR^3}{EI}\left[\left(1 + \frac{1}{4}(-1)\right) - \left(0 + \frac{1}{4}\right)\right]$$

$$= \frac{WR^3}{EI}\left[1 - \frac{1}{4} - \frac{1}{4}\right]$$

i.e. $\boldsymbol{\delta_{HORIZ} = \dfrac{WR^3}{2EI}}$ which is identical to δH in Problem 6, page 253.

The unit load method is particularly useful for solving deflections of pin jointed frameworks, or trusses. In a plane of pin jointed framework consisting of bars (members) pinned at the ends, only axial forces exist.

The deflection, $\delta = \int_0^L \frac{N\,n}{A\,E}dx = \frac{N\,n\,L}{A\,E}$

For several components: $\delta = \sum \frac{N\,n\,L}{A\,E} \qquad (11.56)$

Tabular format facilitates this procedure of listing the information about each member of the truss, as shown in Problem 19.

Problem 19. A pin joined framework ABCD, is pin jointed at C and guided at D. It carries a horizontal load, N, at B as shown in the Figure 11.37. All horizontal and vertical members are of length L and remain elastic. They all have the same Young's modulus and the same cross-sectional area. Find the horizontal deflection of joint A using the unit load method.

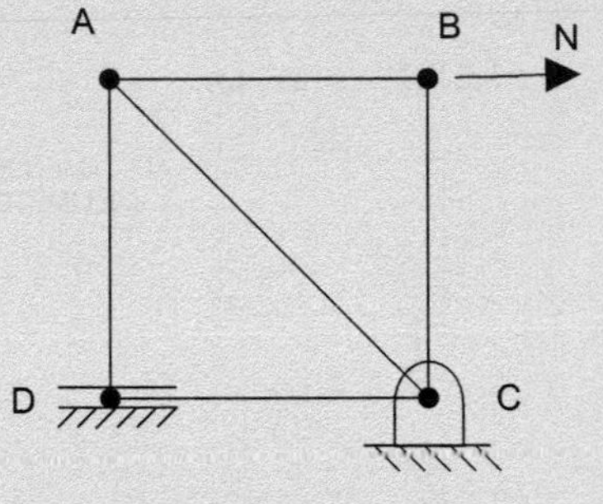

Figure 11.37

The equilibrium of pin joined frameworks is discussed in Section 3.4, page 56. A free body diagram is drawn in Figure 11.38.

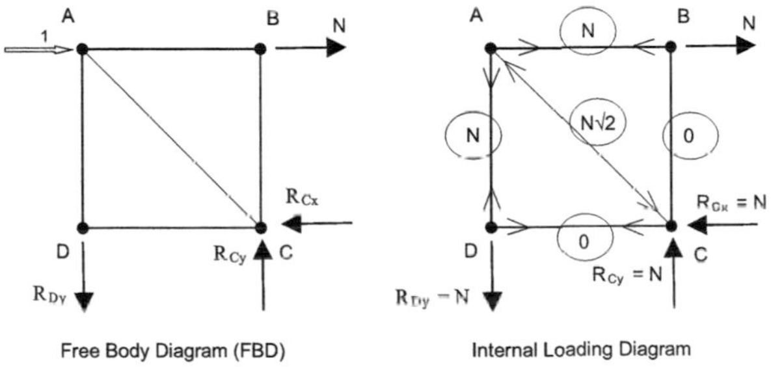

Figure 11.38

From figure 11.38:

Consideration of the horizontal equilibrium of the framework gives $R_{Cx} = N$

Taking moments about C, gives the value of $R_{Dy} = N$ (as all members have equal length).

Consideration of vertical equilibrium of the framework gives $R_{Cy} = N$

From the internal loading diagram and using the method of joints:

Considering Joint B gives $AB = N$ and $BC = 0$

Considering Joint D gives $AD = N$ and $CD = 0$

Considering joint A:

$[\Sigma F_Y = 0]$ ↑+ve −AD − AC cos 45° = 0

from which, $AC = \dfrac{-AD}{\cos 45^\circ} = -N\sqrt{2}$

Internal force AB must point away from joint B, and so the member is in tension and hence POSITIVE (see Section 3.4).

This data is then put in tabular format, along with the loads due to the unit load, which follow a similar pattern.

$$\delta_{EACH\ MEMBER} = \int_0^L \frac{N\,n}{AE}dx = \frac{N\,n\,L}{AE}$$

$$\delta_{B\,HORIZ} = \sum \frac{N\,n\,L}{AE}$$

Member	Length	N	n	NnL
AB	L	+N	0	0
BC	L	0	0	0
AD	L	+N	+1n	NLn
CD	L	0	0	0
AC	√2L	−N√2	−n√2	2√2LNn

$\Sigma = 3.828$ NnL

However, $n = 1$

Hence, $\delta_{B\,Horiz} = \dfrac{\mathbf{3.828}\,NL}{\mathbf{AE}}$

Now try the following Practice Exercise

Practice Exercise 49. Unit load

1. Using the unit load method, determine the maximum vertical deflection of the light cantilever AB at point A, shown in Figure 11.39. The beam is loaded at point A with a moment M at the free end, as shown. It is made from a material of Young's modulus E and shear modulus G. The second moment of area and polar second moment of area are I and G respectively.

   Assume that $E = 200$ GPa, $I = 20 \times 10^{-9}$ m^4, $M = 150$ Nm, $L = 1$ m

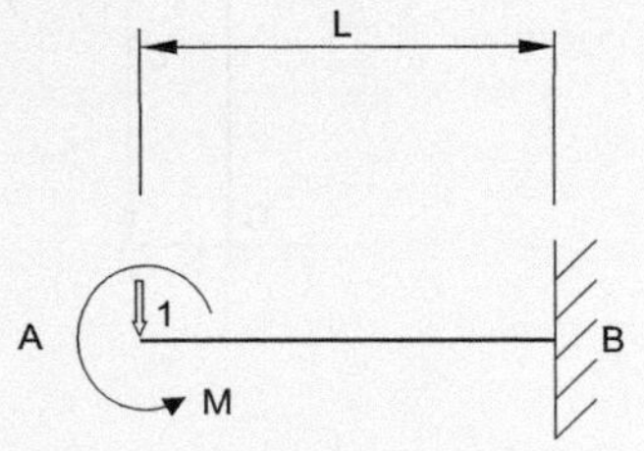

Figure 11.39

Hint: Put unit load at free end of cantilever.

$$\left[\delta = \frac{ML^2}{2EI} = 18.8\text{mm}\right]$$

2. Force P acts vertically downwards at point A on the light curved beam shown in Figure 11.40. Using the unit load method, find the maximum HORIZONTAL deflection of point A, in terms of the second moment of area I and Young's modulus E.

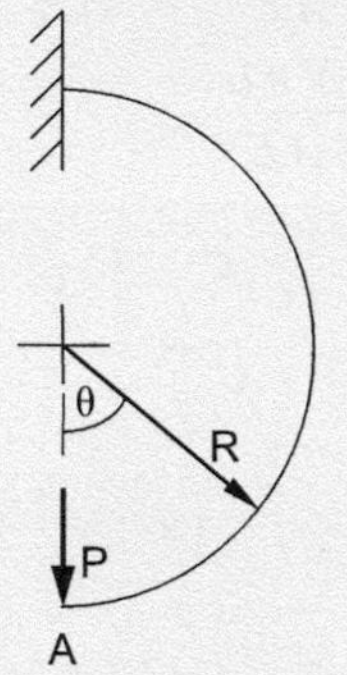

Figure 11.40

$$\left[\delta = \frac{2PR^3}{EI}\right]$$

3 Force P acts horizontally at point A on the light framework shown in Figure 11.41. All sections have a second moment of area, I and Young's modulus E. Using the unit load method, derive an expression for maximum horizontal deflection of point A. It can be assumed that any deflection due to direct compression is small and the self-weight of the beam can be ignored.

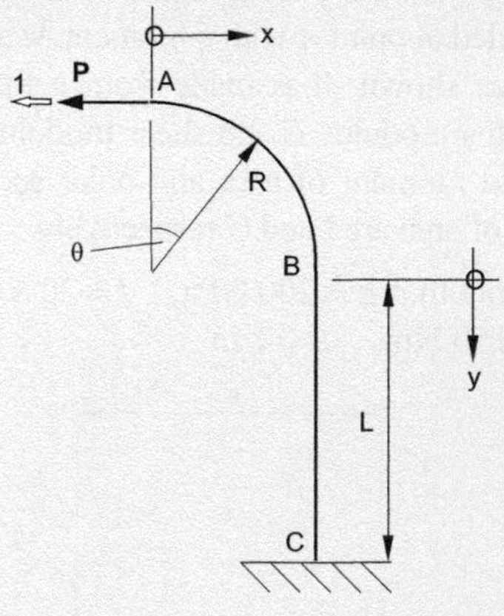

Figure 11.41

$$\left[\delta = \frac{1}{EI}\left\{PR^3\left(\frac{3\pi}{4} - 2\right) + PL\left(R^2 + RL + \frac{L^2}{3}\right)\right\}\right]$$

4. Use the unit load method to calculate the maximum deflection of the cantilever loaded as shown in Figure 11.42 ignoring its self-weight. Take the load P = 180 N and the Young's modulus, E = 200 GPa.

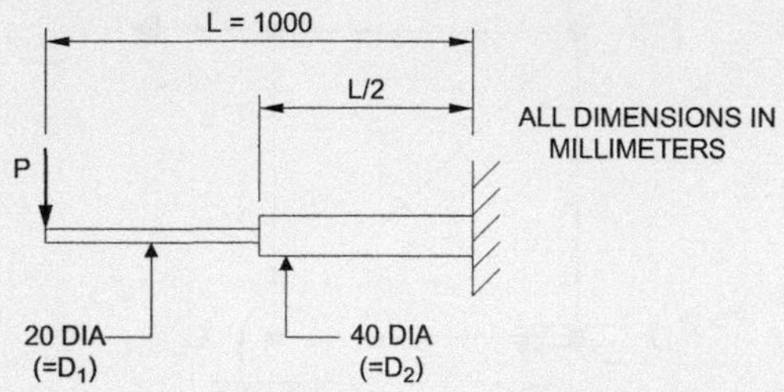

Figure 11.42

$[\delta = 6.86 \text{ mm}]$

5. In Figure 11.43, find the vertical deflection of point A by means of the unit load method, assuming deflection due to direct compression to be small. Assume the second moment of area is I and Young's modulus is E.

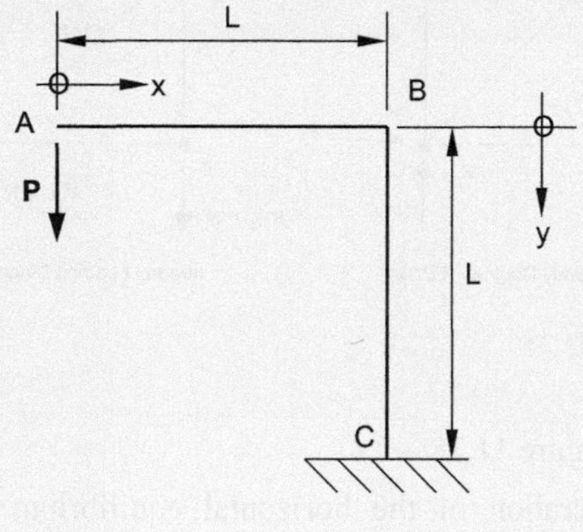

Figure 11.43

$$\left[\delta = \frac{4PL^3}{3EI}\right]$$

11.12 Suddenly applied and impact loads

When structures are subjected to suddenly applied or impact loads, the resulting stresses can be considerably greater than those which will occur if these same loads are gradually applied.

Using a theoretical solution, it will be proven later that if a stationary load were suddenly applied to a structure, at zero velocity, then the stresses set up in the structure would have twice the magnitude that they would have had if the load were gradually applied.

The importance of stresses due to impact will be demonstrated by a number of worked examples, but prior to this, it will be necessary to make a few definitions.

11.13 Resilience

This is a term that is often used when considering elastic structures under impact. The resilience of an elastic body is a measurement of the amount of elastic strain energy that the body will store under a given load.

Problem 20. A vertical rod, of cross-sectional area A, is secured firmly at its top end and has an inextensible collar of negligible mass attached firmly to its bottom end, as shown in Figure 11.44. If a mass of magnitude M is dropped onto the collar from a distance h above it, determine the maximum values of stress and deflection that will occur owing to this impact. Neglect energy and other losses, and assume that the mass is in the form of an annular ring.

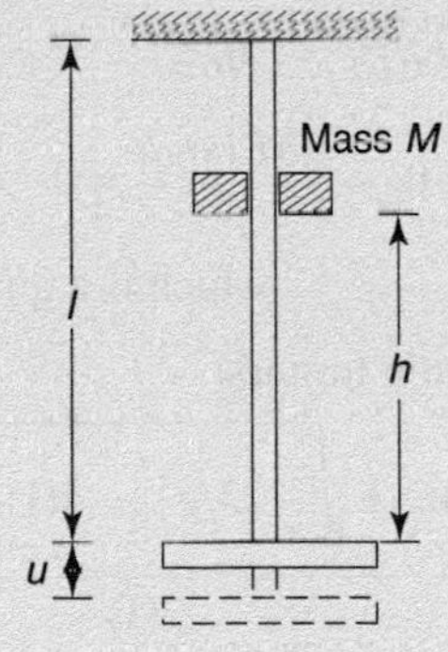

Figure 11.44 Rod under impact

Energy and friction losses are neglected, so that the rod is assumed to absorb all the energy on impact. From the point of view of the structural analyst, this assumption is reasonable, as it simplifies the solution and the errors are on the so-called 'safe side'.

Let u be the maximum deflection of the rod on impact and

if PE = potential energy and σ = the maximum stress in the rod due to impact

then PE of mass $= Mg(h+u)$

= strain energy stored in the rod

$= \frac{\sigma^2}{2E} \times$ volume of rod, from equation (11.15)

i.e. $$Mg(h+u) = \frac{\sigma^2}{2E} \times \text{volume of rod} \qquad (11.57)$$

However $u = \varepsilon l = \frac{\sigma l}{E}$

Substituting into equation (11.57) gives:

$$Mg\left(h + \frac{\sigma l}{E}\right) = \frac{\sigma^2}{2E} \times Al$$

Transposing gives $$\frac{2MgE}{Al}\left(h + \frac{\sigma l}{E}\right) = \sigma^2$$

or $$\sigma^2 - \frac{2\sigma Mg}{A} - \frac{2MghE}{Al} = 0$$

i.e. $$\sigma^2 - \left(\frac{2Mg}{A}\right)\sigma - \frac{2MghE}{Al} = 0$$

which is a quadratic equation in σ

Using the quadratic formula

$$\sigma = \frac{-\left(-\frac{2Mg}{A}\right) \pm \sqrt{\left[\left(-\frac{2Mg}{A}\right)^2 - 4(1)\left(-\frac{2MghE}{Al}\right)\right]}}{2(1)}$$

i.e. $$= \frac{\left(\frac{2Mg}{A}\right) \pm \sqrt{4}\sqrt{\left[\left(\frac{Mg}{A}\right)^2 + \left(\frac{2MghE}{Al}\right)\right]}}{2(1)}$$

i.e. $$\boldsymbol{\sigma = \frac{Mg}{A} + \sqrt{\left[\frac{M^2g^2}{A^2} + \frac{2MghE}{Al}\right]}} \qquad (11.58)$$

and $$\boldsymbol{u = \frac{\sigma l}{E}} \qquad (11.59)$$

Suddenly applied load

If the mass M in Figure 11.44 were *just above* the top surface of the collar and *suddenly released,* so that $h = 0$ in equation (11.58) then

$$\sigma = \frac{Mg}{A} + \sqrt{\left[\frac{M^2g^2}{A^2}\right]} = \frac{Mg}{A} + \frac{Mg}{A}$$

i.e. $\sigma = \dfrac{2Mg}{A}$

That is, *a suddenly applied load*, at zero velocity, *induces twice the stress* of a gradually applied load of the same magnitude.

If, in Problem 20 above, the following applied, then the maximum stress on impact and, also, the maximum stress if the same load were gradually applied to the top of the collar can be determined:

$$M = 1\text{ kg},\ A = 400\text{ mm}^2,\ l = 1\text{ m},\ h = 0.5\text{ m}$$

$$E = 2\times 10^{11}\text{N/m}^2,\ g = 9.81\text{ m/s}^2$$

From equation (11.58) the maximum stress on impact is:

$$\sigma = \frac{1\times 9.81}{400\times 10^{-6}} + \sqrt{\left[\frac{1^2\times 9.81^2}{\left(400\times 10^{-6}\right)^2} + \frac{2\times 1\times 9.81\times 0.5\times 2\times 10^{11}}{400\times 10^{-6}\times 1}\right]}$$

$$= 0.025\times 10^6 + 70.04\times 10^6 = 70.07\times 10^6$$

i.e. $\boldsymbol{\sigma = 70.07\ \mathrm{MN/m^2}}$

If the load were gradually applied to the top of the collar

$$\text{maximum static stress} = 0.025\ \text{MN/m}^2$$

i.e. *the impact stress* is 70.07/0.025, *i.e. about* 2800 *times greater than the static stress.*

Problem 21. A simply supported beam of uniform section is subjected to an impact load at mid-span, as shown in Figure 11.45. Determine an expression for the maximum deflection of the beam.

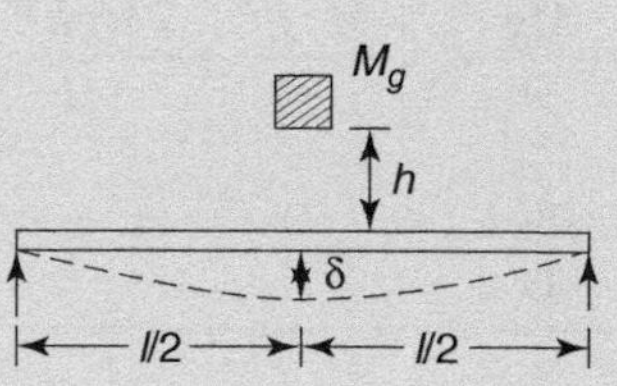

Figure 11.45 Beam under impact

Let δ = the maximum central deflection of the beam due to impact

W_e = equivalent static load to cause the deflection δ

$$PE \text{ of load} = Mg(h+\delta) \tag{11.60}$$

From equation (11.30),

$$U_b = \frac{W_e^2 l^3}{96EI} \tag{11.61}$$

Equating equations (11.60) and (11.61) gives:

$$Mg(h+\delta) = \frac{W_e^2 l^3}{96EI} \tag{11.62}$$

However, from Problem 3 on pages 248 and 249,

$$\delta = \frac{W_e l^3}{48EI}$$

which on substitution into equation (11.62) gives:

$$Mg\left(h + \frac{W_e l^3}{48EI}\right) = \frac{W_e^2 l^3}{96EI}$$

i.e. $$Mgh + \frac{MgW_e l^3}{48EI} = \frac{W_e^2 l^3}{96EI}$$

i.e. $$\frac{W_e^2 l^3}{96EI} - \frac{MgW_e l^3}{48EI} - Mgh = 0$$

or $$W_e^2 - \frac{\dfrac{MgW_e l^3}{48EI}}{\dfrac{l^3}{96EI}} - \frac{Mgh}{\dfrac{l^3}{96EI}} = 0$$

i.e. $$W_e^2 - 2MgW_e - \frac{96EIMgh}{l^3} = 0$$

which is a quadratic equation

Using the quadratic formula

$$W_e = \frac{--2Mg \pm \sqrt{(-2Mg)^2 - 4(1)\left(-\dfrac{96EIMgh}{l^3}\right)}}{2(1)}$$

$$= \frac{2Mg \pm \sqrt{4M^2g^2 + 4\left(\dfrac{96EIMgh}{l^3}\right)}}{2(1)}$$

$$= \frac{2Mg \pm \sqrt{4}\sqrt{M^2g^2 + \left(\dfrac{96EIMgh}{l^3}\right)}}{2(1)}$$

i.e. $$\boldsymbol{W_e = Mg + \sqrt{M^2g^2 + \left(\frac{96EIMgh}{l^3}\right)}} \tag{11.63}$$

If Mg were suddenly applied, at zero velocity, so that $h = 0$ in equation (11.63), then

$$W_e = Mg + \sqrt{M^2 g^2} = Mg + Mg$$

i.e. $\quad \boldsymbol{W_e = 2Mg}$

i.e. once again, it can be seen that a suddenly applied load, at zero velocity, has twice the value of a gradually applied load.

Hence, **maximum deflection of the beam,**

$$\boldsymbol{\delta} = \frac{W_e l^3}{48EI} = \frac{2Mgl^3}{48EI} = \boldsymbol{\frac{Mgl^3}{24EI}}$$

Problem 22. A uniform-section beam, which is initially horizontal, is simply supported at one end, and is supported at the other end by an elastic wire, as shown in Figure 11.46. Assuming that a mass of magnitude 2 kg, which is situated at a height of 0.2 m above the beam, is dropped onto the mid-span of the beam, determine the central deflection of the beam and the maximum stress induced in the wire. The beam and wire may be assumed to have negligible masses, and the following may also be assumed to apply:

$E = 2\times10^{11}\,\text{N/m}^2$ (for both beam and wire)

I = second moment of area of the beam section about a horizontal plane $= 2\times10^{-8}\,\text{m}^4$

A = cross-sectional area of wire $= 1.3\times10^{-6}\,\text{m}^2$

L = length of beam = 2 m

l = length of wire = 0.9 m $\quad h = 0.4$ m

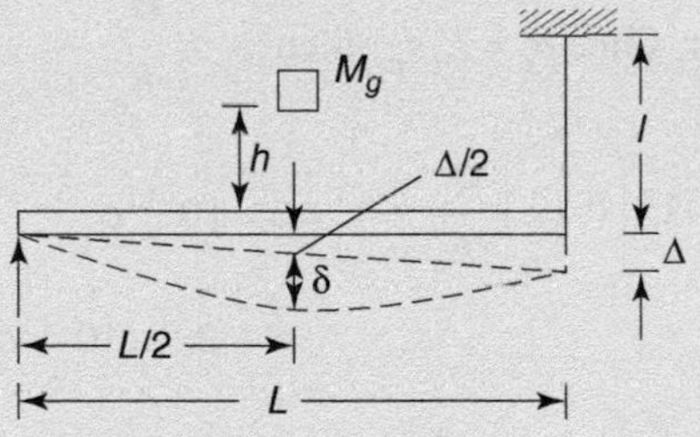

Figure 11.46 Impact on a beam

Let W_e = equivalent static load to cause the deflections δ and Δ

Δ = deflection of wire due to impact, as shown in Figure 11.46

δ = maximum central deflection of beam due to flexure alone, on impact, as shown in Figure 11.46

i.e. the maximum central deflection of the beam due to impact is $\delta + \Delta/2$

From equilibrium considerations, the maximum force in the wire is $W_e/2$ or

$$\Delta = \frac{W_e l}{2AE} \tag{11.64}$$

Now, from equation (11.30)

$$U_b = \frac{W_e^2 L^3}{96EI} \tag{11.65}$$

From equation (11.15), the strain energy in the wire is

$$U_e = \frac{\sigma^2}{2E} \times Al$$

However

$$\sigma = E\varepsilon = \frac{E\Delta}{l} = \frac{E}{l}\left(\frac{W_e l}{2AE}\right) = \frac{W_e}{2A}$$

Hence $\quad U_e = \dfrac{\left(\dfrac{W_e}{2A}\right)^2}{2E} \times Al = \dfrac{W_e^2}{2E\left(4A^2\right)} \times Al$

i.e.

$$U_e = \frac{W_e^2 \times l}{8AE} \tag{11.66}$$

so that the total strain energy is

$$U = \frac{W_e^2 L^3}{96EI} + \frac{W_e^2 \times l}{8AE} \tag{11.67}$$

PE of mass $= Mg(h + \delta + \Delta/2)$ from Figure 11.46 (11.68)

However

$$\delta = \frac{W_e l^3}{48EI} \tag{11.69}$$

Equating equations (11.67) and (11.68), and by substituting equations (11.64) and (11.69), the following relationship is obtained:

$$\frac{W_e^2 L^3}{96EI} + \frac{W_e^2 \times l}{8AE} = Mg(h + \delta + \Delta/2)$$

$$= Mg\left(h + \frac{W_e l^3}{48EI} + \frac{\frac{W_e l}{2AE}}{2}\right)$$

i.e.

$$\frac{W_e^2 L^3}{96EI} + \frac{W_e^2 \times l}{8AE} = Mg\left(h + \frac{W_e L^3}{48EI} + \frac{W_e l}{4AE}\right) \tag{11.70}$$

Substituting the appropriate values into equation (11.70) gives:

$$\frac{W_e^2 2^3}{96\times2\times10^{11}\times2\times10^{-8}} + \frac{W_e^2 \times 0.9}{8\times1.3\times10^{-6}\times2\times10^{11}}$$

$$= 2 \times 9.81 \times \left(0.4 + \frac{W_e \times 2^3}{48 \times 2 \times 10^{11} \times 2 \times 10^{-8}} + \frac{W_e \times 0.9}{4 \times 1.3 \times 10^{-6} \times 2 \times 10^{11}} \right)$$

i.e. $W_e^2 \left(2.083 \times 10^{-5} + 4.327 \times 10^{-7}\right) = 7.848 + W_e\left(8.175 \times 10^{-4} + 1.698 \times 10^{-5}\right)$

i.e. $2.126 \times 10^{-5} W_e^2 - 8.345 \times 10^{-4} W_e - 7.848 = 0$ which is a quadratic equation

Using a CASIO fx-991ES PLUS calculator, or using the quadratic formula, the solution is

$\mathbf{W_e = 607.8}$ **N** (taking the positive answer only)

and **the maximum stress in the wire due to impact**

$$= \frac{W_e}{2A} = \frac{607.8}{2 \times 1.3 \times 10^{-6}} = \mathbf{233.8\ MN/m^2}$$

Problem 23. A concrete pillar of length 4 m has a cross-sectional area of 0.25 m^2, where 10% is composed of steel reinforcement and the remainder of concrete. If a mass of 10 kg is dropped onto the top of the concrete pillar, from a height of 0.12 m above it, determine the maximum stresses in the steel and the concrete due to impact. Assume that:

E_s = elastic modulus of steel = 2×10^{11} N/m^2

E_c = elastic modulus of concrete = 1.5×10^{10} N/m^2

$g = 9.81$ m/s^2

Let δ = the maximum deflection of the column under impact, as shown in Figure 11.47

Work done, WD = $Mg(h + \delta)$ (11.71)

$$\text{Strain energy} = \frac{\sigma_c^2}{2E_c} \times (0.90 \times 0.25) \times 4 + \sigma_s^2 \times \frac{0.10 \times 0.25 \times 4}{2E_s}$$

$$= \frac{\sigma_c^2}{2E_c} \times 0.225 \times 4 + \sigma_s^2 \times \frac{0.025 \times 4}{2E_s} \quad (11.72)$$

where σ_c = maximum stress in the concrete due to impact $= E_c \varepsilon_c$

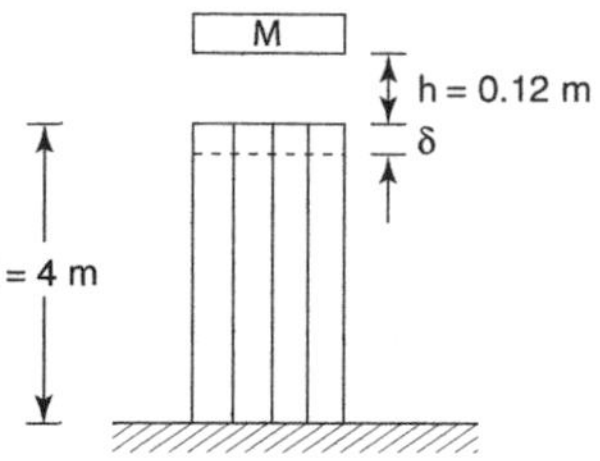

Figure 11.47 Reinforced concrete column

σ_s = maximum stress in the steel due to impact $= E_s \varepsilon_s$

ε_c and ε_s = maximum strain in concrete and steel, respectively

Change in length, δ = strain × original length $= \varepsilon_c \times 4\text{m}$ in this case

Equating equations (11.71) and (11.72) gives:

$$Mg(h + \delta) = \frac{\sigma_c^2}{2E_c} \times 0.225 \times 4 + \sigma_s^2 \times \frac{0.025 \times 4}{2E_s}$$

i.e. $10 \times 9.81(0.12 + \varepsilon_c \times 4)$

$$= \frac{\sigma_c^2}{2 \times 1.5 \times 10^{10}} \times 0.225 \times 4 + \sigma_s^2 \times \frac{0.025 \times 4}{2 \times 2 \times 10^{11}}$$

i.e. $98.1\left(0.12 + \varepsilon_c \times 4\right) = 3 \times 10^{-11} \sigma_c^2 + 2.5 \times 10^{-13} \sigma_s^2$

However, since $\sigma_c = E_c \varepsilon_c$, then $\varepsilon_c = \dfrac{\sigma_c}{E_c}$

$$\text{Hence } 98.1\left(0.12 + \frac{\sigma_c}{E_c} \times 4\right) = 3 \times 10^{-11} \sigma_c^2 + 2.5 \times 10^{-13} \sigma_s^2$$

$$\text{i.e. } 98.1\left(0.12 + \frac{\sigma_c}{1.5 \times 10^{10}} \times 4\right) = 3 \times 10^{-11} \sigma_c^2 + 2.5 \times 10^{-13} \sigma_s^2$$

i.e. $11.772 + 2.616 \times 10^{-8} \sigma_c = 3 \times 10^{-11} \sigma_c^2 + 2.5 \times 10^{-13} \sigma_s^2$ (11.73)

Now $\varepsilon_s = \varepsilon_c$

and therefore $\dfrac{\sigma_s}{E_s} = \dfrac{\sigma_c}{E_c}$

from which $\sigma_s = \dfrac{\sigma_c E_s}{E_c} = \dfrac{\sigma_c\left(2\times10^{11}\right)}{1.5\times10^{10}}$

$$= 13.333\sigma_c \qquad (11.74)$$

Substituting equation (11.74) into equation (11.73) gives:

$$11.772 + 2.616\times10^{-8}\sigma_c = 3\times10^{-11}\sigma_c^{\,2} + 2.5\times10^{-13}\left(13.333\sigma_c\right)^2$$

i.e. $11.772 + 2.616\times10^{-8}\sigma_c = 3\times10^{-11}\sigma_c^{\,2} + 4.444\times10^{-11}\sigma_c^{\,2}$

i.e. $11.772 + 2.616\times10^{-8}\sigma_c = 7.444\times10^{-11}\sigma_c^{\,2}$

or $7.444\times10^{-11}\sigma_c^{\,2} - 2.616\times10^{-8}\sigma_c - 11.772 = 0$

which is a quadratic equation

Using a CASIO fx-991ES PLUS calculator, mode 5, choice 3, (or using the quadratic formula) the solution is $\sigma_c = 398745$ N/m^2 = **0.399 MN/m²** (11.75)

and from equation (11.65)

$$\sigma_s = 13.333(0.399) = \mathbf{5.320\ MN/m^2}$$

Now try the following Practice Exercise

Practice Exercise 50. Suddenly applied and impact loads

1. A rod, composed of two elements of different cross-sectional areas, is firmly fixed at its top end and has an inextensible collar, firmly secured to its bottom end, as shown in Figure 11.48. If a mass of magnitude 2 kg is dropped from a height of 0.4 m above the top of the collar, determine the maximum deflection of the rod, and also the maximum stress. Assume that:

   $E = 2\times10^{11}$ N/m^2 and $g = 9.81$ m/s^2

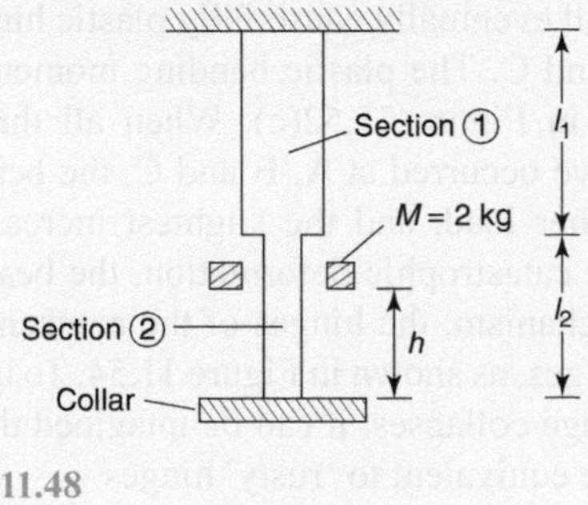

Figure 11.48

For section 1: $A_1 = 500$ mm^2 and $l_1 = 1.2$ m

For section 2: $A_2 = 300$ mm^2 and $l_2 = 0.8$ m

[0.631 mm, 83.04 MN/m²]

2. A reinforced concrete pillar, of length 3 m, is fixed firmly at its base and is free at the top, on to which a 20 kg mass is dropped, as shown in Figure 11.49. Determine the maximum stresses in the steel and the concrete, assuming the following apply:

   A_s = cross-sectional area of steel reinforcement = 0.01 m^2

   E_s = elastic modulus in steel = 2×10^{11} N/m^2

   A_c = cross-sectional area of concrete reinforcement = 0.2 m^2

   E_c = elastic modulus in concrete = 1.4×10^{10} N/m^2

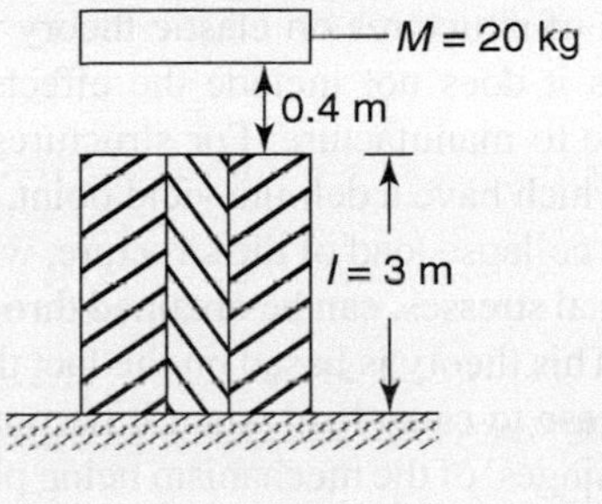

Figure 11.49 Reinforced concrete pillar

$\left[\sigma_{steel} = -20.89\ \text{MN/m}^2,\ \sigma_{concrete} = -1.46\ \text{MN/m}^2\right]$

3. An initially horizontal beam, of length 2 m, is supported at its ends by two wires, one made from aluminium alloy and the other from steel, as shown in Figure 11.50. If a mass of 203.87 kg is dropped a distance of 0.1 m above the mid-span of the beam, determine the maximum stresses in the aluminium and steel wires and the deflection of the beam under this load. The following may be assumed to apply:

   For steel: $E_s = 2\times10^{11}$ N/m^2, $l_s = 2$ m, $A_s = 2\times10^{-4}$ m^2

   For aluminium alloy: $E_{Al} = 7\times10^{10}$ N/m^2, $l_{Al} = 1$ m, $A_{Al} = 4\times10^{-4}$ m^2

   EI (for beam) = 2×10^6 N/m^2

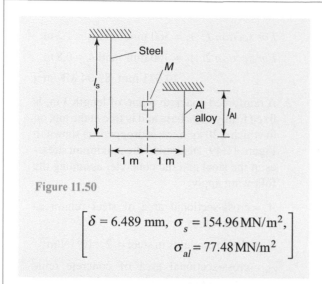

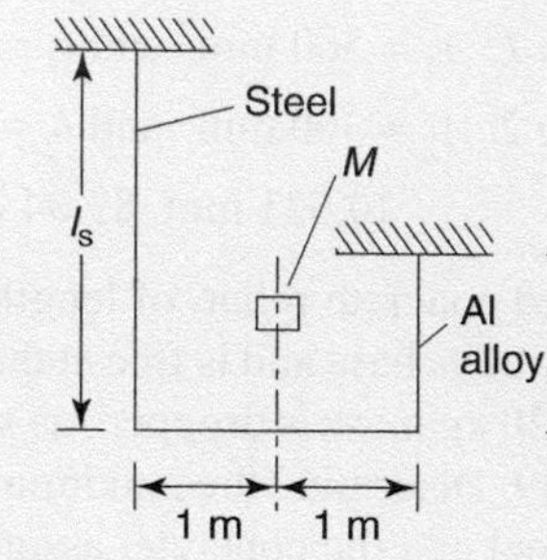

Figure 11.50

$$\left[\begin{array}{l}\delta = 6.489 \text{ mm}, \ \sigma_s = 154.96 \text{ MN/m}^2, \\ \qquad \sigma_{al} = 77.48 \text{ MN/m}^2\end{array}\right]$$

11.14 Plastic collapse of beams

The design of structures on elastic theory is somewhat illogical, as it does not include the effects of residual stresses due to manufacture. For structures made from materials which have a definite yield point, a better estimate of the collapse load of the structure, whether or not it has residual stresses, can be obtained through the plastic theory. This theory is based on the fact that the structure will cease to carry load when it becomes a mechanism, the 'hinges' of the mechanism being plastic hinges. In the plastic hinge theory, the material is assumed to be ideally elastic-plastic, as shown in Figure 11.51.

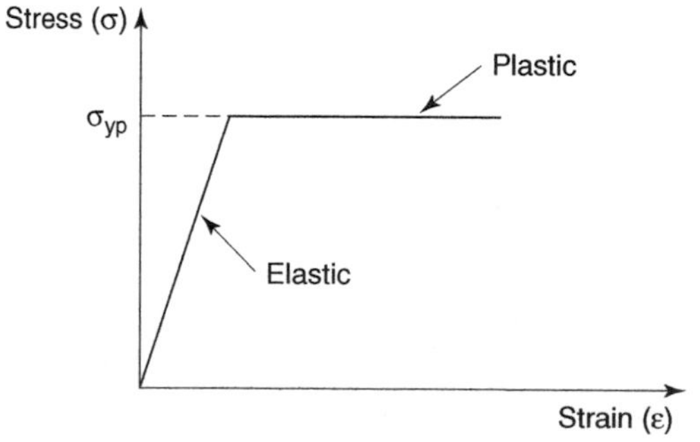

Figure 11.51 Ideally elastic–plastic material

To demonstrate the plastic theory, consider the encastré beam of Figure 11.52(a). Within the elastic limit, the bending moment diagram will be as in Figure 11.52(b), and the stress diagram across a cross-section will be as in Figure 11.53(b). If the load W is increased, the beam will yield at B, assuming that $a > b$; the stress diagram will now take the form of Figure 11.53(c), i.e., although yield has taken place, the beam will not fail, because the cross-section at B still has the ability to resist deformation. Further increase of the load W may cause other sections to become elastic-plastic, especially at A and C, and the plastic penetration at B will become even deeper.

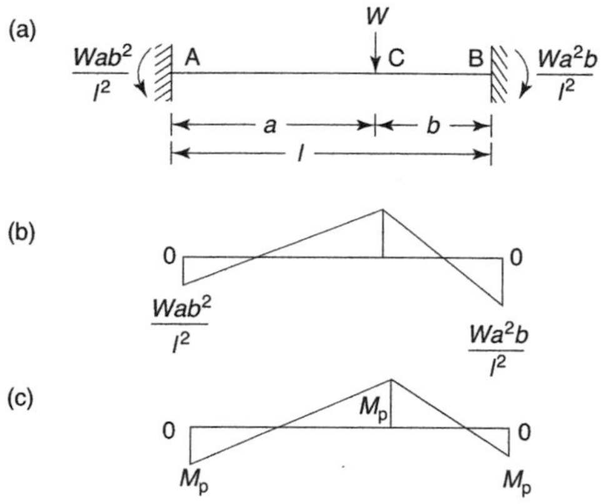

Figure 11.52 Encastré beam (a): (b) elastic bending moment diagram; (c) plastic bending moment diagram

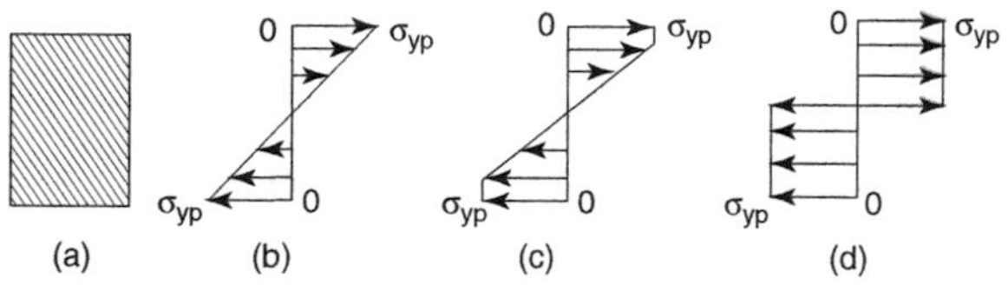

Figure 11.53 Stress diagrams across the cross-section: (a) section; (b) stress diagram up to elastic limit; (c) elastic–plastic stress diagram; (d) fully plastic stress diagram

Eventually at a certain load, the section at B becomes fully plastic, as shown in Figure 11.53(d). When this section at B becomes fully plastic, this part of the beam cannot withstand further load, as a plastic hinge has formed here; the beam simply rotates at B and transforms the bending resistance of the beam, due to an increasing load, to sections A and C. Further increase of load will eventually cause fully plastic hinges to occur at A and C. The plastic bending moment diagram is shown in Figure 11.52(c). When all three plastic hinges have occurred at A, B and C, the beam cannot resist further load, and the slightest increase in load will cause catastrophic deformation, the beam turning into a mechanism, the hinges of the mechanism being plastic hinges, as shown in Figure 11.54. To understand plastic hinge collapses, it can be imagined that **plastic hinges** are equivalent to 'rusty' hinges.

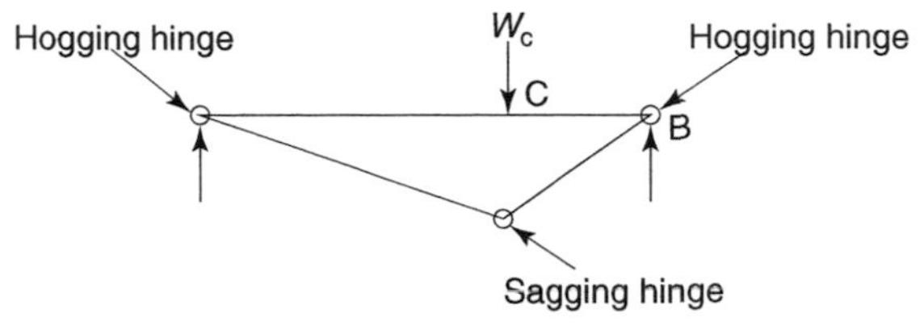

Figure 11.54 Plastic hinges in the beam

Plastic neutral axis

The plastic neutral axis lies at the centre of area of the section or the central axis, (see Section 5.2, page 112), because the stress distribution is constant across the beam's section.

From Section 11.3, page 244, it can be seen that to determine the horizontal central axis of a beam, the area below the central axis must be equal to the area above the central axis.

Load factor (λ)

The safety factor for the plastic hinge theory is called the load factor, where

$$\text{load factor, } \lambda = \frac{\text{plastic collapse load}}{\text{working load}}$$

Shape factor (S)

The shape factor is a measure of the geometry of the beam's section, where

$$\text{shape factor, } S = \frac{M_p}{M_{yp}}$$

where M_p = plastic moment of resistance of the section

and M_{yp} = elastic moment of resistance of the section up to the yield point

Shape factor for a rectangular section

Consider the rectangular section of Figure 11.55(a).

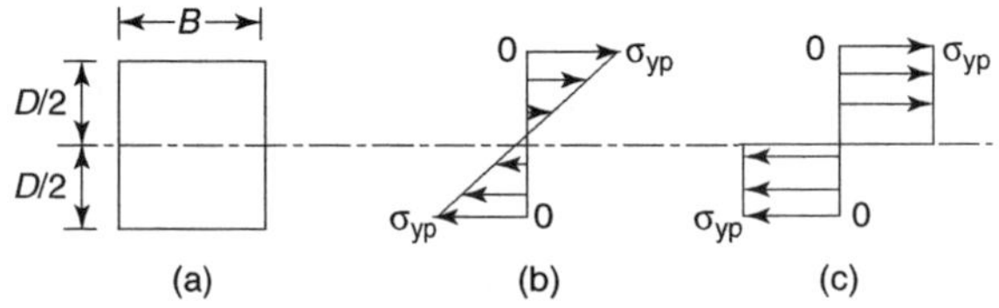

Figure 11.55 Stress diagrams: (a) cross-section; (b) stress diagram up to yield point; (c) fully plastic stress diagram

To determine M_{yp}, we can use the elastic theory of equation (6.1), i.e.

$$\frac{\sigma}{y} = \frac{M}{I}$$

i.e. $$\frac{\sigma_{yp}}{D/2} = \frac{M_{yp}}{\frac{BD^3}{12}}$$

from which $$M_{yp} = \frac{\sigma_{yp}}{D/2} \times \frac{BD^3}{12} = \sigma_{yp}\left(\frac{BD^2}{6}\right) \quad (11.76)$$

which is the *elastic moment of resistance of the section* up to the yield point.

From the stress diagram of Figure 11.55(c)

$$M_p = \text{plastic moment of resistance}$$

$$= \sigma_{yp} \times B \times \frac{D}{2} \times \frac{D}{4} \times 2 \quad (11.77)$$

i.e. $$M_p = \sigma_{yp}\left(\frac{BD^2}{4}\right)$$

Thus the shape factor is:

$$S = \frac{M_p}{M_{yp}} = \frac{\sigma_{yp}\left(\frac{BD^2}{4}\right)}{\sigma_{yp}\left(\frac{BD^2}{6}\right)} = \frac{6}{4} = 1.5$$

For a rolled steel joist, $S = 1.14$ and for a circular cross-section, $S = 1.7$

Problem 24. Using the plastic hinge theory, design a suitable sectional modulus for the encastré beam of Figure 11.56, given that: $\lambda = 3$, $S = 1.14$ and $\sigma_{yp} = 300$ MPa

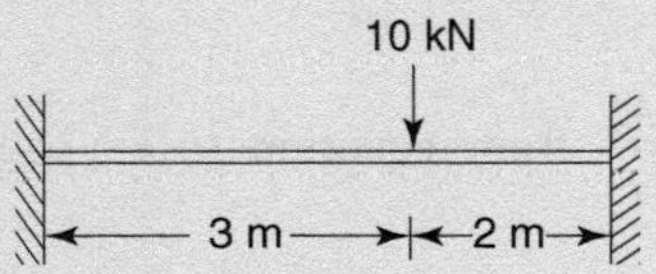

Figure 11.56 Encastré beam

The beam will collapse plastically as shown in Figure 11.57.

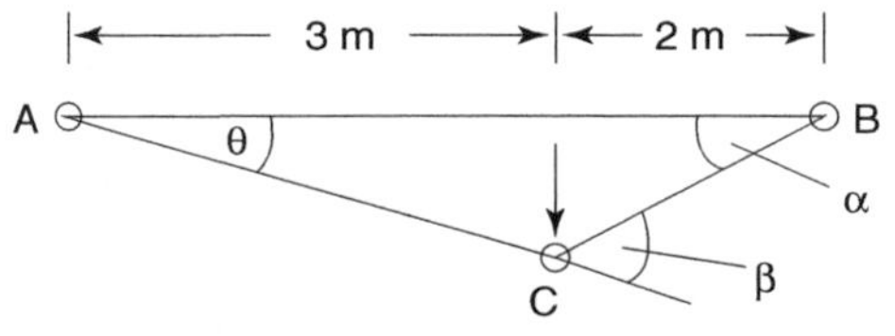

Figure 11.57 Beam mechanism

When the beam collapses, it can be assumed that the beam element remains straight, as shown in Figure 11.57. The reason for the assumption is that, although the beam will be curved, as the bending moments at A, B and C cannot change; the curvature of the beam will remain essentially unchanged during failure.

From small-deflection theory, and Figure 11.57,

$$3\theta = 2\alpha$$

from which $\alpha = \dfrac{3\theta}{2}$ (11.78)

Also $\beta = \alpha + \theta = \dfrac{3\theta}{2} + \theta$ from Figure 11.57

i.e. $\beta = \dfrac{5\theta}{2}$ (11.79)

From the principle of virtual work,

virtual work done by the load = virtual work done by the hinges during rotation

i.e. $10 \times 3\theta = M_p\theta + M_p\beta + M_p\alpha$

i.e. $30\theta = M_p\theta + M_p\left(\dfrac{5\theta}{2}\right) + M_p\left(\dfrac{3\theta}{2}\right)$

i.e. $30\theta = 5M_p\theta$

from which $M_p = \dfrac{30}{5}$

i.e. $M_p = 6$ kN m

Design $M_p = M_p \times \lambda = 6 \times 3 = 18$ kN m

Shape factor, $S = \dfrac{M_p}{M_{yp}}$ therefore, $M_{yp} = \dfrac{M_p}{S}$

i.e. $M_{yp} = \dfrac{18}{1.14} = 15.79$ kN m

Since $\dfrac{\sigma}{y} = \dfrac{M}{I}$ from equation (6.3), page 136, then

$\dfrac{I}{y} = \dfrac{M}{\sigma} = Z$, the sectional modulus

Hence, sectional modulus, $Z = \dfrac{M_{yp}}{\sigma_{yp}} = \dfrac{15.79 \times 10^3}{300 \times 10^6}$

i.e. $\mathbf{Z = 5.263 \times 10^{-5}\,m^2}$

Problem 25. Using the plastic hinge theory, determine a suitable sectional modulus for the propped cantilever of Figure 11.58, given that: $\lambda = 3$, $S = 1.14$ and $\sigma_{yp} = 300$ MPa

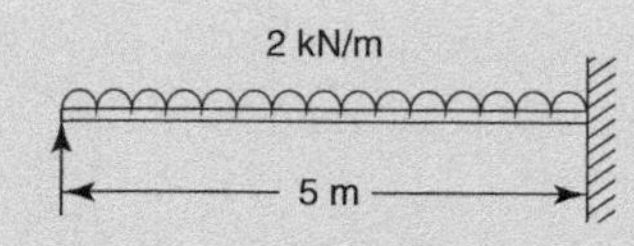

Figure 11.58 Propped cantilever

In this case it will be necessary for two plastic hinges to occur, as the beam is statically indeterminate to the first degree.

The collapse mechanism is shown in Figure 11.59 where it can be seen that the sagging hinge is assumed to occur at a distance X to the right of mid-span.

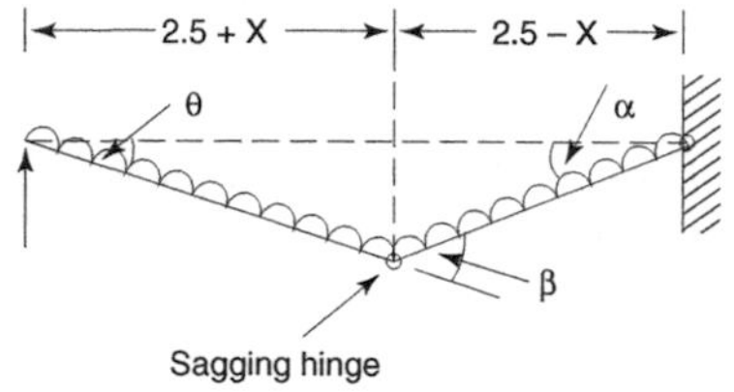

Figure 11.59 Collapse mechanism

Assuming θ is small, then from Figure 11.59,

$$(2.5 + X)\theta = (2.5 - X)\alpha$$

or $\alpha = \dfrac{(2.5 + X)}{(2.5 - X)}\theta$ (11.80)

Now $\beta = \alpha + \theta = \dfrac{(2.5 + X)}{(2.5 - X)}\theta + \theta$ from Figure 11.59

i.e. $= \dfrac{(2.5 + X)\theta + (2.5 - X)\theta}{(2.5 - X)}$

i.e. $\beta = \dfrac{5}{(2.5 - X)}\theta$ (11.81)

Virtual work done by falling load = virtual work done by plastic hinges resisting rotation

$$\frac{2 \times (2.5 + X)}{2} \times (2.5 + X)\theta + \frac{2 \times (2.5 - X)}{2} \times (2.5 - X)\,\alpha = M_p\beta + M_p\alpha$$

Therefore $(2.5 + X)^2\theta + (2.5 - X)^2\dfrac{(2.5 + X)}{(2.5 - X)}\theta$

$= M_p\dfrac{5}{(2.5 - X)}\theta + M_p\dfrac{(2.5 + X)}{(2.5 - X)}\theta$

from equations (11.80) and (11.81),

from which

$$\left(6.25+5X+X^2\right)\theta+\left(6.25-X^2\right)\theta=\left(\frac{7.5+X}{2.5-X}\right)\theta M_p$$

i.e. $$\left(12.5+5X\right)=\left(\frac{7.5+X}{2.5-X}\right)M_p$$

Hence $$M_p=\frac{\left(12.5+5X\right)\left(2.5-X\right)}{\left(7.5+X\right)}$$

$$=\frac{31.25-12.5X+12.5X-5X^2}{\left(7.5+X\right)}$$

i.e. $$M_p=\frac{31.25-5X^2}{\left(7.5+X\right)} \qquad (11.82)$$

For maximum M_p, $$\frac{dM_p}{dX}=0$$

Equation (11.82) is a quotient, hence the quotient rule of differentiation is used, i.e.

if $$M_p=\frac{u}{v} \text{ then } \frac{dM_p}{dX}=\frac{v\frac{du}{dX}-u\frac{dv}{dX}}{v^2}$$

Hence, from equation (11.82)

$$\frac{dM_p}{dX}=\frac{(7.5+X)(-10X)-\left(31.25-5X^2\right)(1)}{\left(7.5+X\right)^2}$$

$$=0 \text{ for maximum } M_p$$

i.e. $$(7.5+X)(-10X)-\left(31.25-5X^2\right)(1)=0$$

i.e. $$-75X-10X^2-31.25+5X^2=0$$

i.e. $$-5X^2-75X-31.25=0$$

or $$5X^2+75X+31.25=0$$

which is a quadratic equation

Solving, either by using a calculator or using the quadratic formula, gives:

$X=-0.429$ or -14.571 (which is impossible)

Hence $$\mathbf{X=-0.429\ m} \qquad (11.83)$$

Substituting equation (11.83) into equation (11.82) gives:

$$M_p=\frac{31.25-5X^2}{\left(7.5+X\right)}=\frac{31.25-5\left(-0.429\right)^2}{7.5+-0.429}$$

$$=\frac{30.330}{7.071}=4.289 \text{ kN m}$$

Design $M_p=\lambda\times M_p=3\times 4.289=12.867$ kN m

Shape factor, $S=\dfrac{M_p}{M_{yp}}$ therefore, $M_{yp}=\dfrac{M_p}{S}$

i.e. $$M_{yp}=\frac{12.867}{1.14}=11.287 \text{ kN m}$$

Sectional modulus, $$Z=\frac{M_{yp}}{\sigma_{yp}}=\frac{11.287\times10^3}{300\times10^6}$$

i.e. $$\mathbf{Z=3.762\times10^{-5}\,m^2}$$

Now try the following Practice Exercise

Practice Exercise 51. Plastic collapse of beams

1. Determine the shape factors for the cross-sections shown in Figures 11.60(a) to (c).

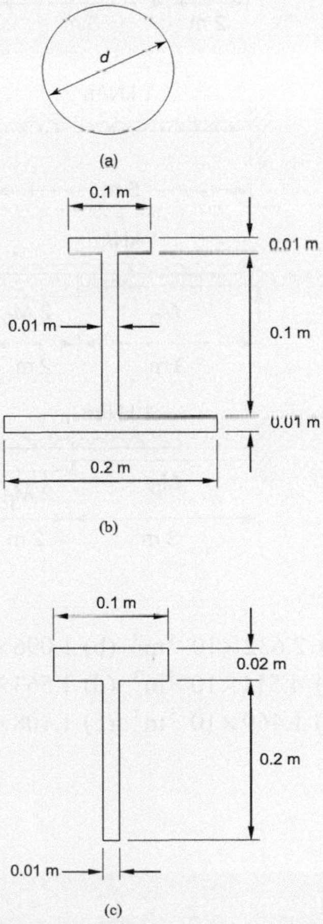

Figure 11.60 Cross sections: (a) solid circular section; (b) RSJ; (c) tee beam

[(a) 1.70 (b) 1.327 (c) 1.811]

2. Using the plastic hinge theory, determine suitable sectional moduli for the beams of Figure 11.61, given that: $\lambda = 3$, $S = 1.14$ and $\sigma_{yp} = 300\,\text{MPa}$

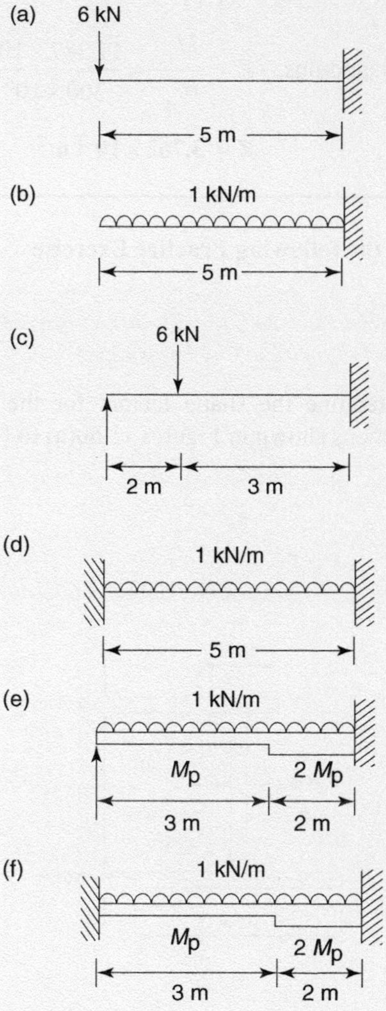

Figure 11.61

$\big[$(a) $2.632 \times 10^{-4}\,\text{m}^3$ (b) $1.096 \times 10^{-4}\,\text{m}^3$
(c) $4.511 \times 10^{-5}\,\text{m}^3$ (d) $1.563 \times 10^{-5}\,\text{m}^3$
(e) $1.469 \times 10^{-5}\,\text{m}^3$ (f) $1.108 \times 10^{-5}\,\text{m}^3\big]$

11.15 Residual stresses in beams

In this section the theory will be based on the material behaving ideally elastic-plastic, as shown in Figure 11.51 on page 274.

The following example will demonstrate how to calculate the residual stresses that remain when a load is removed, after this load has caused the beam to become partially plastic.

Problem 26. A steel beam of constant rectangular section, of depth 0.2 m and width 0.1 m, is subjected to four-point loading, as shown in Figure 11.62. Determine:

(a) the depth of plastic penetration at mid-span
(b) the central deflection
(c) the length of the beam over which yield takes place.

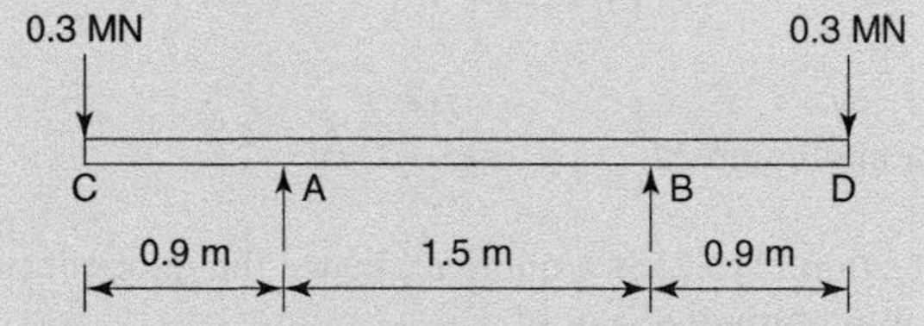

Figure 11.62 Beam under four-point loading

If the above load is removed:

(d) determine the residual central deflection
(e) plot the residual stress distribution at mid-span.

The following may be assumed to apply:

$\sigma_{yp} = 350\,\text{MN/m}^2 \qquad E = 2 \times 10^{11}\,\text{N/m}^2$

(a) To determine the depth of plastic penetration

At mid-span and throughout the length AB, the bending moment is:

$$M = 0.3\,\text{MN} \times 0.9\,\text{m} = \mathbf{0.27\,MN\,m}$$

Owing to this bending moment, the stress distribution will be as in Figure 11.63.

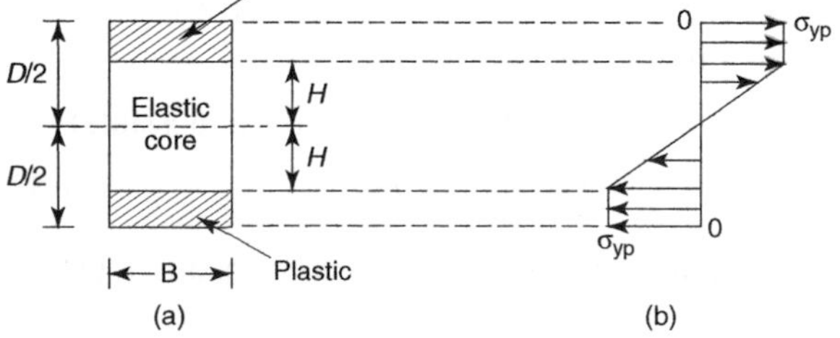

Figure 11.63 Elastic–plastic deformation of beam section: (a) section; (b) stress distribution

Let M_e = moment of resistance of the elastic portion of the beam's section

M'_p = moment of resistance of the plastic portion of the beam's section

I_e = second moment of area of the elastic portion of the beam's section

$$= \frac{B\times(2H)^3}{12} = 0.6667BH^3$$

From elementary elastic theory, equation (6.1), page 134

$$\frac{\sigma}{y} = \frac{M}{I} = \frac{E}{R} \qquad (11.84)$$

where σ = stress at any distance y from the neutral axis (NA)
M = bending moment
I = second moment of area about NA
R = radius of curvature of NA

Applying equation (11.84) to the elastic portion of the beam's section

$$M_e = \frac{\sigma_{yp}\times I_e}{H}$$

$$= \frac{350\times10^6\times0.6667\times0.1\times H^3}{H}$$

i.e. $\mathbf{M_e = 23.33\times10^6 H^2}$ (11.85)

From Figure 11.59 (on page 276)

$$M'_p = 2\times\sigma_{yp}\times B\times\left(\frac{D}{2}-H\right)\times\frac{\left(\frac{D}{2}+H\right)}{2}$$

$$= 2\times350\times10^6\times0.1\times\left(\frac{0.2}{2}-H\right)\times\frac{\left(\frac{0.2}{2}+H\right)}{2}$$

$$= 35\times10^6\times(0.1-H)\times(0.1+H)$$

$$= 35\times10^6\times(0.01-H^2)$$

i.e. $M'_p = (0.35 - 35H^2)\times10^6$ (11.86)

Now $M = M_e + M'_p$

i.e. $0.27\times10^6 = 23.33\times10^6 H^2 + (0.35-35H^2)\times10^6$

or $0.27 = 23.33H^2 + 0.35 - 35H^2$

i.e. $35H^2 - 23.33H^2 = 0.35 - 0.27$

i.e. $11.67H^2 = 0.08$

from which $H = \sqrt{\frac{0.08}{11.67}}$

i.e. $\mathbf{H = 0.0828}$ **m** (11.87)

Therefore, the *depth of plastic penetration* at the top and the bottom of the beam is:

$(0.1 - 0.0828) = 0.0172$ m = **1.72 cm**

(b) To determine the central deflection (δ)

This can be calculated by substituting the value of M_e into equation (11.84), i.e.

$$\frac{M_e}{I_e} = \frac{E}{R}$$

where $I_e = 0.6667BH^3 = 0.6667\times0.1\times0.0828^3$

$= \mathbf{3.785\times10^{-5}\,m^4}$

and $M_e = 23.33\times10^6 H^2 = 23.33\times10^6\times0.0828^2$

$=$ **0.160 MN m**

Since $\frac{M_e}{I_e} = \frac{E}{R}$ then $R = \frac{EI_e}{M_e} = \frac{2\times10^{11}\times3.785\times10^{-5}}{0.16\times10^6}$

i.e. $\mathbf{R = 47.31}$ **m**

The central deflection (δ) can be calculated from the properties of a circle, as shown in Figure 11.64.

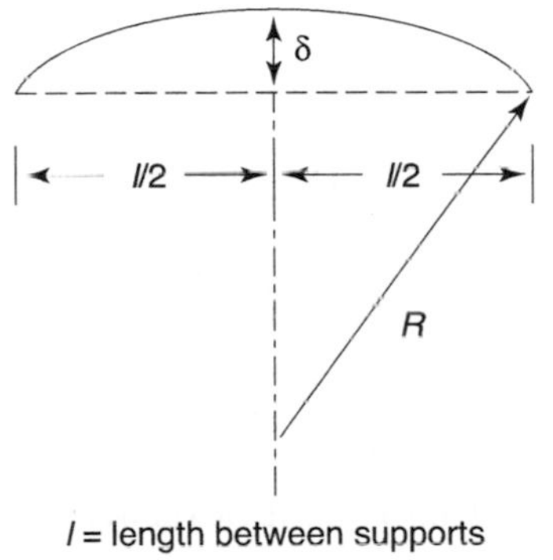

Figure 11.64 Deflected form of beam

Thus, from the properties of a chord of a circle

$$\delta(2R-\delta) = \left(\frac{l}{2}\right)^2 \qquad (11.88)$$

i.e. $\delta(2\times47.31-\delta) = \left(\frac{1.5}{2}\right)^2$

i.e. $94.62\delta - \delta^2 = 0.5625$

and $\delta^2 - 94.62\delta + 0.5625 = 0$ which is a quadratic equation

Using a calculator or the quadratic formula:
$\delta = 5.945\times10^{-3}$ m or 94.61 m (which is ignored)

Hence, **the central deflection, δ = 5.945 mm**

(c) To determine the length of the beam over which yield takes place

Let M_{yp} = moment required just to cause first yield

$$= \frac{\sigma_{yp} \times I}{(D/2)} = \frac{\sigma_{yp} \times \left(\frac{BD^3}{12}\right)}{(D/2)} = \frac{\sigma_{yp} \times BD^2}{6}$$

$$= \frac{350 \times 10^6 \times 0.1 \times 0.2^2}{6}$$

$$= 0.2333 \times 10^6$$

i.e. $\mathbf{M_{yp} = 0.2333\ MN\ m}$

From Figure 11.65

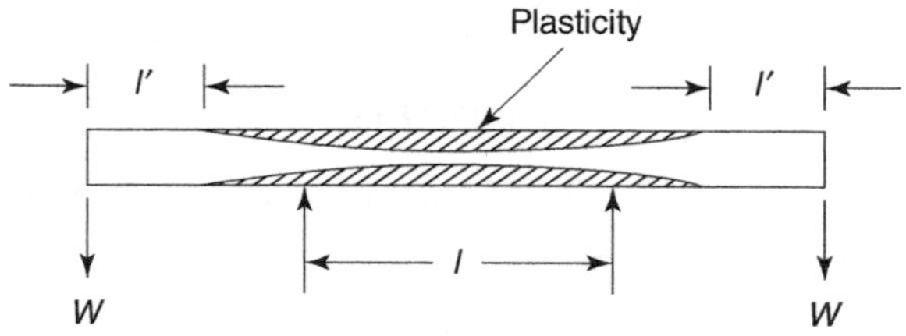

Figure 11.65 Plastic region on beam

$$Wl' = M_{yp}$$

i.e. $0.3 \times 10^6\, l' = 0.2333 \times 10^6$

Therefore $l' = \frac{0.2333}{0.3} = \mathbf{0.778\ m}$

Therefore, **the length of the beam over which plasticity occurs is:**

$$1.8 + 1.5 - 0.778 \times 2 = \mathbf{1.744\ m}$$

(d) To determine the residual central deflection (δ_R)

On unloading the beam, it behaves elastically, as shown by Figure 11.66

i.e. since $\frac{M_e}{I_e} = \frac{E}{R}$ then $R = \frac{EI}{M} = \frac{E\left(\frac{BD^3}{12}\right)}{M}$

$$= \frac{2 \times 10^{11} \times \left(\frac{0.1 \times 0.2^3}{12}\right)}{0.27 \times 10^6}$$

from which $\mathbf{R = 49.383\ m}$

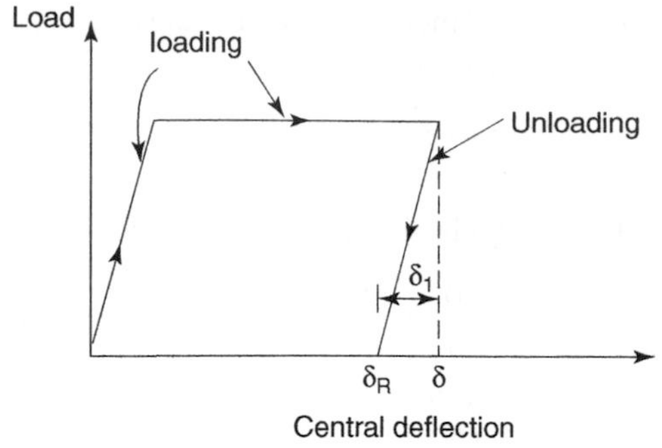

Figure 11.66 Load–deflection relationship

From equation (11.88)

$$\delta(2R - \delta) = \left(\frac{l}{2}\right)^2$$

i.e. $\delta_1(2 \times 49.383 - \delta_1) = \left(\frac{1.5}{2}\right)^2$

i.e. $98.765\delta_1 - \delta_1^2 = 0.5625$

or $\delta_1^2 - 98.765\delta_1 + 0.5625 = 0$ which is a quadratic equation

Using a calculator or the quadratic formula, $\delta_1 = 5.696 \times 10^{-3}$ m or 98.759 m (which is ignored)

Therefore, the residual deflection is:

$$\delta_R = 5.945 - 5.696 = \mathbf{0.249\ mm}$$

(e) To determine the residual stress distribution

This can be obtained by superimposing the elastic-plastic stress distribution, on loading, with the elastic stress distribution, on unloading, as shown in Figures 11.67(a) to (c).

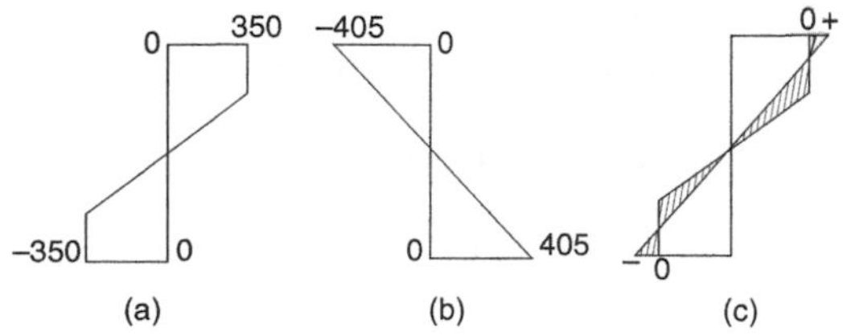

Figure 11.67 Stress distribution across section: (a) elastic–plastic stress distribution, on loading; (b) elastic stress distribution, on unloading; (c) residual stress distribution (shown shaded)

Now try the following Practice Exercise

Practice Exercise 52. Residual stresses in beams

1. A uniform-section tee beam, with the sectional properties shown in Figure 11.68(a), is subjected to four-point bending, as shown in Figure 11.68(b). Determine the central deflection on application of the load, and the residual central deflection on its removal. Assume that: $\sigma_{yp} = 300\,\text{MN/m}^2$ and $E = 2 \times 10^{11}\,\text{N/m}^2$

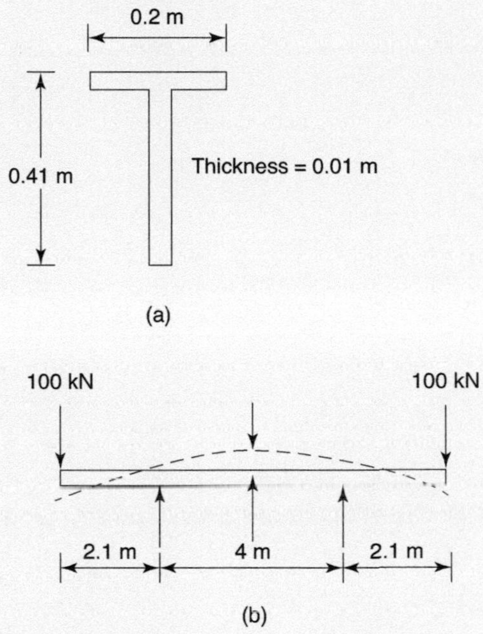

Figure 11.68 Beam under four-point loading

[0.0247 m, 0.0055 m]

For fully worked solutions to each of the problems in Exercises 47 to 52 in this chapter, go to the website:
www.routledge.com/cw/bird

Chapter 12

Theories of elastic failure

Why it is important to understand: **Theories of elastic failure**

For a structural material which suffers elastic failure, there are many different theories of elastic failure. Unfortunately, not all these theories make the same predictions. Moreover, some of these theories are good for materials such as mild steel, which has a definite yield point, while others are more satisfactory for brittle materials, such as cast iron. Also, for materials such as composites, the theories of elastic failure are very much more complex. The theories can also be affected by the complex combination of stresses that occur on the structure. For these reasons, a study of the theories of elastic failure is important, and many of them are presented in this chapter. The theories of elastic failure for composites are given in Chapter 19.

At the end of this chapter you should be able to:

- appreciate the many theories as to which stress or strain causes the onset of yield
- state and use the maximum principal stress theory – Rankine
- state and use the maximum principal strain theory – St Venant
- state and use the total strain energy theory – Beltrami and Haigh
- state and use the maximum shear stress theory – Tresca
- state and use the maximum shear strain energy theory – Hencky and von Mises
- plot and interpret yield loci
- calculate pressures and torques based on the five major theories of elastic failure

Mathematical references

In order to understand the theory involved in this chapter on **theories of elastic failure**, knowledge of the following mathematical topics is required: ***algebraic manipulation and solving equations.*** If help/revision is needed in these areas, see page 50 for some textbook references.

12.1 Introduction

There are many theories as to which stress or strain, or any combination of these, causes the onset of yield, but in this chapter we will restrict ourselves to a consideration of the five major theories, namely:

(a) Maximum principal stress theory
(b) Maximum principal strain theory
(c) Total strain energy theory
(d) Maximum shear stress theory
(e) Shear strain energy theory.

The combination of stresses and strains that causes the onset of yield is of much importance in stress analysis, as the onset of yield is very often related to the ultimate failure of the structure. For convenience, the five major theories are related to a triaxial principal stress system, where

$$\sigma_1 > \sigma_2 > \sigma_3$$

σ_1 = maximum principal stress
σ_2 = minimax principal stress*
σ_3 = minimum principal stress

*where minimax is between minimum and maximum.

The reason for choosing the triaxial principal stress system to investigate yield criteria is that such a system describes the complete stress situation at a point, without involving the complexities caused by the shear stresses on six planes, which would have resulted if a different three-dimensional co-ordinate system were adopted. The five major theories of elastic failure, together with some of the reasons why they have gained popularity, are described in the following sections.

*William John Macquorn Rankine, (5 July 1820 – 24 December 1872) was a Scottish civil engineer, physicist and mathematician. He was a founding contributor, with Rudolf Clausius and William Thomson (Lord Kelvin), to the science of thermodynamics, particularly focusing on the first of the three thermodynamic laws. To find out more about Rankine go to www.routledge.com/cw/bird

12.2 Maximum principal stress theory (Rankine)*

This theory states that yield will occur when

$$\sigma_1 = \sigma_{yp} \tag{12.1}$$

or, if the material is in compression, when

$$\sigma_3 = \sigma_{yp} \tag{12.2}$$

where σ_{yp} = yield stress in tension, obtained in the simple uniaxial tensile test

and σ_{ypc} = yield stress in compression, obtained in the simple uniaxial compression test

12.3 Maximum principal strain theory (St Venant)

Experimental tests have revealed that comparison between the maximum principal stress theory and experiment was very often found to be poor, and St Venant suggested that perhaps yield occurred owing to the maximum principal strain, rather than the maximum principal stress, as the former involved all three principal stresses and Poisson's ratio, whilst the latter did not.

The maximum principal strain theory states that yield will occur when

$$\varepsilon_1 = \frac{1}{E}\left[\sigma_1 - \nu\left(\sigma_2 + \sigma_3\right)\right] = \frac{\sigma_{yp}}{E}$$

or $$\sigma_1 - \nu\left(\sigma_2 + \sigma_3\right) = \sigma_{yp} \quad (12.3)$$

or, if the material is in compression,

$$\sigma_3 - \nu\left(\sigma_1 + \sigma_2\right) = \sigma_{ypc} \quad (12.4)$$

where ε_1 is the maximum principal strain.
In *two dimensions,* $\sigma_3 = 0$; therefore equations (12.3) and (12.4) become of the simpler forms:

$$\sigma_1 - \nu\sigma_2 = \sigma_{yp} \quad (12.5)$$

and $$-\nu\left(\sigma_1 + \sigma_2\right) = \sigma_{ypc} \quad (12.6)$$

For convenience, for the remaining three theories, the assumption will be made that σ_{yp} is of the same magnitude as σ_{ypc}. If the magnitude of σ_{ypc} is less than that of σ_{yp}, the theories can be easily modified to take this into account.

12.4 Total strain energy theory (Beltrami* and Haigh)

This theory states that elastic failure will occur when the total strain energy per unit volume, at a point, reaches the total strain energy per unit volume of a specimen made from the same material, when it is subjected to a simple uniaxial test.
Now for a three-dimensional stress system, the total strain energy per unit volume, in terms of the three principal stresses, is

$$\frac{U_T}{vol} = \frac{1}{2}\sigma_1\varepsilon_1 + \frac{1}{2}\sigma_2\varepsilon_2 + \frac{1}{2}\sigma_3\varepsilon_3$$

but from section 9.10, page 216

$$\varepsilon_1 = \frac{1}{E}\left[\sigma_1 - \nu\left(\sigma_2 + \sigma_3\right)\right] = \text{maximum principal strain}$$

$$\varepsilon_2 = \frac{1}{E}\left[\sigma_2 - \nu\left(\sigma_1 + \sigma_3\right)\right] = \text{minimax principal strain}$$

* Eugenio Beltrami (16 November 1835 – 4 June 1899) was an Italian mathematician notable for his work concerning differential geometry and mathematical physics. To find out more about Beltrami go to www.routledge.com/cw/bird

$$\varepsilon_3 = \frac{1}{E}\left[\sigma_3 - \nu\left(\sigma_1 + \sigma_2\right)\right] = \text{minimum principal strain}$$

i.e. $$\frac{U_T}{vol} = \frac{1}{2}\sigma_1\left\{\frac{1}{E}\left[\sigma_1 - \nu\left(\sigma_2 + \sigma_3\right)\right]\right\} + \frac{1}{2}\sigma_2\left\{\frac{1}{E}\left[\sigma_2 - \nu\left(\sigma_1 + \sigma_3\right)\right]\right\} + \frac{1}{2}\sigma_3\left\{\frac{1}{E}\left[\sigma_3 - \nu\left(\sigma_1 + \sigma_2\right)\right]\right\}$$

i.e. $$\frac{U_T}{vol} = \frac{1}{2E}\{\sigma_1^2 - \nu\sigma_1\sigma_2 - \nu\sigma_1\sigma_3 + \sigma_2^2 - \nu\sigma_1\sigma_2 - \nu\sigma_2\sigma_3 + \sigma_3^2 - \nu\sigma_1\sigma_3 - \nu\sigma_2\sigma_3\}$$

$$= \frac{1}{2E}\{\sigma_1^2 + \sigma_2^2 + \sigma_3^2 - 2\nu\sigma_1\sigma_2 - 2\nu\sigma_1\sigma_3 - 2\nu\sigma_2\sigma_3\}$$

i.e. $$\frac{U_T}{vol} = \frac{1}{2E}\left(\sigma_1^2 + \sigma_2^2 + \sigma_3^2\right) - \frac{\nu}{E}\left(\sigma_1\sigma_2 + \sigma_1\sigma_3 + \sigma_2\sigma_3\right) \quad (12.7)$$

In the simple uniaxial test,

$$\sigma_1 = \sigma_{yp} \text{ and } \sigma_2 = \sigma_3 = 0$$

Therefore $$\frac{U_T}{vol} = \frac{\sigma_{yp}^2}{2E} \quad (12.8)$$

Equating equations (12.7) and (12.8) gives:

$$\frac{1}{2E}\left(\sigma_1^2 + \sigma_2^2 + \sigma_3^2\right) - \frac{\nu}{E}\left(\sigma_1\sigma_2 + \sigma_1\sigma_3 + \sigma_2\sigma_3\right) = \frac{\sigma_{yp}^2}{2E}$$

i.e. $$\left(\sigma_1^2 + \sigma_2^2 + \sigma_3^2\right) - 2\nu\left(\sigma_1\sigma_2 + \sigma_1\sigma_3 + \sigma_2\sigma_3\right) = \sigma_{yp}^2 \quad (12.9)$$

This theory, like the principal strain theory, also involves all three principal stresses and Poisson's ratio. In *two dimensions,* $\sigma_3 = 0$; therefore equation (12.9) becomes

$$\sigma_1^2 + \sigma_2^2 - 2\nu\sigma_1\sigma_2 = \sigma_{yp}^2 \quad (12.10)$$

12.5 Maximum shear stress theory (Tresca)*

The problem with the three previous theories is that they all fail in the case of hydrostatic stress. Experiments have shown that whether or not a solid piece of material is soft and ductile or hard and brittle when it is subjected to a large uniform external hydrostatic pressure, then despite the fact that the yield stress is grossly exceeded, the material does not suffer elastic breakdown in this condition. For example, lumps of chalk or similar low-strength substances can survive intact at great depths in the oceans.

In such cases, all the principal stresses are equal to the water pressure P, so that

*Henri Édouard Tresca (12 October 1814 – 21 June 1885) was a French mechanical engineer, and a professor at the Conservatoire National des Arts et Métiers in Paris. Tresca was also among the designers of the prototype metre bar that served as the first standard of length for the metric system. To find out more about Tresca go to www.routledge.com/cw/bird

$$\sigma_1 = \sigma_2 = \sigma_3 = -P \quad (12.11)$$

From equation (12.11), it can be seen that there are no shear stresses in a hydrostatic stress condition, and this is why low-strength materials survive intact under large values of water pressure.

An argument held against this hypothesis is that in a normal hydrostatic stress condition, the stresses are all compressive, and this is the reason why failure does not take place. However, a Russian scientist carried out tests on a piece of glass, which by a process of heating and cooling was believed to be subjected to a hydrostatic tensile stress of about 6.895×10^9 Pa (1×10^6 lbf/in^2) at a certain point in the material, but inspection revealed that there were no signs of cracking at this point.

Thus, it can be concluded that for elastic failure to take place, the material must distort (change shape), and for this to occur, it is necessary for shear stress to exist.

The maximum shear stress theory states that elastic failure will take place when the maximum shear stress at a point equals the maximum shear stress obtained in a specimen, made from the same material, in the simple uniaxial test,

i.e. $$\sigma_1 - \sigma_3 = \sigma_{yp} \quad (12.12)$$

In two dimensions, $\sigma_3 = 0$, and equation (12.12) becomes:

$$\sigma_1 - \sigma_2 = \sigma_{yp} \quad (12.13)$$

12.6 Maximum shear strain energy theory (Hencky and von Mises)*

The problem with the maximum shear stress theory is that it states that:

$$\tau_{yp} = 0.5\sigma_{yp}$$

where τ_{yp} = shear stress at yield

*Richard Edler von Mises (19 April 1883 – 14 July 1953) was a German scientist and mathematician who worked on solid mechanics, fluid mechanics, aerodynamics, aeronautics, statistics and probability theory. To find out more about von Mises go to www.routledge.com/cw/bird

However, torsional tests on mild steel specimens have found that:

$$\tau_{yp} = \pm \mathbf{0.577}\sigma_{yp}$$

and this implies that the maximum shear stress theory is not always suitable. The reason for this may be due to the fact that it ignores the effects of σ_2 and v.

A theory, therefore, that takes into consideration all these factors, and does not fail under the hydrostatic stress condition, is the shear strain energy theory, which states that elastic failure takes place when the shear strain energy per unit volume, at a point, equals the shear strain energy per unit volume in a specimen of the same material, in the simple uniaxial test.

Now, the shear strain energy(SSE)/vol is:

total strain energy/vol – hydrostatic strain energy/vol

i.e. $\text{SSE} = U_T - U_H$

where, from equation (12.7),

$$U_T = \text{total strain energy} = \frac{1}{2E}\left[\left(\sigma_1^2+\sigma_2^2+\sigma_3^2\right) - \frac{v}{E}\left(\sigma_1\sigma_2+\sigma_1\sigma_3+\sigma_2\sigma_3\right)\right] \times \text{volume}$$

$$= \frac{1}{2E}\left[\left(\sigma_1^2+\sigma_2^2+\sigma_3^2\right) - 2v\left(\sigma_1\sigma_2+\sigma_1\sigma_3+\sigma_2\sigma_3\right)\right] \times \text{volume}$$

$$P = \text{hydrostatic stress} = \frac{\left(\sigma_1+\sigma_2+\sigma_3\right)}{3}$$

$$U_H = \text{hydrostatic strain energy}$$

$$= \frac{1}{2E}\left(3\times P^2 - 2v\times 3P^2\right) \times \text{volume}$$

$$= \frac{1}{2E}\left(3P^2(1-2v)\right) \times \text{volume}$$

and U_H is obtained by substituting P for σ_1, σ_2 and σ_3 into U_T.

Therefore, $\text{SSE/vol} = U_T - U_H$

i.e.

$$\text{SSE/vol} = \frac{1}{2E}\left[\left(\sigma_1^2+\sigma_2^2+\sigma_3^2\right) - 2v\left(\sigma_1\sigma_2+\sigma_1\sigma_3+\sigma_2\sigma_3\right) - \left(3\left(\frac{\left(\sigma_1+\sigma_2+\sigma_3\right)^2}{9}\right)(1-2v)\right)\right]$$

$$= \frac{1}{2E}\left[\left(\sigma_1^2+\sigma_2^2+\sigma_3^2\right) - 2v\left(\sigma_1\sigma_2+\sigma_1\sigma_3+\sigma_2\sigma_3\right) - \left(\left(\frac{\sigma_1^2+\sigma_1\sigma_2+\sigma_1\sigma_3+\sigma_1\sigma_2+\sigma_2^2+\sigma_2\sigma_3+\sigma_1\sigma_3+\sigma_2\sigma_3+\sigma_3^2}{3}\right)(1-2v)\right)\right]$$

$$= \frac{1}{6E}\left[3\left(\sigma_1^2+\sigma_2^2+\sigma_3^2\right) - 6v\left(\sigma_1\sigma_2+\sigma_1\sigma_3+\sigma_2\sigma_3\right) - \left(\sigma_1^2+2\sigma_1\sigma_2+2\sigma_1\sigma_3+\sigma_2^2+2\sigma_2\sigma_3+\sigma_3^2\right)(1-2v)\right]$$

$$= \frac{1}{6E}\left[3\sigma_1^2+3\sigma_2^2+3\sigma_3^2-6v\sigma_1\sigma_2-6v\sigma_1\sigma_3-6v\sigma_2\sigma_3 - \sigma_1^2-2\sigma_1\sigma_2-2\sigma_1\sigma_3-\sigma_2^2-2\sigma_2\sigma_3-\sigma_3^2 + 2v\sigma_1^2+4v\sigma_1\sigma_2+4v\sigma_1\sigma_3+2v\sigma_2^2+4v\sigma_2\sigma_3+2v\sigma_3^2\right]$$

$$= \frac{1}{6E}\left[\left(\sigma_1^2-2\sigma_1\sigma_2+\sigma_2^2\right)+\left(\sigma_1^2-2\sigma_1\sigma_3+\sigma_3^2\right)+\left(\sigma_2^2-2\sigma_2\sigma_3+\sigma_3^2\right) + v\left(-2\sigma_1\sigma_2-2\sigma_1\sigma_3-2\sigma_2\sigma_3+2\sigma_1^2+2\sigma_2^2+2\sigma_3^2\right)\right]$$

$$= \frac{1}{6E}\begin{bmatrix} \left(\sigma_1^2 - 2\sigma_1\sigma_2 + \sigma_2^2\right) + \left(\sigma_1^2 - 2\sigma_1\sigma_3 + \sigma_3^2\right) + \left(\sigma_2^2 - 2\sigma_2\sigma_3 + \sigma_3^2\right) \\ + v\left[\left(\sigma_1^2 - 2\sigma_1\sigma_2 + \sigma_2^2\right) + \left(\sigma_1^2 - 2\sigma_1\sigma_3 + \sigma_3^2\right) + \left(\sigma_2^2 - 2\sigma_2\sigma_3 + \sigma_3^2\right)\right] \end{bmatrix}$$

$$= \frac{(1+v)}{6E}\left[\left(\sigma_1 - \sigma_2\right)^2 + \left(\sigma_1 - \sigma_3\right)^2 + \left(\sigma_2 - \sigma_3\right)^2\right]$$

However, $G = \frac{E}{\left[2(1+v)\right]}$ from Section 4.9, from which $\frac{(1+v)}{E} = \frac{1}{2G}$

Hence

$$\textbf{SSE/vol} = \frac{\mathbf{1}}{\mathbf{12G}}\left[\left(\sigma_1 - \sigma_2\right)^2 + \left(\sigma_1 - \sigma_3\right)^2 + \left(\sigma_2 - \sigma_3\right)^2\right] \tag{12.14}$$

The shear strain energy/vol for a specimen in the uniaxial tensile test can be obtained from equation (12.14) by substituting $\sigma_1 = \sigma_{yp}$ and $\sigma_2 = \sigma_3 = 0$

giving

$$\textbf{SSE/vol} = \frac{1}{12G}\left[\left(\sigma_{yp}\right)^2 + \left(\sigma_{yp}\right)^2\right] = \frac{\boldsymbol{\sigma_{yp}}^{\mathbf{2}}}{\mathbf{6G}} \tag{12.15}$$

Equating equations (12.14) and (12.15), the criterion for yielding, according to the shear strain energy theory, is

$$\frac{1}{12G}\left[\left(\sigma_1 - \sigma_2\right)^2 + \left(\sigma_1 - \sigma_3\right)^2 + \left(\sigma_2 - \sigma_3\right)^2\right] = \frac{\sigma_{yp}^{\ 2}}{6G}$$

i.e.

$$\left(\sigma_1 - \sigma_2\right)^2 + \left(\sigma_1 - \sigma_3\right)^2 + \left(\sigma_2 - \sigma_3\right)^2 = 2\,\sigma_{yp}^{\ 2} \tag{12.16}$$

In *two dimensions*, $\sigma_3 = 0$, and equation (12.16) reduces to:

$$\left(\sigma_1 - \sigma_2\right)^2 + \left(\sigma_1\right)^2 + \left(\sigma_2\right)^2 = 2\,\sigma_{yp}^{\ 2}$$

i.e. $\sigma_1^2 - 2\sigma_1\sigma_2 + \sigma_2^2 + \sigma_1^2 + \sigma_2^2 = 2\,\sigma_{yp}^{\ 2}$

i.e. $2\sigma_1^2 - 2\sigma_1\sigma_2 + 2\sigma_2^2 = 2\,\sigma_{yp}^{\ 2}$

or

$$\sigma_1^{\ 2} + \sigma_2^{\ 2} - \sigma_1\sigma_2 = \sigma_{yp}^{\ 2} \tag{12.17}$$

Another interpretation of equation (12.16) is that elastic failure takes place when the *von Mises stress* reaches yield, where

von Mises stress

$$= \frac{\sqrt{\left(\sigma_1 - \sigma_2\right)^2 + \left(\sigma_1 - \sigma_3\right)^2 + \left(\sigma_2 - \sigma_3\right)^2}}{\sqrt{2}} \tag{12.18}$$

or in *two dimensions*, $\sigma_3 = 0$

and von Mises stress

$$= \frac{\sqrt{\left(\sigma_1 - \sigma_2\right)^2 + \left(\sigma_1\right)^2 + \left(\sigma_2\right)^2}}{\sqrt{2}}$$

$$= \frac{\sqrt{\sigma_1^2 - 2\sigma_1\sigma_2 + \sigma_2^2 + \sigma_1^2 + \sigma_2^2}}{\sqrt{2}}$$

$$= \frac{\sqrt{2\sigma_1^2 - 2\sigma_1\sigma_2 + 2\sigma_2^2}}{\sqrt{2}}$$

$$= \frac{\sqrt{2}\sqrt{\sigma_1^2 - \sigma_1\sigma_2 + \sigma_2^2}}{\sqrt{2}}$$

$$= \sqrt{\sigma_1^2 + \sigma_2^2 - \sigma_1\sigma_2}$$

12.7 Yield loci

In two dimensions, equations (12.1), (12.2), (12.5), (12.6), (12.10), (12.13) and (12.17) can be expressed graphically, as shown in Figure 12.1.

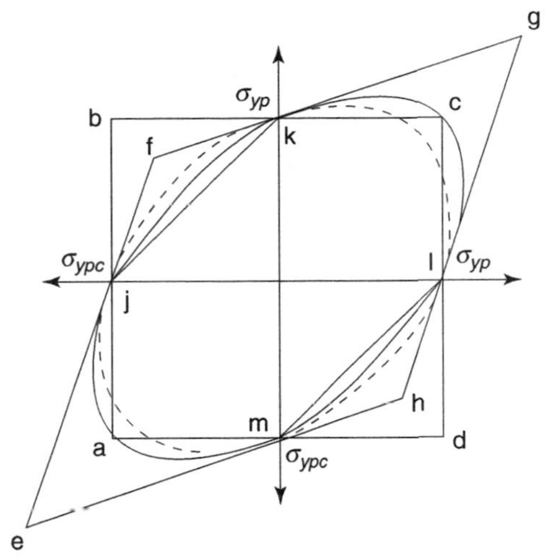

Figure	Theory enclosed by figure
abcd	Maximum principal stress theory
efgh	Maximum principal strain theory
ajkclm	Maximum shear stress theory
- - - - ellipse	Total strain energy theory
____ ellipse	Distortion of shear strain energy theory

Figure 12.1 Yield loci for a two-dimensional stress system

The figures are obtained by plotting the above equations, using σ_1 as the horizontal co-ordinate and σ_2 as the vertical co-ordinate, and the interpretation of each figure is that yield will not occur according to the theory under consideration if the point described by the values of σ_1 and σ_2 does not fall outside the appropriate figure.

Problem 1. A long thin-walled cylinder of wall thickness 2 cm and internal diameter 10 m is subjected to a uniform internal pressure. Determine the pressure that will cause yield, based on the five major theories of elastic failure, and given the following:

$$\sigma_{yp} = \sigma_{ypc} = 300 \text{ MN/m}^2 \quad v = 0.3$$

From Chapter 10,

$$\sigma_1 = \text{hoop stress} = \frac{PR}{t} = \frac{P \times 5}{2 \times 10^{-2}} = 250P \text{ from equation (10.1)}$$

σ_2 = longitudinal stress

$$= \frac{PR}{2t} = \frac{P \times 5}{2 \times 2 \times 10^{-2}} = 125P \text{ from equation (10.3)}$$

σ_3 = radial stress on inside surface of cylinder wall = $-P$

where R = internal radius of cylinder
t = wall thickness
and P = internal pressure – to be determined

(1) Maximum principal stress theory

$$\sigma_1 = \sigma_{yp}$$

Hence, $250P = 300$ MPa

from which **P = 1.2 MPa**

(2) Maximum principal strain theory

$$\sigma_1 - v(\sigma_2 + \sigma_3) = \sigma_{yp} \quad \text{from equation (12.3)}$$

i.e. $250P - 0.3(125P - P) = 300 \times 10^6$

i.e. $250P - 37.5P + 0.3P = 300 \times 10^6$

i.e. $212.8P = 300 \times 10^6$

and $P = \dfrac{300 \times 10^6}{212.8}$

i.e. **$P = 1.41 \times 10^6$ Pa = 1.41 MPa**

(3) Total strain energy theory

$$\left(\sigma_1^2 + \sigma_2^2 + \sigma_3^2\right) - 2v\left(\sigma_1\sigma_2 + \sigma_1\sigma_3 + \sigma_2\sigma_3\right) = \sigma_{yp}^{\ 2}$$

from equation (12.9)

Hence $\left((250P)^2 + (125P)^2 + (-P)^2\right)$

$$-2(0.3)\left[(250P)(125P) + (250P)(-P) + (125P)(-P)\right] = \left(300 \times 10^6\right)^2$$

$[62500 + 15625 + 1 - 0.6(31250 - 250 - 125]\, P^2 = 9 \times 10^{16}$

i.e. $[78126 - 0.6(30875)]\, P^2 = 9 \times 10^{16}$

i.e. $59601\, P^2 = 9 \times 10^{16}$

and $P = \sqrt{\dfrac{9 \times 10^{16}}{59601}}$

$= 1.229 \times 10^6$ Pa

= 1.229 MPa

(4) Maximum shear stress theory

$$\sigma_1 - \sigma_3 = \sigma_{yp} \quad \text{from equation (12.12)}$$

i.e. $250P - -P = 300 \times 10^6$

i.e. $251P = 300 \times 10^6$

from which $P = \dfrac{300 \times 10^6}{251}$

$= 1.195 \times 10^6$ Pa = **1.195 MPa**

(5) Shear strain energy theory

$(\sigma_1 - \sigma_2)^2 + (\sigma_1 - \sigma_3)^2 + (\sigma_2 - \sigma_3)^2 = 2\,\sigma_{yp}^{\,2}$ from equation (12.16)

i.e. $(250P - 125P)^2 + (250P - -P)^2 + (125P - -P)^2$

$$= 2\left(300 \times 10^6\right)^2$$

i.e. $[15625 + 63001 + 15876]\,P^2 = 2\left(300 \times 10^6\right)^2$

i.e. $[94502]\,P^2 = 18 \times 10^{16}$

and $P = \sqrt{\dfrac{18 \times 10^{16}}{94501}}$

$= 1.380 \times 10^6$ Pa

= 1.380 MPa

From the calculations in this problem, it can be seen that the maximum principal strain theory is the most optimistic, and the maximum shear stress theory is the most pessimistic. As expected, the shear strain energy theory is more optimistic than the maximum shear stress theory.

Problem 2. A solid circular-section shaft of diameter 0.1 m is subjected to a bending moment of 15 kN m. Determine the required torque to cause yield, based on the five major theories of elastic failure, given the following:
$\sigma_{yp} = 300$ MN/m^2 $v = 0.3$

This is a two-dimensional system of stress; hence, it is necessary to use equations (12.1), (12.5), (12.10), (12.13) and (12.17).
For a circular shaft,

$$I = \frac{\pi \times D^4}{64} = \frac{\pi \times 0.1^4}{64} = 4.909 \times 10^{-6}\ \text{m}^4$$

$$\bar{y} = \frac{D}{2} = \frac{0.1}{2} = 0.05\ \text{m}$$

Since $\dfrac{\sigma}{y} = \dfrac{M}{I}$ then $\sigma_x = \dfrac{M \times \bar{y}}{I}$ = maximum stress in axial direction

i.e. $\sigma_x = \dfrac{15 \times 10^3 \times 0.05}{4.909 \times 10^6}$

i.e. $\boldsymbol{\sigma_x = 152.8}$ **MPa**

By inspection $\boldsymbol{\sigma_y = 0}$

Now from Chapter 8,

$\dfrac{\tau}{r} = \dfrac{T}{J}$ from which, $T = \dfrac{\tau_{xy} \times J}{r}$

$$= \frac{\tau_{xy} \times \dfrac{\pi D^4}{32}}{r} = \frac{\tau_{xy} \times \dfrac{\pi(0.1)^4}{32}}{0.05}$$

$$= \frac{\tau_{xy} \times 9.817 \times 10^{-6}}{0.05}$$

i.e. $\boldsymbol{T = 1.9634 \times 10^{-4}\,\tau_{xy}}$

Also, from Chapter 9, equation (9.6),

$$\sigma_1 = \frac{1}{2}(\sigma_x + \sigma_y) + \frac{1}{2}\sqrt{(\sigma_x - \sigma_y)^2 + 4\tau_{xy}^{\,2}}$$

$$= \frac{1}{2}(152.8 \times 10^6 + 0) + \frac{1}{2}\sqrt{(\sigma_x - \sigma_y)^2 + 4\tau_{xy}^{\,2}}$$

$$= 76.4 \times 10^6 + k \text{ where } k = \frac{1}{2}\sqrt{(\sigma_x - \sigma_y)^2 + 4\tau_{xy}^{\,2}}$$

Also $\sigma_2 = \dfrac{1}{2}(\sigma_x + \sigma_y) - \dfrac{1}{2}\sqrt{(\sigma_x - \sigma_y)^2 + 4\tau_{xy}^{\,2}}$

i.e. $\sigma_2 = 76.4 \times 10^6 - k$

Thus $\boldsymbol{\sigma_1 = 76.4 \times 10^6 + k}$ and $\boldsymbol{\sigma_2 = 76.4 \times 10^6 - k}$

where $k = \dfrac{1}{2}\sqrt{(\sigma_x - \sigma_y)^2 + 4\tau_{xy}^{\,2}}$

(1) Maximum principal stress theory

$$\sigma_1 = \sigma_{yp}$$

Hence $76.4 \times 10^6 + k = 300 \times 10^6$

from which $k = 300 - 76.4 = 223.6 \times 10^6$

Thus $223.6 \times 10^6 = \dfrac{1}{2}\sqrt{(\sigma_x - \sigma_y)^2 + 4\tau_{xy}^{\,2}}$

$$223.6 \times 10^6 = \frac{1}{2}\sqrt{(152.8 \times 10^6 - 0)^2 + 4\tau_{xy}^{\,2}}$$

$$2 \times 223.6 \times 10^6 = \sqrt{(152.8 \times 10^6 - 0)^2 + 4\tau_{xy}^{\,2}}$$

$$\left(2 \times 223.6 \times 10^6\right)^2 = \left(152.8 \times 10^6\right)^2 + 4\tau_{xy}^{\,2}$$

$$\left(2\times 223.6\times 10^6\right)^2 - \left(152.8\times 10^6\right)^2 = 4\tau_{xy}^{\ 2}$$

Hence $1.7664 \times 10^{17} = 4\tau_{xy}^{\ 2}$

and $\tau_{xy} = \sqrt{\dfrac{1.7664\times 10^{17}}{4}}$

i.e. $\boldsymbol{\tau_{xy}} = 210.1 \times 10^6$ Pa = **210.1 MPa**

However, from above,

$$T = 1.9634\times 10^{-4}\ \tau_{xy}$$

i.e. $= 1.9634\times 10^{-4} \times 210.1\times 10^6$

i.e. **torque, T** $= 41.25 \times 10^3$ N m = **41.25 kN m**

(2) Maximum principal strain theory

From equation (12.5) $\sigma_1 - v\sigma_2 = \sigma_{yp}$

i.e. $76.4\times 10^6 + k - 0.3(76.4\times 10^6 - k) = 300\times 10^6$

i.e. $76.4\times 10^6 + k - 22.92\times 10^6 + 0.3k = 300\times 10^6$

i.e. $53.48\times 10^6 + 1.3k = 300\times 10^6$

$1.3k = 300\times 10^6 - 53.48\times 10^6$

and $k = \dfrac{300\times 10^6 - 53.48\times 10^6}{1.3} = 189.6\times 10^6$

Hence $189.6\times 10^6 = \dfrac{1}{2}\sqrt{\left(\sigma_x - \sigma_y\right)^2 + 4\tau_{xy}^{\ 2}}$

$$= \frac{1}{2}\sqrt{\left(152.8\times 10^6\right)^2 + 4\tau_{xy}^2}$$

and $2\times 189.6\times 10^6 = \sqrt{\left(152.8\times 10^6\right)^2 + 4\tau_{xy}^{\ 2}}$

Thus $\left(2\times 189.6\times 10^6\right)^2 = \left(152.8\times 10^6\right)^2 + 4\tau_{xy}^{\ 2}$

and $\left(2\times 189.6\times 10^6\right)^2 - \left(152.8\times 10^6\right)^2 = 4\tau_{xy}^{\ 2}$

from which $1.2045\times 10^{17} = 4\tau_{xy}^{\ 2}$

and $\tau_{xy} = \sqrt{\dfrac{1.2045\times 10^{17}}{4}}$

i.e. $\boldsymbol{\tau_{xy}} = 173.5 \times 10^6$ Pa = **173.5 MPa**

However, from above, $T = 1.9634\times 10^{-4}\ \tau_{xy}$

i.e. $= 1.9634\times 10^{-4} \times 173.5\times 10^6$

i.e. **torque, T** $= 34.06 \times 10^3$ N m = **34.06 kN m**

(3) Total strain energy theory

$$\sigma_1^2 + \sigma_2^2 - 2v\ \sigma_1\sigma_2 = \sigma_{yp}^{\ 2}$$

$$\left(76.4\times 10^6 + k\right)^2 + \left(76.4\times 10^6 - k\right)^2$$

$$-2(0.3)\left(76.4\times 10^6 + k\right)\left(76.4\times 10^6 - k\right) = \left(300\times 10^6\right)^2$$

i.e.

$$\left[\left(76.4\times 10^6\right)^2 + 2k\left(76.4\times 10^6\right) + k^2\right]$$

$$+\left[\left(76.4\times 10^6\right)^2 - 2k\left(76.4\times 10^6\right) + k^2\right]$$

$$-2(0.3)\left[\left(76.4\times 10^6\right)^2 - k^2\right] = 9\times 10^{16}$$

i.e. $2\left(76.4\times 10^6\right)^2 + 2k^2$

$$-3.502\times 10^{15} + 0.6k^2 = 9\times 10^{16}$$

i.e. $2.6k^2 + 8.172\times 10^{15} = 9\times 10^{16}$

i.e. $2.6k^2 = 9\times 10^{16} - 8.172\times 10^{15} = 8.183\times 10^{16}$

from which $k = \sqrt{\dfrac{8.183\times 10^{16}}{2.6}}$

$= 177.4 \times 10^6$

Hence $177.4 \times 10^6 = \dfrac{1}{2}\sqrt{\left(152.8\times 10^6\right)^2 + 4\tau_{xy}^{\ 2}}$

and $2 \times 177.4 \times 10^6 = \sqrt{\left(152.8\times 10^6\right)^2 + 4\tau_{xy}^{\ 2}}$

$$\left(2\times 177.4\times 10^6\right)^2 = \left(152.8\times 10^6\right)^2 + 4\tau_{xy}^{\ 2}$$

$$\left(2\times 177.4\times 10^6\right)^2 - \left(152.8\times 10^6\right)^2 = 4\tau_{xy}^{\ 2}$$

i.e. $1.0254\times 10^{17} = 4\tau_{xy}^{\ 2}$

and $\tau_{xy} = \sqrt{\dfrac{1.0254\times 10^{17}}{4}}$

i.e. $\boldsymbol{\tau_{xy}} = 160.1 \times 10^6$ Pa = **161.1MPa**

From above $T = 1.9634\times 10^{-4}\ \tau_{xy}$

i.e. $= 1.9634\times 10^{-4} \times 161.1\times 10^6$

i.e. **torque, T** $= 31.63 \times 10^3$ N m = **31.63 kN m**

(4) Maximum shear stress theory

$\sigma_1 - \sigma_2 = \sigma_{yp}$ from equation (12.13)

i.e. $\left(76.4\times 10^6 + k\right) - \left(76.4\times 10^6 - k\right) = 300\times 10^6$

i.e. $2k = 300 \times 10^6$

from which $k = 150 \times 10^6$

Hence $150 \times 10^6 = \frac{1}{2}\sqrt{\left(152.8\times10^6\right)^2 + 4\tau_{xy}^{\ 2}}$

and $2\times150 \times 10^6 = \sqrt{\left(152.8\times10^6\right)^2 + 4\tau_{xy}^{\ 2}}$

$$\left(2\times150\ \times10^6\right)^2 = \left(152.8\times10^6\right)^2 + 4\tau_{xy}^{\ 2}$$

$$\left(2\times150\ \times10^6\right)^2 - \left(152.8\times10^6\right)^2 = 4\tau_{xy}^{\ 2}$$

i.e. $6.6652\times10^{16} = 4\tau_{xy}^{\ 2}$

and $\tau_{xy} = \sqrt{\frac{6.6652\times10^{16}}{4}}$

i.e. $\boldsymbol{\tau_{xy}} = 129.1 \times 10^6$ Pa = **129.1 MPa**

From above $T = 1.9634\times10^{-4}\ \tau_{xy}$

i.e. $= 1.9634\times10^{-4} \times 129.1\times10^6$

i.e. **torque, *T*** = 25.35×10^3 N m = **25.35 kN m**

(5) Shear strain energy theory

$\sigma_1^2 + \sigma_2^2 - \sigma_1\sigma_2 = \sigma_{yp}^{\ 2}$ from equation (12.17)

i.e. $\left(76.4\times10^6 + k\right)^2 + \left(76.4\times10^6 - k\right)^2$

$$-\left(76.4\times10^6 + k\right)\left(76.4\times10^6 - k\right) = \left(300\times10^6\right)^2$$

i.e. $\left[\left(76.4\times10^6\right)^2 + 2k\left(76.4\times10^6\right) + k^2\right]$

$$+\left[\left(76.4\times10^6\right)^2 - 2k\left(76.4\times10^6\right) + k^2\right]$$

$$-\left[\left(76.4\times10^6\right)^2 - k^2\right] = 9\times10^{16}$$

i.e. $2\left(76.4\times10^6\right)^2 + 2k^2 - 5.837\times10^{15} + k^2 = 9\times10^{16}$

i.e. $3k^2 + 5.837\times10^{15} = 9\times10^{16}$

i.e. $3k^2 = 9\times10^{16} - 5.837\times10^{15} = 8.416\times10^{16}$

from which $k = \sqrt{\frac{8.416\times10^{16}}{3}} = 167.5 \times 10^6$

Hence $167.5 \times 10^6 = \frac{1}{2}\sqrt{\left(152.8\times10^6\right)^2 + 4\tau_{xy}^{\ 2}}$

and $2\times167.5 \times 10^6 = \sqrt{\left(152.8\times10^6\right)^2 + 4\tau_{xy}^{\ 2}}$

$$\left(2\times167.5\ \times10^6\right)^2 = \left(152.8\times10^6\right)^2 + 4\tau_{xy}^{\ 2}$$

$$\left(2\times167.5\ \times10^6\right)^2 - \left(152.8\times10^6\right)^2 = 4\tau_{xy}^{\ 2}$$

i.e. $8.8877\times10^{16} = 4\tau_{xy}^{\ 2}$

and $\tau_{xy} = \sqrt{\frac{8.8877\times10^{16}}{4}}$

i.e. $\boldsymbol{\tau_{xy}} = 149.1\times10^6$ Pa = **149.1 MPa**

From above $T = 1.9634\times10^{-4}\ \tau_{xy}$

i.e. $= 1.9634\times10^{-4} \times 149.1\times10^6$

i.e. **torque, *T*** = $29.27\ \times10^3$ N m = **29.27 kN m**

The calculations above have shown that for this problem, the maximum principal stress theory is the most optimistic, and the maximum shear stress theory is the most pessimistic, where the ratio of the former to the latter is about 1.6: 1.

Now try the following Practice Exercise

Practice Exercise 53. Theories of elastic failure

1. A submarine pressure hull, which may be assumed to be a long thin-walled circular cylinder, of external diameter 10 m and wall thickness 5 cm, is constructed from high-tensile steel. Assuming that buckling does not occur, determine the maximum permissible diving depths that the submarine can achieve, without suffering elastic failure, based on the five major theories of yield and given the following:

   $\sigma_{yp} = -\sigma_{ypc} = 400$ MN/m² $\quad v = 0.3$
   Density of water = 1020 kg/m³
   $g = 9.81$ m/s²

   [400 m, 472.3 m, 412.3 m, 404 m, 466.5 m]

2. A circular-section torsion specimen, of diameter 2 cm, yields under a pure torque of 0.25 kN m. What is the shear stress due to yield? What is the yield stress according to (a) Tresca, (b) Hencky-von Mises? What is the ratio τ_{yp}/σ_{yp} according to these two theories?

   [159.1 MPa (a) 318.3 MPa (b) 275.6 MPa; 0.50; 0.577]

3. A shaft of diameter 0.1 m is found to yield under a torque of 30 kN m. Determine the pure bending moment that will cause a similar shaft, with no torque applied to it, to yield, assuming that the Tresca theory applies. What would be the bending moment to cause yield if the Hencky-von Mises theory applied?

[30 kN m, 30 kN m]

12.8 Conclusions

The above worked problems have shown that the predictions from various yield criteria can be very different. Furthermore, from the heuristic arguments of Section 12.5, it would appear that the only two theories that do not fail the hydrostatic stress condition are the maximum shear stress theory and the shear strain energy theory, and in any case, when materials such as mild steel are tested to destruction in tension, the characteristic 'cup and cone' failure mode (Figure 12.2) indicates the importance that shear stress plays in elastic failure. Because of this, many structural designers often prefer to use the maximum shear stress and the shear strain energy theories when designing structures involving two- and three-dimensional stress systems, where the former often lends itself to neat mathematical computations.

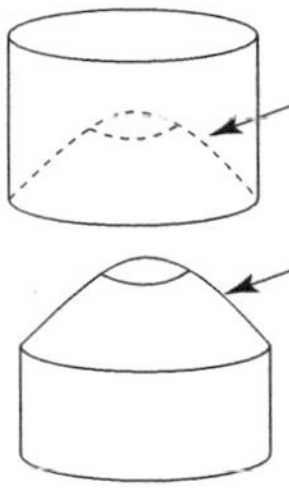

Figure 12.2 Cup and cone failure, indicating the importance of shear stress

Finally, it can be concluded from the calculations for both worked problems in this chapter that in certain two- and three-dimensional stress systems, some of the theories of elastic failure can be dangerous when applied to practical cases.

For fully worked solutions to each of the problems in Exercise 53 in this chapter, go to the website:
www.routledge.com/cw/bird

Chapter 13

Thick cylinders and spheres

Why it is important to understand: **Thick cylinders and spheres**

Thick-walled circular cylinders and spheres and rotating discs and rings are very important in a number of different branches of engineering, including ocean engineering, nuclear engineering and automobile and aeronautical engineering. In ocean engineering they appear as thick-walled submarines and drilling devices for both military and commercial uses. In military uses, C.T.F. Ross has applied them to a 'star wars underwater system' where these submarines are used as a major war deterrent. In commercial uses, he has adapted these structures to mine deep-sea methane hydrates. In the deep oceans, some 10,000 billion tonnes of methane hydrates are buried underneath the sea floor, where the hydrostatic pressure is equivalent to a 7 ton double-decker London bus acting on an area of only about one square inch. Moreover, in monetary terms, these methane hydrates are worth about £1.5 million per person on Earth! These vessels are also important in the design of nuclear reactors and underwater pipes, and for the barrels of guns. The thick-walled theory can be extended to deal with the shattering of rings and discs due to rotation, of much importance in automobile and aeronautical engineering.

At the end of this chapter you should be able to:

- derive hoop and radial stress equations for a thick-walled cylinder
- understand and interpret Lamé lines
- perform calculations on thick-walled cylinders
- perform calculations on compound cylinders
- appreciate plastic yielding of thick tubes
- perform calculations on plastic yielding of thick tubes
- derive equations for stresses in thick spherical shells
- derive equations for stresses in rotating discs
- perform calculations involving stresses in rotating discs
- derive an equation for the angular velocity required to fracture a rotating ring

Mathematical references

In order to understand the theory involved in this chapter on **thick cylinders and spheres**, knowledge of the following mathematical topics is required: ***algebraic manipulation/transposition, differentiation, straight line graphs, trigonometry, integration and integration by parts.*** If help/revision is needed in these areas, see page 50 for some textbook references.

13.1 Introduction

If the thickness to radius ratio of a shell exceeds 1: 30, the theory for the thin shells starts to break down. The reason for this is that for thicker shells, the radius of the shell changes appreciably over its thickness, so that the membrane strain can no longer be assumed to be constant over the thickness. Thick shells are of great importance in ocean engineering, civil engineering, nuclear engineering and so on.

In this chapter, in addition to considering thick shells under pressure, considerations will also be made of the plastic collapse of thick cylinders and discs. The collapse of discs and rings due to high-speed rotation will also be discussed.

For practical video demonstrations of the collapse of submarine pressure hulls, the buckling of a rail car tank under external pressure, bottle buckling – simulating submarine pressure hull collapse, and recent advances in submarine pressure hull design, go to www.routledge.com/cw/bird and click on the menu for 'Mechanics of Solids 3rd Edition'

13.2 Derivation of the hoop and radial stress equations for a thick-walled cylinder

The following convention will be used, where all the stresses and strains are assumed to be positive if they are tensile. At any radius r (see Figure 13.1),

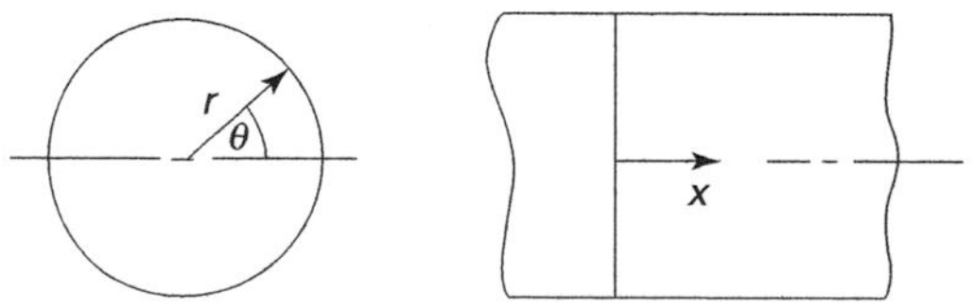

Figure 13.1 Thick cylinder

σ_θ = hoop stress ε_θ = hoop strain

σ_r = radial stress ε_r = radial strain

σ_x = longitudinal stress ε_x = longitudinal strain (assumed to be constant)

w = radial deflection

From Figure 13.2, it can be seen that at any radius r

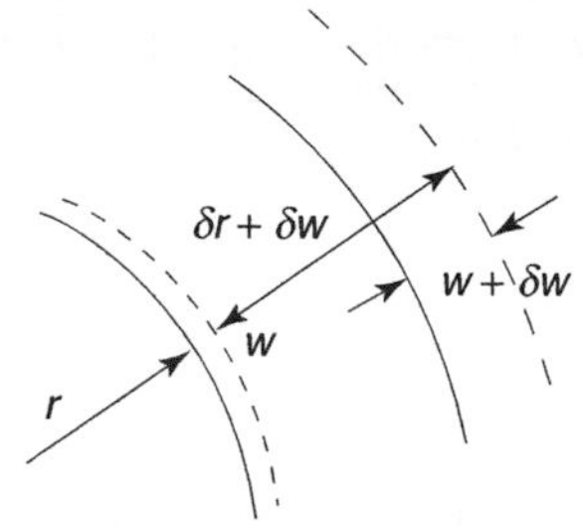

Figure 13.2 Deformation at any radius r

$$\varepsilon_\theta = \frac{2\pi(r+w) - 2\pi r}{2\pi r}$$

or $$\varepsilon_\theta = \frac{w}{r} \quad (13.1)$$

Similarly $$\varepsilon_r = \frac{\delta w}{\delta r} = \frac{dw}{dr} \quad (13.2)$$

From the standard stress-strain relationships (see Chapter 4),

$$E\varepsilon_x = \sigma_x - \nu\sigma_\theta - \nu\sigma_r = \text{a constant}$$

$$E\varepsilon_\theta = \frac{Ew}{r} = \sigma_\theta - \nu\sigma_x - \nu\sigma_r \quad (13.3)$$

$$E\varepsilon_r = E\frac{dw}{dr} = \sigma_r - \nu\sigma_\theta - \nu\sigma_x \quad (13.4)$$

Multiplying equation (13.3) by r gives:

$$Ew = \sigma_\theta \times r - \nu\sigma_x \times r - \nu\sigma_r \times r \quad (13.5)$$

and differentiating equation (13.5) w.r.t. r using the product rule, gives:

$$E\frac{dw}{dr} = \sigma_\theta + r\frac{d\sigma_\theta}{dr} - \left(\nu\sigma_x + \nu r\frac{d\sigma_x}{dr}\right) - \left(\nu\sigma_r + \nu r\frac{d\sigma_r}{dr}\right)$$

$$E\frac{dw}{dr} = \sigma_\theta - \nu\sigma_x - \nu\sigma_r + r\left(\frac{d\sigma_\theta}{dr} - \nu\frac{d\sigma_x}{dr} - \nu\frac{d\sigma_r}{dr}\right) \quad (13.6)$$

Subtracting equation (13.4) from equation (13.6) gives:

$$\sigma_\theta - \nu\sigma_x - \nu\sigma_r + r\left(\frac{d\sigma_\theta}{dr} - \nu\frac{d\sigma_x}{dr} - \nu\frac{d\sigma_r}{dr}\right)$$

$$- \sigma_\theta - -\nu\sigma_x - -\nu\sigma_r = 0$$

i.e.

$$\sigma_\theta + \nu\sigma_\theta - \nu\sigma_x + \nu\sigma_x - \sigma_r - \nu\sigma_r$$

$$+ r\left(\frac{d\sigma_\theta}{dr} - \nu\frac{d\sigma_x}{dr} - \nu\frac{d\sigma_r}{dr}\right) = 0$$

i.e.

$$\sigma_\theta(1+\nu) - \sigma_x(1+\nu)$$

$$+ r\left(\frac{d\sigma_\theta}{dr} - \nu\frac{d\sigma_x}{dr} - \nu\frac{d\sigma_r}{dr}\right) = 0$$

i.e.

$$(\sigma_\theta - \sigma_r)(1+\nu) + r\frac{d\sigma_\theta}{dr} - \nu r\frac{d\sigma_x}{dr}$$

$$- \nu r\frac{d\sigma_r}{dr} = 0 \quad (13.7)$$

Since ε_x is constant,

$$\sigma_x - \nu\sigma_\theta - \nu\sigma_r = \text{a constant} \quad (13.8)$$

Differentiating equation (13.8) w.r.t. r gives:

$$\frac{d\sigma_x}{dr} - \nu\frac{d\sigma_\theta}{dr} - \nu\frac{d\sigma_r}{dr} = 0$$

or

$$\frac{d\sigma_x}{dr} = \nu\left(\frac{d\sigma_\theta}{dr} + \frac{d\sigma_r}{dr}\right) \quad (13.9)$$

Substituting equation (13.9) into equation (13.7) gives:

$$(\sigma_\theta - \sigma_r)(1+\nu) + r\frac{d\sigma_\theta}{dr}$$

$$- \nu^2 r\left(\frac{d\sigma_\theta}{dr} + \frac{d\sigma_r}{dr}\right) - \nu r\frac{d\sigma_r}{dr} = 0$$

i.e.

$$(\sigma_\theta - \sigma_r)(1+\nu) + r(1-\nu^2)\frac{d\sigma_\theta}{dr}$$

$$-\nu r(1+\nu)\frac{d\sigma_r}{dr} = 0 \quad (13.10)$$

and dividing through by $(1 + \nu)$ gives:

$$(\sigma_\theta - \sigma_r) + r\frac{(1-\nu)(1+\nu)}{(1+\nu)}\frac{d\sigma_\theta}{dr} - \nu r\frac{d\sigma_r}{dr} = 0$$

i.e.

$$(\sigma_\theta - \sigma_r) + r(1-\nu)\frac{d\sigma_\theta}{dr} - \nu r\frac{d\sigma_r}{dr} = 0 \quad (13.11)$$

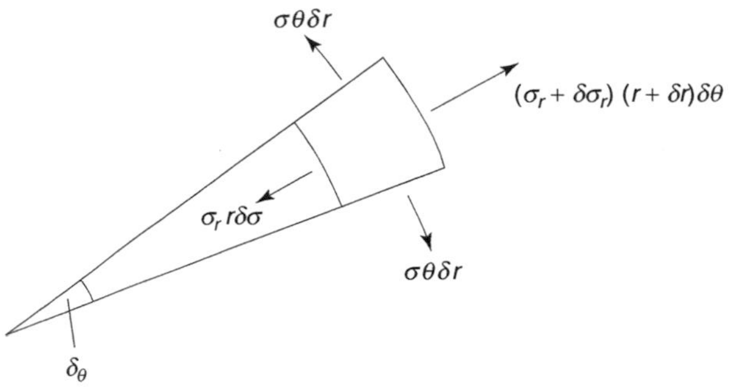

Figure 13.3 Shell element

Considering now the equilibrium of an element of the shell, as shown in Figure 13.3,

$$2\sigma_\theta\,\delta r\sin\left(\frac{\delta\theta}{2}\right) + \sigma_r\,r\,\delta\theta - (\sigma_r + \delta\sigma_r)(r+\delta r)\delta\theta = 0$$

In the limit, when δr and $\delta\theta \to 0$,

$$2\sigma_\theta\frac{\delta r}{2} + \sigma_r r - (\sigma_r + \delta\sigma_r)(r+\delta r) = 0$$

i.e.

$$\sigma_0\,\delta r + \sigma_r r - \sigma_r r - \sigma_r\delta r - r\delta\sigma_r - d\sigma_r\delta r = 0$$

i.e.

$$\sigma_\theta\,\delta r - r d\sigma_r - \sigma_r\delta r = 0$$

or

$$\sigma_\theta - \sigma_r - r\frac{d\sigma_r}{dr} = 0 \quad (13.12)$$

Subtracting equation (13.11) from equation (13.12) gives

$$\left\{\sigma_\theta - \sigma_r - r\frac{d\sigma_r}{dr}\right\}$$

$$- \left\{(\sigma_\theta - \sigma_r) + r(1-\nu)\frac{d\sigma_\theta}{dr} - \nu r\frac{d\sigma_r}{dr}\right\} = 0$$

i.e.

$$-r\frac{d\sigma_r}{dr} - r(1-\nu)\frac{d\sigma_\theta}{dr} + \nu r\frac{d\sigma_r}{dr} = 0$$

i.e.

$$-\frac{d\sigma_r}{dr} - \frac{d\sigma_\theta}{dr} + \nu\frac{d\sigma_\theta}{dr} + \nu\frac{d\sigma_r}{dr} = 0$$

i.e.

$$(\nu-1)\frac{d\sigma_r}{dr} + (\nu-1)\frac{d\sigma_\theta}{dr} = 0$$

Dividing throughout by $(\nu - 1)$ gives:

$$\frac{d\sigma_r}{dr} + \frac{d\sigma_\theta}{dr} = 0 \quad (13.13)$$

Therefore, on integrating w.r.t. 'r',

$$\sigma_\theta + \sigma_r = \text{a constant, say, } 2A \qquad (13.14)$$

Subtracting equation (13.12) from equation (13.14) gives:

$$(\sigma_\theta + \sigma_r) - \left(\sigma_\theta - \sigma_r - r\frac{d\sigma_r}{dr}\right) = 2A$$

i.e. $$2\sigma_r + r\frac{d\sigma_r}{dr} = 2A \qquad \text{(i)}$$

Using the product rule for $\frac{d(\sigma_r r^2)}{dr}$, with $u = \sigma_r$ and $v = r^2$ gives: $\sigma_r(2r) + r^2\frac{d\sigma_r}{dr}$

and $$\frac{1}{r}\frac{d(\sigma_r r^2)}{dr} = 2\sigma_r + r\frac{d\sigma_r}{dr}$$

Comparing this with equation (i) leads to:

$$\frac{1}{r}\frac{d(\sigma_r r^2)}{dr} = 2A$$

or $$\frac{d(\sigma_r r^2)}{dr} = 2Ar$$

Integrating w.r.t. r gives $\int\left(\frac{d(\sigma_r r^2)}{dr}\right)dr = \int 2Ar\,dr$

i.e. $$\sigma_r r^2 + B = 2A\left(\frac{r^2}{2}\right) \quad \text{where } B = \text{constant}$$

or $$\sigma_r r^2 = Ar^2 - B$$

or $$\sigma_r = A - \frac{B}{r^2} \qquad (13.15)$$

From equation (13.14)

$$\sigma_\theta + \left(A - \frac{B}{r^2}\right) = 2A$$

i.e. $$\sigma_\theta = 2A - \left(A - \frac{B}{r^2}\right)$$

i.e. $$\sigma_\theta = A + \frac{B}{r^2} \qquad (13.16)$$

13.3 Lamé* line

Equations (13.15) and (13.16) can be represented by a single straight line if they are plotted against $\frac{1}{r^2}$ as shown in Figure 13.4, where σ_r is on the left of the diagram and σ_θ is on the right.

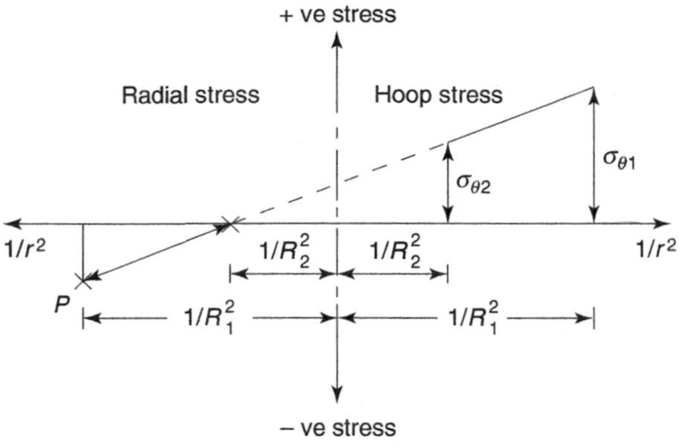

Figure 13.4 Lamé line for the case of internal pressure

In Figure 13.4, we see the case of a thick-walled cylinder of internal radius R_1 and external radius R_2, subjected to an internal pressure P. Now two points are known on this straight line: the radial stresses on the internal and external surfaces, which are $-P$ and zero, respectively (shown by a 'X'). Hence, the straight line can be drawn, and any stress calculated throughout the thickness of the walls by equating similar triangles.

In Figure 13.4

$\sigma_{\theta 1}$ = internal hoop stress, which can be seen to be a maximum stress

and $\sigma_{\theta 2}$ = external hoop stress

Furthermore, from Figure 13.4, it can be seen that both $\sigma_{\theta 1}$ and $\sigma_{\theta 2}$ are tensile, and that σ_r is compressive.

To calculate $\sigma_{\theta 1}$ and $\sigma_{\theta 2}$

Equating similar triangles in Figure 13.4 gives:

$$\frac{\sigma_{\theta 1}}{\left(\frac{1}{R_1^2} + \frac{1}{R_2^2}\right)} = \frac{P}{\left(\frac{1}{R_1^2} - \frac{1}{R_2^2}\right)}$$

from which $$\sigma_{\theta 1} = \frac{P\left(\frac{1}{R_1^2} + \frac{1}{R_2^2}\right)}{\left(\frac{1}{R_1^2} - \frac{1}{R_2^2}\right)} = \frac{P\left(\frac{R_1^2 + R_2^2}{R_1^2 R_2^2}\right)}{\left(\frac{R_2^2 - R_1^2}{R_1^2 R_2^2}\right)}$$

*Gabriel Léon Jean Baptiste Lamé (22 July 1795 – 1 May 1870) was a French mathematician who contributed to the theory of partial differential equations by the use of curvilinear co-ordinates, and the mathematical theory of elasticity. His most significant contribution to engineering was to accurately define the stresses and capabilities of a press fit joint. To find out more information about Lamé go to www.routledge.com/cw/bird

i.e. $$\boldsymbol{\sigma_{\theta 1} = \frac{P\left(R_1^2 + R_2^2\right)}{R_2^2 - R_1^2}}$$

Similarly $$\frac{\sigma_{\theta 2}}{\left(\frac{1}{R_1^2} + \frac{1}{R_2^2}\right)} = \frac{P}{\left(\frac{1}{R_1^2} - \frac{1}{R_2^2}\right)}$$

from which

$$\sigma_{\theta 2} = \frac{P\left(\frac{2}{R_2^2}\right)}{\left(\frac{1}{R_1^2} - \frac{1}{R_2^2}\right)} = \frac{P\left(\frac{2}{R_2^2}\right)}{\left(\frac{R_2^2 - R_1^2}{R_1^2 R_2^2}\right)}$$

$$= P\left(\frac{2}{R_2^2}\right)\left(\frac{R_1^2 R_2^2}{R_2^2 - R_1^2}\right)$$

i.e. $$\boldsymbol{\sigma_{\theta 2} = \frac{2PR_1^2}{R_2^2 - R_1^2}}$$

Problem 1. A thick-walled cylinder of internal diameter 0.1 m is subjected to an internal pressure of 50 MPa. If the maximum permissible stress in the cylinder is limited to 200 MPa, determine the minimum possible external diameter.

Let d_2 be the external diameter – to be determined

The Lamé line for this case would appear as shown in Figure 13.5, where it can be seen that the Lamé line is obtained by knowing two values of radial stress. These are the radial stresses on the internal and external surfaces of the cylinder, which are –50 MPa and zero, respectively. However, as d_2 is unknown, a third point is required, which in this case is the maximum stress (i.e. $\sigma_{\theta 1}$); this is 200 MPa.

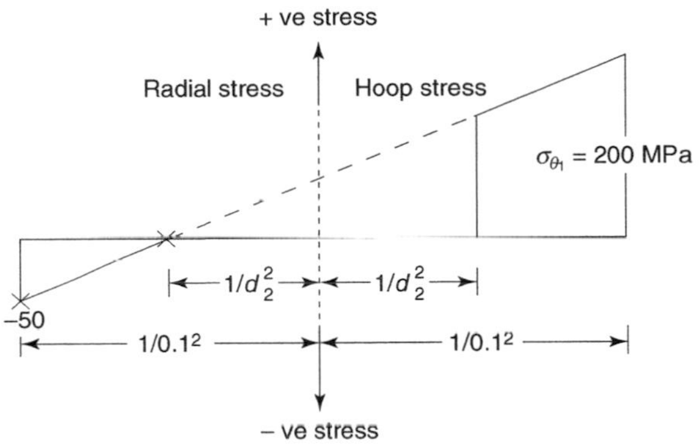

Figure 13.5 Lamé line for thick cylinder

Equating similar triangles in Figure 13.5

$$\frac{50\times10^6}{\left(\frac{1}{0.1^2} - \frac{1}{d_2^2}\right)} = \frac{200\times10^6}{\left(\frac{1}{0.1^2} + \frac{1}{d_2^2}\right)}$$

from which $$\frac{\left(\frac{1}{0.1^2} + \frac{1}{d_2^2}\right)}{\left(\frac{1}{0.1^2} - \frac{1}{d_2^2}\right)} = \frac{200}{50} = 4$$

i.e. $$\frac{\left(\frac{d_2^2 + 0.1^2}{0.1^2 d_2^2}\right)}{\left(\frac{d_2^2 - 0.1^2}{0.1^2 d_2^2}\right)} = \frac{d_2^2 + 0.1^2}{d_2^2 - 0.1^2} = 4$$

and $\quad d_2^2 + 0.1^2 = 4\left(d_2^2 - 0.1^2\right)$

i.e. $\quad d_2^2 + 0.1^2 = 4d_2^2 - 4 \times 0.1^2$

i.e. $\quad 5 \times 0.1^2 = 3d_2^2$

from which $\quad d_2 = \sqrt{\dfrac{5 \times 0.1^2}{3}}$

i.e. $\quad \mathbf{d_2 = 0.129\ m}$

Problem 2. If the vessel of Problem 1 is subjected to an external pressure of 50 MPa, determine the maximum value of stress that would occur in this vessel.

The Lamé line for this case is as shown in Figure 13.6, where it can be seen that two values of radial stress are known. These are the radial stresses on the internal and external surfaces, which are zero and −50 MPa, respectively. Let $\sigma_{\theta 1}$ = hoop stress on the internal surface, which from Figure 13.6 can be seen to have the largest magnitude.

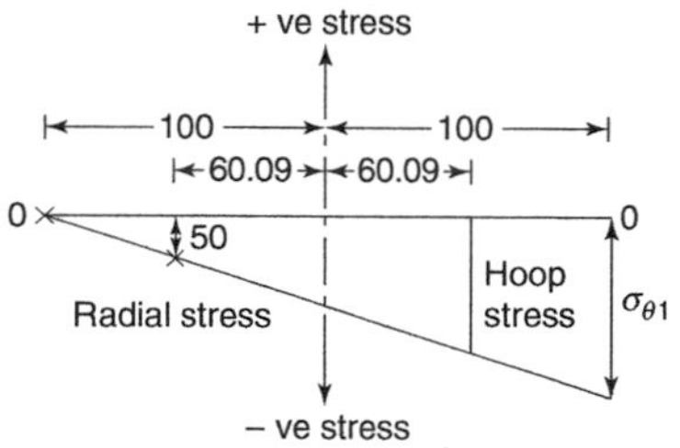

Figure 13.6 Lamé line for external pressure case

(Note that in Figure 13.6, $\dfrac{1}{d_2^2}$ is $\dfrac{1}{0.129^2} = 60.09$)

By equating similar triangles

$$\frac{\sigma_{\theta 1}}{(100+100)} = \frac{-50 \times 10^6}{(100-60.09)}$$

from which,

$$\sigma_{\theta 1} = \frac{-50 \times 10^6 (100+100)}{(100-60.09)} = \frac{-50 \times 10^6 \times 200}{39.91}$$

i.e. $\quad \boldsymbol{\sigma_{\theta 1} = -250.6}$ **MPa**

NB It should be noted in Problems 1 and 2 that the maximum stress for both cases was the internal hoop stress.

Problem 3. A steel ring of external diameter 10 cm and internal diameter 5 cm is to be shrunk into a solid steel shaft of diameter 5 cm, where all the dimensions are nominal. If the interference fit at the common surface between the ring and the shaft is 0.01 cm, based on a diameter, determine the maximum stress in the material. Assume that: $E = 2 \times 10^{11}\ \mathrm{N/m^2}$ and $\nu = 0.3$

Consider first the steel ring

The Lamé line for the steel ring will be as shown in Figure 13.7, where the radial stress on its outer surface will be zero, and that on its internal surface will be P_c (the radial pressure at the common surface between the steel ring and the shaft).

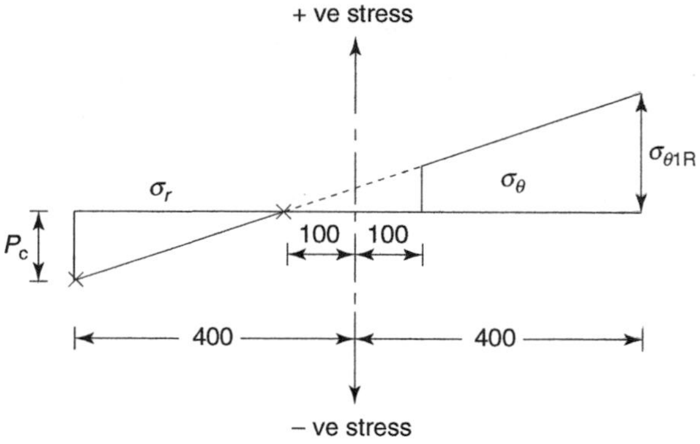

Figure 13.7 Lamé line for steel ring

Let $\quad \sigma_{\theta 1R}$ = hoop stress (maximum stress) on the internal surface of the ring

$\quad \sigma_{r1R}$ = radial stress on the internal surface of the ring

(Note that in Figure 13.7, $\dfrac{1}{d_1^2}$ is $\dfrac{1}{0.05^2} = 400$, and $\dfrac{1}{d_2^2}$ is $\dfrac{1}{0.1^2} = 100$)

Equating similar triangles

$$\frac{\sigma_{\theta 1R}}{(400+100)} = \frac{P_c}{(400-100)}$$

from which $\quad \sigma_{\theta 1} = \dfrac{P_c(400+100)}{(400-100)} = \dfrac{P_c \times 500}{300}$

i.e. $\quad \boldsymbol{\sigma_{\theta 1} = 1.667 P_c} \qquad (13.17)$

Consider now the shaft

For this case, the Lamé line will be horizontal, because if it had any slope at all, the stresses at the centre would be infinite for a finite value of P_c, which is impossible.

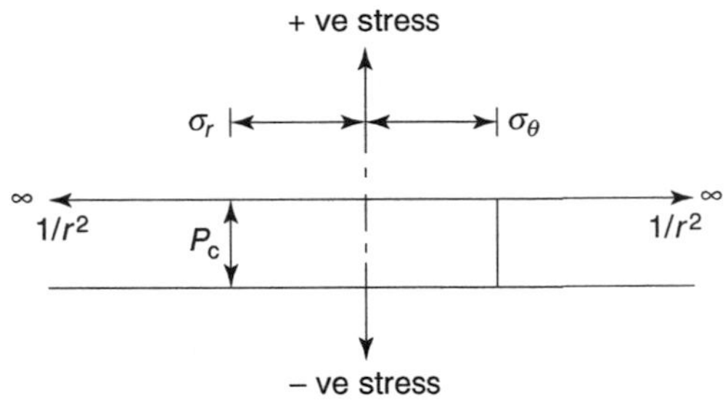

Figure 13.8 Lamé line for a solid shaft

From Figure 13.8, it can be seen that for a solid shaft, under an axisymmetric external pressure of P_c

$$\sigma_r = \sigma_\theta = -P_c \text{ (everywhere)} \qquad (13.18)$$

Let w_R = increase in the radius of the ring at its inner surface

w_s = increase in the radius of the shaft at its outer surface

Now, applying the expression

$$E\varepsilon_\theta = \frac{Ew}{r} = \sigma_\theta - \nu\sigma_r - \nu\sigma_x$$

to the inner surface of the ring gives:

$$\frac{Ew_R}{2.5\times10^{-2}} = \sigma_{\theta 1R} - \nu\sigma_{r1R}$$

However $\sigma_{r1R} = -P_c$

Therefore $\dfrac{2\times10^{11}\times w_R}{2.5\times10^{-2}} = 1.667P_c - -0.3P_c$

from which $w_R = \dfrac{2.5\times10^{-2}}{2\times10^{11}}\left(1.667P_c + 0.3P_c\right)$

i.e. $\mathbf{w_R = 2.459\times10^{-13}\,P_c}$ (13.19)

Similarly for the shaft,

$$\frac{Ew_s}{2.5\times10^{-2}} = \sigma_{\theta s} - \nu\sigma_{rs}$$

However $\sigma_{\theta s} = \sigma_{rs} = -P_c$

Therefore $\dfrac{2\times10^{11}\times w_s}{2.5\times10^{-2}} = -P_c - -0.3P_c$

from which $w_s = \dfrac{2.5\times10^{-2}}{2\times10^{11}}\left(-0.7P_c\right)$

i.e. $\mathbf{w_s = -8.75\times10^{-14}\,P_c}$ (13.20)

Now the interference fit on the diameters is 0.01×10^{-2} m. Therefore the interference fit on the radii is 5×10^{-5} m $= w_R - w_s$ i.e. half the interference fit on the diameters (i.e. $\frac{1}{2}\times0.01\times10^{-2} = 5\times10^{-5}$ m)

Hence $w_R - w_s = 5\times10^{-5}$

i.e. $2.459\times10^{-13}P_c - -8.75\times10^{-14}P_c = 5\times10^{-5}$

from which $3.334\times10^{-13}P_c = 5\times10^{-5}$

and $P_c = \dfrac{5\times10^{-5}}{3.334\times10^{-13}} = 149.97\times10^6$ Pa

i.e. $\mathbf{P_c = 149.97}$ **MPa**

From equation (13.17)

the maximum stress, $\sigma_{\theta 1} = 1.667\,P_c$

$= 1.667\times149.97$ MPa

$= \mathbf{250}$ **MPa**

Now try the following Practice Exercise

Practice Exercise 54. Thick-walled cylinders

1. Determine the maximum permissible internal pressure that a thick-walled cylinder of internal diameter 0.2 m and wall thickness 0.1 m can be subjected to, if the maximum permissible stress in this vessel is not to exceed 250 MPa. [150 MPa]
2. Determine the maximum permissible internal pressure that a thick-walled cylinder of internal diameter 0.2 m and wall thickness 0.1 m can be subjected to, if the cylinder is also subjected to an external pressure of 20 MPa. Assume that $\sigma_{yp} = 300$ MPa. [212 MPa]

3. A steel ring of 9 cm external diameter and 5 cm internal diameter is to be shrunk into a solid bronze shaft, where the interference fit is 0.005×10^{-2} m, based on the diameter. Determine the maximum tensile stress that is set up in the material given that:

For steel $E_s = 2 \times 10^{11} \text{N/m}^2$ and $v_s = 0.3$

For bronze $E_b = 1 \times 10^{11} \text{N/m}^2$ and $v_b = 0.35$

[$P_c = 57.25$ MPa, $\hat{\sigma}_\theta = 108.3$ MPa]

13.4 Compound cylinders

A cylinder made from two different materials is sometimes found useful in engineering when one material is suitable for resisting corrosion in a certain environment, but because this material is expensive or weak, another material is used to strengthen it.

Problem 4. An aluminium-alloy disc of constant thickness, and of internal and external radii R_1 and R_2 respectively, is shrunk onto a solid steel shaft of external radius $R_1 + \delta$. Show that the maximum stress ($\hat{\sigma}$) in the disc is given by:

$$\hat{\sigma} = \frac{\delta}{\left\{R_1\left[\left(\frac{1-v_s}{E_s} + \frac{v_a}{E_a}\right)\left(\frac{R_2^2 - R_1^2}{R_2^2 + R_1^2}\right) + \frac{1}{E_s}\right]\right\}}$$

where

E_s = elastic modulus of steel

E_a = elastic modulus of aluminium alloy

v_s = Poisson's ratio for steel

v_a = Poisson's ratio for aluminium alloy

Let P_c be the radial pressure at the common surface – to be determined. Now the radial stress for the disc on the external surface is zero, and the radial stress on the internal surface of the disc is $-P_c$; hence, the Lamé line will take the form shown in Figure 13.9.

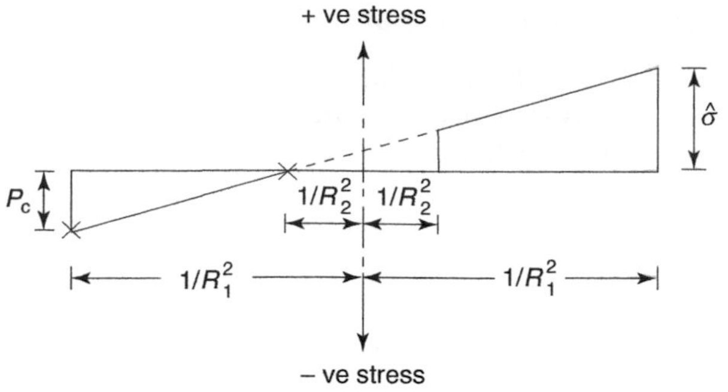

Figure 13.9 Lamé line for the disc

Equating similar triangles in Figure 13.9 gives:

$$\frac{P_c}{\left(\frac{1}{R_1^2} - \frac{1}{R_2^2}\right)} = \frac{\hat{\sigma}}{\left(\frac{1}{R_1^2} + \frac{1}{R_2^2}\right)}$$

from which $$P_c = \frac{\hat{\sigma}\left(\frac{1}{R_1^2} - \frac{1}{R_2^2}\right)}{\left(\frac{1}{R_1^2} + \frac{1}{R_2^2}\right)} = \frac{\hat{\sigma}\left(\frac{R_2^2 - R_1^2}{R_1^2 R_2^2}\right)}{\left(\frac{R_2^2 + R_1^2}{R_1^2 R_2^2}\right)}$$

i.e. $$P_c = \hat{\sigma} \times \frac{\left(R_2^2 - R_1^2\right)}{\left(R_2^2 + R_1^2\right)} \qquad (13.21)$$

Now for the steel shaft, the Lamé line must be horizontal, otherwise the stresses at the centre of the shaft will be infinite for a finite value of P_c

i.e. $\sigma_{\theta s}$ = hoop stress in the shaft

= radial stress in the shaft (σ_{rs})

= $-P_c$ (everywhere in the shaft)

Let w_a = increase in radius of the aluminium disc at its internal surface

and w_s = increase in radius of the steel shaft on its external surface

Applying the expression

$$E\varepsilon_\theta = \frac{Ew}{r} = \sigma_\theta - v\sigma_r - v\sigma_x$$

to the inner surface of the aluminium disc gives:

$$\frac{E_a w_a}{R_1} = \hat{\sigma} + v_a P_c$$

or $$w_a = \frac{R_1}{E_a}\left(\hat{\sigma} + v_a P_c\right) \quad (13.22)$$

Similarly, for the steel shaft,

$$\frac{E_s w_s}{R_1} = -P_c + v_s P_c$$

or $$w_s = \frac{-R_1}{E_s} \times P_c\left(1 - v_s\right) \quad (13.23)$$

However $\delta = w_a - w_s$

$$= \frac{R_1}{E_a}\left(\hat{\sigma} + v_a P_c\right) - \frac{-R_1}{E_s} \times P_c\left(1 - v_s\right)$$

$$= \frac{R_1}{E_a}\left(\hat{\sigma} + v_a P_c\right) + \frac{R_1}{E_s} \times P_c\left(1 - v_s\right) \quad (13.24)$$

Substituting equation (13.21) into equation (13.24) gives:

$$\delta = \frac{R_1}{E_a}\left(\hat{\sigma} + v_a\left(\hat{\sigma} \times \frac{\left(R_2^2 - R_1^2\right)}{\left(R_2^2 + R_1^2\right)}\right)\right) + \frac{R_1}{E_s} \times \hat{\sigma} \times \frac{\left(R_2^2 - R_1^2\right)}{\left(R_2^2 + R_1^2\right)}\left(1 - v_s\right)$$

i.e. $$\delta = \hat{\sigma}\left\{\begin{array}{l}\frac{R_1}{E_a}\left(1 + \frac{v_a\left(R_2^2 - R_1^2\right)}{\left(R_2^2 + R_1^2\right)}\right) \\ + \frac{R_1}{E_s} \times \frac{\left(R_2^2 - R_1^2\right)}{\left(R_2^2 + R_1^2\right)}\left(1 - v_s\right)\end{array}\right\}$$

i.e. $$\delta = \hat{\sigma} R_1\left\{\begin{array}{l}\left(\frac{1}{E_a} + \frac{v_a\left(R_2^2 - R_1^2\right)}{E_a\left(R_2^2 + R_1^2\right)}\right) \\ + \frac{\left(R_2^2 - R_1^2\right)}{E_s\left(R_2^2 + R_1^2\right)}\left(1 - v_s\right)\end{array}\right\}$$

i.e. $$\delta = \hat{\sigma} R_1\left\{\begin{array}{l}\left(\frac{1}{E_a} + \frac{v_a\left(R_2^2 - R_1^2\right)}{E_a\left(R_2^2 + R_1^2\right)}\right) \\ + \frac{\left(R_2^2 - R_1^2\right)}{E_s\left(R_2^2 + R_1^2\right)} - \frac{v_s\left(R_2^2 - R_1^2\right)}{E_s\left(R_2^2 + R_1^2\right)}\end{array}\right\}$$

i.e. $$\delta = \hat{\sigma}\left\{R_1\left[\left(\frac{1 - v_s}{E_s} + \frac{v_a}{E_a}\right)\left(\frac{R_2^2 - R_1^2}{R_2^2 + R_1^2}\right) + \frac{1}{E_s}\right]\right\}$$

from which

$$\hat{\sigma} = \frac{\delta}{\left\{R_1\left[\left(\frac{1 - v_s}{E_s} + \frac{v_a}{E_a}\right)\left(\frac{R_2^2 - R_1^2}{R_2^2 + R_1^2}\right) + \frac{1}{E_s}\right]\right\}}$$

Problem 5. A steel cylinder with external and internal diameters of 10 cm and 8 cm, respectively, is shrunk onto an aluminium-alloy cylinder with internal and external diameters of 5 cm and 8 cm, respectively, where all the dimensions are nominal. Find the radial pressure at the common surface due to shrinkage alone, so that when there is an internal pressure of 150 MPa, the maximum hoop stress in the inner cylinder is 110 MPa. Determine, also, the maximum hoop stress in the outer cylinder, and plot the distributions across the sections. Assume that:

For steel $E_s = 2 \times 10^{11}$ N/m^2 and $v_s = 0.3$

For aluminium alloy $E_a = 6.7 \times 10^{10}$ N/m^2 and $v_a = 0.32$

Let P_c^S = the radial pressure at the common surface due to shrinkage alone

σ_θ^S = the hoop stress due to shrinkage alone

σ_θ^P = the hoop stress due to pressure alone

$\sigma_{\theta,10s}$ = hoop stress in the steel at its 10 cm diameter

$\sigma_{\theta,8s}$ = hoop stress in the steel at its 8 cm diameter

$\sigma_{r,10s}$ = radial stress in the steel at its 10 cm diameter

$\sigma_{r,8s}$ = radial stress in the steel at its 8 cm diameter

$\sigma_{\theta,8a}$ = hoop stress in the aluminium alloy at its 8 cm diameter

$\sigma_{r,8a}$ = radial stress in the aluminium alloy at its 8 cm diameter

$\sigma_{\theta,5a}$ = hoop stress in the aluminium alloy at its 5 cm diameter

$\sigma_{r,5a}$ = radial stress in the aluminium alloy at its 5 cm diameter

Consider first the stress due to shrinkage alone

For the aluminium-alloy tube, the Lamé line due to shrinkage alone is shown in Figure 13.10.

(Note that in Figure 13.10, $\frac{1}{d_1^2}$ is $\frac{1}{0.05^2} = 400$, and

$\frac{1}{d_2^2}$ is $\frac{1}{0.08^2} = 156.25$)

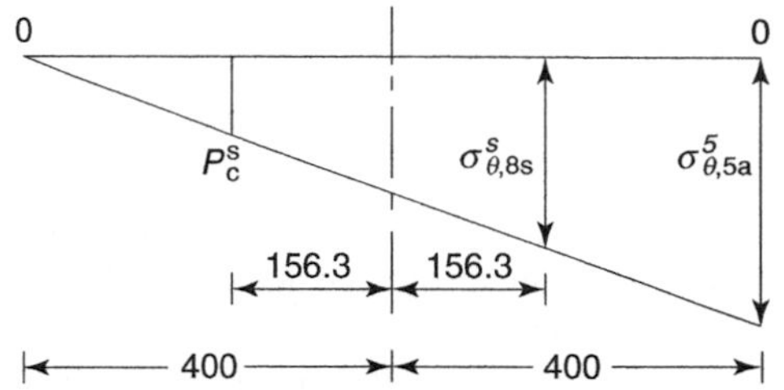

Figure 13.10 Lamé line for aluminium-alloy tube

Equating similar triangles in Figure 13.10 gives:

$$\frac{\sigma^s{}_{\theta,5a}}{(400+400)} = \frac{-P_c^s}{(400-156.25)}$$

Therefore $$\sigma^s{}_{\theta,5a} = \frac{-P_c^s(400+400)}{(400-156.25)}$$

i.e. $$\boldsymbol{\sigma^s_{\theta,5a} = -3.282 P_c^s} \qquad (13.25)$$

Similarly $$\frac{\sigma^s{}_{\theta,8a}}{(400+156.25)} = \frac{-P_c^s}{(400-156.25)}$$

Therefore $$\sigma^s{}_{\theta,8a} = \frac{-P_c^s(400+156.25)}{(400-156.25)}$$

i.e. $$\boldsymbol{\sigma^s{}_{\theta,8a} = -2.282 P_c^s} \qquad (13.26)$$

For the steel tube, the Lamé line due to shrinkage alone will be as shown in Figure 13.11.
Equating similar triangles in Figure 13.11 gives:

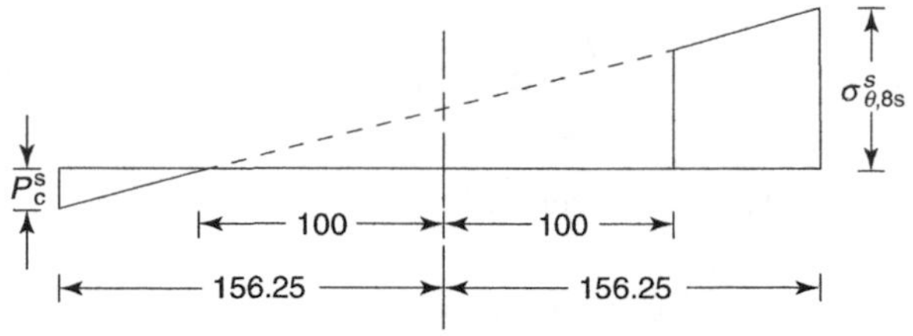

Figure 13.11 Lamé line for steel tube due to shrinkage

$$\frac{\sigma^s{}_{\theta,8s}}{(400+100)} = \frac{P_c^s}{(156.25-100)}$$

Therefore $$\sigma^s{}_{\theta,8s} = \frac{P_c^s(156.25+100)}{(156.25-100)}$$

i.e. $$\boldsymbol{\sigma^s{}_{\theta,8s} = 4.556 P_c^s} \qquad (13.27)$$

Consider now the stresses due to pressure alone

Let P = internal pressure

P_c^p = pressure at the common surface due to pressure alone

For the aluminium-alloy tube, the Lamé line due to pressure alone will be as shown in Figure 13.12.

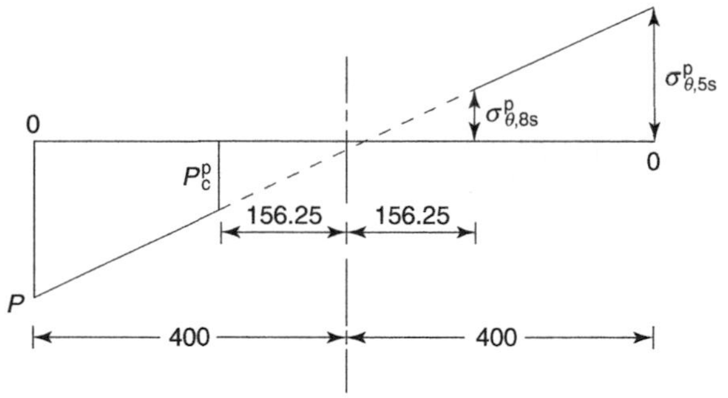

Figure 13.12 Lamé line in aluminium alloy due to pressure alone

Equating similar triangles in Figure 13.12 gives:

$$\frac{P-P_c^p}{(400-156.25)} = \frac{\sigma^p{}_{\theta,8a}+P}{(400+156.25)}$$

i.e. $$\frac{150-P_c^p}{243.75} = \frac{\sigma^p{}_{\theta,8a}+150}{556.25}$$

i.e. $556.25\left(150 - P_c^p\right) = 243.75\left(\sigma^p{}_{\theta,8a} + 150\right)$

i.e. $83437.5 - 556.25P_c^p = 243.75\sigma^p{}_{\theta,8a} + 36562.5$

i.e. $243.75\sigma^p{}_{\theta,8a} = 83437.5 - 36562.5 - 556.25P_c^p$

i.e. $243.75\sigma^p{}_{\theta,8a} = 46875 - 556.25P_c^p$

and $$\sigma^p{}_{\theta,8a} = \frac{46875}{243.75} - \frac{556.25P_c^p}{243.75}$$

i.e. $$\boldsymbol{\sigma^p{}_{\theta,8a} = 192.3 - 2.282P_c^p} \qquad (13.28)$$

Similarly $$\frac{P - P_c^p}{(400 - 156.25)} = \frac{\sigma^p{}_{\theta,5a} + P}{(400 + 400)}$$

i.e. $$\frac{150 - P_c^p}{243.75} = \frac{\sigma^p{}_{\theta,5a} + 150}{800}$$

i.e. $800\left(150 - P_c^p\right) = 243.75\left(\sigma^p{}_{\theta,5a} + 150\right)$

i.e. $120000 - 800P_c^p = 243.75\sigma^p{}_{\theta,5a} + 36562.5$

i.e. $243.75\sigma^p{}_{\theta,5a} = 120000 - 36562.5 - 800P_c^p$

i.e. $243.75\sigma^p{}_{\theta,5a} = 83437.5 - 800P_c^p$

and $$\sigma^p{}_{\theta,5a} = \frac{83437.5}{243.75} - \frac{800P_c^p}{243.75}$$

i.e. $$\boldsymbol{\sigma^p{}_{\theta,5a} = 342.3 - 3.282P_c^p} \qquad (13.29)$$

For the steel tube due to pressure alone, the Lamé line is as shown in Figure 13.13.

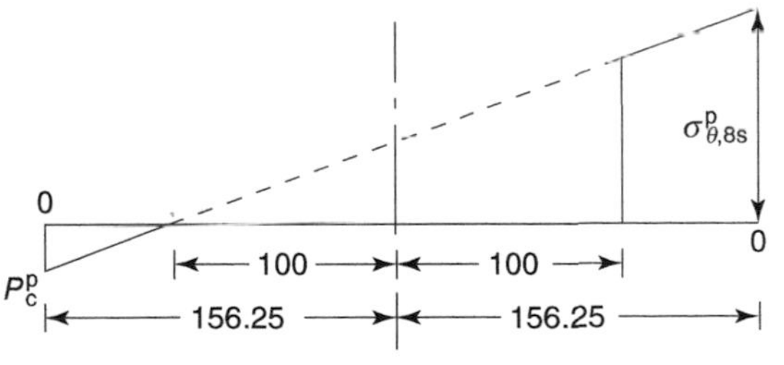

Figure 13.13 Lamé line for steel due to pressure alone

Equating similar triangles in Figure 13.13 gives:

$$\frac{\sigma^p{}_{\theta,8s}}{(156.25 + 100)} = \frac{P_c^p}{(156.25 - 100)}$$

Therefore $$\sigma^p{}_{\theta,8s} = \frac{P_c^p(156.25 + 100)}{(156.25 - 100)}$$

i.e. $$\boldsymbol{\sigma^p{}_{\theta,8s} = 4.556P_c^p} \qquad (13.30)$$

Owing to pressure alone, there is no interference fit, so that:

$$w_s^p = w_a^p$$

Now $$\frac{E_s \times w_s^p}{r_s} = \sigma^p{}_{\theta,8s} + v_s P_c^p$$

Therefore $$w_s^p = \frac{r_s}{E_s}\left(4.556P_c^p + v_s P_c^p\right)$$

$$= \frac{4 \times 10^{-2}}{2 \times 10^{11}}(4.556 + 0.3)P_c^p$$

i.e. $$\boldsymbol{w_s^p = 9.712 \times 10^{-13} P_c^p} \qquad (13.31)$$

Similarly $$\frac{E_a \times w_a^p}{r_a} = \sigma^p{}_{\theta,8a} + v_a P_c^p$$

Therefore $$w_a^p = \frac{r_a}{E_a}\left(\left(192.3 - 2.282P_c^p\right) + 0.32P_c^p\right)$$

$$= \frac{4 \times 10^{-2}}{6.7 \times 10^{10}}\left(192.3 - 1.962P_c^p\right)$$

i.e. $$\boldsymbol{w_a^p = 1.148 \times 10^{-10} - 1.171 \times 10^{-12} P_c^p} \qquad (13.32)$$

Equating equations (13.31) and (13.32) gives:

$$9.712 \times 10^{-13} P_c^p = 1.148 \times 10^{-10} - 1.171 \times 10^{-12} P_c^p$$

i.e.
$$9.712 \times 10^{-13} P_c^p + 1.171 \times 10^{-12} P_c^p = 1.148 \times 10^{-10}$$

i.e. $$2.1422 \times 10^{-12} P_c^p = 1.148 \times 10^{-10}$$

and $$P_c^p = \frac{1.148 \times 10^{-10}}{2.1422 \times 10^{-12}}$$

i.e. $$P_c^p = 53.59 \text{ MPa} \qquad (13.33)$$

Substituting equation (13.33) into equations (13.28) and (13.29) gives:

$$\sigma^p{}_{\theta,8a} = 192.3 - 2.282 \times 53.59 = 70 \text{ MPa}$$

$$\sigma^p_{\theta,5a} = 342.3 - 3.282 \times 53.59 = 166.4 \text{ MPa}$$

Now the maximum hoop stress in the inner tube lies either on its outer surface or on its inner surface, so that:

either $$\sigma^p_{\theta,8a} + \sigma^s_{\theta,8a} = 110 \quad (13.34)$$

or $$\sigma^p_{\theta,5a} + \sigma^s_{\theta,5a} = 110 \quad (13.35)$$

From equation (13.34),

$$(70) + (-2.282 P_c^s) = 110 \quad \text{from equation (13.26)}$$

i.e. $$P_c^s = \frac{70 - 110}{2.282} = -17.5 \text{ MPa}$$

and from equation (13.35),

$$166.4 + \sigma^s_{\theta,5a} = 110$$

i.e. $$166.4 + (-3.282 P_c^s) = 110 \text{ from equation (13.25)}$$

i.e. $$P_c^s = \frac{166.4 - 110}{3.282} = -17.18 \text{ MPa}$$

i.e. the required $\mathbf{P_c^s = 17.18}$ **MPa**

Hence, owing to the *combined effects of pressure and shrinkage*

$$\mathbf{P_c = P_c^p + P_c^s = 53.59 + 17.18 = 70.77 \text{ MPa}}$$

The resultant hoop stress in the steel tube on its 8 cm diameter is

$$\sigma_{\theta,8s} = 4.556 \times (P_c^s + P_c^p) = 4.556 \times (17.18 + 53.59)$$
$$= 4.556 \times 70.77 = \mathbf{322.4 \text{ MPa}}$$

Similarly, the resultant hoop stress in the aluminium-alloy tube on its 8 cm diameter is

$$\sigma_{\theta,8a} = 192.3 - 2.282 \times (P_c^s + P_c^p)$$
$$= 192.3 - 2.282 \times (17.18 + 53.59)$$
$$= 192.3 - 161.5 = \mathbf{30.8 \text{ MPa}}$$

and

$$\sigma_{\theta,5a} = 342.3 - 3.282 \times (P_c^s + P_c^p)$$
$$= 342.3 - 3.282 \times (17.18 + 53.59)$$
$$= 342.3 - 232.3 = \mathbf{110 \text{ MPa}}$$

The hoop stress distribution through the two walls is shown in Figure 13.14.

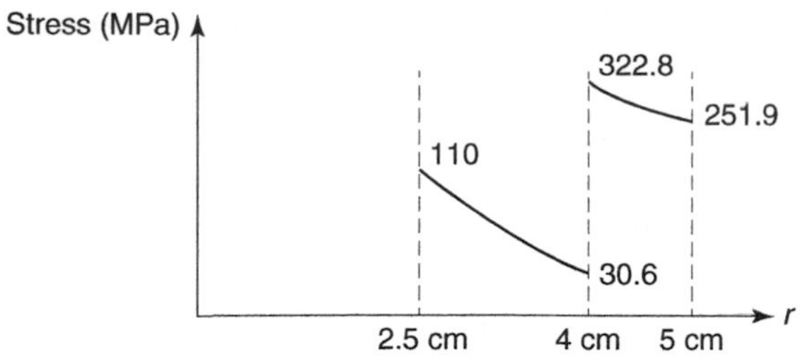

Figure 13.14 Stress distribution across the compound cylinder

Now try the following Practice Exercise

Practice Exercise 55. Compound cylinders

1. A compound cylinder is manufactured by shrinking a steel cylinder of external diameter 22 cm and internal diameter 18 cm onto another steel cylinder of internal diameter 14 cm, the dimensions being nominal. If the maximum tensile stress in the outer cylinder is 100 MPa, determine the radial compressive stress at the common surface and the interference fit at the common diameter. Determine, also the maximum stress in the inner cylinder. Assume that:

   $E_s = 2 \times 10^{11}$ N/m^2 and $\nu_s = 0.3$

   [19.8 MPa, $\delta = 0.16$ mm, −100 MPa]

2. If the inner cylinder of Problem 1 were made from bronze, what would be the value of δ?

   For bronze: $E_b = 1 \times 10^{11}$ N/m^2 and $\nu_b = 0.4$

   [$\delta = 0.226$ mm]

3. If the compound cylinder of Problem 1 were subjected to an internal pressure of 50 MPa, what would be the value of the maximum resultant stress?

   [227.9 MPa]

4. If the compound cylinder of Problem 2 were subjected to an internal pressure of 50 MPa, what would be the value of the maximum resultant stress?

   [241.1 MPa]

13.5 Plastic yielding of thick tubes

The following assumptions are made in this theory:

1. The tube is constructed from an ideally elastic-plastic material.
2. The longitudinal stress is the 'minimax' stress.
3. Yield occurs according to Tresca's criterion.

For this case, the equilibrium considerations of equation (13.12) apply

i.e. $$\sigma_\theta - \sigma_r - r\frac{d\sigma_r}{dr} = 0 \qquad (13.36)$$

Now, Tresca's criterion is that

$$\sigma_\theta - \sigma_r = \sigma_{yp}$$

i.e. $$\sigma_\theta = \sigma_{yp} + \sigma_r \qquad (13.37)$$

Substituting equation (13.37) into equation (13.36) gives:

$$\sigma_{yp} + \sigma_r - \sigma_r - r\frac{d\sigma_r}{dr} = 0$$

i.e. $$\sigma_{yp} - r\frac{d\sigma_r}{dr} = 0$$

i.e. $$\sigma_{yp} = r\frac{d\sigma_r}{dr}$$

and $$\frac{d\sigma_r}{dr} = \frac{\sigma_{yp}}{r}$$

Integrating both sides gives:

$$\int \frac{d\sigma_r}{dr}dr = \int \frac{\sigma_{yp}}{r}dr = \sigma_{yp}\int \frac{1}{r}dr$$

i.e. $$\sigma_r = \sigma_{yp}\ln r + C \qquad (13.38)$$

For the case of a partially plastic cylinder, as shown in Figure 13.15,

$$\text{at } r = r_2,\ \sigma_r = -P_2$$

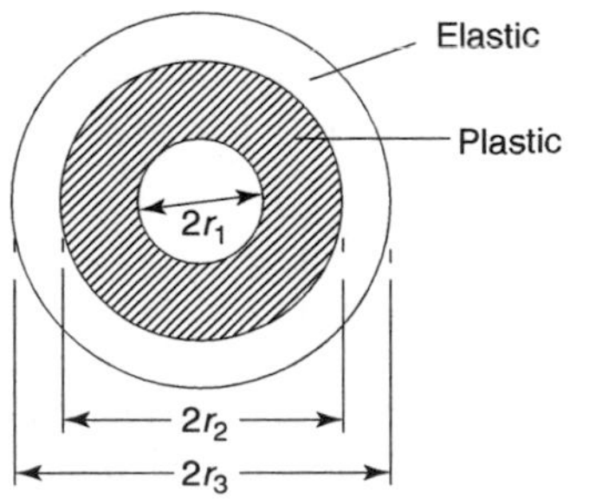

Figure 13.15 Partially plastic cylinder

Substituting the above boundary value into equation (13.38) gives:

$$-P_2 = \sigma_{yp}\ln r_2 + C$$

Therefore $$C = -\sigma_{yp}\ln r_2 - P_2$$

and from equation (13.38)

$$\sigma_r = \sigma_{yp}\ln r + \left(-\sigma_{yp}\ln r_2 - P_2\right)$$

i.e. $$\sigma_r = \sigma_{yp}\ln r - \sigma_{yp}\ln r_2 - P_2$$

i.e. $$\sigma_r = \sigma_{yp}\left(\ln r - \ln r_2\right) - P_2$$

i.e. $$\boldsymbol{\sigma_r = \sigma_{yp}\ln\left(\frac{r}{r_2}\right) - P_2} \qquad (13.39)$$

Similarly, from equation (13.37) $\sigma_\theta = \sigma_{yp} + \sigma_r$

$$= \sigma_{yp} + \sigma_{yp}\ln\left(\frac{r}{r_2}\right) - P_2$$

i.e. $$\boldsymbol{\sigma_\theta = \sigma_{yp}\left[1 + \ln\left(\frac{r}{r_2}\right)\right] - P_2} \qquad (13.40)$$

where r_1 = internal radius
r_2 = outer radius of plastic section of cylinder
r_3 = external radius
P_1 = internal pressure
P_2 = radial pressure at outer radius of plastic zone

The tube can be assumed to behave as a compound cylinder, with the internal portion behaving plastically and the external portion elastically.

The Lamé line for the elastic portion of the cylinder is shown in Figure 13.16.

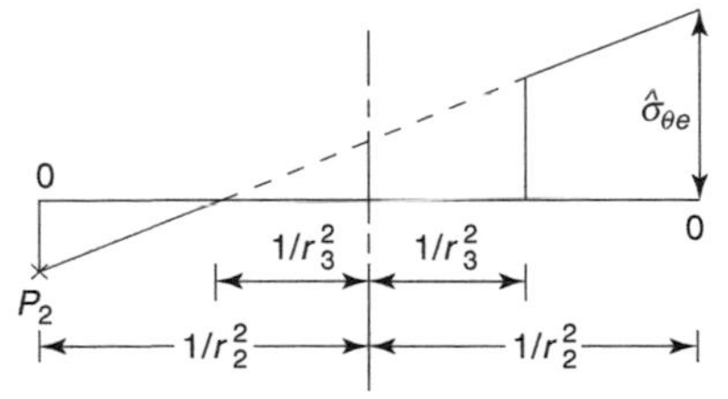

Figure 13.16 Lamé line for elastic zone

In Figure 13.16, $\hat{\sigma}_{\theta e}$ is the elastic hoop stress at $r = r_2$, so that according to Tresca's criterion, on this radius

$$\sigma_{yp} = \hat{\sigma}_{\theta e} + P_2 \qquad (13.41)$$

From Figure 13.16

$$\frac{P_2}{\left(\frac{1}{r_2^2}-\frac{1}{r_3^2}\right)}=\frac{\hat{\sigma}_{\theta e}}{\left(\frac{1}{r_2^2}+\frac{1}{r_3^2}\right)}$$

Therefore $$\hat{\sigma}_{\theta e}=\frac{P_2\left(\frac{1}{r_2^2}+\frac{1}{r_3^2}\right)}{\left(\frac{1}{r_2^2}-\frac{1}{r_3^2}\right)}=\frac{P_2\left(\frac{r_3^2+r_2^2}{r_2^2 r_3^2}\right)}{\left(\frac{r_3^2-r_2^2}{r_2^2 r_3^2}\right)}$$

i.e. $$\hat{\sigma}_{\theta e}=\frac{P_2\left(r_3^2+r_2^2\right)}{\left(r_3^2-r_2^2\right)} \tag{13.42}$$

Substituting equation (13.42) into equation (13.41) gives:

$$\sigma_{yp}=\frac{P_2\left(r_3^2+r_2^2\right)}{\left(r_3^2-r_2^2\right)}+P_2=P_2\left(1+\frac{\left(r_3^2+r_2^2\right)}{\left(r_3^2-r_2^2\right)}\right)$$

$$=P_2\left(\frac{\left(r_3^2-r_2^2\right)+\left(r_3^2+r_2^2\right)}{\left(r_3^2-r_2^2\right)}\right)$$

$$=P_2\left(\frac{2r_3^2}{\left(r_3^2-r_2^2\right)}\right)$$

Hence $$P_2=\sigma_{yp}\left(\frac{\left(r_3^2-r_2^2\right)}{2r_3^2}\right) \tag{13.43}$$

Consider now the portion of the cylinder that is plastic. Substituting equation (13.43) into equations (13.39) and (13.40), the stress distributions in the plastic zone are given by:

$$\sigma_r=\sigma_{yp}\ln\left(\frac{r}{r_2}\right)-\sigma_{yp}\left(\frac{\left(r_3^2-r_2^2\right)}{2r_3^2}\right)$$

i.e. $$\sigma_r=-\sigma_{yp}\ln\left(\frac{r_2}{r}\right)-\sigma_{yp}\left(\frac{\left(r_3^2-r_2^2\right)}{2r_3^2}\right)$$

i.e. $$\boldsymbol{\sigma_r=-\sigma_{yp}\left[\ln\left(\frac{r_2}{r}\right)+\left(\frac{\left(r_3^2-r_2^2\right)}{2r_3^2}\right)\right]} \tag{13.44}$$

and $$\sigma_\theta=\sigma_{yp}\left[1+\ln\left(\frac{r}{r_2}\right)\right]-\sigma_{yp}\left(\frac{\left(r_3^2-r_2^2\right)}{2r_3^2}\right)$$

i.e. $$\sigma_\theta=\sigma_{yp}\left\{1-\ln\left(\frac{r_2}{r}\right)+\frac{\left(r_2^2-r_3^2\right)}{2r_3^2}\right\}$$

i.e. $$\sigma_\theta=\sigma_{yp}\left\{\frac{2r_3^2}{2r_3^2}-\ln\left(\frac{r_2}{r}\right)+\frac{\left(r_2^2-r_3^2\right)}{2r_3^2}\right\}$$

i.e. $$\sigma_\theta=\sigma_{yp}\left\{-\ln\left(\frac{r_2}{r}\right)+\frac{2r_3^2+\left(r_2^2-r_3^2\right)}{2r_3^2}\right\}$$

i.e. $$\boldsymbol{\sigma_\theta=\sigma_{yp}\left[\frac{\left(r_3^2+r_2^2\right)}{2r_3^2}-\ln\left(\frac{r_2}{r}\right)\right]} \tag{13.45}$$

To find the *pressure just to cause yield,* put

$$\sigma_r=-P_1 \text{ at } r=r_1$$

where P_1 is the internal pressure that causes the onset of yield.

Therefore, from equation (13.44)

$$\boldsymbol{P_1=\sigma_{yp}\left[\ln\left(\frac{r_2}{r_1}\right)+\left(\frac{\left(r_3^2-r_2^2\right)}{2r_3^2}\right)\right]} \tag{13.46}$$

but if yield is only on the inside surface, $r_1=r_2$ in equation (13.46), so that:

$$P_1=\sigma_{yp}\left[\ln(1)+\left(\frac{\left(r_3^2-r_2^2\right)}{2r_3^2}\right)\right]$$

i.e. $$\boldsymbol{P_1=\sigma_{yp}\left[\left(\frac{\left(r_3^2-r_1^2\right)}{2r_3^2}\right)\right]} \tag{13.47}$$

To determine the *plastic collapse pressure* P_p, put $r_2=r_3$ in equation (13.46), giving:

$$P_p=\sigma_{yp}\left[\ln\left(\frac{r_3}{r_1}\right)+\left(\frac{\left(r_3^2-r_3^2\right)}{2r_3^2}\right)\right]$$

i.e. $$\boldsymbol{P_p=\sigma_{yp}\ln\left(\frac{r_3}{r_1}\right)} \tag{13.48}$$

To determine the hoop stress distribution in the plastic zone, $\sigma_{\theta p}$, it must be remembered that:

$$\sigma_{yp}=\sigma_\theta-\sigma_r$$

Therefore $$\sigma_{\theta p} = \sigma_{yp}\left[1+\ln\left(\frac{r_3}{r_1}\right)\right] \qquad (13.49)$$

Plots of the stress distributions in a partially plastic cylinder, under internal pressure, are shown in Figure 13.17.

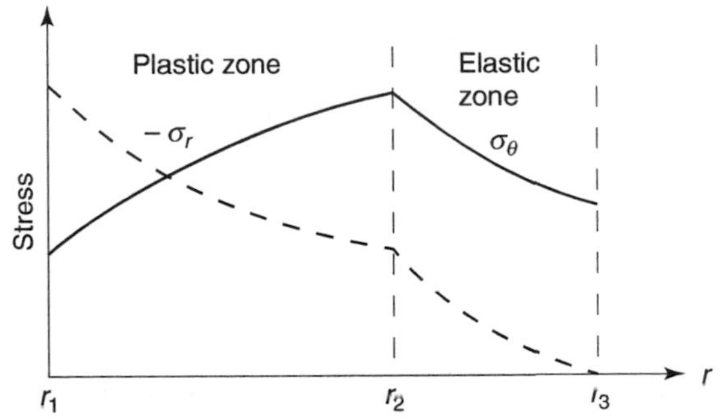

Figure 13.17 Stress distributions in a partially plastic cylinder

Problem 6. A high-tensile steel cylinder of 0.5 m outer diameter and 0.4 m inner diameter is shrunk onto a mild-steel cylinder of 0.3 m bore. If the interference fit is such that when the internal pressure is 100 MN/m^2, the inner face of the inner cylinder is on the point of yielding, determine the internal pressure which will cause plastic penetration through half the thickness of the inner cylinder. The material of the outer cylinder may be assumed to be of a higher quality, so that it does not yield, and the inner cylinder material is perfectly elastic-plastic, yielding at a constant shear stress of 140 MN/m^2. Both materials may be assumed to have the same elastic modulus and Poisson's ratio.

The Lamé line for the compound cylinder at the onset of yield is shown in Figure 13.18.

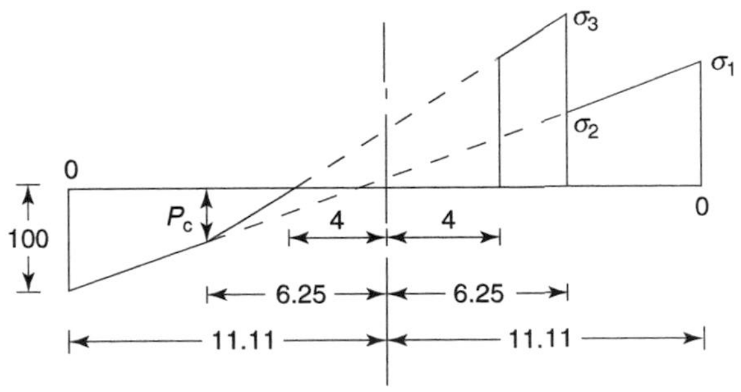

Figure 13.18 Lamé line for compound cylinder

In Figure 13.18

σ_1 = hoop stress on inner surface of inner cylinder

σ_2 = hoop stress on outer surface of inner cylinder

σ_3 = hoop stress on inner surface of outer cylinder

As yield occurs on the inner surface of the inner cylinder,

$$\frac{\sigma_1 - (-100)}{2} = \tau_{yp} = 140$$

Therefore $\sigma_1 = 140\times 2 - 100 = 180$ MPa

where τ_{yp} is the yield stress in shear.

(Note that in Figure 13.18, $\frac{1}{d_1^2}$ is $\frac{1}{0.3^2} = 11.11$, and $\frac{1}{d_2^2}$ is $\frac{1}{0.4^2} = 6.25$)

Equating similar triangles in Figure 13.18 gives:

$$\frac{\sigma_1 + 100}{11.11+11.11} = \frac{100 - P_c}{11.11 - 6.25}$$

Hence $$\frac{180+100}{22.22} = \frac{100-P_c}{4.86}$$

and $$4.86\times 280 = 22.22(100 - P_c)$$

i.e. $$1360.8 = 2222 - 22.22\,P_c$$

and $$22.22\,P_c = 2222 - 1360.8 = 861.2$$

from which $$P_c = \frac{861.2}{22.22} = \mathbf{38.76\ MPa}$$

Similarly from Figure 13.18

$$\frac{\sigma_2 + 100}{11.11+6.25} = \frac{\sigma_1+100}{11.11+11.11}$$

i.e. $$\frac{\sigma_2+100}{17.36} = \frac{280}{22.22}$$

i.e. $$22.22(\sigma_2 + 100) = 17.36\times 280$$

i.e. $$22.22\sigma_2 + 2222 = 4860.8$$

i.e. $$\sigma_2 = \frac{2638.8}{22.22} = \mathbf{118.8\ MPa} \qquad (13.50)$$

Also from Figure 13.18

$$\frac{\sigma_3}{6.25+4} = \frac{P_c}{6.25-4}$$

However, from above, $P_c = 38.76$

Hence
$$\frac{\sigma_3}{6.25+4} = \frac{38.76}{6.25-4}$$

i.e.
$$\frac{\sigma_3}{10.25} = \frac{38.76}{2.25}$$

and
$$2.25\sigma_3 = 10.25 \times 38.76$$

from which $\sigma_3 = \dfrac{10.25 \times 38.76}{2.25} = \mathbf{176.6\ MPa}$ (13.51)

Consider, now, plastic penetration of the inner cylinder to a diameter of 0.35 m. The Lamé line in the elastic zones will be as shown in Figure 13.19.

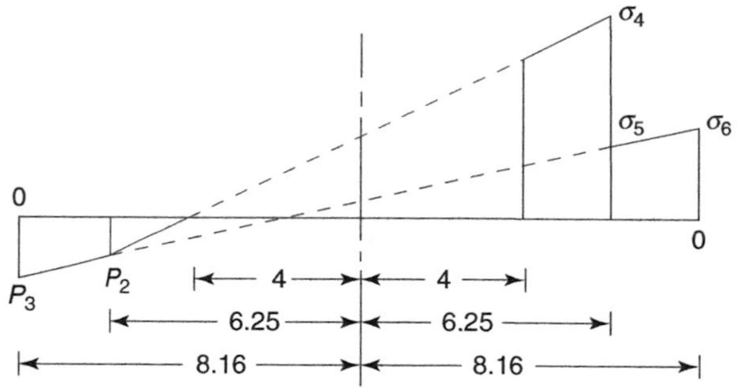

Figure 13.19 Lamé line in elastic zones

From this figure

$$\frac{\sigma_6 + P_3}{2} = 140$$

Therefore
$$\sigma_6 = 280 - P_3 \qquad (13.52)$$

(Note that in Figure 13.19, $\dfrac{1}{d_1^2}$ is $\dfrac{1}{0.35^2} = 8.16$, and $\dfrac{1}{d_2^2}$ is $\dfrac{1}{0.4^2} = 6.25$)

Similarly
$$\frac{P_3 - P_2}{8.16-6.25} = \frac{\sigma_6 + P_3}{8.16 \times 2} = \frac{(280 - P_3) + P_3}{16.32}$$

i.e.
$$\frac{P_3 - P_2}{1.91} = \frac{280}{16.32}$$

from which
$$P_3 - P_2 = \frac{280 \times 1.91}{16.32} = 32.77$$

or
$$P_2 = P_3 - 32.77 \qquad (13.53)$$

Also from Figure 13.19

$$\frac{\sigma_4}{6.25+4} = \frac{P_2}{6.25-4}$$

i.e.
$$\frac{\sigma_4}{10.25} = \frac{P_2}{2.25}$$

and
$$2.25\sigma_4 = 10.25P_2$$

from which
$$\sigma_4 = \frac{10.25}{2.25}P_2$$

i.e.
$$\boldsymbol{\sigma_4 = 4.556P_2} \qquad (13.54)$$

Substituting equation (13.53) into equation (13.54) gives:

$$\sigma_4 = 4.556(P_3 - 32.77)$$

i.e.
$$\boldsymbol{\sigma_4 = 4.556P_3 - 149.3} \qquad (13.55)$$

and
$$\frac{\sigma_5 + P_3}{8.16+6.25} = \frac{280}{16.32}$$

i.e.
$$16.32(\sigma_5 + P_3) = 280(8.16+6.25) = 4034.8$$

i.e.
$$16.32\sigma_5 + 16.32P_3 = 4034.8$$

and
$$\sigma_5 = \frac{4034.8}{16.32} - \frac{16.32}{16.32}P_3$$

i.e.
$$\boldsymbol{\sigma_5 = 247.2 - P_3} \qquad (13.56)$$

Consider strains during the additional pressurisation. Now,

$$w = \frac{r}{E}(\sigma_\theta - \nu\sigma_r)$$

which will be the same for both cylinders at the common surface,

i.e.
$$\frac{1}{E}\left[(\sigma_5 - \sigma_2) - \nu(P_2 - P_c)\right] = \left(\frac{1}{E}(\sigma_4 - \sigma_3) - \nu(P_2 - P_c)\right)$$

or
$$(\sigma_5 - \sigma_2) = (\sigma_4 - \sigma_3) \qquad (13.57)$$

Substituting equations (13.50), (13.51), (13.55) and (13.56) into equation (13.57) gives:

$$247.2 - P_3 - 118.8 = 4.556P_3 - 149.3 - 176.6$$

i.e. $247.2 - 118.8 + 149.3 + 176.6 = 5.556P_3$

i.e. $454.3 = 5.556P_3$

and $\mathbf{P_3 = \dfrac{454.3}{5.556} = 81.77\ MPa}$

Consider the *yielded portion.*

Now, $\sigma_r = \sigma_{yp} \ln r + C$ from equation (13.38)

and $\sigma_{yp} = 280$ MPa

At $r = 0.175$ $\sigma_r = -P_3 = -81.77$

Hence $-81.77 = 280 \ln 0.175 + C$

i.e. $-81.77 = -488.03 + C$

from which $C = 406.26$

At $r = 0.15$ $\sigma_r = -P$

Hence $-P = 280 \ln(0.15) + 406.3$

i.e. $-P = -531.19 + 406.3$

from which $\mathbf{P = 124.9\ MPa}$

i.e. **pressure to cause plastic penetration = 124.9 MPa**

Problem 7. Determine the internal pressure that will cause total plastic failure of the compound vessel of Problem 6, given that σ_{yp} for the outer cylinder is 600 MPa and that E and v are the same for both cylinders.

Now $P_p = \sigma_{yp} \ln\left(\dfrac{r_3}{r_1}\right)$

$$= 280 \ln\left(\frac{0.2}{0.15}\right) + 600 \ln\left(\frac{0.25}{0.2}\right)$$

$$= 80.55 + 133.89$$

i.e. $\mathbf{P_p = 214.44\ MPa}$

i.e. **plastic collapse pressure = 214.44 MPa**

Now try the following Practice Exercise

Practice Exercise 56. Plastic yielding of thick tubes

1. A thick compound cylinder consists of a brass cylinder of internal diameter 0.1 m and external diameter 0.2 m, surrounded by a steel cylinder of external diameter 0.3 m and of the same length. If the compound cylinder is subjected to a compressive axial load of 5 MN, and the axial strain is constant for both cylinders, determine the pressure at the common surface and the longitudinal stresses in the two cylinders due to this load.

The following assumptions may be made:

(a) σ_{Ls} = a longitudinal stress in steel cylinder = a constant

(b) σ_{Lb} = longitudinal stress in brass cylinder = a constant

For steel $E_s = 2 \times 10^{11}\ \mathrm{N/m^2}$ and $v_s = 0.3$

For brass $E_b = 1 \times 10^{11}\ \mathrm{N/m^2}$ and $v_b = 0.4$

[2.20 MPa; $\sigma_{Ls} = -96.52$ MPa; $\sigma_{Lb} = -51.14$ MPa]

13.6 Thick spherical shells

Consider a hemispherical element of a thick spherical shell at any radius r under a compressive radial stress P, as shown in Figure 13.20.

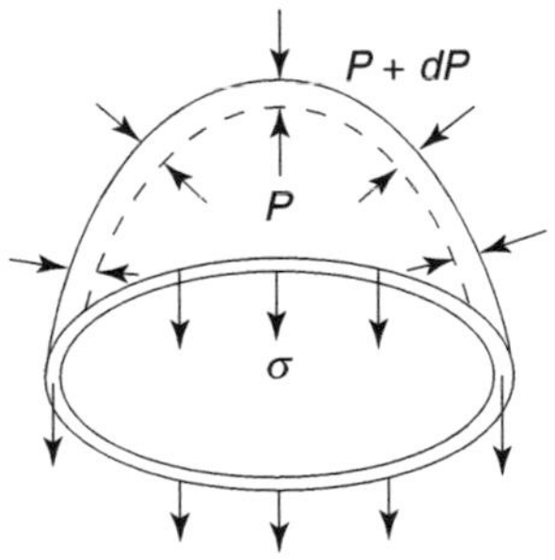

Figure 13.20 Hemispherical shell element

Let w be the radial deflection at any radius r so that

hoop strain = w/r

and radial strain = dw/dr

From three-dimensional stress-strain relationships

$$E\frac{w}{r} = \sigma - v\sigma + vP \quad (13.58)$$

and

$$E\frac{dw}{dr} = -P - v\sigma - v\sigma$$

$$= -P - 2v\sigma \quad (13.59)$$

Multiplying equation (13.58) by r gives:

$$Ew = \sigma r - v\sigma r + v\,Pr$$

and differentiating w.r.t. r gives:

$$E\frac{dw}{dr} = \sigma + r\frac{d\sigma}{dr} - v\sigma - vr\frac{d\sigma}{dr} + vP + vr\frac{dP}{dr}$$

$$= (1-v)\left(\sigma + r\frac{d\sigma}{dr}\right) + v\left(P + r\frac{dP}{dr}\right) \quad (13.60)$$

Equating equations (13.59) and (13.60) gives:

$$-P - 2v\sigma = (1-v)\left(\sigma + r\frac{d\sigma}{dr}\right) + v\left(P + r\frac{dP}{dr}\right)$$

or $$-P - 2v\sigma = \sigma + r\frac{d\sigma}{dr} - v\sigma - vr\frac{d\sigma}{dr} + vP + vr\frac{dP}{dr}$$

or $$0 = P + 2v\sigma + \sigma + r\frac{d\sigma}{dr} - v\sigma - vr\frac{d\sigma}{dr} + vP + vr\frac{dP}{dr}$$

i.e. $$(1+v)(\sigma + P) + r(1-v)\frac{d\sigma}{dr} + vr\frac{dP}{dr} = 0 \quad (13.61)$$

Considering now the equilibrium of the hemispherical shell element

$$\sigma \times 2\pi r \times dr = P \times \pi r^2 - (P + dP) \times \pi \times (r + dr)^2 \quad (13.62)$$

Neglecting higher-order terms, equation (13.62) becomes:

$$\sigma + P = \left(-\frac{r}{2}\right)\frac{dP}{dr} \quad (13.63)$$

Substituting equation (13.63) into equation (13.61) gives:

$$\left(-\frac{r}{2}\right)\frac{dP}{dr}(1+v) + r(1-v)\frac{d\sigma}{dr} + vr\frac{dP}{dr} = 0$$

or $$r(1-v)\frac{d\sigma}{dr} = -vr\frac{dP}{dr} + \left(\frac{r}{2}\right)\frac{dP}{dr}(1+v)$$

i.e. $$r(1-v)\frac{d\sigma}{dr} = -vr\frac{dP}{dr} + \frac{r}{2}\frac{dP}{dr} + \frac{vr}{2}\frac{dP}{dr}$$

i.e. $$r(1-v)\frac{d\sigma}{dr} = \frac{r}{2}\frac{dP}{dr} - \frac{vr}{2}\frac{dP}{dr}$$

i.e. $$r(1-v)\frac{d\sigma}{dr} = \frac{r}{2}(1-v)\frac{dP}{dr}$$

i.e. $$\frac{d\sigma}{dr} = \frac{1}{2}\frac{dP}{dr}$$

or $$\frac{d\sigma}{dr} - \frac{1}{2}\frac{dP}{dr} = 0 \quad (13.64)$$

which on integrating becomes:

$$\int \frac{d\sigma}{dr}dr - \frac{1}{2}\int \frac{dP}{dr}dr = 0$$

i.e. $$\sigma - \frac{1}{2}P = A \quad (13.65)$$

Substituting equation (13.65) into equation (13.63) gives:

$$\frac{1}{2}P + A + P = \left(-\frac{r}{2}\right)\frac{dP}{dr}$$

or $$\frac{3}{2}P + A = \left(-\frac{r}{2}\right)\frac{dP}{dr}$$

or $$3P + 2A = -r\frac{dP}{dr}$$

or $$-r\frac{dP}{dr} - 3P = 2A$$

The left-hand side of this equation may be shown to be the same as: $-\frac{1}{r^2}\frac{d(Pr^3)}{dr}$

i.e. using the product rule of differentiation:

$$-\frac{1}{r^2}\frac{d(Pr^3)}{dr} = -\frac{1}{r^2}\left[P(3r^2) + r^3\frac{dP}{dr}\right] = -3P - r\frac{dP}{dr}$$

Hence $$-\frac{1}{r^2}\frac{d(Pr^3)}{dr} = 2A$$

or $$\frac{d\left(Pr^3\right)}{dr} = -2Ar^2$$

which on integrating becomes:

$$\int \frac{d\left(Pr^3\right)}{dr} dr = \int -2Ar^2\, dr$$

i.e. $$Pr^3 = -2A\frac{r^3}{3} + B$$

or $$P = -\frac{2}{3}A + \frac{B}{r^3} \qquad (13.66)$$

and from equation (13.65) $\sigma = \frac{1}{2}P + A$

i.e. $$\sigma = \frac{1}{2}\left(-\frac{2}{3}A + \frac{B}{r^3}\right) + A$$

i.e. $$\sigma = -\frac{1}{3}A + \frac{B}{2r^3} + A$$

i.e. $$\boldsymbol{\sigma = \frac{2}{3}A + \frac{B}{2r^3}} \qquad (13.67)$$

13.7 Rotating discs

These are a common feature in engineering, which from time to time suffer failure due to high-speed rotation. In this section, equations will be obtained for calculating hoop and radial stresses, and the theory will be extended for calculating angular velocities to cause plastic collapse of rotating discs and rings.

Consider a uniform thickness disc, of density ρ, rotating at a constant angular velocity ω.

From Section 13.2

$$E\frac{dw}{dr} = \sigma_r - \nu\sigma_\theta \qquad (13.68)$$

and $$E\frac{w}{r} = \sigma_\theta - \nu\sigma_r \qquad (13.69)$$

or $$Ew = \sigma_\theta \times r - \nu\sigma_r \times r \qquad (13.70)$$

Differentiating equation (13.70) w.r.t. r using the product rule, gives:

$$E\frac{dw}{dr} = \sigma_\theta + r\frac{d\sigma_\theta}{dr} - \nu\sigma_r - \nu r\frac{d\sigma_r}{dr} \qquad (13.71)$$

Equating equations (13.68) and (13.71) gives:

$$\sigma_r - \nu\sigma_\theta = \sigma_\theta + r\frac{d\sigma_\theta}{dr} - \nu\sigma_r - \nu r\frac{d\sigma_r}{dr}$$

and $$-\sigma_r + \nu\sigma_\theta + \sigma_\theta + r\frac{d\sigma_\theta}{dr} - \nu\sigma_r - \nu r\frac{d\sigma_r}{dr} = 0$$

from which

$$\left(\sigma_\theta - \sigma_r\right)\left(1+\nu\right) + r\frac{d\sigma_\theta}{dr} - \nu r\frac{d\sigma_r}{dr} = 0 \qquad (13.72)$$

Considering equilibrium of an element of the disc, as shown in Figure 13.21

$$2\sigma_\theta \times dr \times \sin\left(\frac{d\theta}{2}\right) + \sigma_1 \times r \times d\theta$$
$$-\left(\sigma_r + d\sigma_r\right)\left(r + dr\right)d\theta = \rho \times \omega^2 \times r^2 \times dr \times d\theta$$

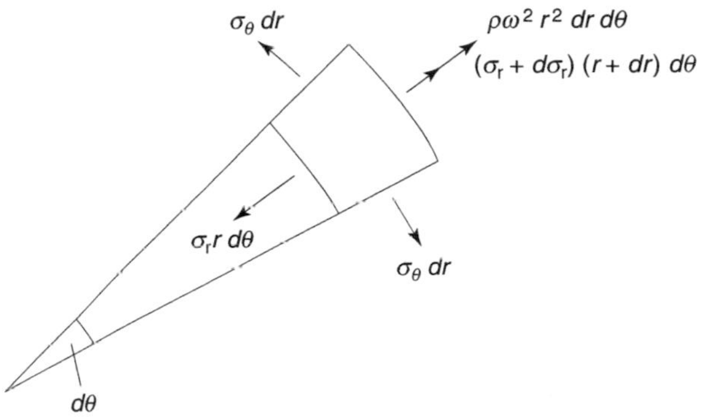

Figure 13.21 Element of disc

In the limit, this reduces to:

$$\sigma_\theta - \sigma_r - r\frac{d\sigma_r}{dr} = \rho\omega^2 r^2$$

or $$\left(\sigma_\theta - \sigma_r\right) = r\frac{d\sigma_r}{dr} + \rho\omega^2 r^2 \qquad (13.73)$$

Substituting equation (13.73) into equation (13.72) gives:

$$\left(r\frac{d\sigma_r}{dr} + \rho\omega^2 r^2\right)\left(1+\nu\right) + r\frac{d\sigma_\theta}{dr} - \nu r\frac{d\sigma_r}{dr} = 0$$

i.e. $$r\frac{d\sigma_r}{dr} + \rho\omega^2 r^2 + \nu r\frac{d\sigma_r}{dr} + \nu\rho\omega^2 r^2$$
$$+ r\frac{d\sigma_\theta}{dr} - \nu r\frac{d\sigma_r}{dr} = 0$$

i.e. $$\frac{d\sigma_r}{dr} + \rho\omega^2 r + \nu\rho\omega^2 r + \frac{d\sigma_\theta}{dr} = 0$$

and $$\frac{d\sigma_\theta}{dr} + \frac{d\sigma_r}{dr} = -\rho\omega^2 r - v\rho\omega^2 r$$

or $$\frac{d\sigma_\theta}{dr} + \frac{d\sigma_r}{dr} = -\rho\omega^2 r(1+v)$$

which on integrating becomes:

$$\sigma_\theta + \sigma_r = -\left(\frac{\rho\omega^2 r^2}{2}\right)(1+v) + 2A \quad (13.74)$$

Subtracting equation (13.73) from equation (13.74) gives:

$$\sigma_\theta + \sigma_r - (\sigma_\theta - \sigma_r) = -\left(\frac{\rho\omega^2 r^2}{2}\right)(1+v) + 2A - \left(r\frac{d\sigma_r}{dr} + \rho\omega^2 r^2\right)$$

i.e. $$2\sigma_r = -\left(\frac{\rho\omega^2 r^2}{2}\right)(1+v) + 2A - r\frac{d\sigma_r}{dr} - \rho\omega^2 r^2$$

i.e. $$2\sigma_r + r\frac{d\sigma_r}{dr} = -\frac{\rho\omega^2 r^2}{2} - \rho\omega^2 r^2 - \frac{v\rho\omega^2 r^2}{2} + 2A$$

i.e. $$2\sigma_r + r\frac{d\sigma_r}{dr} = -\frac{3\rho\omega^2 r^2}{2} - \frac{v\rho\omega^2 r^2}{2} + 2A$$

i.e. $$2\sigma_r + r\frac{d\sigma_r}{dr} = -\left(\frac{\rho\omega^2 r^2}{2}\right)(3+v) + 2A$$

The left-hand side may be written as: $\frac{1}{r}\frac{d(\sigma_r \times r^2)}{dr}$

since by the product rule of differentiation

$$\frac{1}{r}\frac{d(\sigma_r \times r^2)}{dr} = \frac{1}{r}\left(\sigma_r(2r) + r^2\frac{d\sigma_r}{dr}\right) = 2\sigma_r + r\frac{d\sigma_r}{dr}$$

Hence $$\frac{1}{r}\frac{d(\sigma_r \times r^2)}{dr} = -\frac{\rho\omega^2 r^2(3+v)}{2} + 2A$$

and $$\frac{d(\sigma_r \times r^2)}{dr} = -\frac{\rho\omega^2 r^3(3+v)}{2} + 2rA$$

which on integrating becomes:

$$\sigma_r \times r^2 = -\left(\frac{\rho\omega^2 r^4}{8}\right)(3+v) + Ar^2 - B$$

or $$\sigma_r = A - \frac{B}{r^2} - (3+v)\left(\frac{\rho\omega^2 r^2}{8}\right) \quad (13.75)$$

If instead of subtracting equation (13.73) from equation (13.74), the two equations are added, then the following may be obtained:

$$\sigma_\theta = A + \frac{B}{r^2} - (1+3v)\left(\frac{\rho\omega^2 r^2}{8}\right) \quad (13.76)$$

Problem 8. Obtain an expression for the radial variation in the thickness of a disc, so that it will be of constant strength when it is rotated at an angular velocity ω.

Let t_0 = thickness at centre

t = thickness at a radius r

$t + dt$ = thickness at a radius $r + dr$

σ = stress = constant (everywhere)

Consider the equilibrium of an element of this disc at any radius r, as shown in Figure 13.22.

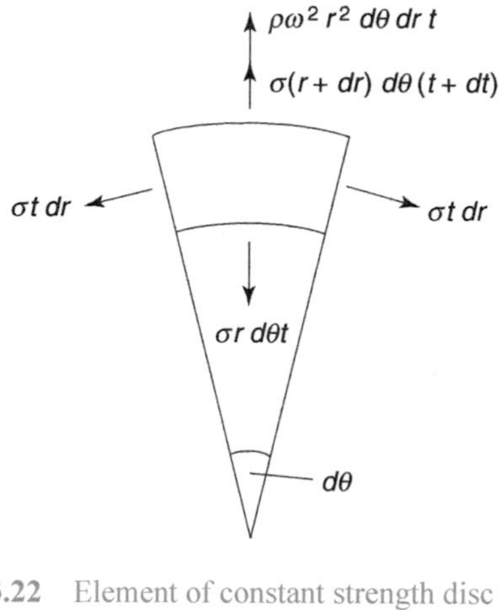

Figure 13.22 Element of constant strength disc

Resolving radially in Figure 13.22 gives:

$$2\sigma \times t \times dr\, sin\left(\frac{d\theta}{2}\right) + \sigma \times t \times r \times d\theta$$
$$= \sigma(r+dr)(t+dt)d\theta + \rho \times \omega^2 \times r^2 \times t \times d\theta \times dr$$

In the limit, this reduces to:

$$\sigma\, t\, dt = \sigma\, r\, dt + \sigma\, t\, dr + \rho\omega^2 rtdr$$

or $$\frac{dt}{dr} = -\frac{\rho\omega^2 rt}{\sigma}$$

which on integrating becomes:

$$\int \frac{dt}{dr}dr = \int -\frac{\rho\omega^2 r t}{\sigma}dr$$

i.e. $$t = -\frac{\rho\omega^2 r^2 t}{2\sigma} + K$$

If $K = \ln C$, then $$t = -\frac{\rho\omega^2 r^2 t}{2\sigma} + \ln C$$

or $$t = Ce^{-\frac{\rho\omega^2 r^2 t}{2\sigma}}$$

Now, at $r = 0$, $t = t_0 = C$,

therefore, $$\boldsymbol{t = t_0 e^{-\frac{\rho\omega^2 r^2 t}{2\sigma}}} \quad (13.77)$$

13.8 Plastic collapse of discs

Assume that $\sigma_\theta > \sigma_r$, and that plastic collapse occurs when

$$\sigma_\theta = \sigma_{yp}$$

Let R be the external radius of the disc. From *equilibrium considerations*

$$\sigma_{yp} - \sigma_r - r\frac{d\sigma_r}{dt} = \rho\omega^2 r^2$$

i.e. $$r\frac{d\sigma_r}{dt} = \sigma_{yp} - \sigma_r - \rho\omega^2 r^2$$

or $$\int r\,d\sigma_r = \int\left(\sigma_{yp} - \sigma_r - \rho\omega^2 r^2\right)dr$$

Integrating (the left-hand side using integration by parts) gives:

$$r\sigma_r - \int \sigma_r\,dr = \sigma_{yp} r - \int \sigma_r\,dr - \frac{\rho\omega^2 r^3}{3} + A$$

Therefore $$r\sigma_r = \sigma_{yp} r - \frac{\rho\omega^2 r^3}{3} + A$$

and $$\boldsymbol{\sigma_r = \sigma_{yp} - \frac{\rho\omega^2 r^2}{3} + \frac{A}{r}} \quad (13.78)$$

For a *solid disc*, at $r = 0$, $\sigma_r \neq \infty$

Therefore $\boldsymbol{A = 0}$

and $$\sigma_r = \sigma_{yp} - \frac{\rho\omega^2 r^2}{3}$$

At $r = R$, $\sigma_r = 0$

Therefore $$0 = \sigma_{yp} - \frac{\rho\omega^2 R^2}{3}$$

i.e. $$\sigma_{yp} = \frac{\rho\omega^2 R^2}{3}$$

and $$\omega = \sqrt{\frac{3\sigma_{yp}}{\rho R^2}}$$

i.e. $$\boldsymbol{\omega = \frac{1}{R}\sqrt{\frac{3\sigma_{yp}}{\rho}}} \quad (13.79)$$

where ω is the angular velocity of the disc, which causes plastic collapse.

For an *annular disc* of internal radius R_1 and external radius R_2, suitable boundary values for equation (13.78) are as follows:

At $r = R_1$, $\sigma_r = 0$

Therefore $$0 = \sigma_{yp} - \frac{\rho\omega^2 R_1^2}{3} + \frac{A}{R_1}$$

i.e. $$\frac{A}{R_1} = \frac{\rho\omega^2 R_1^2}{3} - \sigma_{yp}$$

and $$A = \left(\frac{\rho\omega^2 R_1^2}{3} - \sigma_{yp}\right)R_1$$

Thus, from equation (13.78)

$$\sigma_r = \sigma_{yp} - \frac{\rho\omega^2 r^2}{3} + \left(\frac{\rho\omega^2 R_1^2}{3} - \sigma_{yp}\right)\left(\frac{R_1}{r}\right) \quad (13.80)$$

At $r = R_2$, $\sigma_r = 0$

Therefore $$0 = \sigma_{yp} - \frac{\rho\omega^2 R_2^2}{3} + \left(\frac{\rho\omega^2 R_1^2}{3} - \sigma_{yp}\right)\left(\frac{R_1}{R_2}\right)$$

i.e. $$0 = \sigma_{yp} - \frac{\rho\omega^2 R_2^2}{3} + \frac{\rho\omega^2 R_1^3}{3R_2} - \sigma_{yp}\left(\frac{R_1}{R_2}\right)$$

i.e. $$\frac{\rho\omega^2 R_2^2}{3} - \frac{\rho\omega^2 R_1^3}{3R_2} = \sigma_{yp} - \sigma_{yp}\left(\frac{R_1}{R_2}\right)$$

and $$\omega^2\left(\frac{\rho R_2^2}{3} - \frac{\rho R_1^3}{3R_2}\right) = \sigma_{yp}\left(1 - \frac{R_1}{R_2}\right)$$

from which $$\omega^2\rho\left(\frac{R_2^3 - R_1^3}{3R_2}\right) = \sigma_{yp}\left(\frac{R_2 - R_1}{R_2}\right)$$

and

$$\omega^2 = \frac{3R_2\sigma_{yp}}{\rho}\left(\frac{R_2 - R_1}{R_2\left(R_2^3 - R_1^3\right)}\right) = \frac{3\sigma_{yp}}{\rho}\left(\frac{R_2 - R_1}{R_2^3 - R_1^3}\right)$$

from which $$\omega = \sqrt{\frac{3\sigma_{yp}}{\rho}\left(\frac{R_2 - R_1}{R_2^3 - R_1^3}\right)} \qquad (13.81)$$

13.9 Rotating rings

Consider the equilibrium of the semi-circular ring element shown in Figure 13.23.

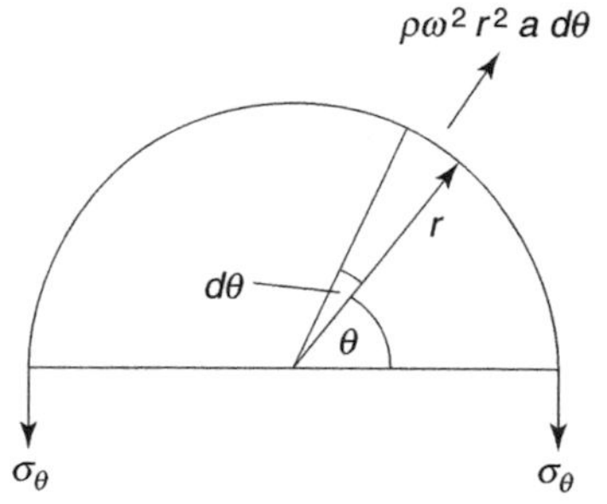

Figure 13.23 Ring element

Let a be the cross-sectional area of the ring. Then, *resolving vertically,*

$$\sigma_\theta \times a \times 2 = \int_0^\pi \rho\omega^2 r^2 a\, d\theta \sin\theta$$

$$= \rho\omega^2 r^2 a\left[-\cos\theta\right]_0^\pi$$

$$= \rho\omega^2 r^2 a\left[(-\cos\pi) - (-\cos 0)\right]$$

$$= \rho\omega^2 r^2 a\left[(--1) - (-1)\right]$$

i.e. $$\sigma_\theta \times a \times 2 = 2\rho\omega^2 r^2 a$$

Therefore $$\sigma_\theta = \rho\omega^2 r^2$$

and $$\omega = \sqrt{\frac{\sigma_\theta}{\rho r^2}} = \frac{1}{r}\sqrt{\frac{\sigma_\theta}{\rho}}$$

At collapse $$\sigma_\theta = \sigma_{yp}$$

Therefore $$\omega = \frac{1}{r}\sqrt{\frac{\sigma_{yp}}{\rho}} \qquad (13.82)$$

where ω is the angular velocity required to fracture the ring.

13.10 Design of the 'Trieste' to conquer the Mariana Trench

Problem 9. August Picard's 'Trieste', a spherical pressure vessel, had an internal diameter of 6 feet (i.e. 1.829 m) and a wall thickness of 6 inches (i.e. 15.24 cm). Determine the maximum stress in her when she conquered the Mariana Trench where depth = 7.16 miles (i.e. 11.523 km), density, σ = 1020 kg/m³, and acceleration due to gravity, g = 9.81 m/s²

Now $$p = -\frac{2A}{3} + \frac{B}{r^3} \qquad (1)$$

and $$\sigma = \frac{2A}{3} + \frac{B}{2r^3} \qquad (2)$$

On the outer surface, $p = -\rho g h$

$$= 1020 \times 9.81 \times 11.523 \times 10^3$$
$$= -115.3 \text{ MPa} \qquad (3)$$

On the inner surface, $p = 0$ (4)

Substituting equation (3) into equation (1) gives:

$$-115.3\times10^6 = -\frac{2A}{3} + \frac{B}{\left(\frac{1.829}{2} + 0.152\right)^3}$$

$$= -0.667\,A + \frac{B}{1.0665^3}$$

i.e. $$-115.3\times10^6 = -0.667A + 0.824B \qquad (5)$$

Substituting equation (4) into equation (1) gives:

$$0 = -0.667A + \frac{B}{(1.829/2)^3}$$

i.e. $$0 = -0.667\,A + 1.308\,B \qquad (6)$$

Equation (5) – equation (6) gives:

$$-115.3\times 10^6 = (0.824 - 1.308)B = -0.484\,B$$

from which $B = \dfrac{-115.3\times 10^6}{-0.484} = 238.2\times 10^6$ (7)

Equation (5) + equation (6) gives:

$$-115.3\times 10^6 = -1.332A + 2.132B \quad (8)$$

Substituting equation (7) into equation (8) gives:

$$-115.3\times 10^6 = -1.332A + 2.132\times 238.2\times 10^6$$

from which

$$-115.3\times 10^6 = -1.332A + 507.84\times 10^6$$

and $-115.3\times 10^6 - 507.84\times 10^6 = -1.332A$

i.e. $-623.14\times 10^6 = -1.332A$

and $\dfrac{-623.14\times 10^6}{-1.332} = A$

i.e. $A = 467.82\times 10^6$

Maximum, stress, $\hat{\sigma}$ occurs on the inside, i.e. $r = 1.829/2 = 0.9145$ m

Hence, from equation (1),

$$\hat{\sigma} = \frac{2A}{3} + \frac{B}{2r^3}$$

i.e. $\hat{\sigma} = \dfrac{2\times 467.82\times 10^6}{3} + \dfrac{238.2\times 10^6}{2(0.9145)^3}$

$$= 311.88\times 10^6 + 155.73\times 10^6$$

i.e. maximum stress, $\hat{\sigma} = 467.55\times 10^6$ Pa

= 467.55 MPa

= 4675.5 bar

= 67796.25 lbf/in²

= 30.3 tonf/in²

For fully worked solutions to each of the problems in Exercise 54 to 56 in this chapter, go to the website:

www.routledge.com/cw/bird

Chapter 14

The buckling of struts

Why it is important to understand: **The buckling of struts**

In a complex structure, analysis is very often carried out by computer methods using small-deflection elastic theory. In general, this results in many of the components of the structure being in axial compression. If these elements of the structure are slim, then there is a possibility of these components buckling at very small axial loads. Very often the axial stresses may be small, but because they are in compression, the structural components catastrophically fail by buckling or instability. If the component is long and slender, then it can fail elastically, but if it is short, it can fail inelastically and in these cases, the elastic theory has to be modified to cater for inelastic instability. Both elastic instability and inelastic instability are considered in this chapter.

At the end of this chapter you should be able to:

- define struts, beam-ties and beam-columns
- define very short struts, very long and slender struts and intermediate struts
- appreciate the elastic instability of very long slender struts
- determine the Euler buckling load for axially loaded struts
- understand struts with various boundary conditions
- appreciate the limit of application of Euler theory
- appreciate and use the Rankine-Gordon formula
- state the Johnson parabolic formula
- understand eccentrically loaded struts and calculate maximum permissible eccentrity
- appreciate struts with initial curvature
- determine the maximum deflection and bending moment for a strut with initial sinusoidal curvature
- derive the Perry-Robertson formula
- calculate the maximum permissible load for a strut using BS449

Mathematical references

In order to understand the theory involved in this chapter on **the buckling of struts**, knowledge of the following mathematical topics is required: ***differentiation, second order differential equations, trigonometry, double angles and quadratic equations.*** If help/revision is needed in these areas, see page 50 for some textbook references.

14.1 Introduction

A *strut* in its most usual form, can be described as a column under axial compression, as shown in Figure 14.1.

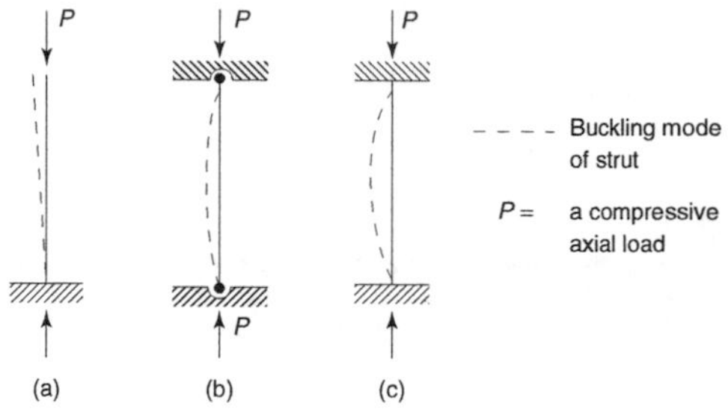

Figure 14.1 Some axially loaded struts: (a) one end free and the other fixed; (b) both ends pinned; (c) both ends clamped

From Figure 14.1, it can be seen that despite the fact that struts are loaded axially, they fail owing to bending moments caused by lateral movement of the struts. The reason for this is that under increasing axial compression, the bending resistance of the strut decreases until a point is reached where the bending stiffness of the strut is so small that the slightest offset of load, or geometrical imperfection of the strut, causes catastrophic buckling failure (or *instability*).

Similarly for a beam (or a length of rubber) under increasing axial tension, its bending stiffness increases until a point is reached where failure occurs. Beams or rods under tension are called *beam-ties* or *ties,* respectively, and beams under compression are often called *beam-columns;* in this chapter, considerations will only be made of the latter.

Struts appear in many and various forms, from the pillars in the hold of a ship to the forks of a bicycle, and from pillars supporting the roofs of ancient monuments to those supporting the roofs of modern football stadiums. For some struts, the design against failure is one of guarding against instability, whilst for others, it is one of determining the stresses due to the combined action of axial and lateral loading.

For practical video demonstrations of the collapse of submarine pressure hulls, the buckling of a rail car tank under external pressure, bottle buckling – simulating submarine pressure hull collapse and recent advances in submarine pressure hull designs go to www.routledge.com/cw/bird and click on the menu for 'Mechanics of Solids 3rd Edition'

14.2 Axially loaded struts

Initially 'straight' struts, which are subjected to axial compression but without additional lateral loading, as shown in Figure 14.1, are classified under the following three broad headings:

Very short struts, as shown in Figure 14.2, which fail owing to excessive stress values. For such cases, instability is not of any importance.

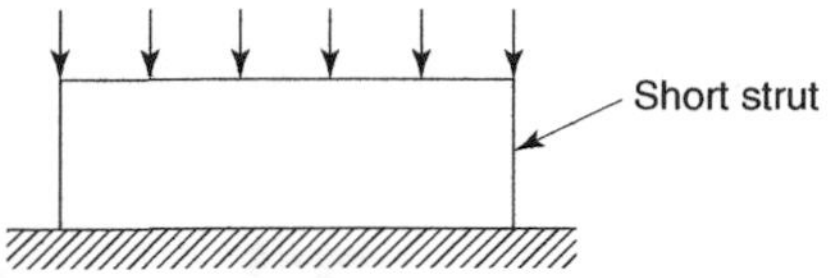

Figure 14.2 Very short strut

Very long and slender struts, which fail owing to elastic instability. For such cases, the calculation of stresses is not important, and the only material properties of interest are elastic modulus and in some cases Poisson's ratio.

Intermediate struts, whose slenderness is somewhere between the previous two extremes. Most struts tend to fall into this category, where both the elastic properties of the material and its failure stress are important. Another feature of much importance for this class of strut is its initial geometrical imperfections. The initial geometrical imperfections of a strut can cause it to fail at a buckling load, which may be a small fraction of the elastic buckling load, and the difference between these two buckling loads is *plastic knockdown,* sometimes called ***elastic knockdown.***

The buckling behaviour of intermediate struts is sometimes termed ***inelastic instability.***

For a practical video demonstration of the buckling of axially loaded struts in compression go to www.routledge.com/cw/bird and click on the menu for 'Mechanics of Solids 3rd Edition'

14.3 Elastic instability of very long slender struts

This theory is due to Euler,* and it breaks down the following classes of strut:

(a) Very short struts
(b) Intermediate struts
(c) Eccentrically loaded struts
(d) Struts with initial curvature
(e) Laterally loaded struts

Despite the fact that the theory presented in this section is only applicable to very long slender struts, which are made from homogeneous, isotropic and elastic materials, the theory can be extended to cater for axially loaded intermediate struts. In addition to this, the differential equation describing the behaviour of this class of strut can be extended to deal with eccentrically loaded, initially curved and laterally loaded struts.

The Euler theory for the elastic instability of a number of initially straight struts will now be considered.

Problem 1. Determine the Euler buckling load for an axially loaded strut, pinned at its ends, as shown in Figure 14.3. It may be assumed that the ends of the strut are free to move axially towards each other.

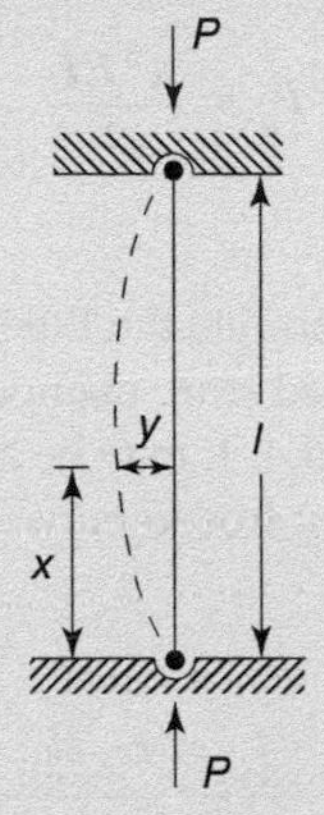

Figure 14.3 Axially loaded strut, pinned at its ends

*Leonhard Euler (15 April 1707 – 18 September 1783) was a pioneering Swiss mathematician and physicist who made important discoveries in infinitesimal calculus and graph theory. He also introduced much of the modern mathematical terminology and notation. To find out more about Euler go to www.routledge.com/cw/bird

In the case of this strut, lateral movement of the strut is prevented at its ends but the strut is free to rotate at these points (i.e. it is 'position fixed' at both ends). Experiments have shown that such struts tend to have a buckling mode, as shown by the dashed line of Figure 14.3.

Just prior to instability, let the lateral deflection of the strut be y at a distance x from its base. Now

$$EI\frac{d^2y}{dx^2} = M = -Py \qquad (14.1)$$

For struts, it is necessary for the sign convention for bending moment to be such that the *bending moment due to the product Py is always negative* and vice versa for a tie. That is, it is assumed that positive bending moment produces negative deflection and vice versa.

The mathematical reasons why the product P_y must have a negative sign for a strut and why it must have a positive sign for a tie are dealt with in a number of texts and will not be dealt with here, as this book is concerned with applying mathematics to some problems in engineering. The student must, however, remember to define the sign convention for bending moment, based on the assumption that the product Py causes a negative bending moment for all struts.

Equation (14.1) can now be rewritten in the form:

$$\frac{d^2y}{dx^2} + \frac{Py}{EI} = 0 \qquad (14.2)$$

Let $\alpha^2 = \frac{P}{EI}$ so that equation (14.2) becomes:

$$\frac{d^2y}{dx^2} + \alpha^2 y = 0 \qquad (14.3)$$

[Note regarding solving second order differential equations:

If $\quad a\frac{d^2y}{dx^2} + b\frac{dy}{dx} + cy = 0$ then the auxiliary equation is: $am^2 + bm + c = 0$

which is a quadratic equation. If the roots are complex, say, $m = p \pm jq$, then the solution of the differential equation is always: $y = e^{px}\left\{A\cos qx + B\sin qx\right\}$

See *Bird's Higher Engineering Mathematics*, 9th Edition, Routledge]

Let $\qquad y = Ae^{\lambda x} \qquad (14.4)$

Thus $\quad \frac{dy}{dx} = \lambda Ae^{\lambda x}$ and $\frac{d^2y}{dx^2} = \lambda^2 Ae^{\lambda x}$

i.e. from equation (14.4) $$\frac{d^2y}{dx^2} = \lambda^2 y \qquad (14.5)$$

Substituting equations (14.4) and (14.5) into equation (14.3) gives:

$$\lambda^2 y + \alpha^2 y = 0$$

i.e. $$\lambda^2 = -\alpha^2$$

Therefore $\lambda = \sqrt{-\alpha^2} = \sqrt{(-1)\alpha^2} = \sqrt{(-1)}\,\alpha = \pm j\alpha$

since $j = \sqrt{-1}$

i.e. $$\lambda = 0 \pm j\alpha$$

and the solution of the differential equation is:

$$y = e^0\left\{A\cos\alpha x + B\sin\alpha x\right\}$$

i.e. $$y = A\cos\alpha x + B\sin\alpha x \qquad (14.6)$$

As there are two constants, namely A and B, it will be necessary to apply two boundary conditions to equation (14.6).

By inspection it can be seen that

$$\text{at } x = 0, y = 0 \text{ and at } x = l, y = 0 \qquad (14.7)$$

From equation (14.6) and (14.7),

$$0 = A\cos 0 + B\sin 0$$

i.e. $$A = 0$$

and $$0 = A\cos\alpha l + B\sin\alpha l$$

Since $A = 0$, $$B\sin\alpha l = 0 \qquad (14.8)$$

Now the condition $B = 0$ in equation (14.8) is not of practical interest, as the strut will not suffer lateral deflections if this were so; therefore, the only possibility for equation (14.8) to apply is that:

$$\sin\alpha l = 0$$

or $$\alpha l = 0, \pi, 2\pi, 3\pi, \text{ and so on}$$

The value of αl that is of interest for the buckling of struts will be the lowest positive value, as nature has shown that the strut of Figure 14.3 will buckle by this mode,

i.e. $$\alpha l = \pi$$

Since $\alpha^2 = \frac{P}{EI}$, then $\alpha = \sqrt{\frac{P}{EI}}$

Hence $$\sqrt{\frac{P}{EI}}\, l = \pi$$

from which $$P = \frac{\pi^2 EI}{l^2} \qquad (14.9)$$

Equation (14.9) is often written in the form of equation (14.10), where P_e, the Euler buckling load of a strut pinned at its ends, is given by:

$$\boldsymbol{P_e = \frac{\pi^2 EI}{l^2}} \qquad (14.10)$$

Problem 2. Determine the Euler buckling load for an axially loaded strut, clamped at its ends, as shown in Figure 14.4. It may be assumed that the ends of the strut are free to move axially towards each other.

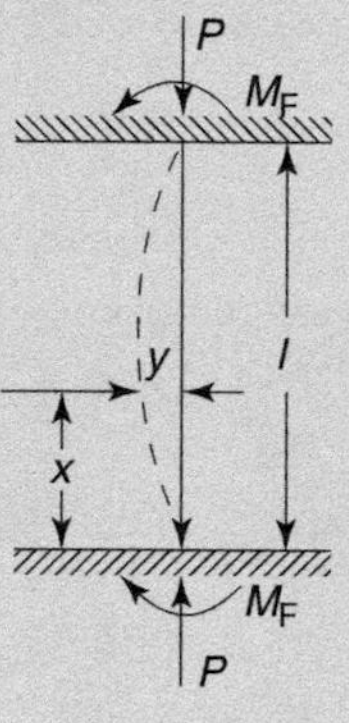

Figure 14.4 Axially loaded strut, clamped at its ends

At both its ends, the strut is prevented from lateral and rotational movement, where the latter is achieved through the reaction moments M_F (i.e. it is 'direction fixed' at both ends). Experiments have shown that the strut will have the buckled form shown by the dashed line. Just prior to buckling, let the deflection at a distance x from the base of the strut be given by y so

that: $$EI\frac{d^2y}{dx^2} = -Py + M_F \qquad (14.11)$$

NB As the product Py is a negative bending moment, M_F will be a positive one.

Equation (14.11) can be rewritten in the form

$$\frac{d^2y}{dx^2} = -\frac{P}{EI}y + \frac{M_F}{EI}$$

i.e. $$\frac{d^2y}{dx^2}+\frac{P}{EI}y=\frac{M_F}{EI}$$

$$\frac{d^2y}{dx^2}+\alpha^2y=\frac{M_F}{EI} \quad (14.12)$$

where $\alpha^2=\dfrac{P}{EI}$

From Problem 1, it can be seen that the complementary function is

$$u=A\cos\alpha x+B\sin\alpha x \quad (14.13)$$

To obtain the particular integral, v, let

$$D=\frac{dy}{dx}=\text{operator 'D'}$$

so that equation (14.12) becomes:

$$\left(D^2+\alpha^2\right)v=\frac{M_F}{EI}$$

Since the term on the right-hand side is a constant, let the particular integral be k

then $$\left(D^2+\alpha^2\right)k=\frac{M_F}{EI}$$

i.e. $$\alpha^2k=\frac{M_F}{EI} \text{ since } D(k)=0 \text{ and } D^2(k)=0$$

Hence, the particular integral $v=k=\dfrac{M_F}{EI\alpha^2}$ (14.14)

From equations (14.13) and (14.14), the complete solution is:

$$y=\text{complementary function}+\text{particular integral}$$

i.e. $y=u+v$

i.e. $$y=A\cos\alpha x+B\sin\alpha x+\frac{M_F}{EI\alpha^2} \quad (14.15)$$

and $$\frac{dy}{dx}=-\alpha A\sin\alpha x+\alpha\beta\cos\alpha x$$

Now there are three unknowns, namely A, B and M_F; hence, it will be necessary to use three boundary conditions, as follows:

(i) When $x=0$, $\dfrac{dy}{dx}=0$, thus $0=-\alpha A\sin 0+\alpha B\cos 0$

from which **$B=0$**

(ii) When $x=0$, $y=0$, thus from equation (14.15)

$$0=A\cos 0+B\sin 0+\frac{M_F}{EI\alpha^2}$$

and as $B=0$, $$0=A+\frac{M_F}{EI\alpha^2}$$

and $$A=-\frac{M_F}{EI\alpha^2}$$

(iii) When $x=l$, $y=0$, thus from equation (14.15)

$$0=A\cos\alpha l+B\sin\alpha l+\frac{M_F}{EI\alpha^2}$$

i.e. $$0=\left(-\frac{M_F}{EI\alpha^2}\right)\cos\alpha l+\frac{M_F}{EI\alpha^2}$$

since $B=0$ and

$$A=-\frac{M_F}{EI\alpha^2}$$

or $$\frac{M_F}{EI\alpha^2}(1-\cos\alpha l)=0$$

i.e. $\cos\alpha l=1$ or $\alpha l=0, 2\pi, 4\pi, 6\pi$, and so on

Nature has shown that the value of interest is $\alpha l=2\pi$ (the lowest positive root).

Hence $$\alpha=\frac{2\pi}{l} \text{ and } \alpha^2=\left(\frac{2\pi}{l}\right)^2=\frac{4\pi^2}{l^2}$$

Since $\alpha^2=\dfrac{P}{EI}$ then $\dfrac{P}{EI}=\dfrac{4\pi^2}{l^2}$

from which $$P_e=\frac{4\pi^2EI}{l^2} \quad (14.16)$$

where P_e is the Euler buckling load for clamped ends.

14.4 Struts with various boundary conditions

By a similar process to above, it can be seen that the Euler buckling load for a strut fixed at one end and free at the other, as shown in Figure 14.5, is given by:

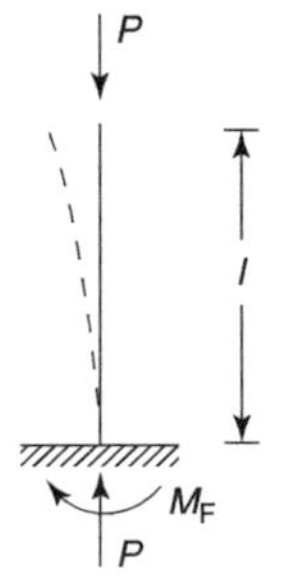

Figure 14.5 Axially loaded strut, fixed at one end and free at the other

$$P_e = \frac{\pi^2 EI}{4l^2} \tag{14.17}$$

Comparing equations (14.10), (14.16) and (14.17), it can be seen that the assumed end conditions play a significant role. Indeed, the 'clamped ends' strut has a Euler buckling load four times greater than the 'pinned ends' strut and 16 times greater than the strut of Figure 14.5.

In practice, however, it is very difficult to obtain the completely clamped condition of Figure 14.4, as there is usually some rotation due to elasticity etc. at the ends and the effect of this is to reduce drastically the predicted value for P_e from equation (14.16).

From Section 14.3, it can be seen that the Euler buckling load can be represented by the equation

$$P_{cr} = \frac{\pi^2 EI}{L_0^{\,2}} \tag{14.18}$$

where L_0 is the effective length of the strut, dependent on the end conditions and end flexibility, as shown in Table 14.1, and l is the actual length of the strut.

The differences between the effective lengths for the Euler theory and BS449 are because the latter allows for flexibility at the end fixings.

Table 14.1 Effective lengths of struts (L_0)

Type of strut	Euler	BS449
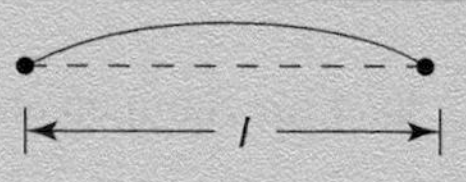 Figure 14.6	$L_0 = l$	$L_0 = l$
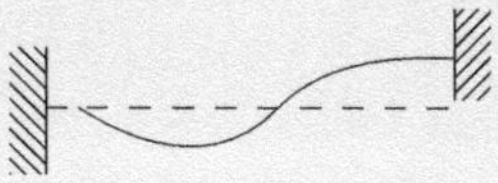 Figure 14.7	$L_0 = l$	l
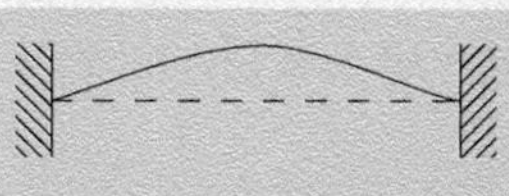 Figure 14.8	$L_0 = 0.5l$	$0.7l$
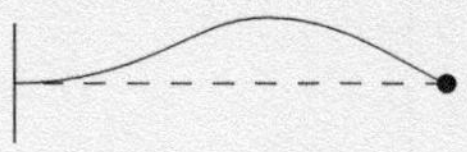 Figure 14.9	$L_0 = 0.7l$	$0.85l$
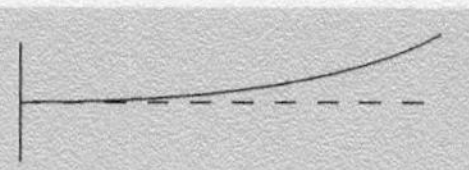 Figure 14.10	$L_0 = 2l$	$2l$

14.5 Limit of application of Euler theory

The Euler formulae are obviously inapplicable where they predict elastic instability loads greater than the crushing strength of the strut material.

For example, assuming that a 'pinned ends' strut is made from mild steel, with a crushing stress of 300 MN/m^2, the lower limit of the Euler theory is obtained, as follows:

$$P_e = \frac{\pi^2 EI}{l^2} = \frac{\pi^2 EAk^2}{l^2} \tag{14.19}$$

where $I = Ak^2$, A = cross-sectional area, k = least radius of gyration

and $$\text{yield load} = 300\ \frac{\text{MN}}{\text{m}^2} \times A \tag{14.20}$$

Equating equations (14.19) and (14.20) gives:

$$\frac{\pi^2 EAk^2}{l^2} = 300A$$

or $$\frac{\pi^2 EA}{\left(\frac{l}{k}\right)^2} = 300A \tag{14.21}$$

Let $E = 2\times10^{11}\,\text{N/m}^2 = 2\times10^5\,\text{MN/m}^2$, and substituting this into (14.21) gives:

$$\left(\frac{l}{k}\right)^2 = \frac{\pi^2 EA}{300A} = \frac{\pi^2\left(2\times10^5\right)}{300} = 6579.7$$

and $$\left(\frac{l}{k}\right) = \sqrt{6579.7}$$

i.e. $$\left(\frac{\boldsymbol{l}}{\boldsymbol{k}}\right) = \mathbf{81} = \text{slenderness ratio} \tag{14.22}$$

From equation (14.22), it can be seen that, for mild steel, the Euler theory is obviously inapplicable where the *slenderness ratio* is less than 80, and in practice, the theory needs correction for some values of slenderness ratios greater than this.

14.6 Rankine-Gordon formula for struts buckling inelastically

The Rankine*-Gordon formula is applicable to intermediate struts, and extends the Euler formulae to take

*William John Macquorn Rankine, (5 July 1820 – 24 December 1872) was a Scottish civil engineer, physicist and mathematician. He was a founding contributor, with Rudolf Clausius and William Thomson (Lord Kelvin), to the science of thermodynamics, particularly focusing on the first of the three thermodynamic laws. To find out more about Rankine go to www.routledge.com/cw/bird

account of the crushing stress of the strut material by a semi-empirical approach, as follows.

For a very short strut, the crushing load

$$P_c = \sigma_c A \qquad (14.23)$$

where A = cross section and σ_c = crushing stress.

Then, to obtain the inelastic buckling load, P_R

let $$\frac{1}{P_R} = \frac{1}{P_e} + \frac{1}{P_c}$$

$$= \frac{L_0^2}{\pi^2 EI} + \frac{1}{\sigma_{yc} \times A}$$

$$= \frac{L_0^2}{\pi^2 EAk^2} + \frac{1}{\sigma_{yc} \times A}$$

$$= \frac{L_0^2 \sigma_{yc} + \pi^2 Ek^2}{\pi^2 EAk^2 \sigma_{yc}}$$

or $$P_R = \frac{\pi^2 EAk^2 \sigma_{yc}}{L_0^2 \sigma_{yc} + \pi^2 Ek^2}$$

i.e. $$= \frac{\sigma_{yc}}{\dfrac{L_0^2 \sigma_{yc}}{\pi^2 EAk^2} + \dfrac{\pi^2 Ek^2}{\pi^2 EAk^2}}$$

i.e. $$P_R = \frac{\sigma_{yc} \times A}{\dfrac{L_0^2 \sigma_{yc}}{\pi^2 Ek^2} + 1} = \frac{\sigma_{yc} \times A}{\left(\dfrac{\sigma_{yc}}{\pi^2 E}\right)\left(\dfrac{L_0}{k}\right)^2 + 1}$$

Let $$a = \frac{\sigma_{yc}}{\pi^2 E} \qquad (14.24)$$

Then $$P_R = \frac{\sigma_{yc} \times A}{a\left(\dfrac{L_0}{k}\right)^2 + 1} \qquad (14.25)$$

where a is the denominator constant in the Rankine-Gordon formula, which is dependent on the boundary conditions and material properties.

A comparison of the Rankine-Gordon and Euler formulae, for geometrically perfect struts, is given in Figure 14.11. Some typical values for $1/a$ and σ_{yc} are given in Table 14.2.

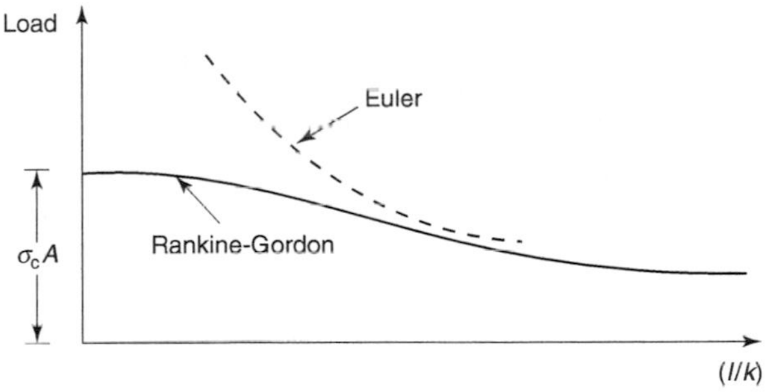

Figure 14.11 Comparison of Euler and Rankine–Gordon formulae

Table 14.2 Rankine constants

Material	1/a	σ_{yc} (MPa)
Mild steel	7500	300
Wrought iron	8000	250
Cast iron	18000	560
Timber	1000	35

Note that in the Rankine-Gordon formula, **'a' is the denominator constant** and **'σ_{yc}' is the numerator constant.**

Now try the following Practice Exercise

Practice Exercise 57. Buckling of struts

1. Determine the Euler buckling load for the axially loaded strut of Figure 14.5.
$$\left[P_e = \frac{\pi^2 EI}{4l^2}\right]$$
2. Determine the Euler buckling load for an initially straight axially loaded strut which is pinned at one end and fixed at the other.
$$\left[P_e = \frac{20.2EI}{l^2}\right]$$
3. Find the Euler crushing load for a hollow cylindrical cast-iron column of 0.15 m external diameter and 20 mm thick, if it is 6 m long and hinged at both ends. Assume that $E = 75\times10^9$ N/m^2.
Compare this load with that given by the Rankine formula using constants of 540 MN/m^2 and 1/1600. For what length of column would these two formulae give the same crushing load?
[363.1 kN, 386.7 kN, 4.55 m]
4. A short steel tube of 0.1 m outside diameter, when tested in compression, was found to fail under an axial load of 800 kN. A 15 m length of the same tube when tested as a pin-jointed strut failed under a load of 30 kN. Assuming that the Euler and Rankine-Gordon formulae apply to the strut, calculate (a) the tube inner diameter, and (b) the denominator constant in the Rankine-Gordon formula. Assume that $E = 196.5$ GN/m^2
[(a) 0.0734 m (b) 1/9117]
5. A steel pipe of 36 mm inner diameter, 6 mm thick and 1 m long is supported so that the ends are hinged, but all expansion is prevented. The pipe is unstressed at 0°C. Calculate the temperature at which buckling will occur. Assume the following: $\sigma_c = 325$ MN/m^2, $a = 1/7500$, $E = 200$ GN/m^2 and $\alpha = 11.1\times10^{-6}$ /°C
[68.4°C]
6. The table below shows the results of a series of buckling tests carried out on a steel tube of external diameter 35 mm and internal diameter 25 mm. Assuming the Rankine-Gordon formula to apply, determine the numerator and denominator constants for this tube.

l (mm)	600	1000	1400	1800
P_R (kN)	150	125	110	88

[344 MN/m^2, 1/36200]

7. The result of two tests on steel struts with pinned ends were found to be:

Test number	1	2
Slenderness ratio	50	80
Average stress at failure (MN/m^2)	266.7	194.4

$A = 1$ m^2

(a) Assuming that the Rankine-Gordon formula applies to both struts, determine the numerator and denominator constants of the Rankine-Gordon formula.
(b) If a steel bar of rectangular section 0.06 m × 0.019 m and of length 0.4 m is used as a strut with both ends clamped, determine the safe load using the constants derived in (a) and employing a safety factor of 4.
[(a) 350 MN/m^2, 1/8000 (b) 342 kN]

14.7 Effects of geometrical imperfections

For intermediate struts with geometrical imperfections, the buckling load is further decreased, as shown in Figure 14.12.

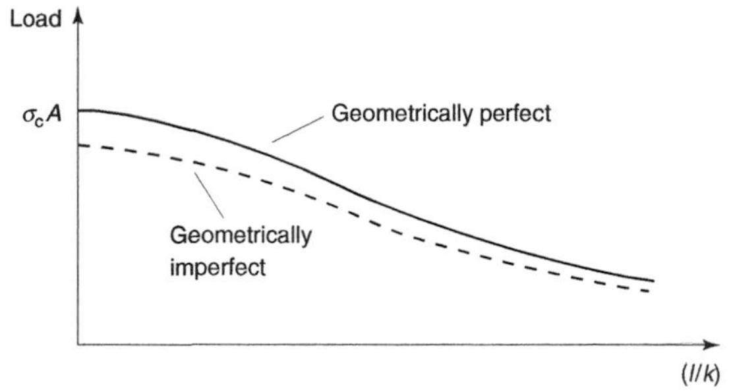

Figure 14.12 Rankine–Gordon loads for perfect and imperfect struts

Johnson's parabolic formula

This is a simplified version of the Rankine-Gordon expression, and it proved to be popular in design offices prior to the invention of the hand-held calculator:

Johnson buckling stress $= \sigma_c - b\left(\frac{l}{k}\right)^2$ (14.26)

where b is a constant depending on end conditions, material properties, and so on.

14.8 Eccentrically loaded struts

Struts in this category, examples of which are shown in Figure 14.13, frequently occur in practice.

For such problems, the main object in stress analysis is to determine the stresses due to the combined effects of bending and axial load, unlike the struts of Sections 14.3 and 14.4, where the main object was to obtain a crippling load.

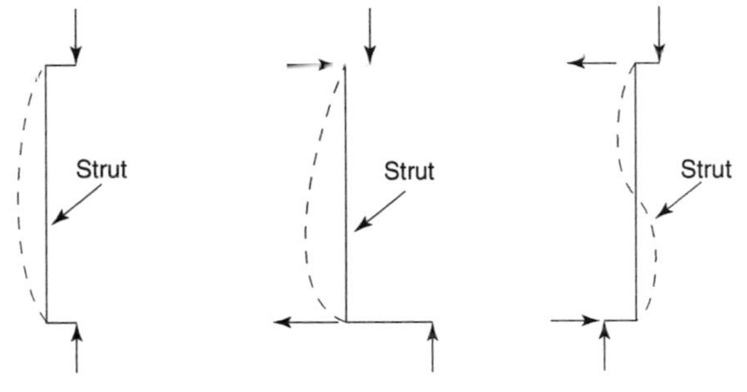

Figure 14.13 Eccentrically loaded struts: (a) eccentricity equal and on the same side of the strut; (b) eccentricity unequal and on the same side of the strut; (c) eccentricity on opposite sides of the strut

Problem 3. A pillar in the hold of a ship is in the form of a tube of external diameter 25 cm, internal diameter 22.5 cm and length 10 m. If the pillar is subjected to an eccentric load of 20 tonnes, as shown in Figure 14.14, calculate the maximum permissible eccentricity if the maximum permissible stress is 75 MN/m^2 It may be assumed that $E = 2 \times 10^{11}$ N/m^2.

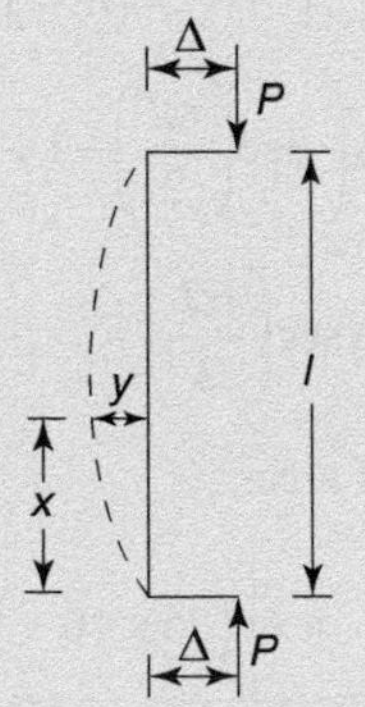

Figure 14.14 Eccentrically loaded strut

At any distance from x the base,

$$EI\frac{d^2y}{dx^2} = M = -P(y+\Delta)$$

or
$$\frac{d^2y}{dx^2} + \frac{P}{EI}y = -\frac{P}{EI}\Delta \qquad (14.27)$$

Let
$$\alpha^2 = \frac{P}{EI}$$

so that equation (14.27) becomes:

$$\frac{d^2y}{dx^2} + \alpha^2 y = -\alpha^2\,\Delta \qquad (14.28)$$

The complementary function of equation (14.28) is:

$$u = A\cos\alpha x + B\sin\alpha x \qquad (14.29)$$

and the particular integral of equation (12.28) is obtained as follows:

$$\left(D^2+\alpha^2\right)v = -\alpha^2\Delta$$

Since the term on the right-hand side is a constant, let the particular integral be k,

then
$$\left(D^2+\alpha^2\right)k = -\alpha^2\Delta$$

i.e. $\alpha^2 k = -\alpha^2\,\Delta$ since $D(k) = 0$ and $D^2(k) = 0$

Hence, the particular integral $v = k = -\Delta$ (14.30)

From equations (14.29) and (14.30), the complete solution is:

y = complementary function + particular integral

i.e. $y = u + v$

i.e. $y = A\cos\alpha x + B\sin\alpha x - \Delta$ (14.31)

Now there are two unknowns, namely A and B; hence, it will be necessary to use two boundary conditions, as follows:

(i) when $x = 0$, $y = 0$ and (ii) when $x = l$, $y = 0$ (14.32)

(i) When $x = 0$, $y = 0$, thus $0 = A\cos 0 + B\sin 0 - \Delta$

from which, $\mathbf{A = \Delta}$ (14.33)

(ii) When $x = l$, $y = 0$, thus $0 = A\cos\alpha l + B\sin\alpha l - \Delta$

i.e. $\Delta - A\cos\alpha l = B\sin\alpha l$

from which $B = \dfrac{\Delta - A\cos\alpha l}{\sin\alpha l}$

i.e. $$B = \frac{\Delta(1 - \cos \alpha l)}{\sin \alpha l}$$ since $A = \Delta$ from equation (14.33)

Now, from double angles (see page 6),
$\sin 2x = 2 \sin x \cos x$

hence $$\sin x = 2 \sin\frac{x}{2}\cos\frac{x}{2}$$

Also $$\cos 2x = 1 - 2\sin^2 x$$

hence $$\cos x = 1 - 2\sin^2 \frac{x}{2}$$

Therefore $$B = \frac{\Delta(1 - \cos \alpha l)}{\sin \alpha l}$$

becomes $$B = \frac{\Delta\left(1 - \left(1 - 2\sin^2\left(\frac{\alpha l}{2}\right)\right)\right)}{2\sin\left(\frac{\alpha l}{2}\right)\cos\left(\frac{\alpha l}{2}\right)}$$

i.e. $$B = \frac{\Delta\left(2\sin^2\left(\frac{\alpha l}{2}\right)\right)}{2\sin\left(\frac{\alpha l}{2}\right)\cos\left(\frac{\alpha l}{2}\right)}$$

i.e. $$B = \frac{\Delta\left(\sin\left(\frac{\alpha l}{2}\right)\right)}{\cos\left(\frac{\alpha l}{2}\right)}$$

i.e. $$\boldsymbol{B = \Delta \tan\left(\frac{\alpha l}{2}\right)} \qquad (14.34)$$

Substituting equations (14.33) and (14.34) into equation (14.31), the deflected form of the strut, at any distance x, is given by:

$$y = \Delta \cos \alpha x + \left[\Delta \tan\left(\frac{\alpha l}{2}\right)\right] \sin \alpha x - \Delta$$

i.e. $$\boldsymbol{y = \Delta\left[\cos \alpha x + \tan\left(\frac{\alpha l}{2}\right)\sin \alpha x - 1\right]} \qquad (14.35)$$

The maximum deflection δ occurs at $x = l/2$

i.e. $$\delta = \Delta\left[\cos\left(\frac{\alpha l}{2}\right) + \tan\left(\frac{\alpha l}{2}\right)\sin\left(\frac{\alpha l}{2}\right) - 1\right]$$

i.e. $$\delta = \Delta\left[\cos\left(\frac{\alpha l}{2}\right) + \frac{\sin\left(\frac{\alpha l}{2}\right)}{\cos\left(\frac{\alpha l}{2}\right)}\sin\left(\frac{\alpha l}{2}\right) - 1\right]$$ since $\tan x = \frac{\sin x}{\cos x}$

i.e. $$\delta = \Delta\left[\cos\left(\frac{\alpha l}{2}\right) + \frac{\sin^2\left(\frac{\alpha l}{2}\right)}{\cos\left(\frac{\alpha l}{2}\right)} - 1\right]$$

i.e. $$\delta = \Delta\left[\cos\left(\frac{\alpha l}{2}\right) + \frac{\left[1 - \cos^2\left(\frac{\alpha l}{2}\right)\right]}{\cos\left(\frac{\alpha l}{2}\right)} - 1\right]$$

since $\sin^2 x = 1 - \cos^2 x$

i.e. $$\delta = \Delta\left[\cos\left(\frac{\alpha l}{2}\right) + \frac{1}{\cos\left(\frac{\alpha l}{2}\right)} - \cos\left(\frac{\alpha l}{2}\right) - 1\right]$$

i.e. $$\boldsymbol{\delta = \Delta\left[\sec\left(\frac{\alpha l}{2}\right) - 1\right]} \qquad (14.36)$$

since $\sec x = \frac{1}{\cos x}$

By inspection, the maximum bending moment M is given by:

$$\widehat{M} = P(\delta + \Delta)$$

i.e. $$\widehat{M} = P\left(\Delta\left[\sec\left(\frac{\alpha l}{2}\right) - 1\right] + \Delta\right)$$

i.e. $$\widehat{M} = P\Delta\left(\sec\left(\frac{\alpha l}{2}\right) - 1 + 1\right)$$

i.e. $$\boldsymbol{\widehat{M} = P\Delta \sec\left(\frac{\alpha l}{2}\right)} \qquad (14.37)$$

Second moment of area,

$$I = \frac{\pi\left(D_2{}^4 - D_1{}^4\right)}{64} = \frac{\pi \times \left[\left(25\times10^{-2}\right)^4 - \left(22.5\times10^{-2}\right)^4\right]}{64}$$

i.e. $\boldsymbol{I = 6.594 \times 10^{-5}\,m^4}$

Cross-sectional area

$$A = \frac{\pi\left(D_2^2 - D_1^2\right)}{4} = \frac{\pi \times \left[\left(25\times10^{-2}\right)^2 - \left(22.5\times10^{-2}\right)^2\right]}{4}$$

i.e. $= \mathbf{9.327 \times 10^{-3}\,m^2}$

Since $\alpha^2 = \dfrac{P}{EI}$ then $\alpha = \sqrt{\dfrac{P}{EI}}$

Hence $\dfrac{\alpha l}{2} = \sqrt{\dfrac{P}{EI}}\left(\dfrac{l}{2}\right) = \sqrt{\dfrac{20\times10^3\times9.81}{2\times10^{11}\times6.594\times10^{-5}}}\left(\dfrac{10}{2}\right)$

i.e. $\boldsymbol{\dfrac{\alpha l}{2} = 0.6099\text{ rad} = 34.94^\circ}$

From equation (14.37)

$$\widehat{M} = P\Delta\sec\left(\frac{\alpha l}{2}\right) - P\Delta \times \frac{1}{\cos\left(\dfrac{\alpha l}{2}\right)}$$

$$= 20000\times9.81\times\Delta\times\frac{1}{\cos 34.94^\circ}$$

i.e. $\boldsymbol{\widehat{M} = 239341\,\Delta}$

Now $\widehat{\sigma} = \sigma\text{ (direct)} + \sigma\text{ (bending)}$

i.e. $\widehat{\sigma} = \dfrac{P}{A} \pm \dfrac{\widehat{M}\times y}{I}$

i.e. $-75\times10^6 = \dfrac{-20\times10^3\times9.81}{9.327\times10^{-3}} - \dfrac{239341\Delta\times12.5\times10^{-2}}{6.594\times10^{-5}}$

(minus because it is compressive)

$$= -2.104\times10^7 - 4.537\times10^8\Delta$$

Hence $4.537\times10^8\Delta = 75\times10^6 - 2.104\times10^7$

$= 5.396\times10^7$

i.e. $\Delta = \dfrac{5.396\times10^7}{4.537\times10^8}$

i.e. $\Delta = \mathbf{0.1189\ m = 11.9\ cm}$

Problem 4. If the pillar of Problem 3 above were subjected to the eccentric loading of Figure 14.15, determine the maximum permissible value of Δ.

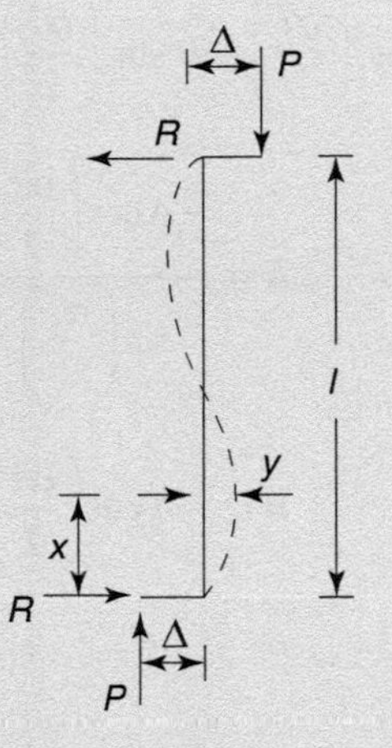

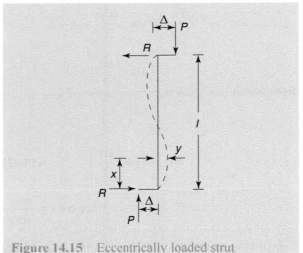

Figure 14.15 Eccentrically loaded strut

The horizontal reactions R are required to achieve equilibrium, and the relationship between them and P can be obtained by taking moments, as follows:

$$R\,l = P\times2\Delta$$

or $$R = \frac{2P\Delta}{l} \qquad (14.38)$$

At any distance from the base

$$EI\frac{d^2y}{dx^2} = -P\left(y+\Delta\right) + Rx$$

or $$\frac{d^2y}{dx^2} + \frac{P}{EI}y = \frac{-P\Delta}{EI} + \frac{Rx}{EI}$$

If $\alpha^2 = \dfrac{P}{EI}$ then $\dfrac{d^2y}{dx^2} + \alpha^2 y = \dfrac{R\alpha^2 x}{P} - \alpha^2\Delta$ (14.39)

The complete solution of equation (14.39) – see Reference on page 50 for help – is:

$$\boldsymbol{y = A\cos\alpha x + B\sin\alpha x + \frac{Rx}{P} - \Delta} \qquad (14.40)$$

There are two unknowns; therefore two boundary conditions are required, as follows:

At $x = 0$, $y = 0$, therefore $\mathbf{A = \Delta}$ (14.41)

At $x = l/2$, $y = 0$, therefore

$$0 = A\cos\left(\frac{\alpha l}{2}\right) + B\sin\left(\frac{\alpha l}{2}\right) + \left(\frac{Rl}{2P}\right) - \Delta$$

and from equations (14.38) and (14.41)

$$0 = \Delta\cos\left(\frac{\alpha l}{2}\right) + B\sin\left(\frac{\alpha l}{2}\right) + \frac{\left(\dfrac{2P\Delta}{l}\right)l}{2P} - \Delta$$

i.e. $0 = \Delta\cos\left(\frac{\alpha l}{2}\right) + B\sin\left(\frac{\alpha l}{2}\right) + \Delta - \Delta$

from which $B = \frac{-\Delta\cos\left(\frac{\alpha l}{2}\right)}{\sin\left(\frac{\alpha l}{2}\right)}$

i.e. $\boldsymbol{B = -\Delta\cot\left(\frac{\alpha l}{2}\right)}$ (14.42)

since $\cot x = \frac{\cos x}{\sin x}$

Substituting equations (14.38), (14.41) and (14.42) into equation (14.40) gives:

$$y = \Delta\cos\alpha x - \Delta\cot\left(\frac{\alpha l}{2}\right)\sin\alpha x + \frac{\left(\frac{2P\Delta}{l}\right)x}{P} - \Delta$$

i.e. $y = \Delta\cos\alpha x - \Delta\cot\left(\frac{\alpha l}{2}\right)\sin\alpha x + \left(\frac{2\Delta x}{l}\right) - \Delta$

i.e. $\boldsymbol{y = \Delta\left[\cos ax - \cot\left(\frac{\alpha l}{2}\right)\sin ax + \frac{2x}{l} - 1\right]}$ (14.43)

Now $\frac{dy}{dx} = \Delta\left[-a\sin ax - \cot\left(\frac{\alpha l}{2}\right)a\cos ax + \frac{2}{l}\right]$

and $\frac{d^2y}{dx^2} = \Delta\left[-a^2\cos ax + \cot\left(\frac{\alpha l}{2}\right)a^2\sin ax\right]$

$$= \Delta\alpha^2\left[-\cos ax + \cot\left(\frac{\alpha l}{2}\right)\sin ax\right]$$

Now $M = EI\frac{d^2y}{dx^2}$

$$= EI\Delta\alpha^2\left[-\cos ax + \cot\left(\frac{\alpha l}{2}\right)\sin ax\right]$$

Since $\alpha^2 = \frac{P}{EI}$ then

$$M = EI\Delta\left(\frac{P}{EI}\right)\left[-\cos ax + \cot\left(\frac{\alpha l}{2}\right)\sin ax\right]$$

$$= -\Delta P\left[\cos ax - \cot\left(\frac{\alpha l}{2}\right)\sin ax\right]$$

$$= -\Delta P\left[\cos ax - \frac{\cos\left(\frac{\alpha l}{2}\right)}{\sin\left(\frac{\alpha l}{2}\right)}\sin ax\right]$$

since $\cot = \frac{\cos x}{\sin x}$

$$= -\Delta P\left[\frac{\cos ax\sin\left(\frac{\alpha l}{2}\right) - \cos\left(\frac{\alpha l}{2}\right)\sin ax}{\sin\left(\frac{\alpha l}{2}\right)}\right]$$

$$= -\Delta P\left[\frac{\sin\left(\frac{\alpha l}{2}\right)\cos ax - \cos\left(\frac{\alpha l}{2}\right)\sin ax}{\sin\left(\frac{\alpha l}{2}\right)}\right]$$

i.e. $M = -\Delta P\left[\frac{\sin\left(\frac{\alpha l}{2} - ax\right)}{\sin\left(\frac{\alpha l}{2}\right)}\right]$ since $\sin(A - B) = \sin A\cos B - \cos A\sin B$

The maximum bending moment $\widehat{M}$ occurs when

$$\left(\frac{\alpha l}{2} - ax\right) = \pm\frac{\pi}{2}$$

i.e. $\widehat{M} = -\Delta P\left[\frac{\sin\left(\pm\frac{\pi}{2}\right)}{\sin\left(\frac{\alpha l}{2}\right)}\right] = \pm\Delta P\left[\frac{1}{\sin\left(\frac{\alpha l}{2}\right)}\right]$

i.e. $\boldsymbol{\widehat{M} = \pm\Delta P\,\mathrm{cosec}\left(\frac{\alpha l}{2}\right)}$ since $\mathrm{cosec}\,x = \frac{1}{\sin x}$

Now $\hat{\sigma} = \sigma\text{ (direct)} \pm \sigma\text{ (bending)}$

i.e. $\hat{\sigma} = \frac{P}{A} \pm \frac{\widehat{M}\times y}{I} = \frac{P}{A} \pm \frac{\left[\Delta P\,\mathrm{cosec}\left(\frac{\alpha l}{2}\right)\right]\times y}{I}$

i.e. $-75\times 10^6 = \frac{-20\times 10^3\times 9.81}{9.327\times 10^{-3}} - \frac{20000\times 9.81\times\mathrm{cosec}(34.94°)\times\Delta\times 12.5\times 10^{-2}}{6.594\times 10^{-5}}$

(minus because it is compressive)

$$= -2.104\times 10^7 - 649.4\times 10^6\Delta$$

Hence $649.4\times10^6\Delta = 75\times10^6 - 2.104\times10^7 = 5.396\times10^7$

i.e. $\Delta = \dfrac{5.396\times10^7}{649.4\times10^6}$

i.e. $\Delta = 0.0831\ \mathbf{m} = \mathbf{8.31\ cm}$

14.9 Struts with initial curvature

Struts in this category are usually assumed to have an initial sinusoidal or parabolic shape of the form shown in Figure 14.16. As in Section 14.8, stresses are of more importance for this class of strut than are crippling loads.

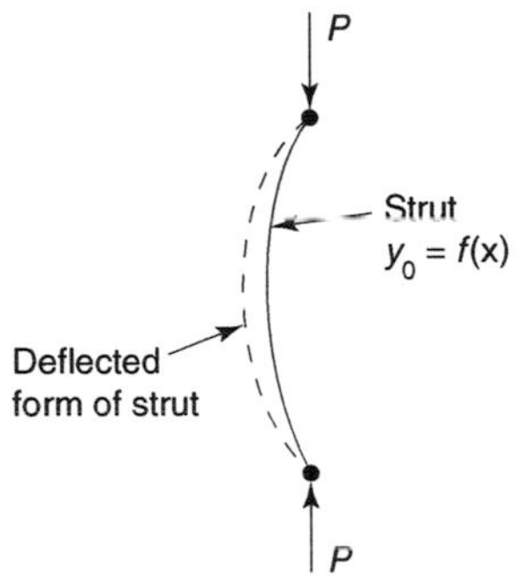

Figure 14.16 Initially curved strut

Let $y_0 = f(x)$ = initial equation of strut

and R_0 = initial radius of curvature of strut of x

$$= \frac{d^2y_0}{dx^2}$$

From the relationship $EI\left(\dfrac{1}{R} - \dfrac{1}{R_0}\right) = M$ (from Section 7.1)

$$EI\left(\frac{d^2y}{dx^2} - \frac{d^2y_0}{dx^2}\right) = M \qquad (14.44)$$

Normally, solution of equation (14.44) is achieved by assuming the initial shape of the strut y_0 to be sinusoidal or parabolic or circular.

Problem 5. Determine the maximum deflection and bending moment for a strut with pinned ends, which has an initial sinusoidal curvature of the form

$$y_0 = \Delta \sin\left(\frac{\pi x}{l}\right) \qquad (14.45)$$

where Δ is a small initial central deflection.

From equation (14.45) $y_0 = \Delta \sin\left(\dfrac{\pi x}{l}\right)$

and $$\frac{dy_0}{dx} = \Delta\left(\frac{\pi}{l}\right)\cos\left(\frac{\pi x}{l}\right)$$

and $$\frac{d^2y_0}{dx^2} = -\Delta\left(\frac{\pi}{l}\right)^2 \sin\left(\frac{\pi x}{l}\right)$$

Substituting the second derivative of equation (14.45) into equation (14.44), the following is obtained:

$$EI\left[\frac{d^2y}{dx^2} - \left[-\Delta\left(\frac{\pi}{l}\right)^2 \sin\left(\frac{\pi x}{l}\right)\right]\right] = M$$

i.e. $$EI\left[\frac{d^2y}{dx^2} + \left(\frac{\pi^2}{l^2}\right)\Delta \sin\left(\frac{\pi x}{l}\right)\right] = -Py$$

i.e. $$\frac{d^2y}{dx^2} + \left(\frac{\pi^2}{l^2}\right)\Delta \sin\left(\frac{\pi x}{l}\right) = \frac{-P}{EI}y$$

If $\alpha^2 = \dfrac{P}{EI}$ then $$\frac{d^2y}{dx^2} + \alpha^2 y = -\left(\frac{\pi^2}{l^2}\right)\Delta \sin\left(\frac{\pi x}{l}\right) \qquad (14.46)$$

The complementary function of equation (14.46) is:

$$u = A\cos\alpha x + B\sin\alpha x \qquad (14.47)$$

Let the particular integral be: $v = A\sin\left(\dfrac{\pi x}{l}\right) + B\cos\left(\dfrac{\pi x}{l}\right)$

Then $$\left(D^2 + \alpha^2\right)\left[A\sin\left(\frac{\pi x}{l}\right) + B\cos\left(\frac{\pi x}{l}\right)\right] = -\left(\frac{\pi^2}{l^2}\right)\Delta \sin\left(\frac{\pi x}{l}\right)$$

$$D\left[A\sin\left(\frac{\pi x}{l}\right) + B\cos\left(\frac{\pi x}{l}\right)\right] = \left[\left(\frac{\pi}{l}\right)A\cos\left(\frac{\pi x}{l}\right) - \left(\frac{\pi}{l}\right)B\sin\left(\frac{\pi x}{l}\right)\right]$$

$$D^2\left[A\sin\left(\frac{\pi x}{l}\right) + B\cos\left(\frac{\pi x}{l}\right)\right] = D\left[\left(\frac{\pi}{l}\right)A\cos\left(\frac{\pi x}{l}\right) - \left(\frac{\pi}{l}\right)B\sin\left(\frac{\pi x}{l}\right)\right]$$

$$= \left[-\left(\frac{\pi}{l}\right)^2 A \sin\left(\frac{\pi x}{l}\right) - \left(\frac{\pi}{l}\right)^2 B \cos\left(\frac{\pi x}{l}\right) \right]$$

Hence $\left(D^2 + \alpha^2\right)\left[A \sin\left(\frac{\pi x}{l}\right) + B \cos\left(\frac{\pi x}{l}\right) \right]$

$$= -\left(\frac{\pi^2}{l^2}\right) \Delta \sin\left(\frac{\pi x}{l}\right)$$

becomes $\left[-\left(\frac{\pi}{l}\right)^2 A \sin\left(\frac{\pi x}{l}\right) - \left(\frac{\pi}{l}\right)^2 B \cos\left(\frac{\pi x}{l}\right) \right]$

$$+ \left[\alpha^2 A \sin\left(\frac{\pi x}{l}\right) + \alpha^2 B \cos\left(\frac{\pi x}{l}\right) \right] = -\left(\frac{\pi^2}{l^2}\right) \Delta \sin\left(\frac{\pi x}{l}\right)$$

Equating coefficients of $\sin\left(\frac{\pi x}{l}\right)$ gives:

$$-\left(\frac{\pi}{l}\right)^2 A + \alpha^2 A = -\left(\frac{\pi^2}{l^2}\right) \Delta$$

from which $A\left[-\left(\frac{\pi}{l}\right)^2 + \alpha^2 \right] = -\left(\frac{\pi^2}{l^2}\right) \Delta$

and $$A = \frac{-\left(\frac{\pi^2}{l^2}\right) \Delta}{\left[-\left(\frac{\pi}{l}\right)^2 + \alpha^2 \right]}$$

Equating coefficients of $\cos\left(\frac{\pi x}{l}\right)$ gives:

$$-\left(\frac{\pi}{l}\right)^2 B + \alpha^2 B = 0 \text{ from which, } B = 0$$

Hence, the particular integral,

$$= \frac{-\left(\frac{\pi^2}{l^2}\right) \Delta \sin\left(\frac{\pi x}{l}\right)}{\left[\left(-\frac{\pi^2}{l^2}\right) + \alpha^2 \right]} \qquad (14.48)$$

From equations (14.47) and (14.48), the complete solution is:

$$y = u + v$$

i.e. $$\boldsymbol{y = A \cos \alpha x + B \sin \alpha x - \frac{\left(\frac{\pi^2}{l^2}\right) \Delta \sin\left(\frac{\pi x}{l}\right)}{\left[\left(-\frac{\pi^2}{l^2}\right) + \alpha^2 \right]}} \qquad (14.49)$$

Two suitable boundary conditions are:

At $x = 0$, $y = 0$, therefore $\boldsymbol{A = 0}$

$$\frac{dy}{dx} = -A \sin \alpha x + B\alpha \cos \alpha x - \frac{\left(\frac{\pi^3}{l^3}\right) \Delta \cos\left(\frac{\pi x}{l}\right)}{\left[\left(-\frac{\pi^2}{l^2}\right) + \alpha^2 \right]}$$

At $x = l/2$, $\frac{dy}{dx} = 0$, therefore, $0 = -(0) \sin\left(\frac{al}{2}\right)$

$$+ Ba \cos\left(\frac{al}{2}\right) - \frac{\left(\frac{\pi^3}{l^3}\right) \Delta \cos\left(\frac{\pi l}{2l}\right)}{\left[\left(-\frac{\pi^2}{l^2}\right) + \alpha^2 \right]}$$

i.e. $0 = 0 + Ba \cos\left(\frac{al}{2}\right)$

from which $\boldsymbol{B = 0}$

Hence, since $A = B = 0$, equation (14.41) becomes:

$$y = -\frac{\left(\frac{\pi^2}{l^2}\right) \Delta \sin\left(\frac{\pi x}{l}\right)}{\left[\left(-\frac{\pi^2}{l^2}\right) + \alpha^2 \right]} = \frac{\left(\frac{\pi^2}{l^2}\right) \Delta \sin\left(\frac{\pi x}{l}\right)}{\left[\left(\frac{\pi^2}{l^2}\right) - \alpha^2 \right]}$$

Since $\alpha^2 = \frac{P}{EI}$ $$y = \frac{\left(\frac{\pi^2}{l^2}\right) \Delta \sin\left(\frac{\pi x}{l}\right)}{\left[\left(\frac{\pi^2}{l^2}\right) - \left(\frac{P}{EI}\right) \right]}$$

$$= \frac{\left(\frac{\pi^2}{l^2}\right) \Delta \sin\left(\frac{\pi x}{l}\right)}{\left[\frac{\pi^2 EI - l^2 P}{l^2 EI} \right]}$$

$$= \frac{\left(\frac{\pi^2 EI}{l^2}\right)\Delta \sin\left(\frac{\pi x}{l}\right)}{\left[\frac{\pi^2 EI - l^2 P}{l^2}\right]}$$

If $P_e = \frac{\pi^2 EI}{l^2}$ then $y = \frac{P_e \Delta \sin\left(\frac{\pi x}{l}\right)}{\left[P_e - P\right]}$ (14.50)

The maximum deflection δ occurs at $x = l/2$:

i.e. $\delta = \frac{P_e \Delta \sin\left(\frac{\pi l}{2l}\right)}{\left[P_e - P\right]} = \frac{\Delta P_e}{\left[P_e - P\right]}$ (14.51)

and $\widehat{M} = \frac{\Delta P P_e}{\left[P_e - P\right]}$ (14.52)

Problem 6. Determine the maximum permissible value of Δ for an initially curved strut of sinusoidal curvature, given the following:
(a) sectional and material properties are as in Problem 4,
(b) length of strut is 10 m
(c) the strut is pinned at its ends.

Now $\hat{\sigma} = \sigma\text{ (direct)} \pm \sigma\text{(bending)}$

$$P_e = \frac{\pi^2 EI}{l^2} = \frac{\pi^2 \times 2 \times 10^{11} \times 6.594 \times 10^{-5}}{10^2} = 1.302 \text{ MN}$$

$$P = 20000 \times 9.81 = 0.196 \text{ MN}$$

Hence $\hat{\sigma} = \frac{P}{A} \pm \frac{\widehat{M} \times y}{I} = \frac{P}{A} \pm \frac{\left[\frac{\Delta P P_e}{\left[P_e - P\right]}\right] \times y}{I}$

i.e. $-75 \times 10^6 = \frac{-20 \times 10^3 \times 9.81}{9.327 \times 10^{-3}}$

$$-\Delta\left(\frac{1.302 \times 0.196 \times 10^6 \times 12.5 \times 10^{-2}}{1.106 \times 6.594 \times 10^{-5}}\right)$$

(minus because it is compressive)

$$= -21.04 \times 10^6 - 437.49 \times 10^6 \Delta$$

i.e. $437.49 \times 10^6 \Delta = 75 \times 10^6 - 21.04 \times 10^6$

$$= 53.94 \times 10^6$$

and $\boldsymbol{\Delta} = \frac{53.94}{437.49} = \mathbf{0.123\ m = 12.3\ cm}$

14.10 Perry-Robertson formula

From Section 14.9, it can be seen that the maximum stress for a strut with initial sinusoidal curvature is given by:

$$\sigma_c = \frac{P}{A} + \frac{\Delta P P_e \bar{y}}{\left(P_e - P\right) I} \quad (14.53)$$

Since $I = Ak^2$ $\sigma_c = \frac{P}{A} + \frac{\Delta P P_e \bar{y}}{\left(P_e - P\right) Ak^2}$

where k is the least radius of gyration of the strut's cross-section

Putting $\gamma = \frac{\Delta \bar{y}}{k^2}$ and $\sigma_e = \frac{P_e}{A}$ or $P_e = \sigma_e A$ (14.54)

and substituting equations (14.54) into equation (14.53) gives:

$$\sigma_c = \frac{P}{A} + \frac{\gamma P P_e}{\left(P_e - P\right) A}$$

$$= \frac{P}{A} + \frac{\gamma P \sigma_e A}{\left(\sigma_e A - P\right) A}$$

i.e. $\sigma_c = \frac{P}{A} + \frac{\gamma P \sigma_e}{\left(\sigma_e A - P\right)}$

$$= \frac{P}{A} + \frac{\gamma P \sigma_e}{\left(\sigma_e - P/A\right) A}$$

i.e. $\sigma_c = \frac{P}{A} + \frac{\gamma (P/A) \sigma_e}{\left(\sigma_e - P/A\right)}$

Therefore $\sigma_c\left(\sigma_e - \frac{P}{A}\right) = \frac{P}{A}\left(\sigma_e - \frac{P}{A}\right) + \gamma\left(\frac{P}{A}\right)\sigma_e$

or $\sigma_c \sigma_e - \sigma_c \frac{P}{A} = \frac{P}{A}\sigma_e - \left(\frac{P}{A}\right)^2 + \gamma\left(\frac{P}{A}\right)\sigma_e$

i.e. $\left(\frac{P}{A}\right)^2 + \sigma_c \sigma_e - \sigma_c \frac{P}{A} - \frac{P}{A}\sigma_e - \gamma\left(\frac{P}{A}\right)\sigma_e = 0$

i.e. $\left(\frac{P}{A}\right)^2 - \left[\sigma_c + \sigma_e + \gamma\, \sigma_e\right]\frac{P}{A} + \sigma_c \sigma_e = 0$

i.e. $$\left(\frac{P}{A}\right)^2 - \left[\sigma_c + (\gamma+1)\sigma_e\right]\frac{P}{A} + \sigma_c\sigma_e = 0$$

which is a quadratic equation.

Hence $\left(\frac{P}{A}\right) =$

$$\frac{--\left[\sigma_c+(\gamma+1)\sigma_e\right] \pm \sqrt{\left[\sigma_c+(\gamma+1)\sigma_e\right]^2 - 4(1)(\sigma_c\sigma_e)}}{2(1)}$$

i.e. $$\left(\boldsymbol{\frac{P}{A}}\right) = \frac{\mathbf{1}}{\mathbf{2}}\left[\sigma_c + (\gamma+\mathbf{1})\sigma_e\right] - \frac{\mathbf{1}}{\mathbf{2}}\sqrt{\left[\sigma_c+(\gamma+\mathbf{1})\sigma_e\right]^2 - \mathbf{4}\sigma_c\sigma_e} \qquad (14.55)$$

minus before the root sign, because it is compressive

From BS449 $$\gamma = 0.003\left(\frac{L_0}{k}\right)$$

Problem 7. Calculate the maximum permissible value for P for the strut of Problem 6, using γ from BS449.

$$k = \sqrt{\frac{I}{A}} = \frac{6.594\times10^{-5}}{9.327\times10^{-3}} = 0.084 \text{ and } \frac{L_0}{k} = 119.05$$

Therefore $$\gamma = 0.003\left(\frac{L_0}{k}\right) = 0.003\times119.05 = 0.357$$

From equation (14.55)

$$\frac{P}{A} = \frac{1}{2}\left[75\times10^6 + 1.357\left(\frac{1.302\times10^6}{9.327\times10^{-3}}\right)\right] - \frac{1}{2}\sqrt{\left[75\times10^7 + 1.357\left(\frac{1.302\times10^6}{9.327\times10^{-3}}\right)\right]^2 - 4\times75\times10^6\times\left(\frac{1.302\times10^6}{9.327\times10^{-3}}\right)}$$

i.e. $$\frac{P}{A} = \frac{1}{2}\left[264.43\times10^6\right] - \frac{1}{2}\sqrt{\left[6.9923\times10^{16}\right] - 4.1878\times10^{16}}$$

$$\frac{P}{A} = 132.22\times10^6 - 83.733\times10^6$$

$$= 48.49\times10^6\,\text{N/m}^2$$

and $$P = 48.49\times10^6\,\text{N/m}^2\times A$$

$$= 48.49\times10^6\,\text{N/m}^2\times9.327\times10^{-3}\,\text{m}^2$$

i.e. $\boldsymbol{P} = 452.3\times10^3\,\text{N} = \mathbf{452.3\ kN} = \mathbf{0.4523\ MN}$

Now try the following Practice Exercise

Practice Exercise 58. Buckling of struts

1. A long slender strut of length L is encastré at one end and pin jointed at the other. At its pinned end, it carries an axial load P, together with a couple M. Show that the magnitude of the couple at the clamped end is given by the expression

$$M\left(\frac{\alpha L - \sin\alpha L}{\alpha L\cos\alpha L - \sin\alpha L}\right)$$

Determine the value of this couple if P is one quarter of the Euler buckling load for this class of strut.

[– 0.672 m]

2. A long strut, initially straight, securely fixed at one end and free at the other, is loaded at the free end with an eccentric load whose line of action is parallel to the original axis. Deduce an expression for the deviation of the free end from its original position.

$$\left[\Delta(\sec\alpha L - 1) \text{ where } \alpha = \sqrt{\frac{P}{EI}} \text{ and } \Delta = \text{eccentricity}\right]$$

3. A tubular steel strut of 70 mm external diameter and 50 mm internal diameter is 3.25 m long. The line of action of the compressive forces is parallel to, but eccentric from, the axis of the tube, as shown in Figure 14.17.

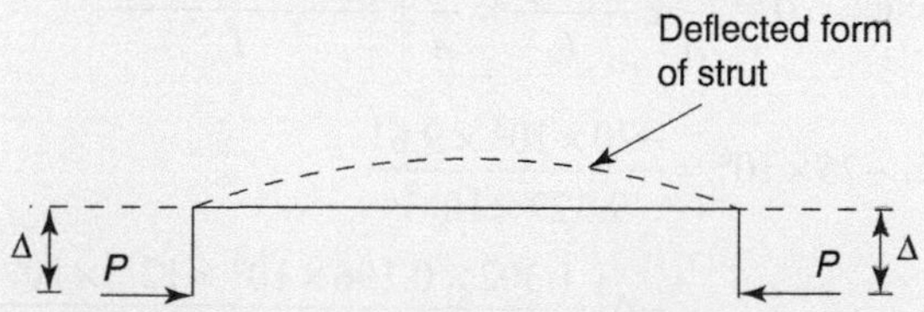

Figure 14.17

Find the maximum allowable eccentricity of these forces if the maximum permissible deflection (total) is not greater than 15 mm. Assume that: $E = 2\times10^{11}\,\text{N/m}^2$ and $P = 114.7$ kN

[Δ = 5 mm]

4. The eccentrically loaded strut of Figure 14.18 is subjected to a compressive load P. If $EI = 20000$ Nm^2, determine the position and value of the maximum deflection assuming the following data apply: $P = 5000$ N, $l = 3$ m and $\Delta = 0.01$ m

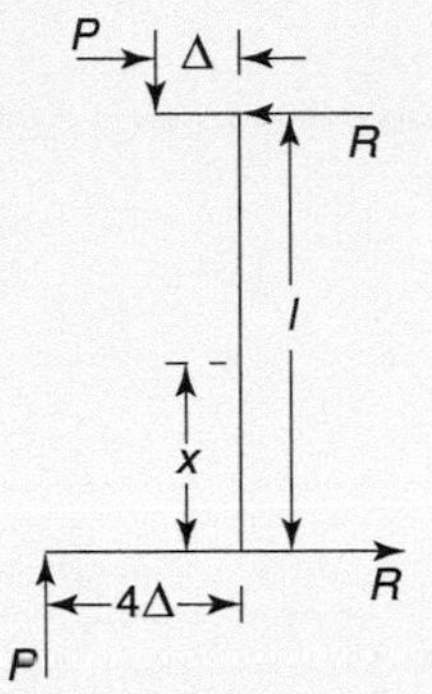

Figure 14.18

[$x = 1.38$ m, $\delta = 9.21 \times 10^{-3}$ m]

5. Show that for the eccentrically loaded strut of Figure 14.19 the bending moment at any distance x is given by:

$$M = P\Delta\left(-2\cos\alpha x + \frac{(1+2\cos\alpha l)}{\sin\alpha l}\sin\alpha x\right)$$

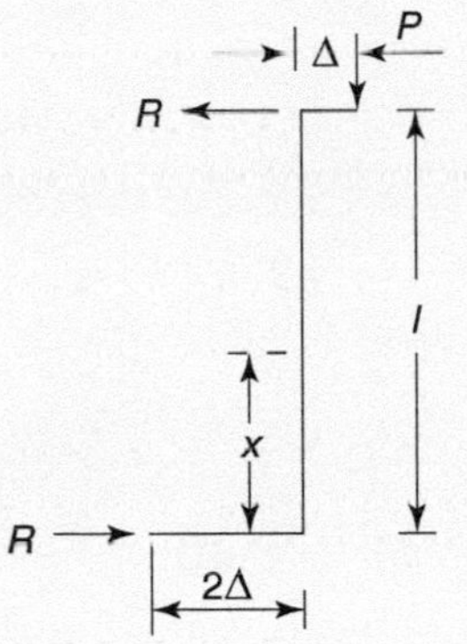

Figure 14.19

6. An initially curved strut, whose initial deflected form is small and parabolic, is symmetrical about its mid-point. If the strut is subjected to a compressive axial load P at its pinned ends, show that the maximum compressive stress is given by:

$$\frac{P}{A}\left[1+\frac{\Delta\hat{y}}{k^2}\frac{8EI}{Pl^2}\left(\sec\frac{\alpha l}{2}-1\right)\right]$$

where Δ = initial central deflection and k = least radius of gyration

Determine Δ for such a strut, assuming the geometrical and material properties of Problem 6 (on page 333) apply. [12.3 cm]

7. An initially curved strut, whose initial deflected form is small and circular, is subjected to a compressive axial load P at its pinned ends. Show that the total deflection y at any distance x is given by:

$$y = -\frac{8\Delta}{\alpha^2 l^2}\left[\cos\alpha x + \tan\frac{\alpha l}{2}\sin\alpha x - 1\right]$$

where Δ is the initial central deflection.

Determine Δ for such a strut, assuming the geometrical and material properties of Problem 6 apply. [12.3 cm]

14.11 Dynamic instability

The theory covered in this chapter has been based on static stability, but it must be pointed out that for struts subjected to compressive periodic axial forces, there is a possibility that dynamic instability can occur when the lateral critical frequency of the beam-column is reached.

The study of dynamic stability is beyond the scope of this book, but is dealt with in much detail in Reference 30 (see page 497).

For fully worked solutions to each of the problems in Exercise 57 and 58 in this chapter, go to the website:
www.routledge.com/cw/bird

Chapter 15

Asymmetrical bending of beams

Why it is important to understand: **Asymmetrical bending of beams**

The asymmetrical bending of beams is a common feature in structural mechanics. Sometimes this feature occurs when a beam of a symmetrical cross-section, such as '*I*' beams, or tee beams, or beams of rectangular cross-section, are subjected to asymmetrical loads to their cross-sections, usually the loads being perpendicular to each other. Other forms of asymmetrical beams occur when beams of unsymmetrical cross-sections are transversely loaded. The cross-sections of asymmetrical beams are usually of the form of angle bars and channel bars.

At the end of this chapter you should be able to:

- understand the most common forms of asymmetrical bending of straight beams
- calculate the maximum and minimum second moments of area of beams of asymmetrical cross-section
- determine stresses in symmetrical-section beams loaded asymmetrically
- determine stresses in beams of asymmetrical cross-sections
- determine deflections of beams of symmetrical cross-sections, loaded asymmetrically
- determine deflections of asymmetrical section beams

Mathematical references

In order to understand the theory involved in this chapter on **asymmetrical bending of beams**, knowledge of the following mathematical topics is required: ***integration, double integrals, trigonometric identities and double angles.*** If help/revision is needed in these areas, see page 50 for some textbook references.

15.1 Introduction

The two most common forms of asymmetrical (or unsymmetrical) bending of straight beams are as follows:

(a) when symmetrical-section beams are subjected to asymmetrical loads, as shown by Figure 15.1, and
(b) when asymmetrical-section beams are subjected to either symmetrical or asymmetrical loading, as shown in Figure 15.2.

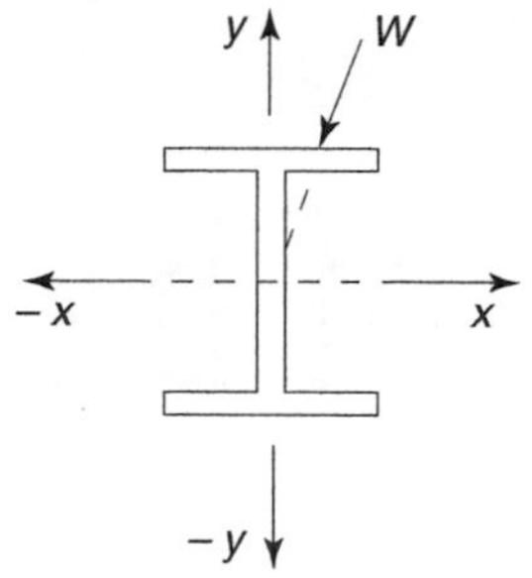

Figure 15.1 A beam of symmetrical section, subjected to an asymmetrical load

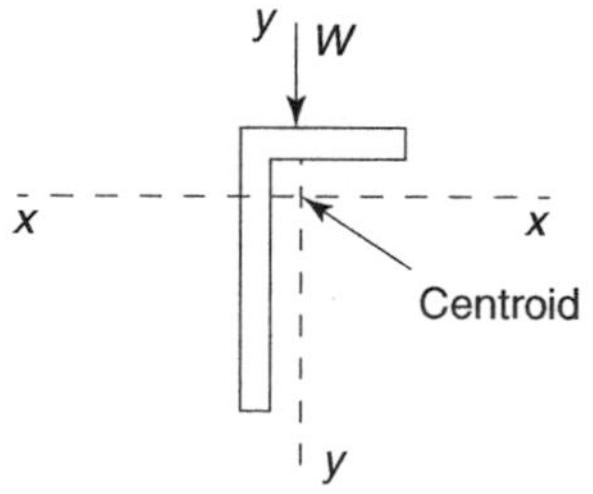

Figure 15.2 A beam of asymmetrical section, subjected to a vertical load

15.2 Symmetrical-section beams loaded asymmetrically

In the case of the symmetrical-section beam which is loaded asymmetrically, the skew load of Figure 15.3(a) can be resolved into two components mutually perpendicular to each other and acting along the axes of symmetry, as shown in Figures 15.3(b) and (c). Assuming that the beam behaves in a linear elastic manner, the effects of bending can be considered separately, about each of the two axes of symmetry, namely *xx* and *yy*, and later the effects of each component of W can be superimposed to give the resultant stresses and deflections. To demonstrate the process, let us assume that the beam of Figure 15.3(a) is of length l, and is simply supported at its ends, with W at mid-span.

It can readily be seen that the components of W are $W \cos \alpha$, acting along the y axis, and $W \sin \alpha$, acting along the x axis, where the former causes the beam to bend about its x-x axis, and the latter causes bending about the y-y axis.

The effect of $W \cos \alpha$ will be to cause the stress in the flange AB, namely $\sigma_{y(AB)}$, to be compressive, whilst the stress in the flange CD, namely $\sigma_{y(CD)}$ will be tensile, so that:

$$\sigma_{y(AB)} = -\frac{(W\cos\alpha)l\bar{y}}{4I_{xx}} \tag{15.1}$$

and

$$\sigma_{y(CD)} = \frac{(W\cos\alpha)l\bar{y}}{4I_{xx}} \tag{15.2}$$

Similarly, owing to $W \sin \alpha$, the stress on the flange edges B and D, namely $\sigma_{x(BD)}$, will be compressive, whilst the stress on the flange edges A and C, namely $\sigma_{x(AC)}$, is tensile, so that:

$$\sigma_{x(BD)} = -\frac{(W\sin\alpha)l\bar{x}}{4I_{yy}} \tag{15.3}$$

and

$$\sigma_{x(AC)} = \frac{(W\sin\alpha)l\bar{x}}{4I_{yy}} \tag{15.4}$$

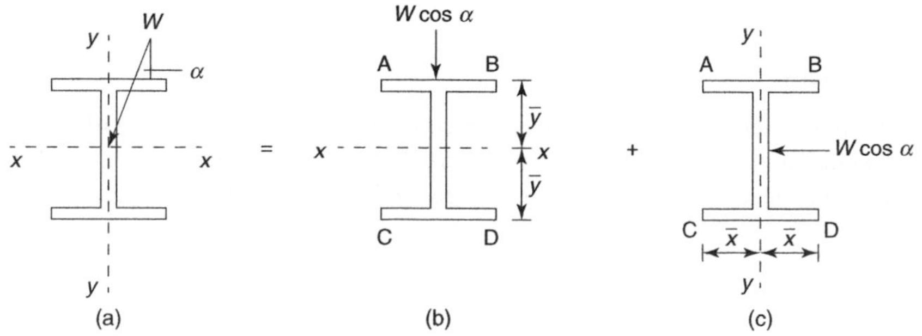

Figure 15.3 Symmetrical beam, loaded asymmetrically

The combined effects of $W \cos \alpha$ and $W \sin \alpha$ will be such that the magnitude of the maximum stresses will be largest at the points B and C, where at point B the stress is compressive, whilst at point C the stress is tensile and of the same magnitude as the stress at point B. At the points A and D, the effects of $W \cos \alpha$ will be to cause stresses of opposite sign to the stresses caused by $W \sin \alpha$, so that the magnitude of these stresses will be less than those at the points B and C.

Thus, in general, the stress at any point in the section of the beam is given by:

$$\sigma = \frac{M(\cos\alpha)y}{I_{xx}} + \frac{M(\sin\alpha)x}{I_{yy}} \qquad (15.5)$$

where x and y are perpendicular distances of the outermost fibres from yy and xx, respectively, and

I_{xx} = second moment of area of section about xx

and I_{yy} = second moment of area of section about yy

M is a bending moment due to W and l, and x and y are as defined in Figure 15.1.

Problem 1. A cantilever of length 2 m and of rectangular section 0.1 m × 0.05 m is subjected to a skew load of 5 kN, at its free end, as shown in Figure 15.4. Determine the stresses at A, B, C and D at its fixed end.

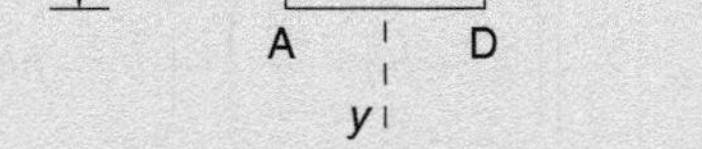

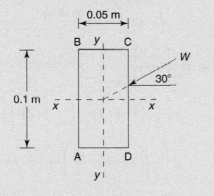

Figure 15.4 Rectangular-section beam

M_y = bending moment about xx

$= W \sin 30° \times 2 = 5\times10^3 \times \sin30° \times 2 = 5$ kN m

M_x = bending moment about yy

$= W \cos 30° \times 2 = 5\times10^3 \times \cos30° \times 2 = 8.66$ kN m

I_{xx} = second moment of area of the cross-section about xx

$$= \frac{BD^3}{12} = \frac{0.05\times0.1^3}{12} = 4.167\times10^{-6}\,\text{m}^4$$

I_{yy} = second moment of area of the cross-section about yy

$$= \frac{DB^3}{12} = \frac{0.1\times0.05^3}{12} = 1.042\times10^{-6}\,\text{m}^4$$

σ_A = maximum stress at the point A

i.e. $$\sigma_A = -\frac{5\times10^{-3}\times0.05}{4.167\times10^{-6}} - \frac{8.66\times10^{-3}\times0.025}{1.042\times10^{-6}}$$

$$= -60\times10^6 - 207.8\times10^6 = \mathbf{-267.8\ MN/m^2}$$

$$\sigma_B = +60\times10^6 - 207.8\times10^6 = \mathbf{-147.8\ MN/m^2}$$

$$\sigma_C = +60\times10^6 + 207.8\times10^6 = \mathbf{267.8\ MN/m^2}$$

$$\sigma_D = -60\times10^6 + 207.8\times10^6 = \mathbf{147.8\ MN/m^2}$$

15.3 Asymmetrical sections

To demonstrate the more complex problem of the bending of an asymmetrical section, consider the cantilever of Figure 15.5, which is of 'Z' section.

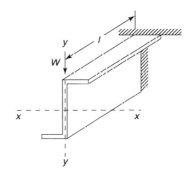

Figure 15.5 Cross-section that is asymmetrical about yy

If the symmetrical theory of bending is applied to the cantilever of Figure 15.5, then at the built-in end, the stress in the top flange would be uniform and tensile, whilst the stress in the bottom flange would be uniform and compressive.

Thus, according to the theory of bending of symmetrical sections, the resisting couple due to the stresses would balance the bending moment about xx due to W, but if such a stress system existed, it would also cause a resisting couple about yy, which is impossible, as there is not applied bending moment about yy.

It is evident, therefore, that the theory of bending for symmetrical sections cannot be applied to asymmetrical sections, and the mathematical explanation for this is as follows:

If M_y = bending moment about yy

and M_x = bending moment about xx

then according to simple bending theory,

$$M_x = \sum \sigma \times y \times \delta a \qquad (15.6)$$

and

$$M_y = \sum \sigma \times x \times \delta a \qquad (15.7)$$

where σ = stress due to bending

$$= \frac{Ey}{R} = \text{constant} \times y \qquad (15.8)$$

and δa = elemental area

Substituting equation (15.8) into equations (15.6) and (15.7) gives:

$$M_x = \sum \text{constant } \times y^2 \times \delta a = \text{constant} \times \sum y^2 \times \delta a$$

$$= \text{constant } \times I_{xx} \qquad (15.9)$$

and

$$M_y = \sum \text{constant } \times x \times y \times \delta a = \text{constant} \times \sum xy \times \delta a$$

$$= \text{constant } \times I_{xy} \qquad (15.10)$$

where I_{xy} = the product of inertia

However, from the heuristic arguments of Chapter 9, page 202

$$M_y = 0$$

Thus the only way that simple bending theory can be satisfied for asymmetrical sections is for the beam to bend about those two mutually perpendicular axes where the product of inertia, namely I_{xy}, is zero.

It will now be shown in the following sections that these two axes are in fact the *principal axes of bending* of the section, rather similar to the axes of principal stresses and principal strains, as discussed in Chapter 9.

For a practical video demonstration of asymmetrical bending of beams go to www.routledge.com/cw/bird and click on the menu for 'Mechanics of Solids 3rd Edition'

15.4 Calculation of I_{xy}

Two elements will be considered in this section, namely a rectangle and the quadrant of a circle.

Consider a rectangle of area A in the positive quadrant of the Cartesian coordinate system of Figure 15.6.

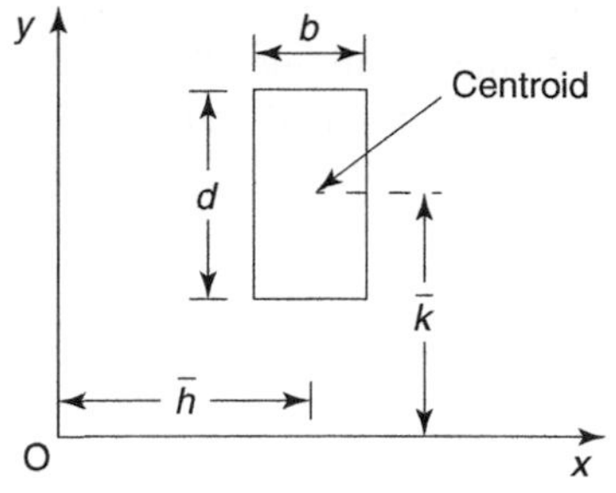

Figure 15.6 Rectangular element

Let $\bar{h}$ = distance of the centroid of the rectangular element from Oy

$\bar{k}$ = distance of the centroid of the rectangular element from Ox

Now, from equation (15.10),

$$I_{xy} = \int xy \, da$$

which for the rectangular element for Figure 15.6 becomes:

$$I_{xy} = \int\int (dx \times dy) x \times y$$

$$= \int_{\bar{k}-d/2}^{\bar{h}+d/2} \int_{\bar{h}-b/2}^{\bar{h}+b/2} xy \, dx \, dy$$

$$\int_{\bar{h}-b/2}^{\bar{h}+b/2} xy \, dx = \left[\frac{yx^2}{2}\right]_{\bar{h}-b/2}^{(\bar{h}+b/2)}$$

$$= \left[\frac{y}{2}\left\{\left(\bar{h}+b/2\right)^2 - \left(\bar{h}-b/2\right)^2\right\}\right]$$

$$= \left[\frac{y}{2}\left\{\left(\bar{h}^2 + b\bar{h} + \frac{b^2}{4}\right) - \left(\bar{h}^2 - b\bar{h} + \frac{b^2}{4}\right)\right\}\right]$$

$$= \left[\frac{y}{2}\left\{2b\bar{h}\right\}\right] = b\bar{h}y$$

Hence $$I_{xy} = \int_{\bar{k}-d/2}^{\bar{k}+d/2} \left(b\bar{h}y\right) dy = \left[\frac{b\bar{h}y^2}{2}\right]_{\bar{k}-d/2}^{\bar{k}+d/2}$$

$$= \left[\frac{b\bar{h}}{2}\left\{\left(\bar{k}+d/2\right)^2 - \left(\bar{k}-d/2\right)^2\right\}\right]$$

$$=\left[\frac{b\bar{h}}{2}\left\{\left(\bar{k}^2+d\bar{k}+\frac{d^2}{4}\right)-\left(\bar{k}^2-b\bar{k}+\frac{d^2}{4}\right)\right\}\right]$$

$$=\left[\frac{b\bar{h}}{2}\left\{2d\bar{k}\right\}\right]$$

i.e. $I_{xy} = bd\bar{h}\bar{k}$

Therefore $\boldsymbol{I_{xy} = A\bar{h}\bar{k}}$ (15.11)

where A is the area of the rectangle. Hence, for a built-up section, consisting of n rectangles, equation (15.11) becomes:

$$I_{xy} = \sum_{i=1}^{n} A_i\bar{h}_i\bar{k}_i \quad (15.12)$$

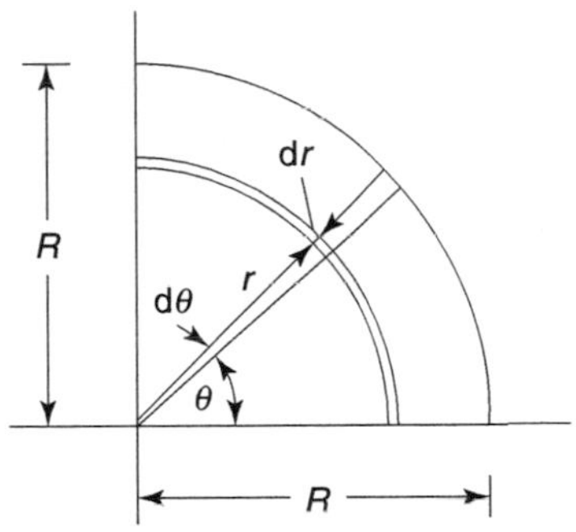

Figure 15.7 Quadrant of a circle

For the *quadrant of the circle* in Figure 15.7

$$I_{xy} = \int xy \times da$$

$$= \int_0^{\pi/2}\int_0^R r\cos\theta \times r\sin\theta \times dr \times r\,d\theta$$

$$= \int_0^{\pi/2}\int_0^R r^3\cos\theta\sin\theta\,dr\,d\theta$$

$$\int_0^R r^3(\cos\theta\sin\theta)dr = (\cos\theta\sin\theta)\left[\frac{r^4}{4}\right]_0^R = \frac{R^4}{4}(\cos\theta\sin\theta)$$

Hence

$$I_{xy} = \int_0^{\pi/2}\frac{R^4}{4}(\cos\theta\sin\theta)\,d\theta$$

$$= \int_0^{\pi/2}\frac{R^4}{4}\,\frac{\sin 2\theta}{2}\,d\theta \text{ since}$$

$\sin 2\theta = 2\sin\theta\cos\theta$

$$= \frac{R^4}{4}\left[-\frac{\cos 2\theta}{4}\right]_0^{\pi/2} = \frac{R^4}{4}\left[\left(-\frac{\cos 2(\pi/2)}{4}\right)-\left(-\frac{\cos 0}{4}\right)\right]$$

$$= \frac{R^4}{4}\left[\left(--\frac{1}{4}\right)-\left(-\frac{1}{4}\right)\right] = \frac{R^4}{4}\left(\frac{1}{2}\right)$$

i.e. $\boldsymbol{I_{xy} = \frac{R^4}{8}}$

15.5 Principal axes of bending

Let OU and OV be the principal axes, and Ox and Oy be the reference (or global) axes, as shown in Figure 15.8.

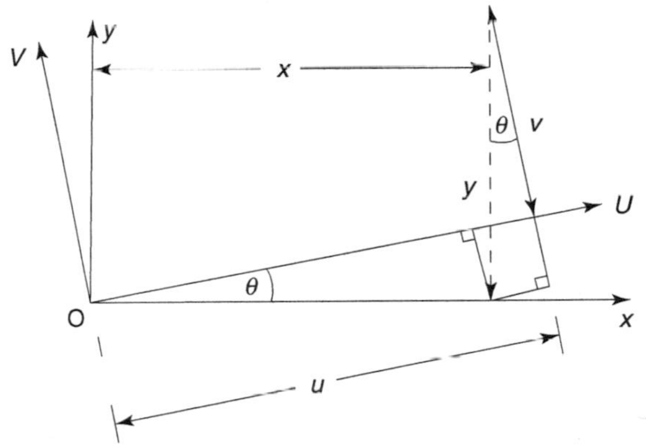

Figure 15.8 Principal and reference axes

From Figure 15.8, it can be seen that:

$$u = x\cos\theta + y\sin\theta \quad (15.13)$$
$$v = y\cos\theta - x\sin\theta \quad (15.14)$$

Equations (15.13) and (15.14) are in fact **co-ordinate transformations**, as these equations transform one set of orthogonal co-ordinates to another set of orthogonal coordinates.

Now to satisfy simple bending theory and equilibrium, it is necessary for the product of inertia to be zero, with reference to the principal axes of bending, namely the axes OU and OV, i.e.

$$I_{uv} = \int uv\,da$$

or $\int(x\cos\theta + y\sin\theta)\times(y\cos\theta - x\sin\theta)\times da = 0$

i.e. $\int xy\cos^2\theta\,da - \int x^2\cos\theta\sin\theta\,da + \int y^2\sin\theta\cos\theta\,da - \int xy\sin^2\theta\,da = 0$

i.e. $$\left(\cos^2\theta - \sin^2\theta\right)\int xy\,da + \cos\theta\sin\theta\left(\int y^2\,da - \int x^2\,da\right) = 0$$

i.e. $$\cos 2\theta \int xy\,da + \frac{1}{2}\sin 2\theta\left(\int y^2\,da - \int x^2\,da\right) = 0$$

i.e. $$\cos 2\theta\, I_{xy} + \frac{1}{2}\sin 2\theta\left(I_x - I_y\right) = 0$$

Therefore, $$\cos 2\theta\, I_{xy} = -\frac{1}{2}\sin 2\theta\left(I_x - I_y\right)$$

i.e. $$\cos 2\theta\, I_{xy} = \frac{1}{2}\sin 2\theta\left(I_y - I_x\right)$$

and $$\frac{2I_{xy}}{\left(I_y - I_x\right)} = \frac{\sin 2\theta}{\cos 2\theta}$$

from which $$\mathbf{\tan 2\theta = \frac{2I_{xy}}{\left(I_y - I_x\right)}} \qquad (15.15)$$

To determine I_U and I_V, the principal second moments of area

Now $$I_U = \int v^2 da \qquad (15.16)$$

Substituting equation (15.14) into (15.16) gives:

$$I_U = \int \left(y\cos\theta - x\sin\theta\right)^2 da$$
$$= \int \left(y^2\cos^2\theta - 2xy\sin\theta\cos\theta + x^2\sin^2\theta\right)da$$
$$= \cos^2\theta \int y^2\,da + \sin^2\theta \int x^2\,da - 2\cos\theta\sin\theta \int xy\,da$$
$$= \cos^2\theta\, I_x + \sin^2\theta\, I_y - 2\cos\theta\sin\theta\, I_{xy}$$

From equation (15.16), $I_{xy} = \left(\frac{I_y - I_x}{2}\right)\tan 2\theta$

Hence

$$I_U = \cos^2\theta\, I_x + \sin^2\theta\, I_y - \sin 2\theta\left(\frac{I_y - I_x}{2}\right)\tan 2\theta$$

and

$$I_U = \left(\frac{1+\cos 2\theta}{2}\right)I_x + \left(\frac{1-\cos 2\theta}{2}\right)I_y - \sin 2\theta\left(\frac{I_y - I_x}{2}\right)\tan 2\theta$$

$$= \frac{1}{2}I_x + \frac{1}{2}\cos 2\theta\, I_x + \frac{1}{2}I_y - \frac{1}{2}\cos 2\theta\, I_y - \frac{1}{2}\sin 2\theta\tan 2\theta\, I_y + \frac{1}{2}\sin 2\theta\tan 2\theta\, I_x$$

$$= \frac{1}{2}\left(I_x + I_y\right) + \frac{1}{2}\left(\cos 2\theta + \sin 2\theta\tan 2\theta\right)I_x - \frac{1}{2}\left(\cos 2\theta + \sin 2\theta\tan 2\theta\right)I_y$$

$$= \frac{1}{2}\left(I_x + I_y\right) + \frac{1}{2}\left(\cos 2\theta + \sin 2\theta\frac{\sin 2\theta}{\cos 2\theta}\right)I_x - \frac{1}{2}\left(\cos 2\theta + \sin 2\theta\frac{\sin 2\theta}{\cos 2\theta}\right)I_y$$

$$= \frac{1}{2}\left(I_x + I_y\right) + \frac{1}{2}\left(\frac{\cos^2 2\theta + \sin^2 2\theta}{\cos 2\theta}\right)I_x - \frac{1}{2}\left(\frac{\cos^2 2\theta + \sin^2 2\theta}{\cos 2\theta}\right)I_y$$

$$= \frac{1}{2}\left(I_x + I_y\right) + \frac{1}{2}\left(\frac{1}{\cos 2\theta}\right)I_x - \frac{1}{2}\left(\frac{1}{\cos 2\theta}\right)I_y$$

$$= \frac{1}{2}\left(I_x + I_y\right) + \frac{1}{2}\sec 2\theta\, I_x - \frac{1}{2}\sec 2\theta\, I_y$$

i.e. $$\mathbf{I_U = \frac{1}{2}\left(I_x + I_y\right) + \frac{1}{2}\left(I_x - I_y\right)\sec 2\theta} \qquad (15.17)$$

Similarly

$$I_V = \int u^2 da = \int \left(x\cos\theta + y\sin\theta\right)^2 da$$
$$= \int \left(x^2\cos^2\theta + 2xy\sin\theta\cos\theta + y^2\sin^2\theta\right)da$$
$$= \cos^2\theta \int x^2\,da + \sin^2\theta \int y^2\,da + 2\cos\theta\sin\theta \int xy\,da$$

i.e.

$$I_V = \cos^2\theta\, I_y + \sin^2\theta\, I_x + \sin 2\theta\, I_{xy}$$

By similar working to above

$$\mathbf{I_V = \frac{1}{2}\left(I_x + I_y\right) - \frac{1}{2}\left(I_x - I_y\right)\sec 2\theta} \qquad (15.18)$$

If equations (15.17) and (15.18) are added together, it can be seen that:

$$I_U + I_V = \frac{1}{2}\left(I_x + I_y\right) + \frac{1}{2}\left(I_x - I_y\right)\sec 2\theta$$
$$+ \frac{1}{2}\left(I_x + I_y\right) - \frac{1}{2}\left(I_x - I_y\right)\sec 2\theta$$

i.e. $\boldsymbol{I_U + I_V = I_x + I_y}$ (15.19)

Equation (15.19) is known as the ***invariant of inertia.***

15.6 Mohr's circle of inertia

Equations (15.15), (15.17) and (15.18) can be represented by a circle of inertia, as shown by Figure 15.9, rather similar to Mohr's circles* for stress and strain, as discussed in Chapter 9.

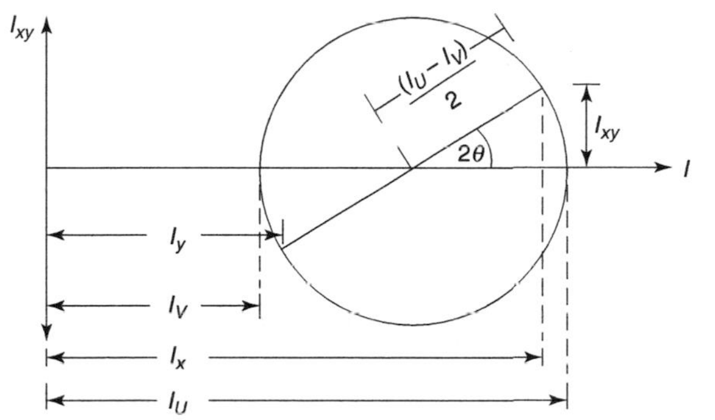

Figure 15.9 Mohr's circle of inertia

From Figure 15.9 it can readily be seen that I_U and I_V are maximum and minimum values of second moments of area, and this is why I_U and I_V are called principal second moments of area.

Plots of the variation of the radius of gyration in any direction are called *momental ellipses,* and typical momental ellipses are shown in Figure 15.10.

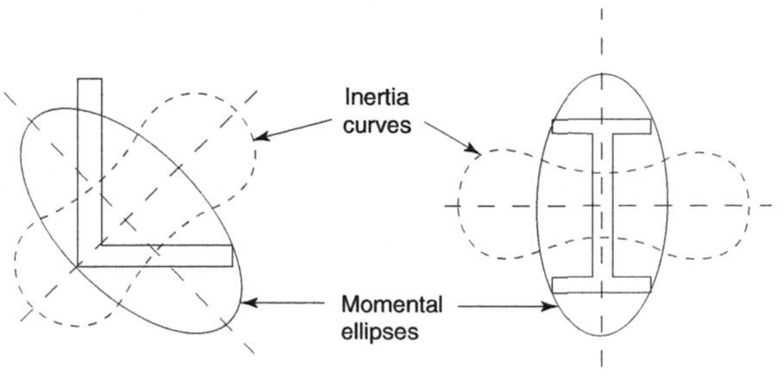

Figure 15.10 Momental ellipses and inertia curves

Structural engineers sometimes find another set of curves useful, which are known as *inertia curves.* The inertia curve of a section is obtained by plotting a radius vector equal to the second moment of area of the section at any angle θ, as shown by the dashed lines of Figure 15.10. From this figure, it can be seen that the inertia curves are effectively plotted in a direction perpendicular to the momental ellipses.

Problem 2. Determine the direction and value of the principal second moments of area of the asymmetrical section of Figure 15.11.

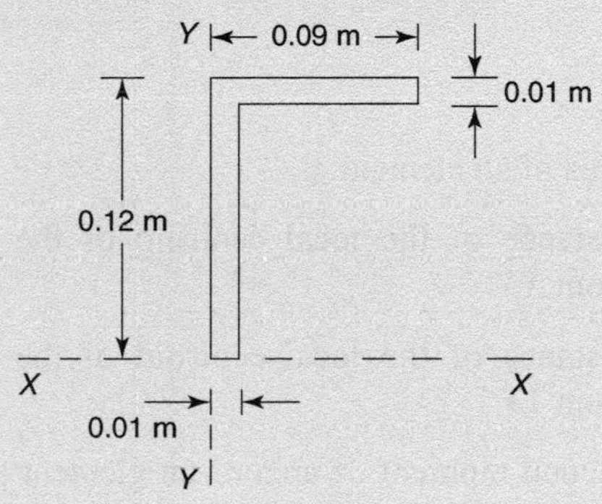

Figure 15.11

Consider the angle bar of Figure 15.11 and assume that its centroid is at O, as shown in Figure 15.12. For convenience, the angle bar can be assumed to be composed of two rectangles, and the geometrical properties of the angle bar can be calculated with the aid of Tables 15.1 and 15.2, where

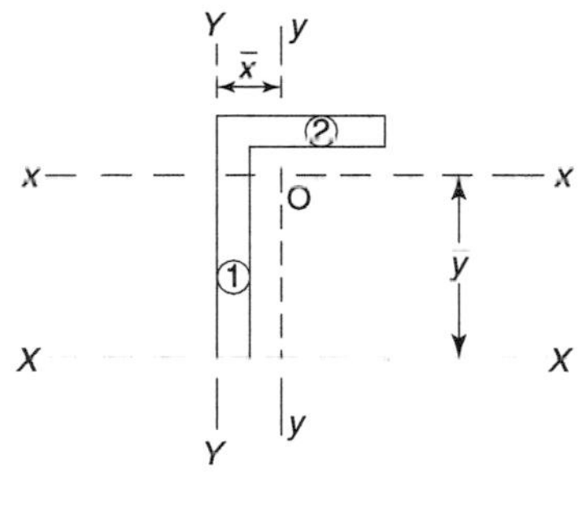

Figure 15.12

*Christian Otto Mohr (8 October 1835 – 2 October 1918) was a German civil engineer. In 1882, he famously developed the graphical method for analysing stress known as Mohr's circle and used it to propose an early theory of strength based on shear stress. To find out more about Mohr go to www.routledge.com/cw/bird

Table 15.1 Calculation of I_{xx}

Section	a	y	ay	ay^2	i_H
1	1.2×10^{-3}	0.06	7.2×10^{-5}	4.32×10^{-6}	$\frac{0.01\times0.12^3}{12}=1.44\times10^{-6}$
2	8×10^{-4}	0.115	9.2×10^{-5}	1.058×10^{-5}	$\frac{0.08\times0.01^3}{12}=6.7\times10^{-9}$
Σ	2×10^{-3}	–	1.64×10^{-4}	1.49×10^{-5}	1.447×10^{-6}

a = area of an element

y = distance of the local centroid of the element from XX

x = distance of the local centroid of the element from YY

i_H = second moment of area of an element about an axis passing through its local centroid and parallel to XX

i_V = second moment of area of an element about an axis passing through its local centroid and parallel to YY

ay = the product $a\times y$ ay^2 = the product $a\times y\times y$

ax = the product $a\times x$ ax^2 = the product $a\times x\times x$

$\sum$ = summation of column, where appropriate

From Table 15.1,

$$\bar{y}=\frac{\sum ay}{\sum a}=\frac{1.64\times10^{-4}}{2\times10^{-3}}=\mathbf{0.082\ m}$$

$$I_{XX}=\sum i_H+\sum ay^2=1.447\times10^{-6}+1.49\times10^{-5}$$

i.e. $\boldsymbol{I_{XX}=1.635\times10^{-5}\ m^4}$

$$I_{xx}=I_{XX}-(\bar{y})^2\sum a=1.635\times10^{-5}-0.082^2\times2\times10^{-3}$$

i.e. $\boldsymbol{I_{xx}=2.902\times10^{-6}\ m^4}$ (15.20)

From Table 15.2,

$$\bar{x}=\frac{\sum ax}{\sum a}=\frac{4.6\times10^{-5}}{2\times10^{-3}}=\mathbf{0.023\ m}$$

$$I_{YY}=\sum i_V+\sum ax^2=4.37\times10^{-7}+2.03\times10^{-6}$$

i.e. $\boldsymbol{I_{YY}=2.467\times10^{-6}\ m^4}$

$$I_{yy}=I_{YY}-(\bar{x})^2\sum a=2.467\times10^{-6}-0.023^2\times2\times10^{-3}$$

i.e. $\boldsymbol{I_{yy}=1.409\times10^{-6}\ m^4}$ (15.21)

To calculate I_{xy}

From equation (15.12)

$$I_{xy}=\sum_{i=1}^{n}A_i\bar{h}_i\bar{k}_i$$

$$=1.2\times10^{-3}\times\left[-(\bar{x}-0.005)\right]\times\left[-(\bar{y}-0.06)\right]+8\times10^{-4}\times(0.05-\bar{x})\times(0.115-\bar{y})$$

$$=1.2\times10^{-3}\times\left[-(0.023-0.005)\right]\times\left[-(0.082-0.06)\right]+8\times10^{-4}\times(0.05-0.023)\times(0.115-0.082)$$

$$=1.2\times10^{-3}\times-0.018\times-0.022+8\times10^{-4}\times0.027\times0.033$$

i.e. $\boldsymbol{I_{xy}=1.188\times10^{-6}\ m^4}$ (15.22)

Table 15.2 Calculation of I_{yy}

Section	a	x	ax	ax^2	i_Y
1	1.2×10^{-3}	0.005	6×10^{-6}	3×10^{-8}	$\frac{0.12\times0.01^3}{12}=1\times10^{-8}$
2	8×10^{-4}	0.05	4×10^{-5}	2×10^{-6}	$\frac{0.01\times0.08^3}{12}=4.27\times10^{-7}$
Σ	2×10^{-3}	–	4.6×10^{-5}	2.03×10^{-6}	4.37×10^{-7}

To calculate θ

Substituting the appropriate values from equations (15.20) to (15.22) into equation (15.15) gives:

$$\tan 2\theta = \frac{2I_{xy}}{(I_y - I_x)} = \frac{2\times1.188\times10^{-6}}{(1.409\times10^{-6}-2.902\times10^{-6})}$$

$$=-1.59143$$

and $$2\theta = \tan^{-1}(-1.59143) = -57.86°$$

from which $$\theta = \frac{-57.86°}{2} = \mathbf{-28.93°} \quad (15.23)$$

where θ is shown in Figure 15.13.

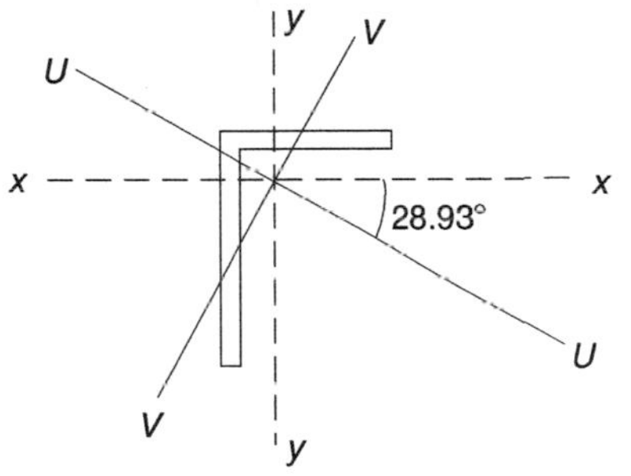

Figure 15.13 Directions of the principal axes

Now $$I_U = \frac{1}{2}(I_x + I_y) + \frac{1}{2}(I_x - I_y)\sec 2\theta$$

from equation (15.17)

and $$I_V = \frac{1}{2}(I_x + I_y) - \frac{1}{2}(I_x - I_y)\sec 2\theta$$

from equation (15.18)

Hence, by substituting the values for I_x, I_y, I_{xy} and θ into these equations gives:

$$I_U = \frac{1}{2}(2.902+1.409)\times10^{-6} + \frac{1}{2}(2.902-1.409)\times10^{-6}\times\left(\frac{1}{\cos(-57.86°)}\right)$$

$$=2.156\times10^{-6}+7.465\times10^{-7}\times1.88$$

i.e. $\mathbf{I_U = 3.56\times10^{-6}\ m^4}$ (15.24)

and

$$I_V = \frac{1}{2}(2.902+1.409)\times10^{-6} - \frac{1}{2}(2.902-1.409)\times10^{-6}\times\left(\frac{1}{\cos(-57.86°)}\right)$$

$$=2.156\times10^{-6}-7.465\times10^{-7}\times1.88$$

i.e. $\mathbf{I_V = 7.53\times10^{-7}\ m^4}$ (15.25)

Problem 3. Determine the direction and value of the principal second moments of area of the asymmetrical section of Figures 15.14.

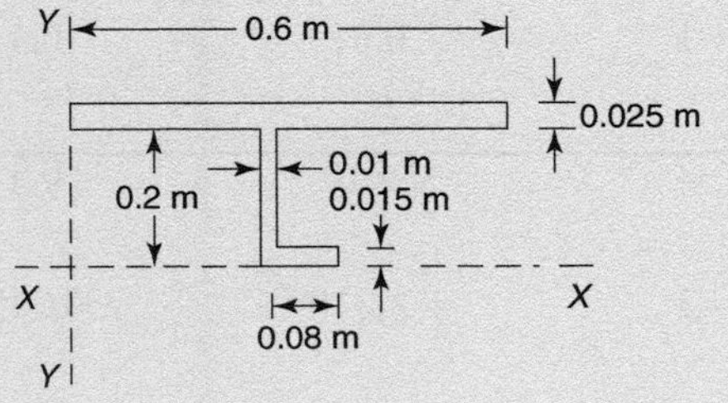

Figure 15.14

Let the centroid of the section in Figure 15.14 be at O, as shown in Figure 15.15.

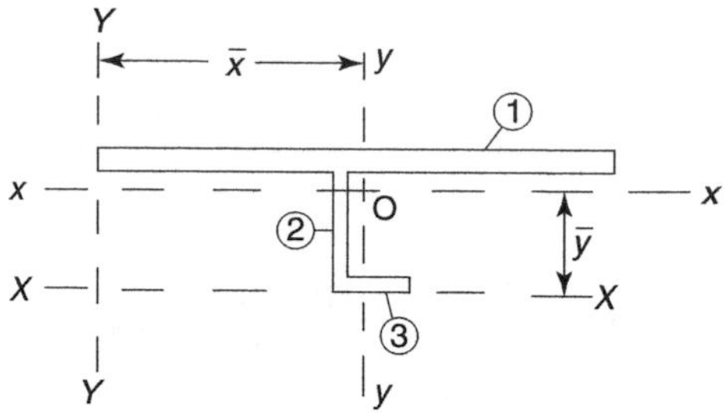

Figure 15.15 Cross-section of beam

Tables 15.3 and 15.4 show the calculations for determining I_{xx} and I_{yy} of this section, where the symbols have the same meanings as in Problem 2.

From Table 15.3

$$\bar{y} = \frac{\sum ay}{\sum a} = \frac{3.397\times10^{-3}}{0.0182} = \mathbf{0.1866\ m}$$

$$I_{XX} = \sum i_H + \sum ay^2 = 7.471\times10^{-6} + 6.974\times10^{-4}$$

i.e. $\mathbf{I_{XX} = 7.049\times10^{-4}\ m^4}$

$$I_{xx} = I_{XX} - (\bar{y})^2 \sum a = 7.049\times10^{-4} - 0.1866^2\times0.0182$$

i.e. $\mathbf{I_{xx} = 7.118\times10^{-5}\ m^4}$ (15.26)

From Table 15.4

$$\bar{x} = \frac{\sum ax}{\sum a} = \frac{5.514\times10^{-3}}{0.0182} = \mathbf{0.303\ m}$$

$$I_{YY} = \sum i_V + \sum ax^2 = 4.51\times10^{-4} + 1.673\times10^{-3}$$

i.e. $\mathbf{I_{YY} = 2.124\times10^{-3}\ m^4}$

$$I_{yy} = I_{YY} - (\bar{x})^2 \sum a = 2.124\times10^{-3} - 0.303^2\times0.0182$$

Table 15.3 Calculation of I_{xx}

Section	a	y	ay	ay^2	i_H
1	0.015	0.2125	3.188×10^{-3}	6.773×10^{-4}	7.81×10^{-7}
2	2×10^{-3}	0.1	2×10^{-4}	2×10^{-5}	6.667×10^{-6}
3	1.2×10^{-3}	7.5×10^{-3}	9×10^{-6}	6.75×10^{-8}	2.25×10^{-8}
Σ	0.0182	–	3.397×10^{-3}	6.974×10^{-4}	7.471×10^{-6}

Table 15.4 Calculation of I_{yy}

Section	a	x	ax	ax^2	i_V
1	0.015	0.3	4.5×10^{-3}	1.35×10^{-3}	4.5×10^{-4}
2	2×10^{-3}	0.3	6×10^{-4}	1.8×10^{-4}	1.67×10^{-8}
3	1.2×10^{-3}	0.345	4.14×10^{-4}	1.428×10^{-4}	6.4×10^{-7}
Σ	0.0182	–	5.514×10^{-3}	1.673×10^{-3}	4.51×10^{-4}

i.e. $$\boldsymbol{I_{yy} = 4.531 \times 10^{-4}\ \mathrm{m}^4} \qquad (15.27)$$

$$I_{xy} = \sum_{i=1}^{n} A_i \overline{h_i}\,\overline{k_i}$$

$$= 0.015 \times (0.3 - \bar{x}) \times (0.2125 - \bar{y}) + 2 \times 10^{-3} \times (0.3 - \bar{x}) \times (0.1 - \bar{y}) + 1.2 \times 10^{-3} \times (0.345 - \bar{x}) \times (7.5 \times 10^{-3} - \bar{y})$$

$$= 0.015 \times (0.3 - 0.303) \times (0.2125 - 0.1866) + 2 \times 10^{-3} \times (0.3 - 0.303) \times (0.1 - 0.1866) + 1.2 \times 10^{-3} \times (0.345 - 0.303) \times (7.5 \times 10^{-3} - 0.1866)$$

$$= 0.015 \times (-0.003) \times (0.0259) + 2 \times 10^{-3} \times (-0.003) \times (-0.0866) + 1.2 \times 10^{-3} \times (0.042) \times (-0.1791)$$

i.e. $$\boldsymbol{I_{xy} = -9.673 \times 10^{-6}\ \mathrm{m}^4} \qquad (15.28)$$

Substituting the appropriate values from equations (15.20) to (15.22) into equation (15.15) gives:

$$\tan 2\theta = \frac{2I_{xy}}{(I_y - I_x)} = \frac{2 \times -9.673 \times 10^{-6}}{(4.531 \times 10^{-4} - 7.188 \times 10^{-5})} = -0.050745$$

and $$2\theta = \tan^{-1}(-0.05075) = -2.905°$$

from which $$\theta = \frac{-2.905°}{2} = \mathbf{-1.45°}$$

Now, $$I_U = \frac{1}{2}(I_x + I_y) + \frac{1}{2}(I_x - I_y)\sec 2\theta$$

from equation (15.17)

and $$I_V = \frac{1}{2}(I_x + I_y) - \frac{1}{2}(I_x - I_y)\sec 2\theta$$

from equation (15.18)

Hence, by substituting the values for I_x, I_y, I_{xy} and θ into these equations gives:

$$I_U = \frac{1}{2}(7.118 \times 10^{-5} + 4.531 \times 10^{-4}) + \frac{1}{2}(7.118 \times 10^{-5} - 4.531 \times 10^{-4}) \times \left(\frac{1}{\cos(-2.905°)}\right)$$

$$= 2.6214 \times 10^{-4} - 1.9096 \times 10^{-4} \times 1.0013$$

i.e. $$\boldsymbol{I_U = 7.093 \times 10^{-5}\ \mathrm{m}^4}$$

and

$$I_V = \frac{1}{2}(7.118 \times 10^{-5} + 4.531 \times 10^{-4}) - \frac{1}{2}(7.118 \times 10^{-5} - 4.531 \times 10^{-4}) \times \left(\frac{1}{\cos(-2.905°)}\right)$$

$$= 2.6214 \times 10^{-4} + 1.9096 \times 10^{-4} \times 1.0013$$

i.e. $$\boldsymbol{I_V = 4.533 \times 10^{-4}\ \mathrm{m}^4}$$

Now try the following Practice Exercise

Practice Exercise 59. Asymmetrical bending of beams

1. Determine the direction and magnitudes of the principal second moments of area of the angle bar shown in Figure 15.16.

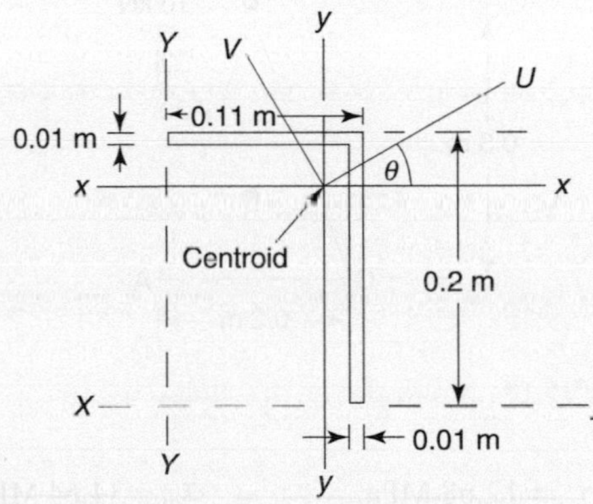

Figure 15.16

$[I_{xx} = 1.268 \times 10^{-5}\,\mathrm{m}^4,\ I_{yy} = 2.849 \times 10^{-6}\,\mathrm{m}^4,$

$I_{xy} = -3.483 \times 10^{-6}\,\mathrm{m}^4, \qquad \theta = 17.64°,$

$I_{UU} = 1.381 \times 10^{-5}\,\mathrm{m}^4,\ I_{VV} = 1.741 \times 10^{-6}\,\mathrm{m}^4]$

2. Determine the direction and magnitudes of the principal second moments of area of the section of Figure 15.17.

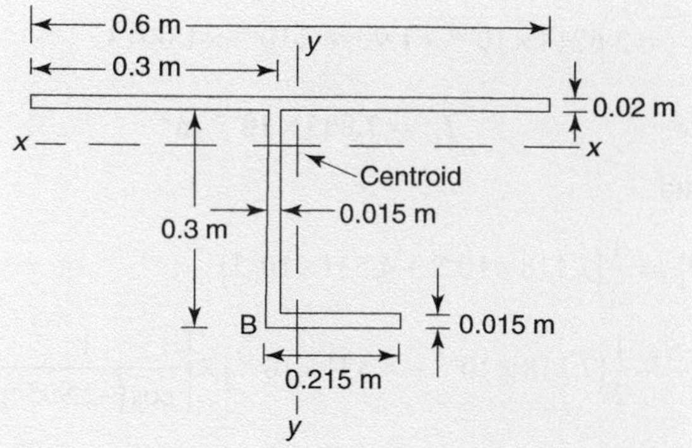

Figure 15.17

$[I_{xx} = 2.844\times10^{-4}\,m^4,\quad I_{yy} = 3.994\times10^{-4}\,m^4,$

$I_{xy} = -7.098\times10^{-5}\,m^4,\quad \theta = -25.49°,$

$I_{UU} = 1.592\times10^{-4}\,m^4,\quad I_{VV} = 5.246\times10^{-4}\,m^4]$

3. Determine the stresses at the corners and the maximum deflection of a cantilever of length 3 m, loaded at its free end with a concentrated load of 10 kN, as shown in Figure 15.18.

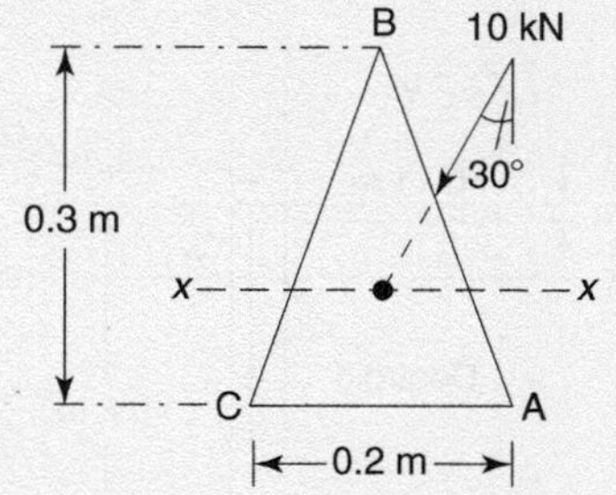

Figure 15.18 Cross-section of cantilever

$[\sigma_A = 12.68$ MPa, $\sigma_B = 34.64$ MPa,

$\sigma_C = -47.32$ MPa, neutral axis is 60° clockwise from xx]

15.7 Stresses in beams of asymmetrical section

Equation (15.5) can be extended to the bending of asymmetrical sections, as shown by equation (15.29) which gives the value of bending stress at any point P in the positive quadrant for the orthogonal axes OU and OV, where θ, u and v are defined in Figure 15.19:

$$\sigma = \frac{M\cos\theta\, v}{I_{UU}} + \frac{M\sin\theta\, u}{I_{VV}} \qquad (15.29)$$

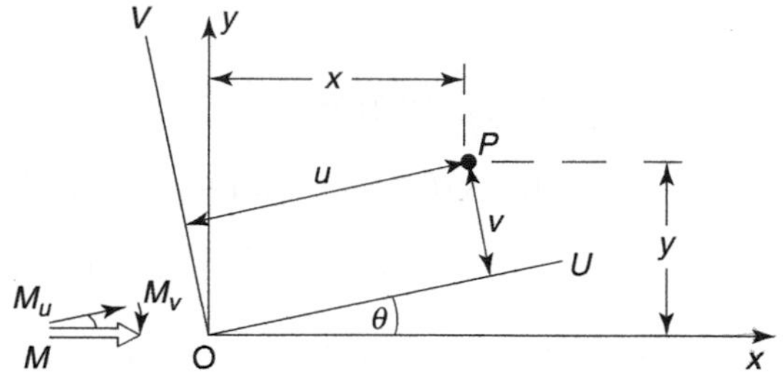

Figure 15.19 Principal axes of bending, Ou and Ov

From equation (15.29), it can be seen that M is due to a load that acts perpendicular to the Ox axis, where in Figure 15.19, M is shown according to the right-hand screw rule. The components of M, namely M_U and M_V, cause bending about the principal axes OU and OV, respectively, where

$$M_U = M\cos\theta$$

and

$$M_V = M\sin\theta$$

Problem 4. A cantilever of length 2 m is subjected to a concentrated load of 5 kN at its free end, as shown in Figure 15.20. Assuming that the cantilever's cross-section is as shown in Figure 15.11 on page 343, determine the direction of its neutral axis and the position and magnitude of the maximum stress.

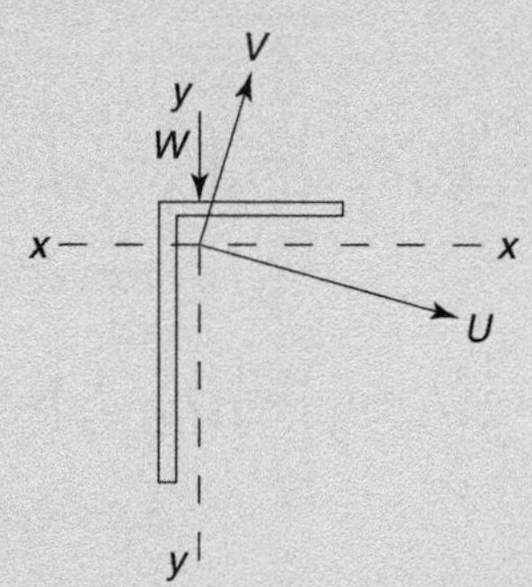

Figure 15.20 Cross-section of cantilever

The maximum bending moment $(\widehat{M})$ occurs at the built-in end, where

$$\widehat{M} = Wl = 5\text{ kN} \times 2\text{ m} = 10\text{ kN m} \qquad (15.30)$$

At the neutral axis, the bending stress is zero,

i.e. $$\frac{M\cos\theta\, v}{I_{UU}} + \frac{M\sin\theta\, u}{I_{VV}} = 0 \qquad (15.31)$$

or $$\frac{M\cos\theta\, v}{I_{UU}} = -\frac{M\sin\theta\, u}{I_{VV}}$$

and $$\frac{v}{u} = -\frac{I_{UU}\sin\theta}{I_{VV}\cos\theta} = -\frac{I_{UU}}{I_{VV}}\tan\theta = \tan\beta$$

where β is defined in Figure 15.21.

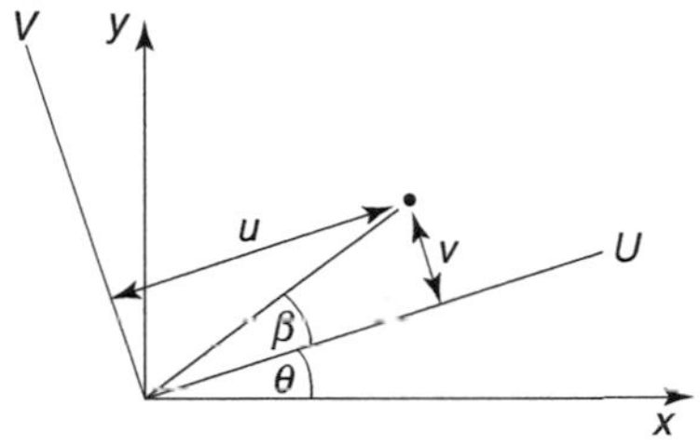

Figure 15.21 Definition of β

Therefore $$\boldsymbol{\beta = \tan^{-1}\left(-\frac{I_{UU}\tan\theta}{I_{VV}}\right)} \qquad (15.32)$$

Substituting the appropriate values from Problem 2, page 343, into equation (15.32) gives:

$$\beta = \tan^{-1}\left(-\frac{I_{UU}\tan\theta}{I_{VV}}\right) = \tan^{-1}\left(-\frac{3.56\times10^{-6}\times\tan(-28.93^\circ)}{7.53\times10^{-7}}\right)$$

i.e. $$\boldsymbol{\beta = 69.06^\circ}$$

The direction of the neutral axis is shown in Figure 15.22, and the largest stress due to bending will occur at a point on the section which is at the furthest perpendicular distance from the neutral axis (NA).

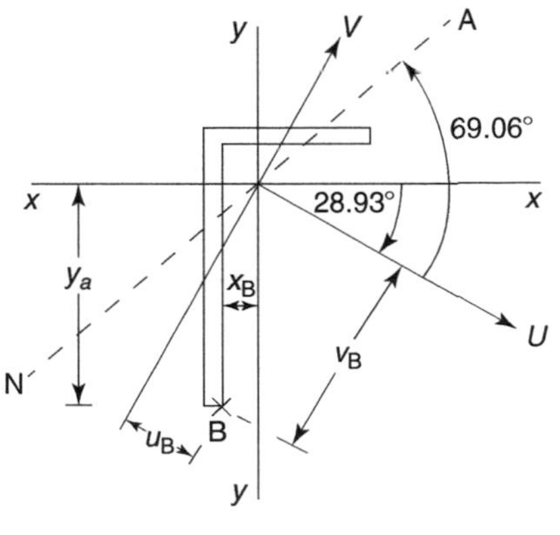

Figure 15.22 Position of neutral axis

From Figure 15.22, it can be seen that the furthest perpendicular distance from NA is at the point B.

Let u_B = deflection of 'B' in the 'u' direction,

and v_B = deflection of 'B' in the 'v' direction,

then from equation (15.13),

$$u_B = x_B\cos\theta + y_B\sin\theta$$
$$= x_B\cos(-28.93^\circ) + y_B\sin(-28.93^\circ)$$
$$= -0.013\times0.875 - 0.083\times(-0.483)$$

i.e. $\boldsymbol{u_B = 0.029\ \text{m}}$

and from equation (15.14)

$$v_B = y_B\cos\theta - x_B\sin\theta$$
$$= -0.083\times0.875 - 0.013\times0.483$$

i.e. $\boldsymbol{v_B = -0.0789\ \text{m}}$

where u_B, v_B, etc., are defined in Figure 15.22.

From equation (15.29) σ_B = stress at the point B

i.e. $$\sigma_B = \frac{M\cos\theta\, v_B}{I_{UU}} + \frac{M\sin\theta\, u_B}{I_{VV}}$$

$$= \frac{10\times10^3\,\text{N m}\times\cos(-28.93^\circ)\times(-0.0789)\,\text{m}}{3.56\times10^{-6}\,\text{m}^4}$$
$$+ \frac{10\times10^3\,\text{N m}\times\sin(-28.93^\circ)\times(0.029)\,\text{m}}{7.53\times10^{-7}\,\text{m}^4}$$
$$= -193.97\times10^6 - 183.3\times10^6$$

i.e. $\sigma_B = -380.3\times10^6 = \boldsymbol{-380.3\ \text{MN/m}^2}$ **(compressive)**

Problem 5. If an encastré beam of length 2 m and with a cross-section as in Figure 15.14 on page 345 is subjected to the uniformly distributed load shown in Figure 15.23, determine the position and value of the maximum stress.

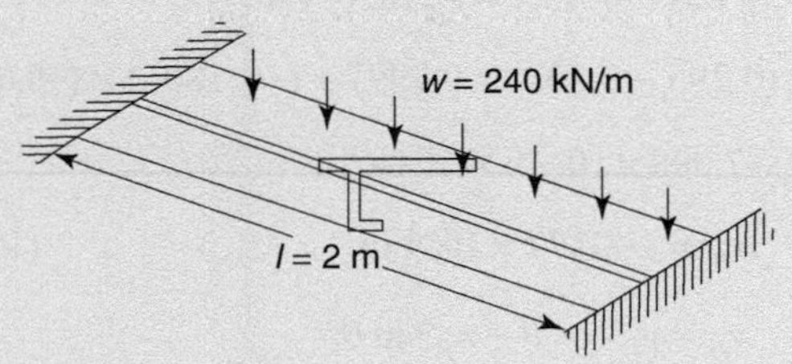

Figure 15.23 Encastré beam with a uniformly distributed load

The maximum value for bending moment $(\widehat{M})$ occurs at the ends, and is given by:

$$\widehat{M}=\frac{wl^2}{12}=\frac{240\times 2^2}{12}=\mathbf{80\ kN\ m}$$

From equation (15.32)

$$\beta=\tan^{-1}\left(-\frac{I_{UU}\tan\theta}{I_{VV}}\right)=\tan^{-1}\left(-\frac{7.095\times10^{-5}\tan(-1.45°)}{4.533\times10^{-4}}\right)$$

i.e. $\beta = \mathbf{0.227°}$ (15.33)

The position of the neutral axis (NA) is shown in Figure 15.24.

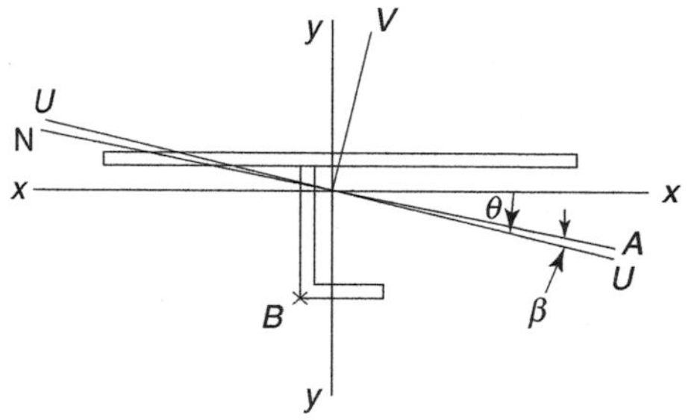

Figure 15.24 Position of neutral axis for encastré beam

By inspection, it can be seen that the point of maximum stress occurs at B which is the furthest perpendicular distance from the neutral axis.

Let u_B = deflection of the point 'B' in the 'u' direction,

and v_B = deflection of the point 'B' in the 'v' direction,

Prior to calculating σ_B, the stress at the point B, the distances u_B and v_B have to be determined, as follows:

$$u_B=x_B\cos\theta+y_B\sin\theta$$
$$=x_B\cos(-1.45°)+y_B\sin(-1.45°)$$
$$=(0.295-0.303)\times0.9997+(-0.1866)\times(-0.0253)$$
$$=-7.998\times10^{-3}+4.721\times10^{-3}$$

i.e. $\mathbf{u_B=-3.277\times10^{-3}\ m}$ (15.34)

and $v_B=y_B\cos\theta-x_B\sin\theta$

$$=-8\times10^{-3}\times(-0.0253)-0.1866\times0.9997$$

i.e. $\mathbf{v_B=-0.1863\ m}$ (15.35)

where u_B, v_B, x_B and y_B etc., are co-ordinates of the point B as defined in Figure 15.22.

From equation (15.29),

$$\sigma_B=\frac{M\cos\theta\, v_B}{I_{UU}}+\frac{M\sin\theta\, u_B}{I_{VV}}$$

$$=\left(80\times10^3\,\mathrm{N\,m}\right)\left(\frac{-0.1867\times0.9997\,\mathrm{m}}{7.095\times10^{-5}\,\mathrm{m}^3}+\frac{-0.0253\times3.277\times10^{-3}\,\mathrm{m}}{4.533\times10^{-4}\,\mathrm{m}^4}\right)$$

$$=\left(80\times10^3\,\mathrm{N\,m}\right)\left(-2630.6+0.183\right)/\mathrm{m}^3$$

i.e. $\sigma_B=\mathbf{-210.5\ MN/m^2}$ **(compressive)**

Problem 6. Determine the end deflection of the cantilever of Problem 4 on page 348, given that $E=2\times10^{11}\ \mathrm{N/m^2}$.

In this case, the components of load can be resolved along the two principal axes of bending, and each component of deflection can then be calculated. Hence, the resultant deflection can be obtained.

Now, from Chapter 7, the maximum deflection (δ) of an end-loaded cantilever is given by:

$$\delta=\frac{Wl^3}{3EI}$$

where W = load
l = length of cantilever
E = elastic modulus
I = second moment of area of cantilever section about its axis of bending

For the present problem,

δ_U = deflection under load in the u direction

$$=-\frac{W\sin\theta\, l^3}{3EI_V}=-\frac{10\times10^3\sin(-28.93°)\times2^3}{3\times2\times10^{11}\times7.53\times10^{-7}}$$

$$=\mathbf{0.0857\ m}$$

and δ_V = deflection under load in the v direction

$$=-\frac{W\cos\theta\, l^3}{3EI_U}=-\frac{10\times10^3\cos(-28.93°)\times2^3}{3\times2\times10^{11}\times3.56\times10^{-6}}$$

$$=\mathbf{-0.0328\ m}$$

These two components of deflection can be drawn as shown in Figure 15.25. Hence, from Pythagoras's theorem and from elementary trigonometry,

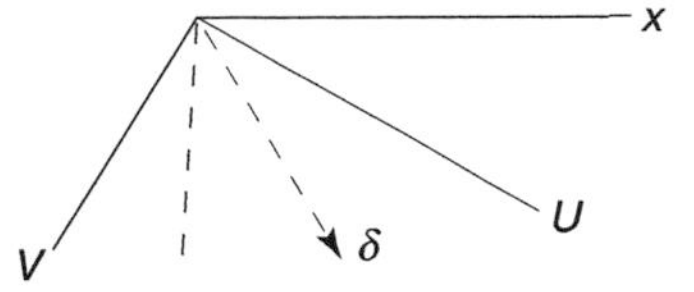

Figure 15.25

$$\delta = \sqrt{\left(\delta_U{}^2 + \delta_V{}^2\right)} = \sqrt{\left(0.0857^2 + (-0.0328)^2\right)}$$

$$= \mathbf{0.0918}$$

at an angle of $\tan^{-1}\left(\dfrac{\delta_V}{\delta_U}\right) = \tan^{-1}\left(\dfrac{-0.0328}{0.0857}\right)$

$= \mathbf{-20.94°}$ from the OU axis

Problem 7. Determine the central deflection of the encastré beam of Problem 5 on page 349, given that $E = 2 \times 10^{11}$ N/m^2.

From Chapter 7, the maximum deflection of an encastré beam with a uniformly distributed load is given by:

$$\delta = \frac{wl^4}{384EI} \qquad \text{from equation (7.19)}$$

where w = load/unit length
l = length of cantilever
E = Young's modulus
I = second moment of area

For the present problem,

δ_U = central deflection in the u direction

$$= \frac{w\sin\theta\, l^4}{384EI_V} = \frac{240\times10^3\sin(-1.45°)\times2^4}{384\times2\times10^{11}\times4.533\times10^{-4}}$$

$$= -2.791\times10^{-6}\text{ m} = \mathbf{-2.791\times10^{-3}\,mm}$$

and δ_V = deflection under load in the v direction

$$= -\frac{w\cos\theta\, l^4}{384EI_U} = -\frac{240\times10^3\cos(-1.45°)\times2^4}{384\times2\times10^{11}\times7.095\times10^{-5}}$$

$$= -7.045\times10^{-4} = -0.7045\times10^{-3}\text{ m}$$

$$= \mathbf{-\,0.7045\ mm}$$

The resultant deflection is:

$$\delta = \sqrt{\left(\delta_U{}^2 + \delta_V{}^2\right)} = \sqrt{\left(-2.791\times10^{-3}\right)^2 + \left(-0.7045\right)^2}$$

$$= \mathbf{0.7045\ mm}$$

at an angle of: $\mathbf{\tan^{-1}}\left(\dfrac{\delta_V}{\delta_U}\right) = \tan^{-1}\left(\dfrac{-0.7045}{-2.791\times10^{-3}}\right)$

$= 89.77° = \mathbf{0.23°}$ **clockwise from the vertical**

NB The theory in this chapter does not include the additional effects of shear stresses due to torsion and bending that occur when asymmetrical beams are loaded through their centroids, but these theories are dealt with in some detail in Chapters 8 and 16.

Now try the following Practice Exercise

Practice Exercise 60. Asymmetrical bending of beams

1. If a cantilever of length 3 m, and with a cross-section as shown in Figure 15.26, is subjected to a vertically applied downward load at its free end, of magnitude 4 kN, determine the position and value of the maximum bending stress.

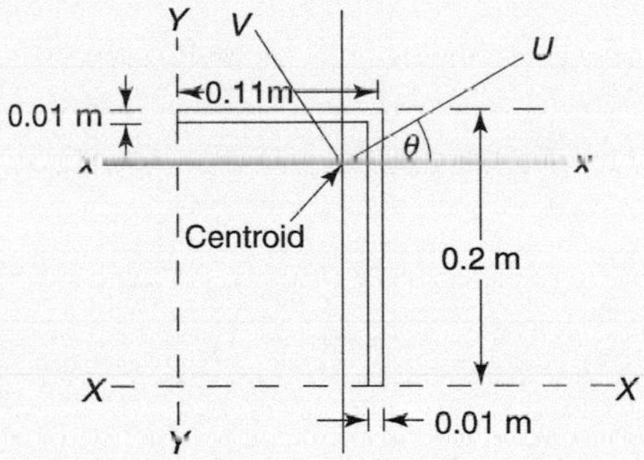

Figure 15.26

[NA is –68.38° from UU, σ_B = –163.63 MPa]

2. A simply supported beam of length 4 m, and with a cross-section as shown in Figure 15.27, is subjected to a centrally placed concentrated load of 20 kN, acting perpendicularly to the xx axis, and through the centroid of the beam. Determine the stress at mid-span at the point B in the cross-section of the beam.

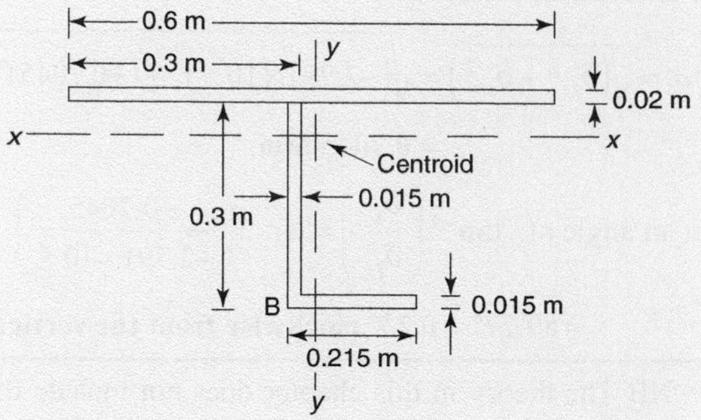

Figure 15.27

[NA is 8.23° from UU, $\sigma_B = -19.82$ MPa]

3. Determine the components of deflection under the load in the directions of the principal axes of bending for the beam of Problem 1 above, given that $E = 2 \times 10^{11}\,\text{N/m}^2$

   [$\delta_U = -0.031$ m, $\delta_V = -0.0124$ m]

4. Determine the components of the central deflection in the directions of the principal axes of bending for the beam of Problem 2 above, given that $E = 1 \times 10^{11}\,\text{N/m}^2$

   [$\delta_U = 2.19 \times 10^{-4}$ m, $\delta_V = -1.51 \times 10^{-3}$ m]

For fully worked solutions to each of the problems in Exercise 59 and 60 in this chapter, go to the website:
www.routledge.com/cw/bird

Chapter 16

Shear stresses in bending and shear deflections

Why it is important to understand: **Shear stresses in bending and shear deflections**

Shearing stresses due to bending occur in a structure when the bending moment changes. For example, if the bending moment changes along the length of a beam, then this will result in shearing forces occurring, as described in Chapter 3. Thus, if the structure is subjected to a lateral shearing force, shearing stresses will occur as a result of this, and structural failure may occur, due to large shearing stresses. Moreover, in the case of beams with thin-walled cross-sections, such as channel and angle bars, the shearing stresses due to bending can cause the beam to twist, because the applied load of the laterally loaded beam is not placed through the structure's shear centre, as described in Section 16.4. If the lateral load is not applied through the beam's shear centre, then shearing stresses due to torsion can occur, in addition to shearing stresses due to bending – where the former can be very large.

At the end of this chapter you should be able to:

- understand shearing stresses due to bending
- understand and calculate 'vertical' shearing stresses
- understand and calculate 'horizontal' shearing stresses
- understand and calculate shearing stress distribution in thin-walled beams, such as channel bars and angles bars
- understand and determine shear centre
- understand and calculate shearing stresses due to torsion
- understand and calculate shear deflections
- understand warping of thin-walled sections

Mathematical references

In order to understand the theory involved in this chapter on **shear stresses in bending and shear deflections**, knowledge of the following mathematical topics is required: ***integration, integration using algebraic substitutions, trigonometry, double integrals, differentiation and the Newton-Raphson method.*** If help/revision is needed in these areas, see page 50 for some textbook references.

16.1 Introduction

If a horizontal beam is subjected to transversely applied vertical loads, so that the bending moment changes, then there will be vertical shearing forces at every point along the length of the beam where the bending moment changes (see Chapter 3).

Furthermore, if the cross-section of the beam is of a built-up section, such as a rolled steel joist (RSJ), or a tee beam or a channel section or an angle bar, then there will be horizontal shearing stresses in addition to the vertical shearing stresses, where both are caused by the same vertically applied shearing forces. Vertical shearing stresses due to bending are those that act in a vertical plane, and horizontal shearing stresses are those that act in a horizontal plane. Later in this chapter, it will be shown that for curved sections such as split tubes etc., the shearing stresses due to bending act in the planes of the curved sections. Similar arguments apply to horizontal beams subjected to laterally applied horizontal shearing forces.

To demonstrate the concept of shearing stresses due to bending, the variation of vertical shearing stresses across the section of a beam will be first considered, followed by a consideration of the variation of horizontal shearing stresses due to the same vertical shearing forces.

16.2 Vertical shearing stresses

Consider a beam subjected to a system of vertical loads, so that the bending moment changes along the length of the beam, as shown in Figure 16.1.

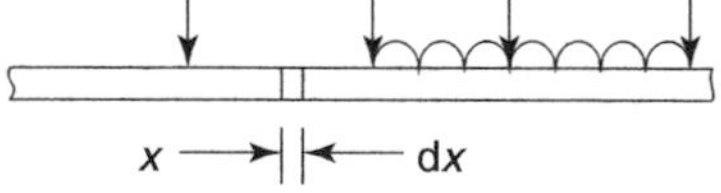

Figure 16.1 Beam subjected to vertical loads

Consider an elemental length dx of the beam, as shown in Figure 16.2(a). Consider the sub-element of Figure 16.2(b), and also the sub-sub-element in the same figure, which is shown by the heavy line. Let the stress on the left of the sub-sub-element of Figure 16.2(b) be σ, and the stress on the right of the sub-sub-element be $\sigma + d\sigma$, due to M and $M + dM$, respectively.

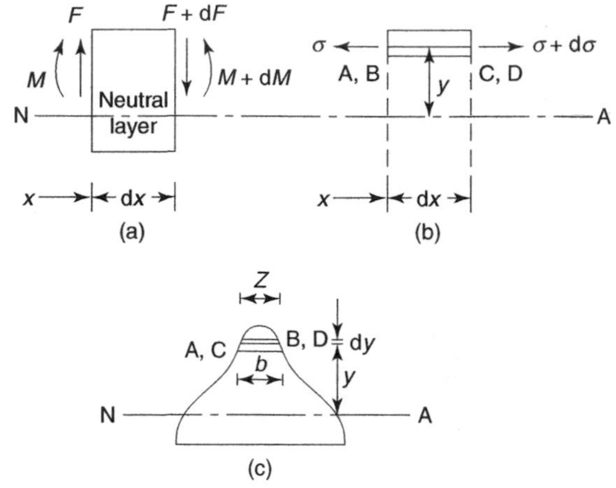

(c)

Figure 16.2 Beam element under shearing force: (a) element; (b) sub-element; (c) cross-section

From Figure 16.2(b), it can be seen that there is an apparent unbalanced force in the x direction, of magnitude $d\sigma \times Z \times dy$, but as the sub-sub-element is in equilibrium, this is not possible, i.e. the only way equilibrium can be achieved is for a longitudinal horizontal stress to act tangentially along the rectangular face ABCD. It is evident that as this stress acts tangentially along the face ABCD, it must be a shearing stress.

Let τ be the shearing stress on the face ABCD. Resolving horizontally gives:

$$\tau \times b \times dx = \int d\sigma \times dA$$

$$= \int \frac{dM \times y \times dA}{I} \quad \text{since } \frac{\sigma}{y} = \frac{M}{I}$$

where $\quad dA = Z \times dy$

Hence $\quad \tau = \frac{dM}{dx}\frac{1}{bI}\int y\,dA$

However $\quad \frac{dM}{dx} = F = \text{shearing force at } x$

Therefore $$\tau = \frac{F}{bI}\int y\,dA \tag{16.1}$$

where, in this case, $\int y\,dA$ is the first moment of area of the section *above the plane* ABCD, and about the neutral axis, NA.

Owing to complementary shearing stresses (see Reference 1 on page 497), the shearing stress on the face ABCD will be accompanied by three other shearing stresses of the same magnitude, as shown in Figure 16.3.

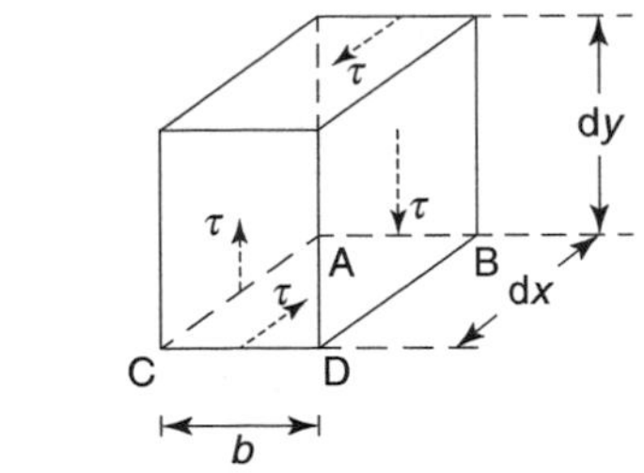

Figure 16.3 Complementary shearing stresses in a vertical plane

From Figure 16.3, it can be seen that these four complementary shearing stresses act together in a vertical plane, and this is why these shearing stresses are called vertical shearing stresses.

16.3 Horizontal shearing stresses

Consider a horizontal top flange, of thickness t, on a horizontal beam under the action of vertical shearing forces, as shown in Figure 16.4.

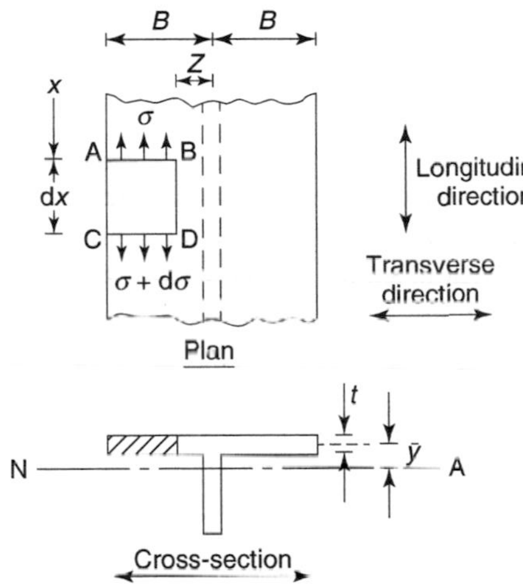

Figure 16.4 Horizontal flange on a beam

Let M = bending moment at AB
$M + dM$ = bending moment at CD
σ = bending stress in flange at AB
$\sigma + d\sigma$ = bending stress in flange at CD

Consider the equilibrium of an element of the flange, ABCD, in the x direction. From the plan view of Figure 16.4, it can be seen that the apparent unbalanced force in the longitudinal direction on this element is $d\sigma \times t \times (B - Z)$. However, as the beam is in equilibrium, no such unbalanced force can exist, and the only way that equilibrium can be achieved in this flange element is for a horizontal shearing stress to act tangentially along BD.

Let τ be the shearing stress on the face BD. Then, considering horizontal equilibrium of the element ABCD in the x direction,

$$d\sigma \times t \times (B - Z) = \tau \times dx \times t$$

or
$$\frac{dM \times \bar{y} \times t \times (B - Z)}{I} = \tau \times dx \times t$$

since
$$d\sigma = \frac{dM \times \bar{y}}{I}$$

However
$$\frac{dM}{dx} = F = \text{the vertical shearing force}$$

Therefore
$$\tau = \frac{F(B - Z)\bar{y}}{I} \tag{16.2}$$

Equation (16.2) can be obtained directly from equation (16.1), as follows:

$$\tau = \frac{F \int y\,dA}{bI}$$

which, for the horizontal flange element ABCD of Figure 16.4, becomes:

$$\tau = \frac{F}{tI} \times (B - Z) \times t \times \bar{y} = \frac{F(B - Z)\bar{y}}{I}$$

which is identical to equation (16.2).

It should be noted that, in this case, $\int y\,dA$ was the first moment of area of the flange element ABCD about NA, and b was, in fact, the flange thickness t. The reason for applying equation (16.1) in this way was because the shearing stresses in the flange act in a horizontal plane, as shown in Figure 16.5.

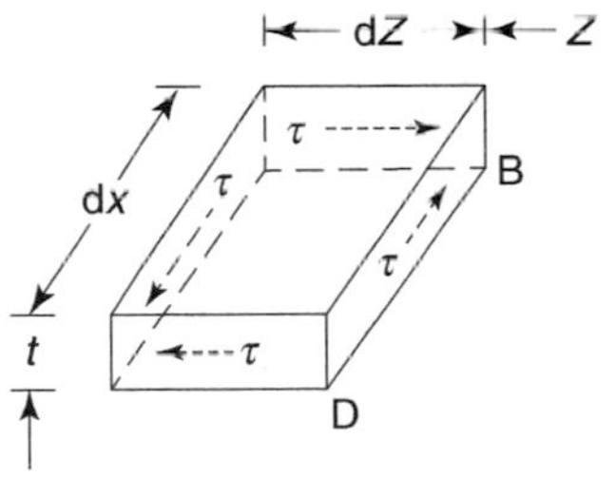

Figure 16.5 Complementary shearing stresses in a horizontal plane

The horizontal shearing stress distribution of equation (16.2) can be seen to vary linearly along the width of the flange, having a maximum value at the centre of the flange, and zero values at the flange edges.

In fact, owing to the effects of complementary shearing stresses, this shearing stress will be accompanied by three other shearing stresses, where all four shearing stresses act together in a horizontal plane, as shown in Figure 16.5.

Problem 1. Determine and sketch the vertical shearing stress distribution in a horizontal beam of rectangular section, subjected to a transverse vertical shearing force F, as shown in Figure 16.6.

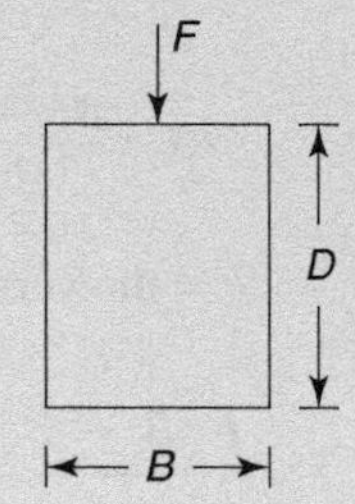

Figure 16.6 Rectangular section under a shearing force F

From equation (16.1), τ the vertical shearing stress at any point y from NA (see Figure 16.7), is given by:

$$\tau = \frac{F \int y\,dA}{bI}$$

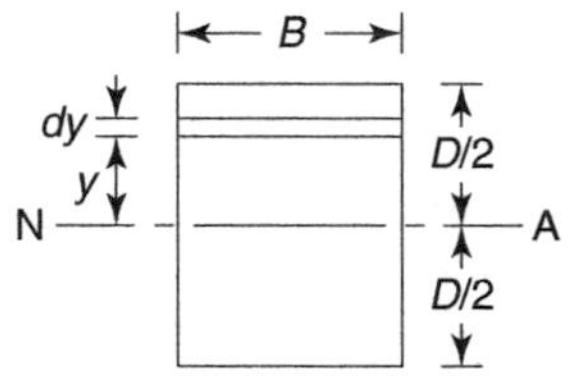

Figure 16.7 Rectangular section

Now, second moment of area $I = \dfrac{BD^3}{12}$

Therefore $$\tau = \frac{F\int_y^{D/2} B \times dy \times y}{B\left(\dfrac{BD^3}{12}\right)} = \frac{12F}{BD^3}\int_y^{D/2} y\,dy$$

$$= \frac{12F}{BD^3}\left[\frac{y^2}{2}\right]_y^{D/2}$$

i.e. $$\tau = \frac{12F}{BD^3}\left[\frac{(D/2)^2}{2} - \frac{y^2}{2}\right]$$

i.e. $$\boldsymbol{\tau = \frac{6F}{BD^3}\left(\frac{D^2}{4} - y^2\right)} \qquad (16.3)$$

Equation (16.3) can be seen to vary parabolically, having a maximum value at NA, and being zero at the top and bottom.

Let $\hat{\tau}$ be the maximum shearing stress, i.e. when $y = 0$,

then, $$\hat{\tau} = \frac{6F}{BD^3}\left(\frac{D^2}{4}\right)$$

i.e. $$\boldsymbol{\hat{\tau} = \frac{1.5F}{BD}} \qquad (16.4)$$

Equation (16.4) shows that the maximum value of shearing stress ($\hat{\tau}$) is 50% greater than the average value (τ_{av}) where

$$\boldsymbol{\tau_{av} = \frac{F}{BD}}$$

A plot of the variation of vertical shearing stress for this section is shown in Figure 16.8.

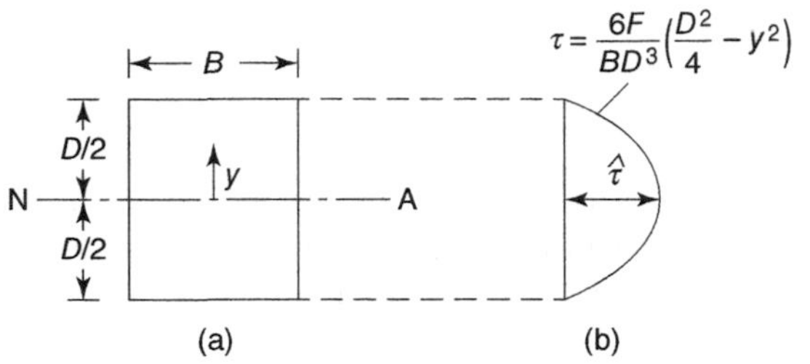

Figure 16.8 Shearing stress distribution in a rectangular section: (a) section; (b) plot of τ

Problem 2. Determine and sketch the vertical shearing stress distribution in a horizontal beam of circular section, subjected to the shearing force F, shown in Figure 16.9.

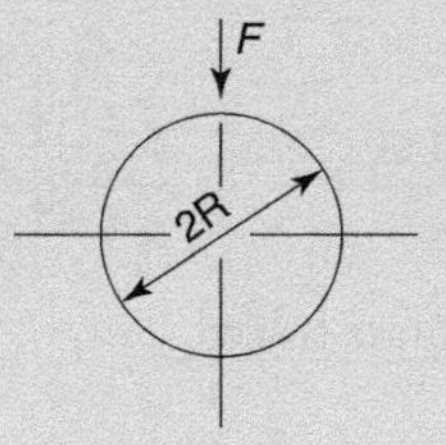

Figure 16.9 Circular section

Let τ be the vertical shearing stress at any distance y from NA (see Figure 16.10).

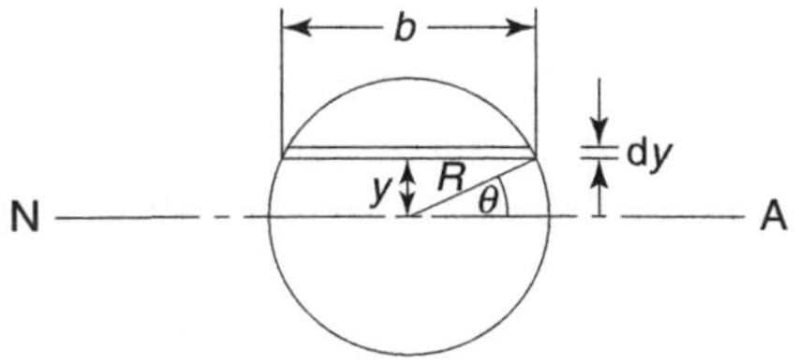

Figure 16.10 Circular section

Then $\tau = \dfrac{F}{bI}\int y\,dA$ (i)

i.e. $\tau = \dfrac{F}{bI}\int_y^R b\,dy\,y$

However $I = \dfrac{\pi R^4}{4}$, $b = 2R\cos\theta$

and $y = R\sin\theta$

By differentiation $\dfrac{dy}{d\theta} = R\cos\theta$

from which $dy = R\cos\theta\, d\theta$

Therefore τ

$$= \frac{F}{(2R\cos\theta)\left(\frac{\pi R^4}{4}\right)}\int_\theta^{\pi/2}(2R\cos\theta)(R\sin\theta)R\cos\theta\,d\theta \quad \text{(ii)}$$

Note that in equation (i), 'b' does not appear within the integral and therefore $b = 2R\cos\theta$ should not be 'cancelled' with the $2R\cos\theta$ within the integral of equation (ii).

Hence $\tau = \dfrac{4F}{2\pi R^2\cos\theta}\displaystyle\int_\theta^{\pi/2}(2\cos^2\theta)(\sin\theta)\,d\theta$

$$= \frac{4F}{\pi R^2\cos\theta}\int_\theta^{\pi 2}\cos^2\theta\sin\theta\,d\theta$$

$$= \frac{4F}{\pi R^2\cos\theta}\left[\frac{-\cos^3\theta}{3}\right]_\theta^{\pi/2}$$

using an algebraic substitution $u = \cos\theta$

$$= \frac{-4F}{3\pi R^2\cos\theta}\left[(\cos(\pi/2))^3 - (\cos\theta)^3\right]$$

$$= \frac{-4F}{3\pi R^2\cos\theta}\left[-\cos^3\theta\right]$$

$$= \frac{4F\cos^2\theta}{3\pi R^2}$$

$$= \frac{4F}{3\pi R^2}\left(1-\sin^2\theta\right)$$

and since $\sin\theta = \dfrac{y}{R}$

$$\tau = \frac{\mathbf{4F}}{\mathbf{3\pi R^2}}\left[\mathbf{1}-\left(\frac{\mathbf{y}}{\mathbf{R}}\right)^2\right] \quad (16.5)$$

which, again, is a parabolic distribution, as shown in Figure 16.11.

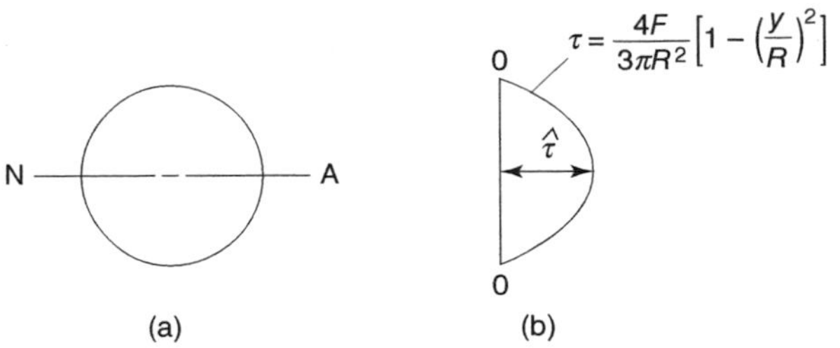

Figure 16.11 Shear stress distribution in a circular section: (a) section; (b) shear stress distribution

Let $\hat{\tau}$ be the maximum value of shearing stress, i.e. when $y = 0$ in equation (16.5)

then $\hat{\tau} = \dfrac{4F}{3\pi R^2}$

which can be seen to be about 33.3% larger than the average shearing stress τ_{av} where

$$\tau_{av} = \frac{\mathbf{F}}{\mathbf{\pi R^2}}$$

Problem 3. Determine and sketch the distribution of vertical shearing stress in the triangular section of Figure 16.12.

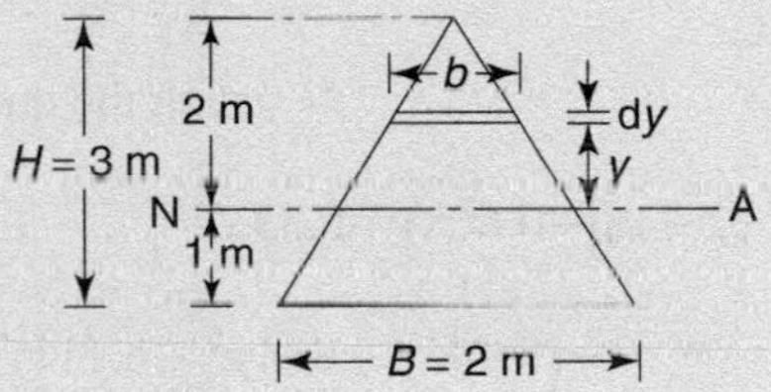

Figure 16.12 Triangular section

At any distance y above NA in the triangular section of Figure 16.12

$$\tau = \frac{F}{bI}\int y\,dA = \frac{F}{bI}\int b\,dy\,y$$

However, from Table 5.3, page 120, the second moment of area through the centroid parallel to the base is:

$$I = \frac{BH^3}{36} = \frac{2\times 3^3}{36} = 1.5$$

and $b = B\left(\dfrac{2}{3}-\dfrac{y}{H}\right) = 2(0.667 - 0.333y)$

Therefore τ

$$= \frac{F}{(2(0.667-0.333y))(1.5)}\int_y^2 2(0.667-0.333y)\,y\,dy$$

$$= \frac{2F}{(2-y)}\int_y^2 \left(0.667y - 0.333y^2\right)dy$$

$$= \frac{2F}{(2-y)}\left[\frac{0.667y^2}{2} - \frac{0.333y^3}{3}\right]_y^2$$

$$= \frac{2F}{(2-y)}\left[\left(\frac{0.667(2)^2}{2} - \frac{0.333(2)^3}{3}\right) - \left(\frac{0.667y^2}{2} - \frac{0.333y^3}{3}\right)\right]$$

$$= \frac{2F}{(2-y)}\left(1.334 - 0.889\right) - \left(0.334y^2 - 0.111y^3\right)$$

i.e. $$\tau = \frac{2F}{(2-y)}\left[0.445 - 0.334y^2 + 0.111y^3\right] \quad \text{(i)}$$

For maximum τ $$\frac{d\tau}{dy} = 0$$

i.e. $$2F\left[\frac{(2-y)\left(-0.668y + 0.333y^2\right) - \left(0.445 - 0.334y^2 + 0.111y^3\right)(-1)}{(2-y)^2}\right] = 0$$

using the quotient rule

i.e. $$(2-y)\left(-0.668y + 0.333y^2\right) - \left(0.445 - 0.334y^2 + 0.111y^3\right)(-1) = 0$$

and $$-1.336y + 0.666y^2 + 0.668y^2 - 0.333y^3 + 0.445 - 0.334y^2 + 0.111y^3 = 0$$

i.e. $$-0.222y^3 + y^2 - 1.336y + 0.445 = 0$$

or $$0.222y^3 - y^2 + 1.336y - 0.445 = 0 \quad (16.7)$$

Equation (16.7) has three roots, but for this case the root of interest is the lowest positive root, which can be obtained by the Newton-Raphson* iterative process (see reference on page 50), as follows:

Let $$f(y) = 0.222y^3 - y^2 + 1.336y - 0.445$$

$$f'(y) = 0.666y^2 - 2y + 1.336$$

* Who were Newton and Raphson? **Go to** www.routledge.com/cw/bird

Let $r_1 = 0$

then $$r_2 = r_1 - \frac{f(r_1)}{f'(r_1)}$$

i.e. $$r_2 = 0 - \frac{-0.455}{1.336} = 0.3406$$

$$r_3 = r_2 - \frac{f(r_2)}{f'(r_2)} = 0.3406 - \frac{-0.097195}{0.732062} = 0.4734$$

$$r_4 = r_3 - \frac{f(r_3)}{f'(r_3)} = 0.4734 - \frac{-0.013093}{0.5384556} = 0.4977$$

$$r_5 = r_4 - \frac{f(r_4)}{f'(r_4)} = 0.4977 - \frac{-0.0004093}{0.5055717} = 0.4985$$

Since the last two values are same, when correct to 2 decimal places, $\boldsymbol{y = 0.50}$ **m**
(NB an fx-991ES PLUS calculator will compute the above, giving an answer of 0.4985)
In this case, the average shear stress, from equation (i), is:

$$\tau = \frac{2F}{(2-y)}\left[0.445 - 0.334y^2 + 0.111y^3\right]$$

$$= \frac{2F}{(2-0.50)}\left[0.445 - 0.334(0.50)^2 + 0.111(0.50)^3\right]$$

$$= \frac{2F}{(2-0.50)}\left[0.375375\right]$$

i.e. $\boldsymbol{\tau = 0.5\,F}$
At NA (i.e. when $y = 0$)

$$\tau = \frac{2F}{(2-0)}\left[0.445 - 0.334(0)^2 + 0.111(0)^3\right]$$

i.e. $\boldsymbol{\tau = 0.445\,F}$
In this case, the average shear stress is:

$$\tau_{av} = \frac{F}{A} = \frac{F}{\left(\frac{1}{2}\times 2\times 3\right)} = 0.333F$$

A sketch of the distribution of vertical shearing stress across the section is shown in Figure 16.13.

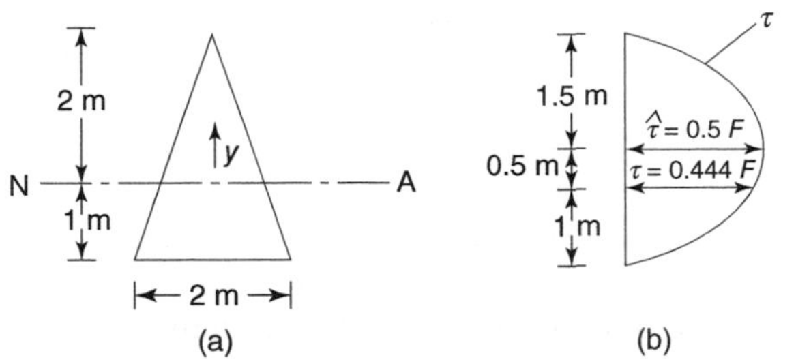

Figure 16.13 Stress distribution in a triangular section: (a) section; (b) plot of τ

NB In practice, shearing stresses in horizontal beams due to transverse vertical shearing forces will not have the distributions assumed in Problems 2 and 3 above and as shown in Figure 16.14, but the more complex forms of Figure 16.15. The determination of the correct forms of Figure 16.15 is beyond the scope of this book, and the reader is referred to more advanced works on this topic (see Reference 31 on page 497). In both figures, the size of the arrows is related to the magnitude of the shearing stresses.

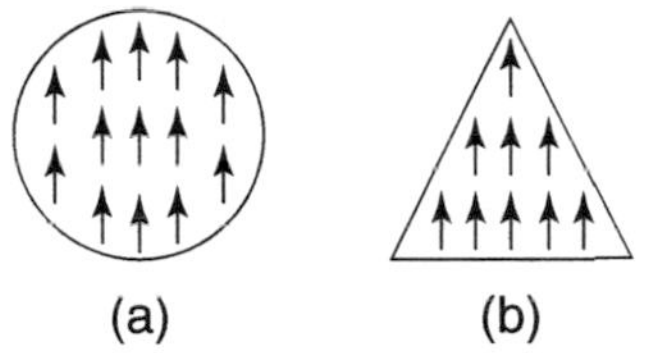

Figure 16.14 Incorrect shear stress distributions due to bending

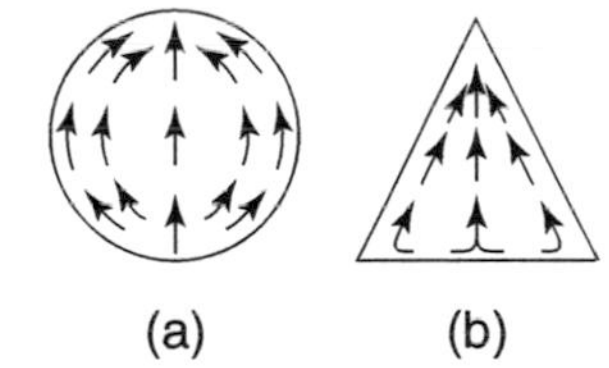

Figure 16.15 Correct shear stress distributions due to bending

Problem 4. A beam of the section in Figure 16.16 is subjected to a bending moment of 0.5 MN m, so that the section bends about a horizontal plane NA. If the maximum principal stress due to this bending moment lies on the axis AA, and it is not to exceed 80% of the greatest bending stress, determine the value of the shearing force that acts on this section.

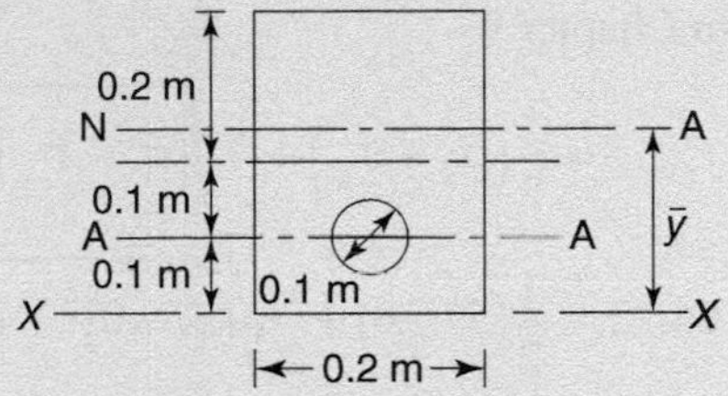

Figure 16.16 Complex section

To determine $\bar{y}$, I, etc., Table 16.1 is used, where the symbols are as defined in Chapter 15.

$$\bar{y} = \frac{\sum ay}{\sum a} = \frac{0.0152}{0.0721} = \mathbf{0.211\ m}$$

$$I_{XX} = \sum ay^2 + \sum i_H = 3.121 \times 10^{-3} + 1.062 \times 10^{-3}$$

i.e. $$\boldsymbol{I_{XX} = 4.183 \times 10^{-3}\ m^4}$$

$$I_{xx} = I_{XX} - \left(\bar{y}\right)^2 \sum a = 4.183 \times 10^{-3} - 0.211^2 \times 0.0721$$

i.e. $$\boldsymbol{I_{xx} = 9.730 \times 10^{-4}\ m^4}$$

Maximum stress

$$\hat{\sigma} = \frac{M \times \bar{y}}{I} = \frac{0.5 \times 10^6 \times 0.211}{9.730 \times 10^{-4}} = 108.4 \times 10^6\ \text{N/m}^2$$

$$= 108.4\ \text{MN/m}^2$$

Table 16.1

Section	a	y	ay	ay^2	i_H
1	0.08	0.2	0.016	3.2×10^{-3}	$\frac{0.2 \times 0.4^3}{12} = 1.0667 \times 10^{-3}$
2	-7.854×10^{-3}	0.1	-7.854×10^{-4}	-7.854×10^{-5}	$\frac{\pi \times 0.1^4}{64} = -4.909 \times 10^{-6}$
$\sum$	0.0721	–	0.0152	3.121×10^{-3}	1.062×10^{-3}

At AA, $\sigma_1 = 0.8 \times 108.4$ = maximum principal stress

i.e. $\boldsymbol{\sigma_1 = 86.72\ \mathrm{MN/m^2}}$

At AA, $\sigma_X = \dfrac{0.5\times10^6 \times (0.211-0.1)}{9.730\times10^{-4}} = 57\times10^6\ \mathrm{N/m^2}$

$= \mathbf{57\ MN/m^2}$

and $\boldsymbol{\sigma_Y = 0}$

Now, from Chapter 9

$$= \frac{1}{2}\left(\sigma_x+\sigma_y\right)+\frac{1}{2}\sqrt{\left(\sigma_x-\sigma_y\right)^2+4\tau_{xy}^2}$$

i.e. $86.72 = \dfrac{1}{2}(57+0)+\dfrac{1}{2}\sqrt{(57-0)^2+4\tau_{xy}^2}$

i.e. $86.72 = 28.5+\dfrac{1}{2}\sqrt{3249+4\tau_{xy}^2}$

Hence $(86.72 - 28.5) \times 2 = \sqrt{3249+4\tau_{xy}^2}$

i.e. $116.44 = \sqrt{3249+4\tau_{xy}^2}$

and $(116.44)^2 = 3249+4\tau_{xy}^2$

from which $(116.44)^2 - 3249 = 4\tau_{xy}^2$

and $\tau_{xy}^2 = \dfrac{116.44^2-3249}{4}$

i.e. $\tau_{xy} = \sqrt{\dfrac{116.44^2-3249}{4}}$

i.e. $\boldsymbol{\tau_{xy} = 50.77\ \mathrm{MN/m^2}}$ at AA

Now $\tau = \dfrac{F}{bI}\int y\,dA$

which, when applied to AA, becomes:

$$50.77 = \frac{F}{0.1\times9.73\times10^{-4}}\int y\,dA$$

where

$$\int y\,dA = 0.211\times0.2\times\frac{0.211}{2}-\frac{\pi\times0.1^2}{4}\times(0.211-0.1)$$

$$= 4.4521\times10^{-3}-8.7179\times10^{-4} = 3.58\times10^{-3}$$

Hence $50.77 = \dfrac{F}{0.1\times9.73\times10^{-4}}\left(3.58\times10^{-3}\right)$

from which $F = \dfrac{50.77\times0.1\times9.73\times10^{-4}}{\left(3.58\times10^{-3}\right)} = 1.38$ MN

i.e. **the vertical shearing force at the section is: F = 1.38 MN**

Problem 5. Calculate and sketch the distribution of vertical and horizontal shearing stresses due to bending, which occur on a beam with the cross-section shown in Figure 16.17, when it is subjected to a vertical shearing force of 30 kN through its centroid.

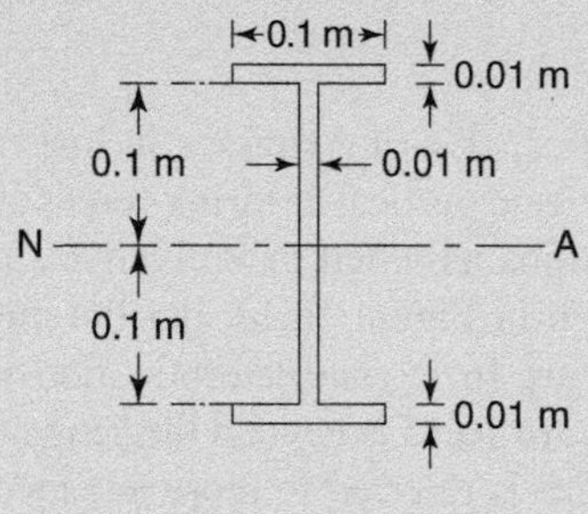

Figure 16.17 Beam cross-section

I = second moment of area about NA

$$= \frac{0.1\times0.22^3}{12}-\frac{0.09\times0.2^3}{12}$$

i.e. $\boldsymbol{I = 2.873\times10^{-5}\,\mathrm{m}^4}$

Vertical shearing stress

At the *top of the flange* $\int y\,dA = 0$

Therefore $\tau_1 = 0$

At the *bottom of the flange*

$$\int y\,dA = 0.1\times0.01\times0.105 = 1.05\times10^{-4}\ \mathrm{m}^2$$

Therefore $\tau_2 = \dfrac{F\int y\,dA}{bI} = \dfrac{30\times10^3\times1.05\times10^{-4}}{0.1\times2.873\times10^{-5}}$

i.e. $\boldsymbol{\tau_2 = 1.096\,\mathrm{MN/m^2}}$

At the *top of the web* $\int y\,dA = 1.05\times10^{-4}$

Therefore $\tau_3 = \dfrac{F\int y\,dA}{bI} = \dfrac{30\times10^3\times1.05\times10^{-4}}{0.01\times2.873\times10^{-5}}$

i.e. $\boldsymbol{\tau_3 = 10.96\,\mathrm{MN/m^2}}$

From equation (16.1), it can be seen that the maximum shear stress, namely $\hat{\tau}$, occurs where $\dfrac{\int y\,dA}{b}$ is a maximum, which in this case is at NA.

At NA $\int y\,dA = 1.05\times10^{-4} + 0.1\times0.01\times0.05$

$$= 1.55\times10^{-4}\,\text{m}^3$$

Therefore $\hat{\tau} = \dfrac{30\times10^3\times1.55\times10^{-4}}{0.01\times2.873\times10^{-5}}$

i.e. $\hat{\tau} = \mathbf{16.19\ MN/m^2}$

Horizontal shearing stress

From equation (16.2), it can be seen that the horizontal shearing stress varies linearly along the flanges of the RSJ, from zero at the free edges to a maximum value $\hat{\tau}_F$ at the centre:

$$\tau_F = \frac{F(B-Z)\bar{y}}{I}$$

and when $Z = 0$ $\tau_F = \dfrac{FB\bar{y}}{I} = \dfrac{30\times10^3\times0.05\times0.105}{2.873\times10^{-5}}$

i.e. $\hat{\tau}_F = \mathbf{5.48\ MN/m^2}$

A plot of the vertical and horizontal shearing stresses is shown in Figure 16.18.

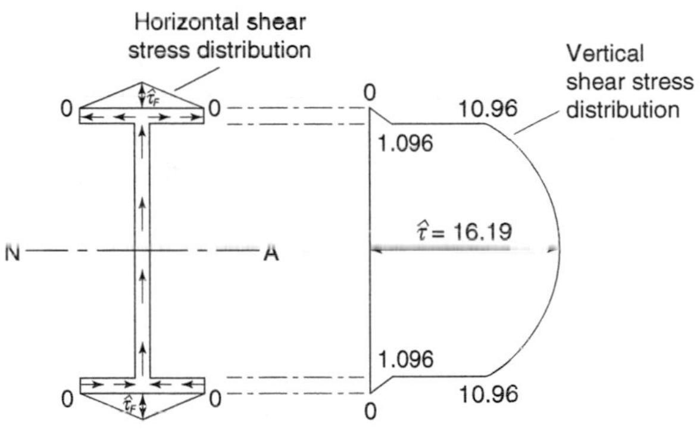

Figure 16.18 Vertical and horizontal shearing stress distributions (MN/m²)

Problem 6. Calculate and sketch the distribution of vertical and horizontal shear stress in the channel bar of Figure 16.19, when it is subjected to a vertical shearing force of 30 kN.

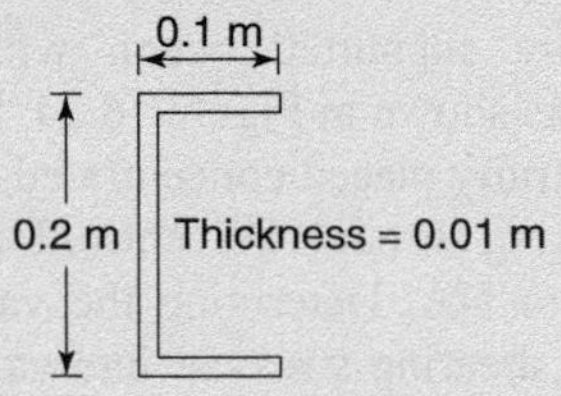

Figure 16.19 Channel section

Second moment of area,

$$I = \frac{0.1\times0.2^3}{12} - \frac{0.09\times0.18^3}{12} = 2.293\times10^{-5}\,\text{m}^4$$

Vertical shearing stress

At the *bottom of the flange*

$$\int y\,dA = 0.1\times0.01\times0.095 = 9.5\times10^{-5}\,\text{m}^3$$

and $\tau_1 = \dfrac{F\int y\,dA}{bI} = \dfrac{30\times10^3\times9.5\times10^{-5}}{0.1\times2.293\times10^{-5}}$

i.e. $\tau_1 = \mathbf{1.24\ MN/m^2}$

At the *top of the web* $\tau_2 = \dfrac{30\times10^3\times9.5\times10^{-5}}{0.01\times2.293\times10^{-5}}$

i.e. $\tau_2 = \mathbf{12.43\,MN/m^2}$

The maximum shear stress, $\hat{\tau}$, occurs at NA, because this is the point at which $\dfrac{\int y\,dA}{b}$ is a maximum.

At NA $\int y\,dA = 9.5\times10^{-5} + 0.095\times0.01\times0.095/2$

$$= 1.40\times10^{-4}\,\text{m}^3$$

and $\hat{\tau} = \dfrac{30\times10^3\times1.40\times10^{-4}}{0.01\times2.293\times10^{-5}}$

i.e. $\hat{\tau} = \mathbf{18.32\ MN/m^2}$

Horizontal shearing stress

From equation (16.2), it can be seen that the horizontal shearing stress varies linearly along the flange width, from zero at the right edge to a maximum value of $\hat{\tau}_F$ at the intersection between the flange and the web.

From equation (16.2) $\tau = \dfrac{F(B-Z)\bar{y}}{I}$

and, when Z = 0, $\tau = \dfrac{FB\bar{y}}{I} = \dfrac{30\times10^3\times0.09\times0.095}{2.293\times10^{-5}}$

i.e. $\hat{\tau} = \mathbf{11.19\ MN/m^2}$

A plot of the vertical and horizontal shearing stress distributions is shown in Figure 16.20, where the arrows are used to indicate the direction and magnitude of the vertical and horizontal shearing stresses acting on the beam's cross-section.

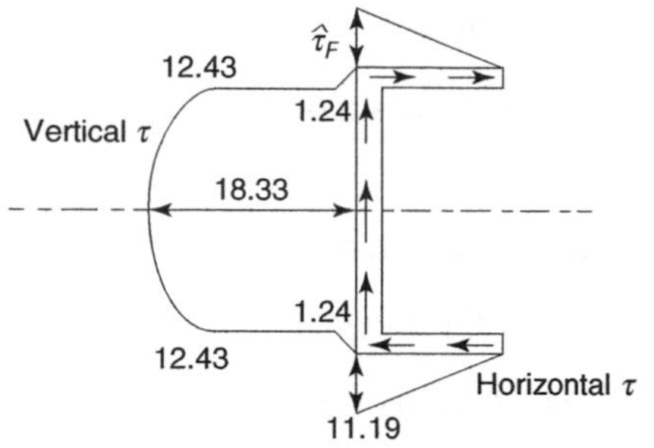

Figure 16.20 Plot of vertical and horizontal shear stress distributions

Now try the following Practice Exercise

Practice Exercise 61. Shear stresses in bending

1. A beam of length 3 m is simply supported at its ends and subjected to a uniformly distributed load of 200 kN/m, spread over its entire length. If the beam has a uniform cross-section of depth 0.2 m and width 0.1 m, determine the position and value of the maximum shearing stress due to bending. What will be the value of the maximum shear stress at mid-span?

   [22.5 MPa at the NA at the ends; at mid-span, stress = 0]

2. Determine the maximum values of shear stress due to bending in the web and flanges of the sections of Figure 16.21 when they are subjected to vertical shearing forces of 100 kN.

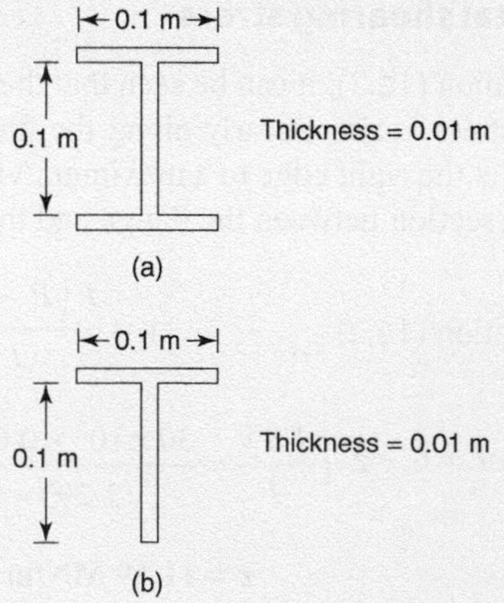

Figure 16.21 Symmetrical sections subjected to vertical shearing forces: (a) RSJ; (b) tee section

   [(a) 97.83 MPa, 35.87 MPa (b) 127.36 MPa, 52.48 MPa]

3. Determine an expression for the maximum shearing stress due to bending for the section of Figure 16.22, assuming that it is subjected to a shearing force of 0.5 MN acting through its centroid and in a perpendicular direction to NA.

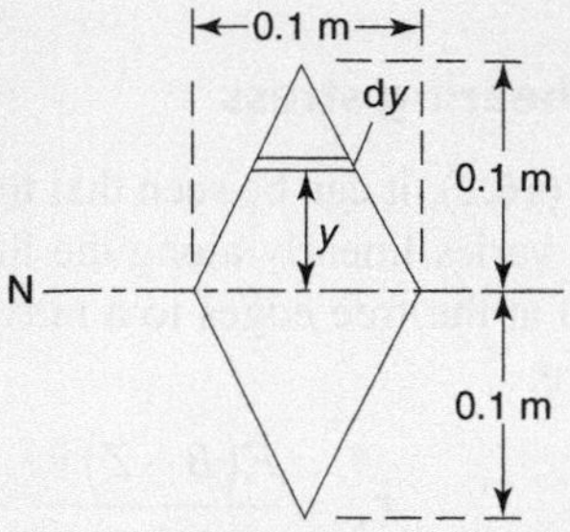

Figure 16.22

   $$\left[\tau = \frac{5000[0.01-(3y^2-20y^3)}{(1-10y)}\right.$$

   $$\left.\hat{\tau} = \pm 56.25 \text{ MPa at } y = \pm 0.025 \text{ m}\right]$$

4. Determine the value of the maximum shear stress for the cross-section of Figure 16.23, assuming that it is subjected to a shearing force of magnitude 0.5 MN acting through its centroid and in a perpendicular direction to NA.

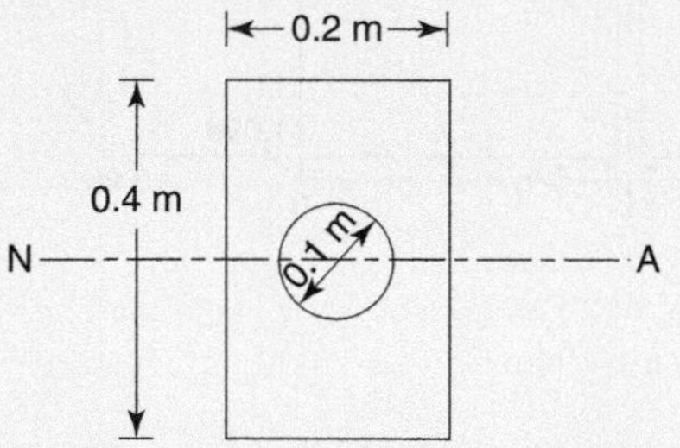

Figure 16.23 Rectangular section with a hole

   [18.44 MPa]

5. A simply supported beam, with a cross-section as shown in Figure 16.24, is subjected to a centrally placed concentrated load of 100 MN, acting through its centroid and perpendicular to NA. Determine the values of the vertical shearing stress at intervals of 0.1 m from NA.

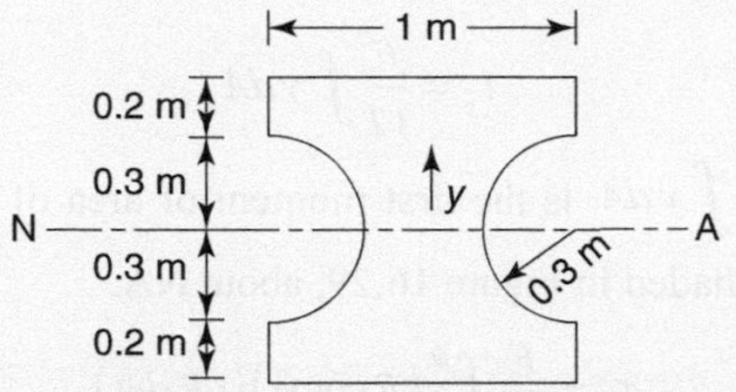

Figure 16.24 Complex cross-section

[At $y = 0$, $\tau_0 = 173.7$ MPa, $\tau_{0.1} = 156.95$ MPa, $\tau_{0.2} = 114.49$ MPa, $\tau_{0.3} = 51.95$ MPa, $\tau_{0.4} = 29.22$ MPa; At $y = 0.5$, $\tau_{0.5} = 0$ MPa]

16.4 Shear centre

When slender symmetrical-section cantilevers are subjected to transverse loads in a laboratory, good agreement is usually found between the predictions of simple bending theory and experimental observations. When, however, similar tests are carried out on cantilevers with unsymmetrical sections, such as channel bars, angle irons, etc., comparison between theoretical predictions and experimental observations are usually poor. The explanation for this can be obtained by considering the channel section of Figure 16.25, and assuming that the shearing force F is applied through its centroid.

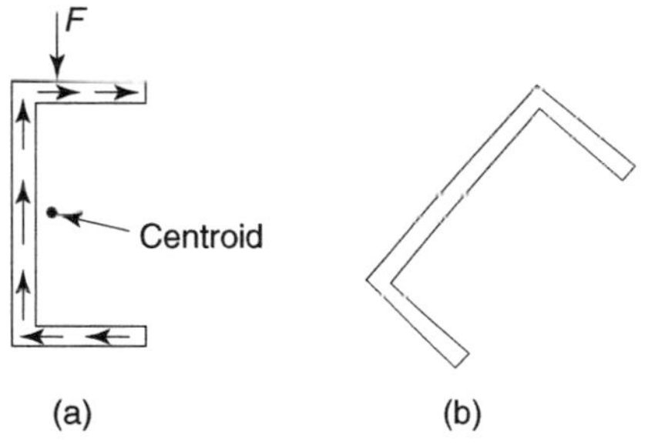

Figure 16.25 Shear flow in the cross-section of a channel bar; (a) shear flow; (b) twisting

From Figure 16.20 on page 362, the shear stress distribution due to bending will be as shown in Figure 16.25(a), where the magnitude and direction of the arrows are used to indicate the magnitude and direction of the shearing stresses due to bending.

From Figure 16.25(a), it can be seen that horizontal equilibrium is achieved by the horizontal shearing force in the top flange being equal, but opposite, to the horizontal shearing force in the bottom flange.

Similarly, vertical equilibrium is achieved by the internal resisting vertical shearing force in the web being equal and opposite to the applied external vertical shearing force F. However, from Figure 16.25(a), it can be seen that rotational equilibrium is not achieved, so that if F is applied through the centroid of the section, the beam will twist, as shown in Figure 16.25(b), and as a result, shearing stresses due to torsion will occur in addition to shearing stresses due to bending.

In general, it is advisable to eliminate shearing stresses due to torsion, as these can be relatively large, and to achieve this, it is necessary to ensure that F acts at a point where rotational equilibrium is achieved. This point is called the shear centre, and some typical positions of shear centre are shown in Figure 16.26, together with distributions of shear stress due to bending caused by F.

NB The term *shear flow* has been introduced in this section, where *shear flow* q = *shear stress* × *wall thickness* = τt.

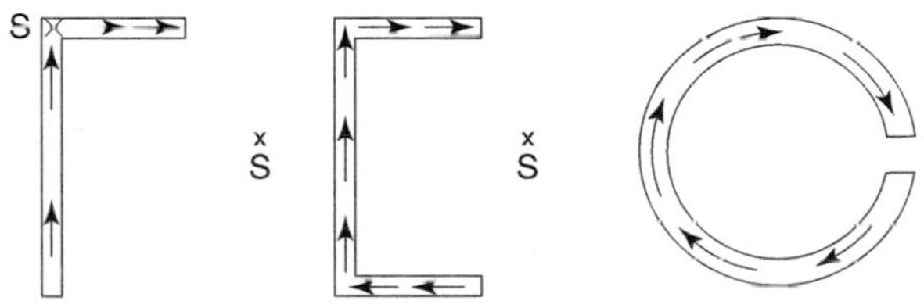

Figure 16.26 Some shear centre positions 'S'

In certain cases, when the shear stress varies inversely with the wall thickness, so that the shear flow q is constant, it may be found convenient to carry out the calculation using q instead of τ.

Problem 7. Determine the shear centre position for the channel section in Figure 16.27.

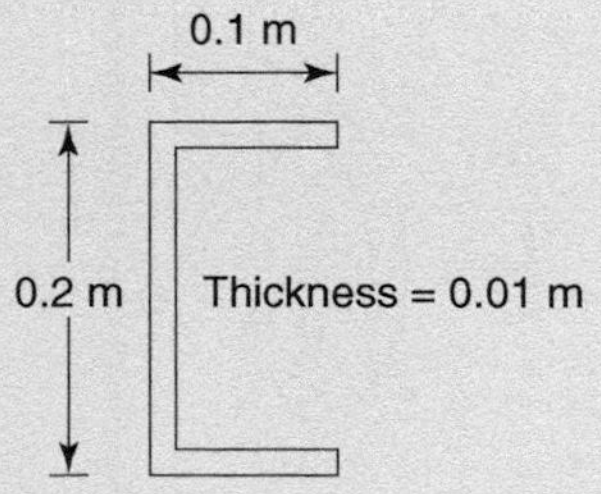

Figure 16.27 Channel section

From Figure 16.20 on page 362, it can be seen that the horizontal shearing stress distribution in the flange is linear, so that the average shear stress in the flange is:

$$\tau_{av} = 11.19/2 = 5.595 \text{ MN/m}^2$$

Let F_F be the resisting shearing force in the flange.

Then $F_F = 5.595 \times 0.01 \times 0.095 = 5.315 \times 10^{-3}$ MN

i.e. $\mathbf{F_F = 5.315}$ **kN**

Now from vertical equilibrium considerations, the resisting shearing force in the web is:

$$\mathbf{F = 30 \text{ kN}}$$

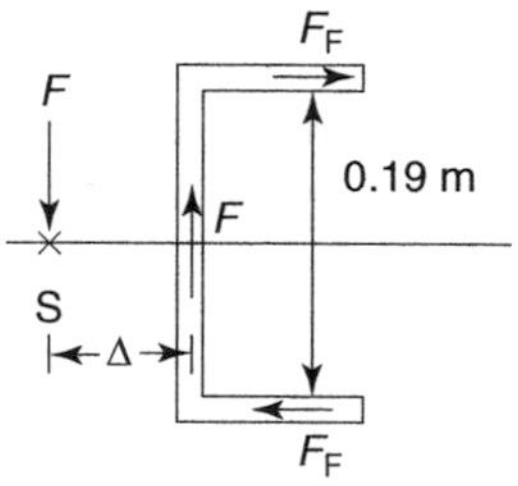

Figure 16.28 Shearing forces on channel section

Let the shear centre position be at S in the beam section, as shown in Figure 16.28, and by taking moments about S

$$F \times \Delta = F_F \times 0.19$$

i.e. $30 \times 10^3 \times \Delta = 5.315 \times 10^3 \times 0.19$

from which $\boldsymbol{\Delta} = \dfrac{5.315 \times 0.19}{30} = 0.0337 \text{ m} = \mathbf{3.37 \text{ cm}}$

Problem 8. Determine the shear centre position for the thin-walled curved section shown in Figure 16.29, which is of constant thickness t.

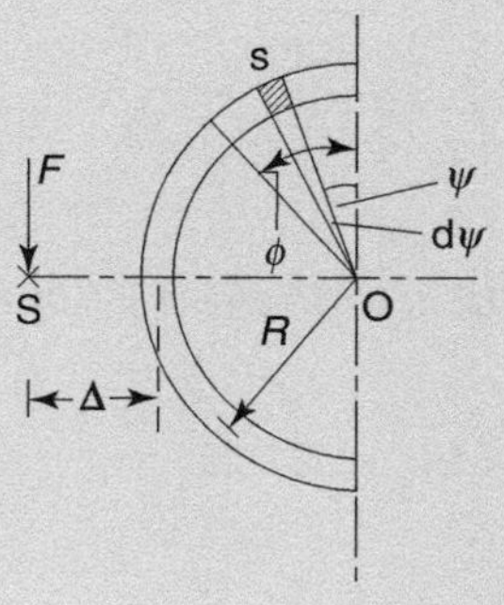

Figure 16.29 Thin-walled curved section

At any distance s, the shear stress due to bending is:

$$\tau_s = \frac{F}{tI}\int y\,dA$$

where $\int y\,dA$ is the first moment of area of the element, shaded in Figure 16.29, about NA,

i.e. $\tau_s = \dfrac{F}{tI}\displaystyle\int_0^{\phi}(R\cos\psi)(tR\,d\psi)$

i.e. $\tau_s = \dfrac{FR^2}{I}\displaystyle\int_0^{\phi}\cos\psi\,d\psi$

i.e. $\tau_s = \dfrac{FR^2}{I}\left[\sin\psi\right]_0^{\phi} = \dfrac{FR^2}{I}\left[\sin\phi - \sin 0\right]$

i.e. $$\boldsymbol{\tau_s = \frac{FR^2\sin\phi}{I}} \qquad (16.8)$$

The shear flow in the section is shown in Figure 16.30, where the magnitude and directions of the arrows are intended to give a measure of the magnitude and direction of the internal resisting shearing stresses.

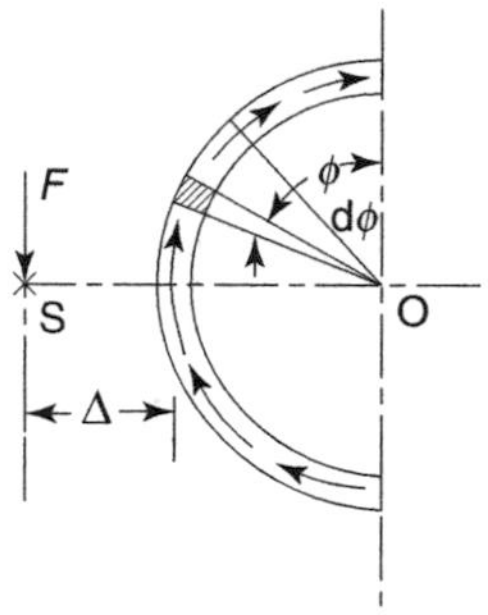

Figure 16.30 Shear flow in thin-walled curved section

To determine Δ, take moments about O,

i.e. $F(\Delta + R) = \displaystyle\int_0^{\pi}\tau_s \times (R\,d\phi)\times t \times R$

$$= \int_0^{\pi}\left(\frac{FR^2\sin\phi}{I}\right)\times(R\,d\phi)\times t\times R$$

from equation (16.8)

$$= \frac{FR^4t}{I}\int_0^{\pi}\sin\phi\,d\phi$$

$$= \frac{FR^4t}{I}\left[-\cos\phi\right]_0^{\pi} = \frac{FR^4t}{I}\left[(-\cos\pi)-(-\cos 0)\right]$$

$$= \frac{FR^4t}{I}\left[(--1)-(-1)\right]$$

i.e. $$F(\Delta + R) = \frac{2FR^4t}{I}$$

or $$F\Delta + FR = \frac{2FR^4 t}{I}$$

and $$F\Delta = \frac{2FR^4 t}{I} - FR$$

i.e. $$\boldsymbol{\Delta = \frac{2R^4 t}{I} - R} \tag{16.9}$$

To determine *I*

From Chapter 5 $I = \int y^2\, dA$

$$= \int_0^{\pi} (R\cos\phi)^2\, t\, R\, d\phi$$

$$= t R^3 \int_0^{\pi} \cos^2\phi\, d\phi$$

$$= t R^3 \int_0^{\pi} \left(\frac{1+\cos 2\phi}{2}\right) d\phi$$

$$= \frac{tR^3}{2} \int_0^{\pi} (1+\cos 2\phi)\, d\phi$$

$$= \frac{tR^3}{2}\left[\phi + \frac{\sin 2\phi}{2}\right]_0^{\pi}$$

$$= \frac{tR^3}{2}\left[\left(\pi + \frac{\sin 2\pi}{2}\right) - \left(0 + \frac{\sin 0}{2}\right)\right]$$

i.e. $$\boldsymbol{I = \frac{\pi R^3 t}{2}} \tag{16.10}$$

Substituting equation (16.10) into equation (16.9) gives:

$$\Delta = \frac{2R^4 t}{\left(\frac{\pi R^3 t}{2}\right)} - R$$

i.e. $$\Delta = \frac{4R}{\pi} - R$$

i.e. $$\boldsymbol{\Delta = R\left(\frac{4}{\pi} - 1\right) = 0.273\, R}$$

Problem 9. Determine the shear centre position for the thin-walled section of Figure 16.31, which is of uniform thickness *t*.

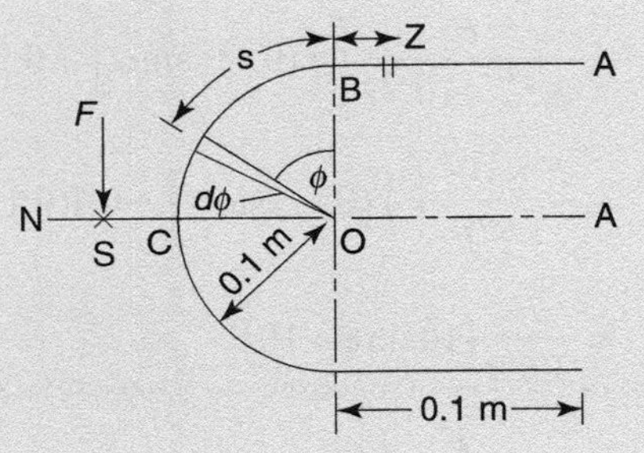

Figure 16.31 Curved section with flanges

Let I be the second moment of area of the section about NA, which can be obtained with the assistance of equation (16.10). Then

$$I = \frac{\pi R^3 t}{2} + 2\left(\frac{0.1t^3}{12} + t \times 0.1 \times 0.1^2\right)$$

but as the section is thin, higher-order terms involving t can be ignored,

therefore $$I = \frac{\pi \times 0.1^3 t}{2} + 2 \times 0.001t$$

i.e. $$\boldsymbol{I = 3.57 \times 10^{-3}\, t}$$

Consider the top flange

τ_F = shear stress in the top flange at Z

$$= \frac{F(B-Z)\bar{y}}{I}$$

At $Z = 0.1,\ \tau_A = 0$ (16.11)

At $Z = 0,\ \tau_B = \dfrac{FB\bar{y}}{I} = \dfrac{F \times 0.1 \times 0.1}{3.57 \times 10^{-3} t}$

i.e. $$\boldsymbol{\tau_B = \frac{2.8F}{t}} \tag{16.12}$$

Consider the curved section

At s, $$\tau_s = \frac{F}{tI}\int y\, dA$$

$$= \frac{F}{t \times 3.57 \times 10^{-3} t}\left(\int_0^{\phi} (R\cos\psi\, t\, R\, d\psi) + 0.1 \times t \times R\right)$$

$$= \frac{F}{3.57 \times 10^{-3} t}\left(\int_0^{\phi} (R^2\cos\psi\, d\psi) + 0.1R\right)$$

$$= \frac{F}{3.57 \times 10^{-3} t}\left(\left[R^2 \sin\psi\right]_0^{\phi} + 0.1R\right)$$

$$= \frac{F}{3.57 \times 10^{-3} t}\left(\left[R^2 \sin\phi\right] + 0.1R\right)$$

$$= \frac{F}{3.57\times10^{-3}t}\left(\left[(0.1)^2 \sin\phi\right] + 0.1(0.1)\right)$$

$$= \frac{F\times10^3}{3.57t}\left(\left[(0.1)^2 \sin\phi\right] + 0.1(0.1)\right)$$

$$= \frac{F}{3.57t}\left(10\sin\phi + 10\right)$$

i.e. $$\boldsymbol{\tau_s = \frac{F}{0.357t}\left(1+\sin\phi\right)} \qquad (16.13)$$

At $\phi = 0$, $\tau_B = \dfrac{F}{0.357t}\left(\sin 0 + 1\right) = \dfrac{F}{0.357t} = \dfrac{2.8F}{t}$ as required (see equation (16.12))

At $\phi = 90°$, $\tau_C = \dfrac{F}{0.357t}\left(1+\sin 90°\right) = \dfrac{2F}{0.357t} = \dfrac{5.6F}{t}$ = maximum shear stress due to bending

To calculate the shear centre position

Let F_F be the resisting shearing force in the flange. Then

$$F_F = \left(\frac{\tau_A + \tau_B}{2}\right)\times 0.1t$$

$$= \left(\frac{0+\dfrac{2.8F}{t}}{2}\right)\times 0.1t$$

i.e. $$\boldsymbol{F_F = 0.14F}$$

To obtain Δ, consider rotational equilibrium about the point O of the section, as shown in Figure 16.32.

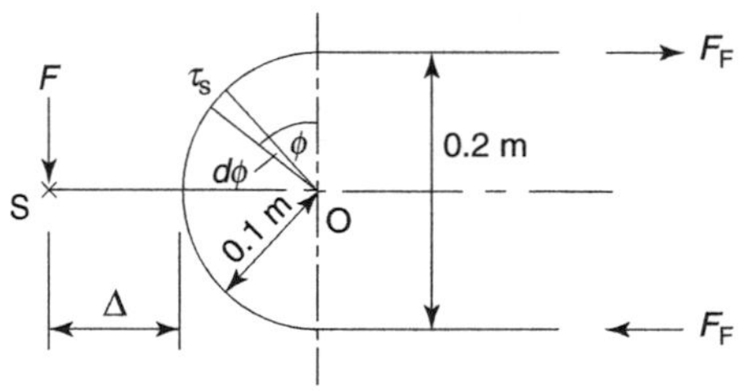

Figure 16.32 Thin-walled open section

$$F\left(\Delta + 0.1\right) = F_F \times 0.2 + \int_0^{\pi} \tau_s \times R \times d\phi \times t \times R$$

$$= 0.14F\times0.2 + \int_0^{\pi}\left[\frac{F}{0.357t}\left(\sin\phi + 1\right)\right]\times 0.1^2 \times t\,d\phi$$

$$= 0.028F + \left(\frac{F\times0.1^2\times t}{0.357t}\right)\int_0^{\pi}\left[\left(\sin\phi+1\right)\right]d\phi$$

$$= 0.028F + \left(\frac{F\times0.1^2}{0.357}\right)\left[-\cos\phi + \phi\right]_0^{\pi}$$

$$= 0.028F + 0.028F\left[\left(-\cos\pi + \pi\right) - \left(-\cos 0 + 0\right)\right]$$

$$= 0.028F + 0.028F\left[\left(--1+\pi\right) - \left(-1+0\right)\right]$$

$$= 0.028F + 0.028F\left[2+\pi\right]$$

$$= 0.028F + 0.144F$$

$$F\left(\Delta + 0.1\right) = 0.172F$$

i.e. $F\Delta + 0.1F = 0.172F$

i.e. $F\Delta = 0.172F - 0.1F = 0.072F$

and **shear centre position, Δ = 0.072 m or 72 mm**

Now try the following Practice Exercise

Practice Exercise 62. Shear centre

1. Determine the position of the shear centre for the thin-walled section of Figure 16.33.

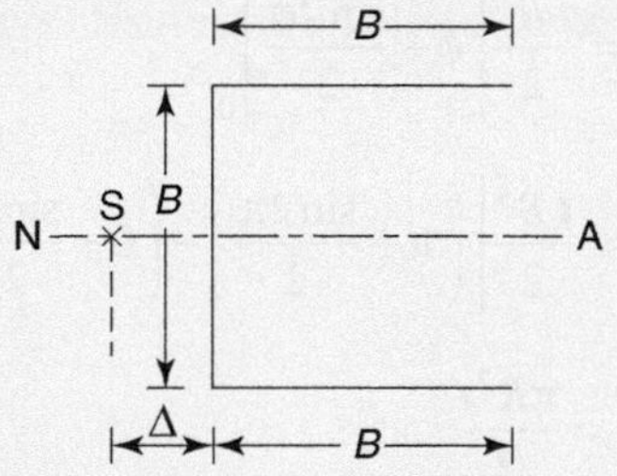

Figure 16.33 Thin-walled open section: channel section

[$\Delta = 0.429B$]

2. Determine the position of the shear centre for the thin-walled section of Figure 16.34.

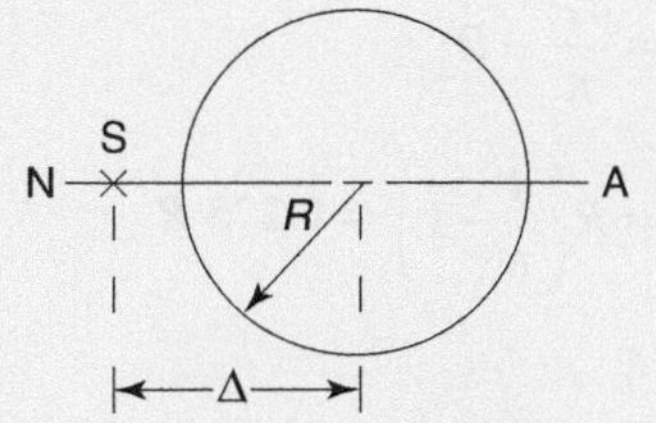

Figure 16.34 Split tube

[$\Delta = 2R$]

3. Determine the shear centre position for the thin-walled section of Figure 16.35.

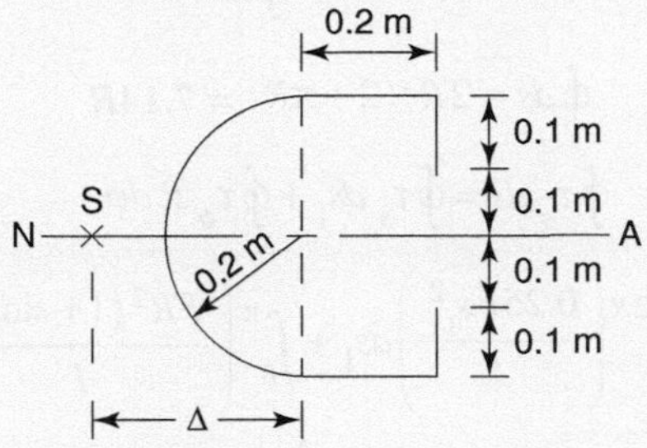

Figure 16.35 Thin-walled complex open section

[Δ = 0.396 m]

4. Determine the shear centre position for the thin-walled section of Figure 16.36.

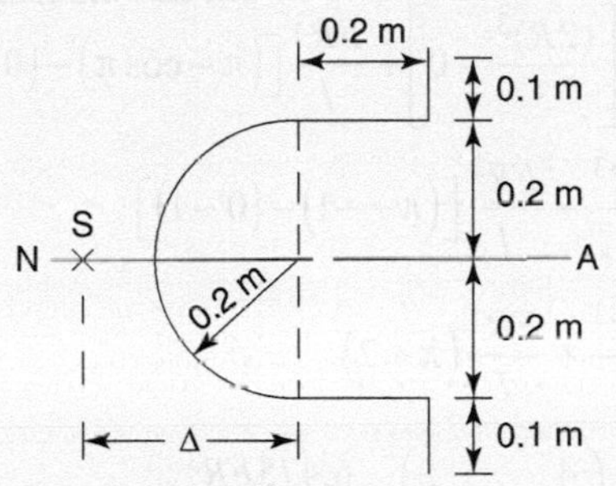

Figure 16.36 Thin-walled complex open section

[Δ = 2.83 m]

16.5 Shear centre positions for closed thin-walled tubes

The main problem of determining the shear stress due to bending in closed thin-walled tubes is that, initially, the shear stress is not known at any point. To overcome this difficulty, the assumption is made that the shear stress due to bending at a certain point in the section has an unknown value of τ_0, as shown in Figure 16.37.

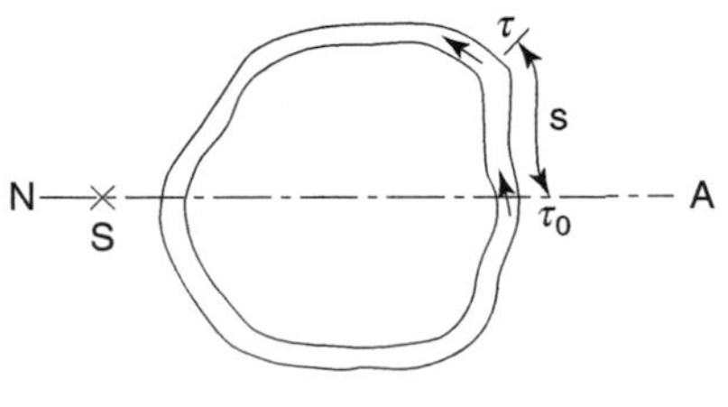

Figure 16.37 Thin-walled closed tube

Megson (see Reference 29 on page 497) has shown that for a thin-walled closed tube, the relationship between the twist and the shear stress is given by:

$$\oint \frac{\tau ds}{G} = 2A\frac{d\theta}{dz}$$

where

A = enclosed area of the cross-section of the tube

$\frac{d\theta}{dz}$ = twist/unit length

θ = angle of twist

z = distance along the axis of the tube

τ = shearing stress due to bending at any distance s from the 'starting' point

$= \tau_0 + \tau_s$

τ_0 = shearing stress due to bending at the 'starting' point

τ_s = shearing stress due to bending at any distance s for an equivalent *open* tube

ds = elemental length

G = rigidity modulus

If F is the shearing force applied through the shear centre S, then there will be no twist

i.e. $$\frac{d\theta}{dz} = 0 = \oint \frac{\tau ds}{G}$$

or $$\oint \frac{(\tau_0 + \tau_s) ds}{G} = 0$$

i.e. $$\oint (\tau_0 + \tau_s) ds = 0$$

However τ_0 = constant

Therefore $$\tau_0 \oint ds = -\oint \tau_s \, ds$$

i.e. $$\tau_0 = -\frac{\oint \tau_s \, ds}{\oint ds} \qquad (16.14)$$

Once τ_0 is determined from equation (16.14), τ can be found.

Problem 10. Determine the shear centre position for the thin-walled closed tube of Figure 16.38, which is of uniform thickness t.

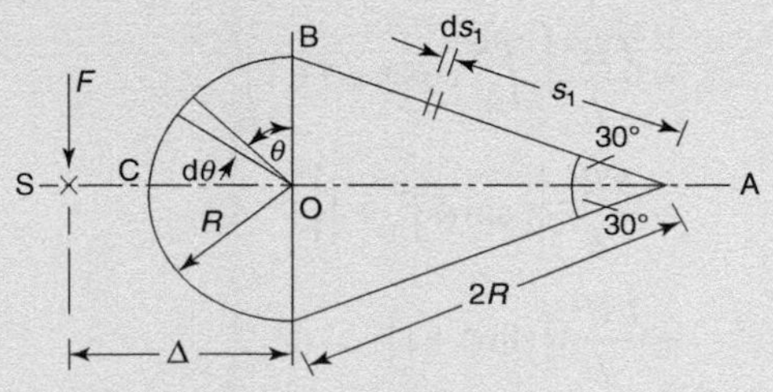

Figure 16.38 Thin-walled closed tube

Let τ_0 be the shear stress due to bending at the point A. Then

$$I = 2\int_0^{2R} (s_1 \sin 30°)^2 \times (t ds_1) + \int_0^{\pi} (R\cos\phi)^2 \times (t R d\phi)$$

$$= 0.5t\int_0^{2R} s_1^2 ds_1 + tR^3 \int_0^{\pi} \cos^2\phi \, d\phi$$

$$= 0.5t\left[\frac{s_1^3}{3}\right]_0^{2R} + tR^3 \int_0^{\pi}\left(\frac{1+\cos 2\phi}{2}\right) d\phi$$

$$= 0.5t\left[\frac{(2R)^3}{3} - 0\right] + \frac{tR^3}{2}\left[\phi + \frac{\sin 2\phi}{2}\right]_0^{\pi}$$

$$= \frac{4R^3 t}{3} + \frac{tR^3}{2}\left[\left(\pi + \frac{\sin 2\pi}{2}\right) - \left(0 + \frac{\sin 0}{2}\right)\right]$$

$$= \frac{4R^3 t}{3} + \frac{\pi t R^3}{2}$$

i.e. $$\boldsymbol{I = R^3 t\left(\frac{4}{3} + \frac{\pi}{2}\right) = 2.904R^3 t}$$

Consider AB

At any distance s_1

$$\tau_{s_1} = \frac{F}{tI}\int_0^{s_1} (s_1 \sin 30°) \times (t\, ds_1)$$

$$= \frac{0.5F}{I}\int_0^{s_1} s_1 \, ds_1 = \frac{0.5F}{I}\left[\frac{s_1^2}{2}\right]_0^{s_1}$$

$$= \frac{0.5F}{I}\left[\frac{s_1^2}{2} - 0\right]$$

i.e. $$\tau_{s_1} = \frac{0.25Fs_1^2}{I} \quad (16.15)$$

Consider BC

At any distance φ_1

$$\tau_\phi = \frac{F}{tI}\left(\int_0^{\phi} (R\cos\phi) \times (tR\, d\phi) + (2Rt) \times \left(\frac{R}{2}\right)\right)$$

$$= \frac{FR^2}{I}\left(\int_0^{\phi} \cos\phi \, d\phi + 1\right)$$

$$= \frac{FR^2}{I}\left\{\left[\sin\phi\right]_0^{\phi} + 1\right\}$$

$$= \frac{FR^2}{I}\{\sin\phi + 1\}$$

i.e. $$\boldsymbol{\tau_\phi = \frac{FR^2(1+\sin\phi)}{I}} \quad (16.16)$$

Now from equation (16.14)

$$\tau_0 = -\frac{\oint \tau_s \, ds}{\oint ds}$$

However $\oint ds = 2R \times 2 + \pi R = \mathbf{7.14R}$ (16.17)

and $$\oint \tau_s \, ds = \oint \tau_{s_1} ds_1 + \oint \tau_\phi R\, d\phi$$

$$= 2\int_0^{2R}\left(\frac{0.25Fs_1^2}{I}\right) ds_1 + \int_0^{\pi}\left(\frac{FR^2(1+\sin\phi)}{I}\right) R\, d\phi$$

$$= \frac{0.5F}{I}\int_0^{2R} s_1^2 \, ds_1 + \frac{FR^3}{I}\int_0^{\pi}(1+\sin\phi)\, d\phi$$

$$= \frac{0.5F}{I}\left[\frac{s_1^3}{3}\right]_0^{2R} + \frac{FR^3}{I}\left[\phi - \cos\phi\right]_0^{\pi}$$

$$= \frac{0.5F}{I}\left[\frac{(2R)^3}{3} - 0\right] + \frac{FR^3}{I}\left[(\pi - \cos\pi) - (0 - \cos 0)\right]$$

$$= \frac{4FR^3}{3I} + \frac{FR^3}{I}\left[(\pi - -1) - (0 - 1)\right]$$

$$= \frac{4FR^3}{3I} + \frac{FR^3}{I}(\pi + 2)$$

$$= \frac{FR^3}{I}\left(\frac{4}{3} + \pi + 2\right) = \frac{6.475FR^3}{I}$$

$$= \frac{6.475FR^3}{2.904R^3 t}$$

i.e. $$\oint \tau_s \, ds = \frac{2.23F}{t} \quad (16.18)$$

Hence, from equations (16.14), (16.18) and (16.17)

$$\tau_0 = -\frac{\oint \tau_s \, ds}{\oint ds} = -\frac{\frac{2.23F}{t}}{7.14R}$$

i.e. $$\boldsymbol{\tau_0 = -\frac{0.312F}{Rt}} \quad (16.19)$$

so that if τ_{AB} is the shear stress at any point between A and B, then:

$$\tau_{AB} = \tau_0 + \tau_{s_1} = -\frac{0.312F}{Rt} + \frac{0.25Fs_1^2}{I}$$

$$= -\frac{0.312F}{Rt} + \frac{0.25Fs_1^2}{2.904R^3 t}$$

i.e. $$\boldsymbol{\tau_{AB} = -\frac{0.312F}{Rt} + \frac{0.0861Fs_1^2}{R^3 t}} \quad (16.20)$$

If τ_{BC} is the shear stress at any point between B and C, then:

$$\tau_{BC} = \tau_0 + \tau_\phi = -\frac{0.312F}{Rt} + \frac{FR^2(1+\sin\phi)}{I}$$

$$= -\frac{0.312F}{Rt} + \frac{FR^2(1+\sin\phi)}{2.904R^3t}$$

$$= -\frac{0.312F}{Rt} + \frac{F(1+\sin\phi)}{2.904Rt}$$

$$= -\frac{0.312F}{Rt} + \frac{F}{2.904Rt} + \frac{F\sin\phi}{2.904Rt}$$

i.e. $$\boldsymbol{\tau_{BC} = \frac{F}{Rt}\left(0.0324 + \frac{\sin\phi}{2.904}\right)} \quad (16.21)$$

Taking moments about O gives:

$$F = \int_0^\pi \tau_{BC} \times (tR\,d\phi) \times R + 2\int_0^{2R} \tau_{AB} \times (t\,ds_1) \times R\cos 30°$$

$$= \int_0^\pi \left(\frac{F}{Rt}\left(0.0324 + \frac{\sin\phi}{2.904}\right)\right) \times (tR\,d\phi) \times R$$

$$+ 2\int_0^{2R}\left(-\frac{0.312F}{Rt} + \frac{0.0861Fs_1^2}{R^3t}\right) \times (t\,ds_1) \times R\cos 30°$$

from equations (16.20) and (16.21)

$$= \frac{FR^2t}{Rt}\int_0^\pi \left(\frac{0.0941 + \sin\phi}{2.904}\right) d\phi$$

$$+1.732Rt\int_0^{2R}\left(-\frac{0.312F}{Rt} + \frac{0.0861Fs_1^2}{R^3t}\right) ds_1$$

$$= \frac{FR}{2.904}\left[0.0941\phi - \cos\phi\right]_0^\pi$$

$$+1.732FRt\left[-\frac{0.312s_1}{Rt} + \frac{0.0861s_1^3}{3R^3t}\right]_0^{2R}$$

$$= \frac{FR}{2.904}\left[(0.0941\pi - \cos\pi) - (0 - \cos 0)\right]$$

$$+1.732FRt\left[\left(-\frac{0.312(2R)}{Rt} + \frac{0.0861(2R)^3}{3R^3t}\right) - (0)\right]$$

$$= \frac{FR}{2.904}\left[(0.0941\pi - 1) - (-1)\right]$$

$$+1.732FRt\left[\left(-\frac{0.624R}{Rt} + \frac{0.6888R^3}{3R^3t}\right)\right]$$

$$= \frac{FR}{2.904}\left[2.2956\right] + 1.732FR\left[(-0.624 + 0.2296)\right]$$

$$= 0.7905FR - 0.6831FR$$

$$= 0.107FR$$

Hence $F\Delta = 0.107FR$

and **the shear centre position, Δ = 0.107R**

Now try the following Practice Exercise

Practice Exercise 63. Shear centre positions for closed thin-walled tubes

1. Determine the shear centre position for the thin-walled closed tube of Figure 16.39 which is of uniform thickness.

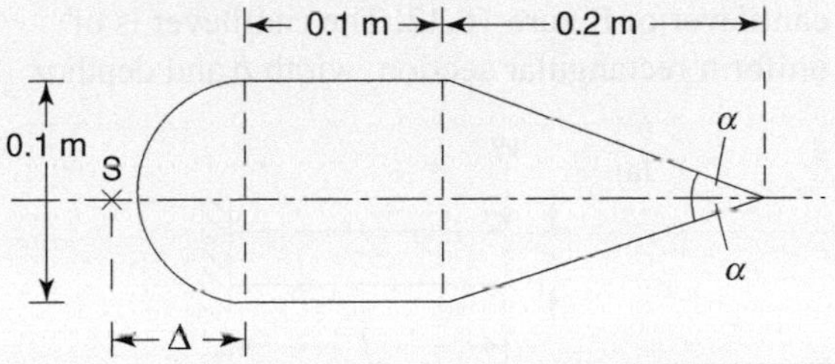

Figure 16.39 Thin-walled closed tube

[Δ = −0.032 m]

2. Determine the shear centre position for the thin-walled closed tube of Figure 16.40, which is of uniform thickness.

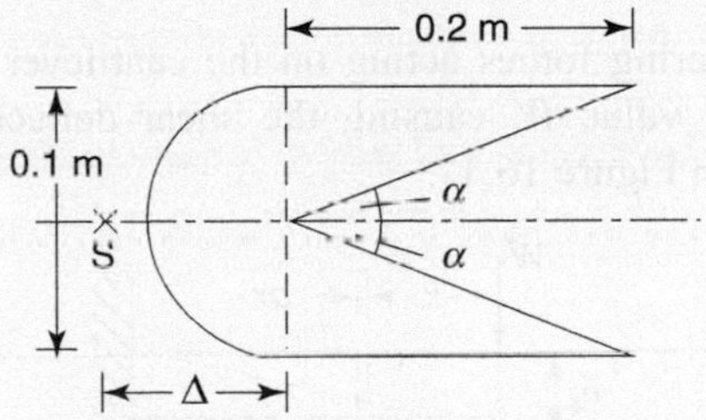

Figure 16.40 Thin-walled closed tube

[Δ = 0.0792 m]

16.6 Shear deflections

Deflections of beams usually consist of two components, namely deflections due to bending and deflections due to shear. If a beam is long and slender, then deflections due to shear are small compared with deflections due to bending. If, however, the beam is short and stout, then deflections due to shear cannot be neglected.

Deflections due to shear are caused by shearing action alone, where each element of the beam tends to change shape, as shown in Figure 16.41, where F is the shearing force acting on a typical element.

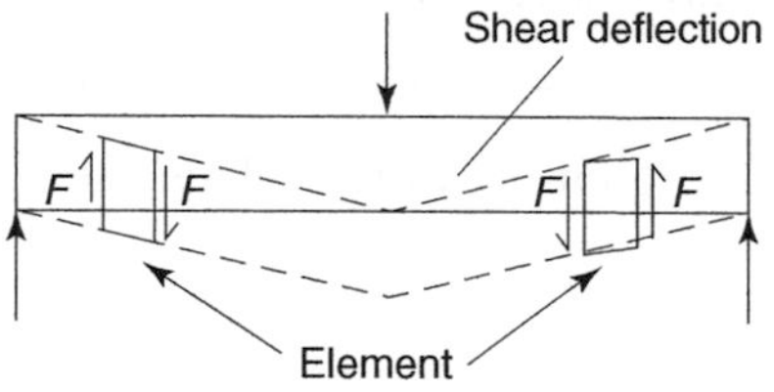

Figure 16.41 Shear deflection of a beam

Problem 11. Determine the value of the maximum deflection due to shear for the end-loaded cantilever of Figure 16.42. The cantilever is of uniform rectangular section, width b and depth d.

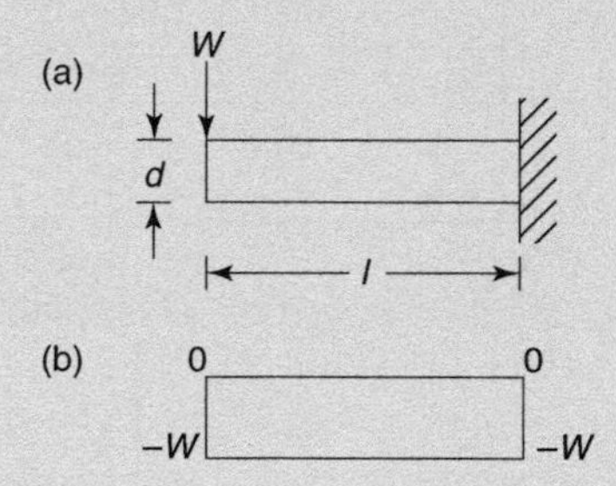

Figure 16.42 Shearing force distribution in an end-loaded cantilever: (a) cantilever; (b) shearing force diagram

The shearing forces acting on the cantilever are of a constant value W, causing the shear deflected form shown in Figure 16.43.

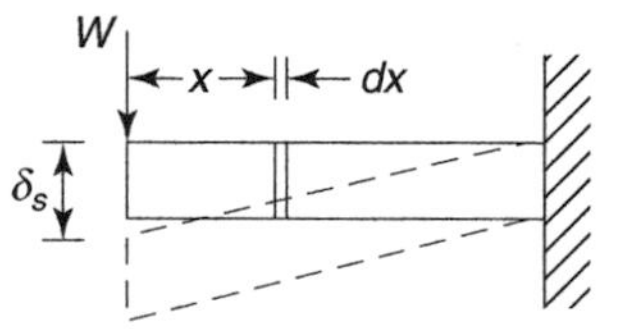

Figure 16.43 Shear deflected form of cantilever

From equation (16.3), the value of shear stress at any distance y from the neutral axis is:

$$\tau = \frac{6W}{bd^3}\left(\frac{d^2}{4} - y^2\right)$$

From Chapter 11, the total shear strain energy of the cantilever is:

$$\text{SSE} = \int \frac{\tau^2}{2G}\, d(\text{vol})$$

$$= 2\int_0^{d/2} \frac{\left(\frac{6W}{bd^3}\left(\frac{d^2}{4} - y^2\right)\right)^2}{2G}\, l\, b\, dy$$

$$= \frac{36W^2 lb}{Gb^2 d^6}\int_0^{d/2}\left(\frac{d^2}{4} - y^2\right)^2 dy$$

$$= \frac{36W^2 l}{Gbd^6}\int_0^{d/2}\left(\frac{d^4}{16} - \frac{d^2}{2}y^2 + y^4\right)dy$$

$$= \frac{36W^2 l}{Gbd^6}\left[\frac{d^4 y}{16} - \frac{d^2 y^3}{6} + \frac{y^5}{5}\right]_0^{d/2}$$

$$= \frac{36W^2 l}{Gbd^6}\left[\left(\frac{d^4(d/2)}{16} - \frac{d^2(d/2)^3}{6} + \frac{(d/2)^5}{5}\right) - (0)\right]$$

$$= \frac{36W^2 l}{Gbd^6}\left(\frac{d^5}{32} - \frac{d^5}{48} + \frac{d^5}{160}\right)$$

i.e. $$\text{SSE} = \frac{3W^2 l}{5Gbd} \qquad (16.22)$$

If δ_s is the maximum deflection due to shear, then the work done by the load W is:

$$WD = \frac{1}{2}W\delta_s \qquad (16.23)$$

Equating equations (16.22) and (16.23) gives:

$$\frac{3W^2 l}{5Gbd} = \frac{1}{2}W\delta_s$$

from which, maximum deflection due to shear,

$$\boldsymbol{\delta_s = \frac{6Wl}{5Gbd}} \qquad (16.24)$$

Problem 12. Determine the value of the maximum deflection due to shear for a cantilever, assuming that it is subjected to a uniformly distributed load w, as shown in Figure 16.44.

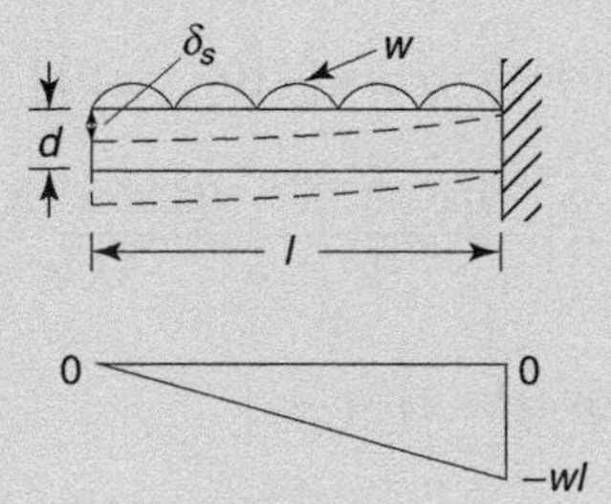

Figure 16.44 Cantilever with a uniformly distributed load: (a) cantilever; (b) shearing force diagram

In this case, as the shearing force varies linearly along the length of the cantilever, the shear deflection will vary parabolically.

At any distance x from the free end, the shearing force on an element of the beam of length dx is:

$$F = wx$$

From equation (16.24), the shear deflection of this element is:

$$\frac{6wx\,dx}{5Gbd}$$

Therefore **the total shear deflection of the cantilever is:**

$$\delta_s = \int_0^l \left(\frac{6wx}{5Gbd}\right) dx$$

$$= \left[\frac{6w}{5Gbd}\frac{x^2}{2}\right]_0^l = \left[\left(\frac{6w}{5Gbd}\frac{l^2}{2}\right) - (0)\right]$$

i.e. $$\boldsymbol{\delta_s = \frac{3wl^2}{5Gbd}}$$

Total deflection of an end-loaded cantilever

From Chapter 7, the maximum deflection due to bending of an end-loaded cantilever is:

$$\delta_b = \frac{Wl^3}{3EI}$$

which for a rectangular section of width b and depth d becomes:

$$\delta_b = \frac{Wl^3}{3E\left(\frac{bd^3}{12}\right)}$$

i.e. $$\boldsymbol{\delta_b = \frac{4Wl^3}{Ebd^3}}$$

Now, the total deflection is:

$$\delta = \delta_b + \delta_s = \frac{4Wl^3}{Ebd^3} + \frac{3Wl}{5Gbd}$$

From Section 4.9, if it is assumed that $E = 2.5G$ then

$$G = \frac{E}{2.5} = \frac{E}{5/2} = \frac{2E}{5}$$

i.e. $$\delta = \frac{4Wl^3}{Ebd^3} + \frac{3Wl}{5\left(\frac{2E}{5}\right)bd}$$

$$= \frac{4Wl^3}{Ebd^3} + \frac{3Wl}{2Ebd}$$

$$\delta = \frac{4Wl^3}{Ebd^3}\left[1 + \frac{3}{4}\left(\frac{d}{l}\right)^2\right] \qquad (16.25)$$

where the second term in the brackets represents the component of deflection due to shear.

From equation (16.25), it can be seen that the deflection due to shear is important when $\frac{d}{l}$ becomes relatively large.

Now try the following Practice Exercise

Practice Exercise 64. Shear deflections

1. Determine the maximum deflection due to shear for the simply supported beam of Figure 16.45. It may be assumed that the beam cross-section is rectangular, of constant width b and of constant depth d.

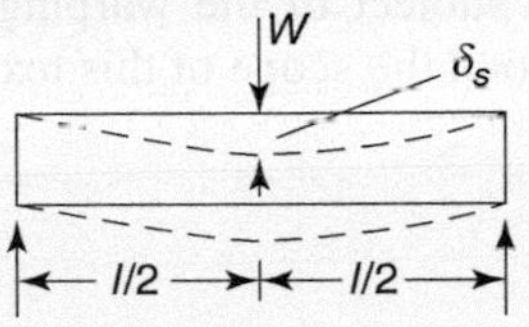

Figure 16.45 Shear deflections of simply-supported beam-centrally loaded

$$\left[\delta_s = \frac{3Wl}{10Gbd}\right]$$

2. Determine the maximum deflection due to shear for the simply supported beam of Figure 16.46. It may be assumed that the beam cross-section is rectangular, of constant width b and of constant depth d.

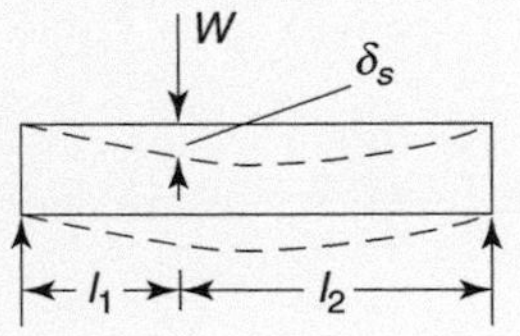

Figure 16.46 Shear deflections of simply-supported beam-off-centre load

$$\left[\delta_s = \frac{6Wl_1l_2}{5Gbd}\right]$$

3. Determine the maximum deflection due to shear for the simply supported beam of Figure 16.47. It may be assumed that the beam cross-section is rectangular, of constant width b and of constant depth d.

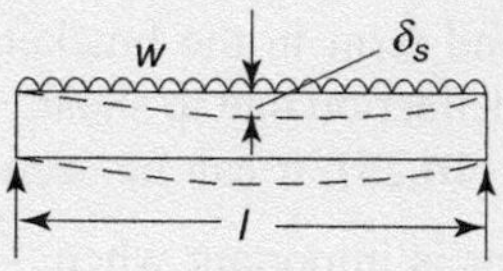

Figure 16.47 Shear deflections of simply supported beam: uniformly distributed load

$$\left[\delta_s = \frac{3wl^2}{20Gbd}\right]$$

16.7 Warping

The effects of warping have not been included in this chapter as the subject of the warping of thin-walled sections is beyond the scope of this text.

Megson (Reference 29 on page 497) describes warping as out-of-plane deformation of a cross-section, particularly when an unsymmetrical section is not loaded through its shear centre, as shown in Figure 16.48.

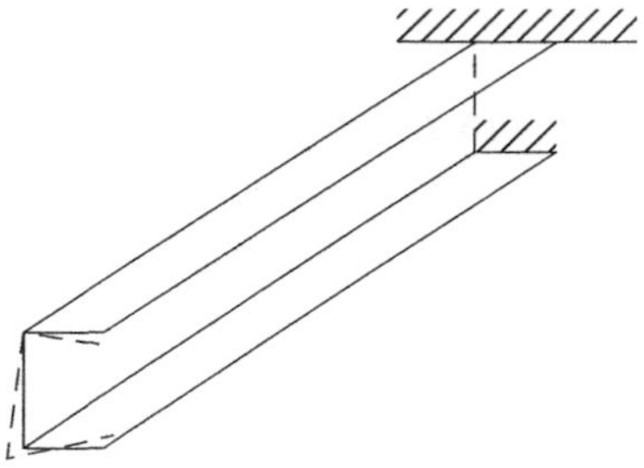

Figure 16.48 Warping of a cross-section

The longitudinal direct stresses caused by warping are of particular importance when the beam is restrained from axial movement. Warping is not of importance in solid circular-section beams and in thin-walled tubes of circular and square cross-section.

For fully worked solutions to each of the problems in Exercises 61 to 64 in this chapter, go to the website:
www.routledge.com/cw/bird

Chapter 17

Experimental strain analysis

Why it is important to understand: **Experimental strain analysis**

Normally, before a structure is built, it is usually convenient and cheaper to design the structure using the theory of structural analysis. For large or unusual structures an additional safeguard of the structure is to determine the stresses in the structure to ensure that it will be safe. Electrical resistance strain gauges, together with other experimental methods of determining strains, are usually added insurance that the structure is going to be safe. Moreover, sometimes a structure may have stress concentrations and suffer dynamic stresses due to sudden gusts of wind or sea waves, and for these structures, experimental back up is essential. This chapter explains a number of different methods in experimental strain analysis, in particular those in electrical resistance strain gauges and photoelasticity.

At the end of this chapter you should be able to:

- understand the construction and principle of operation of an electrical resistance strain gauge, including temperature and pressure compensation and null and deflection methods
- understand the principle of operation of the Wheatstone bridge
- understand foil gauge and foil gauge type of electrical resistance strain gauge, including advantages and disadvantages and gauge materials
- appreciate different types of gauge adhesives, including cellulose acetates, epoxy resins, cyanoacrylates and Norton-Rockide
- understand some methods of water-proofing
- understand combined bending and torsion of circular section shafts, developing formulae for M and T
- understand combined bending and axial strains and full bridge for measuring M
- appreciate photoelasticity, birefringence, circularly polarised light, experimental photoelasticity, material fringe value, stress trajectories, three-dimensional photoelasticity
- understand Moire fringes, brittle lacquer techniques
- appreciate semi-conductor and acoustical strain gauges
- appreciate the use of lacquers to determine experimental strains

17.1 Introduction

In this chapter, a brief description will be given of some of the major methods in experimental strain analysis, and in particular, to the use of electrical resistance strain gauges and photoelasticity.

The aim in this chapter is to expose the reader to various methods of experimental strain analysis, and to encourage him or her to consult other publications which cover this topic in a more comprehensive manner (see References 23 to 25 on page 497).

17.2 Electrical resistance strain gauges

The elastic strain in most structures, constructed from steel, aluminium alloy, etc., seldom exceeds 0.1%, and it is evident that such small magnitudes of strain will need considerable magnification to record them precisely. This feature presented a major problem to structural engineers in the past, and it was not until the 1930s that this problem was resolved, when the electrical resistance strain gauge was invented. This gauge was invented as a direct result of requiring lighter aircraft structures, although the principle that the electrical resistance strain gauge is based on was discovered as early as 1856, by Lord Kelvin,* when he observed that the electrical resistance of copper and iron wires varied with strain. Lord Kelvin's discovery was that, when a length of wire is strained, its electrical resistance changes and, within certain limits, this relationship is linear, and can be expressed as follows:

Strain ∝ change of electrical resistance; or strain

$$(\varepsilon) = K\frac{\Delta R}{R} \qquad (17.1)$$

where K is known as the **gauge factor**, and is dependent on the material of construction (i.e. it is a material constant);

for an ordinary Cu/Ni gauge, K is about 2

R is the electrical resistance of the strain gauge (ohms)

ΔR is the change of electrical resistance of the strain gauge (ohms) due to ε

As K is known, and R and ΔR can be measured, ε can be readily obtained from equation (17.1).

Temperature compensation

If a strain gauge is subjected to a change of temperature, it very often suffers a larger change in length than that caused by external loadings. To overcome this deficiency, it is necessary to attach another strain gauge (called a 'dummy' gauge) to a piece of material with the same properties as the structure itself, and to subject this piece of material to the same temperature changes as the structure, but not to 'constrain' the 'dummy' gauge, or allow it to undergo any external loading.

The dummy gauge should have identical properties to the active gauge, and for static analysis, the two gauges should be connected together in the form of a Wheatstone bridge, as shown in Figure 17.1.

*William Thomson, 1st Baron Kelvin (26 June 1824 – 17 December 1907) did important work in the mathematical analysis of electricity and formulation of the first and second laws of thermodynamics, and did much to unify the emerging discipline of physics in its modern form. He also had a career as an electric telegraph engineer and inventor, which propelled him into the public eye and ensured his wealth, fame and honour. To find out more about Kelvin go to www.routledge.com/cw/bird

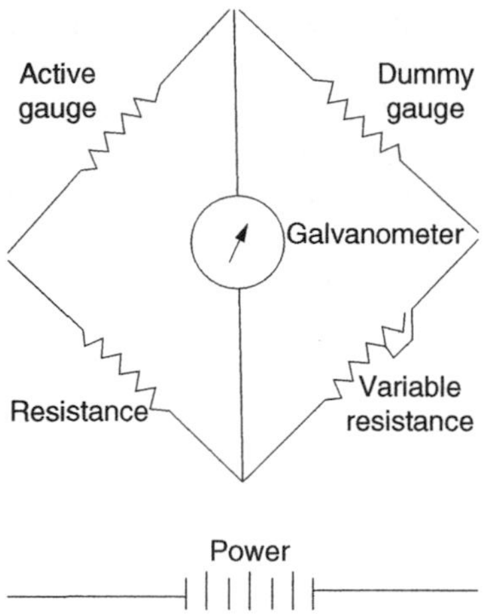

Figure 17.1 The 'null' method of measuring strains

Pressure compensation

If an electrical resistance strain gauge is subjected to fluid pressure, its electrical resistance changes owing to the Poisson effect of the gauge material. To overcome this deficiency, it is necessary to subject the strain gauge to the same pressure medium conditions as the active gauge.

The 'null' method of measuring strain

Figure 17.1 shows the circuitry for this method of strain measurement, which is suitable for static analysis. Although the active gauge and the dummy gauge may have the same initial electrical resistance, after attaching these gauges to their respective surfaces, there may be small differences in their electrical resistances, so that the galvanometer will become unbalanced. Thus, prior to loading the structure, it will be necessary to balance the galvanometer, by suitably changing the electrical resistance on the variable resistance arm. Once the structure is loaded, and the gauge experiences strain, the galvanometer will once again become unbalanced, and by balancing the galvanometer, the change of electrical resistance due to this load can be measured. Normally, the strain gauge equipment has facilities which allow it to directly record strain, providing the gauge factor is pre-set for the particular batch of gauges.

Further loading of the structure will cause the galvanometer to become unbalanced, once again, and by rebalancing the galvanometer, the strain can be recorded for this particular loading condition.

The 'deflection' method of measuring strain

It is evident from above that the 'null' method is only suitable for static analysis, where there is a sufficient time available to record each individual strain. Thus, for dynamic analysis, or where many measurements of static strain are required in quick succession, the 'null' method is unsuitable.

One method of overcoming this problem is to use the 'deflection' method of strain analysis, where the strain gauge circuit of Figure 17.2 is used.

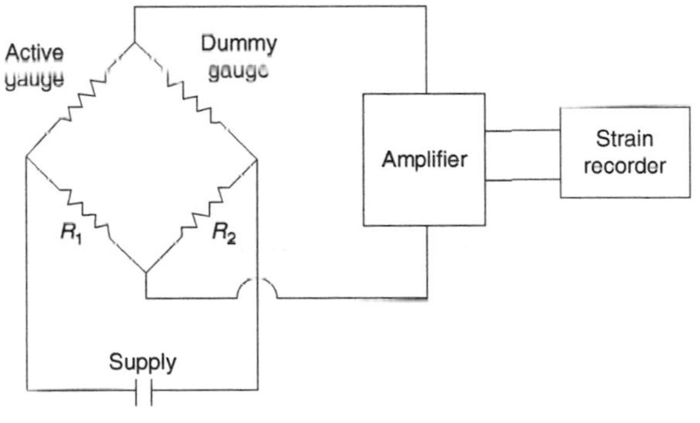

Figure 17.2 Strain gauge circuit for the 'deflection' method of strain measurement

In this case, as the measured strains are very small, they need to be amplified prior to being recorded. The strain recorder can take many forms, varying from chart recorders and storage oscilloscopes to computers. Thus, it is evident that this method of measuring strain is much more expensive than the null method, because of its requirements for large-scale amplification and sensitive strain recorders, but, nevertheless, it has been successfully employed for dynamic strain recording. The electrical supply shown in Figure 17.2 can be either a.c. or d.c.

17.3 Types of electrical resistance strain gauge

The two most common types of electrical resistance strain gauge are the wire gauge and the foil gauge, and these gauges are now described.

Wire gauges

One of the simplest forms of the wire gauge is the *zigzag* gauge, where the diameter of its wire can be as small as 0.025 mm, as shown in Figure 17.3. The reason why this gauge wire is of zigzag form is because a minimum length of wire is required, so that the supplied power can be sufficiently small to prevent heating of the gauge itself. The gauge is constructed by sandwiching a length of high-resistance wire in zigzag form, between two pieces of paper.

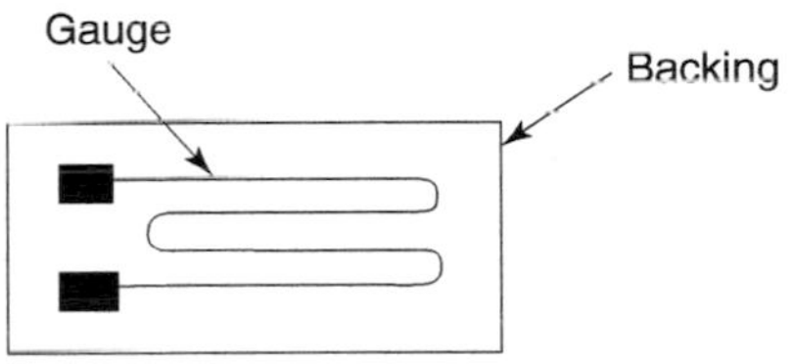

Figure 17.3 A linear zigzag wire gauge

It should be noted that the gauge is attached to the structure via its backing, and that the latter is required as an electrical insulation. The problem with the zigzag gauge is that it is prone to cross-sensitivity.

Cross-sensitivity is an undesirable property of a strain gauge, as the high-resistance wire, which is perpendicular to the axis of the gauge, measures erroneous strains in this direction, in addition to the required strains which lie along the axis of the gauge.

To some extent, cross-sensitivity can be reduced by employing the strain gauge of Figure 17.4. This gauge is constructed by placing thin strands of high-resistance wire parallel to each other, and connecting them together by low-resistance welded crossbars. Thus, as the electrical resistance of the crossbars is small compared with the electrical resistance of the wires, the effects of cross-sensitivity are much smaller than for zigzag wire gauges.

The backing material for these gauges can be either paper or plastic.

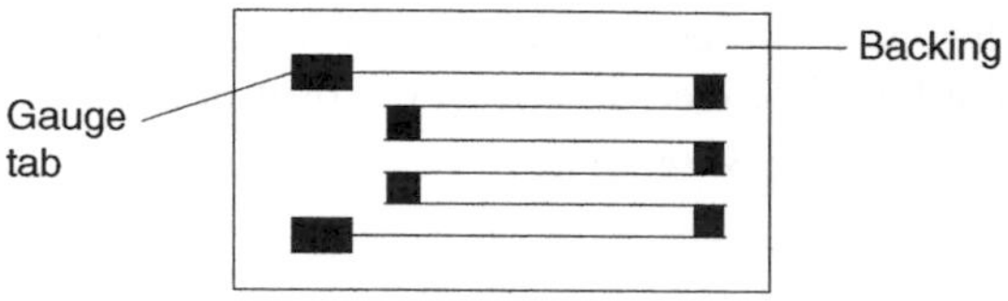

Figure 17.4 A linear wire gauge, with welded cross-bars

Foil gauges

Another popular strain gauge is the foil gauge of Figure 17.5, which was invented by Saunders-Roe in 1952. This gauge is etched out of flat metal foil, so that its electrical resistance along its axis can be very large compared with its electrical resistance perpendicular to its axis. This property allows successful construction of such gauges to be as small as 0.2 mm (1/128 in) or as long as 2.54 m (100 in), with negligible cross-sensitivity effects. In the cases of these extreme lengths of strain gauge, considerable expertise is required to attach these gauges successfully to the structure. Small strain gauges are required for strain investigations in regions of stress concentrations, and strain gauges of length 2.54 m are used to investigate the longitudinal strength characteristics of ship structures.

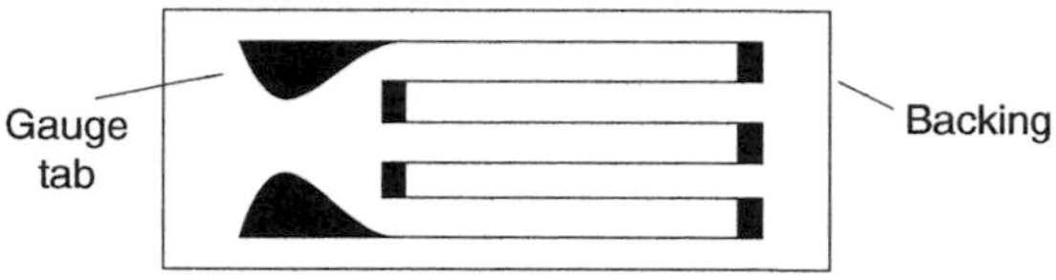

Figure 17.5 A linear foil gauge

In general, the most popular lengths of linear gauges vary between about 5 and 10 mm. The main advantage of the foil gauge is that because it is etched out of flat metal foil, it and its backing can be very thin, so that the gauge can be in a more intimate contact with the structure than can a wire gauge.

Disadvantages of foil gauges are that they are generally more expensive than wire gauges, and that for very thin foil gauges, they can easily fracture, if roughly handled. Foil gauges are usually plastic backed.

17.4 Gauge material

Most electrical resistance strain gauges of normal resistance (about 100 ohms) are constructed from a copper/nickel alloy, where the proportion of Cu:Ni usually varies from about 55:45 to 60:40. For such gauges, the gauge factor is about 2, and depends on the Cu:Ni ratio.

A popular alloy for high-resistance gauges (about 1000 ohms) is called *Nichrome,* which consists of 75% Ni, 12% Fe, 11% Cr and 2% Mn.

17.5 Gauge adhesives

In the electrical resistance strain gauge technique, adhesives are one of the most important considerations. It is important that the experimentalist uses a satisfactory adhesive, as an unsatisfactory adhesive can cause hysteresis or zero drift. In addition to this, another important consideration is the preparation of the surface to which the gauge is to be attached.

Prior to attaching the gauge, it should be ensured that the surface is clean and free from rust, paint, dirt, grease, etc. This surface should not be highly polished; if it is, it should be roughened a little with a suitable abrasive, and then it should be degreased with a suitable solvent, such as acetone, alcohol, etc. After application of the strain gauge cement to the appropriate part of the structure's surface, and in some cases to the back of the gauge itself, the gauge should be gently placed onto its required position.

To ensure that no air bubbles become trapped between the gauge and the surface of the structure, the thumb, or finger, should be pressed firmly onto the surface of the gauge, and gently rolled to and fro, until any excess strain gauge cement is squeezed out.

The adhesion time of the gauge varies from a few minutes to several days, depending on the type of adhesive used and the environment that the adhesive is to be exposed to.

Gauge adhesives are generally either organic based or ceramic based, the former being satisfactory for temperatures below 260°C and the latter for temperatures in excess of this. A brief description of some of the different types of gauge adhesive will now be given.

Cellulose acetates

These are among the most common forms of adhesive (e.g. Durofix) that can be bought from high street shops. They adhere by evaporation, and take from 24 hours to 3 days to gain full strength, depending on the surrounding temperature. They are usually used for paper-backed gauges.

Epoxy resins

These require the mixing of a resin with a hardener. A popular combination is to use Araldite strain gauge cement, with either Araldite hardener HY951 or

Araldite hardener HY956, in the following proportions by weight:

Araldite strain gauge cement 100
with either HY951 hardener 4 to 4.5
or HY956 hardener 8 to 10

Another epoxy resin adhesive is the M-Bond epoxy supplied by Omega Products.

These cements adhere by curing, where the time taken to gain full strength can take 24 hours, but in the case of Araldite, the curing time can be accelerated by exposing the cement to ultra-violet radiation. These adhesives are suitable for either plastic-backed or paper-backed gauges.

Cyanoacrylates

These are pressure adhesives, which adhere by applying pressure to the cement, via the gauge. One of the earliest of these adhesives is called Eastman-Kodak 910, which allows a strain gauge to be used within a few minutes of attaching it to the structure.

Another more recent cyanoacrylate is M-Bond 200, which is supplied by Omega Products.

Norton-Rokide

This adhesive is suitable for high-temperature work, and because of this, the strain gauge does not normally have any backing. Instead, the cement is first sprayed onto the surface, and then the gauge is firmly pressed down, but care has to be taken to ensure that some cement lies between the gauge and the structure, so that its resistance to earth does not break down. After the wires have been attached to the gauge tabs, further cement is sprayed over the gauge and the surrounding surface, so that it is encapsulated.

17.6 Water-proofing

If a strain gauge is exposed to water or damp conditions, its electrical resistance to earth will break down, and render the gauge useless. Thus, if a gauge is likely to be exposed to such an environment, it is advisable to water-proof the gauge and its wiring.

Some methods of water-proofing strain gauges are discussed below, but prior to using any of these methods, it should be assumed that the gauges and their surrounding surfaces are free from water.

Di-jell is a micro-crystalline wax which has the appearance of a jelly-like substance. In general, it is only suitable for damp-proofing or for water-proofing when the water is stationary. To water-proof the gauge, the Di-jell is simply applied to the gauge and its surrounding surface.

Silicone greases and petroleum jelly (Vaseline) can also be used for damp-proofing, but care should be taken to ensure that these substances are not subjected to temperatures which will melt them.

A more robust and permanent method of water-proofing strain gauges is that recommended by Omega Products. This consists of painting M-Coat A or D over the gauge and its adhesive, and then covering the M-Coat with aluminium foil. Finally, the whole surface is covered with M-Coat G, the electrical leads being first covered with M-Coat B, as shown in Figure 17.6.

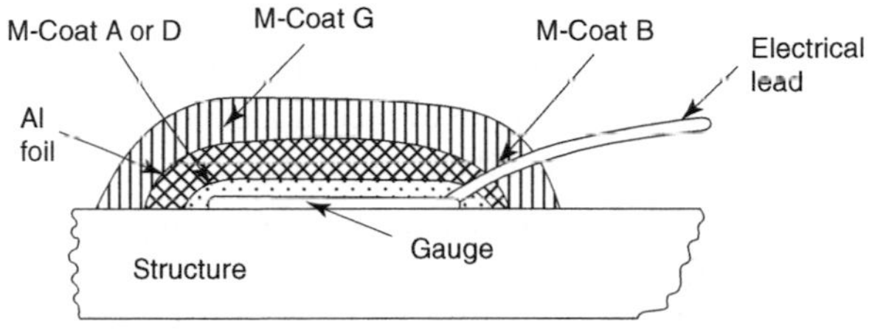

Figure 17.6 Omega Products method of water-proofing

Prior to the development of the Omega Products method of water-proofing strain gauges, the Saunders-Roe technique proved popular. This method consisted of applying successive coats of expoxy resin and glass cloth, and covering the whole with an impervious rubber-based solution.

Other methods of water-proofing consist simply of covering the gauge and its surrounding surface with various types of sealant, including automobile underbody sealants.

Line (see Reference 26 on page 497) carried out an investigation on the water-proofing qualities of a number of sealants, as shown in Table 17.1. Line used foil gauges of dimensions 2.54 cm length, 1.02 cm width. He measured the electrical resistance of these strain gauges in air, before immersion into water, and then took these measurements again, 1 hour and also 3 weeks later, after continuous immersion in water. He also measured the electrical resistances of these gauges at water pressures of 3.45 and 6.9 MPa. He found that all the sealants were satisfactory under test, and he recorded the following observations:

(a) The underbody seal appeared to have softened and, although it was still water-proof, it could easily be chipped off by hand.
(b) Under the M-Coat A, it was possible to see that the metal was completely rust-free.

Table 17.1 Effect of pressure and water immersion on the electrical resistance of strain gauges

	Electrical resistance (ohms)				
Sealant	**Before**	**1 hour in water**	**3 weeks in water**	**3.45 MPa**	**6.9 MPa**
Bostik 6	55	54	55	55	55
Underwater seal	56	56	55	56	56
M-coat A	56	56	55	56	56
M-coat D	56	55	55	55	55
Di-Jell	52	52	52	–	–

(c) The pressures were too low and the instruments were too insensitive to record a change of resistance due to the effects of pressure.

Note: Dally and Riley (see Reference 25 on page 497) report on tests by Milligan (see Reference 27), and by Brace (see Reference 28), who found that the effect of pressure caused a strain of about 0.58×10^{-6} per MPa of pressure, and they concluded that for most problems, the effects of pressure on strain gauges can be ignored at pressures below 20.7 MPa (3000 lbf/in^2).

17.7 Other strain gauges

Other forms of gauge include *shear pairs* and *strain rosettes,* as described in Chapter 9, together with *crack measuring* and *diaphragm* gauges. Diaphragm gauges consist of a combination of radial and circumferential gauges, and crack measuring gauges consist of several parallel strands of wire and tabs.

17.8 Gauge circuits

Skilful use of strain gauge circuits can eliminate the use of dummy gauges and provide increased sensitivity.

Combined bending and torsion of circular section shafts

For circular-section shafts, under the effects of combined bending and torsion, it is convenient to use two pairs of 'shear pairs' as shown in Figure 17.7. By fitting two pairs of shear pairs, it is possible to record either the bending moment M or the torque T, depending on the circuit used.

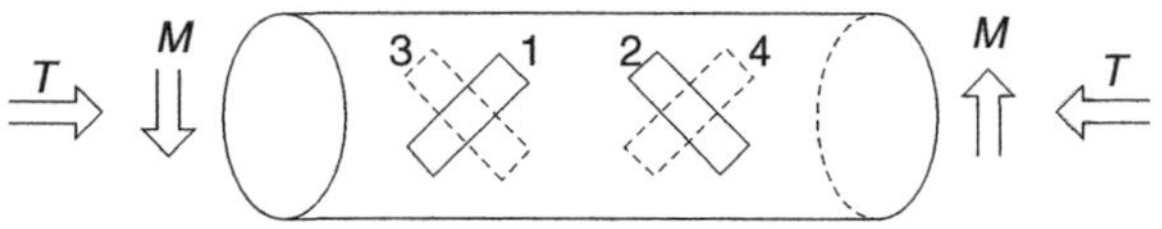

Figure 17.7 Shaft under combined bending and torque

The shear pairs must be fitted at 45° to the axis of the shaft, and a typical shear pair of strain gauges is shown in Figure 17.8. Shear pairs are so called because, under pure torque, the maximum principal stresses in a circular-section shaft, which are numerically equal to the maximum shear stress, lie at 45° to the axis of the shaft (see Chapter 9).

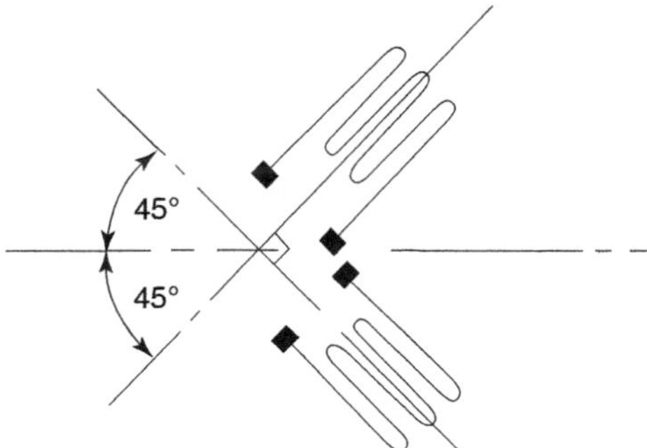

Figure 17.8 A shear pair of strain gauges

To determine moment *M*

Let ε_{45B} = direct strain due to M, which lies at 45° to the axis of the shaft

ε_{45T} = direct strain due to T, which lies at 45° to the axis of the shaft

$$= \frac{\gamma}{2} \text{ (see Chapter 9)}$$

γ = maximum shear strain due to T

If the gauges are connected together in the form of a full Wheatstone bridge, as shown in Figure 17.9, γ will be eliminated. From Figure 17.9, the output will be:

$$(\text{gauge }1 - \text{gauge }3) - (\text{gauge }4 - \text{gauge }2)$$

$$= \left(\varepsilon_{45B} + \frac{\gamma}{2} + \varepsilon_{45B} - \frac{\gamma}{2}\right) - \left(-\varepsilon_{45B} - \frac{\gamma}{2} - \varepsilon_{45B} + \frac{\gamma}{2}\right)$$

$$= 4\varepsilon_{45B}$$

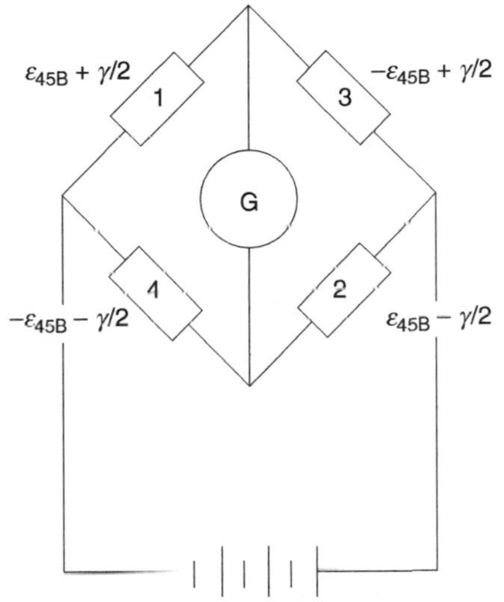

Figure 17.9 Circuit for measuring M

or $\qquad \text{output} = 4\varepsilon_{45B} = \dfrac{\sigma_{45B}}{E}(1-\nu)\times 4$

$$= \frac{\sigma_B}{2E}(1-\nu)\times 4$$

$$= \frac{2\sigma_B(1-\nu)}{E}$$

Therefore $\qquad \sigma_B = \dfrac{\text{output}}{2(1-\nu)} \times E = \text{bending stress}$

However $\qquad M = \sigma_B \times \dfrac{\pi d^4}{64} \times \dfrac{2}{d}$

i.e. $\qquad \boldsymbol{M = \dfrac{\pi d^3 \sigma_B}{32}} \qquad (17.2)$

To determine torque *T*

By adopting the circuit of Figure 17.10, ε_{45B} can be eliminated. The output from the circuit of Figure 17.10 will be:

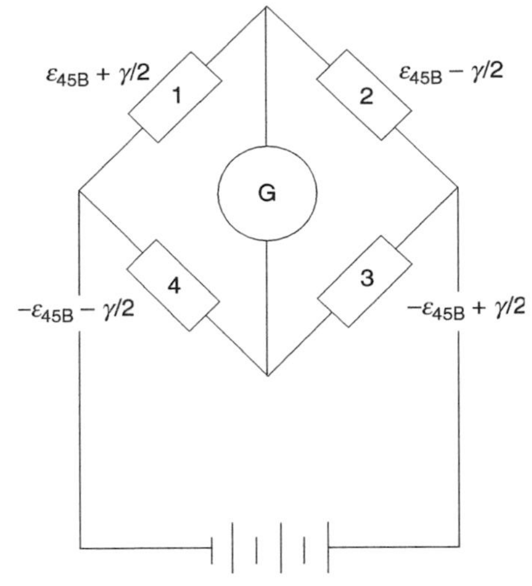

Figure 17.10 Circuit for measuring T

$$(\text{gauge }1 - \text{gauge }2) - (\text{gauge }4 - \text{gauge }3)$$

$$= \left(\varepsilon_{45B} + \frac{\gamma}{2} - \varepsilon_{45B} + \frac{\gamma}{2}\right) - \left(-\varepsilon_{45B} - \frac{\gamma}{2} + \varepsilon_{45B} - \frac{\gamma}{2}\right)$$

i.e. $\qquad \text{output} = 2\gamma$

or $\qquad \gamma = \dfrac{\text{output}}{2}$

However, torque, $T = r \times \dfrac{\pi d^4}{32} \times \dfrac{2}{d}$

i.e. $\qquad \boldsymbol{T = \dfrac{\pi G \gamma d^3}{16}} \qquad (17.3)$

NB Strains due to axial loads, including *thermal effects*, will automatically be eliminated when the circuits of Figures 17.9 and 17.10 are adopted.

Combined bending and axial strains

Consider a length of beam under combined bending and axial load, as shown in Figure 17.11.

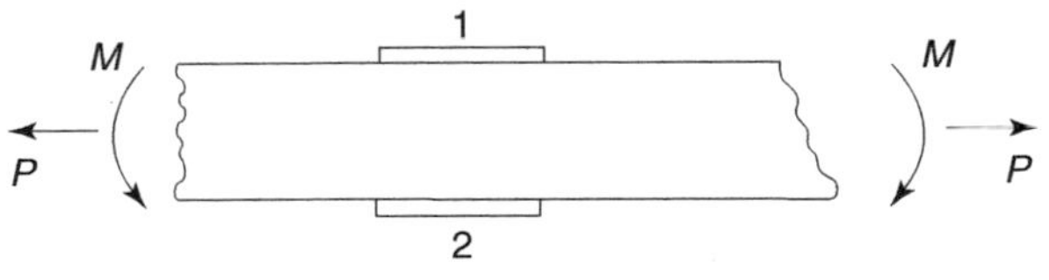

Figure 17.11 Combined M and P

Let $\qquad \varepsilon_B$ = bending strain due to M

and $\qquad \varepsilon_D$ = direct strain due to P, plus thermal effects

The bending moment M can be obtained by adopting the circuit of Figure 17.12.

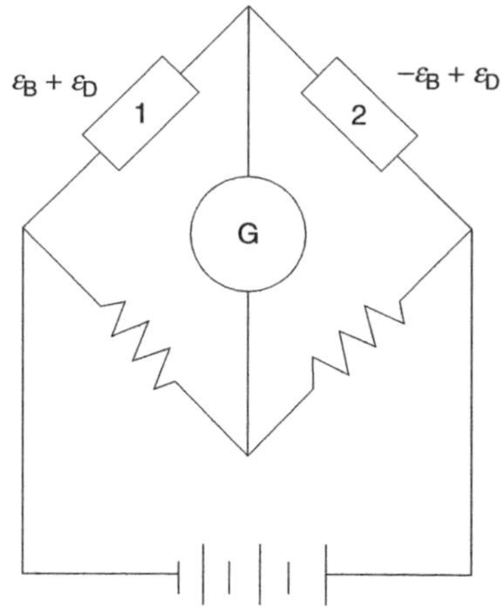

Figure 17.12 Circuit to determine M

The output from the circuit of Figure 17.12 is

gauge 1 – gauge 2 = $2\varepsilon_B$

or $\varepsilon_B = \dfrac{\text{output}}{2} = \dfrac{\varepsilon_B}{E}$

Hence, M can be obtained.

Full bridge for measuring *M*

output = (gauge 1 – gauge 2) – (gauge 4 – gauge 3)

$= \mathbf{4}\,\varepsilon_B$

i.e. the circuit of Figure 17.13 will give four times the sensitivity of a single strain gauge, and twice the sensitivity of the circuit of Figure 17.12. It will also automatically eliminate thermal strains.

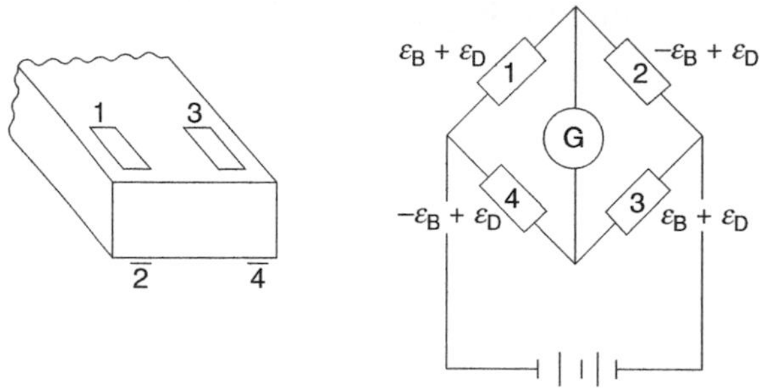

Figure 17.13 Full bridge for determining M

Half-bridge for measuring axial strains

See Figure 17.14.

$$\textbf{Output} = 2\varepsilon_D = \mathbf{2} \times \frac{\sigma_D}{E}$$

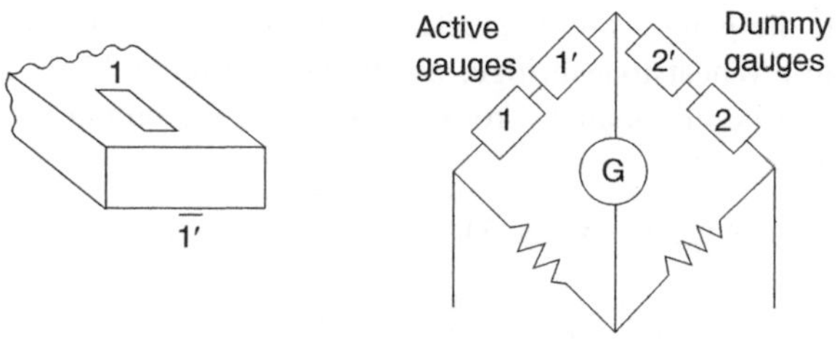

Figure 17.14 Circuit for eliminating bending and thermal strains

Hence, the axial load can be obtained.

17.9 Photoelasticity

The practice of photoelasticity is dependent on the shining of light through transparent or translucent materials, but prior to discussing the method, it will be necessary to make some definitions.

A beam of ordinary light vibrates in many planes, transverse to its axis of propagation. It is evident that, as light vibrates in so many different planes, it will be necessary for a successful photoelastic analysis to allow only those components of a beam of light to vibrate in a known plane.

This is achieved by the use of a *polariser,* which, in general, only allows those components of light to pass through it that vibrate in a vertical plane. Polarisers are made from Polaroid sheet, and apart from photoelastic analysis, they are used for sunglasses ('shades' in the USA).

A polariser which has a horizontal axis of transmission is called an *analyzer.* Thus, if the axes of transmission of a polariser and an analyzer are at 90° to each other (crossed), as shown in Figure 17.15, no light will emerge from the analyzer.

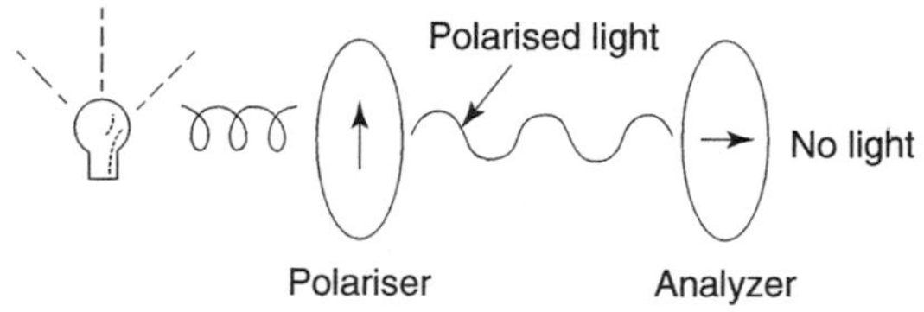

Figure 17.15 Polariser and analyzer

Birefringence

The practice of photoelasticity is dependent on birefringence or double refraction. Birefringence can be described as the property possessed by some transparent and translucent materials, whereby a single ray of polarised light is split into two rays on emerging from the birefringent material, where the two rays are perpendicular to each other. In general, the two emerging rays are out of phase, and this is called the relative retardation. Some materials are permanently birefringent; others, such as that from which the models are constructed are only birefringent under stress. In the case of the latter, a single ray of polarised light is split into two rays on emerging from the model, where the directions of the two rays lie along the planes of the principal stresses. The relative retardation of the two rays, namely R_t, is proportional to the magnitude of the principal stresses and the thickness of the model, as follows:

$$R_t + C(\sigma_1 - \sigma_2)h \tag{17.4}$$

where C = the stress optical coefficient (i.e. it is a material constant)

σ_1 = maximum principal stress

σ_2 = minimum principal stress

and h = model thickness

Typical materials used for photoelastic models include epoxy resin (e.g. Araldite), polycarbonate (e.g. Lexan or Makrolan), urethane rubber, etc.

A *plane polariscope* is one of the simplest pieces of equipment used for photoelasticity. It consists of a light source, a polariser, a model, an analyzer and a screen, as shown in Figure 17.16.

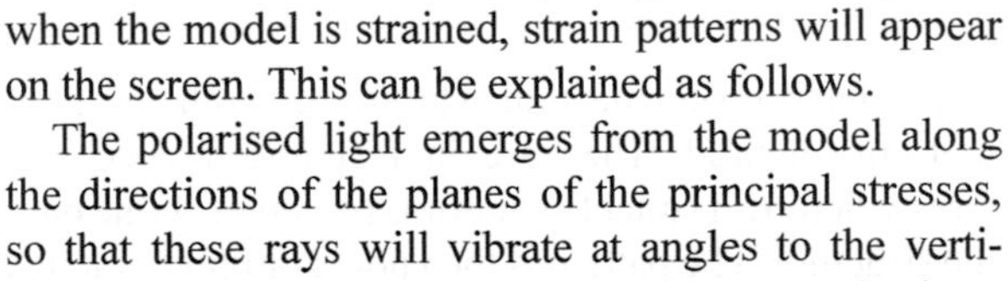

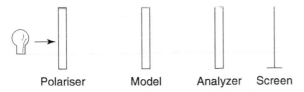

Figure 17.16 A plane polariscope

When the model is unstrained, no light will emerge, but when the model is strained, strain patterns will appear on the screen. This can be explained as follows.

The polarised light emerges from the model along the directions of the planes of the principal stresses, so that these rays will vibrate at angles to the vertical plane. On emerging from the analyzer, only those components of light which vibrate in a horizontal plane will be displayed on the screen, to give a measure of the stress distribution in the model. If daylight or white light is used in a plane polariscope, the stress patterns will involve all the colours of the rainbow. The reason for this is that daylight or white light is composed of all the colours of the rainbow, each colour vibrating at a different frequency. The approximate frequency of deep red light is about 390×10^{12} Hz, and the approximate frequency of deep violet light is 770×10^{12} Hz, the speed of light in a vacuum being about 2.998×10^8 m/s 186,282 miles/s). The difference in frequencies between different coloured lights is one of the reasons why violet is on the inside of the rainbow and red is on the outside, and these differences probably account for why we can distinguish between different colours.

However, whereas the stress patterns are quite spectacular when daylight is used, it is difficult to analyse the model. For this reason, it is preferable to use *monochromatic light.*

Monochromatic light is light of one wavelength only, and when it is used, the stress patterns consist of dark lines against a light background. Typical lamps used to produce monochromatic light include mercury vapour and sodium.

The use of monochromatic light in a plane polariscope will produce *isoclinics* and *isochromatics,* which are defined as follows:

Isoclinic fringe patterns are lines of constant stress direction, and occur when monochromatic light is used. They are useful for determining the principal stress directions.

Isochromatic fringe patterns are lines of constant maximum shear stress, i.e. lines of constant $(\sigma_1 - \sigma_2)$ and occur when monochromatic light is used.

Isotropic points are points on the model where $\sigma_1 - \sigma_2$, and ***singular points*** are points on the model where $\sigma_1 = \sigma_2 = 0$

Circularly polarised light

One of the problems with using a plane polariscope is that both isochromatic and isoclinic fringes appear together, and much difficulty is experienced in distinguishing between the two. The problem can be overcome by using a circular polariscope, which can extinguish the isoclinics. A plane polariscope can be converted to a circular polariscope by inserting *quarter-wave plates,* as shown in Figure 17.17.
Quarter-wave plates are constructed from permanently birefringent materials, which have a relative retardation of $\lambda/4$, and to extinguish the isoclinics, they are placed with their fast axes at 45° to the planes of polarisation of the polariser and the analyzer, and at 90° to each other, where λ is the wavelength of the selected light.

Some notes on experimental photoelasticity

To obtain the isoclinics, remove the quarter-wave plates and rotate the polariser and the analyzer by the same angle and in the same direction. This process will

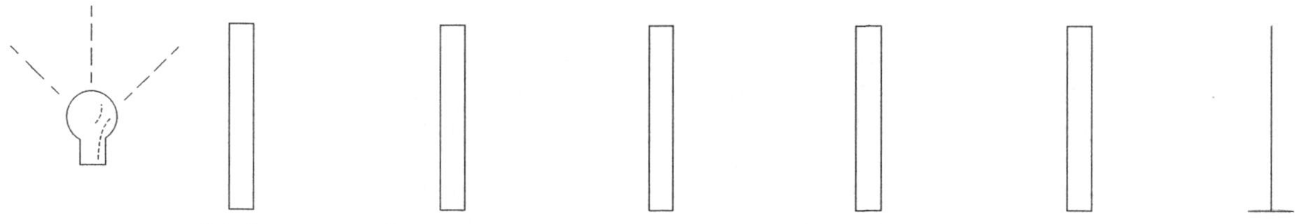

Figure 17.17 Circular polariscope

make the isoclinic fringes move with the change of angle, but will not affect the isochromatic fringes.

To obtain *isochromatic half-fringes* in a circular polariscope, place the axes of polarisation of the polariser in the same plane as the axis of polarisation of the analyzer.

The *isochromatic fractional fringes* can be obtained by appropriately placing the plane of the polariser at various angles, other than 0° and 90°, to the plane of polarisation of the analyzer.

Material fringe value *f*

The material fringe value for any given photoelastic material is given by:

$$f = \frac{\lambda}{2C} \text{ (kN/m fringe)} \tag{17.5}$$

where C = stress optical coefficient and λ = wavelength of the selected light

If there are *n isochromatic fringes,* then from equations (12.12) and (12.13),

$$\sigma_1 - \sigma_2 = 2\hat{\tau} = \frac{nf}{h} \tag{17.6}$$

where h = model thickness and $\hat{\tau}$ = maximum shear stress at the *n*th fringe

Thus, at the *i*th fringe, the maximum shear stress τ, is given by

$$\hat{\tau}_i = i \times \frac{f}{2h}$$

Typical values of *f*, under mercury green light, together with other material constants for some photoelastic materials, are given in Table 17.2.

Stress trajectories

These are a family of lines which lie orthogonally to each other. They are lines of constant principal stress, and they can be constructed from isoclinics.

Three-dimensional photoelasticity

One popular method in three-dimensional photoelasticity is to load the model whilst it is being subjected to a temperature between 120°C and 180°C. The model is then allowed to cool slowly to room temperature, but the load is not removed during this process. On removing the load, when the model is at room temperature, the stress system will be 'frozen' into the model, and then by taking thin two-dimensional slices from the model, a photoelastic investigation can be carried out–sometimes by immersing the slices into a liquid of the same refractive index as the model's material.

There are many other methods in three-dimensional photoelasticity, but descriptions of these methods are beyond the scope of this text.

Table 17.2 Some material constants for photoelastic materials

Material	*f* (kN/m fringe)	*E* (MPa)	*v*	UTS* (MPa)
Araldite B	10.8	2760	0.38	70
Makralon	7.0	2930	0.38	62
Urethane rubber	0.2	3	0.46	0.15*

*This quoted value is not the ultimate tensile strength (UTS) of the material but its elastic limit

17.10 Moire fringes

These have no connection whatsoever with the photoelastic method. They are in fact interference patterns whereby a suitable pattern is placed or shone onto the structure. The pattern is then noted or photographed before and after deformation, and the two patterns are superimposed to produce 'fringes'. Examination of these fringes by the use of a comparator or a microscope can be made to determine the experimental strains.

The patterns can consist of lines, grids or dots, etc., which can be parallel, radial or concentric, depending on the shape of the structure to be analysed.

17.11 Brittle lacquer techniques

This is another method of experimental strain analysis, where a thin coating of a brittle lacquer is sprayed or painted onto the surface of the structure before it is loaded. After loading the structure, the lacquer will be found to have cracked patterns, and these patterns will be related to the direction and magnitude of the maximum tensile stresses. One of the most popular brittle lacquers is called 'stresscoat', which is manufactured by the Magnaflux Corporation of the USA.

One of the problems with using brittle lacquer is that the cracks only occur under tension, so that if regions of compressive stresses are required to be examined, the lacquer must be applied to the surface of the structure while it is under load. On removing the load, the brittle lacquer will crack in those zones where the structure was in compression when loaded.

17.12 Semiconductor strain gauges

These are constructed from a single crystal of silicon, and each gauge takes the form of a short rectangular filament.

Semiconductor strain gauges are very sensitive to strain, where their change of resistance for a given strain can be about 100 times the value of the change of resistance of an electrical resistance strain gauge. Semiconductor strain gauges are usually small (e.g. 0.5 mm), and because of this and their high sensitivity, they are particularly useful for experimental strain analysis in regions of high stress concentration. Their main disadvantage is that they are much more expensive than electrical resistance strain gauges. They are usually manufactured from either a P-type material or an N-type material. The gauges can be made to be self-temperature compensating, by constructing the gauge from a P-type element together with an N-type element, the two being connected in one half of a Wheatstone bridge.

17.13 Acoustical gauges

These gauges are based on measuring the magnitude of the resonating frequency of a piece of wire stretched between two knife edges. Initially, the wire must be in tension, so that any compressive strains it is likely to receive under load will not cause it to lose its bending stiffness. The wire is 'plucked' by an electromagnet, and another electromagnet, called the 'pick-up', receives the signals from the resonating wire. The signal received by the pick-up magnet is then amplified, and apart from the signal being sent to an oscilloscope, it is used further to excite the 'plucking' magnet, so that the amplitude of the resonating wire will be maximised.

One knife edge of the acoustical gauge is fixed, and the other knife edge is movable, so that the latter can transmit strain to the resonating wire. Change of length of the wire will cause its resonant frequency to change, which can then be compared with a reference acoustical gauge. This gauge can then be adjusted with the aid of a micrometer screw gauge attached to it, so that its resonant frequency is the same as the gauge under test, and, hence, the strain recorded.

Acoustical gauges vary in length from 2.54 cm (1 in) to 15.24 cm (6 in), and because of their bulk, they are generally only preferred to electrical resistance strain gauges in special circumstances, where their inherent robustness and long-term stability characteristics are considered to be of prime importance.

Revision Test 4 Specimen examination questions for Chapters 9 to 17

This Revision Test covers the material in Chapters 9 to 17. *The marks for each question are shown in brackets at the end of each question.*

1. Determine the end fixing moments and reactions for the uniform section encastré beam of Figure RT4.1. (28)

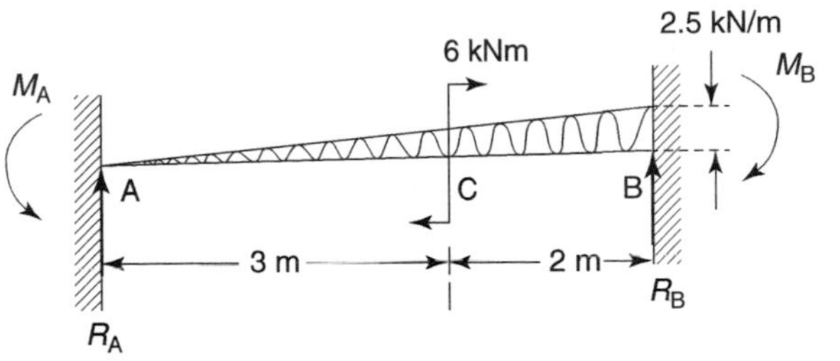

Figure RT4.1

2. Determine the end fixing moments and reactions for the uniform section encastré beam of Figure RT4.2. (28)

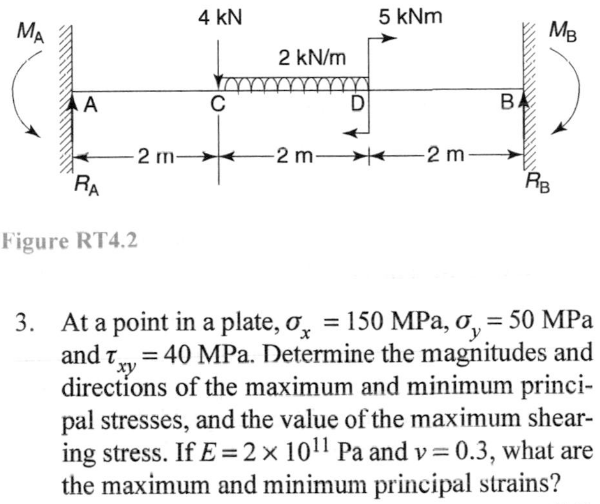

Figure RT4.2

3. At a point in a plate, $\sigma_x = 150$ MPa, $\sigma_y = 50$ MPa and $\tau_{xy} = 40$ MPa. Determine the magnitudes and directions of the maximum and minimum principal stresses, and the value of the maximum shearing stress. If $E = 2 \times 10^{11}$ Pa and $v = 0.3$, what are the maximum and minimum principal strains? (16)

4. At a point in a plate, $\sigma_x = 55$ MPa, $\sigma_y = -120$ MPa and $\tau_{xy} = 60$ MPa. Determine the magnitudes and directions of the maximum and minimum principal stresses, and the value and direction of the maximum shearing stress. If $E = 1 \times 10^{11}$ Pa and $v = 0.33$, what are the maximum and minimum principal strains? (16)

5. The stresses $\sigma_x = 90$ MPa, $\tau_{xy} = 50$ MPa and $\sigma_1 = 220$ MPa are shown on the triangular element abc of Figure RT4.3. Determine the minimum principal stress, and also σ_y and α. (19)

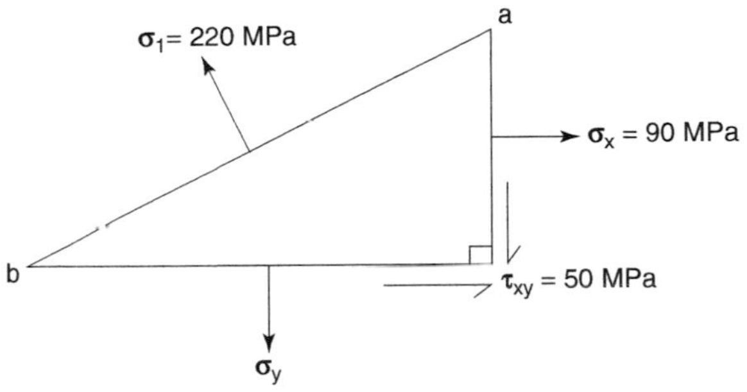

Figure RT4.3

6. At a point in a two-dimensional stress system, the known stresses are as shown in Figure RT4.4. Determine the stresses σ_y and τ_{xy}. (19)

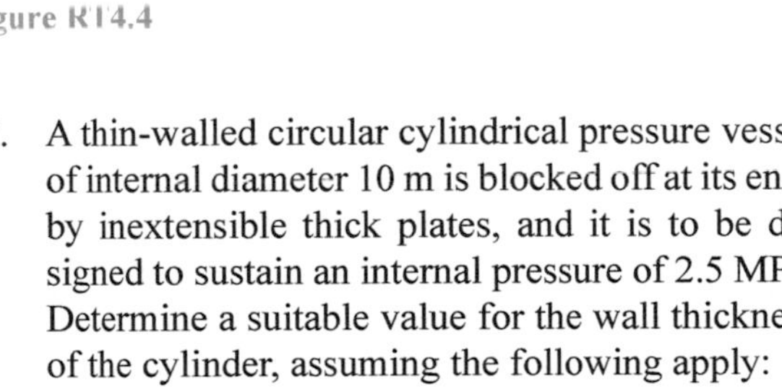

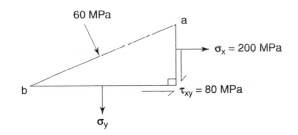

Figure RT4.4

7. A thin-walled circular cylindrical pressure vessel of internal diameter 10 m is blocked off at its ends by inextensible thick plates, and it is to be designed to sustain an internal pressure of 2.5 MPa. Determine a suitable value for the wall thickness of the cylinder, assuming the following apply:

longitudinal joint efficiency = 46%
circumferential joint efficiency = 98%
maximum permissible stress = 120 MPa (7)

8. A thin-walled circular cylinder of internal diameter 8 m and wall thickness 2.5 cm, and length 4 m, is just filled with water. Determine the additional water that is required to be pumped into the vessel to raise its pressure by 0.8 MPa. Assume the following:

   Young's modulus, $E = 2 \times 10^{11}$ Pa
   Poisson's ratio, $v = 0.3$
   the bulk modulus of water, $k = 2 \times 10^9$ Pa (15)

9. Determine an expression for the deflection at the free end 'A' of the thin section beam shown in Figure RT4.5. (20)

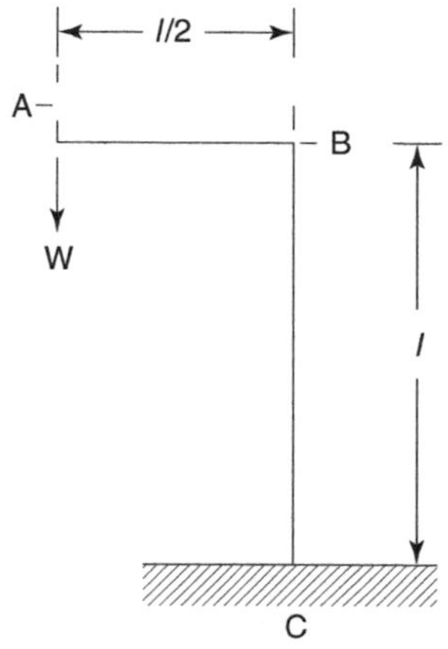

Figure RT4.5

10. Determine an expression for the deflection at the free end 'A' of the thin uniform circular section beam shown in Figure RT4.6, which has an out-of-plane concentrated load 'W' applied to its free end. (8)

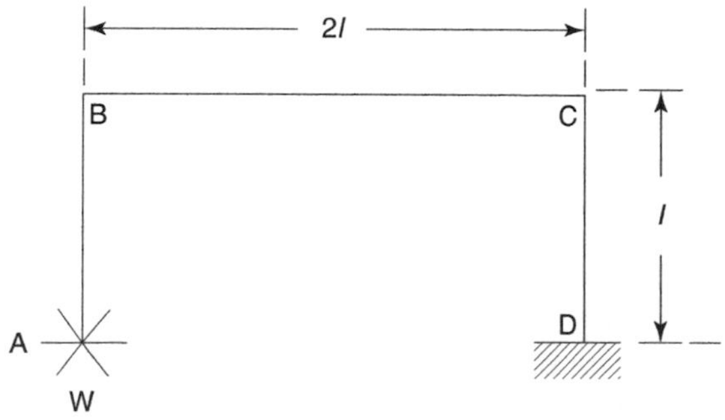

Figure RT4.6

11. A rod composed of 2 elements of different cross-sectional areas, A_1 and A_2, where A_1 is for the top element, is firmly fixed at the top and has an inextensible collar firmly fixed at its bottom end, as shown in Figure RT4.7. Determine expressions for the maximum deflection and maximum stress in the rod, if a mass of 1.5 kg falls a distance 'h', as shown.

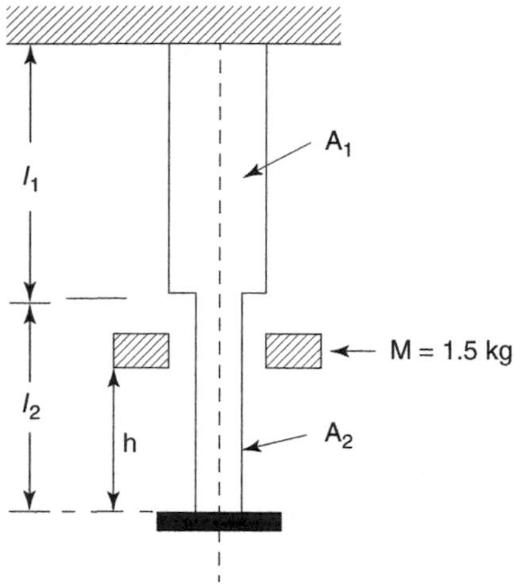

Figure RT4.7

Assume the following: $E = 2 \times 10^{11}$ Pa, $g = 9.81$ m/s^2, $h = 0.1$ m
For section 1: A_1 = cross-sectional area $= 0.04$ m^2 and $l_1 = 1.3$ m
For section 2: A_2 = cross-sectional area $= 0.02$ m^2 and $l_2 = 0.9$ m (19)

12. A long thin-walled circular cylinder of wall thickness 3 cm and of internal diameter 8 m is subjected to uniform internal pressure. Determine the pressure that will cause yield, based on the five major theories of elastic failure, and given the following:

    $\sigma_{yp} = 200$ MPa $E = 2 \times 10^{11}$ Pa, $v = 0.3$ (25)

13. (a) Determine the Euler crushing load for a hollow cylindrical steel column of 20 cm external diameter, 25 mm thick and of length 7 m, and pinned-hinged at both ends. Assume that $E = 2 \times 10^{11}$ Pa.

    (b) Compare the load in part (a) with that given by the Rankine-Gordon formula, using constants of 300 MPa and 1/7500.

    (c) For what length of column would the two formulae of parts (a) and (b) have the same crushing load? (16)

14. A very short steel tube of 0.1 m external diameter, when tested in compression, failed under an axial load of 1 MN. A 12 m length of the same column, pin-jointed at its ends, failed under an axial compressive load of 40 kN. Assuming that both the Euler and Rankine-Gordon formulae apply to this strut, calculate (a) the tube's internal diameter, and (b) the denominator constant in the Rankine-Gordon formula. Assume that $E = 200$ GPa. (14)

For lecturers/instructors/teachers, fully worked solutions to each of the problems in Revision Test 4, together with a full marking scheme, are available at the website:
www.routledge.com/cw/bird

Chapter 18

An introduction to matrix algebra

Why it is important to understand: **An introduction to matrix algebra**

Matrices are used to solve problems in statics, quantum mechanics, electronics, optics, robotics, linear programming, optimisation, genetics, and much more. Matrix calculus is a mathematical tool used in connection with linear equations, linear transformations, systems of differential equations, and so on, and is vital for calculating forces, vectors, tensions, masses, loads and a lot of other factors that must be accounted for in engineering to ensure safe and resource-efficient structures. Mechanical and electrical engineers, chemists, biologists and scientists all need knowledge of matrices to solve problems. In computer graphics, matrices are used to project a three-dimensional image on to a two-dimensional screen, and to create realistic motion. Matrices are therefore very important in solving engineering problems.

In the main, matrices and determinants are used to solve a system of simultaneous linear equations. The simultaneous solution of multiple equations finds its way into many common engineering problems, in fact, modern structural engineering analysis techniques are all about solving systems of equations simultaneously.

At the end of this chapter you should be able to:

- understand the terms: matrix, column, row, elements and order of matrices
- appreciate types of matrices: column, row, square, null, diagonal and scalar matrices
- transpose a matrix
- add and subtract matrices
- multiply two matrices
- calculate determinants of 2 by 2 and 3 by 3 matrices
- determine the inverse for 2 by 2 and 3 by 3 matrices

18.1 Introduction

This introductory chapter on matrix algebra is included in this text because the following chapters need a knowledge of matrices, and it may be that many studying this text will benefit by a revision of the salient points.

18.2 Elementary matrix algebra

A matrix in its most usual form is an array of scalar quantities arranged as a rectangular table of m rows and n columns, as shown in (18.1). This compact method of representing scalar quantities allows matrices to be particularly suitable for analysing complicated physical problems with the aid of computers.

$$[A] = \begin{bmatrix} A_{11} & A_{12} & A_{13} & \cdots & A_{1n} \\ A_{21} & A_{22} & A_{23} & \cdots & A_{2n} \\ A_{31} & A_{32} & A_{33} & \cdots & A_{3n} \\ \cdot & \cdot & \cdot & & \cdot \\ \cdot & \cdot & \cdot & & \cdot \\ \cdot & \cdot & \cdot & & \cdot \\ A_{m1} & A_{m2} & A_{m3} & \cdots & A_{mn} \end{bmatrix} \tag{18.1}$$

A ***row*** is defined as a horizontal line of numbers and a ***column*** as a vertical line of numbers.
The numbers A_{11}, A_{12}, and so on, are called the ***elements*** of a matrix.
The ***order*** of a matrix is defined by the number of its rows and columns. For example, the matrix in (18.1) is said to be of order m × n.
The matrix shown in (18.2) is of order 2 × 3, usually expressed as 2 by 3, since it has 2 rows and 3 columns.

$$\begin{pmatrix} 2 & 7 & 6 \\ 4 & -1 & 3 \end{pmatrix} \tag{18.2}$$

A ***column matrix*** is where n = 1, as in (18.3).

$$[A] = \begin{bmatrix} A_{11} \\ A_{21} \\ \cdot \\ \cdot \\ \cdot \\ A_{m1} \end{bmatrix} \tag{18.3}$$

A ***row matrix*** is where m = 1, as in (18.4).

$$[A] = [A_{11}\ A_{12}\ \cdots\ A_{1n}] \tag{18.4}$$

A ***square matrix*** is where m = n, as in (18.5).

$$[A] = \begin{bmatrix} A_{11} & A_{12} & \cdots & A_{1n} \\ A_{21} & A_{22} & \cdots & A_{2n} \\ \cdot & \cdot & \cdot & \cdot \\ \cdot & \cdot & \cdot & \cdot \\ \cdot & \cdot & \cdot & \cdot \\ \cdot & \cdot & \cdot & \cdot \\ A_{m1} & A_{m2} & \cdots & A_{mn} \end{bmatrix} \tag{18.5}$$

The square matrix of (18.5) is said to be of order n.
The ***transpose*** of a matrix is obtained by interchanging its rows and its columns. For example,

if $[A] = \begin{bmatrix} 1 & 2 \\ -4 & 5 \end{bmatrix}$ then the transpose of [A] is

given by: $[A]^T = \begin{bmatrix} 1 & -4 \\ 2 & 5 \end{bmatrix}$

A ***null-matrix*** is one which has all its elements equal to zero.
A ***diagonal matrix*** is a square matrix where all the elements except those of the main diagonal are zero, as in (18.6).

$$[A] = \begin{bmatrix} A_{11} & 0 & 0 & \cdot & \cdot & \cdot & 0 \\ 0 & A_{22} & & & & & \\ 0 & 0 & A_{33} & & & & \\ \cdot & \cdot & & \cdot & & & \\ \cdot & \cdot & & & \cdot & & \\ \cdot & \cdot & & & & \cdot & \\ 0 & 0 & & & & & A_{mn} \end{bmatrix} \tag{18.6}$$

A ***scalar matrix*** is a diagonal matrix where all the diagonal elements are equal to the same scalar quantity. When the scalar quantity is unity, the matrix is called the ***unit*** or ***identity matrix*** and is denoted by [I]
For example, a 3 by 3 unit matrix is shown in (18.7).

$$\begin{pmatrix} 1 & 0 & 0 \\ 0 & 1 & 0 \\ 0 & 0 & 1 \end{pmatrix} \tag{18.7}$$

18.3 Addition and subtraction of matrices

If $[A] = \begin{bmatrix} 3 & 1 \\ -1 & 2 \end{bmatrix}$ and $[B] = \begin{bmatrix} -1 & 2 \\ 3 & -2 \end{bmatrix}$

then [A] + [B] is obtained by adding together the corresponding elements as in (18.8).

$$\text{i.e. } [A] + [B] = \begin{bmatrix} 3+-1 & 1+2 \\ -1+3 & 2+-2 \end{bmatrix} = \begin{bmatrix} \mathbf{2} & \mathbf{3} \\ \mathbf{2} & \mathbf{0} \end{bmatrix} \quad (18.8)$$

$$\text{Similarly, } [A] - [B] = \begin{bmatrix} 3--1 & 1-2 \\ -1-3 & 2--2 \end{bmatrix} = \begin{bmatrix} \mathbf{4} & \mathbf{-1} \\ \mathbf{-4} & \mathbf{4} \end{bmatrix}$$

If $[A] = 3\begin{bmatrix} 1 & 5 \\ -3 & 2 \end{bmatrix}$ then $[A] = \begin{bmatrix} 3 & 15 \\ -9 & 6 \end{bmatrix}$ i.e. each element of the original matrix has been multiplied by 3. This is called **scalar multiplication**.

Now try the following Practice Exercise

Practice Exercise 65. Further problems on addition and subtraction of matrices

Matrices A to F are as follows:

$A = \begin{pmatrix} 3 & -1 \\ -4 & 7 \end{pmatrix}$ $B = \begin{pmatrix} 5 & 7 \\ -4 & -2 \end{pmatrix}$

$C = \begin{pmatrix} -6 & 2 \\ 3 & -9 \end{pmatrix}$ $D = \begin{pmatrix} 4 & -7 & 3 \\ -2 & 4 & 0 \\ 5 & 7 & -4 \end{pmatrix}$

$E = \begin{pmatrix} 3 & 6 & 1 \\ 5 & -2 & 7 \\ -1 & 0 & 4 \end{pmatrix}$ $F = \begin{pmatrix} 3 & 4 & 6 \\ -1 & 8 & -11 \\ 5 & 3 & -2 \end{pmatrix}$

In Problems 1 to 6, perform the matrix operation stated.

1. A + B $\left[\begin{pmatrix} 8 & 6 \\ -8 & 5 \end{pmatrix}\right]$
2. D + E $\left[\begin{pmatrix} 7 & -1 & 4 \\ 3 & 2 & 7 \\ 4 & 7 & 0 \end{pmatrix}\right]$
3. A − B $\left[\begin{pmatrix} -2 & -8 \\ 0 & 9 \end{pmatrix}\right]$
4. A + B − C $\left[\begin{pmatrix} 14 & 4 \\ -11 & 14 \end{pmatrix}\right]$
5. 5A + 6B $\left[\begin{pmatrix} 45 & 37 \\ -44 & 23 \end{pmatrix}\right]$
6. 2D + 3E − 4F $\left[\begin{pmatrix} 5 & -12 & -15 \\ 15 & -30 & 65 \\ -13 & 2 & 12 \end{pmatrix}\right]$

18.4 Matrix multiplication

In the relationship [A][B] = [C]

[A] is known as the ***premultiplier***, [B] the ***postmultiplier*** and [C] the ***product***.

Furthermore, if [A] is of order m × n and [B] is of order n × p, then [C] is of order m × p

It should be noted that [B] must always have its number of rows equal to the number of columns in [A]

Let $[A] = \begin{bmatrix} 2 & 1 & 3 \\ -1 & 2 & -2 \\ 4 & 0 & 5 \end{bmatrix}$ and $[B] = \begin{bmatrix} 4 & 1 \\ 0 & -6 \\ 3 & 2 \end{bmatrix}$

The product [C] is obtained by multiplying the columns of [B] by the rows of [A] as in (18.9).

Hence,

$$[C] = \begin{bmatrix} (2\times4+1\times0+3\times3) & (2\times1+1\times-6+3\times2) \\ (-1\times4+2\times0+-2\times3) & (-1\times1+2\times-6+-2\times2) \\ (4\times4+0\times0+5\times3) & (4\times1+0\times-6+5\times2) \end{bmatrix}$$

$$= \begin{bmatrix} \mathbf{17} & \mathbf{2} \\ \mathbf{-10} & \mathbf{-17} \\ \mathbf{31} & \mathbf{14} \end{bmatrix} \quad (18.9)$$

If $[A] = \begin{bmatrix} 2 & 1 & 3 \\ -1 & 2 & -2 \\ 4 & 0 & 2 \end{bmatrix}$ and $[B] = \begin{bmatrix} 4 \\ 0 \\ 5 \end{bmatrix}$ then

$$[A][B] = \begin{bmatrix} (2\times4+1\times0+3\times5) \\ (-1\times4+2\times0+-2\times5) \\ (4\times4+0\times0+2\times5) \end{bmatrix} = \begin{bmatrix} \mathbf{23} \\ \mathbf{-14} \\ \mathbf{26} \end{bmatrix}$$

If $[A] = \begin{bmatrix} 1 & 5 & -2 & -1 \\ 4 & -1 & 2 & 3 \\ -3 & 6 & -4 & 2 \\ -5 & 1 & 0 & 7 \end{bmatrix}$ and $[B] = \begin{bmatrix} 3 \\ -2 \\ 4 \\ -1 \end{bmatrix}$

$$\text{then } [A][B] = \begin{bmatrix} (1\times3+5\times-2+-2\times4+-1\times-1) \\ (4\times3+-1\times-2+2\times4+3\times-1) \\ (-3\times3+6\times-2+-4\times4+2\times-1) \\ (-5\times3+1\times-2+0\times4+7\times-1) \end{bmatrix} = \begin{bmatrix} \mathbf{-14} \\ \mathbf{19} \\ \mathbf{-39} \\ \mathbf{-24} \end{bmatrix}$$

If $[A] = \begin{bmatrix} 2 & 1 & 3 \\ -1 & 2 & -2 \\ 4 & 0 & 5 \end{bmatrix}$ and $[B] = \begin{bmatrix} 3 & 0 & 0 \\ 0 & 3 & 0 \\ 0 & 0 & 3 \end{bmatrix}$ (18.10)

$$\text{then } [A][B] = \begin{bmatrix} 6 & 3 & 9 \\ -3 & 6 & -6 \\ 12 & 0 & 15 \end{bmatrix} \quad (18.11)$$

It should be noted that (18.11) can be obtained simply by multiplying [A] by the scalar number 3, and this is why the matrix [B] of (18.10) is called a scalar matrix. In general, with matrices, **[A][B] ≠ [B][A]**

Now try the following Practice Exercise

Practice Exercise 66. Further problems on multiplication of matrices

Matrices A to L are:

$A = \begin{pmatrix} 3 & -1 \\ -4 & 7 \end{pmatrix}$ $B = \begin{pmatrix} 6 & 2 \\ -3 & -5 \end{pmatrix}$

$C = \begin{pmatrix} -10 & 7 \\ 9 & -9 \end{pmatrix}$ $D = \begin{pmatrix} 4 & -7 & 6 \\ -2 & 4 & 0 \\ 5 & 7 & -4 \end{pmatrix}$

$E = \begin{pmatrix} 3 & 6 & 2 \\ 5 & -3 & 7 \\ -1 & 0 & 5 \end{pmatrix}$ $F = \begin{pmatrix} 3 & 2 & 6 \\ -5 & 8 & -9 \\ 6 & 4 & -4 \end{pmatrix}$

$G = \begin{pmatrix} -2 \\ 5 \end{pmatrix}$ $H = \begin{pmatrix} 4 \\ -11 \\ 7 \end{pmatrix}$ $J = \begin{pmatrix} 1 & 0 \\ 0 & 1 \\ 1 & 0 \end{pmatrix}$

$K = \begin{bmatrix} 2 & 3 & -4 & 1 \\ 0 & 5 & 7 & -3 \\ 4 & 1 & 0 & 2 \\ -3 & 2 & -1 & 4 \end{bmatrix}$ $L = \begin{bmatrix} 2 \\ 5 \\ 1 \\ 7 \end{bmatrix}$

In Problems 1 to 7, perform the matrix operation stated.

1. A × G $\left[\begin{pmatrix} -11 \\ 43 \end{pmatrix}\right]$
2. A × B $\left[\begin{pmatrix} 21 & 11 \\ -45 & -43 \end{pmatrix}\right]$
3. A × C $\left[\begin{pmatrix} -39 & 30 \\ 103 & -91 \end{pmatrix}\right]$
4. D × H $\left[\begin{pmatrix} 135 \\ -52 \\ -85 \end{pmatrix}\right]$

5. $E \times J$ $\left[\begin{pmatrix} 5 & 6 \\ 12 & -3 \\ 4 & 0 \end{pmatrix}\right]$

6. $D \times F$ $\left[\begin{pmatrix} 83 & -24 & 63 \\ -26 & 28 & -48 \\ -44 & 50 & -17 \end{pmatrix}\right]$

7. $K \times L$ $\left[\begin{pmatrix} 22 \\ 11 \\ 27 \\ 31 \end{pmatrix}\right]$

8. Show that $A \times C \neq C \times A$

$\left[A \times C = \begin{pmatrix} -39 & 30 \\ 103 & -91 \end{pmatrix} \right.$

$C \times A = \begin{pmatrix} -58 & 59 \\ 63 & -72 \end{pmatrix}$

$\left. \text{Hence, } A \times C \neq C \times A \right]$

18.5 Two by two determinants

The determinant of a **2 by 2 matrix**,

$$D = \begin{vmatrix} A_{11} & A_{12} \\ A_{21} & A_{22} \end{vmatrix} = A_{11}A_{22} - A_{12}A_{21}$$

Note that the elements of the determinant of a matrix are written between vertical lines.

For example,

$$\begin{vmatrix} 3 & 4 \\ -1 & 2 \end{vmatrix} = 3 \times 2 - 4 \times -1 = 6 + 4 = \mathbf{10}$$

The **inverse** of a 2 by 2 matrix A is A^{-1} such that $A \times A^{-1} = I$, the unit matrix.

For any matrix $\begin{pmatrix} A_{11} & A_{12} \\ A_{21} & A_{22} \end{pmatrix}$ one method of obtaining the inverse is to:

(i) interchange the positions of A_{11} and A_{22}
(ii) change the signs of A_{12} and A_{21}
(iii) multiply the new matrix by the reciprocal of the determinant of $\begin{pmatrix} A_{11} & A_{12} \\ A_{21} & A_{22} \end{pmatrix}$

For example, the inverse of $\begin{pmatrix} 3 & 1 \\ 4 & 2 \end{pmatrix}$ is

$$\frac{1}{3 \times 2 - 1 \times 4}\begin{pmatrix} 2 & -1 \\ -4 & 3 \end{pmatrix} = \frac{1}{2}\begin{pmatrix} 2 & -1 \\ -4 & 3 \end{pmatrix} = \begin{pmatrix} 1 & -0.5 \\ -2 & 1.5 \end{pmatrix}$$

Now try the following Practice Exercise

Practice Exercise 67. Further problems on 2 by 2 determinants

1. Calculate the determinant of $\begin{pmatrix} 3 & -1 \\ -4 & 7 \end{pmatrix}$ [17]

2. Calculate the determinant of $\begin{pmatrix} 1 & 3 \\ -5 & -4 \end{pmatrix}$ [11]

3. Calculate the determinant of $\begin{vmatrix} -11 & 7 \\ 2 & -5 \end{vmatrix}$ [41]

4. Determine the inverse of $\begin{pmatrix} 3 & -1 \\ -4 & 7 \end{pmatrix}$

$\left[\frac{1}{17}\begin{pmatrix} 7 & 1 \\ 4 & 3 \end{pmatrix}\right]$

5. Determine the inverse of $\begin{pmatrix} 2 & 3 \\ -1 & -5 \end{pmatrix}$

$\left[-\frac{1}{7}\begin{pmatrix} -5 & -3 \\ 1 & 2 \end{pmatrix}\right]$

6. Determine the inverse of $\begin{pmatrix} -3 & 7 \\ 2 & -9 \end{pmatrix}$

$\left[\frac{1}{13}\begin{pmatrix} -9 & -7 \\ -2 & -3 \end{pmatrix} \text{ or } -\frac{1}{13}\begin{pmatrix} 9 & 7 \\ 2 & 3 \end{pmatrix}\right]$

18.6 Three by three determinants

The **minor** of an element of a 3 by 3 matrix is the value of the 2 by 2 determinant obtained by covering up the row and column containing that element. For example, for the matrix

$$\begin{pmatrix} 1 & 2 & 3 \\ 4 & 5 & 6 \\ 7 & 8 & 9 \end{pmatrix} \text{ the minor of element 4 is } \begin{vmatrix} 2 & 3 \\ 8 & 9 \end{vmatrix}$$

i.e. $18 - 24 = \mathbf{-6}$

Similarly, the minor of element 9 is $\begin{vmatrix} 1 & 2 \\ 4 & 5 \end{vmatrix}$

i.e. $5 - 8 = \mathbf{-3}$

The sign of a minor depends on its position within the matrix, the sign pattern being $\begin{pmatrix} + & - & + \\ - & + & - \\ + & - & + \end{pmatrix}$

Thus, the signed minor of element 4 in the matrix

$$\begin{pmatrix} 1 & 2 & 3 \\ 4 & 5 & 6 \\ 7 & 8 & 9 \end{pmatrix} \text{ is } -\begin{vmatrix} 2 & 3 \\ 8 & 9 \end{vmatrix} = -(18-24) = -(-6) = \mathbf{6}$$

The **signed-minor** of an element is called the **cofactor** of the element.

The cofactor of element 7 is $+\begin{vmatrix} 2 & 3 \\ 5 & 6 \end{vmatrix} = 12 - 15 = \mathbf{-3}$

The value of a 3 by 3 determinant is the sum of the products of the elements and their cofactors of any row or any column of the corresponding 3 by 3 matrix.

Using the first row, the determinant of the 3 by 3 matrix, $\begin{pmatrix} A_{11} & A_{12} & A_{13} \\ A_{21} & A_{22} & A_{23} \\ A_{31} & A_{32} & A_{33} \end{pmatrix}$ is evaluated as follows:

$$D = A_{11}\begin{vmatrix} A_{22} & A_{23} \\ A_{32} & A_{33} \end{vmatrix} - A_{12}\begin{vmatrix} A_{21} & A_{23} \\ A_{31} & A_{33} \end{vmatrix} + A_{13}\begin{vmatrix} A_{21} & A_{22} \\ A_{31} & A_{32} \end{vmatrix}$$

For example, $\begin{vmatrix} 2 & 4 & -3 \\ -5 & 2 & 6 \\ -1 & 3 & 5 \end{vmatrix}$

$$= 2\begin{vmatrix} 2 & 6 \\ 3 & 5 \end{vmatrix} - 4\begin{vmatrix} -5 & 6 \\ -1 & 5 \end{vmatrix} + -3\begin{vmatrix} -5 & 2 \\ -1 & 3 \end{vmatrix}$$

$$= 2(10 - 18) - 4(-25 - -6) - 3(-15 - -2)$$
$$= 2(-8) - 4(-19) - 3(-13) = -16 + 76 + 39 = \mathbf{99}$$

Using a calculator to evaluate a determinant

Evaluating the 3 by 3 determinant $\begin{vmatrix} 2 & 4 & -3 \\ -5 & 2 & 6 \\ -1 & 3 & 5 \end{vmatrix}$ can also be achieved using a **calculator.**

Here is the procedure for the above determinant using a CASIO 991 ES PLUS calculator:

1. Press 'Mode' and choose '6'
2. Select Matrix A by entering '1'
3. Select 3 × 3 by entering '1'
4. Enter in the nine numbers of the above matrix in this order, following each number by pressing '='; i.e. 2 =, 4 =, –3 =, –5 =, 2 =, and so on
5. Press 'ON' and a zero appears
6. Press 'Shift' and '4' and 8 choices appear
7. Select choice '7' and 'det(' appears
8. Press 'Shift' and '4' and choose Matrix A by entering '3'
9. Press '=' and the answer **99** appears

The **adjoint** of a matrix A is obtained by:
(i) forming a matrix B of the cofactors of A, and (ii) transposing matrix B to give B^T (i.e. where B^T is the matrix obtained by writing the rows of B as the columns). Then **adj A = $\mathbf{B^T}$**

The **inverse of matrix A**, A^{-1} is given by: $\mathbf{A^{-1} = \dfrac{adj\,A}{|A|}}$

where adj A is the adjoint of A and $|A|$ is the determinant of matrix A.

For example, to find the inverse of the matrix

$$\begin{pmatrix} 3 & 4 & -1 \\ 2 & 0 & 7 \\ 1 & -3 & -2 \end{pmatrix}$$

(i) the matrix of the cofactors of the elements is

$$\begin{pmatrix} 21 & 11 & -6 \\ 11 & -5 & 13 \\ 28 & -23 & -8 \end{pmatrix}$$

(ii) the transpose of the matrix of cofactors is

$$\begin{pmatrix} 21 & 11 & 28 \\ 11 & -5 & -23 \\ -6 & 13 & -8 \end{pmatrix}$$

(iii) the determinant of the matrix $\begin{pmatrix} 3 & 4 & -1 \\ 2 & 0 & 7 \\ 1 & -3 & -2 \end{pmatrix}$ is given by:

$$\begin{vmatrix} 3 & 4 & -1 \\ 2 & 0 & 7 \\ 1 & -3 & -2 \end{vmatrix} =$$

$$3\begin{vmatrix} 0 & 7 \\ -3 & -2 \end{vmatrix} - 4\begin{vmatrix} 2 & 7 \\ 1 & -2 \end{vmatrix} + -1\begin{vmatrix} 2 & 0 \\ 1 & -3 \end{vmatrix}$$

$$= 3(21) - 4(-11) - (-6) = 113$$

Hence, the inverse of $\begin{pmatrix} 3 & 4 & -1 \\ 2 & 0 & 7 \\ 1 & -3 & -2 \end{pmatrix}$

$$A^{-1} = \frac{\text{adj}A}{|A|} = \frac{\begin{pmatrix} 21 & 11 & 28 \\ 11 & -5 & -23 \\ -6 & 13 & -8 \end{pmatrix}}{113}$$

$$= \frac{1}{113}\begin{pmatrix} 21 & 11 & 28 \\ 11 & -5 & -23 \\ -6 & 13 & -8 \end{pmatrix}$$

Now try the following Practice Exercises

Practice Exercise 68. Further problems on 3 by 3 determinants

1. Evaluate $\begin{vmatrix} 8 & -2 & -10 \\ 2 & -3 & -2 \\ 6 & 3 & 8 \end{vmatrix}$ [−328]

2. Calculate the determinant of $\begin{pmatrix} 3 & 4 & 6 \\ -1 & 8 & -9 \\ 5 & 4 & -2 \end{pmatrix}$

   [−392]

3. For the matrix $A = \begin{pmatrix} 4 & -7 & 6 \\ -2 & 4 & 0 \\ 5 & 7 & -4 \end{pmatrix}$

   determine:

   (a) the matrix of minors,
   (b) the matrix of cofactors,
   (c) the transpose of the matrix of cofactors,
   (d) the determinant of matrix A,
   (e) the inverse of matrix A

   [(a) $\begin{pmatrix} -16 & 8 & -34 \\ -14 & -46 & 63 \\ -24 & 12 & 2 \end{pmatrix}$

   (b) $\begin{pmatrix} -16 & -8 & -34 \\ 14 & -46 & -63 \\ -24 & -12 & 2 \end{pmatrix}$

   (c) $\begin{pmatrix} -16 & 14 & -24 \\ -8 & -46 & -12 \\ -34 & -63 & 2 \end{pmatrix}$

   (d) -212

   (e) $-\frac{1}{212}\begin{pmatrix} -16 & 14 & -24 \\ -8 & -46 & -12 \\ -34 & -63 & 2 \end{pmatrix}$]

4. For the matrix $B = \begin{pmatrix} 1 & -2 & 4 \\ 6 & -5 & 0 \\ 0 & 1 & -3 \end{pmatrix}$

determine

(a) the matrix of minors,
(b) the matrix of cofactors,
(c) the transpose of the matrix of cofactors,
(d) the determinant of matrix B,
(e) the inverse of matrix B

[(a) $\begin{pmatrix} 15 & -18 & 6 \\ 2 & -3 & 1 \\ 20 & -24 & 7 \end{pmatrix}$

(b) $\begin{pmatrix} 15 & 18 & 6 \\ -2 & -3 & -1 \\ 20 & 24 & 7 \end{pmatrix}$

(c) $\begin{pmatrix} 15 & -2 & 20 \\ 18 & -3 & 24 \\ 6 & -1 & 7 \end{pmatrix}$

(d) 3

(e) $\frac{1}{3}\begin{pmatrix} 15 & -2 & 20 \\ 18 & -3 & 24 \\ 6 & -1 & 7 \end{pmatrix}$]

5. For the matrix $C = \begin{pmatrix} 3 & 6 & 1 \\ 5 & -2 & 7 \\ -1 & 0 & 3 \end{pmatrix}$

determine

(a) the matrix of minors,
(b) the matrix of cofactors,
(c) the transpose of the matrix of cofactors,
(d) the determinant of matrix C,
(e) the inverse of matrix C

[(a) $\begin{pmatrix} -6 & 22 & -2 \\ 18 & 10 & 6 \\ 44 & 16 & -36 \end{pmatrix}$

(b) $\begin{pmatrix} -6 & -22 & -2 \\ -18 & 10 & -6 \\ 44 & -16 & -36 \end{pmatrix}$

(c) $\begin{pmatrix} -6 & -18 & 44 \\ -22 & 10 & -16 \\ -2 & -6 & -36 \end{pmatrix}$

(d) -152

(e) $-\frac{1}{152}\begin{pmatrix} -6 & -18 & 44 \\ -22 & 10 & -16 \\ -2 & -6 & -36 \end{pmatrix}$ or $\frac{1}{152}\begin{pmatrix} 6 & 18 & -44 \\ 22 & -10 & 16 \\ 2 & 6 & 36 \end{pmatrix}$]

For fully worked solutions to each of the problems in Exercises 65 to 68 in this chapter, go to the website:
www.routledge.com/cw/bird

Multiple-Choice Questions Test 5

This test covers the material in Chapter 18 on matrices and determinants. All questions have only one correct answer (answers on page 496).

1. The vertical lines of elements in a matrix are called:

   (a) rows (b) a row matrix
   (c) columns (d) a column matrix

2. If a matrix P has the same number of rows and columns, then matrix P is called a:

   (a) row matrix (b) square matrix
   (c) column matrix (d) rectangular matrix

3. A matrix of order $m \times 1$ is called a:

   (a) row matrix (b) column matrix
   (c) square matrix (d) rectangular matrix

4. $3\begin{pmatrix} 4 & -2 \\ 1 & 0 \end{pmatrix} - 2\begin{pmatrix} -3 & 0 \\ 5 & 1 \end{pmatrix}$ is equal to:

   (a) $\begin{pmatrix} 6 & -4 \\ -7 & 2 \end{pmatrix}$ (b) $\begin{pmatrix} 18 & -8 \\ 7 & -2 \end{pmatrix}$

   (c) $\begin{pmatrix} 18 & -6 \\ -7 & -2 \end{pmatrix}$ (d) $\begin{pmatrix} 6 & -6 \\ -4 & -1 \end{pmatrix}$

5. The value of $\begin{vmatrix} 2 & -3 \\ 5 & -4 \end{vmatrix}$ is:

   (a) -23 (b) -7

   (c) $\frac{1}{7}\begin{pmatrix} 4 & 3 \\ -5 & -2 \end{pmatrix}$ (d) 7

6. The matrix product $\begin{pmatrix} 2 & 3 \\ -1 & 4 \end{pmatrix}\begin{pmatrix} 1 & -5 \\ -2 & 6 \end{pmatrix}$ is equal to:

   (a) $\begin{pmatrix} -13 \\ 26 \end{pmatrix}$ (b) $\begin{pmatrix} 3 & -2 \\ -3 & 10 \end{pmatrix}$

   (c) $\begin{pmatrix} 1 & -2 \\ -3 & -2 \end{pmatrix}$ (d) $\begin{pmatrix} -4 & 8 \\ -9 & 29 \end{pmatrix}$

7. If $\begin{vmatrix} 2k & -1 \\ 4 & 2 \end{vmatrix} = \begin{vmatrix} 3 & 0 \\ 2 & 1 \end{vmatrix}$ then the value of k is:

   (a) $\frac{2}{3}$ (b) 3

   (c) $-\frac{1}{4}$ (d) $\frac{3}{2}$

8. The value of $\begin{vmatrix} 6 & 0 & -1 \\ 2 & 1 & 4 \\ 1 & 1 & 3 \end{vmatrix}$ is:

   (a) -7 (b) 8
   (c) 7 (d) 10

9. Solving $\begin{vmatrix} (2k+5) & 3 \\ (5k+2) & 9 \end{vmatrix} = 0$ for k gives:

   (a) $-\frac{17}{11}$ (b) -17

   (c) 17 (d) -13

10. $\begin{vmatrix} j2 & j \\ j & -j \end{vmatrix}$ is equal to:

    (a) 5 (b) 4
    (c) 3 (d) 2

11. If $P = \begin{pmatrix} 5 & 3 & 2 \\ 0 & 4 & 1 \\ 0 & 0 & 3 \end{pmatrix}$ then $|P|$ is equal to:

    (a) 30 (b) 40
    (c) 50 (d) 60

12. If the order of a matrix P is $a \times b$ and the order of a matrix Q is $b \times c$ then the order of $P \times Q$ is:

    (a) $a \times c$ (b) $c \times a$
    (c) $c \times b$ (d) $a \times b$

13. The value of $\begin{vmatrix} (1+j2) & j \\ j & -j \end{vmatrix}$ is:

(a) $-3-j$ (b) $-1-j$

(c) $3-j$ (d) $1-j$

14. The inverse of the matrix $\begin{pmatrix} 5 & -3 \\ -2 & 1 \end{pmatrix}$ is:

(a) $\begin{pmatrix} -5 & -3 \\ 2 & -1 \end{pmatrix}$ (b) $\begin{pmatrix} -1 & -3 \\ -2 & -5 \end{pmatrix}$

(c) $\begin{pmatrix} -1 & 3 \\ 2 & -5 \end{pmatrix}$ (d) $\begin{pmatrix} 1 & 3 \\ 2 & 5 \end{pmatrix}$

15. The value of the determinant $\begin{vmatrix} 2 & -1 & 4 \\ 0 & 1 & 5 \\ 6 & 0 & -1 \end{vmatrix}$

is:

(a) -56 (b) 52

(c) 4 (d) 8

16. The value of $\begin{vmatrix} j2 & -(1+j) \\ (1-j) & 1 \end{vmatrix}$ is:

(a) $-j2$ (b) $2(1+j)$

(c) 2 (d) $-2+j2$

17. The inverse of the matrix $\begin{pmatrix} 2 & -3 \\ 1 & -4 \end{pmatrix}$ is:

(a) $\begin{pmatrix} 0.8 & 0.6 \\ -0.2 & -0.4 \end{pmatrix}$ (b) $\begin{pmatrix} -4 & 3 \\ -1 & 2 \end{pmatrix}$

(c) $\begin{pmatrix} -0.4 & -0.6 \\ 0.2 & 0.8 \end{pmatrix}$ (d) $\begin{pmatrix} 0.8 & -0.6 \\ 0.2 & -0.4 \end{pmatrix}$

18. Matrix A is given by: $A = \begin{pmatrix} 3 & -2 \\ 4 & -1 \end{pmatrix}$

The determinant of matrix A is:

(a) -11 (b) 5

(c) 4 (d) $\begin{pmatrix} -1 & 2 \\ -4 & 3 \end{pmatrix}$

19. The inverse of the matrix $\begin{pmatrix} -1 & 4 \\ -3 & 2 \end{pmatrix}$ is:

(a) $\begin{pmatrix} \frac{1}{10} & -\frac{3}{10} \\ \frac{2}{5} & -\frac{1}{5} \end{pmatrix}$ (b) $\begin{pmatrix} -\frac{1}{7} & \frac{2}{7} \\ -\frac{3}{14} & \frac{1}{14} \end{pmatrix}$

(c) $\begin{pmatrix} \frac{1}{5} & -\frac{2}{5} \\ \frac{3}{10} & -\frac{1}{10} \end{pmatrix}$ (d) $\begin{pmatrix} \frac{1}{14} & -\frac{3}{14} \\ \frac{2}{7} & -\frac{1}{7} \end{pmatrix}$

20. The value(s) of λ given

$$\begin{vmatrix} (2-\lambda) & -1 \\ -4 & (-1-\lambda) \end{vmatrix} = 0 \text{ is:}$$

(a) -2 and $+3$ (b) $+2$ and -3

(c) -2 (d) 3

Chapter 19

Composites

Why it is important to understand: **Composites**

The design engineer's interest in composites derives from the properties shown in the table in Section 19.1. Carbon fibre reinforced plastic (CFRP) can be seen to have similar properties to steel along the fibres in terms of Young's modulus and strength, but with approximately one fifth of the density. By careful design, the fibres can be laid in the direction of the load paths, which results in the possibility of extremely efficient structures compared with metallic equivalents.

Composites are very important in structural engineering, because composite materials have a much higher strength to weight ratio than metals. For example, if it is required to design a submarine pressure hull in high-tensile steel, of internal diameter 10 m, to dive to the bottom of the Mariana Trench, a pressure hull of thickness of about 2.3 m would be required! Such a vessel will have no reserve buoyancy and will sink like a stone and not be able to re-surface! If, however, the pressure hull were made from a high-tensile composite, its thickness need only be about 0.5 m – and even thinner if it is made with carbon nanotubes or graphene! However, composites are not usually isotropic materials, and because of this the equivalent equations for stress analysis and material failure are explained in this chapter. Composites are also important in other branches of engineering, including aeronautical and aerospace engineering, where the strength to weight ratios of the material of construction is very important. Composites are also used in automobile engineering, especially for racing cars – together with numerous artefacts in many sports.

At the end of this chapter you should be able to:

- understand the difference between an isotropic material and an orthotropic material
- appreciate what a plane stress is for a composite material
- appreciate what a plane strain is for a composite material
- appreciate what an anisotropic material is
- appreciate what a ply or sheet or layer is
- understand and develop a stiffness matrix for a composite
- understand and develop a compliance matrix for a composite
- appreciate what a Reuter's matrix is
- understand the relationships between stress and strain for composites
- appreciate some failure criteria for composites

Mathematical references

In order to understand the theory involved in this chapter on **composites**, knowledge of the following mathematical topics is required: ***matrices and integration.*** If help/revision is needed in these areas, see chapter 18 and page 50 for some textbook references.

19.1 A comparison of mechanical properties of materials

Material	Density (kg/m^3)	Young's Modulus (GPa)	Compressive Strength (MPa)
Steel	7800	207	275–3340
GFRP (Epoxy/S-glass unidirectional)	2100	65	1200
GFRP (Epoxy/S-glass filament wound)	2100	50	1000
CFRP (Epoxy/HS unidirectional)	1600	210	1200
CFRP (Epoxy/HS filament wound)	1600	170	1000
Metal Matrix Composite, MMC (6061Al/SiC fibre UD)	2700	140	3000
Metal Matrix Composite, MMC (6061Al/Alumina Fibre UD)	3100	190	3100
Carbon Nanotubes (CNTs)	1600	270–950	11000–63000
Graphene	Approx. 2000	500	130000

19.2 Matrix equations for composites

Consider a piece of rectangular bar made from an isotropic material, fixed at one end and with a tensile load F_1 in direction 1 and a tensile load F_2 in direction 2, as shown in Figure 19.1.

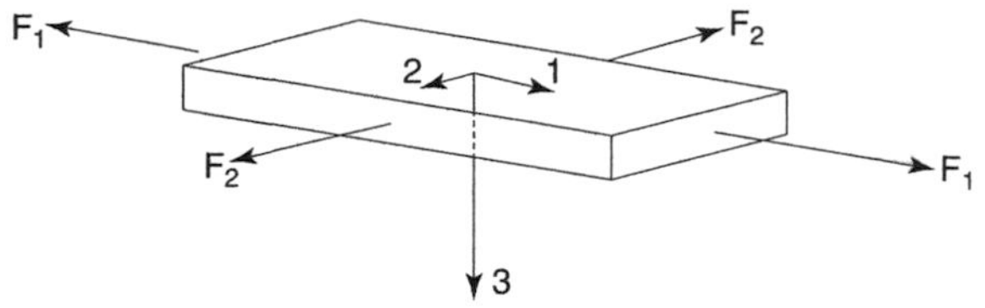

Figure 19.1

Now, stress, $\sigma = \varepsilon \times E$

where $\varepsilon_1 = \frac{\Delta L_1}{L_1}$ and $\varepsilon_2 = \frac{\Delta L_2}{L_2}$; L_1 and L_2 refer to the lengths in directions 1 and 2, and ΔL_1 and ΔL_2 refer to the displacements in directions L_1 and L_2 respectively.

$\upsilon = -\frac{\varepsilon_2}{\varepsilon_1}$ due to stress in direction 1

From Section 4.9, direct strain is given by:

$$\varepsilon_1 = \frac{\sigma_1}{E} - \upsilon\frac{\sigma_2}{E} \text{ and } \varepsilon_2 = \frac{\sigma_2}{E} - \upsilon\frac{\sigma_1}{E}$$

From Section 4.9, shear strain, $\gamma_{12} = \frac{\tau}{G}$

For an isotropic material, it can be shown that:

$$\text{shear modulus, } G = \frac{E}{2(1+\upsilon)}$$

In matrix notation, the above equations can be written as:

$$(\varepsilon) = (S)(\sigma)$$

where matrix (S) is known as the **compliance matrix**,

$$\text{or} \begin{pmatrix} \varepsilon_1 \\ \varepsilon_2 \\ \gamma_{12} \end{pmatrix} = \begin{pmatrix} 1/E & -\upsilon/E & 0 \\ -\upsilon/E & 1/E & 0 \\ 0 & 0 & 1/G \end{pmatrix} \begin{pmatrix} \sigma_1 \\ \sigma_2 \\ \tau_{12} \end{pmatrix} \quad (19.1)$$

For an isotropic block of material, in plane stress, the above equations are written as:

$$\begin{pmatrix} \varepsilon_1 \\ \varepsilon_2 \\ \varepsilon_3 \\ \gamma_{23} \\ \gamma_{31} \\ \gamma_{12} \end{pmatrix} = \begin{pmatrix} 1/E & -\upsilon/E & -\upsilon/E & 0 & 0 & 0 \\ -\upsilon/E & 1/E & -\upsilon/E & 0 & 0 & 0 \\ -\upsilon/E & -\upsilon/E & 1/E & 0 & 0 & 0 \\ 0 & 0 & 0 & 1/G & 0 & 0 \\ 0 & 0 & 0 & 0 & 1/G & 0 \\ 0 & 0 & 0 & 0 & 0 & 1/G \end{pmatrix} \begin{pmatrix} \sigma_1 \\ \sigma_2 \\ \sigma_3 \\ \tau_{23} \\ \tau_{31} \\ \tau_{12} \end{pmatrix} \quad (19.2)$$

where suffixes: i – force applied in 'i' direction

j – force acting on area normal to 'j' direction

Hence

$$\begin{pmatrix} \varepsilon_1 \\ \varepsilon_2 \\ \varepsilon_3 \\ \varepsilon_4 \\ \varepsilon_5 \\ \varepsilon_6 \end{pmatrix} = \begin{pmatrix} S_{11} & S_{12} & S_{13} & 0 & 0 & 0 \\ S_{12} & S_{22} & S_{23} & 0 & 0 & 0 \\ S_{13} & S_{23} & S_{33} & 0 & 0 & 0 \\ 0 & 0 & 0 & S_{44} & 0 & 0 \\ 0 & 0 & 0 & 0 & S_{55} & 0 \\ 0 & 0 & 0 & 0 & 0 & S_{66} \end{pmatrix} \begin{pmatrix} \sigma_1 \\ \sigma_2 \\ \sigma_3 \\ \sigma_4 \\ \sigma_5 \\ \sigma_6 \end{pmatrix} \quad (19.3)$$

Note: (S) is symmetrical, since $S_{12} = S_{21}$ and so on

Most applications of composite materials can be considered as plane stress, as in equation (19.4). This applies to curved surfaces too, as long as the radius (R) to thickness (t) ratio, $R/t > 10$

$$\begin{pmatrix} \varepsilon_1 \\ \varepsilon_2 \\ \varepsilon_6 \end{pmatrix} = \begin{pmatrix} S_{11} & S_{12} & 0 \\ S_{21} & S_{22} & 0 \\ 0 & 0 & S_{66} \end{pmatrix} \begin{pmatrix} \sigma_1 \\ \sigma_2 \\ \sigma_6 \end{pmatrix} \quad (19.4)$$

It is, however, usual for double suffix notation to be retained for shear, as shown in equation (19.5).

$$\begin{pmatrix} \varepsilon_1 \\ \varepsilon_2 \\ \gamma_{12} \end{pmatrix} = \begin{pmatrix} S_{11} & S_{12} & 0 \\ S_{21} & S_{22} & 0 \\ 0 & 0 & S_{66} \end{pmatrix} \begin{pmatrix} \sigma_1 \\ \sigma_2 \\ \tau_{12} \end{pmatrix} \quad (19.5)$$

We can invert the compliance matrix (S) to give the stiffness matrix (Q):

$$(\sigma) = (Q)(\varepsilon) \text{ where } (Q) = (S)^{-1} \quad (19.6)$$

19.3 Derivation of the stiffness matrix (Q) and $(S)^{-1}$ for isotropic materials

(Q), the stiffness matrix, is obtained by inverting the compliance matrix (S), as follows:

$$(Q) = (S)^{-1} = \frac{(S)^a}{|S|} = \frac{(S)^{C^T}}{|S|} \quad (19.7)$$

where the superscripts of the matrices, namely, 'a' and 'C' represent the adjoint and cofactor matrices (see chapter 18 for more on matrices).

$$(S) = \begin{pmatrix} 1/E & -\upsilon/E & 0 \\ -\upsilon/E & 1/E & 0 \\ 0 & 0 & 1/G \end{pmatrix}$$

The matrix of cofactors for (S) is given by:

$$(S)^C = \begin{pmatrix} \frac{1}{EG} & \frac{\upsilon}{EG} & 0 \\ \frac{\upsilon}{EG} & \frac{1}{EG} & 0 \\ 0 & 0 & \frac{1}{E^2} - \frac{\upsilon^2}{E^2} \end{pmatrix}$$

The adjoint matrix for $(S)^C$ is $(S)^a$, which is:

$$(S)^{C^T} = \begin{pmatrix} \frac{1}{EG} & \frac{\upsilon}{EG} & 0 \\ \frac{\upsilon}{EG} & \frac{1}{EG} & 0 \\ 0 & 0 & \frac{1}{E^2} - \frac{\upsilon^2}{E^2} \end{pmatrix} \quad (19.8)$$

The determinant of S, $|S| = \frac{1}{E}\left(\frac{1}{EG}\right) - \left(-\frac{\upsilon}{E}\right)\left(-\frac{\upsilon}{EG}\right) + 0$

i.e. $$|S| = \frac{1}{E^2G} - \frac{\upsilon^2}{E^2G} = \frac{1}{E^2G}\left(1-\upsilon^2\right)$$

from which $$\frac{1}{|S|} = \left(\frac{1}{1-\upsilon^2}\right)E^2G \quad (19.9)$$

From equations (19.8) and (19.9)

the stiffness matrix $(Q) = (S)^{-1} = \frac{(S)^a}{|S|} = \frac{(S)^{C^T}}{|S|}$ (19.10)

$$=\left(\frac{1}{1-\upsilon^2}\right)E^2G\begin{pmatrix}\frac{1}{EG} & \frac{\upsilon}{EG} & 0\\ \frac{\upsilon}{EG} & \frac{1}{EG} & 0\\ 0 & 0 & \frac{1}{E^2}-\frac{\upsilon^2}{E^2}\end{pmatrix}$$

$$=\left(\frac{1}{1-\upsilon^2}\right)\begin{pmatrix}E & \upsilon E & 0\\ \upsilon E & E & 0\\ 0 & 0 & \left(1-\upsilon^2\right)G\end{pmatrix} \qquad (19.11)$$

It is conventional to work on the basis of forces and moments per unit width.

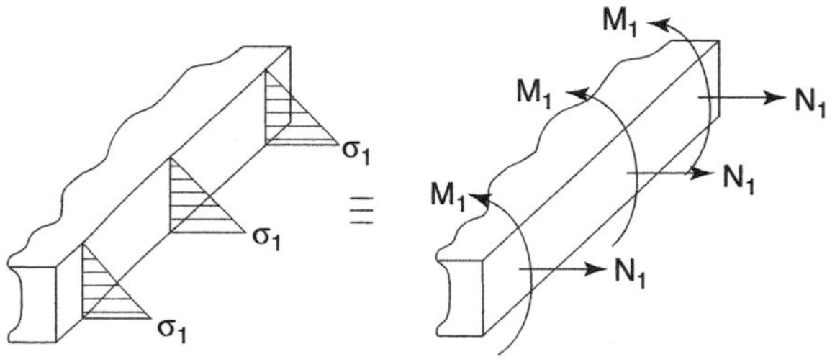

Figure 19.2

For a sheet of thickness h, under stress σ_1, force per unit width.

$$\left(N\right)=\int_{-h/2}^{+h/2}\left(\sigma\right)dz \qquad (19.12)$$

where (σ) is shown on the right-hand side of equation (19.2)

The 3 force resultants (N_1, N_2, N_{12}) acting on a sheet in plane stress can be written as:

$$\left(N\right)=\begin{pmatrix}N_1\\ N_2\\ N_{12}\end{pmatrix}=\int_{-h/2}^{+h/2}\left(\sigma\right)dz \qquad (19.13)$$

Similarly, the moment resultants (M_1, M_2, M_{12}) acting on a sheet can be written as:

$$\left(M\right)=\begin{pmatrix}M_1\\ M_2\\ M_{12}\end{pmatrix}\int_{-h/2}^{+h/2}\left(\sigma\right)z\,dz \qquad (19.14)$$

Strain profile through the thickness of the ply or sheet is given by:

$$\left(\varepsilon(z)\right)=\left(\varepsilon^o\right)+z\left(\kappa\right)$$

where 'κ' is the curvature at mid-plane, (i.e. 1 divided by the bending radius of the 'beam') and 'z' is through the thickness of the sheet or ply.

This is the same in Beam Theory, Chapter 6, where:

$$\varepsilon=\frac{y}{R}$$

Stress profile through the thickness of the ply or sheet is given by:

$$\left(\sigma(z)\right)=\left(Q\right)\left(\varepsilon^o\right)+z\left(Q\right)\left(\kappa\right) \qquad \text{since } \left(\sigma\right)=\left(Q\right)\left(\varepsilon(z)\right)$$

From equation (19.12)

$$\left(N\right)=\int_{-h/2}^{+h/2}\left(\sigma\right)dz$$

$$=\int_{-h/2}^{+h/2}\left(Q\right)\left(\varepsilon^o\right)dz+\int_{-h/2}^{+h/2}z\left(Q\right)\left(\kappa\right)dz \qquad (19.15)$$

(1) For a thin sheet or ply, (ε^0) and (κ) relate to mid-point only, and are not dependent on z.
(2) (Q) is a set of elastic constants for the material, and is not dependent on z.

Therefore, from equation (19.15)

$$\left(N\right)=\left(Q\right)\left(\varepsilon^o\right)\int_{-h/2}^{+h/2}dz+\left(Q\right)\left(\kappa\right)\int_{-h/2}^{+h/2}z\,dz$$

$$=\left(Q\right)\left(\varepsilon^o\right)\left[z\right]_{-h/2}^{+h/2}+\left(Q\right)\left(\kappa\right)\left[\frac{z^2}{2}\right]_{-h/2}^{+h/2}$$

$$=\left(Q\right)\left(\varepsilon^o\right)\left[\frac{h}{2}-\left(-\frac{h}{2}\right)\right]$$

$$+\left(Q\right)\left(\kappa\right)\left[\left(\frac{\left(h/2\right)^2}{2}\right)-\left(\frac{\left(h/2\right)^2}{2}\right)\right]$$

$$=\left(Q\right)\left(\varepsilon^o\right)h+\left(Q\right)\left(\kappa\right)\left[\frac{h^2}{8}-\frac{h^2}{8}\right]$$

i.e. $\left(N\right)=\left(A\right)\left(\varepsilon^o\right)+\left(B\right)\left(\kappa\right)$ (19.16)

$\left(A\right)=\left(Q\right)h$ and is known as the **'extensional stiffness matrix'** (19.17)

$\left(B\right)=\left(Q\right)\left(\frac{h^2}{8}-\frac{h^2}{8}\right)$ and is known as the **'bending extension coupling matrix'** (19.18)

Note that $(B) = 0$ for a single ply, but is non-zero for asymmetric multilayered composites.
Similarly for (M), from equation (19.14):

$$(M)=\int_{-h/2}^{+h/2}(\sigma)z\,dz$$

$$=\int_{-h/2}^{+h/2}(Q)(\varepsilon^o)z\,dz+\int_{-h/2}^{+h/2}z^2(Q)(\kappa)dz$$

$$=(Q)(\varepsilon^o)\int_{-h/2}^{+h/2}z\,dz+(Q)(\kappa)\int_{-h/2}^{+h/2}z^2\,dz$$

$$=(Q)(\varepsilon^o)\left[\frac{z^2}{2}\right]_{-h/2}^{+h/2}+(Q)(\kappa)\left[\frac{z^3}{3}\right]_{-h/2}^{+h/2}$$

$$=(Q)(\varepsilon^o)\left[\frac{(h/2)^2}{2}-\frac{(h/2)^2}{2}\right]+(Q)(\kappa)\left[\left(\frac{(h/2)^3}{3}\right)-\left(\frac{(-h/2)^3}{3}\right)\right]$$

$$=(Q)(\varepsilon^o)\left[\frac{h^2}{8}-\frac{h^2}{8}\right]+(Q)(\kappa)\left[\left(\frac{h^3}{24}\right)-\left(\frac{-h^3}{24}\right)\right]$$

$$=(Q)\left[\frac{h^2}{8}-\frac{h^2}{8}\right](\varepsilon^o)+(Q)\left[\frac{h^3}{12}\right](\kappa)$$

i.e. $$(M)=(B)(\varepsilon^o)+(D)(\kappa) \qquad (19.19)$$

where $(D)=(Q)\,\frac{h^3}{12}\,(\kappa)$ and is known as the 'bending stiffness matrix'

A single sheet or layer or ply, from equations (19.16) and (19.19), can be summed up by the matrix equation:

$$\begin{pmatrix} N \\ M \end{pmatrix}=\begin{pmatrix} A & B \\ B & D \end{pmatrix}\begin{pmatrix} \varepsilon^o \\ \kappa \end{pmatrix} \qquad (19.20)$$

It follows that an inverse matrix can be obtained to define the mid-plane strain and the curvature.
From equation (19.20):

$$\begin{pmatrix} \varepsilon^o \\ \kappa \end{pmatrix}=\begin{pmatrix} A & B \\ B & D \end{pmatrix}^{-1}\begin{pmatrix} N \\ M \end{pmatrix}$$

Let $$\begin{pmatrix} A & B \\ B & D \end{pmatrix}^{-1}=\begin{pmatrix} a & b \\ h & d \end{pmatrix}$$

where $$(a)=(A)^{-1}$$

and $$(d)=(D)^{-1}$$

If $$(B)=0 \text{ then } (b)=(h)=0$$

Note that $(B)=0$ for **all symmetric laminates**. NB A laminate is made from a collection of individual plies, or sheets, or layers or laminae, (where laminae is the plural of lamina).

So, when $(B) = 0$, $$(\varepsilon^o)=(a)(N)$$

and $$(\kappa)=(d)(M)$$

For a given loading defined by (N) and (M), the strain profile can be obtained from:

$$(\varepsilon(z))=(\varepsilon^o)+z(\kappa)$$

Stresses can then be found using the stiffness matrix (Q) for the relevant layer of the laminate.

$$\begin{pmatrix} \sigma_x \\ \sigma_y \\ \tau_{xy} \end{pmatrix}=\begin{pmatrix} Q_{11} & Q_{12} & 0 \\ Q_{21} & Q_{22} & 0 \\ 0 & 0 & Q_{66} \end{pmatrix}\begin{pmatrix} \varepsilon_x \\ \varepsilon_y \\ \gamma_{xy} \end{pmatrix}$$

i.e. $$(\sigma(z))=(Q)(\varepsilon^o)+(Q)z(\kappa)$$

19.4 Compliance matrix (*S*) for an orthotropic ply or sheet or layer

Now, from equation (19.7)

$$\begin{pmatrix} \varepsilon_1 \\ \varepsilon_2 \\ \gamma_{12} \end{pmatrix}=\begin{pmatrix} S_{11} & S_{12} & 0 \\ S_{21} & S_{22} & 0 \\ 0 & 0 & S_{66} \end{pmatrix}\begin{pmatrix} \sigma_1 \\ \sigma_2 \\ \tau_{12} \end{pmatrix} \qquad (19.21)$$

To find S_{11}

Let $$\sigma_2=\tau_{12}=0$$

hence, from equation (19.21)

$$\varepsilon_1=S_{11}\sigma_1$$

Therefore $$S_{11}=\frac{\varepsilon_1}{\sigma_1}=\frac{\varepsilon_1}{\varepsilon_1 E_1}$$

i.e. $$S_{11}=\frac{1}{E_1}$$

To find S_{22}

Let $$\sigma_1=\tau_{12}=0$$

hence, from equation (19.21)

$$\varepsilon_2 = S_{22}\sigma_2$$

Therefore $$S_{22} = \frac{\varepsilon_2}{\sigma_2} = \frac{\varepsilon_2}{\varepsilon_2 E_2}$$

i.e. $$S_{22} = \frac{1}{E_2}$$

To find S_{66}

Let $$\sigma_1 = \sigma_2 = 0$$

hence, from equation (19.21)

$$\gamma_{12} = S_{66}\tau_{12}$$

Therefore $$S_{66} = \frac{\gamma_{12}}{\tau_{12}} = \frac{\gamma_{12}}{\gamma_{12}G_{12}}$$

i.e. $$S_{66} = \frac{1}{G_{12}}$$

To find S_{21}

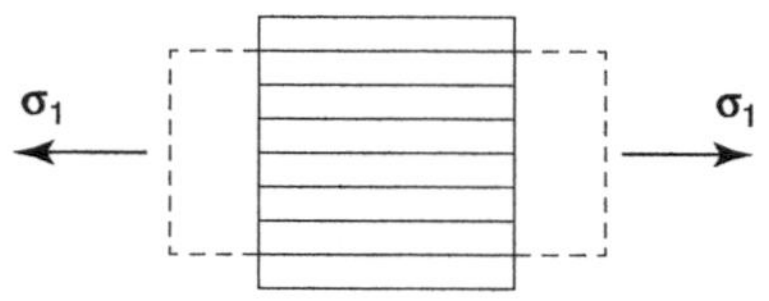

Figure 19.3

Let $$\sigma_2 = \tau_{12} = 0$$

hence, from equation (19.21),

$$\varepsilon_2 = S_{21}\sigma_1$$

Therefore $$S_{21} = \frac{\varepsilon_2}{\sigma_1}$$

Using $$\upsilon_{12} = -\frac{\varepsilon_2}{\varepsilon_1}$$ where υ_{12} is the major Poisson's ratio

from which $$\varepsilon_2 = -\upsilon_{12}\varepsilon_1$$

Now $$S_{21} = \frac{\varepsilon_2}{\sigma_1} = \frac{-\upsilon_{12}\varepsilon_1}{\sigma_1}$$

However $$\frac{\varepsilon_1}{\sigma_1} = \frac{1}{E_1} \quad (\text{as } \sigma_1 = \varepsilon_1 E_1)$$

Thus, $$S_{21} = \frac{-\upsilon_{12}}{E_1}$$

To find S_{12}

Let $$\sigma_1 = \tau_{12} = 0$$

hence, from equation (19.21),

$$\varepsilon_1 = S_{12}\sigma_2$$

Therefore $$S_{12} = \frac{\varepsilon_1}{\sigma_2}$$

Using $$\upsilon_{21} = -\frac{\varepsilon_1}{\varepsilon_2}$$ where υ_{21} is the minor Poisson's ratio

then, $$\varepsilon_1 = -\upsilon_{21}\varepsilon_2$$

Now $$S_{12} = \frac{\varepsilon_1}{\sigma_2} = \frac{-\upsilon_{21}\varepsilon_2}{\sigma_2}$$

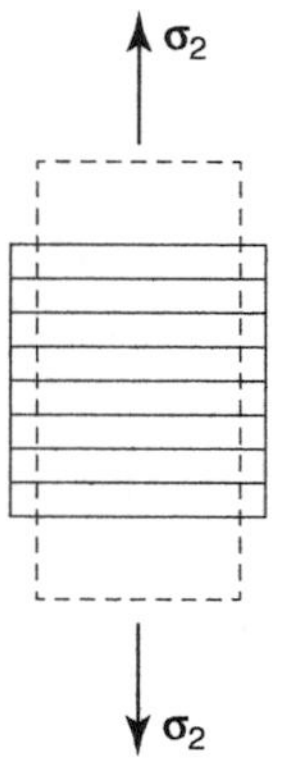

Figure 19.4

However $$\frac{\varepsilon_2}{\sigma_2} = \frac{1}{E_2}$$

Thus $$S_{12} = \frac{-\upsilon_{21}}{E_2}$$

It can be shown from strain energy arguments that the (*S*) or compliance matrix is symmetrical.

$$(S) = \begin{pmatrix} 1/E_1 & -\upsilon_{21}/E_2 & 0 \\ -\upsilon_{12}/E_1 & 1/E_2 & 0 \\ 0 & 0 & 1/G_{12} \end{pmatrix} \quad (19.22)$$

Consequently $$\frac{\upsilon_{12}}{E_1} = \frac{\upsilon_{21}}{E_2}$$

$$\upsilon_{12} = \upsilon_{21}\left(\frac{E_1}{E_2}\right)$$

For carbon fibre $\upsilon_{12} = \upsilon_{21}\left(\frac{180}{10}\right)$

i.e. υ_{12} is larger by a factor approaching 20.

Problem 1. Determine the compliance and the stiffness matrices for a sheet of copper with a Young's modulus of 117 GPa and Poisson's ratio of 0.355.

For an isotropic material, $G = \frac{E}{2(1+\nu)}$

$$G = \frac{117}{2(1+0.355)} = 43.17\text{ GPa} = 43170\text{ MPa}$$

$$(S) = \begin{pmatrix} 1/E & -\upsilon/E & 0 \\ -\upsilon/E & 1/E & 0 \\ 0 & 0 & 1/G \end{pmatrix}$$

$$= \begin{pmatrix} 1/117000 & -0.355/117000 & 0 \\ -0.355/170000 & 1/170000 & 0 \\ 0 & 0 & 1/43200 \end{pmatrix} \text{mm}^2/\text{N}$$

i.e. the compliance matrix,

$$(S) = \begin{pmatrix} \mathbf{8.547} & \mathbf{-3.03} & \mathbf{0} \\ \mathbf{-3.03} & \mathbf{8.547} & \mathbf{0} \\ \mathbf{0} & \mathbf{0} & \mathbf{23.15} \end{pmatrix} \times \mathbf{10^{-6}\ mm^2/N}$$

$$(Q) = \left(\frac{1}{1-\upsilon^2}\right)\begin{pmatrix} E & \upsilon E & 0 \\ \upsilon E & E & 0 \\ 0 & 0 & (1-\upsilon^2)G \end{pmatrix}$$

$$1-\nu^2 = 1-0.355^2 = 0.874$$

$$(Q) = \left(\frac{1}{0.874}\right)\begin{pmatrix} 117000 & 41540 & 0 \\ 41540 & 117000 & 0 \\ 0 & 0 & 0.874(43170) \end{pmatrix} \text{N/mm}^2$$

i.e. the stiffness matrix,

$$(Q) = \begin{pmatrix} \mathbf{133900} & \mathbf{47530} & \mathbf{0} \\ \mathbf{47530} & \mathbf{133900} & \mathbf{0} \\ \mathbf{0} & \mathbf{0} & \mathbf{43170} \end{pmatrix} \mathbf{N/mm^2}$$

Problem 2. Find the ABD matrix for a 5 mm thick sheet of isotropic material, given the stiffness matrix (Q):

$$(Q) = \begin{pmatrix} 133900 & 47530 & 0 \\ 47530 & 133900 & 0 \\ 0 & 0 & 43200 \end{pmatrix} \text{N/mm}^2$$

The **extensional stiffness matrix,** $(A) = h\,(Q)$

$$= 5 \times \begin{pmatrix} 133900 & 47530 & 0 \\ 47530 & 133900 & 0 \\ 0 & 0 & 43200 \end{pmatrix} \text{N/mm}^2$$

$$(A) = \begin{pmatrix} \mathbf{669500} & \mathbf{237700} & \mathbf{0} \\ \mathbf{237700} & \mathbf{669500} & \mathbf{0} \\ \mathbf{0} & \mathbf{0} & \mathbf{216000} \end{pmatrix} \mathbf{N/mm^2}$$

The **bending, extension coupling matrix,** $(B) = 0$ for an isotropic ply,

i.e. $$(\boldsymbol{B}) = \begin{pmatrix} \mathbf{0} & \mathbf{0} & \mathbf{0} \\ \mathbf{0} & \mathbf{0} & \mathbf{0} \\ \mathbf{0} & \mathbf{0} & \mathbf{0} \end{pmatrix} \mathbf{N/mm^2}$$

The bending stiffness matrix, $(D) = (Q)\left[\frac{h^3}{12}\right]$

$$(D) = \frac{5^3}{12} \times \begin{pmatrix} 133900 & 47530 & 0 \\ 47530 & 133900 & 0 \\ 0 & 0 & 43200 \end{pmatrix} \text{N/mm}^2$$

i.e. $$(\boldsymbol{D}) = \begin{pmatrix} \mathbf{1395000} & \mathbf{495100} & \mathbf{0} \\ \mathbf{495100} & \mathbf{1395000} & \mathbf{0} \\ \mathbf{0} & \mathbf{0} & \mathbf{450000} \end{pmatrix} \mathbf{N/mm^2}$$

Now try the following Practice Exercise

Practice Exercise 69. Composites – isotropic plies

1. Calculate the stiffness matrix (Q) and derive the (A), (B) and (D) matrices for a sheet of aluminium 1 mm thick. The following material properties can be assumed:

   $E = 70$ GPa and $\upsilon = 0.35$ and for an isotropic material: $G = \frac{E}{2(1+\upsilon)}$

$$\left[(Q) = \begin{pmatrix} 79.77 & 27.92 & 0 \\ 27.92 & 79.77 & 0 \\ 0 & 0 & 25.93 \end{pmatrix} \text{in GPa}, \; (A) = \begin{pmatrix} 79.77 & 27.92 & 0 \\ 27.92 & 79.77 & 0 \\ 0 & 0 & 25.93 \end{pmatrix} \text{in kN/mm} \quad (D) = \begin{pmatrix} 6.648 & 2.327 & 0 \\ 2.327 & 6.648 & 0 \\ 0 & 0 & 2.161 \end{pmatrix} \text{in kN/mm} \right]$$

Problems 2 to 4 follow on from Problem 1

2. Show that the inverse of (A) in Problem 1 is:

$$(a) = \begin{pmatrix} 1.428\times10^{-5} & -5.0\times10^{-6} & 0 \\ -5.0\times10^{-6} & 1.428\times10^{-5} & 0 \\ 0 & 0 & 3.856\times10^{-5} \end{pmatrix} \text{in mm/N}$$

3. An aluminium sheet is 1 mm thick, 300 mm long and 200 mm wide. The sheet is clamped at one end, and a tensile load of 6 kN is applied across the 200 mm width (i.e. the x direction). Using the stiffness matrix (Q) (for Problem 1 above), determine the strains and hence the stress in the x and y directions, given the matrix (a). For the aluminium assume that $E = 70$ GPa, $\upsilon = 0.35$, and for an isotropic material: $G = \dfrac{E}{2(1+\upsilon)}$

$$\text{Also given: } (a) = \begin{pmatrix} 1.429\times10^{-5} & -5.0\times10^{-6} & 0 \\ -5.0\times10^{-6} & 1.429\times10^{-5} & 0 \\ 0 & 0 & 3.857\times10^{-5} \end{pmatrix} \text{in mm/N}$$

$$\left[(\varepsilon^o) = \begin{pmatrix} 4.287\times10^{-4} \\ -1.5\times10^{-4} \\ 0 \end{pmatrix} \text{in units of strain}, \quad \begin{pmatrix} \sigma_x \\ \sigma_y \\ \tau_{xy} \end{pmatrix} = \begin{pmatrix} 30 \\ 0 \\ 0 \end{pmatrix} \text{MPa} \right]$$

4. The aluminium sheet described in Problem 1 above is 1 mm thick, 300 mm long and 200 mm wide. The sheet is clamped at one end forming a 300 mm long cantilever. A load is applied which produces a pure bending moment of 2000 N mm across the 200 mm width of the sheet (i.e. M_x). Using the stiffness matrix (Q) (for Problem 1 above), determine the strain on the top and bottom of the sheet. Use this to determine the stress on the top and bottom of the sheet and sketch the graph of stresses in the x and y directions, given the matrix (d),

$$(d) = \begin{pmatrix} 0.0001714 & -0.00006 & 0 \\ -0.00006 & 0.0001714 & 0 \\ 0 & 0 & 0.0004628 \end{pmatrix} \text{in units N}^{-1}\text{mm}^{-1}$$

$$\left[(\kappa) = \begin{pmatrix} 0.001714 \\ -0.00006 \\ 0 \end{pmatrix} \text{in units of mm}^{-1}, \; (\sigma)_{TOP} = \begin{pmatrix} -60.0 \\ 0 \\ 0 \end{pmatrix} \text{in MPa}, \; (\sigma)_{BOT} = \begin{pmatrix} 60.0 \\ 0 \\ 0 \end{pmatrix} \text{in MPa} \right]$$

19.5 Derivation of the stiffness matrix (Q) for orthotropic materials

(Q) is obtained by inverting the compliance matrix (S). From equations (19.6) and (19.22),

$$(Q)=(S)^{-1}=\frac{(S)^a}{|S|}=\frac{(S)^{C^T}}{|S|}$$

where $(S)^a$ is the adjoint matrix of (S) and $(S)^C$ is the cofactor matrix.

Now $$(S)=\begin{pmatrix} 1/E_1 & -\upsilon_{21}/E_2 & 0 \\ -\upsilon_{12}/E_1 & 1/E_2 & 0 \\ 0 & 0 & 1/G_{12} \end{pmatrix}$$

from which

$$(S)^C=\begin{pmatrix} \frac{1}{E_2G_{12}} & -\frac{\upsilon_{12}}{E_1G_{12}} & 0 \\ \frac{\upsilon_{21}}{E_2G_{12}} & \frac{1}{E_1G_{12}} & 0 \\ 0 & 0 & \left(\frac{1}{E_1E_2}-\frac{\upsilon_{12}\upsilon_{21}}{E_1E_2}\right) \end{pmatrix}$$

and

$$(S)^{C^T}=\begin{pmatrix} \frac{1}{E_2G_{12}} & \frac{\upsilon_{21}}{E_2G_{12}} & 0 \\ \frac{\upsilon_{12}}{E_1G_{12}} & \frac{1}{E_1G_{12}} & 0 \\ 0 & 0 & \left(\frac{1}{E_1E_2}-\frac{\upsilon_{12}\upsilon_{21}}{E_1E_2}\right) \end{pmatrix} \quad (19.23)$$

The determinant,

$$|S|=\left(\frac{1}{E_1}\right)\left[\left(\frac{1}{E_2}\right)\left(\frac{1}{G_{12}}\right)\right]-\left(-\frac{\upsilon_{21}}{E_2}\right)\left[\left(-\frac{\upsilon_{12}}{E_2}\right)\left(\frac{1}{G_{12}}\right)\right]+0$$

$$=\left(\frac{1}{E_1E_2G_{12}}\right)-\left(\frac{\upsilon_{12}\upsilon_{21}}{E_1E_2G_{12}}\right)$$

i.e. $$|S|=\left(\frac{1}{E_1E_2G_{12}}\right)\left(1-\upsilon_{12}\upsilon_{21}\right)$$

from which $$\frac{1}{|S|}=\left(\frac{1}{1-\upsilon_{12}\upsilon_{21}}\right)\left(E_1E_2G_{12}\right) \quad (19.24)$$

From equations (19.23) and (19.24),

$$(Q)=(S)^{-1}=\frac{(S)^a}{|S|}=\frac{(S)^{C^T}}{|S|}$$

$$=\left(\frac{1}{1-\upsilon_{12}\upsilon_{21}}\right)\left(E_1E_2G_{12}\right)\begin{pmatrix} \frac{1}{E_2G_{12}} & \frac{\upsilon_{21}}{E_2G_{12}} & 0 \\ \frac{\upsilon_{12}}{E_1G_{12}} & \frac{1}{E_1G_{12}} & 0 \\ 0 & 0 & \left(\frac{1}{E_1E_2}-\frac{\upsilon_{12}\upsilon_{21}}{E_1E_2}\right) \end{pmatrix}$$

i.e.

$$(Q)=\left(\frac{1}{1-\upsilon_{12}\upsilon_{21}}\right)\begin{pmatrix} E_1 & \upsilon_{21}E_1 & 0 \\ \upsilon_{12}E_2 & E_2 & 0 \\ 0 & 0 & (1-\upsilon_{12}\upsilon_{21})G_{12} \end{pmatrix}$$

For a ply, or sheet, or layer:

$$(A)=(Q)h$$

$$(B)=(Q)\left(\frac{h^2}{8}-\frac{h^2}{8}\right)$$

$$(D)=(Q)\frac{h^3}{12}$$

19.6 An orthotropic ply with off-axis loading

With reference to Figure 19.5, fibres are at angle θ to global axis (x,y), which is the direction of loading. An anticlockwise rotation is defined as positive.

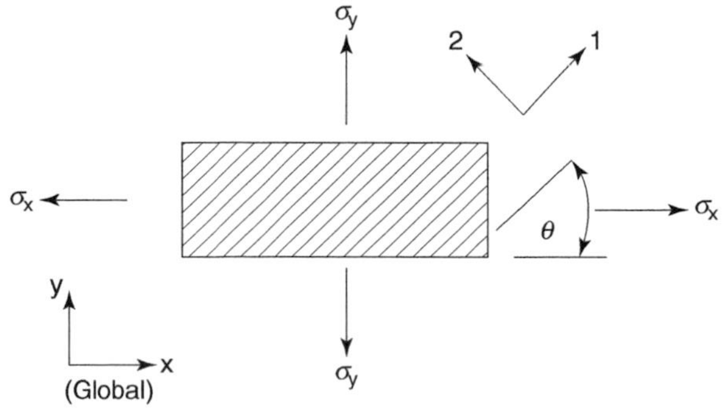

Figure 19.5

Material properties are known along and perpendicular to the fibre direction i.e. axes (1,2), but loads are applied in (x,y).

We wish to obtain (A), (B) and (D) for this 'off axis' loading, and to do this we require a stiffness matrix (Q^*) such that:

$$(\sigma)_{GLOBAL} = (Q^*)(\varepsilon)_{GLOBAL} \quad \text{from equation (19.6)}$$

i.e.

$$(\sigma)_{GLOBAL} \begin{pmatrix} \sigma_x \\ \sigma_y \\ \tau_{xy} \end{pmatrix} = (Q^*) \begin{pmatrix} \varepsilon_x \\ \varepsilon_y \\ \gamma_{xy} \end{pmatrix}$$

The process to find (Q^*) is as follows:

(1) Express stresses in the directions in which the loads are applied (global x,y) to the directions parallel and perpendicular to the fibres (1,2).
(2) Use Hooke's Law to relate the stresses to strains in (1,2) direction.
(3) Transform the strains in the (1,2) directions to strains in (x,y).

A transformation matrix (T) is defined to relate local (fibre) stresses to global stresses.

$$(\sigma)_{FIBRE} = (T)(\sigma)_{GLOBAL}$$

and

$$(T) = \begin{pmatrix} c^2 & s^2 & 2sc \\ s^2 & c^2 & -2sc \\ -sc & sc & c^2 - s^2 \end{pmatrix}$$

where $c = \cos\theta$ and $s = \sin\theta$

Also

$$(\sigma)_{GLOBAL} = (T)^{-1}(\sigma)_{FIBRE}$$

$$(T)^{-1} = \begin{pmatrix} c^2 & s^2 & -2sc \\ s^2 & c^2 & 2sc \\ sc & -sc & c^2 - s^2 \end{pmatrix}$$

Note that these transformation matrices are not symmetrical.

$$\begin{pmatrix} \sigma_1 \\ \sigma_2 \\ \tau_{12} \end{pmatrix} = (T) \begin{pmatrix} \sigma_x \\ \sigma_y \\ \tau_{xy} \end{pmatrix}$$

and hence

$$\begin{pmatrix} \sigma_x \\ \sigma_y \\ \tau_{xy} \end{pmatrix} = (T)^{-1} \begin{pmatrix} \sigma_1 \\ \sigma_2 \\ \tau_{12} \end{pmatrix} \tag{19.25}$$

It can be shown that transformation for strain gives rise to a similar matrix to the matrix (T) for the transformation of stress. The only difference is a factor of 0.5 applied to the shear component. This factor of 2 difference is analogous to the behaviour of the shear strain term in Mohr's circle for strain (in Chapter 9).

The matrix (T) can be used, as long as it is multiplied by another matrix which simply multiplies the shear value.

$$\begin{pmatrix} \varepsilon_1 \\ \varepsilon_2 \\ \frac{\gamma_{12}}{2} \end{pmatrix} = (T) \begin{pmatrix} \varepsilon_x \\ \varepsilon_y \\ \frac{\gamma_{xy}}{2} \end{pmatrix}$$

and

$$\begin{pmatrix} \varepsilon_1 \\ \varepsilon_2 \\ \gamma_{12} \end{pmatrix} = \begin{pmatrix} 1 & 0 & 0 \\ 0 & 1 & 0 \\ 0 & 0 & 2 \end{pmatrix} \begin{pmatrix} \varepsilon_1 \\ \varepsilon_2 \\ \frac{\gamma_{12}}{2} \end{pmatrix} = (R) \begin{pmatrix} \varepsilon_1 \\ \varepsilon_2 \\ \frac{\gamma_{12}}{2} \end{pmatrix}$$

where (R) is the Reuter matrix, defined as:

$$(R) = \begin{pmatrix} 1 & 0 & 0 \\ 0 & 1 & 0 \\ 0 & 0 & 2 \end{pmatrix} \tag{19.26}$$

and thus

$$(R)^{-1} = \begin{pmatrix} 1 & 0 & 0 \\ 0 & 1 & 0 \\ 0 & 0 & 0.5 \end{pmatrix}$$

The Reuter and inverse Reuter matrices simply multiply and divide the shear strain term by 2, avoiding the need for an additional 3×3 transformation matrix and its inverse.

Now

$$\begin{pmatrix} \sigma_x \\ \sigma_y \\ \tau_{xy} \end{pmatrix} = (T)^{-1} \begin{pmatrix} \sigma_1 \\ \sigma_2 \\ \tau_{12} \end{pmatrix} \quad \text{as stated above}$$

$$= (T)^{-1}(Q) \begin{pmatrix} \varepsilon_1 \\ \varepsilon_2 \\ \gamma_{12} \end{pmatrix}$$

However, from equation (19.10), $(Q)=(S)^{-1}$ and

$$\begin{pmatrix} \sigma_1 \\ \sigma_2 \\ \tau_{12} \end{pmatrix} = (Q) \begin{pmatrix} \varepsilon_1 \\ \varepsilon_2 \\ \gamma_{12} \end{pmatrix}$$ where (Q) is the stiffness matrix

and (S) is the compliance matrix

Hence $$\begin{pmatrix} \sigma_x \\ \sigma_y \\ \tau_{xy} \end{pmatrix} = (T)^{-1}(Q)(R) \begin{pmatrix} \varepsilon_1 \\ \varepsilon_2 \\ \frac{\gamma_{12}}{2} \end{pmatrix}$$

$$= (T)^{-1}(Q)(R)(T) \begin{pmatrix} \varepsilon_x \\ \varepsilon_y \\ \frac{\gamma_{xy}}{2} \end{pmatrix}$$

from equation (19.26)

and $$\begin{pmatrix} \sigma_x \\ \sigma_y \\ \tau_{xy} \end{pmatrix} = (T)^{-1}(Q)(R)(T)(R)^{-1} \begin{pmatrix} \varepsilon_x \\ \varepsilon_y \\ \gamma_{xy} \end{pmatrix}$$

from equation (19.25)

This is the equation required to link global stresses to global strains:

$$(\sigma)_{GLOBAL} = (Q^*)(\varepsilon)_{GLOBAL}$$

where $$(Q^*) = (T)^{-1}(Q)(R)(T)(R)^{-1}$$

For a single ply under off axis loading:

$$\begin{pmatrix} N \\ M \end{pmatrix} = \begin{pmatrix} A & B \\ B & D \end{pmatrix} \begin{pmatrix} \varepsilon^o \\ \kappa \end{pmatrix}$$

and so $$(A) = (Q^*)h$$

$$(B) = (Q^*)\left[\frac{h^2}{8} - \frac{h^2}{8}\right]$$

$$(D) = (Q^*)\left[\frac{h^3}{12}\right]$$

19.7 A laminate or ply based on orthotropic plies with off-axis loading

The approach here is to sum the forces applied to each layer to give the resultant for the whole laminate, by integrating through the thickness (depth) of the laminate. See Figure 19.6.

A laminate can be subjected to both in-plane and transverse loading. It will stretch and bend, and both these effects must be considered. The total strain will be composed of in-plane strain ε^0, constant across the thickness of the laminate, and strains due to bending which are linear across the thickness. NB A laminate is made from a collection of individual plies, or sheets, or layers or laminae (where laminae is the plural of lamina).

The bending strains are defined in terms of the plate curvatures κ_x, κ_y and κ_{xy}. Note that z is measured from the **mid-plane** of the laminate.

- There are 'p' plies in the laminate
- Plies are numbered from top down

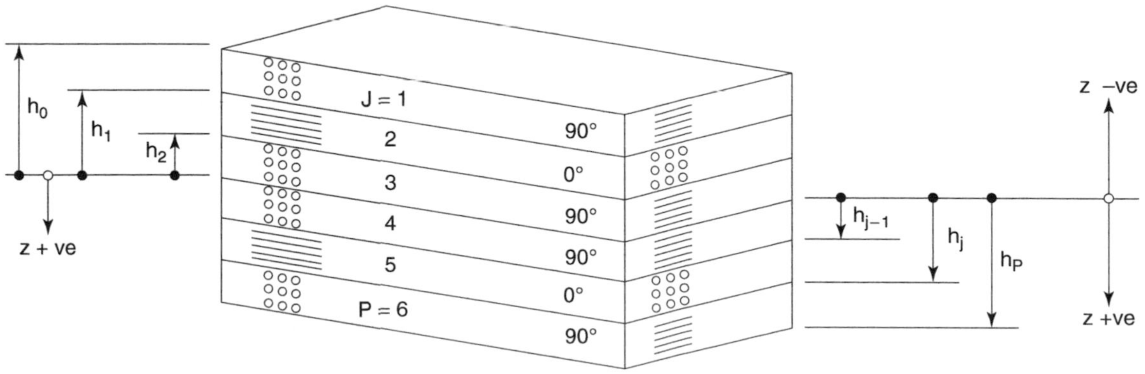

Figure 19.6

- The lower surface of the pth ply is assigned the co-ordinate h_p
- The thickness of a ply is $h_j - h_{j-1}$ where 'j' is the 'jth' ply, counting from the top down to the bottom
- Total thickness of laminate is $-h_0 + h_p$ where 'p' is the total number of plies or laminates.

For a single layer, in global co-ordinates:

$$\begin{pmatrix} \varepsilon_x \\ \varepsilon_y \\ \gamma_{xy} \end{pmatrix} = \begin{pmatrix} \varepsilon_x^{\,o} \\ \varepsilon_y^{\,o} \\ \gamma_{xy}^{\,o} \end{pmatrix} + z \begin{pmatrix} \kappa_x \\ \kappa_y \\ \kappa_{xy} \end{pmatrix}$$

For 'p' layers of the laminate:

$$\left(N^L\right) = \begin{pmatrix} N_x \\ N_y \\ N_{xy} \end{pmatrix}_{\text{GLOBAL}} = \sum_{j=1}^{p} \int_{h_{j-1}}^{h_j} \begin{pmatrix} \sigma_x \\ \sigma_y \\ \tau_{xy} \end{pmatrix}_{\text{GLOBAL}} dz$$

Substituting for stress gives:

$$\begin{pmatrix} N_x \\ N_y \\ N_{xy} \end{pmatrix}_{\text{GLOBAL}} = \sum_{j=1}^{p} \left(\int_{h_{j-1}}^{h_j} (Q^*)(\varepsilon^o)\,dz + \int_{h_{j-1}}^{h_j} (Q^*)(\kappa)\,z\,dz \right)$$

(direct stress + bending stress)

But ε^0 and κ are not functions of z and within a layer, Q* is not a function of z, hence

$$\left(N^L\right) = \sum_{j=1}^{p} \left((Q^*)(\varepsilon^o) \int_{h_{j-1}}^{h_j} dz + (Q^*)(\kappa) \int_{h_{j-1}}^{h_j} z\,dz \right)$$

In addition, ε^0 and κ are not functions of j giving:

$$\left(N^L\right) = \left(\varepsilon^o\right) \sum_{j=1}^{p} (Q^*) \int_{h_{j-1}}^{h_j} dz + (\kappa) \sum_{j=1}^{p} (Q^*) \int_{h_{j-1}}^{h_j} z\,dz$$

or $\left(N^L\right) = (A)^L \left(\varepsilon^o\right) + (B)^L (\kappa)$

where the superscript 'L' represents the whole laminate plate.

$(A)^L$ is the 'extensional stiffness matrix' for the laminate, sometimes called the 'in-plane stiffness matrix'.

$$(A)^L = \sum_{j=1}^{p} (Q^*) [z]_{h_{j-1}}^{h_j}$$

i.e. $$(A)^L = \sum_{j=1}^{p} (Q^*) \left[h_j - h_{j-1} \right]$$

where h_j is the bottom surface of the jth layer, so $h_j - h_{j-1}$ is the layer thickness.

The dimensions for $(A)^L$ are: Force/length, and **the conventional units are:** N/mm.

$(B)^L$ is the 'bending extension coupling matrix', and $(B)^L = 0$ for all **symmetric** laminates.

$$(B)^L = \sum_{j=1}^{p} (Q^*) \int_{h_{j-1}}^{h_j} z\,dz = \sum_{j=1}^{p} (Q^*) \left[\frac{z^2}{2} \right]_{h_{j-1}}^{h_j}$$

i.e. $(B)^L = \frac{1}{2} \sum_{j=1}^{p} (Q^*) \left[h_j^{\,2} - h_{j-1}^{\,2} \right]$ where the superscript 'L' represents the whole laminate plate

The dimensions for $(B)^L$ **are:** Force and **the conventional unit is:** N.

Adopting a similar procedure for the moments gives:

$$\left(M^L\right) = (B)^L \left(\varepsilon^o\right) + (D)^L (\kappa)$$

where the superscript 'L' represents the whole laminate plate and where $(D)^L$ is the 'bending stiffness matrix', given by:

$$(D)^L = \sum_{j=1}^{p} (Q^*) \int_{h_{j-1}}^{h_j} z^2\,dz = \sum_{j=1}^{p} (Q^*) \left[\frac{z^3}{3} \right]_{h_{j-1}}^{h_j}$$

i.e. $$(D)^L = \frac{1}{3} \sum_{j=1}^{p} (Q^*) \left[h_j^{\,3} - h_{j-1}^{\,3} \right]$$

The dimensions for $(D)^L$ **are:** Force × length and **the conventional units are:** N mm.

Combining the A^L, B^L and D^L matrices from above, into a single matrix generates the ABD matrix defining material properties.

$$(ABD)^L = \begin{pmatrix} (A)^L & (B)^L \\ (B)^L & (D)^L \end{pmatrix}$$ where the superscript 'L' represents the whole laminate plate.

19.8 Failure criteria for composite materials

There are a number of criteria used for the prediction of failure of isotropic materials. Each one has its advantages and disadvantages, and these are well documented in undergraduate texts. The Maximum Stress, the Maximum Strain and the von Mises criteria are used with some success for composite materials – see Chapter 12.

The von Mises failure criterion, from Section 12.6, is given by:

$$(\sigma_1-\sigma_2)^2+(\sigma_1-\sigma_3)^2+(\sigma_2-\sigma_3)^2 = 2\left(\sigma_{yp}\right)^2$$

However, orthotropic materials pose considerable difficulties when it come to the prediction of failure. There have been many theories proposed.

It must be stated that most major manufactures involved with the production of composite materials have their own preferred criteria. These criteria are generally particular variations of the Tsai-Hill and the Tsai-Wu failure criteria. The reasons for these variations are largely historical, each one better suited to the company's application, and based on that company's experience and their particular application. The continued existence of custom failure criteria is confirmed by a NAFEMS report (see Reference 31 on page 497) on the finite element analysis of composite materials, where the ability to formulate a failure criterion was described as a very desirable feature of a finite element package.

The Deviatoric strain energy (Tsai-Hill) failure criterion (see Section 12.3)

$$\frac{\sigma_1^2}{F_1^2}+\frac{\sigma_2^2}{F_2^2}+\frac{\tau_{12}^2}{F_{12}^2}-\frac{\sigma_1\sigma_2}{F_1^2}=1$$

where σ_1 = principal longitudinal stresses, i.e. in the direction of the fibres

σ_2 = principal transverse stresses i.e. normal to the fibres

τ_{12} = in-plane shear stress

F_1 = uniaxial longitudinal stress to failure

F_2 = uniaxial transverse stress to failure

F_{12} = in-plane shear stress to failure

The Interactive tensor polynomial (Tsai-Wu) failure criterion (see Section 12.3)

This is reputed to be more accurate than the Tsai-Hill, but it requires considerably more material strength data.

$$f_1\sigma_1+f_2\sigma_2+f_{11}\sigma_1^2+f_{22}\sigma_2^2+f_{66}\,\tau_{12}^2+2f_{12}\sigma_1\sigma_2^2=1$$

where the coefficients are given below.

F_{1T} = uniaxial longitudinal stress to failure in tension

F_{1C} = uniaxial longitudinal stress to failure in compression

F_{2T} = uniaxial transverse stress to failure in tension

F_{2C} = uniaxial transverse stress to failure in compression

F_{12} = in-plane shear stress to failure

$$f_1=\frac{1}{F_{1T}}-\frac{1}{F_{1C}}\quad f_{11}=\frac{1}{F_{1T}F_{1C}}\quad f_2=\frac{1}{F_{2T}}-\frac{1}{F_{2C}}$$

$$f_{22}=\frac{1}{F_{2T}F_{2C}}\qquad f_{66}=\frac{1}{F_{12}{}^2}$$

$$f_{12}=-\frac{1}{2}\left(\frac{1}{F_{1T}F_{1C}F_{2T}F_{2C}}\right)^{\frac{1}{2}}$$

An **anisotropic material** is one that has many different structural properties in many directions. Analysis of anisotropic materials is beyond the scope of the present text and is not dealt with here.

Problem 3. Determine the compliance and the stiffness matrices for a sheet of carbon fibre reinforced PEEK (polyetheretherketone). The following material properties apply:

E_1 = 134 GPa, E_2 = 8.9 GPa, G_{12} = 5.1 GPa

and $\upsilon_{12}=0.28$

$$\upsilon_{21}E_1=\upsilon_{12}E_2$$

$$\text{so } \upsilon_{21}=\frac{E_2}{E_1}\upsilon_{12}=\frac{8.9}{134}(0.28)=0.0186$$

$$(S)=\begin{pmatrix}1/E_1 & -\upsilon_{21}/E_2 & 0\\ -\upsilon_{12}/E_1 & 1/E_2 & 0\\ 0 & 0 & 1/G_{12}\end{pmatrix}$$

$$=\begin{pmatrix}1/134000 & -0.0186/8900 & 0\\ -0.28/134000 & 1/8900 & 0\\ 0 & 0 & 1/5100\end{pmatrix}$$

i.e. **the compliance matrix,**

$$(S)=\begin{pmatrix}7.462 & -2.090 & 0\\ -2.090 & 112.4 & 0\\ 0 & 0 & 196.1\end{pmatrix}\times 10^{-6}\ \text{mm}^2/\text{N}$$

Stiffness matrix

$$(Q)=\left(\frac{1}{1-\upsilon_{12}\,\upsilon_{21}}\right)\begin{pmatrix} E_1 & \upsilon_{21}E_1 & 0 \\ \upsilon_{12}E_2 & E_2 & 0 \\ 0 & 0 & (1-\upsilon_{12}\upsilon_{21})G_{12} \end{pmatrix}$$

$$\frac{1}{1-\upsilon_{12}\,\upsilon_{21}}=\frac{1}{1-(0.28)(0.0186)}=\frac{1}{0.9948}$$

i.e.

$$(Q)=\left(\frac{1}{0.9948}\right)\begin{pmatrix} 134000 & 0.0186(134000) & 0 \\ 0.28(8900) & 8900 & 0 \\ 0 & 0 & 0.9948(5100) \end{pmatrix}$$

i.e. **stiffness matrix,** $(\boldsymbol{Q})=\begin{pmatrix} \mathbf{134700} & \mathbf{2505} & \mathbf{0} \\ \mathbf{2505} & \mathbf{8947} & \mathbf{0} \\ \mathbf{0} & \mathbf{0} & \mathbf{5100} \end{pmatrix}$

Problem 4. Determine the stiffness matrix (Q) of a unidirectional sheet of Kevlar epoxy. Determine also the ABD matrix if the sheet is 4 mm thick. Assume that:

E_1 = 75 GPa, E_2 = 5.6 GPa, G_{12} = 2.2 GPa and υ_{12} = 0.33

$$\upsilon_{21}E_1=\upsilon_{12}E_2 \text{ so } \upsilon_{21}=\frac{E_2}{E_1}\upsilon_{12}=\frac{5.6}{75}(0.33)=0.02464$$

$$(Q)=\left(\frac{1}{1-\upsilon_{12}\,\upsilon_{21}}\right)\begin{pmatrix} E_1 & \upsilon_{21}E_1 & 0 \\ \upsilon_{12}E_2 & E_2 & 0 \\ 0 & 0 & (1-\upsilon_{12}\upsilon_{21})G_{12} \end{pmatrix}$$

$$\frac{1}{1-\upsilon_{12}\,\upsilon_{21}}=\frac{1}{1-(0.33)(0.02464)}=\frac{1}{0.9919}$$

Hence

$$(Q)=\left(\frac{1}{0.9919}\right)\begin{pmatrix} 75000 & 1848 & 0 \\ 1848 & 5600 & 0 \\ 0 & 0 & 0.9919(2.2) \end{pmatrix}$$

i.e.

$$(Q)=\begin{pmatrix} 75610 & 1863 & 0 \\ 1863 & 5646 & 0 \\ 0 & 0 & 2.2 \end{pmatrix}\text{N/mm}^2$$

The extensional stiffness matrix $(A) = h\,(Q)$

$$= 4\times\begin{pmatrix} 75610 & 1863 & 0 \\ 1863 & 5646 & 0 \\ 0 & 0 & 2.2 \end{pmatrix}\text{N/mm}^2$$

i.e.

$$(A)=\begin{pmatrix} \mathbf{302400} & \mathbf{7452} & \mathbf{0} \\ \mathbf{7452} & \mathbf{22580} & \mathbf{0} \\ \mathbf{0} & \mathbf{0} & \mathbf{8.8} \end{pmatrix}\textbf{N/mm}^2$$

The bending, extension coupling matrix, $(B) = 0$ for an isotropic ply,

i.e. $(\boldsymbol{B})=\begin{pmatrix} \mathbf{0} & \mathbf{0} & \mathbf{0} \\ \mathbf{0} & \mathbf{0} & \mathbf{0} \\ \mathbf{0} & \mathbf{0} & \mathbf{0} \end{pmatrix}$ **N/mm²**

The bending stiffness matrix

$$(D)=(Q)\left[\frac{h^3}{12}\right]=(Q)\left[\frac{4^3}{12}\right]=5.333\,(Q)$$

i.e. $$(D)=\frac{4^3}{12}\times\begin{pmatrix} 75610 & 1863 & 0 \\ 1863 & 5646 & 0 \\ 0 & 0 & 2.2 \end{pmatrix}\text{N/mm}^2$$

i.e. **the bending stiffness matrix**

$$(\boldsymbol{D})=\begin{pmatrix} \mathbf{403200} & \mathbf{9935} & \mathbf{0} \\ \mathbf{9935} & \mathbf{30110} & \mathbf{0} \\ \mathbf{0} & \mathbf{0} & \mathbf{11.73} \end{pmatrix}\textbf{N/mm}^2$$

Problem 5. A structural shell is made from 2 mm thick unidirectional glass epoxy. It is loaded along the fibre direction by a tensile force of 100 N/mm and another tensile force 20 N/mm perpendicular to the fibres. Determine the Tsai-Hill load factor for the shell, given the following data for the glass epoxy:

E_1 = 45.6 GPa, E_2 = 10.73 GPa,

G_{12} = 5.14 GPa and υ_{12} = 0.274

F_{1T} = uniaxial longitudinal stress to failure in tension = 1000 MPa

F_{1C} = uniaxial longitudinal stress to failure in compression = −1600 MPa

F_{2T} = uniaxial transverse stress to failure in tension = 100 MPa

F_{2C} = uniaxial transverse stress to failure in compression = −270 MPa

F_{12} = in-plane shear stress to failure = 80 MPa

$$\sigma_1 = \frac{100\,\text{N/mm}}{2\,\text{mm}} = 50\text{ MPa}$$

$$\sigma_2 = \frac{20\,\text{N/mm}}{2\,\text{mm}} = 10\text{ MPa}$$

$$\frac{\sigma_1{}^2}{F_1^2} + \frac{\sigma_2{}^2}{F_2^2} + \frac{\tau_1{}^2}{F_{12}{}^2} - \frac{\sigma_1\sigma_2}{F_1^2} = 1$$

Including the load factor

$$\frac{\sigma_1^2}{F_1^2} + \frac{\sigma_2^2}{F_2^2} + \frac{\tau_1^2}{F_{12}^2} - \frac{\sigma_1\sigma_2}{F_1^2} = \frac{1}{(LF)^2}$$

$$\frac{1}{(LF)^2} = \frac{50^2}{1000^2} + \frac{10^2}{100^2} + \frac{0^2}{80^2} - \frac{50\times10}{1000^2}$$

$$\frac{1}{(LF)^2} = 0.0025 + 0.01 + 0 - 0.0005 = 0.012$$

$$(LF) = \sqrt{\frac{1}{0.012}} = 9.129$$

i.e. **the Tsai-Hill load factor for the shell is 9.129.** Hence a safety factor of nearly 10 before failure.

Now try the following Practice Exercise

Practice Exercise 70. Composites – orthotropic plies

1. A laminate, shown in Figure 19.7, consists of 4 plies of unidirectional high modulus carbon fibres in an epoxy resin, each 0.125 mm thick and arranged as $(0/90)_s$. Given that $E_1 = 180$ GPa, $E_2 = 8$ GPa, $G_{12} = 5$ GPa and $\nu_{12} = 0.3$, calculate the reduced stiffness matrices (Q) for the single plies along the fibres (i.e. at 0°).

$$\left[(Q_0) = (Q_{90}) = \begin{pmatrix} 180.7 & 2.41 & 0 \\ 2.41 & 8.032 & 0 \\ 0 & 0 & 5 \end{pmatrix} \text{ in GPa}\right]$$

2. For the laminate described in Problem 1, write out the transformation matrices (T) and $(T)^{-1}$

$$\left[(T)_0 = (T)_0^{-1} = \begin{pmatrix} 1 & 0 & 0 \\ 0 & 1 & 0 \\ 0 & 0 & 1 \end{pmatrix}\right.$$

$$\left.(T)_{90} = (T)_{90}^{-1} = \begin{pmatrix} 0 & 1 & 0 \\ 1 & 0 & 0 \\ 0 & 0 & -1 \end{pmatrix}\right]$$

3. For the laminate described in Problem 1, calculate the reduced stiffness matrices $(Q^*)_0$ and $(Q^*)_{90}$ for the single plies with respect to the global x-axis (i.e. $\theta = 0°$).

$$\left[(Q^*)_0 = \begin{pmatrix} 180.7 & 2.41 & 0 \\ 2.41 & 8.032 & 0 \\ 0 & 0 & 5.0 \end{pmatrix} \text{ in GPa},\right.$$

$$\left.(Q^*)_{90} = \begin{pmatrix} 8.032 & 2.41 & 0 \\ 2.41 & 180.7 & 0 \\ 0 & 0 & 5.0 \end{pmatrix} \text{ in GPa}\right]$$

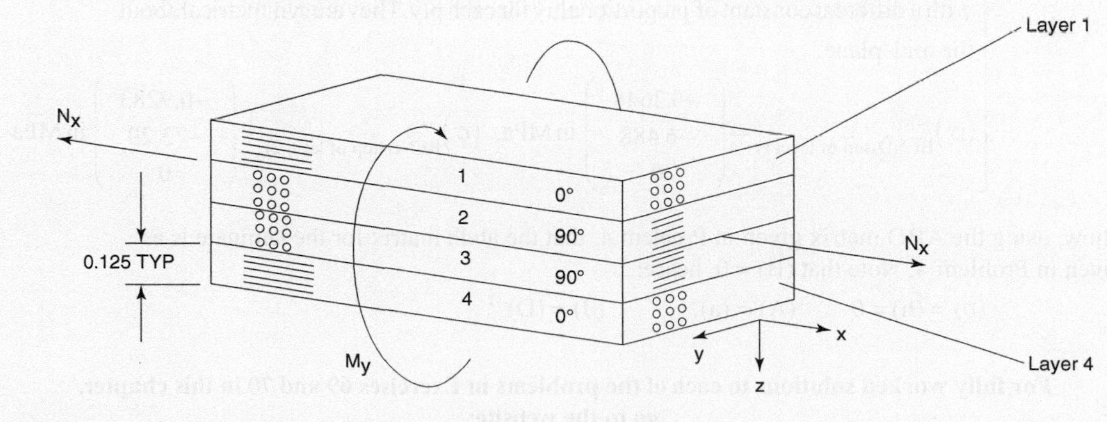

Figure 19.7

4. A load of 1 N/mm is applied in the x-direction (N_x) and a moment of 1 N mm/mm in the y-direction (M_y) to the laminate described in Problem 1 (see Figure 19.7). Using the ABD and the abd matrices for the laminate given below:

   (i) Calculate the direct and the bending strains in the 0° and the 90° layers of the laminate.

   (ii) Sketch the graph of the direct and the bending strains in the x and y directions through each layer of the laminate.

   (iii) Calculate the direct and the bending stresses in the 0° and the 90° layers of the laminate.

   (iv) Sketch the graph of the direct and the bending stresses in the x and y directions through each layer of the laminate.

$$(\text{ABD}) = \begin{pmatrix} 47.19 & 1.205 & 0 & 0 & 0 & 0 \\ 1.205 & 47.19 & 0 & 0 & 0 & 0 \\ 0 & 0 & 2.50 & 0 & 0 & 0 \\ 0 & 0 & 0 & 1.658 & 0.0251 & 0 \\ 0 & 0 & 0 & 0.0251 & 0.3085 & 0 \\ 0 & 0 & 0 & 0 & 0 & 0.05208 \end{pmatrix} \times 10^3 \qquad \begin{matrix} \text{A in N/mm} \\ \text{B in N} \\ \text{D in N mm} \end{matrix}$$

$$(\text{abd}) = \begin{pmatrix} 21.21 & -0.5414 & 0 & 0 & 0 & 0 \\ -0.5414 & 21.21 & 0 & 0 & 0 & 0 \\ 0 & 0 & 400 & 0 & 0 & 0 \\ 0 & 0 & 0 & 604 & -49.14 & 0 \\ 0 & 0 & 0 & -49.14 & 3245 & 0 \\ 0 & 0 & 0 & 0 & 0 & 19200 \end{pmatrix} \times 10^{-6} \qquad \begin{matrix} \text{a in mm/N} \\ \text{b in 1/N} \\ \text{d in 1/(N mm)} \end{matrix}$$

$$\left[(\varepsilon^0)_{0\ \text{DIR}} = (\varepsilon^0)_{90\ \text{DIR}} = \begin{pmatrix} 21.2 \\ -0.5414 \\ 0 \end{pmatrix} \text{in } \mu\varepsilon, \ (\varepsilon)_{\text{BEND (top of layer 1)}} = \begin{pmatrix} +12.48 \\ -811.3 \\ 0 \end{pmatrix} \text{in } \mu\varepsilon, \right.$$

$$(\varepsilon)_{\text{BEND (bottom of layer 4)}} = \begin{pmatrix} -12.48 \\ 811.3 \\ 0 \end{pmatrix} \text{in } \mu\varepsilon$$

Bending strains and bending stresses vary linearly through the depth of the laminate, with a different constant of proportionality for each ply. They are symmetrical about the mid-plane.

$$\left. (\sigma)_{\text{BEND (top of layer 1)}} = \begin{pmatrix} +0.2648 \\ -6.488 \\ 0 \end{pmatrix} \text{in MPa}, \ (\sigma)_{\text{BEND (top of layer 2)}} = \begin{pmatrix} -0.9283 \\ -73.29 \\ 0 \end{pmatrix} \text{in MPa} \right]$$

5. Show, using the ABD matrix given in Problem 4, that the abdh matrix for the laminate is as given in Problem 4. Note that (B) = 0, hence:

   (b) = (h) = 0 (A) = $(\text{a})^{-1}$ (d) = $(\text{D})^{-1}$

For fully worked solutions to each of the problems in Exercises 69 and 70 in this chapter, go to the website:

www.routledge.com/cw/bird

Chapter 20

The matrix displacement method

Why it is important to understand: **The matrix displacement method**

The matrix displacement method is one of the most powerful computer methods for analysing complex structures and associated problems in continuum mechanics, providing that access to the required computer program is available. It is not particularly useful for solving closed loop trivial problems, which can be better solved by the simple solutions described earlier in this book. In this chapter, the matrix displacement method will be used to analyse plane pin-jointed trusses and continuous beams. The method can easily be extended to solve three-dimensional pin-jointed trusses, together with rigid-jointed space frames in both statics and dynamics. These latter structures will not be considered in the present text because they involve too much arithmetic; moreover, they are better solved with computers and their associated software. In this chapter, a description will be given of free software that can be used via SmartPhones or iPads or laptops and so on. It should be noted that with a chromecast device, the screen of the SmartPhone or iPad can be wirelessly projected on to an HD TV screen!

At the end of this chapter you should be able to:

- understand what a load vector is for a truss and beam
- understand what a displacement vector is for a truss and beam
- understand what a stiffness matrix is for a truss and beam
- appreciate what the stiffness matrix of a plane truss element looks like
- appreciate what the stiffness matrix of a continuous beam element looks like
- understand how to analyse statically determinate and statically indeterminate plane pin-jointed trusses
- understand how to analyse statically determinate and statically indeterminate continuous beams under lateral loading

Mathematical references

In order to understand the theory involved in this chapter on **the matrix displacement method**, knowledge of the following mathematical topic is required: ***matrices.*** If help/revision is needed in this area, see chapter 18.

20.1 Introduction

The finite element method (see Reference 5 on page 497) is one of the most powerful methods of solving partial differential equations, particularly if these equations apply over complex shapes. The method consists of subdividing the complex shape into several elements of simpler shape, each of which is more suitable for mathematical analysis. The process then, as far as structural analysis is concerned, is to obtain the elemental stiffnesses of these simpler shapes and then, by considering equilibrium and compatibility at the inter-element boundaries, to assemble all the elements, so that a mathematical model of the entire structure is obtained.

Hence, owing to the application of loads on this mathematical model, the 'deflections' at various points of the structure can be obtained through the solution of the resulting simultaneous equations. Once these 'deflections' are known, the stresses in the structure can be determined through Hookean elasticity.

Each finite element is described by 'nodes' or 'nodal points', and the stiffnesses, displacements, loads, etc., are all related to those nodes. Finite elements vary in shape, depending on the systems they have to describe, and some typical finite elements are shown in Figure 20.1.

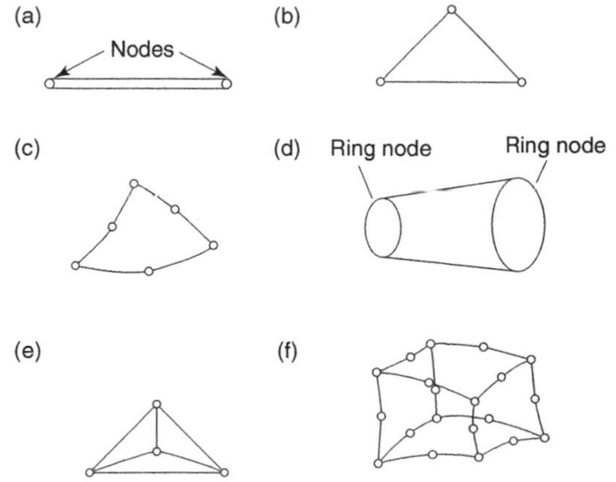

Figure 20.1 Some typical finite elements: (a) one-dimensional rod element, with end nodes; (b) two-dimensional triangular element with corner nodes; (c) curved triangular plate, with additional 'mid-side' nodes; (d) truncated conical element with ring nodes; (e) solid tetrahedral element with corner nodes; (f) 20-node curved brick element

Now, the finite element method is a vast topic, covering problems in structural mechanics, fluid flow, heat transfer, acoustics, etc., and is described in Chapter 21 following. Because the finite element method is based on the matrix displacement method, a brief description of the latter will be given in the present chapter.

20.2 The matrix displacement method

The matrix displacement method is also known as the stiffness method. It is based on obtaining the stiffness of the entire structure by assembling together all the individual stiffnesses of each member or element of the structure. When this is done, the mathematical model of the structure is subjected to the externally applied loads, and by solving the resulting simultaneous equations, the nodal deflections are determined. Once the nodal deflections are known, the stresses in the structure can be obtained through Hookean elasticity.

Now, in the small-deflection theory of elasticity, a structure can be said to behave like a complex spring, where each member or element of the structure can be regarded as an individual spring, with a different type and value of stiffness.

Thus to introduce the method, let us consider the single elemental spring of Figure 20.2.

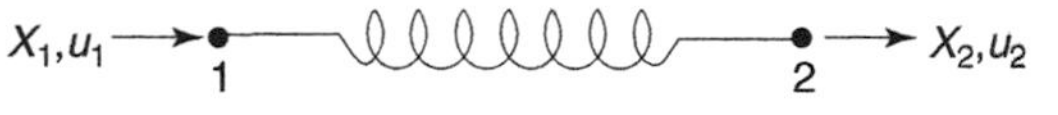

Figure 20.2 Spring element

Let k = stiffness of spring

= slope of load-deflection relationship for the spring

X_1 = axial force at node 1

X_2 = axial force at node 2

u_1 = nodal displacement at node 1 in the direction of X_1

u_2 = nodal displacement at node 2 in the direction of X_2

In Figure 20.2, 1 and 2 are known as the nodes or nodal points.
From Hooke's law

$$X_1 = k(u_1 - u_2) \tag{20.1}$$

and from considerations of equilibrium

$$\begin{aligned} X_2 &= -X_1 \\ &= k(u_2 - u_1) \end{aligned} \tag{20.2}$$

If equations (20.1) and (20.2) are put into matrix form, they appear as follows:

$$\begin{pmatrix} X_1 \\ X_2 \end{pmatrix} = \begin{pmatrix} k & -k \\ -k & k \end{pmatrix} \begin{pmatrix} u_1 \\ u_2 \end{pmatrix} \tag{20.3}$$

or $\quad (P_i) = (k)(u_i)$

where (P_i) = a vector of elemental nodal forces

(u_i) = a vector of elemental nodal displacements

and (k) = elemental stiffness matrix

20.3 The structural stiffness matrix (K)

Consider a simple structure composed of two elemental springs, as shown in Figure 20.3, where

k_a = stiffness of spring 1–2

and k_b = stiffness of spring 2–3

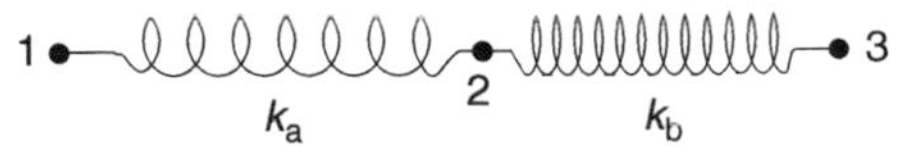

Figure 20.3 Simple structure

To determine (K)

(K) will be of order 3 × 3 because there are three degrees of freedom, namely u_1, u_2 and u_3, and it can be obtained as follows.

For *element* 1–2, from equation (20.3) the elemental stiffness matrix for spring 1–2 is given by (20.4), where the components of stiffness are related to the nodal displacements, u_1 and u_2:

$$(k_{1-2}) = \begin{array}{c} \begin{matrix} u_1 & u_2 \end{matrix} \\ \begin{pmatrix} k_a & -k_a \\ -k_a & k_a \end{pmatrix} \end{array} \begin{matrix} u_1 \\ u_2 \end{matrix} \tag{20.4}$$

Similarly for *element* 2–3

$$(k_{2-3}) = \begin{array}{c} \begin{matrix} u_2 & u_3 \end{matrix} \\ \begin{pmatrix} k_b & -k_b \\ -k_b & k_b \end{pmatrix} \end{array} \begin{matrix} u_2 \\ u_3 \end{matrix} \tag{20.5}$$

Superimposing the components of stiffness corresponding to the displacements, u_1, u_2 and u_3, from (20.4) and (20.5), the stiffness matrix for the entire structure is obtained as follows:

$$(K) = \begin{array}{c} \begin{matrix} u_1 & u_2 & u_3 \end{matrix} \\ \begin{pmatrix} k_a & -k_a & 0 \\ -k_a & (k_a + k_b) & -k_b \\ 0 & -k_b & k_b \end{pmatrix} \end{array} \begin{matrix} u_1 \\ u_2 \\ u_3 \end{matrix} \tag{20.6}$$

Method of solution

The above matrix is singular because free-body displacements have been allowed, i.e. it is necessary to apply boundary conditions. If the system is fixed at node 3, such that the deflection of 3 is zero, and R is the 'reaction', then the equations become:

$$\begin{pmatrix} Q_1 \\ Q_2 \\ R \end{pmatrix} = \begin{pmatrix} k_a & -k_a & 0 \\ -k_a & (k_a + k_b) & -k_b \\ 0 & -k_b & k_b \end{pmatrix} \begin{pmatrix} u_1 \\ u_2 \\ u_3 \end{pmatrix}$$

or

$$\begin{pmatrix} q_F \\ R \end{pmatrix} = \begin{pmatrix} K_{11} & K_{12} \\ K_{21} & K_{22} \end{pmatrix} \begin{pmatrix} u_F \\ 0 \end{pmatrix}$$

i.e. the nodal displacements (u_F) are given by:

$$(u_F) = (K_{11})^{-1}(q_F) \tag{20.7}$$

and the reactions (R) are given by:

$$(R) = (K_{21})(u_F) \tag{20.8}$$

where Q_1 and Q_2 are loads applied to nodes 1 and 2, respectively, and

(q_F) = a vector of externally applied loads corresponding to the free displacements

(K_{11}) = that part of the structural stiffness matrix corresponding to the free displacements, namely, u_1 and u_2

$$= \begin{pmatrix} k_a & -k_a \\ -k_a & (k_a + k_b) \end{pmatrix}$$

(u_F) = a vector of free displacements

$$= \begin{pmatrix} u_1 \\ u_2 \end{pmatrix}$$

For large stiffness matrices of a banded form, (K_{11}) is, in general, not inverted, and solution is carried out by Gaussian elimination or by Choleski's method. (For the Gaussian elimination method, see *Bird's Higher Engineering Mathematics* 9th Edition).

20.4 Elemental stiffness matrix for a plane rod

A rod is defined as a member of a framework which resists its load axially, e.g. a member of a pin-jointed truss.

Under an axial load X, a rod of length l and uniform cross-sectional area A will deflect a distance

$$u = \frac{Xl}{AE}$$

Consider the one-dimensional rod shown in Figure 20.4.

Figure 20.4

$$X_1 = \frac{AE}{l}(u_1 - u_2) = \text{axial force at node 1}$$

$$X_2 = \frac{AE}{l}(u_2 - u_1) = \text{axial force at node 2}$$

or, in matrix form:

$$\begin{pmatrix} X_1 \\ X_2 \end{pmatrix} = \frac{AE}{l}\begin{pmatrix} 1 & -1 \\ -1 & 1 \end{pmatrix}\begin{pmatrix} u_1 \\ u_2 \end{pmatrix}$$

The above can be seen to be of similar form as equation (20.3). Hence, the elemental stiffness matrix for a plane rod is given by:

$$(k) = \frac{AE}{l}\begin{pmatrix} 1 & -1 \\ -1 & 1 \end{pmatrix} \qquad (20.9)$$

Equation (20.9) is the *elemental stiffness matrix for a rod in local co-ordinates*, but in practice it is more useful to obtain the elemental stiffness matrix in global co-ordinates.

Let, O_x^0 and O_y^0 be the global axes and Ox and Oy the local axes, as shown in Figure 20.5.

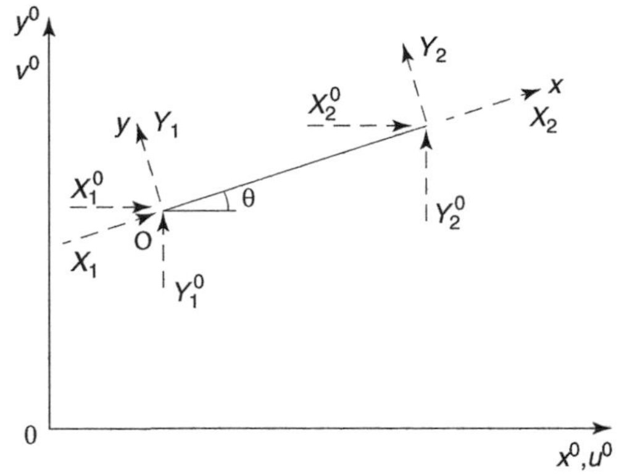

Figure 20.5 Local and global axes

In global co-ordinates, both u and v displacements are important; hence, equation (20.9) must be written as follows:

$$(k) = \frac{AE}{l}\begin{pmatrix} 1 & 0 & -1 & 0 \\ 0 & 0 & 0 & 0 \\ -1 & 0 & 1 & 0 \\ 0 & 0 & 0 & 0 \end{pmatrix} \begin{matrix} u_1 \\ v_1 \\ u_2 \\ v_2 \end{matrix} \qquad (20.10)$$

(columns: $u_1 \; v_1 \; u_2 \; v_2$)

From Figure 20.5, it can be seen that at node 2

$$X_2 = X_2^0 \cos\theta + Y_2^0 \sin\theta$$

$$Y_2 = -X_2^0 \sin\theta + Y_2^0 \cos\theta \qquad (20.11)$$

Similar expressions apply to node 1. Hence, in matrix form

$$\begin{pmatrix} X_1 \\ Y_1 \\ X_2 \\ Y_2 \end{pmatrix} = \begin{pmatrix} \cos\theta & \sin\theta & 0 & 0 \\ -\sin\theta & \cos\theta & 0 & 0 \\ 0 & 0 & \cos\theta & \sin\theta \\ 0 & 0 & -\sin\theta & \cos\theta \end{pmatrix}\begin{pmatrix} X_1^0 \\ Y_1^0 \\ X_2^0 \\ Y_2^0 \end{pmatrix}$$

i.e. $(P) = (\Xi)(P^0)$

where (Ξ) = a matrix of directional cosines

and (P^0) = a vector of elemental nodal forces in global co-ordinates

Now, (Ξ) can be seen to be orthogonal, as $\cos^2\theta + \sin^2\theta = 1$, $\cos\theta \times -\sin\theta + \sin\theta \times \cos\theta = 0$, and so on.

Hence $(\Xi)^{-1} = (\Xi)^T$

Therefore $(P^0) = (\Xi)^T (P^0)$

Similarly $(u) = (\Xi)(u^0)$ (20.12)

Now $(P) = (k)(u)$

Therefore $(\Xi)(P^0) = (k)(\Xi)(u^0)$

and $(P^0) = (\Xi)^T (k)(\Xi)(u^0)$ because (Ξ) is orthogonal, i.e. $(\Xi)^{-1} = (\Xi)^T$

and hence $(k^0) = (\Xi)^T (k)(\Xi)$ (20.13)

Similarly $(K^0) - (\Xi)^T (K)(\Xi)$ (20.14)

Hence, from equations (20.10) and (20.13), the *elemental stiffness matrix for a rod in global co-ordinates* is as follows:

$$\begin{matrix} & u_1^0 & v_1^0 & u_2^0 & v_2^0 & \end{matrix}$$

$$(k^0) = \frac{AE}{l}\begin{pmatrix} C^2 & CS & -C^2 & -CS \\ CS & S^2 & -CS & -S^2 \\ -C^2 & -CS & C^2 & CS \\ -CS & -S^2 & CS & S^2 \end{pmatrix} \begin{matrix} u_1^0 \\ v_1^0 \\ u_2^0 \\ v_2^0 \end{matrix} \quad (20.15)$$

where $C = \cos\theta$, $S = \sin\theta$

Problem 1. Determine the forces in the members of the plane pin-jointed truss of Figure 20.6. It may be assumed that AE is constant for all members.

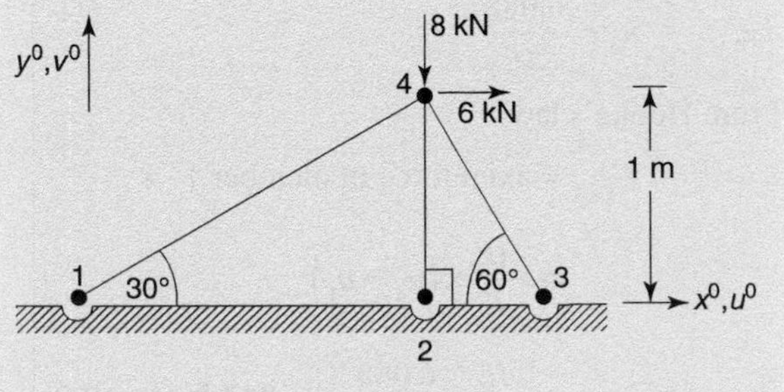

Figure 20.6 Plane pin-jointed truss

Member 1–4

The member points from node 1 to node 4, so that:

$\theta = 30°$, $C = 0.866$, $S = 0.5$ and $\sin 30° = \frac{1}{l}$ hence, $l = \frac{1}{\sin 30°} = 2$ m, where l = length of element 1–4

The member points in the direction from node 1 to node 4 as shown in Figure 20.7. Now as the member is firmly pinned at node 1, it is only necessary to consider the components of the stiffness matrix corresponding to the free displacements u_4^0 and v_4^0

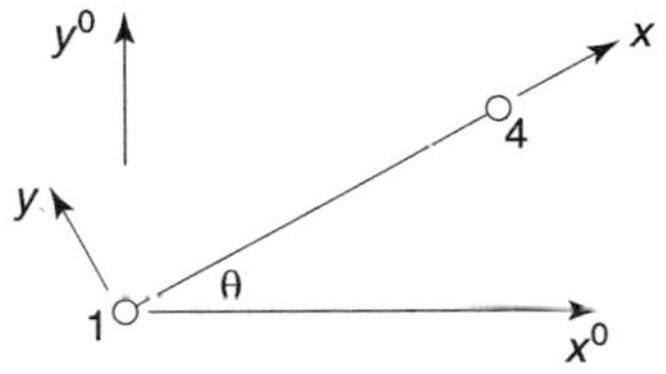

Figure 20.7 Member 1–4

Hence, from equation (20.15)

$$(k^0_{1-4}) - \frac{AE}{l}\begin{pmatrix} C^2 & CS \\ CS & S^2 \end{pmatrix} \begin{matrix} u_4^0 \\ v_4^0 \end{matrix} = \frac{AE}{2}\begin{pmatrix} 0.75 & 0.433 \\ 0.433 & 0.25 \end{pmatrix} \begin{matrix} u_4^0 \\ v_4^0 \end{matrix}$$

(columns: u_4^0, v_4^0)

$$= AE\begin{pmatrix} 0.375 & 0.216 \\ 0.216 & 0.125 \end{pmatrix} \quad (20.16)$$

Member 2–4

$\theta = 90°$, $C = 0$, $S = 1$ and $l = 1$ m, where l = length of element 2–4

The member points in the direction from node 2 to node 4 are as shown in Figure 20.8, so that:

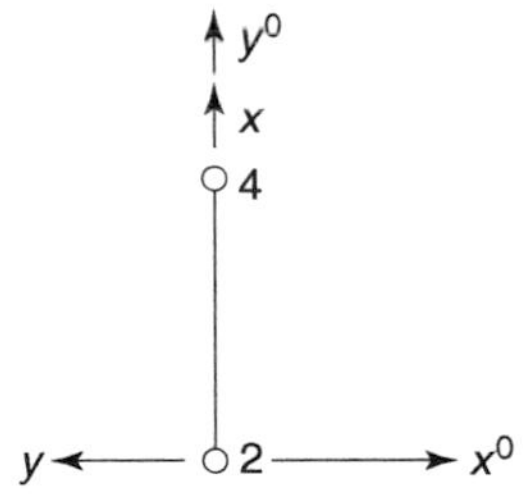

Figure 20.8 Member 2–4

$$(k^0_{2-4}) = \frac{AE}{1}\begin{pmatrix} 0 & 0 \\ 0 & 1 \end{pmatrix} \begin{matrix} u_4^0 \\ v_4^0 \end{matrix} \quad (20.17)$$

(columns: u_4^0, v_4^0)

Member 4–3

$\theta = -60°$, $C = 0.5$, $S = -0.866$ and $l = \dfrac{1}{\sin 60°} = 1.155$ m,

where l = length of element 4–3

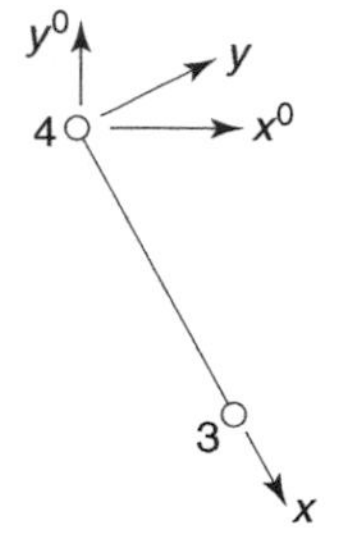

Figure 20.9 Member 4–3

The member points in the direction from node 4 to node 3 are as shown in Figure 20.9, so that:

$$\left(k^0{}_{4-3}\right) = \frac{AE}{1.155}\begin{pmatrix} 0.25 & -0.433 \\ -0.433 & 0.75 \end{pmatrix} \begin{matrix} u_4{}^0 \\ v_4{}^0 \end{matrix} \quad (\text{columns } u_4{}^0, v_4{}^0)$$

$$= AE\begin{pmatrix} 0.216 & -0.375 \\ -0.375 & 0.649 \end{pmatrix} \qquad (20.18)$$

From equations (20.16) to (20.18), the *structural stiffness matrix* (K_{11}) corresponding to the *free displacements* is obtained by assembling equations (20.16) to (20.18)

i.e.

$$(K_{11}) = AE\begin{pmatrix} (0.375+0+0.216) & (0.216+0-0.375) \\ (0.216+0-0.375) & (0.125+1+0.649) \end{pmatrix} \begin{matrix} u_4^0 \\ v_4^0 \end{matrix} \quad (\text{columns } u_4^0, v_4^0)$$

$$= AE\begin{pmatrix} 0.591 & -0.159 \\ -0.159 & 1.744 \end{pmatrix} \begin{matrix} u_4{}^0 \\ v_4{}^0 \end{matrix} \qquad (20.19)$$

The *vector of loads* corresponding to the *free displacements*, namely, u_4^o and v_4^o, is:

$$(q_F) = \begin{pmatrix} 6 \\ -8 \end{pmatrix} \begin{matrix} u_4^0 \\ v_4^0 \end{matrix} \qquad (20.20)$$

where the load 6 kN corresponds to the displacement u_4^0, and the load – 8 kN corresponds to the displacement v_4^0

Substituting equations (20.19) and (20.20) into (20.7) gives:

$$(u_F) = \begin{pmatrix} u_4^0 \\ v_4^0 \end{pmatrix} = \frac{1}{AE}\begin{pmatrix} 0.591 & -0.159 \\ -0.159 & 1.774 \end{pmatrix}^{-1}\begin{pmatrix} 6 \\ -8 \end{pmatrix}$$

$$= \frac{\frac{1}{AE}\begin{pmatrix} 1.774 & 0.159 \\ 0.159 & 0.591 \end{pmatrix}\begin{pmatrix} 6 \\ -8 \end{pmatrix}}{\{(0.591 \times 1.774) - 0.159^2\}}$$

$$= \frac{\frac{1}{AE}\begin{pmatrix} 1.774 & 0.159 \\ 0.159 & 0.591 \end{pmatrix}\begin{pmatrix} 6 \\ -8 \end{pmatrix}}{1.0232}$$

$$= \frac{1}{AE}\begin{pmatrix} 1.734 & 0.155 \\ 0.155 & 0.578 \end{pmatrix}\begin{pmatrix} 6 \\ -8 \end{pmatrix}$$

or $$\boldsymbol{u_4{}^0 = \frac{9.16}{AE}} \qquad (20.21)$$

and $$\boldsymbol{v_4{}^0 = -\frac{3.69}{AE}} \qquad (20.22)$$

The forces in the members of the framework can be obtained from the theory of Hookean elasticity, if the axial extension or contraction of each element is known. This can be achieved by resolving equations (20.21) and (20.22) along the local axis of each element, as follows.

Member 1–4

$$u_1 = 0$$

and u_4 can be obtained from (20.12):

$$u_4 = \begin{pmatrix} C & S \end{pmatrix}\begin{pmatrix} u_4{}^0 \\ v_4{}^0 \end{pmatrix}$$

$$= \begin{pmatrix} 0.866 & 0.5 \end{pmatrix}\frac{1}{AE}\begin{pmatrix} 9.16 \\ -3.69 \end{pmatrix}$$

i.e. $$u_4 = \frac{6.088}{AE}$$

From Hooke's law

F_{1-4} = axial force in member 1–4

$$= \frac{AE}{l} \times (u_4 - u_1)$$

$$= \frac{AE}{2} \times \frac{6.088}{AE} = \textbf{3.04 kN (tensile)}$$

Member 2–4

$$u_2 = 0$$

and $$u_4 = \begin{pmatrix} C & S \end{pmatrix} \begin{pmatrix} u_4^0 \\ v_4^0 \end{pmatrix}$$

$$= \begin{pmatrix} 0 & 1 \end{pmatrix} \frac{1}{AE} \begin{pmatrix} 9.16 \\ -3.69 \end{pmatrix}$$

i.e. $$u_4 = -\frac{3.69}{AE}$$

F_{2-4} = axial force in member 2–4

$$= \frac{AE}{1} \times (u_4 - u_2)$$

$$= \frac{AE}{1} \times \frac{-3.69}{AE} = \mathbf{-3.69\ kN\ (compressive)}$$

Member 4–3

$$u_3 = 0$$

and $$u_4 = \begin{pmatrix} C & S \end{pmatrix} \begin{pmatrix} u_4^0 \\ v_4^0 \end{pmatrix}$$

$$= \begin{pmatrix} +0.5 & -0.866 \end{pmatrix} \frac{1}{AE} \begin{pmatrix} 9.16 \\ -3.69 \end{pmatrix}$$

i.e. $$u_4 = +\frac{7.776}{AE}$$

$$F_{4-3} = \frac{AE}{1.155} \times (u_3 - u_4)$$

$$= \frac{AE}{1.155} \times \left(0 - \frac{7.776}{AE}\right)$$

$$= \mathbf{-6.73\ kN\ (compressive)}$$

The method can be applied to numerous other problems, which are beyond the scope of this book, but if the reader requires a greater depth of coverage, he or she should consult References 16 to 20, and 29, page 497.

Now try the following Practice Exercise

Practice Exercise 71. The matrix displacement method

1. Determine the nodal displacements and member forces in the plane pin-jointed truss of Figure 20.10. For all members, AE is constant.

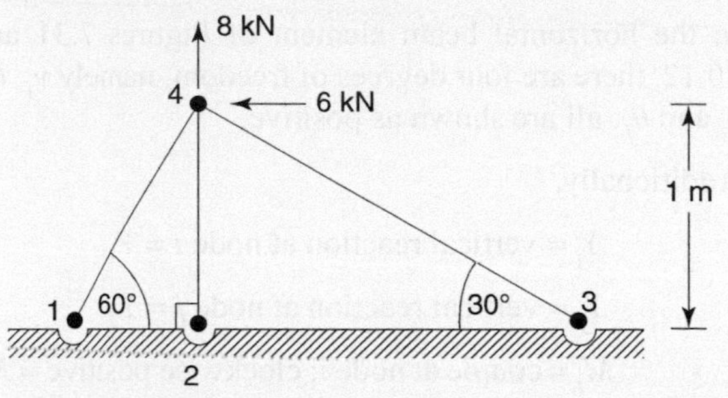

Figure 20.10

[$u_4^0 = -11.63/AE$, $v_4^0 = 5.54/AE$, $F_{1-4} = -0.88$ kN, $F_{2-4} = 5.54$ kN, $F_{3-4} = 6.42$ kN]

2. Determine the nodal displacements and member forces in the plane pin-jointed truss of Figure 20.11. For all members, AE is constant.

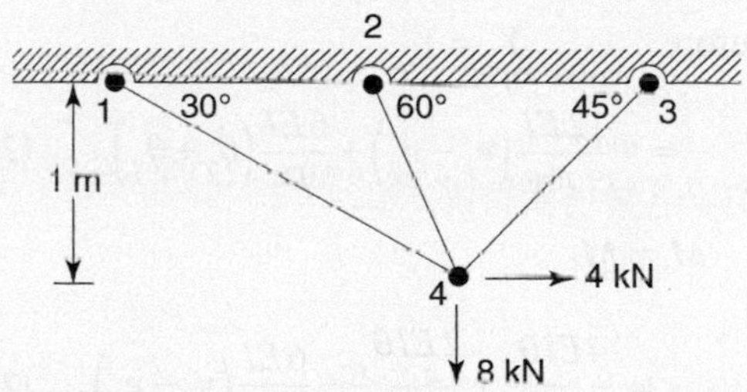

Figure 20.11

[$u_4^0 = 2.584/AE$, $v_4^0 = -6.544/AE$, $F_{1-4} = 2.755$ kN, $F_{2-4} = 6.025$ kN, $F_{3-4} = 1.980$ kN]

20.5 Continuous beams

The elemental stiffness matrix for a beam element can be obtained by considering the beam element of Figure 7.31 (on page 167) and Figure 20.12 below.

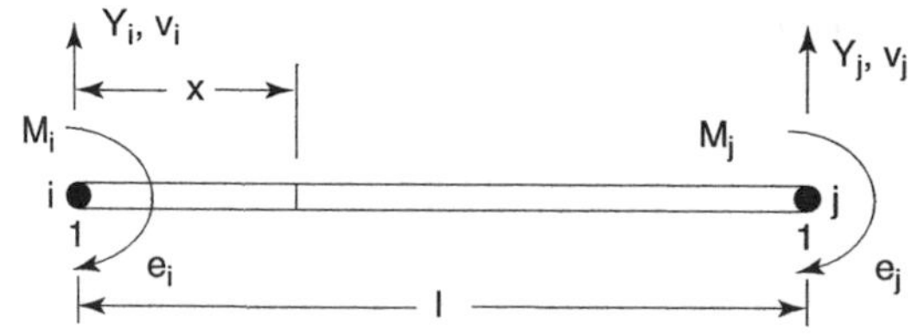

Figure 20.12 Beam element.

In the horizontal beam element of Figures 7.31 and 20.12, there are four degrees of freedom, namely v_i, θ_i, v_j and θ_j; all are shown as positive.

Additionally,

Y_i = vertical reaction at node $i = Y_1$

Y_j = vertical reaction at node $j = Y_2$

M_i = couple at node i; clockwise positive $= M_1$

M_j = couple at node j; clockwise positive $= M_2$

Now, from equation (7.57) on page 167

$$Y_1 = Y_i = \frac{12EI}{l^3}\left(v_i - v_j\right) - \frac{EI}{l^2}\left(6\theta_i + 6\theta_j\right)$$

or
$$Y_i = \frac{12EI}{l^3}\left(v_i - v_j\right) - \frac{6EI}{l^2}\left(\theta_i + \theta_j\right) \quad (20.23)$$

Resolving vertically gives $Y_j = -Y_i = -Y_1 = +Y_2$

Therefore $Y_2 = Y_j$

$$= -\frac{12EI}{l^3}\left(v_i - v_j\right) + \frac{6EI}{l^2}\left(\theta_i + \theta_j\right) \quad (20.24)$$

Also $M_i = M_1$

$$= \frac{4EI\theta_i}{l} + \frac{2EI\theta_j}{l} - \frac{6EI}{l^2}\left(v_i - v_j\right) \quad (20.25)$$

$M_1 = M_j$ and $M_j = M_2$

$$= \frac{2EI\theta_i}{l} + \frac{4EI\theta_j}{l} - \frac{6EI}{l^2}\left(v_i - v_j\right) \quad (20.26)$$

In matrix form, equations (20.23) to (20.26) become:

$$\begin{pmatrix} Y_i \\ M_i \\ Y_j \\ M_j \end{pmatrix} = EI \begin{pmatrix} 12/l^3 & -6/l^2 & -12/l^3 & -6/l^2 \\ -6/l^2 & 4/l & 6/l^2 & 2/l \\ -12/l^3 & 6/l^2 & 12/l^3 & 6/l^2 \\ -6/l^2 & 2/l & 6/l^2 & 4/l \end{pmatrix} \begin{pmatrix} v_i \\ \theta_i \\ v_j \\ \theta_j \end{pmatrix}$$

(20.27)

or $\left(P_i\right) = \left(k\right)\left(u_i\right)$

and $\left(P_i\right)^T$ = vector of elemental forces

$= \left(Y_i, M_i, Y_j, M_j\right)$

(k) = the elemental stiffness matrix of a beam

$$\text{Hence } \left(k\right) = EI \begin{pmatrix} 12/l^3 & -6/l^2 & -12/l^3 & -6/l^2 \\ -6/l^2 & 4/l & 6/l^2 & 2/l \\ -12/l^3 & 6/l^2 & 12/l^3 & 6/l^2 \\ -6/l^2 & 2/l & 6/l^2 & 4/l \end{pmatrix} \begin{matrix} v_i \\ \theta_i \\ v_j \\ \theta_j \end{matrix}$$

(columns: v_i, θ_i, v_j, θ_j)

(20.28)

Problem 2. Determine the nodal displacements and bending moments for the encastré beam of Figure 20.13.

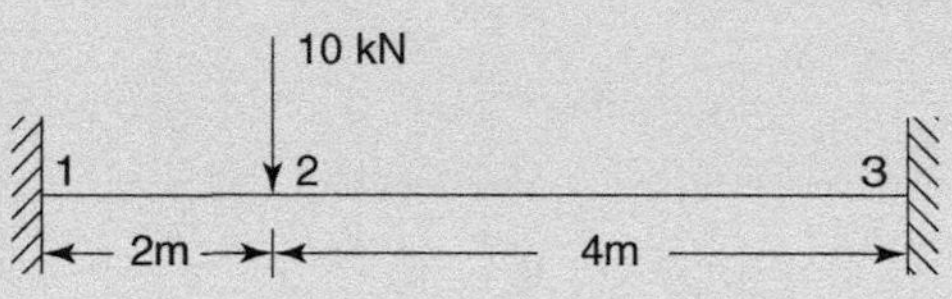

Figure 20.13 Encastré beam.

For beam elements, the elements must be numbered from left to right.

Element 1–2

$$i = 1, j = 2, l = 2 \text{ m}$$

Substituting $l = 2$ into equation (20.28), and removing the columns and rows corresponding to the zero displacements, which in this case are v_1 and θ_1, the following is obtained for the stiffness matrix for element 1–2:

$$\left(k_{1-2}\right) = EI \begin{pmatrix} & & & \\ & & & \\ & & 12/2^3 & 6/2^2 \\ & & 6/2^2 & 4/2 \end{pmatrix} \begin{matrix} v_1 \\ \theta_1 \\ v_2 \\ \theta_2 \end{matrix}$$

(columns: v_1, θ_1, v_2, θ_2)

$$= EI\begin{pmatrix} 1.5 & 1.5 \\ 1.5 & 2 \end{pmatrix}\begin{matrix} v_2 \\ \theta_2 \end{matrix} \qquad (20.29)$$

(columns v_2, θ_2)

Element 2–3

$$i = 2, j = 3, l = 4 \text{ m}$$

Substituting $l = 4$ into equation (20.28), and removing the columns and rows corresponding to the zero displacements, which in this case are v_3 and θ_3, the following is obtained for the stiffness matrix for element 2–3:

$$\left(k_{2-3}\right) = EI\begin{pmatrix} 12/4^3 & -6/4^2 & & \\ -6/4^2 & 4/4 & & \\ & & & \\ & & & \end{pmatrix}\begin{matrix} v_2 \\ \theta_2 \\ v_3 \\ \theta_3 \end{matrix}$$

(columns v_2, θ_2, v_3, θ_3)

$$= EI\begin{pmatrix} 0.1875 & -0.375 \\ -0.375 & 1 \end{pmatrix}\begin{matrix} v_2 \\ \theta_2 \end{matrix} \qquad (20.30)$$

(columns v_2, θ_2)

To obtain that part of the system matrix $\left(K_{11}\right)$ that corresponds to the free displacements v_2 and θ_2, the coefficients of the two elemental stiffness matrices, namely equations (20.29) and (20.30), corresponding to these free displacements, namely v_2 and θ_2, must be added together, as follows:

$$\left(K_{11}\right) = EI\begin{pmatrix} 1.5+0.1875 & 1.5-0.375 \\ 1.5-0.375 & 2+1 \end{pmatrix}\begin{matrix} v_2 \\ \theta_2 \end{matrix}$$

$$= EI\begin{pmatrix} 1.6875 & 1.125 \\ 1.125 & 3 \end{pmatrix}\begin{matrix} v_2 \\ \theta_2 \end{matrix} \qquad (20.31)$$

Now
$$\left(K_{11}\right)^{-1} = \frac{\frac{1}{EI}\begin{pmatrix} 3 & -1.125 \\ -1.125 & 1.6875 \end{pmatrix}}{1.6875\times 3 - 1.125\times 1.125}$$

$$= \frac{\frac{1}{EI}\begin{pmatrix} 3 & -1.125 \\ -1.125 & 1.6875 \end{pmatrix}}{3.7969}$$

i.e.
$$\left(K_{11}\right)^{-1} = \frac{1}{EI}\begin{pmatrix} 0.790 & -0.296 \\ -0.296 & 0.444 \end{pmatrix}$$

and
$$\left(u_F\right) = \begin{pmatrix} v_2 \\ \theta_2 \end{pmatrix} = \left(K_{11}\right)^{-1}\left(q_F\right)$$

where the load vector of loads $\left(q_F\right)$ is obtained by considering the forces and couples in the directions of the free displacements, namely v_2 and θ_2, as follows:

$$\left(q_F\right) = \begin{pmatrix} -10\,\text{kN} \\ 0\,\text{kN m} \end{pmatrix}\begin{matrix} v_2 \\ \theta_2 \end{matrix}$$

Hence
$$\begin{pmatrix} v_2 \\ \theta_2 \end{pmatrix} = \frac{1}{EI}\begin{pmatrix} 0.790 & -0.296 \\ -0.296 & 0.444 \end{pmatrix}\begin{pmatrix} -10 \\ 0 \end{pmatrix}$$

i.e.
$$\begin{pmatrix} v_2 \\ \theta_2 \end{pmatrix} = \frac{1}{EI}\begin{pmatrix} -7.9 \\ 2.96 \end{pmatrix} \qquad (20.32)$$

To determine the nodal moments, the slope-deflection equations, namely equations (20.25) and (20.26), have to be used as follows.

Element 1–2

$$i = 1, j = 2,\ v_1 = \theta_1 = 0,\ v_2 = -\frac{7.9}{EI} \text{ and } \theta_2 = \frac{2.96}{EI}$$

Substituting the above into equation (20.25) gives:

$$M_1 = \frac{4EI\theta_1}{l} + \frac{2EI\theta_2}{l} - \frac{6EI}{l^2}\left(v_1 - v_2\right)$$

$$= \frac{4EI(0)}{2} + \frac{2EI\left(\frac{2.96}{EI}\right)}{2} - \frac{6EI}{2^2}\left(0 - -\frac{7.9}{EI}\right)$$

since $l = 2$ m

$$= 2.96 - \frac{6}{4}(7.9)$$

i.e. $\mathbf{M_1 = -8.890}$ **kN m**

Substituting the above into equation (20.26) gives:

$$M_2 = \frac{2EI\theta_1}{l} + \frac{4EI\theta_2}{l} - \frac{6EI}{l^2}(v_1 - v_2)$$

$$= \frac{2EI(0)}{2} + \frac{4EI\left(\frac{2.96}{EI}\right)}{2} - \frac{6EI}{2^2}\left(0 - -\frac{7.9}{EI}\right)$$

$$= 2(2.96) - \frac{6}{4}(7.9)$$

i.e. $\mathbf{M_2 = -5.930}$ **kN m**

Element 2–3

$$i = 2, j = 3,\ v_2 = -\frac{7.9}{EI},\ \theta_2 = \frac{2.96}{EI} \text{ and } v_3 = \theta_3 = 0$$

Substituting the above into equation (20.25) gives:

$$M_2 = \frac{4EI\theta_2}{l} + \frac{2EI\theta_3}{l} - \frac{6EI}{l^2}(v_2 - v_3)$$

$$= \frac{4EI\left(\frac{2.96}{EI}\right)}{4} + \frac{2EI(0)}{4} - \frac{6EI}{4^2}\left(-\frac{7.9}{EI} - 0\right)$$

since $l = 4$ m

$$= 2.96 + \frac{6}{16}(7.9)$$

i.e. $\mathbf{M_2 = 5.923}$ **kN m**

Substituting the above displacements into the slope-deflection equation (20.26) gives:

$$M_3 = \frac{2EI\theta_2}{l} + \frac{4EI\theta_3}{l} - \frac{6EI}{l^2}(v_2 - v_3)$$

$$= \frac{2EI\left(\frac{2.96}{EI}\right)}{4} + \frac{4EI(0)}{4} - \frac{6EI}{4^2}\left(-\frac{7.9}{EI} - 0\right)$$

since $l = 4$ m

$$= \frac{1}{2}(2.96) + \frac{6}{16}(7.9)$$

i.e. $\mathbf{M_3 = 4.443}$ **kN m**

NB The slight differences between the magnitudes for the two values for M_2 were due to rounding errors.

Problem 3. Determine the nodal displacements and bending moments for the continuous beam of Figure 20.14, which is subjected to a downward uniformly distributed load.

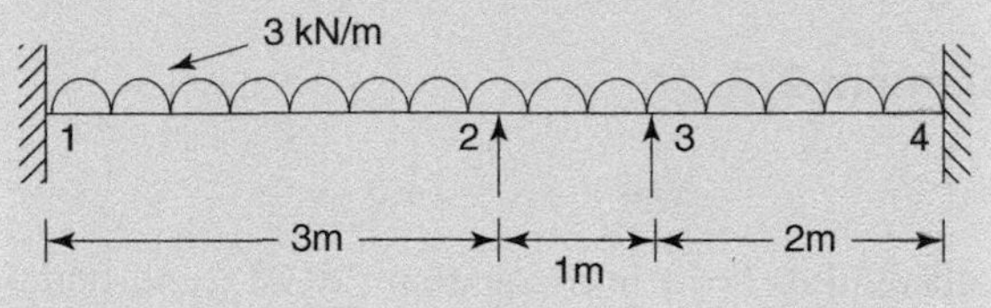

Figure 20.14 Continuous beam.

The main difficulty of analysing the continuous beam of Figure 20.14 is in obtaining the equivalent nodal forces for the distributed load; this is necessary as the finite element method can only cope with nodal forces. To deal with distributed loads on beams and frames, the following process, which is based on superposition, is used.

(a) Fix the beam or frame at all the nodes of joints, and calculate the end fixing 'forces' to achieve this.
(b) The beam or frame in condition (a) is not in equilibrium, hence, to keep it in equilibrium, subject the beam or frame to the negative resultants of the end fixing 'forces' to the appropriate node.
(c) Use the negative resultants of the end fixing 'forces' as the external load factor, and calculate the nodal displacements and bending moments due to this vector.
(d) Now as the beam or frame was fixed at its nodes in condition (a), it will be necessary to superimpose the end fixing moments with the nodal moments calculated in condition (c) to give the final values of the nodal moments.

To calculate (q_F), the beam will be fixed at all nodes from 1 to 4, as follows:

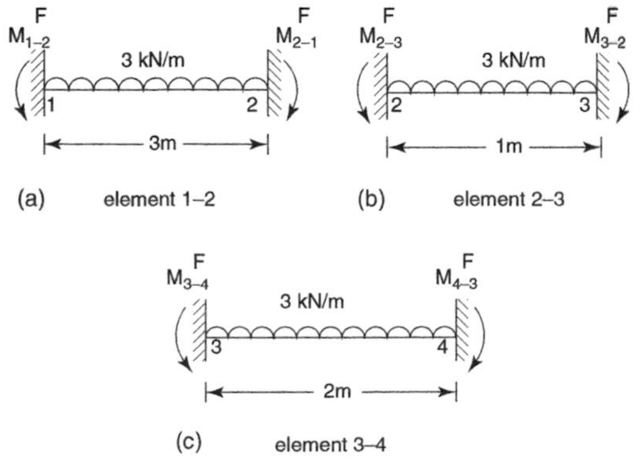

Figure 20.15 Beam elements with end fixing moments.

From Figure 20.15(a) $$M^F_{1-2} = -\frac{wl^2}{12} = -\frac{3\times 3^2}{12}$$

$$= -2.25 \text{ kN m}$$

and $$M^F_{2-1} = \frac{wl^2}{12} = \frac{3\times 3^2}{12} = 2.25 \text{ kN m}$$

From Figure 20.15(b) $$M^F_{2-3} = -\frac{wl^2}{12} = -\frac{3\times 1^2}{12}$$

$$= -0.25 \text{ kN m}$$

and $$M^F_{3-2} = \frac{wl^2}{12} = \frac{3\times 1^2}{12} = 0.25 \text{ kN m}$$

From Figure 20.15(c) $$M^F_{3-4} = -\frac{wl^2}{12} = -\frac{3\times 2^2}{12}$$

$$= -1.0 \text{ kN m}$$

and $$M^F_{4-3} = \frac{wl^2}{12} = \frac{3\times 2^2}{12} = 1.0 \text{ kN m}$$

Now $v_1 = \theta_1 = v_2 = v_3 = v_4 = \theta_4 = 0$, hence, the only unbalanced moments that need to be considered are in the directions θ_2 and θ_3 as shown in Figure 20.16(a); the negative resultants, corresponding to the free displacements θ_2 and θ_3 are shown in Figure 20.16(b).

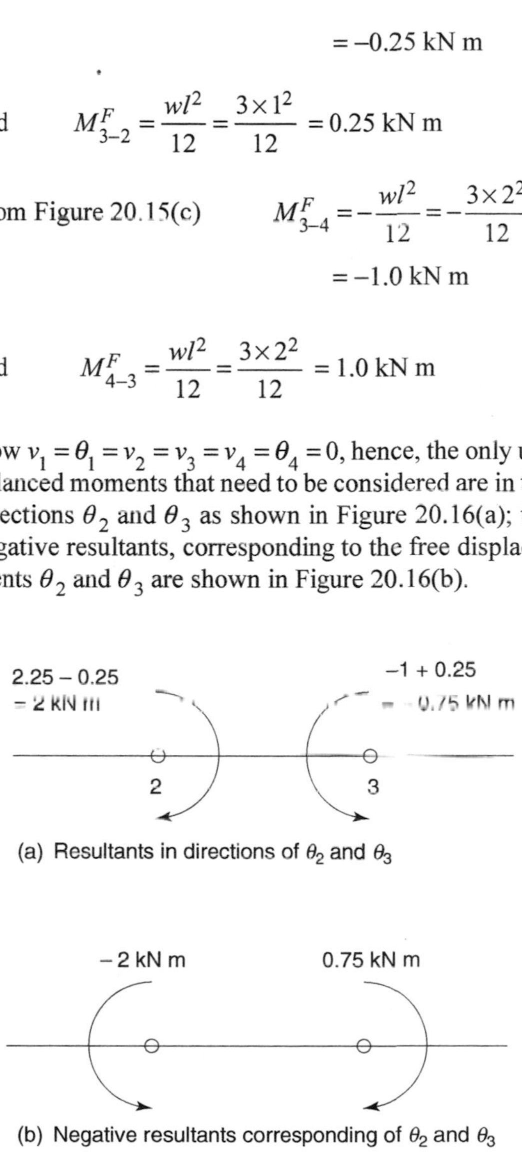

Figure 20.16 Resultants and negative resultants.

From Figure 20.16(b) $$(q_F) = \begin{pmatrix} -2 \\ 0.75 \end{pmatrix} \begin{matrix} \theta_2 \\ \theta_3 \end{matrix} \qquad (20.33)$$

Element 1–2

$$i = 1, j = 2, l = 3 \text{ m}$$

Substituting l = 3 into equation (20.28), and removing the columns and rows corresponding to the zero displacements, which in this case are v_1, θ_1 and v_2, the following is obtained for the stiffness matrix for element 1–2:

$$\begin{matrix} & v_1 & \theta_1 & v_2 & \theta_2 \end{matrix}$$

$$(k_{1-2}) = EI \begin{pmatrix} & & & \\ & & & \\ & & & \\ & & & 4/3 \end{pmatrix} \begin{matrix} v_1 \\ \theta_1 \\ v_2 \\ \theta_2 \end{matrix} \qquad (20.34)$$

$$= EI\,(1.333)\,\theta_2$$

Element 2–3

$$i = 2, j = 3, l = 1 \text{ m}$$

Substituting l = 1 into equation (20.28), and removing the columns and rows corresponding to the zero displacements, which in this case are v_2 and v_3, the following is obtained for the stiffness matrix for element 2–3:

$$\begin{matrix} & v_2 & \theta_2 & v_3 & \theta_3 \end{matrix}$$

$$(k_{2-3}) = EI \begin{pmatrix} & & & \\ & 4/1 & & 2/1 \\ & & & \\ & 2/1 & & 4/1 \end{pmatrix} \begin{matrix} v_2 \\ \theta_2 \\ v_3 \\ \theta_3 \end{matrix} \qquad (20.35)$$

Element 3–4

$$i = 3, j = 4, l = 2 \text{ m}$$

Substituting $l = 2$ into equation (20.28), and removing the columns and rows corresponding to the zero displacements, which in this case are v_3, v_4 and θ_4, the following is obtained for the stiffness matrix for element 3–4:

$$\begin{matrix} & v_3 & \theta_3 & v_4 & \theta_4 \end{matrix}$$

$$(k_{3-4}) = EI \begin{pmatrix} & & & \\ & 4/2 & & \\ & & & \\ & & & \end{pmatrix} \begin{matrix} v_3 \\ \theta_3 \\ v_4 \\ \theta_4 \end{matrix} \qquad (20.36)$$

To obtain that part of the structural stiffness matrix, (K_{11}), that corresponds to the free displacements, namely θ_2 and θ_3, the appropriate coefficients of the elemental stiffness matrices of equations (20.34) to (20.36) must be added together, as follows:

$$\begin{matrix} & \theta_2 & \theta_3 & \\ (K_{11}) = EI & \begin{pmatrix} 1.333+4 & 2 \\ 2 & 4+2 \end{pmatrix} & & \begin{matrix}\theta_2 \\ \theta_3\end{matrix} \end{matrix} \quad (20.37)$$

i.e.

$$\begin{matrix} & \theta_2 & \theta_3 & \\ (K_{11}) = EI & \begin{pmatrix} 5.333 & 2 \\ 2 & 6 \end{pmatrix} & & \begin{matrix}\theta_2 \\ \theta_3\end{matrix} \end{matrix} \quad (20.38)$$

Now $(K_{11})^{-1} = \dfrac{\dfrac{1}{EI}\begin{pmatrix} 6 & -2 \\ -2 & 5.333 \end{pmatrix}}{5.333\times 6 - 2\times 2} = \dfrac{\dfrac{1}{EI}\begin{pmatrix} 6 & -2 \\ -2 & 5.333 \end{pmatrix}}{28}$

i.e. $(K_{11})^{-1} = \dfrac{1}{EI}\begin{pmatrix} 0.214 & -0.0714 \\ -0.0714 & 0.190 \end{pmatrix}$ (20.39)

Now $(u_F) = \begin{pmatrix} \theta_2 \\ \theta_3 \end{pmatrix} = (K_{11})^{-1}(q_F)$ (20.40)

Hence, from equations (20.33) and (20.39)

$$\begin{pmatrix} \theta_2 \\ \theta_3 \end{pmatrix} = \frac{1}{EI}\begin{pmatrix} 0.214 & -0.0714 \\ -0.0714 & 0.190 \end{pmatrix}\begin{pmatrix} -2 \\ 0.75 \end{pmatrix}$$

$$= \frac{1}{EI}\begin{pmatrix} -0.482 \\ 0.285 \end{pmatrix} \quad (20.41)$$

To determine the nodal moments, the nodal displacements will be substituted into equations (20.25) and (20.26), and then the end fixing moments will be added, as follows.

Element 1–2

$$i = 1, j = 2,\ v_1 = \theta_1 = v_2 = 0 \text{ and } \theta_2 = -\frac{0.482}{EI}$$

Substituting the above data into the slope-deflection equation (20.25) gives:

$$M_1 = \frac{6EI}{9}\left[(0-0) + \left(0 - \frac{0.482}{EI}\right)\frac{3}{3}\right] - 2.25$$

i.e. $\mathbf{M_1 = -2.571\ kN\ m}$

Substituting the above into equation (20.26) gives:

$$M_2 = \frac{6EI}{9}\left[(0-0) + \left(0 - 2\times\frac{0.482}{EI}\right)\frac{3}{3}\right] + 2.25$$

i.e. $\mathbf{M_2 = 1.607\ kN\ m}$

Element 2–3

$$i = 2, j = 3,\ v_2 = v_3 = 0,\ \theta_2 = -\frac{0.482}{EI}$$

$$\text{and } \theta_3 = -\frac{0.286}{EI}$$

Substituting the above into equation (20.25) gives:

$$M_2 = \frac{6EI}{1}\left[(0-0) + \left(-2\times\frac{0.482}{EI} + \frac{0.286}{EI}\right)\frac{1}{3}\right] - 0.25$$

i.e. $\mathbf{M_2 = -1.606\ kN\ m}$

Substituting the above displacements into the slope-deflection equation (20.26) gives:

$$M_3 = \frac{6EI}{1}\left[(0-0) + \left(\frac{-0.482}{EI} + 2\times\frac{0.286}{EI}\right)\frac{1}{3}\right] + 0.25$$

i.e. $\mathbf{M_3 = 0.430\ kN\ m}$

Element 3–4

$$i = 3, j = 4,\ v_3 = v_4 = \theta_4 = 0, \text{ and } \theta_3 = \frac{0.286}{EI}$$

Substituting the above into equation (20.25) gives:

$$M_3 = \frac{6EI}{4}\left[(0-0) + \left(2\times\frac{0.286}{EI} + 0\right)\frac{2}{3}\right] - 1.0$$

i.e. $\mathbf{M_3 = -0.428\ kN\ m}$

Substituting the above displacements into the slope-deflection equation (20.26) gives:

$$M_4 = \frac{6EI}{4}\left[(0-0) + \left(\frac{0.286}{EI} + 0\right)\frac{2}{3}\right] + 1.0$$

i.e. $\mathbf{M_4 = 1.286\ kN\ m}$

NB On Carl Ross's YouTube sites there are available a number of computer programs, which can analyse beams, 2D and 3D trusses and frames, on a number of devices, including SmartPhones, iPads, laptops, and so on. The authors believe that the impact of the

SmartPhone on humankind will be greater than was the efforts of Elvis Presley, the Beatles and Bill Gates! For Ross' Finite Element Computer Programs, which can be used on SmartPhones/iPads/ laptops, etc., visit: carltfross YouTube computer programs

Now try the following Practice Exercise

Practice Exercise 72. The matrix displacement method for continuous beams

1. Using the matrix displacement method, determine the nodal displacements and bending moments for the beam of Figure 20.17.

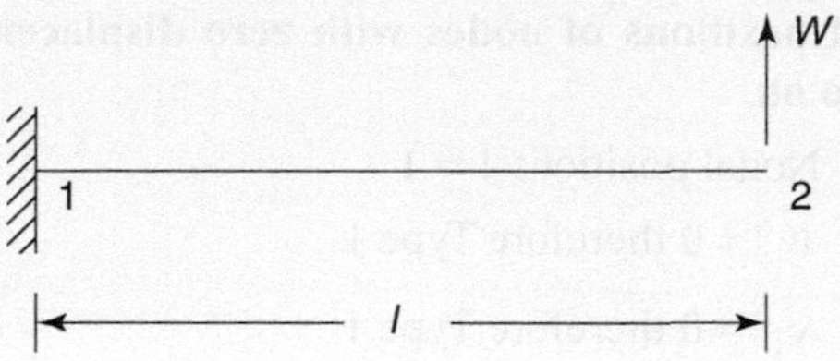

Figure 20.17

$$[v_1 = \theta_1 = 0,\ v_2 = \frac{Wl^3}{3EI},\ \theta_2 = -\frac{Wl^2}{2EI},\ M_1 = Wl]$$

2. Using the matrix displacement method, determine the nodal displacements and bending moments for the beam of Figure 20.18.

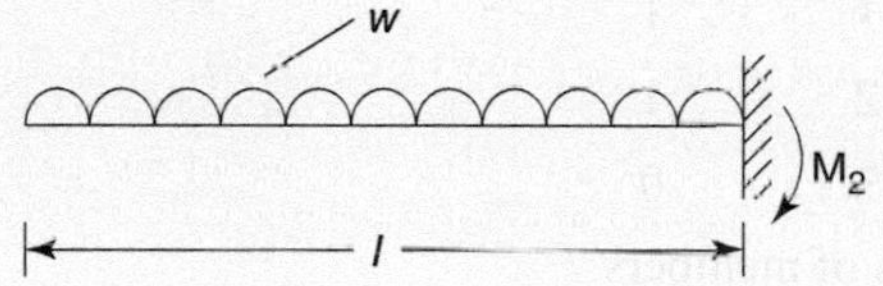

Figure 20.18

$$[v_1 = -\frac{wl^4}{8EI},\ \theta_1 = -\frac{wl^3}{6EI},$$

$$v_2 = \theta_2 = 0,\ M_2 = \frac{wl^2}{2}]$$

3. Using the matrix displacement method, determine the nodal displacements and bending moments for the continuous beam of Figure 20.19.

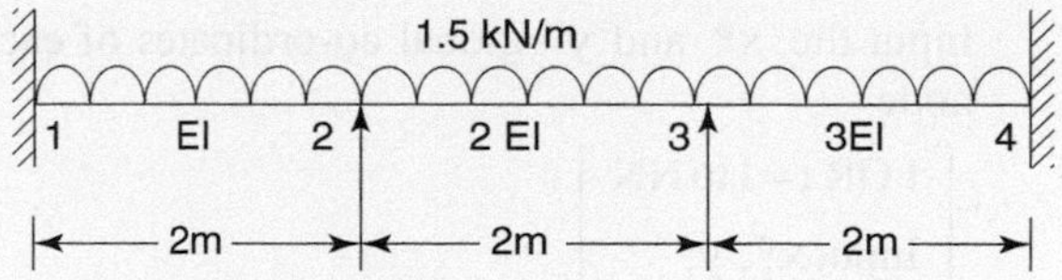

Figure 20.19

$$[v_1 = \theta_1 = v_2 = v_3 = v_4 = \theta_4 = 0,\ \theta_2 = 0,\ \theta_3 = 0,$$
$$M_1 = -0.5 \text{ kN m},\ M_2 = \pm 0.5 \text{ kN m},$$
$$M_3 = \pm 0.5 \text{ kN m},\ M_4 = 0.5 \text{ kN m}]$$

20.6 Analysis of pin-jointed trusses on SmartPhones, tablets and Microsoft computers

Computer programs, namely, TRUSS.EXE and TRUSSF.EXE are available for the analysis of plane pin-jointed trusses. The programs are given away free on Professor Ross's website. Some of the programs can be run on Android SmartPhones, Android tablets, laptops and windows-based computers and laptops. The program TRUSS.EXE is interactive and the program TRUSSF.EXE requires an input file.

(1) Input data for TRUSS.EXE and TRUSSF.EXE

The input data file should be created in the following sequence:

(i) Number the nodal points or pin-joints – NN

(ii) Number of one-dimensional structural members = ES

(iii) Number of nodes that have zero displacements = NF

(iv) **Details of zero displacements**

Input the nodal position of each 'suppressed' node and whether or not the appropriate displacements are zero at these nodes

FOR i = 1 to NF Input NS_i Type 1 if the u° displacement is zero at this node or 0 if it is not after this, type 1 if the v° displacement is zero at this node, or 0 if it is not	where NS_i = a node with one or more zero displacements

(v) **Nodal co-ordinates**
Input the x^o and y^o global co-ordinates of each node

$$\begin{bmatrix} \text{FOR } i = 1 \text{ to NN} \\ \text{Input } x_i^o, y_i^o \end{bmatrix}$$

(vi) **Member details**

$$\begin{bmatrix} \text{FOR EL} = 1 \text{ to ES} \\ \text{Input i} - \text{Input 'i' node for the member} \\ \text{Input j} - \text{Input 'j' node for the member} \\ \text{Input A} - \text{Input cross-sectional area} \\ \qquad \text{for the member} \\ \text{Input E} - \text{Input Young's modulus of} \\ \qquad \text{elasticity for the member} \end{bmatrix}$$

(vii) **Details of nodal forces**
NC = number of nodes with nodal forces

$$\begin{bmatrix} \text{FOR } i = 1 \text{ to NC} \\ \text{Input node i} \\ \text{Input loads in } x^o \text{ and } y^o \text{ directions,} \\ \text{respectively, at node 'i'} \end{bmatrix}$$

A typical **input data file** is shown in section (3) below; this must be created prior to running the program.

(2) Results

The results are output onto a data file which **must not** have the same name as the input data file or the name of any other file that you do not wish to overwrite.
The results are output in the following sequence:
(i) **Vector of nodal displacements**

$$\begin{bmatrix} \text{FOR } i = 1 \text{ to NN} \\ \text{PRINT } u_i^o, v_i^o \end{bmatrix}$$

(ii) **Nodal forces in each member**

$$\begin{bmatrix} \text{FOR } i = 1 \text{ to ES} \\ \text{PRINT nodes defining each member and} \\ \text{the forces in each member (tensile +ve)} \end{bmatrix}$$

(iii) Reactions in the u^o and v^o directions, respectively, at each of the 'suppressed' nodes

Problem 4. Determine the forces in the plane pin-jointed truss shown in Figure 20.20. All members are of constant '*E*', and all members are of constant '*A*', except for member 1–3, which has a cross-sectional area of '2*A*'.

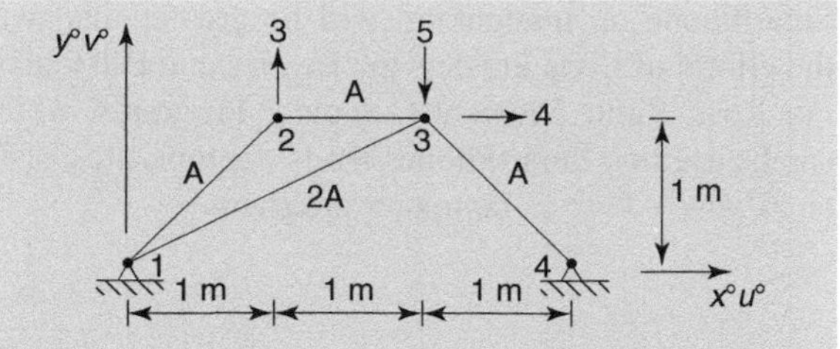

Figure 20.20

As only the forces are required, it will be convenient to assume that $A = E = 1$. If, however, the nodal displacements are required, then it will be necessary to feed in the true values of A and E.

Number of nodes = 4

Number of members = 4

Number of nodes with zero displacement = 2

Nodal positions of nodes with zero displacements, and so on.

Nodal positions 1 = 1

$u_i^o = 0$ therefore Type 1

$v_i^o = 0$ therefore Type 1

Nodal position 2 = 4

$u_4^o = 0$ therefore Type 1

$v_4^o = 0$ therefore Type 1

Nodal co-ordinates

x_i^o	y_i^o
0	0
1	1
2	1
3	0

Details of members

i	j	A	E
1	2	1	1
2	3	1	1
1	3	2	1
3	4	1	1

Nodal forces

NC = 2

Nodal positions 1 = 2 loads = 0 and 3

Nodal position 2 = 3 loads = 4 and – 5

(3) Input data file for TRUSSF.EXE

A typical input data file for this program, namely, TRUSS.DAT, is shown below. This data file was used to solve Problem 4 above.

```
TRUSS.DAT
4
4
2
1,1,1
4,1,1
0,0
1,1
2,1
3,0
1,2,1,1
2,3,1,1
1,3,2,1
3,4,1,1
2
2,0,3
3,4,–5
EOF
```

NB It should be noted that instead of using commas to separate the various items of data, two or more spaces could have been used. Do not use the TAB key to insert spaces! Also, TRUSS.EXE is an interactive computer program, but TRUSSF.EXE requires an input and an output file.

(4) Output file for TRUSSF.EXE

The results for Problem 4 above are given below.

Displacements x AE

Node	u^0	v^0
1	0	0
2	–2.028	10.513
3	0.972	–9.393
4	0	0

Forces in each member

Member	Forces	
1	$F_{1-2} = 4.243$	in tension
2	$F_{2-3} = 3.000$	in tension
3	$F_{1-3} = -2.981$	in compression
4	$F_{3-4} = -5.185$	in compression

The subscripts i–j represent the nodes defining the member, and the negative sign denotes compression.

Reactions

$H_{x1} = 0.333$ $V_{y1} = -1.667$ (Node 1)

See Figure 20.21

$H_{x3} = -3.667$ $V_{y3} = 3.667$ (Node 4)

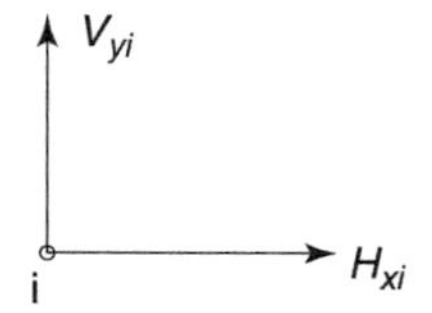

Figure 20.21

(5) Screen dump

A screen dump of the truss and its deflected form is shown in Figure 20.22.

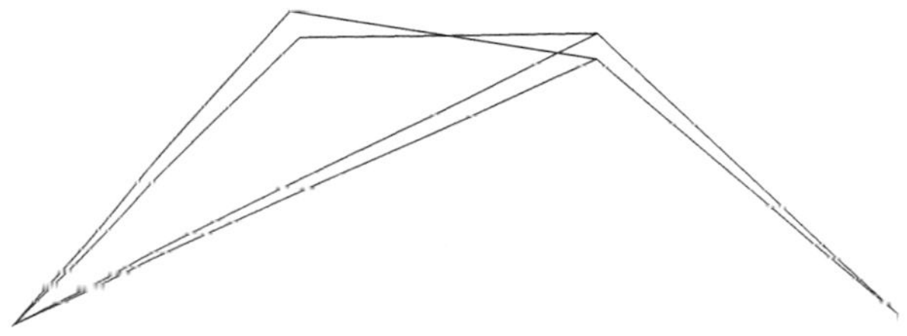

Figure 20.22 Deflected form of truss

20.7 Analysis of continuous beams on SmartPhones, tablets and Microsoft computers

Free computer programs are available on Professor Ross's website for continuous beams. These programs are called BEAMS.EXE, which is an interactive program, and BEAMSF.EXE, which is a program which needs an input file.

The type of beams which can be analysed by these programs on Android SmartPhones and tablets, and Microsoft computers, are shown in Figure 20.23.

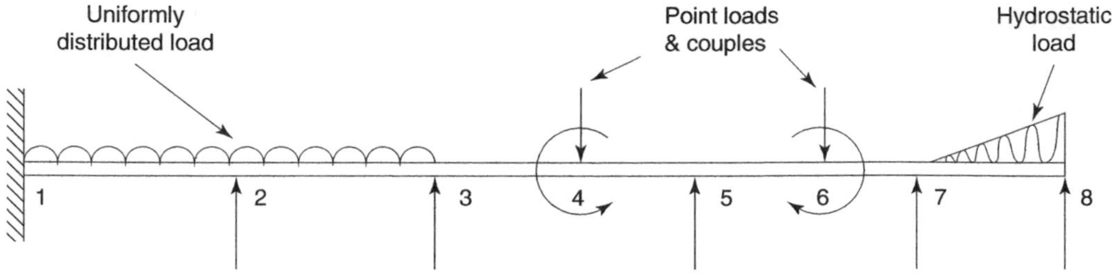

Figure 20.23 Continuous beam under lateral bending

Both programs calculate the nodal displacements, namely v and θ, and then nodal bending moments and 'reactions', but BEAMS.EXE is interactive, and BEAMSF.EXE requires an input and an output file.

The beam element has two degrees of freedom at each node, namely v and θ, as shown by Figure 20.24.

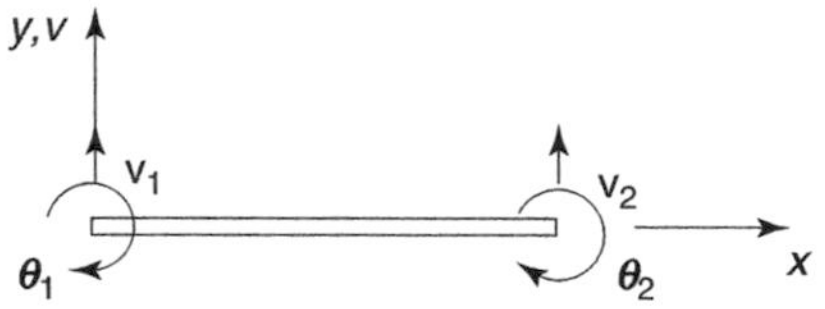

Figure 20.24 Beam element

It must be ensured that the beam lies horizontally, and that the **nodes are numbered in ascending order from the left of the beam to its right.**

(1) Data for BEAMS.EXE and BEAMSF.EXE

The input data file should be created in the following sequence:

(i) Number of elements = LS

(ii) Number of nodes with zero displacements = NF

(iii) Number of nodes with concentrated loads and couples = NC

(iv) **Nodal positions, and so on of zero displacements**

FOR i = 1 to NF

Input NS_i (nodal position of suppressed node and whether or not the displacements are zero at this node)

Type 1 if $v = 0$ at this node, and 0 if it is not zero; immediately after this, type 1 if $\theta = 0$ at this node, and 0 if it is not zero

(v) Elastic modulus = E

(vi) If there is any hydrostatic load feed 1, otherwise feed 0 (if NL = 1 or NL = 0, respectively)

(vii) **Nodal positions and value of concentrated loads and couples**

NC = number of nodes with concentrated loads and couples

FOR i = 1 to NC

Input 'nodal' position of load and couple

Input value of load and couple at this node

(viii) If NL = 1, ignore (viii)

Member details (no hydrostatic load)

FOR I = 1 to LS

(Type in the details of the members/elements from left to right)

Input SA – Input 2nd moment of area for member

Input XL – Input elemental length

Input UD – Input value of distributed load/length (+ve if in 'y' direction)

(ix) If NL = 0, ignore (ix)

Member details (hydrostatic load)

FOR I = 1 to LS

(Type in the details of the members/elements from left to right)

Input SA – Input 2nd moment of area for member (I)

Input XL – Input elemental length (l)

Input WA – Value of distributed load on left

Input WB – Value of distributed load on right

Input AS – The distance of WA from the left node. See Figure 20.25

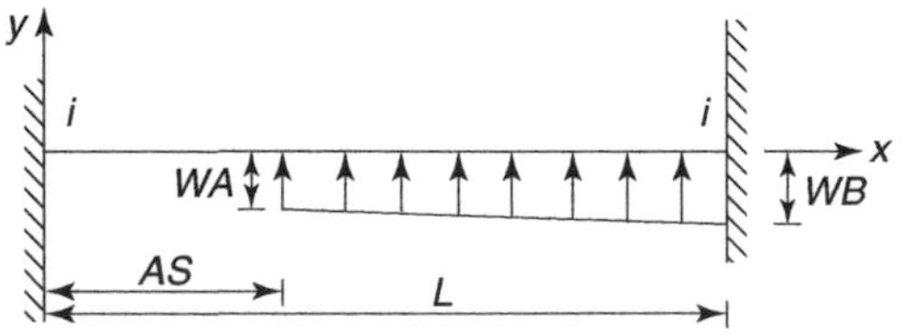

Figure 20.25 Element in Local Co-ordinates

(2) Output

The nodal displacements v_1, θ_1, v_2 … v_n, θ_n where $n = \text{LS} + 1$

The total bending moments

Element	Nodes defining each element		Nodal bending moments
1	i	j	BM_i
	i	j	BM_j
2	j	k	BM_j
	j	k	
	k		BM_k
	k		BM_k
LS			

Problem 5. Determine the nodal bending moments for the beam shown in Figure 20.26. The value of the second moment of area for element 2–3 is twice the value for element 1–2.

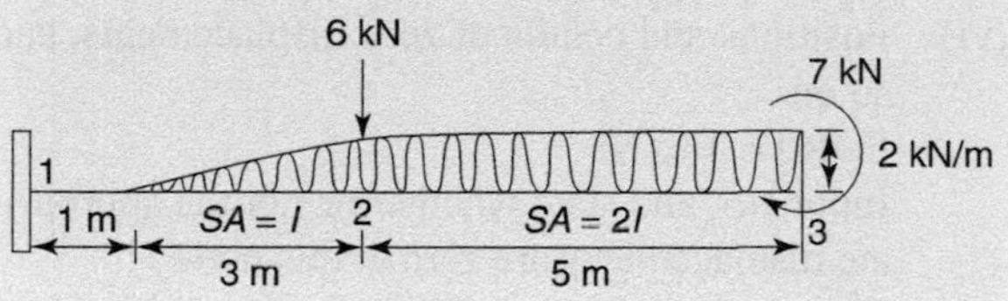

Figure 20.26

As only the moments are required, it will be convenient to assume that $I = E = 1$. If, however, the displacements are required, it will be necessary to feed in the true values for 'E' and 'I'.

The data is as follows:

LS = 2
NF = 2
NC = 2

Details of zero displacements

Nodal position 1 = 1
$v_1 = 0$ therefore, Type 1; $\theta_1 = 0$, therefore, Type 1
Nodal position 2 = 3
$v_3 = 0$ therefore, Type 1; $\theta_3 \neq 0$, therefore, Type 0
E = 1
NL = 1

Nodal positions and values of concentrated loads and couples

Nodal position 1 = 2
Value of load = –6
Value of couple = 0
Nodal position 2 = 3
Value of load = 0
Value of couple = 7 (clockwise +ve)

Member details (hydrostatic load data for this example)

SA	XL	WA	WB	AS
1	4	0	–2	1
2	5	–2	–2	0

A typical input data file for BEAMSF is shown below. The file was used to solve Problem 4 on page 428.

```
Beams.DAT
2
2
2
1,1,1
3,1,0
1
1
2,–6,0
3,0,7
1,4
0,–2,1
2,5
–2,–2,0
EOF
```

The various items of data in this input file can be separated by two spaces instead of commas. Do not use the TAB key for spaces.

(3) Results

The results are given below:

Nodal displacements × EI

Node	v	θ
1	0	0
2	–55.68	4.096
3	0	–16.98

Nodal bending moments (kN m) and reactions (kN)

Node	Moments	Reactions
1	19.45	9.494
2 (left)	–15.53	0
2 (right)	–15.53	0
3	7.00	–16.98

Hogging bending moments are positive.

(4) Screen dumps

Screen dumps of the bending moment and shearing force diagrams for Problem 4 on page 428 are shown in Figure 20.27.

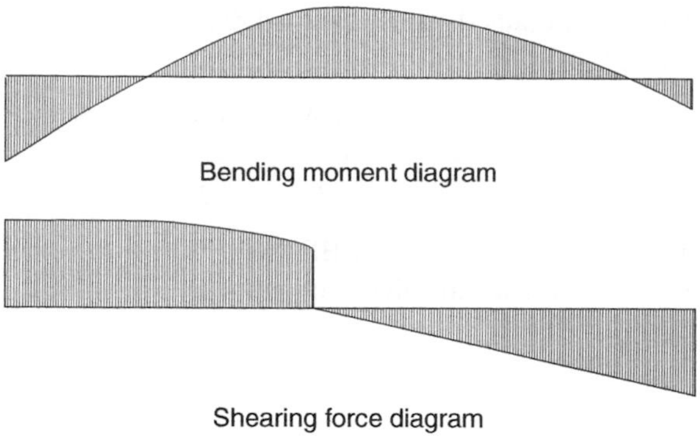

Figure 20.27 Screendumps of the BM & SF diagrams for Problem 4

20.8 Analysis of rigid-jointed plane frames on SmartPhones, tablets and Microsoft computers

Free computer programs, namely FRAME2D.EXE and FRAME2DF.EXE, are available for analysis of regular and skew rigid-jointed plane frames, including the effects of sidesway, on Professor Ross's website. The program FRAME2D.EXE is interactive, and the program FRAME2DF.EXE requires an input and an output file.

For both programs, each element has two end nodes, with three degrees of freedom per node, namely, u^o, v^o and θ as shown in Figure 20.28.

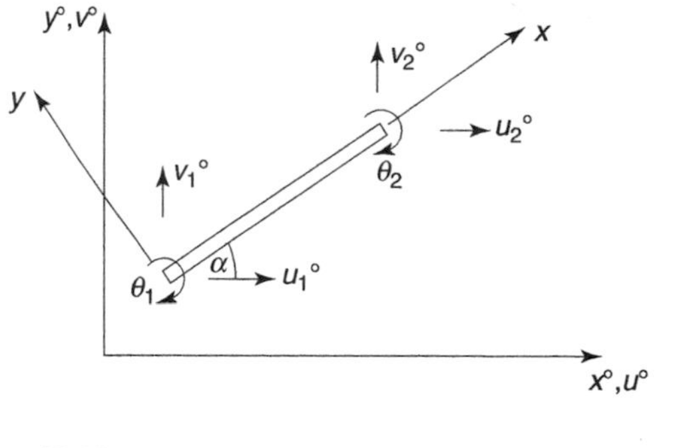

Figure 20.28

Displacement positions are calculated as follows:

Displacement position corresponding to $u_1^o = 3 \times \mathrm{i} - 2$

Displacement position corresponding to $v_1^o = 3 \times \mathrm{i} - 1$

Displacement position corresponding to $\theta_1 = 3 \times \mathrm{i}$

As there are six degrees of freedom per element, the elemental stiffness matrix $\left(k^o\right)$ will be of order 6 × 6. This elemental stiffness matrix consists of an axial component $\left(k_r^o\right)$ and a flexural component $\left(k_b^o\right)$, as shown by equation (20.42).

$$\left(k^o\right) = \left(k_b^o\right) + \left(k_r^o\right) \tag{20.42}$$

(1) Data for FRAME2D.EXE and FRAME2DF.EXE

The input file for FRAME2DF should be created in the following sequence:

(i) Number of nodal points = NJ
(ii) Number of elements = MS
(iii) Number of nodes with zero displacements = NF
(iv) Number of nodes with concentrated loads and couples = NC
(v) Nodal co-ordinates
FOR i = 1 to NJ
Input x_1^o, y_1^o
(vi) Positions and details of zero displacements, and so on.
FOR i = 1 to NF
Input NS_i and state whether or not the appropriate displacements are zero at this node.
Type 1 if $u^o = 0$, or 0 if it is not, then
Type 1 if $v^o = 0$, or 0 if it is not, then
Type 1 in $\theta = 0$, or 0 if it is not for node NS_i
(vii) Elastic modulus = E
(viii) If there is any hydrostatic load, type 1, otherwise type 0, (i.e. HY = 1 or HY = 0, respectively)
(ix) If HY = 1, ignore (ix)
FOR I = 1 to MS
Input i – input 'i' node of element
Input j – input 'j' node of element
Input SA – input 2nd moment of area of element
Input A – input cross-sectional area of element
Input UD – input of uniformly distributed load/length (+ve if in 'y' direction)
(x) If HY = 0, ignore (x)
FOR I = 1 to MS
Input i – input 'i' node of element
Input j – input 'j' node of element
Input SA – input 2nd moment of area of element
Input A – input cross-sectional area of element
Input WA – input value of distributed load on 'left'
Input WB – input value of distributed load on 'right'
Input AS – distance of WA from the left node. See Figure 20.25
(xi) Nodal positions and value of concentrated loads and couples

FOR i = 1 to NC
Input nodal position of load

$$\left.\begin{array}{l}\text{Value of load in } x^o \text{ direction}\\ \text{Value of load in } y^o \text{ direction}\\ \text{Value of couple (clockwise +ve)}\end{array}\right\}\text{ at this node}$$

(2) Output

The nodal displacements $u_1, v_1, \theta_1, u_2^o, v_2^o \ldots u_{NJ}^o, v_{NJ}^o, \theta_{NJ}$

Nodal moments and axial forces

Element	Nodes defining element	Axial force	Moment
1	i–j	F_1	$M_{i\text{-}j}$
	i–j	F_1	$M_{j\text{-}i}$
2	k–	F_2	$M_{j\text{-}k}$
	k–		
MS			

Reaction in x° and y° directions at the suppressed nodes.

Problem 6. Determine the bending moments for the rigid-jointed skew frame shown in Figure 20.29.

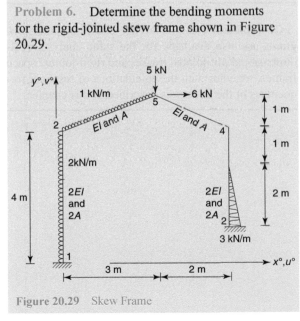

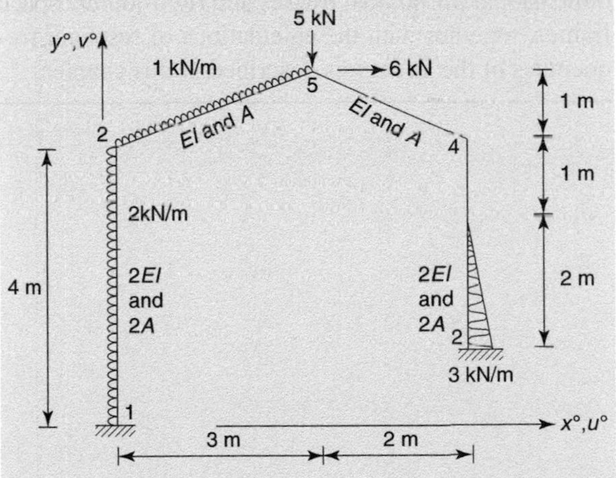

Figure 20.29 Skew Frame

As only the moments are required, it will not be necessary to feed in the true values of *A*, *I* and *E*, where '*I*' is the second moment of area for the inclined members.

As element 2–4 has a hydrostatic load, the variable HY must be made equal to one.

The data file should be created in the following sequence:
NJ = 5
MS = 4
NF = 2
NC = 1

Nodal co-ordinates

x_i^o	y_i^o
0	0
5	1
0	4
5	4
3	5

Nodal details of zero displacements

Nodal position 1 = 1

$u_1^o = 0$, therefore, Type 1;

$v_1^o = 0$, therefore, Type 1;

$\theta_1 = 0$, therefore, Type 1

Nodal position 2 = 2

$u_2^o = 0$, therefore, Type 1;

$v_2^o = 0$, therefore, Type 1;

$\theta_2 = 0$, therefore, Type 1

E = 1

HY = 1

Member details

i	j	SA	A	WA	WB	AS
1	3	2	2000	–2	–2	0
3	5	1	1000	–1	–1	0
5	4	1	1000	0	0	0
4	2	2	2000	0	–3	1

Nodal positions and values of concentrated loads

Nodal position 1 = 5

Load in x° direction = 6

Load in y° direction = –5

Value of couple = 0

(3) Output

The nodal displacements u_1^o, v_1^o, θ_1, u_2^o, v_2^o, θ_2, u_3^o, v_3^o, θ_3, u_4^o, v_4^o, θ_4, u_5^o, v_5^o and θ_5

Moments and forces

Member	Nodes defining member	Axial force	Nodal Moments (kN m)
1	1–3	−0.87	$M_{1-3} = 11.60$
	1–3	−0.87	$M_{3-1} = -2.57$
2	3–5	−0.71	$M_{3-5} = -2.57$
	3–5	−0.71	$M_{5-3} = 0.28$
3	5–4	−9.86	$M_{5-4} = 0.28$
	5–4	−9.86	$M_{4-5} = 7.08$
4	4–2	−7.13	$M_{4-2} = 7.08$
	4–2	−7.13	$M_{2-4} = 13.3$

Reactions

$H_{x1} = -7.54$ $\quad V_{y1} = 0.87$ – Node 1

$H_{x2} = -4.28$ $\quad V_{y2} = 7.13$ – Node 2

(4) 'FRAME2DF'

A typical data file for FRAME2DF is given below.
The data file 'FRAME2D.INP' was used to analyse Problem 6 above, which was for the **hydrostatic load case**.

```
FRAME2DF.INP
5
4
2
1
0,0
5,1
0,4
5,4
3,5
1,1,1,1
2,1,1,1
1
1
1,3,2,2000
–2,–2,0
3,5,1,1000
–1,–1,0
5,4,1,1000
0,0,0
4,2,2,2000
0,–3,1
5,6,–5,0
EOF
```

The various items in the above file can be separated by two or more spaces, instead of commas. **Do not use the TAB key** to create spaces!

On Professor Ross's YouTube sites, computer programs are also available for the static analysis of 3-dimensional pin-jointed trusses and rigid-jointed space frames, together with the calculations of resonant frequencies of the structures described in this chapter.

For fully worked solutions to each of the problems in Exercise 71 and 72 in this chapter, go to the website:
www.routledge.com/cw/bird

Chapter 21

The finite element method

Why it is important to understand: **The finite element method**

The finite element method is one of the most powerful methods for solving differential equations, including partial differential equations, which apply over complex shapes, with complex boundary equations. The method requires powerful computers together with the appropriate computer programs. The method can be used for the statics and dynamics of problems in structural and continuum mechanics. The finite elements used in this analysis include the truss and beam elements described in Chapter 20, and also in-plane and out-of-plane plate elements, together with shell and solid elements. The method described here demonstrates how to determine the stiffness matrix of an in-plane triangular plate element, as shown by Figure 21.2. The same method can be used for the more complex problems involving shell and solid elements. Apart from its use for structural analysis, the finite element method can also be used for vibrations, acoustics, electrostatics, magnetostatics, heat transfer, fluid flow and so on.

At the end of this chapter you should be able to:

- understand what a load vector is for an in-plane plate
- understand what a displacement vector is for an in-plane plate
- understand what an elemental stiffness matrix is for an in-plane plate
- determine a shape function
- determine a load vector for a complex shape
- determine a stiffness matrix for a complex shape, such as a triangle

Mathematical references

In order to understand the theory involved in **the finite element method**, knowledge of the following mathematical topics is required: ***matrices and partial differentiation.*** If help/revision is needed in these areas, see chapter 18 and page 50 for some textbook references.

21.1 Introduction

In this chapter, the finite element method proper will be introduced. The finite element method is based on the matrix displacement method of Chapter 20, in that the unknown nodal displacements have to be calculated prior to determining the elemental stresses.

The main problem, however, with the finite element method, is in determining the stiffness matrices of elements of complex shape. That is, the methods of Chapter 7 cannot be used to determine the stiffness matrices of triangular or quadrilateral plate elements or of thin or thick doubly curved shell elements, because these shapes are too complex for the approaches of Chapter 7.

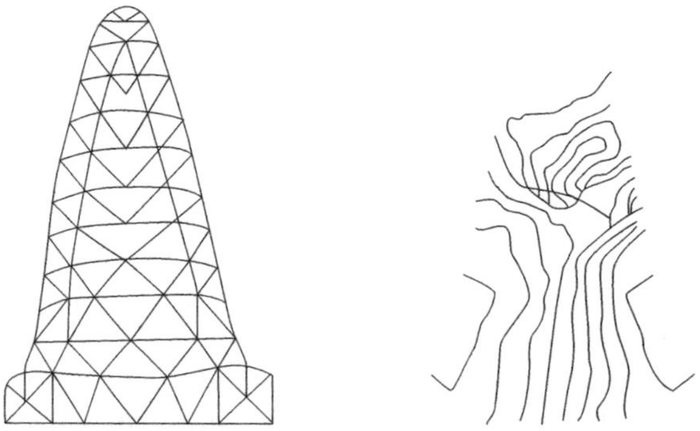

(a) Gear tooth

(b) Maximum shear stress contours

Figure 21.1 Parent shape

The finite element method is particularly useful for solving complex partial differential equations which apply over a complex parent shape, such as shown in Figure 21.1(a). This process is achieved by sub-dividing the complex parent shape of Figure 21.1(a) into many finite elements of smaller shape and determining the nodal displacements and element stresses by solving the complex partial differential equation over each of the finite elements of simpler shape. By considering equilibrium and compatibility at the inter-element boundaries, a large number of simultaneous equations are obtained. Solution of these simultaneous equations results in values of the unknown function; in the present chapter these unknown functions are in fact displacements. Apart from its use for structural analysis, the finite element method can be used for vibrations, acoustics, electrostatics, magnetostatics, heat transfer, fluid flow and so on.

The true finite element method was invented by J. Turner, M.J. Clough, R.W. Martin and M.C. Topp ('Stiffness and deflection analysis of complex structures', 1956, reference 5, page 497), when they presented the three node in-plane triangular plate element of Figure 21.2. The derivation of the stiffness matrix of this element is described in the following section.

21.2 Stiffness matrix for the in-plane triangular element

The in-plane triangular element is described by three corner nodes, namely nodes 1, 2 and 3, as shown in Figure 21.2.

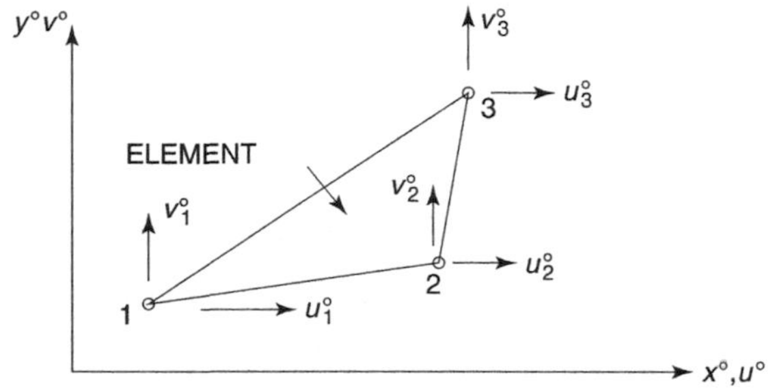

Figure 21.2 In-plane triangular plate element.

The element is useful for mathematically modelling flat plates subjected to in-plane forces. Each node has two degrees of freedom, making a total of 6 degrees of freedom per element; these degrees of freedom, or boundary values, are u_1^0, v_1^0, u_2^0, v_2^0, u_3^0 and v_3^0.

The method of derivation of the stiffness matrix is based on boundary values, and as there are six boundary values, it will be necessary to assume polynomials for the displacements u^0 and v^0, which have a total of six arbitrary constants α_i as shown by equations (21.1) and (21.2).

$$u^0 = \alpha_1 + \alpha_2 x^0 + \alpha_3 y^0 \tag{21.1}$$

$$v^0 = \alpha_4 + \alpha_5 x^0 + \alpha_6 y^0 \tag{21.2}$$

By applying each of the six boundary values to equations (21.1) and (21.2), six simultaneous equations will result, hence, the six α's can be determined, as follows.

(1) Boundary conditions

The six boundary conditions are:

$$\left.\begin{aligned} &\text{At } x^0 = x_1^0 \text{ and } y^0 = y_1^0, u^0 = u_1^0 \text{ and } v^0 = v_1^0 \\ &\text{At } x^0 = x_2^0 \text{ and } y^0 = y_2^0, u^0 = u_2^0 \text{ and } v^0 = v_2^0 \\ &\text{At } x^0 = x_3^0 \text{ and } y^0 = y_3^0, u^0 = u_3^0 \text{ and } v^0 = v_3^0 \end{aligned}\right\} \quad (21.3)$$

Substituting these six boundary values into equations (21.1) and (21.2), the following six simultaneous equations are obtained:

$$\left.\begin{aligned} u_1^0 &= \alpha_1 + \alpha_2 x_1^0 + \alpha_3 y_1^0 \\ u_2^0 &= \alpha_1 + \alpha_2 x_2^0 + \alpha_3 y_2^0 \\ u_3^0 &= \alpha_1 + \alpha_2 x_3^0 + \alpha_3 y_3^0 \end{aligned}\right\} \quad (21.4)$$

$$\left.\begin{aligned} v_1^0 &= \alpha_4 + \alpha_5 x_1^0 + \alpha_6 y_1^0 \\ v_2^0 &= \alpha_4 + \alpha_5 x_2^0 + \alpha_6 y_2^0 \\ v_3^0 &= \alpha_4 + \alpha_5 x_3^0 + \alpha_6 y_3^0 \end{aligned}\right\} \quad (21.5)$$

Rewriting equations (21.4) and (21.5) in matrix form, the following is obtained:

$$\begin{pmatrix} u_1^0 \\ u_2^0 \\ u_3^0 \\ v_1^0 \\ v_2^0 \\ v_3^0 \end{pmatrix} = \begin{pmatrix} 1 & x_1^0 & y_1^0 & & & \\ 1 & x_2^0 & y_2^0 & & 0_3 & \\ 1 & x_3^0 & y_3^0 & & & \\ & & & 1 & x_1^0 & y_1^0 \\ & 0_3 & & 1 & x_2^0 & y_2^0 \\ & & & 1 & x_3^0 & y_3^0 \end{pmatrix} \begin{pmatrix} \alpha_1 \\ \alpha_2 \\ \alpha_3 \\ \alpha_4 \\ \alpha_5 \\ \alpha_6 \end{pmatrix} \quad (21.6)$$

$$= \begin{pmatrix} A & 0_3 \\ 0_3 & A \end{pmatrix} (\alpha_i) \quad (21.7)$$

or

$$(\alpha_i) = \begin{pmatrix} A^{-1} & 0_3 \\ 0_3 & A^{-1} \end{pmatrix} (u_i^0) \quad (21.8)$$

where

$$(A) = \begin{pmatrix} 1 & x_1^0 & y_1^0 \\ 1 & x_2^0 & y_2^0 \\ 1 & x_3^0 & y_3^0 \end{pmatrix}$$

and

$$(A)^{-1} = \frac{\begin{pmatrix} a_1 & a_2 & a_3 \\ b_1 & b_2 & b_3 \\ c_1 & c_2 & c_3 \end{pmatrix}}{|A|} \quad (21.9)$$

$$\left.\begin{aligned} a_1 &= x_2^0 y_3^0 - x_3^0 y_2^0 \\ a_2 &= x_3^0 y_1^0 - x_1^0 y_3^0 \\ a_3 &= x_1^0 y_2^0 - x_2^0 y_1^0 \end{aligned}\right\} \quad (21.10)$$

$$\left.\begin{aligned} b_1 &= y_2^0 - y_3^0 \\ b_2 &= y_3^0 - y_1^0 \\ b_3 &= y_1^0 - y_2^0 \end{aligned}\right\} \quad (21.11)$$

$$\left.\begin{aligned} c_1 &= x_3^0 - x_2^0 \\ c_2 &= x_1^0 - x_3^0 \\ c_3 &= x_2^0 - x_1^0 \end{aligned}\right\} \quad (21.12)$$

$$\text{Determinant } |A| = x_2^0 y_3^0 - y_2^0 y_3^0 - x_1^0\left(y_3^0 - y_2^0\right) + y_1^0\left(x_3^0 - x_2^0\right)$$

$$= 2\Delta$$

where Δ = area of triangular element

$$\left.\begin{aligned} &\text{Hence} \quad u^0 = N_1 u_1^0 + N_2 u_2^0 + N_3 u_3^0 \\ &\text{and} \quad v^0 = N_1 v_1^0 + N_2 v_2^0 + N_3 v_3^0 \end{aligned}\right\} \quad (21.13)$$

or

$$\begin{pmatrix} u^0 \\ v^0 \end{pmatrix} = \begin{pmatrix} N_1 & N_2 & N_3 & 0 & 0 & 0 \\ 0 & 0 & 0 & N_1 & N_2 & N_3 \end{pmatrix} \begin{pmatrix} u_1^0 \\ u_2^0 \\ u_3^0 \\ v_1^0 \\ v_2^0 \\ v_3^0 \end{pmatrix} \quad (21.14)$$

$$= (N)(U_i)$$

where (N) = a matrix of shape functions

$$= \begin{pmatrix} N_1 & N_2 & N_3 & 0 & 0 & 0 \\ 0 & 0 & 0 & N_1 & N_2 & N_3 \end{pmatrix} \quad (21.15)$$

$$\left.\begin{aligned} N_1 &= \frac{1}{2\Delta}\left(a_1 + b_1 x^0 + c_1 y^0\right) \\ N_2 &= \frac{1}{2\Delta}\left(a_2 + b_2 x^0 + c_2 y^0\right) \\ N_3 &= \frac{1}{2\Delta}\left(a_3 + b_3 x^0 + c_3 y^0\right) \end{aligned}\right\} \quad (21.16)$$

(2) The matrix (*B*)

For a two-dimensional system of stress, the strains are given by:

$$\left.\begin{aligned}\varepsilon_x &= \frac{\partial u^0}{\partial x^0} = \text{strain in the } x^0 \text{ direction}\\ \varepsilon_y &= \frac{\partial v^0}{\partial y^0} = \text{strain in the } y^0 \text{ direction}\\ \gamma_{xy} &= \frac{\partial u^0}{\partial y^0} + \frac{\partial v^0}{\partial x^0} = \text{shear strain in the } x^0 - y^0 \text{ direction}\end{aligned}\right\} \quad (21.17)$$

Hence, from equations (21.13) and (21.17),

$$\left.\begin{aligned}\varepsilon_x &= \frac{1}{2\Delta}\left(b_1 u_1{}^0 + b_2 u_2{}^0 + b_3 u_3{}^0\right)\\ \varepsilon_y &= \frac{1}{2\Delta}\left(c_1 v_1{}^0 + c_2 v_2{}^0 + c_3 v_3{}^0\right)\\ \gamma_{xy} &= \frac{1}{2\Delta}\left(\begin{aligned}&c_1 u_1{}^0 + c_2 u_2{}^0 + c_3 u_3{}^0\\ &\quad + b_1 v_1{}^0 + b_2 v_2{}^0 + b_3 v_3{}^0\end{aligned}\right)\end{aligned}\right\} \quad (21.18)$$

or in matrix form:

$$\begin{pmatrix}\varepsilon_x\\ \varepsilon_y\\ \gamma_{xy}\end{pmatrix} = \frac{1}{2\Delta}\begin{pmatrix} b_1 & b_2 & b_3 & 0 & 0 & 0\\ 0 & 0 & 0 & c_1 & c_2 & c_3\\ c_1 & c_2 & c_3 & b_1 & b_2 & b_3\end{pmatrix}\begin{pmatrix}u_1{}^0\\ u_2{}^0\\ u_3{}^0\\ v_1{}^0\\ v_2{}^0\\ v_3{}^0\end{pmatrix} \quad (21.19)$$

$$= (B)(u_i)$$

where (*B*) = a matrix relating co-ordinate strains to nodal displacements

$$= \frac{1}{2\Delta}\begin{pmatrix} b_1 & b_2 & b_3 & 0 & 0 & 0\\ 0 & 0 & 0 & c_1 & c_2 & c_3\\ c_1 & c_2 & c_3 & b_1 & b_2 & b_3\end{pmatrix} \quad (21.20)$$

(3) The matrix (*D*)

Now the stress-strain relationship for an in-plane plate, in **plane stress**, are given by equations (9.43), (9.44) and (9.53) on pages 215 and 216.

i.e.

$$\left.\begin{aligned}\sigma_x &= \frac{E}{\left(1-v^2\right)}\left(\varepsilon_x + v\varepsilon_y\right)\\ \sigma_y &= \frac{E}{\left(1-v^2\right)}\left(\varepsilon_y + v\varepsilon_x\right)\\ \tau_{xy} &= G\gamma_{xy}\end{aligned}\right\} \quad (21.21)$$

where σ_x = normal stress in x^0 direction

σ_y = normal stress in y^0 direction

τ_{xy} = shear stress in the $x^0 - y^0$ plane

E = Young's modulus

G = rigidity modulus = $\dfrac{E}{2(1+v)}$

v = Poisson's ratio

In matrix form, equation (21.21) becomes:

$$\begin{pmatrix}\sigma_x\\ \sigma_y\\ \tau_{xy}\end{pmatrix} = \frac{E}{\left(1-v^2\right)}\begin{pmatrix}1 & v & 0\\ v & 1 & 0\\ 0 & 0 & \dfrac{(1-v)}{2}\end{pmatrix}\begin{pmatrix}\varepsilon_x\\ \varepsilon_y\\ \gamma_{xy}\end{pmatrix} \quad (21.22)$$

or $\quad (\sigma) = (D)(\varepsilon) \quad$ (21.23)

where (*D*) = a matrix of material constants

$$= \frac{E}{\left(1-v^2\right)}\begin{pmatrix}1 & v & 0\\ v & 1 & 0\\ 0 & 0 & \dfrac{(1-v)}{2}\end{pmatrix} \quad (21.24)$$

For **plane strain** (see Section 9.10), which is a two-dimensional system of strain and a three-dimensional system of stress, the matrix of material constants (*D*), is given by:

$$(D) = \frac{E}{(1+v)(1-2v)}\begin{pmatrix}(1-v) & v & 0\\ v & (1-v) & 0\\ 0 & 0 & \dfrac{(1-2v)}{2}\end{pmatrix} \quad (21.25)$$

In general, $\quad (D) = E^1\begin{pmatrix}1 & \mu & 0\\ \mu & 1 & 0\\ 0 & 0 & \gamma\end{pmatrix} \quad$ (21.26) where for **plane stress**, $E^1 = \dfrac{E}{(1-v^2)}$

$$\mu = v \quad \text{and} \quad \gamma = \frac{(1-v)}{2} \qquad (21.27)$$

and for **plane strain**, $E^1 = \dfrac{E}{(1+v)(1-2v)}$

$$\mu = v \quad \text{and} \quad \gamma = \frac{(1-2v)}{2} \qquad (21.28)$$

(4) The stiffness matrix (k^0)

From reference (4), the elastic strain energy stored in a body $= U_e = \dfrac{1}{2}E\int \sigma^2 d(vol)$ (21.29)

However $\dfrac{\sigma}{\varepsilon} = E$ from which $\sigma = E\varepsilon$

Hence $U_e = \dfrac{1}{2E}\int E^2\varepsilon^2\, d(vol) = \dfrac{1}{2}\int E\varepsilon^2\, d(vol)$

(21.30)

In matrix form, equation (21.30) appears as:

$$U_e = \frac{1}{2}\int (\varepsilon)^T (D)(\varepsilon)\, d(vol) \qquad (21.31)$$

It is necessary to write equation (21.31) in the form shown, because if it is multiplied out, the result will be a scalar, which is the correct form for strain energy.

However $(\varepsilon) = (B)(U_i)$ (21.32)

Therefore $U_e = (U_i)^T \int (B)^T (D)(B)(U_i)\, d(vol)$

(21.33)

However, the potential of the external loads

$$= -(U_i)^T (P_i) \qquad (21.34)$$

and the total potential is given by:

$$\pi = U_e - (U_i)^T (P_i) \qquad (21.35)$$

For minimum potential $\dfrac{d\pi_p}{d(U_i)} = 0$ (21.36)

Therefore $(P_i) = \int (B)^T (D)(B)\, d(vol)(U_i)$ (21.37)

$$= (k^o)(U_i)$$

Therefore (k^o) = elemental stiffness matrix in global co-ordinates

$$= \int (B)^T (D)(B)\, d(vol) \qquad (21.38)$$

Substituting equations (21.20) and (21.26) into equation (21.38) gives:

$$(k^o) = \int \frac{1}{4\Delta^2}\begin{pmatrix} b_1 & 0 & c_1 \\ b_2 & 0 & c_2 \\ b_3 & 0 & c_3 \\ 0 & c_1 & b_1 \\ 0 & c_2 & b_2 \\ 0 & c_3 & b_3 \end{pmatrix} E^1 \begin{pmatrix} 1 & \mu & 0 \\ \mu & 1 & 0 \\ 0 & 0 & \gamma \end{pmatrix}\begin{pmatrix} b_1 & b_2 & b_3 & 0 & 0 & 0 \\ 0 & 0 & 0 & c_1 & c_2 & c_3 \\ c_1 & c_2 & c_3 & b_1 & b_2 & b_3 \end{pmatrix} t\, dA$$

(21.39)

As b_1, b_2, b_3, and so on, are constants, $\int dA = \Delta$

therefore, $(k^o) = t\begin{pmatrix} P_{ij} & Q_{ij} \\ Q_{ji} & R_{ij} \end{pmatrix}$ (21.40)

where t = plate thickness

$$P_{ij} = \frac{0.25E^1(b_i b_j + \gamma c_i c_j)}{\Delta}$$

$$Q_{ij} = \frac{0.25E^1(\mu b_i c_j + \gamma c_i b_j)}{\Delta}$$

$$Q_{ji} = \frac{0.25E^1(\mu b_j c_i + \gamma c_j b_i)}{\Delta}$$

$$R_{ij} = \frac{0.25E^1(c_i c_j + \gamma b_i b_j)}{\Delta} \qquad (21.41)$$

where i and j vary from 1 to 3.

NB It should be noted that (k^0) is of order 6, which is the same number as the degrees of freedom for the element.

Now try the following Practice Exercise

Practice Exercise 73. The finite element method

1. Using the finite element method, determine the elemental stiffness matrix for a uniform section rod, of length l and with end nodes.

   $$\left[\frac{AE}{l}\begin{pmatrix} 1 & -1 \\ -1 & 1 \end{pmatrix}\right]$$

2. Determine the stiffness matrix for a rod element whose cross-sectional area varies uniformly from A_1, at node 1 to A_2 at node 2.

   $$\left[\frac{(A_1 + A_2)E}{2l}\begin{pmatrix} 1 & -1 \\ -1 & 1 \end{pmatrix}\right]$$

21.3 Stiffness matrix for a three node rod element

A three-node rod element of uniform cross-section has a mid-side node in addition to the end nodes, as shown in Figure 21.3. Very often, a better stiffness matrix can be obtained if additional mid-side nodes are used.

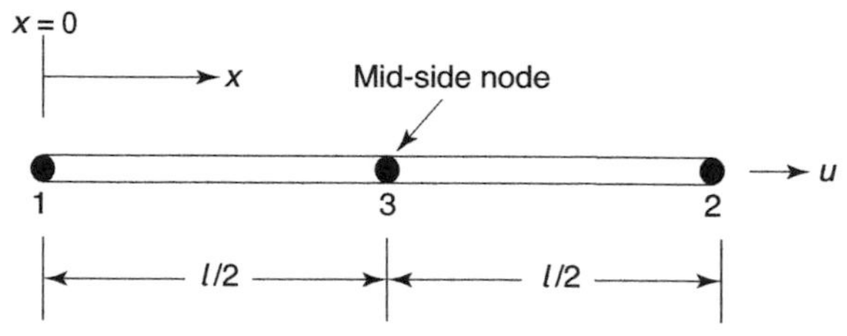

Figure 21.3 Rod element with 3 nodes.

(1) To obtain (*N*)

In this case, there are three nodal displacements, namely u_1, u_2 and u_3, hence, it will be necessary to assume a polynomial with three α's, as shown by equation (21.42).

$$u = \alpha_1 + \alpha_2 x + \alpha_3 x^2 \quad (21.42)$$

The three boundary values are:

at $x = 0$, $u = u_1$

at $x = l$, $u = u_2$

and at $x = l/2$, $u = u_3$ (21.43)

Substituting these three boundary values into equation (21.42), the following three simultaneous equations are obtained:

$$u_1 = \alpha_1 \quad (21.44a)$$

$$u_2 = \alpha_1 + \alpha_2 l + \alpha_3 l^2 \quad (21.44b)$$

$$u_3 = \alpha_1 + \frac{\alpha_2 l}{2} + \frac{\alpha_3 l^2}{4} \quad (21.44c)$$

i.e. $\boldsymbol{\alpha_1 = u_1}$ (21.45)

Hence, from equations (21.44b) and (21.44c),

$$u_2 = u_1 + \alpha_2 l + \alpha_3 l^2$$

i.e. $$u_2 - u_1 = \alpha_2 l + \alpha_3 l^2 \quad (21.46)$$

and $$u_3 = u_1 + \frac{\alpha_2 l}{2} + \frac{\alpha_3 l^2}{4}$$

i.e $$u_3 - u_1 = \frac{\alpha_2 l}{2} + \frac{\alpha_3 l^2}{4} \quad (21.47)$$

To eliminate α_2, multiply equation (21.47) by 2 and subtract from equation (21.46) to give:

$$(u_2 - u_1) - (2u_3 - 2u_1) = (\alpha_2 l + \alpha_3 l^2) - \left(\alpha_2 l + \frac{\alpha_3 l^2}{2}\right)$$

i.e. $$u_2 + u_1 - 2u_3 = \alpha_3 l^2 - \frac{\alpha_3 l^2}{2}$$

i.e. $$u_1 + u_2 - 2u_3 = \frac{\alpha_3 l^2}{2}$$

and $$\alpha_3 = \frac{2u_1 + 2u_2 - 4u_3}{l^2} \quad (21.48)$$

Substituting equation (21.48) into equation (21.46) gives:

$$u_2 - u_1 = \alpha_2 l + \left(\frac{2u_1 + 2u_2 - 4u_3}{l^2}\right) l^2$$

i.e. $$u_2 - u_1 = \alpha_2 l + 2u_1 + 2u_2 - 4u_3$$

i.e. $$u_2 - u_1 - 2u_1 - 2u_2 + 4u_3 = \alpha_2 l$$

and $$\alpha_2 = \frac{-3u_1 - u_2 + 4u_3}{l} \quad (21.49)$$

Substituting equations (21.45), (21.48) and (21.49) into equation (21.42) gives:

$$u = u_1 + \left(\frac{-3u_1 - u_2 + 4u_3}{l}\right) x + \left(\frac{2u_1 + 2u_2 - 4u_3}{l^2}\right) x^2 \quad (21.50)$$

Let $\xi = \frac{x}{l}$

Then $$u = u_1 + (-3u_1 - u_2 + 4u_3)\xi + (2u_1 + 2u_2 - 4u_3)\xi^2 \quad (21.51)$$

i.e. $$u = u_1(1 - 3\xi + 2\xi^2) + u_2(-\xi + 2\xi^2) + u_3(4\xi - 4\xi^2) \quad (21.52)$$

i.e.

$$u = \left((1 - 3\xi + 2\xi^2) \; (-\xi + 2\xi^2) \; (4\xi - 4\xi^2)\right) \begin{pmatrix} u_1 \\ u_2 \\ u_3 \end{pmatrix} \quad (21.53)$$

i.e. $u = (N)(U_i)$

where (N) = the matrix of shape functions

$$= \left(\left(1-3\xi+2\xi^2\right)\ \left(-\xi+2\xi^2\right)\ \left(4\xi-4\xi^2\right)\right) \quad (21.54)$$

(2) To obtain (*D*)

In one dimension, and from Hooke's law,

$$\sigma = E\varepsilon$$

or in matrix form, $(\sigma)=(D)(\varepsilon)$

i.e. $$(D) = (E) \quad (21.55)$$

(3) To obtain (*B*)

In one dimension, ε = axial strain = $\frac{du}{dx}=\frac{du}{l\,d\xi}$

$$= \frac{u_1(-3+4\xi)+u_2(-1+4\xi)+u_3(4-8\xi)}{l}$$

from equation (21.52)

$$= \left((-3+4\xi)\ (-1+4\xi)\ (4-8\xi)\right)\begin{pmatrix}u_1\\u_2\\u_3\end{pmatrix}/l$$

or $u=(B)(U_i)$

$$\text{i.e. } (B)=\frac{\left((-3+4\xi)\ (-1+4\xi)\ (4-8\xi)\right)}{l} \quad (21.56)$$

Substituting equation (21.55) and (21.56) into equation (21.38) gives:

(k) = the stiffness matrix for a 3-node rod element

$$= \frac{1}{2}\int_0^1 \frac{1}{l^2}\begin{pmatrix}(-3+4\xi)\\(-1+4\xi)\\(4-8\xi)\end{pmatrix}E(-3+4\xi)(-1+4\xi)(4-8\xi)\,A\,l\,d\xi$$

$$\text{where } (k)=\begin{pmatrix}k_{11} & k_{12} & k_{13}\\ k_{21} & k_{22} & k_{23}\\ k_{31} & k_{32} & k_{33}\end{pmatrix} \quad (21.57)$$

and $k_{11}=\frac{AE}{l}\int_0^1(-3+4\xi)^2\,d\xi$

$$=\frac{AE}{l}\int_0^1\left(9-24\xi+16\xi^2\right)d\xi \quad (21.58)$$

$$=\frac{AE}{l}\left[9\xi-12\xi^2+\frac{16\xi^3}{3}\right]_0^1$$

$$=\frac{AE}{l}\left[\left(9-12+\frac{16}{3}\right)-(0)\right]=\frac{AE}{l}\left[2\frac{1}{3}\right]$$

i.e. $k_{11}=\frac{7AE}{3l}$

Similarly $k_{12}=k_{21}=\frac{AE}{l}\int_0^1(-3+4\xi)(-1+4\xi)\,d\xi$

$$=\frac{AE}{l}\int_0^1\left(3-12\xi-4\xi+16\xi^2\right)d\xi$$

$$=\frac{AE}{l}\left[3\xi-8\xi^2+\frac{16\xi^3}{3}\right]_0^1 \quad (21.59)$$

$$=\frac{AE}{l}\left[\left(3-8+\frac{16}{3}\right)-(0)\right]=\frac{AE}{l}\left[\frac{1}{3}\right]$$

i.e. $k_{12}=\frac{AE}{3l}$

Similarly $k_{13}=k_{31}=\frac{AE}{l}\int_0^1(-3+4\xi)(4-8\xi)\,d\xi$

$$=\frac{AE}{l}\int_0^1\left(-12+40\xi-32\xi^2\right)d\xi \quad (21.60)$$

$$=\frac{AE}{l}\left[-12\xi+20\xi^2-\frac{32\xi^3}{3}\right]_0^1$$

$$=\frac{AE}{l}\left[\left(-12+20-\frac{32}{3}\right)-(0)\right]=\frac{AE}{l}\left[-2\frac{2}{3}\right]$$

i.e. $k_{13}=-\frac{8AE}{3l}$

Similarly $k_{22}=\frac{AE}{l}\int_0^1(-1+4\xi)(-1+4\xi)\,d\xi$

$$=\frac{AE}{l}\int_0^1\left(1-8\xi+16\xi^2\right)d\xi \quad (21.61)$$

$$=\frac{AE}{l}\left[\xi-4\xi^2+\frac{16\xi^3}{3}\right]_0^1$$

$$=\frac{AE}{l}\left[\left(1-4+\frac{16}{3}\right)-(0)\right]=\frac{AE}{l}\left[2\frac{2}{3}\right]$$

i.e. $k_{22}=\frac{7AE}{3l}$

Similarly $k_{23}=k_{32}=\frac{AE}{l}\int_0^1(-1+4\xi)(4-8\xi)\,d\xi$

$$=\frac{AE}{l}\int_0^1\left(-4+24\xi-32\xi^2\right)d\xi \quad (21.62)$$

$$=\frac{AE}{l}\left[-4\xi+12\xi^2-\frac{32\xi^3}{3}\right]_0^1$$

$$=\frac{AE}{l}\left[\left(-4+12-\frac{32}{3}\right)-(0)\right]=\frac{AE}{l}\left[-2\frac{2}{3}\right]$$

i.e. $k_{23}=k_{32}=-\frac{8AE}{3l}$

Similarly $k_{33} = \frac{AE}{l}\int_0^1 (4-8\xi)(4-8\xi)\,d\xi$

$$= \frac{AE}{l}\int_0^1 \left(16-64\xi+64\xi^2\right)d\xi \qquad (21.63)$$

$$= \frac{AE}{l}\left[16\xi-32\xi^2+\frac{64\xi^3}{3}\right]_0^1$$

$$= \frac{AE}{l}\left[\left(16-32+\frac{64}{3}\right)-(0)\right] = \frac{AE}{l}\left[5\frac{1}{3}\right]$$

i.e. $k_{33} = \frac{16AE}{3l}$

NB It should be noted that (k) is of order 3, which is the same number as the degrees of freedom of the element.

Computer programs

It should be noted that Ross provides a number of free computer programs, which can analyse in-plane plates by a number of different devices, including SmartPhones, iPads, laptops, and so on.

For Ross's free Finite Element Computer Programs, which can be used on SmartPhones/iPads/laptops, etc., go to carltfross YouTube computer programs

Almost any name can be used for the output file, but its length should not exceed 8 characters, because of working in DOS. It should be noted that with a chromecast device, the screen of the SmartPhone or pad can be wirelessly projected on to an HD TV screen!

For fully worked solutions to each of the problems in Exercise 73 in this chapter, go to the website:

www.routledge.com/cw/bird

Revision Test 5: Specimen examination questions for Chapters 19 to 21

This Revision Test covers the material in Chapters 19 to 21. *The marks for each question are shown in brackets at the end of each question.*

1. A thick-walled submarine pressure hull in the form of a circular cylinder, of 10 m external diameter, is to be designed to dive to a depth of 1000 m. Determine its wall thickness, based on a safety factor of 2, and assume the following apply:

   Yield stress = 500 MPa

   Density of salt water = 1020 kg/m^3

   $g = 9.81$ m/s^2 (10)

2. A thick-walled submarine pressure hull, in the form of a circular cylinder, of 2 m external diameter, is to be designed to dive to the bottom of the Mariana Trench. Determine its wall thickness, based on a safety factor of 1.8, and assuming the following apply:

   Yield stress = 900 MPa

   Depth of the Mariana Trench = 11.5 km

   Density of salt water 1020 kg/m^3

   $g = 9.81$ m/s^2 (10)

3. A steel cylinder, with external and internal diameters 12 cm and 10 cm respectively, is shrunk on to an aluminium-alloy cylinder, with internal and external diameters of 8 cm and 10 cm respectively, where all dimensions are nominal. Find the radial pressure at the common surface due to shrinkage alone, so that when there is an internal pressure of 130 MPa, the maximum hoop stress in the inner cylinder is 100 MPa. Determine also the maximum hoop stress in the outer cylinder. Assume the following: $E_s = 2 \times 10^{11}$ N/m^2, $E_{al} = 1 \times 10^{11}$ N/m^2, $\nu_s = 0.3$, $\nu_{al} = 0.33$ (59)

4. A steel ring of 10 cm external diameter and 6 cm internal diameter, is to be shrunk on to a solid bronze shaft, where the interference fit is 0.004×10^{-2} m, based on the diameter. Determine the maximum tensile stress in the material, given that: $E_s = 2 \times 10^{11}$ N/m^2, $\nu_s = 0.3$, $E_b = 1 \times 10^{11}$ N/m^2, $\nu_b = 0.35$ (21)

5. A pillar in the hold of a ship is in the form of a tube of internal diameter 16 cm and of external diameter 23 cm and length 10 m. If the pillar is subjected to an eccentric load of 15 tonnes, as shown in Figure RT5.1, calculate 'Δ', given that maximum permissible stress = 80 MPa and $E_s = 2 \times 10^{11}$ N/m^2 (56)

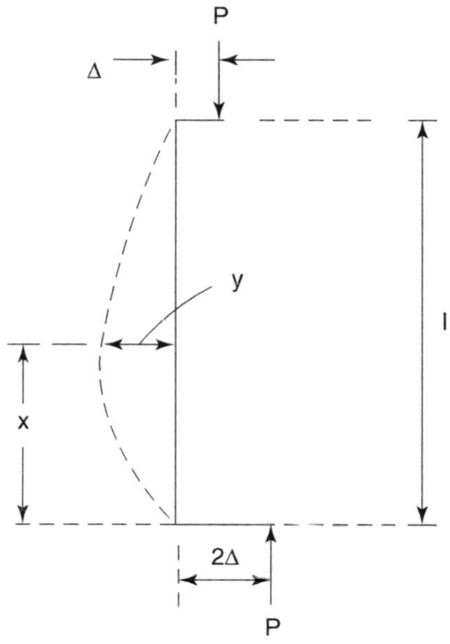

Figure RT5.1

6. Calculate the principal second moments of area, and their directions, for the angle bar section of Figure RT5.2. (38)

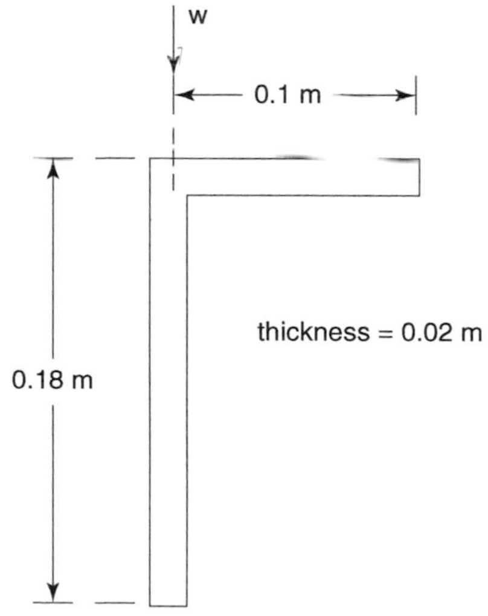

Figure RT5.2

7. Calculate the principal second moments of area, and their directions, for the beam cross-section shown in Figure RT5.3. (46)

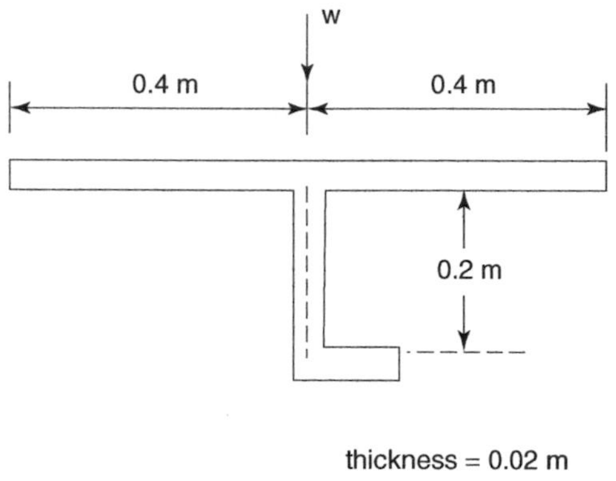

Figure RT5.3

8. If the beam of question 6 were encastré at its ends, and subjected to a downward vertical load of 8 kN at its mid-span, determine the direction of its neutral axis, and its maximum bending stress. The length of the beam is 2 m. (20)

9. If the beam of question 7 was of length 3 m, and encastré at its ends, and was subjected to a downward vertical load of 10 kN at its mid-span, determine its maximum deflection. Assume that $E = 2 \times 10^{11}$ N/m^2 (16)

10. Calculate and sketch the distribution of vertical and horizontal shear stress of the channel section shown in Figure RT5.4, when it is subjected to a vertical shearing force of 50 kN through its shear centre 'S'. Hence, or otherwise, determine the position of its shear centre. (19)

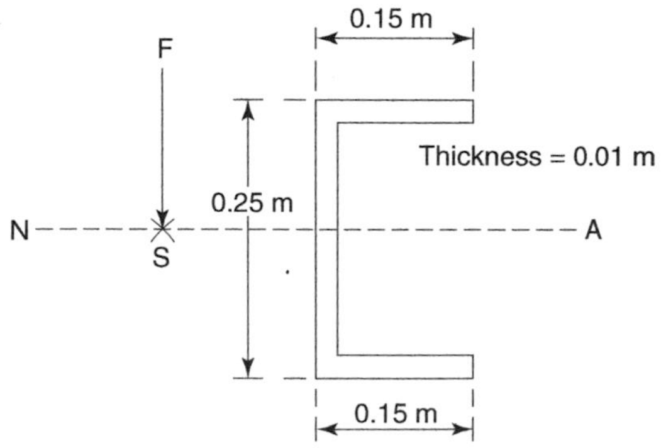

Figure RT5.4

11. Calculate and sketch the distribution of shearing stress of the section shown in Figure RT5.5, when it is subjected to a vertical downward force through its shear centre 'S'. (14)

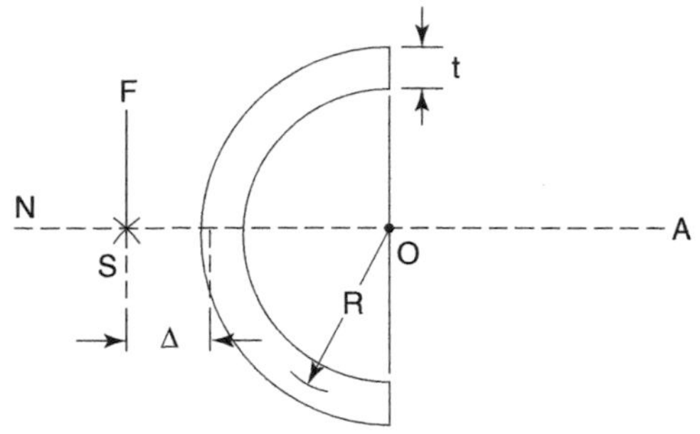

Figure RT5.5

12. Determine the forces in the members of the plane pin-jointed truss of Figure RT5.6, by the matrix displacement method. All members are of the same cross-section and material properties. (35)

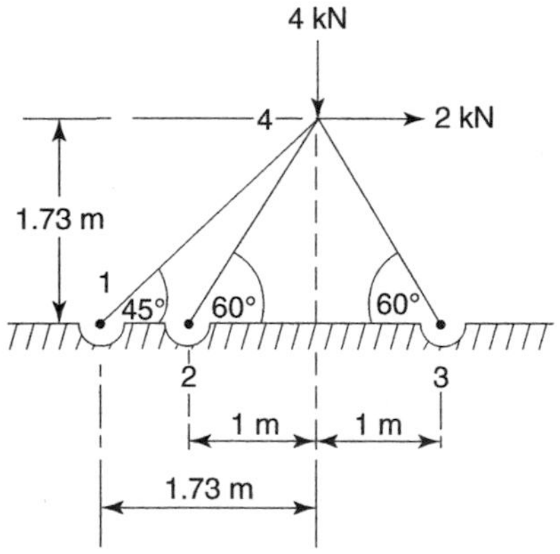

Figure RT5.6

13. Determine the end reactions for the uniform section beam of Figure RT5.7, which is encastré at its ends, by the matrix displacement method. E = constant (34)

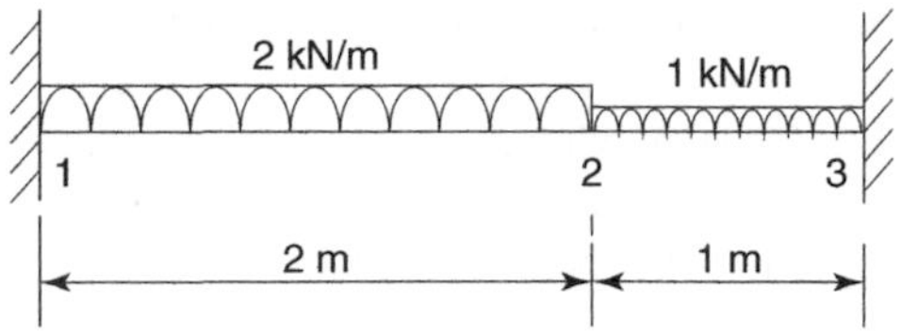

Figure RT5.7

14. Using the finite element method, determine the stiffness matrix a the 2-node rod element, which is of uniform taper, of cross-sectional area ‘A’ at the left end, and ‘2A’ at the right end, as shown in Figure RT5.8. (12)

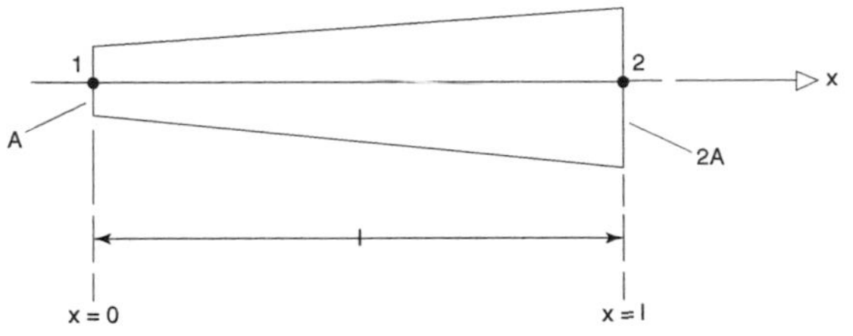

Figure RT5.8

15. Using the finite element method, determine the stiffness matrix for a 3-node rod element, of uniform taper, where the cross-section varies linearly from ‘A’ at the left end to ‘3A’ at the right end, as shown in Figure RT 5.9. (60)

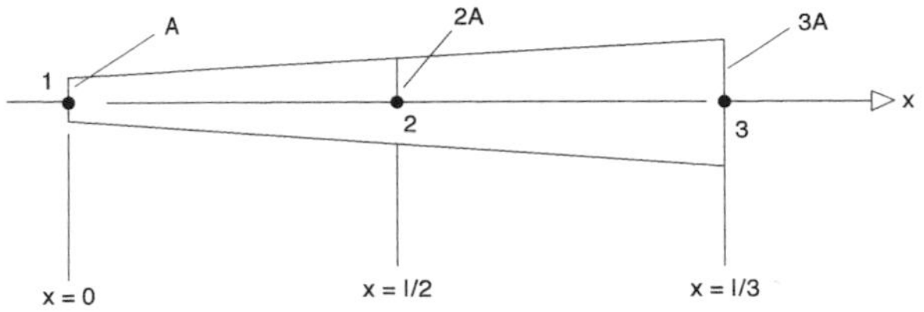

Figure RT5.9

For lecturers/instructors/teachers, fully worked solutions to each of the problems in Revision Test 5, together with a full marking scheme, are available at the website:
www.routledge.com/cw/bird

Chapter 22

An introduction to linear elastic fracture mechanics

Why is it important to understand: **An introduction to linear elastic fracture mechanics**

The majority of topics within mechanics of solids assume homogeneous, fault-free material for analysis. This is a necessary first step to predict the behaviour of materials under load, but professional engineers have to be aware of limitations that occur due to inconsistent material properties. In addition, there are situations where stress levels are increased around features such as existing cracks or around sites from which a crack could grow. Cracks can result from minor problems, many of which are almost inevitable during manufacture, such as inclusions of impurities during melting, or as part of a welding process. However, many problems can be prevented at the design stage with some knowledge of fracture mechanics and attention to detail. In many cases it is a case of accepting the existence of cracks, understanding them and living with them. This involves potential problem areas being identified and then arrangements made to monitor them. To emphasise this point, aircraft have been described as a series of cracks flying in close formation!

At the end of this chapter you should be able to:

- recognise the basis of fracture mechanics theory
- understand strain energy release and crack propagation
- appreciate energy balance and stress intensity approaches
- understand plane stress and plane strain behaviour
- appreciate the allowance for small scale yielding at crack tip
- understand fracture toughness crack tip opening displacement (CTOD)
- apply fracture mechanics to fatigue crack growth
- understand the J-integral and recognise crack extension resistance curves (*R*-curves)

Mathematical references

In order to understand the theory involved in this chapter on **linear fracture mechanics**, knowledge of the following mathematical topics are required: ***algebra, differentiation, integration, partial differentiation and differential equations***. If help/revision is needed in these areas, see page 50 for some textbook references.

22.1 Introduction

Fracture Mechanics is the 'applied mechanics of crack growth', starting from a flaw, and developed from the study of brittle fracture. A brittle fracture is the catastrophic failure of a structure made from a normally ductile material with a load less than that necessary to cause general yielding and with no significant plastic deformation.

These fractures always originate from a crack or flaw, such as a fatigue crack developed during service or a weld flaw introduced during initial fabrication. Examples include the welded Liberty ships during the Second World War, bridges, pressure vessels, pipelines and many more.

Brittle fractures in steels occur particularly at low temperature in thick sections, but metals can suffer brittle fracture in very thin sheets as well.

The toughness of a material can be described as its resistance to crack growth. In the past, measurements of material ductility obtained from the tensile test, such as elongation or reduction in area, have been used for specifying material quality in general terms.

For assessing resistance to fast crack propagation, the Charpy or Izod v-notch tests can be used, where the amount of energy absorbed when a notched specimen is broken under impact conditions is measured. These tests give comparative values of material toughness. They do not provide information to enable the magnitude of stress required to produce fast crack propagation in structures containing cracks varying in size and geometry, to be calculated. Fracture mechanics allows quantitative measurements to be made of a material's resistance to crack propagation, i.e. the toughness of the material.

22.2 Basis of fracture mechanics theory

For crack growth under static loading two conditions are necessary:

(a) Sufficient stress must be available at the crack tip to operate some mechanism of crack growth.

(b) Sufficient energy must flow to the crack tip to supply the work done in the creation of new crack surfaces.

Initially, it was thought that only the first was required.

***Sir C. E. Inglis** (in 1913) derived a mathematical solution for stress concentration at an elliptical hole, and the stress concentration factor, K_T states:

$$K_t = \frac{\sigma_{\max}}{\sigma} = 1 + 2\sqrt{\frac{a}{b}} \quad (22.1)$$

The dimensions a and b are defined in Figure 22.1. Dimension 'b' can be thought of as the crack tip radius.

However, for sharp cracks where b tends to zero, this indicates that the peak stress tends to infinity. If this were completely true, it would mean that a structure with any cracks whatsoever could not support a load!

In his classic fracture theory for glass, **A. A. Griffith** (in 1920) ignored the precise physical condition at the crack tip and showed that a satisfactory fracture theory could be found from energy balance considerations. When the crack length is such that the strain energy release rate equals the surface energy required per unit extension, the crack will grow in a catastrophic manner. The more brittle the material, the shorter the critical crack length. This solution applies to brittle materials.

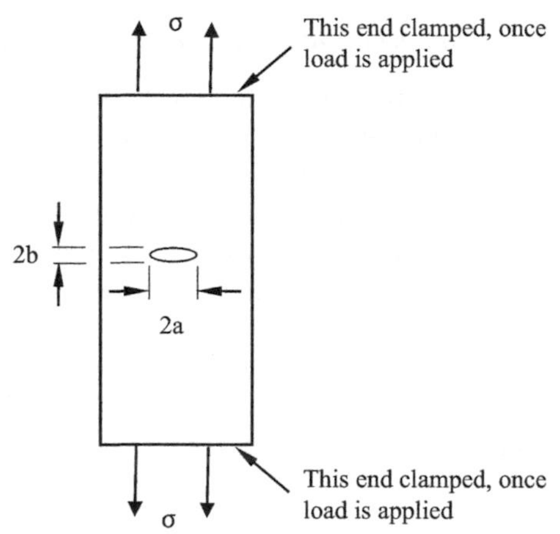

Figure 22.1 A crack in a specimen modelled as an ellipse

***G. R. Irwin** (1957) extended this energy balance approach for brittle materials to ductile materials by allowing for the energy associated with plastic deformation at the crack tip, using the stress intensity approach.

22.3 Strain energy release and crack propagation

Consider a thin sheet of material, with no prior crack, pulled and then maintained in tension. If a small cut is made at the centre of the sheet plate with the cut perpendicular to the direction of the stress, and if the cut is slowly extended further, a critical stage is reached when the crack starts growing on its own.

As the crack advances, the sheet becomes less stiff since the effective width of the material reduces. The ends are rigidly held and so the displacement remains the same as the stiffness is reduced. The energy stored is reduced, and this energy released is available for crack growth.

A detailed calculation of the amount of energy released is a complex problem, but a good approximation can be obtained from a simplified analysis, using Figure 22.2. A load is applied to the plate and when stretched, the ends of the plate are rigidly clamped. The crack is much smaller than the width and the length of the plate so that the stress away from the crack is constant. To ease the analysis, the major area of the plate where its strain energy is released is taken to be a triangle on each side of the crack plane as shown. As the

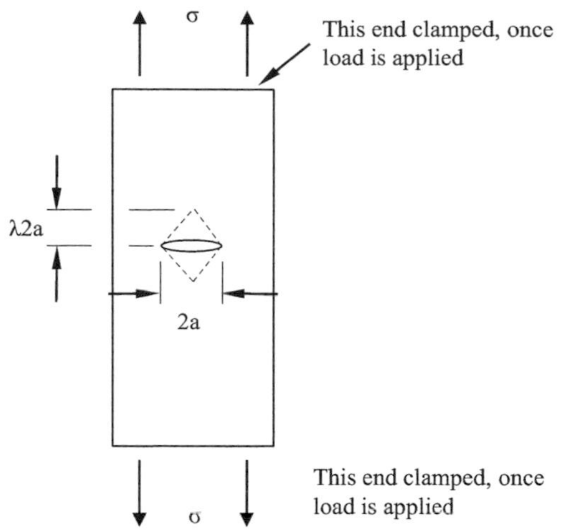

Figure 22.2 An elliptical crack with the area of strain energy release modelled as a triangle

*Sir Charles Edward Inglis (31 July 1875 – 19 April 1952) was a British civil engineer. The son of a doctor, he was educated at Cheltenham College and won a scholarship to King's College, Cambridge, where he would later forge a career as an academic. Inglis spent a two-year period with the engineering firm run by John Wolfe-Barry before he returned to King's College as a lecturer. Working with Professors James Alfred Ewing and Bertram Hopkinson, he made several important studies into the effects of vibration on structures and defects on the strength of plate steel. He has been described as the greatest teacher of engineering of his time and has a building named in his honour at Cambridge University. To find out more about Inglis go to www.routledge.com/cw/bird

*George Rankin Irwin (26 February 1907 – 9 October 1998) was an American scientist in the field of fracture mechanics and strength of materials. He was internationally known for his study of fracture of materials. When with the US Naval Research Laboratory (NRL) he developed methods for determining the penetration force that a projectile exerts on its target. Part of this work led to the development of several nonmetallic armors; this coupled with his observation that thick armor plate made from ductile material (such as steel) failed in a brittle manner during test firings initiated his interest in brittle fracture. To find out more about Irwin go to www.routledge.com/cw/bird

crack length increases, so do the bases and the heights of both triangles.
From Section 11.7, page 246, strain energy per unit volume

$$=\frac{1}{2}\sigma\varepsilon=\frac{\sigma^2}{2E} \qquad (22.2)$$

If B is the thickness of the plate, the strain energy released, U_R is given by:

$$U_R=\left(\frac{\sigma^2}{2E}\right)\times\left(\text{area of both triangles}\right)\times B \qquad (22.3)$$

i.e.

$$U_R=\left(\frac{\sigma^2}{2E}\right)\times 2\left(\frac{1}{2}\times 2a\right)\left(\lambda\times 2a\right)\times B \qquad (22.4)$$

where σ is the tensile stress and E is the Young's modulus.

Rigorous analysis (see Reference 38, page 498) shows that $\lambda=\pi/2$ for thin plates (for plane stress) giving:

$$U_R=\frac{\pi\sigma^2a^2B}{E} \quad \text{for plane stress} \qquad (22.5)$$

and

$$U_R=\left(1-\upsilon^2\right)\frac{\pi\sigma^2a^2B}{E} \quad \text{for plane strain} \qquad (22.6)$$

Plane stress and plane strain are important considerations in solid mechanics generally and fracture mechanics in particular. They are discussed further in Section 22.4.

Energy is required to create each of the two new surfaces of the crack. If γ is the surface energy per unit area of one surface, the total surface energy required, U_S is given by:

$$U_S=2\left(2a\right)B\gamma=4aB\gamma \qquad (22.7)$$

For brittle materials,

$$\gamma=\frac{1}{2}G \qquad (22.8)$$

The symbol G is the energy release rate and is named after ***A. A. Griffith**. Strictly speaking, it is the energy release rate per unit area, but the shorter name is in common use. It incorporates both surfaces of the crack, giving the factor of two. The symbol has widespread use within fracture mechanics, but it is somewhat unfortunate that it is the same symbol, G, often used for the shear modulus (or the modulus of rigidity). Once aware of the potential issue, it is usually very clear from the context.

Substituting equation (22.8) into equation (22.7) gives:

$$U_S=2aBG \qquad (22.9)$$

The surface energy required for propagation (U_S) is proportional to the crack length, whereas the strain energy released by crack growth (U_R) is proportional to the square of the crack length. Griffith proposed that *a crack will become unstable when an increment of growth causes the release of more energy than can be absorbed by the material.* The net energy of the crack is the algebraic sum of the two energies, which is stated mathematically as:

$$U_{Total}=2aBG-\frac{\pi\sigma^2a^2B}{E} \qquad (22.10)$$

A critical condition occurs when the rate of change of this energy = 0

i.e.

$$\frac{d}{da}\left(U_{Total}\right)=\frac{d}{da}\left(2aBG_C-\frac{\pi\sigma^2a^2B}{E}\right)=0 \qquad (22.11)$$

This critical condition occurs when $2BG_C=\dfrac{2\pi\sigma^2aB}{E}$

i.e.

$$BG_C=\frac{\pi\sigma^2aB}{E} \qquad (22.12)$$

and

$$EG_C=\pi\sigma^2a \qquad (22.13)$$

*Alan Arnold Griffith (13 June 1893 – 13 October 1963), son of Victorian science fiction author George Griffith, was an English engineer. Among many other contributions he is best known for his work on stress and fracture in metals that is now known as metal fatigue, as well as being one of the first to develop a strong theoretical basis for the jet engine. Griffith's advanced axial-flow turbojet engine designs were integral in the creation of Britain's first operational axial-flow turbojet engine, the Metropolitan-Vickers F.2 which first ran successfully in 1941. To find out more about Griffith go to www.routledge.com/cw/bird

Also, $(EG_C)^{\frac{1}{2}} = \sigma(\pi a)^{\frac{1}{2}}$ for plane stress (22.14)

Also, $\left(\dfrac{EG_C}{1-\upsilon^2}\right)^{\frac{1}{2}} = \sigma(\pi a)^{\frac{1}{2}}$ for plane strain (22.15)

See Section 22.6 for more information about plane stress and plane strain.

Note that the above equations apply for fast fracture in a brittle material and that G_C is a material property. It is known as the 'critical strain energy release rate' or the 'crack extension force' and has units of J/m². Glass has a low value of G_C, around 10 J/m² making crack propagation easy, whereas the value for alloy steel is around 30,000 J/m².

Problem 1. A large sheet of aluminium alloy carries a tensile load resulting in a stress of 90 MPa. If there is a 45 mm long central crack in the sheet, and the E value for the material is 72 GPa, estimate the value of G_C for the alloy.

$\sigma = 90$ MPa, $a = 45/2 = 22.5$ mm $= 0.0225$ m and

$E = 72$ GPa

From equation (22.14), $(EG_C)^{\frac{1}{2}} = \sigma(\pi a)^{\frac{1}{2}}$ for plane stress

from which, $(EG_C) = \sigma^2(\pi a)$

and $$G_C = \frac{\sigma^2 \pi a}{E} = \frac{(90\times10^6)^2 \pi(0.0225)}{72\times10^9}$$

$$= \mathbf{7952\ J/m^2}$$

Problem 2. Approximate the theoretical tensile strength of flaw-free glass, assuming that any flaw is internal, and that its length will be less than 3.5×10^{-9} m. The Young's modulus of glass is 70 GPa and G_C can be taken to be 10 J/m².

$a = 3.5\times10^{-9}/2 = 1.75\times10^{-9}$ m, $G_C = 10$ J/m²

and E = 70 MPa

From equation (22.14), $(EG_C)^{\frac{1}{2}} = \sigma(\pi a)^{\frac{1}{2}}$

and $$\sigma = \left(\frac{EG_C}{\pi a}\right)^{\frac{1}{2}} = \left(\frac{(70\times10^9)(10)}{\pi(1.75\times10^{-9})}\right)^{\frac{1}{2}}$$

$$= 11.28\times10^9\ \text{Pa}$$

i.e. **the theoretical tensile strength of flaw-free glass, σ = 11.28 GPa**

This theoretical high value of stress can only be reached if no flaw exceeds this very small value.

Problem 3. A part of a machine is to be treated as a sheet for the purposes of analysis. G_c for the material is 28,000 J/m² and the E value for the material is 116 GPa. There is a 50 mm long crack in the centre of the component, aligned normal to the stress field. What would be the expected fracture stress of this sheet?

$a = 50/2 = 25$ mm $= 0.025$ m, $G_C = 28{,}000$ J/m² and

$E = 116$ GPa

From equation (22.14), $(EG_C)^{\frac{1}{2}} = \sigma(\pi a)^{\frac{1}{2}}$ for plane stress

and $$\sigma = \left(\frac{EG_C}{\pi a}\right)^{\frac{1}{2}} = \left(\frac{(116\times10^9)(28000)}{\pi(0.025)}\right)^{\frac{1}{2}}$$

$$= 203.4\times10^6\ \text{Pa}$$

i.e. **the expected fracture stress of the sheet, σ = 203.4 MPa**

22.4 Energy balance approach

When there is work done on a cracked component/test piece, as a result of crack growth, this work done has to be incorporated into the energy sum.

Let U = Strain energy stored in a structure

W_{EXT} = External work done

A = Area of the crack

and G = energy release rate (i.e. energy release per unit increase in area during crack growth) then $$G\Delta A = \Delta W_{\text{EXT}} - \Delta U \qquad (22.16)$$

Dividing each side of equation (22.16) by ΔA and taking the limit as $\Delta A \to 0$ gives:

$$\boldsymbol{G = \frac{d}{dA}\left(\Delta W_{EXT} - \Delta U\right)} \qquad (22.17)$$

This is a powerful equation and Problems 4 and 5 demonstrate how it is used.

Problem 4. An example is given of a rectangular, double cantilever beam (DCB) test. Each cantilever is subjected to a constant load via a pin joint (so that it can rotate). This type of test is a convenient way of determining the energy release rate, G and it is shown diagrammatically in Figure 22.3. If the thickness of the material is B, the height of each cantilever is h and the crack length is 'a', determine an expression for G in terms of F, a, E, B and h.

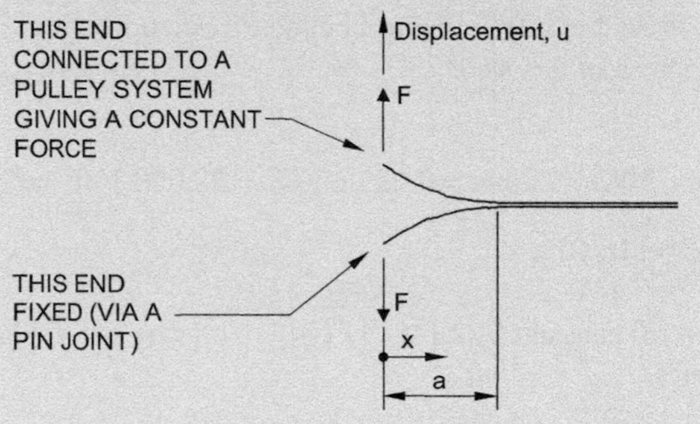

Figure 22.3 A double cantilever beam (DCB) test

From equation (22.16), the energy balance is:

$$G\Delta A = \Delta W_{EXT} - \Delta U$$

Introducing B, the thickness of the plate, and changing the order within the brackets, gives:

$$dA = B\,da$$

$$G = -\frac{1}{B}\frac{d}{da}\left(\Delta U - \Delta W_{EXT}\right) \qquad (22.18)$$

The strain energy, $$\Delta U = \frac{1}{2}Fu \qquad (22.19)$$

The work done, $\Delta W_{EXT} = Fu$

Hence, $$G = -\frac{1}{B}\frac{d}{da}\left(\frac{1}{2}Fu - Fu\right)$$

$$= -\frac{1}{B}\frac{d}{da}\left(-\frac{1}{2}Fu\right) \qquad (22.20)$$

i.e. $$G = \frac{F}{2B}\frac{du}{da} \qquad (22.21)$$

Now, force F can be defined in terms of the compliance, C (the inverse of the stiffness),

i.e. $$F = \frac{1}{C}u \quad \text{or} \quad u = FC \qquad (22.22)$$

Hence, from equation (22.21),

$$G = \frac{F}{2B}\frac{d(FC)}{da} = \frac{F^2}{2B}\frac{dC}{da} \qquad (22.23)$$

The compliance of a cantilever can be obtained from beam theory.
The deflection of a cantilever is a well-known standard form (see Section 7.2, Equation (7.9)).

Maximum deflection, $$\delta = \frac{1}{3}\frac{FL^3}{EI} \qquad (22.24)$$

where E is the Young's modulus, and I is the second moment of area of the section.
For a double cantilever, a factor of 2 is included, giving:

$$u = \frac{2}{3}\frac{Fa^3}{EI} \qquad (22.25)$$

Since $C = \dfrac{u}{F}$ from equation (22.22), then

$$C = \frac{2}{3}\frac{a^3}{EI} \qquad (22.26)$$

The second moment of area of the rectangular section,

$$I = \frac{Bh^3}{12} \qquad (22.27)$$

Hence, from equations (22.26) and (22.27),

$$C = \frac{2}{3}\frac{a^3}{EI} = \frac{2}{3}\frac{a^3}{E\left(\frac{Bh^3}{12}\right)} = \frac{8a^3}{EBh^3} \qquad (22.28)$$

Substituting into equation (22.23) gives:

$$G = \frac{F^2}{2B}\frac{d}{da}\left(\frac{8a^3}{EBh^3}\right) = \frac{F^2}{2B}\left(\frac{24a^2}{EBh^3}\right) \qquad (22.29)$$

i.e. $$\boldsymbol{G = \frac{12F^2a^2}{EB^2h^3}} \qquad (22.30)$$

Problem 5. A rectangular block of aluminium is fixed at one end and the other end is cracked. The block carries equal and opposite pure moments on each of the two cantilevers formed by the crack, tending to open the crack further. Determine an expression for the energy release rate, G, for the edge crack shown in Figure 22.4.

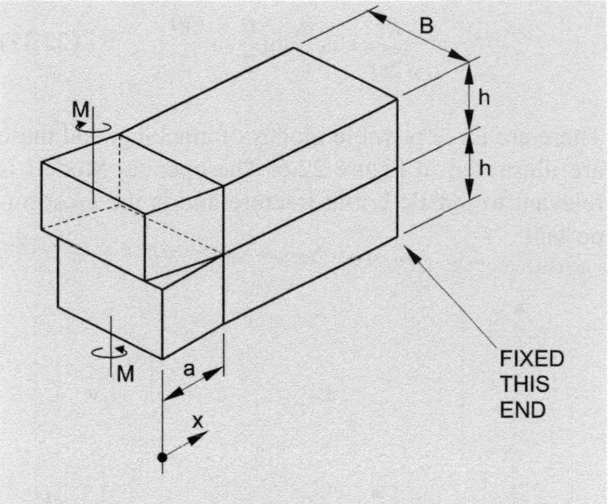

Figure 22.4 A pure moment applied to an edge crack

The moment M, is a constant load and so from equation (22.16) the energy balance is:

$$G\,\Delta A = \Delta W_{EXT} - \Delta U$$

Now A is the area of the crack and B is the thickness of the plate.

So, $\Delta A = B\,da$

From equation (22.18), $G = \dfrac{1}{B}\dfrac{d}{da}\left(\Delta W_{EXT} - \Delta U\right)$

In this case, M is a constant and so the work done is twice the value for a variable moment.

The work done by a single cantilever, $\Delta W_{EXT} = \displaystyle\int \frac{M^2 dx}{EI}$

Incorporating the factor of 2 for both cantilevers, gives:

$$\Delta W_{EXT} = 2\int_0^a \frac{M^2 dx}{EI} = 2\frac{M^2}{EI}\left[x\right]_0^a = 2\frac{M^2 a}{EI}$$

The strain energy, U, stored in a cantilever $= \dfrac{1}{2}\displaystyle\int \frac{M^2 dx}{EI}$

Again, M is a constant, so incorporating the factor of 2 for both cantilevers gives:

$$\Delta U = 2\left(\frac{1}{2}\int_0^a \frac{M^2 dx}{EI}\right) = \frac{M^2 a}{EI}$$

Substituting into equation (22.18) gives:

$$G = -\frac{1}{B}\frac{d}{da}\left(\frac{M^2 a}{EI} - 2\frac{M^2 a}{EI}\right) = +\frac{1}{B}\frac{d}{da}\left(\frac{M^2 a}{EI}\right) \quad (22.31)$$

$$\frac{da}{da} = 1 \text{ and therefore, } \quad G = \frac{1}{B}\left(\frac{M^2}{EI}\right) \quad (22.32)$$

The second moment of area of this rectangular section,

$$I = \frac{hB^3}{12}$$

Hence, from equation (22.32),

$$G = \frac{1}{B}\frac{M^2}{E}\left(\frac{12}{hB^3}\right) = \frac{1}{B}\left(\frac{M^2}{E\left(\frac{hB^3}{12}\right)}\right) \quad (22.33)$$

Hence, **the energy release rate,** $\boldsymbol{G = \dfrac{12M^2}{EB^4 h}}$ (22.34)

Now try the following Practice exercise

Practice Exercise 74. Strain energy release

1. A semi-circular section, double cantilever beam (DCB) test is to be used to derive a value of the energy release rate, G, in a similar manner to Problem 4 on page 452. The diagram of the test set up is the same and is shown in Figure 22.3. The cantilevers are formed from a circular bar of diameter, D, and is subjected to a constant load via a pin joint (so that it can rotate). It is shown diagrammatically in Figure 22.3. Assuming perfect symmetry and material homogeneity and that the crack follows the centre of the beam, determine an expression for the energy release rate, G. The second moment of area of each of the two cantilevers can be taken as: $I = 0.006863\,D^4$

$$\left[G = \frac{145.7\,F^2 a^2}{E\,D^5}\right]$$

2. A double cantilever beam (DCB) type test is to be prepared to investigate a material. It has a Young's modulus of 200 GPa and the expected energy release rate, G is around 20,000 J/m^2. Each cantilever of the test piece is 25 mm wide and has a depth of 8 mm. What length of crack would be expected at an applied load of 10 kN?

[$a = 32.7$ mm]

3. A rectangular-sectioned double cantilever beam (DCB) specimen is loaded in a tensile testing machine. The thickness of the DCB specimen is 25 mm, the depth of each cantilever is 10 mm and the crack length is 45 mm. It is made of steel with a modulus of 205 GPa. If the crack is about to propagate at a tensile load of 14.5 kN, calculate the critical energy release rate of the steel.

$\left[G = 39\,880\,\mathrm{J/m^2}\right]$

22.5 The stress intensity approach

G. R. Irwin showed that the field around any crack tip was always of the same form. This stress field could be characterised by a stress intensity factor (SIF), K, which is a function of the applied load, crack size and the geometry. The coordinate system used around the crack tip is shown in Figure 22.5.

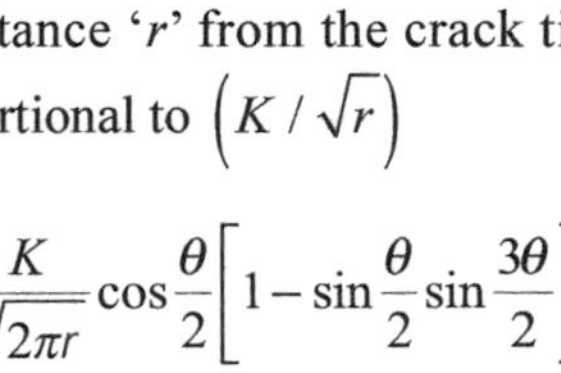

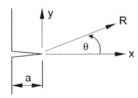

Figure 22.5 Crack tip coordinate system

At a point distance 'r' from the crack tip, the elastic stress is proportional to $\left(K/\sqrt{r}\right)$

$$\sigma_x = \frac{K}{\sqrt{2\pi r}}\cos\frac{\theta}{2}\left[1 - \sin\frac{\theta}{2}\sin\frac{3\theta}{2}\right] \tag{22.35}$$

$$\sigma_y = \frac{K}{\sqrt{2\pi r}}\cos\frac{\theta}{2}\left[1 + \sin\frac{\theta}{2}\sin\frac{3\theta}{2}\right] \tag{22.36}$$

$$\tau_{xy} = \frac{K}{\sqrt{2\pi r}}\cos\frac{\theta}{2}\sin\frac{\theta}{2}\sin\frac{3\theta}{2} \tag{22.37}$$

There are three possible modes of cracking, and these are illustrated in Figure 22.6. The opening Mode I is relevant to tensile brittle fracture and is the most important.

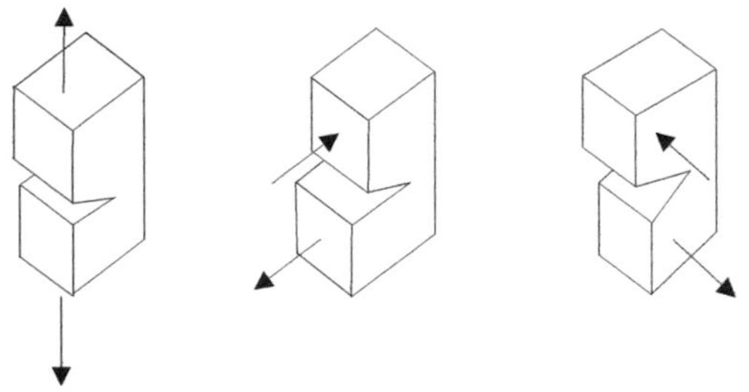

Figure 22.6 The three modes of crack growth

The opening mode stress intensity factor was shown by Irwin to be:

$$K_I = \sigma\sqrt{\pi a}\,\alpha \tag{22.38}$$

where σ is the gross tensile stress perpendicular to the crack,

a is the crack length, and
α is the geometry correction factor, which is approximately 1.

The critical value of K at which unstable crack growth occurs is a property of that material under plane strain conditions (see Section 22.6). Mechanical properties in the presence of a crack can be expressed in terms of stress intensity factor K, in the same way that properties of plain specimens are expressed in terms of stress.

Fracture toughness is the critical stress intensity factor and is designated K_{IC}. It predicts the onset of unstable crack growth in a cracked specimen. It can be used in a similar way to a yield stress (or a 0.2% proof stress) as a critical stress level allowing a specified amount of plastic deformation in a plain specimen. The units used for fracture toughness appear strange initially. They are consistent and logical and become familiar with use. The units are N/m$^{3/2}$.

$$\text{From equation (22.14)}, \quad \left(EG_C\right)^{\frac{1}{2}} = \sigma\left(\pi a\right)^{\frac{1}{2}} \tag{22.39}$$

This equation links the material properties G_C and K_{IC}, such that:

$$(EG_C)^{\frac{1}{2}} = K_{IC} \quad (22.40)$$

Fracture toughness can be regarded as the critical stress intensity factor (or critical toughness for unstable crack growth in a cracked specimen).

22.6 Plane stress and plane strain

For linear isotropic materials deforming elastically, the stress-strain relations are well known (see equations (9.47) to (9.49) on page 216). They are:

$$\varepsilon_x = \frac{1}{E}(\sigma_x - \upsilon\sigma_y - \upsilon\sigma_z) \quad (22.41)$$

$$\varepsilon_y = \frac{1}{E}(\sigma_y - \upsilon\sigma_x - \upsilon\sigma_z) \quad (22.42)$$

$$\varepsilon_z = \frac{1}{E}(\sigma_z - \upsilon\sigma_x - \upsilon\sigma_y) \quad (22.43)$$

Consider a thin plate that is deformed in plane stress. On the free surfaces, the out-of-plane stresses (in the z direction) are zero and they are usually negligible in the interior points of the plate. For plane stress it is assumed:

$$\sigma_z = \tau_{yz} = \tau_{xz} = 0$$

Therefore, the plate carries only in-plane stresses. The stress-strain relations are simplified to:

$$\varepsilon_x = \frac{1}{E}(\sigma_x - \upsilon\sigma_y) \quad (22.44)$$

$$\varepsilon_y = \frac{1}{E}(\sigma_y - \upsilon\sigma_x) \quad (22.45)$$

On the other hand, the plane strain case corresponds to a sufficiently thick plate for which:

(i) displacement in z direction is restricted
(ii) variation in z direction is zero

i.e. $$\varepsilon_z = \varepsilon_{xz} = \varepsilon_{yz} = 0 \quad (22.46)$$

Therefore, for plane strain cases, simplified stress-strain equations are obtained:

$$\sigma_z = \upsilon(\sigma_x + \sigma_y) \quad (22.47)$$

Substituting in equation (22.41) gives:

$$\varepsilon_x = \frac{1}{E}\left(\sigma_x - \upsilon\sigma_y - \upsilon\left[\upsilon(\sigma_x + \sigma_y)\right]\right) \quad (22.48)$$

i.e. $$\varepsilon_x = \frac{1}{E}(\sigma_x - \upsilon\sigma_y - \upsilon^2\sigma_x - \upsilon^2\sigma_y) \quad (22.49)$$

i.e. $$\varepsilon_x = \frac{1}{E}\left(\sigma_x(1-\upsilon^2) - \sigma_y\upsilon(1+\upsilon)\right) \quad (22.50)$$

Rearranging gives:

$$\varepsilon_x = \left(\frac{1-\upsilon^2}{E}\right)\left\{\sigma_x - \sigma_y\upsilon\left(\frac{1+\upsilon}{1-\upsilon^2}\right)\right\} \quad (22.51)$$

However, $$\upsilon\left(\frac{1+\upsilon}{1-\upsilon^2}\right) = \upsilon\left(\frac{1+\upsilon}{(1-\upsilon)(1+\upsilon)}\right) = \frac{\upsilon}{1-\upsilon} \quad (22.52)$$

From equation (22.51),

$$\varepsilon_x = \left(\frac{1-\upsilon^2}{E}\right)\left\{\sigma_x - \sigma_y\left(\frac{\upsilon}{1-\upsilon}\right)\right\} \quad (22.53)$$

Similarly, $$\varepsilon_y = \left(\frac{1-\upsilon^2}{E}\right)\left\{\sigma_y - \sigma_x\left(\frac{\upsilon}{1-\upsilon}\right)\right\} \quad (22.54)$$

This means that the stress-strain equations, equation (22.41) and equation (22.42) for plane stress conditions are also applicable for plane strain if the following changes are made:

$$E^* = E \text{ for plane stress} \quad (22.55)$$

and

$$E^* = \left(\frac{E}{1-\upsilon^2}\right) \text{ and } \upsilon^* = \left(\frac{\upsilon}{1-\upsilon}\right) \text{ for plane strain} \quad (22.56)$$

22.7 Plane stress and plain strain behaviour

Crack tip stress fields are essentially plane stress or plane strain.

A thin plate loaded in uniaxial tension is in a state of **plane stress**. When a crack is introduced, the highly stressed material at the crack tip is constrained by the less highly stressed surrounding material and stresses are introduced in the through thickness direction at the crack tip. This scenario creates a **plane strain** situation.

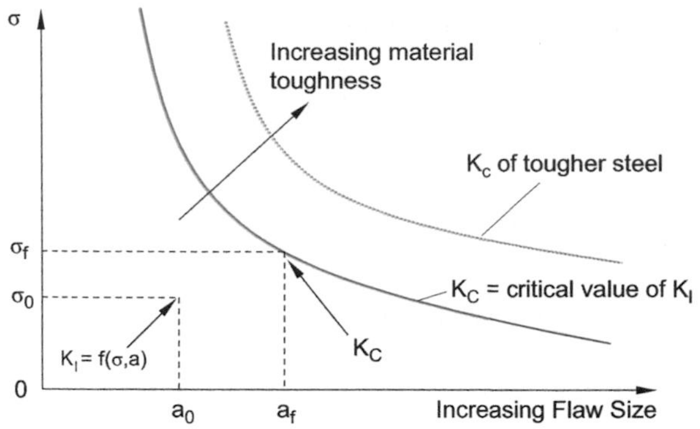

Figure 22.7 Relation between stress, flaw size and material toughness

Figure 22.7 shows the relation between stress, flaw size and material toughness.

22.8 Allowance for small scale yielding at crack tip

From equation (22.38) for an 'infinite' plate ($\alpha = 1$):

$$K = \sigma\sqrt{\pi a} \tag{22.57}$$

For limited plasticity at the crack tip, consider the crack lengthened by twice the radius of the plastic zone at the crack tip (see Figure 22.8).

From equation (22.36),

$$\sigma_y = \frac{K}{\sqrt{2\pi r}}\cos\frac{\theta}{2}\left[1+\sin\frac{\theta}{2}\sin\frac{3\theta}{2}\right] \tag{22.58}$$

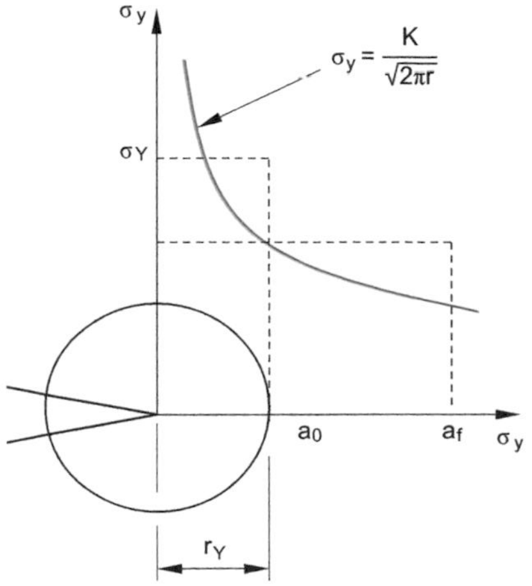

Figure 22.8 Limited plasticity at the crack tip

When $\theta = 0$, $\cos\theta = 1$ and $\sin\theta = 0$

Thus, $$\sigma_y = \frac{K}{\sqrt{2\pi r}} \tag{22.59}$$

σ_y, the stress in the y direction $= \sigma_{yield}$

Therefore, in Figure 22.8, $$r_y = \frac{1}{2\pi}\left(\frac{K}{\sigma_y}\right)^2 \tag{22.60}$$

For plane strain, the plastic zone is less and is given approximately by:

$$r_y = \frac{1}{6\pi}\left(\frac{K}{\sigma_y}\right)^2 \text{ for plane strain} \tag{22.61}$$

When the crack tip plastic zone size becomes comparable to the thickness of the plate, yielding occurs in the through thickness direction, allowing relaxation of through thickness stresses so that the whole plate approaches a state of plane stress.

Figure 22.9 shows plane strain and plane stress in cracked plates, using the rounded factor of 2.5 (≈ 2.65).

Fracture toughness, K_{IC}, is a minimum under plane strain conditions. To obtain these conditions, it is generally accepted that the specimen thickness must be up to about 50 times the radius of the plane strain plastic zone at the crack tip,

i.e. specimen thickness,

$$B = 50r_p = 50\left(\frac{1}{6\pi}\right)\left(\frac{K_I}{\sigma_{yield}}\right)^2 = 2.65\left(\frac{K_I}{\sigma_{yield}}\right)^2 \tag{22.62}$$

The important point is to ensure that testing obtains plane strain conditions, and to ensure this, ASTM test requirements are that:

the specimen thickness, $$B > 2.5\left(\frac{K_I}{\sigma_{yield}}\right)^2 \tag{22.63}$$

Equation (22.63) gives the accepted minimum specimen thickness to ensure strain conditions.

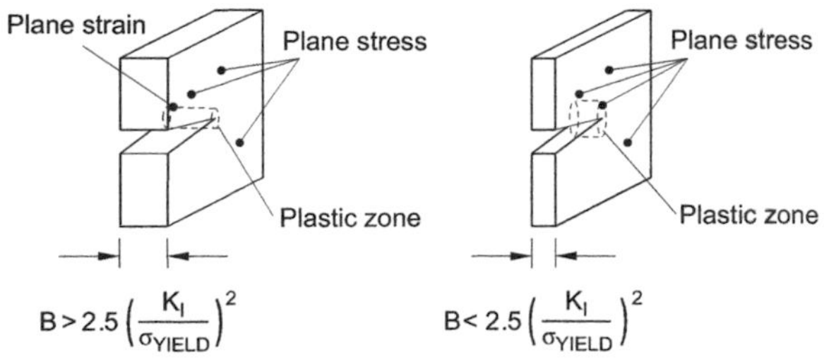

Figure 22.9 Plane strain and plane stress in cracked plates

22.9 Fracture toughness crack tip opening displacement (CTOD)

For high strength steels and other materials, it becomes impossible to achieve plane strain conditions even with very thick specimens. Here another approach such as the 'crack tip opening displacement'(CTOD) approach method has to be used to determine material properties experimentally.

This method of characterising a crack can be used at and beyond general yield. Wells showed that there is a critical value of CTOD at which crack growth starts. Unlike parameters G and K, it can be used for both linear elastic fracture mechanics (LEFM) and elastic-plastic fracture mechanics (EPFM).

Figure 22.10 shows a typical tension crack test specimen and Figure 22.11 shows an assumed plastic zone around a crack tip.

Dugdale showed that:

the crack tip opening displacement (CTOD),

$$\delta = \frac{\pi\sigma^2 a}{E\sigma_y} \qquad (22.64)$$

For a through thickness crack in an infinite plate:

$$K_I = \sigma\sqrt{\pi a} \text{ from which, } \sigma = \frac{K_I}{\sqrt{\pi a}} \qquad (22.65)$$

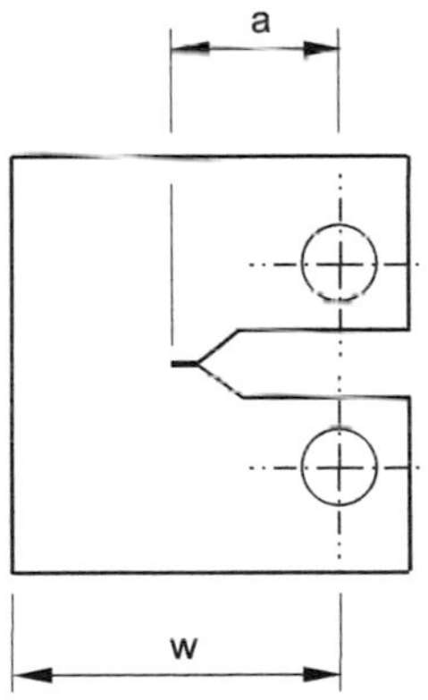

Figure 22.10 A typical tension crack test specimen

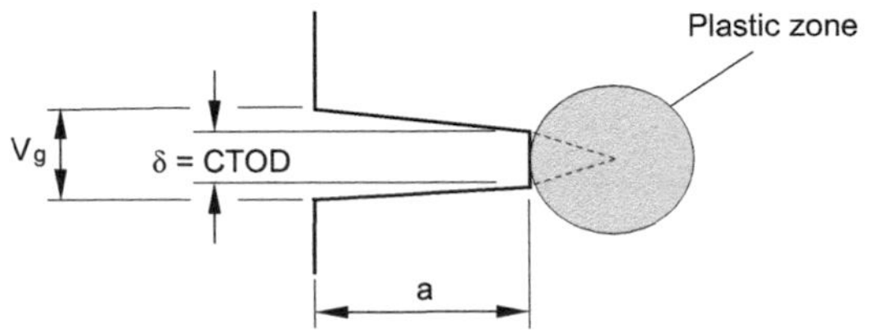

Figure 22.11 Assumed plastic zone around a crack tip

Substituting in equation (22.64) gives:

$$\delta = \frac{\pi\left(\frac{K_I}{\sqrt{\pi a}}\right)^2 a}{E\sigma_y} = \frac{\pi\left(\frac{K_I^2}{\pi a}\right)a}{E\sigma_y}$$

Hence, $(\text{CTOD}) = \delta = \dfrac{K_I^2}{E\sigma_y}$ for plane stress (22.66)

and $K_{IC} = \sqrt{\delta_C E\sigma_y}$ for plane stress (22.67)

Problem 6. The stress intensity expression at a flaw site in a component is $K_I = 0.9\sigma\sqrt{\pi a}$

For a fracture toughness $K_{IC} = 60$ MN/m$^{3/2}$ find:

(a) the critical stress for a flaw where $a = 5$ mm,
(b) the critical flaw size for a stress of 1000 MPa

(a) $K_I = 0.9\sigma\sqrt{\pi a}$

from which, **critical stress**,

$$\sigma = \frac{K_I}{0.9\sqrt{\pi a}} = \frac{60\times10^6}{0.9\sqrt{\pi(0.005)}}$$

$$= 532\times10^6 \text{ Pa} = \mathbf{532\ MPa}$$

(b) Since $K_I = 0.9\sigma\sqrt{\pi a}$ then $K_I^2 = (0.9\sigma)^2\pi a$

and **the critical flaw size,**

$$a = \frac{K_I^2}{0.9^2\sigma^2\pi} = \frac{(60\times10^6)^2}{(0.9)^2(1000\times10^6)^2\pi}$$

$$= \mathbf{1.415\times10^{-3}\ m = 1.415\ mm}$$

Problem 7. A part of a satellite launching system incorporates a casing which is 6.5 m in diameter and has a wall thickness of 18.5 mm. The casing must withstand a test pressure of 6.6 MPa working on the basis that there could be a flaw of up to 5 mm in the wall, normal to the direction of the hoop stress, as shown in Figure 22.12. For this geometry, $K_I = \sigma\sqrt{\pi a}$. The material options are:

(1) A steel where σ_Y = 1650 MPa and K_{IC} = 60.5 MN / m$^{3/2}$

(2) A steel where σ_Y = 1520 MPa and K_{IC} = 165 MN / m$^{3/2}$

Find the maximum stress and determine the acceptable size of flaw for each material.

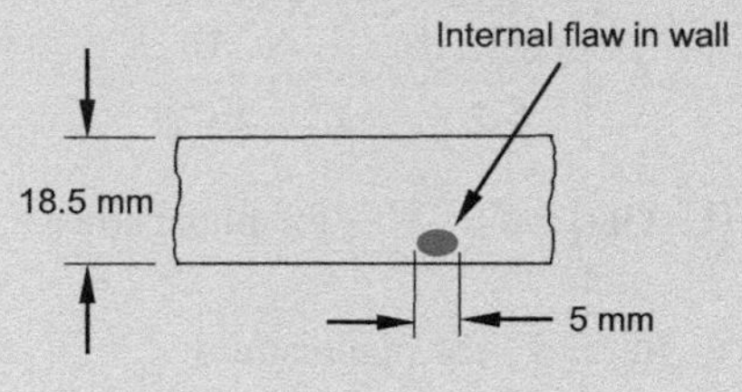

Figure 22.12 Location of flaw in wall

From equation (10.3), page 234,

$$\sigma_{Hoop} = \frac{pd}{2t} = \frac{(6.6\times10^6)(6.5)}{2(0.0185)} = 1159\,\text{MPa}$$

$K_I = \sigma\sqrt{\pi a}$ from which, $K_I^2 = \sigma^2 \pi a$ and

$$a = \frac{K_I^2}{\sigma^2 \pi} = \left(\frac{K_I}{\sigma}\right)^2\left(\frac{1}{\pi}\right)$$

An internal crack means that $a = 2.5$ mm

For Material 1:

$$\textbf{Flaw size, } a = \left(\frac{K_I}{\sigma}\right)^2\left(\frac{1}{\pi}\right) = \left(\frac{60.5\times10^6}{1159\times10^6}\right)^2\left(\frac{1}{\pi}\right)$$

$$= \mathbf{867.3\times10^{-6}\,m} \text{ or } \mathbf{0.8673\,mm}$$

This is less than 2.5 mm minimum value for 'a'.

For Material 2:

$$\textbf{Flaw size, } a = \left(\frac{K_I}{\sigma}\right)^2\left(\frac{1}{\pi}\right) = \left(\frac{165\times10^6}{1159\times10^6}\right)^2\left(\frac{1}{\pi}\right)$$

$$= \mathbf{6.451\times10^{-3}\,m} \text{ or } \mathbf{6.451\,mm}$$

Material 1 would be expected to fail before proof pressure is reached.

Material 2 is satisfactory. It would stand the proof pressure even with a flaw size $2a = 12.9$ mm, which is about 2/3 of the wall thickness.

Figure 22.13 shows graphs of stress versus half crack length for the two materials.

Now try the following Practice exercise

Practice Exercise 75. Applications of stress intensity

1. A large sheet containing a 50 mm long internal crack fracture when loaded to 500 MPa. Determine the fracture load of a similar sheet with a 100 mm long internal crack. Take α to be 1.2 [$\sigma = 353.6$ MPa]

2. (a) Find the critical stress (see Figure 22.14) when the crack length is 20 mm, $\alpha = 1.2$ and the material has a fracture toughness, $K_{IC} = 35\ \text{MN/m}^{3/2}$.

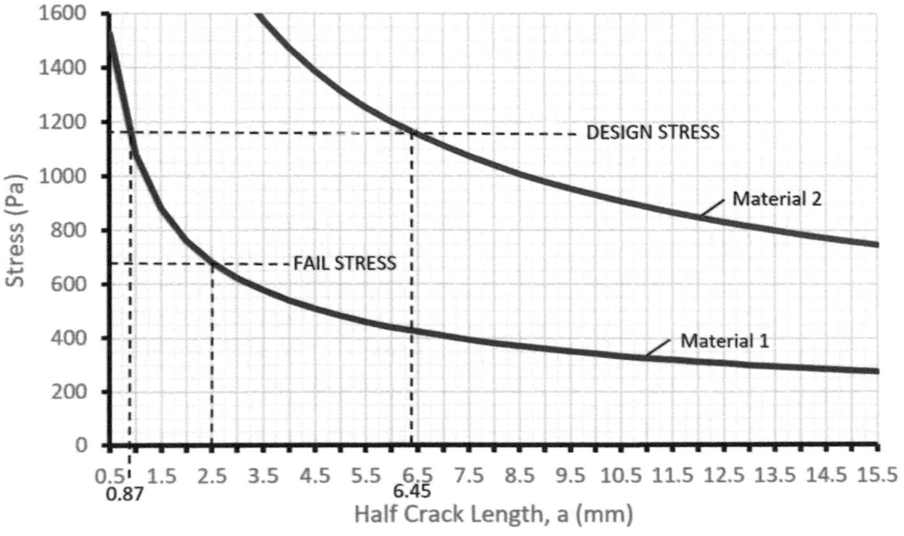

Figure 22.13 Stress versus half crack length for the materials in Problem 7

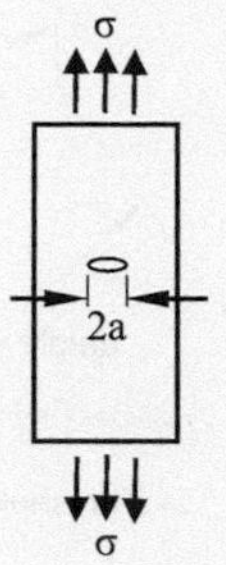

Figure 22.14

(b) What is the critical crack length for the same material at a stress of 550 MN/m$^{3/2}$?

[(a) $\sigma = 164.6$ MPa (b) Critical crack length = 1.79 mm]

3. A plate is made from a steel with a Young's modulus of 920 MPa and it has a critical stress intensity factor, K_{IC}, of 45 MN/m$^{3/2}$. What is the minimum thickness B for this plate to give plane strain behaviour?

[$B = 5.98$ mm]

4. A pressure vessel is made from a material with a 0.2% proof stress of 1300 MPa and a critical stress intensity factor of 55 MN/m$^{3/2}$. The vessel was constructed by welding, and has an outer diameter of 3.950 m and an inner diameter of 3.930 m. The vessel is found to have a welding flaw (edge) of 3.2 mm depth and roughly semi-circular in shape, for which α can be taken to be 0.9

(a) Calculate the stress level at which the vessel will fail.
(b) Determine the permissible flaw size to withstand a hoop stress equal to the proof stress.

[(a) 609.5×10^{6} Pa (b) a = 0.703 mm]

5. A 75 mm diameter internally pressurised polycarbonate pipe with a wall thickness of 3 mm failed in service. It was found to have an elliptical crack which grew from a surface scratch, aligned with the cylindrical axis of the pipe. If the working pressure was 30 bar and the K_{IC} value is 1.8 MN/m$^{3/2}$, estimate the depth of the scratch that would have propagated to cause the failure. Ignore any safety factors that would normally be incorporated and assume α to be 0.9.

[$a_C = 905.4\times10^{-6}$m or 0.905 mm]

6. A sheet of brittle material is 0.5 m wide and 4 mm thick. It is held horizontally on two supports at the extreme ends as shown in Figure 22.15. Calculate the maximum length of the sheet to avoid fracture under its own weight, if there is a surface crack 3 mm deep and 12 mm long on the underside of the sheet. Take the density of the material to be 2500 kg/m^3, $K_{IC} = 0.25$ MN/m$^{3/2}$, and $\alpha = 0.9$

[$L = 0.789$ m]

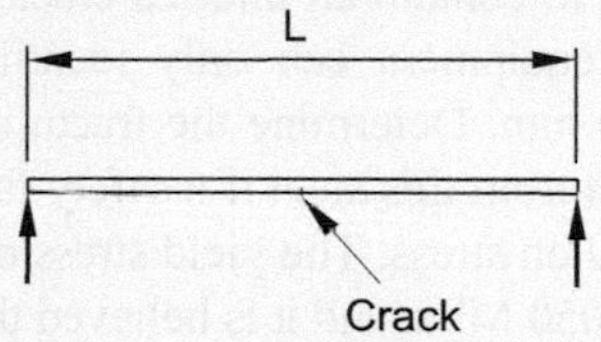

Figure 22.15

7. A large plate carrying a tensile load contains a 6 mm diameter hole with two cracks as shown in Figure 22.16. Each crack is of length $L = 2$ mm, and propagates from the hole in a direction perpendicular to the applied load. The stress, $\sigma = 170$ MPa.

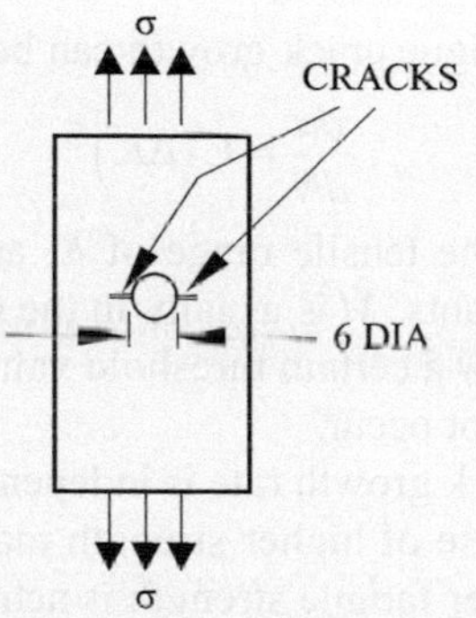

Figure 22.16

Calculate the stress intensity factor (K_{IC}) making the following assumptions:

(a) Each crack is an external crack, where $\alpha = 1.12$
(b) The hole with the cracks has the effect of one continuous crack, where $\alpha = 1.0$

[(a) $K_{IC} = 15.09\times10^{6}\,\mathrm{N/m^{3/2}}$
(b) $K_{IC} = 21.31\times10^{6}\,\mathrm{N/m^{3/2}}$]

8. Engineers are aware that a 20 mm long crack exists in a wide sheet of aluminium alloy, which is approximately 5 mm below the surface. The

fracture stress for the sheet is 180 MPa and the yield stress of the material is 300 MPa. Calculate the fracture toughness of the material using plastic zone correction and compare this to a value obtained using linear elastic fracture mechanics. It can be assumed that α is equal to 1.0

[(a) $K_{IC} = 34.66 \times 10^6 \text{N/m}^{3/2}$

(b) $K_{IC} = 31.90 \times 10^6 \text{N/m}^{3/2}$]

9. Part of a large component of lifting equipment is thought to contain an internal crack. The crack testing equipment can only identify flaws of over 20 mm. Determine the fracture toughness required from this steel if a safety factor of 2 is required on stress. The yield stress of the material is 1050 MPa, and it is believed that this material would exhibit yielding around the crack tip. Take α to be 0.64 for the crack geometry.

[$K_{IC} = 61.06 \times 10^6 \text{N/m}^{3/2}$]

22.10 Application of fracture mechanics to fatigue crack growth

The rate of fatigue crack growth can be expressed as:

$$\frac{da}{dN} = C(\Delta K)^m \qquad (22.68)$$

where ΔK is the tensile range of K, and C and m are material constants. M is usually in the range of 3 to 4. If ΔK is below a certain threshold value, fatigue crack growth does not occur.
As fatigue crack growth rate is independent of material strength, the use of higher strength materials does not mean that better fatigue strength is achieved.
Figure 22.17 shows threshold, with steady and fast crack growth (with logarithmic scales).

From equation (22.38), $\Delta K = \Delta\sigma(\pi a)^{\frac{1}{2}}\alpha$

From equation (22.68),

$$\frac{da}{dN} = C(\Delta K)^m = C\left(\Delta\sigma(\pi a)^{\frac{1}{2}}\alpha\right)^m = C\left(\Delta\sigma\,\pi^{\frac{1}{2}}\alpha\right)^m a^{\frac{m}{2}} \qquad (22.69)$$

An expression for the number of cycles to failure can be obtained by inverting this equation and then integrating both sides,

$$\text{i.e.}\quad \frac{dN}{da} = C^{-1}\left(\Delta\sigma\,\pi^{\frac{1}{2}}\alpha\right)^{-m} a^{-\frac{m}{2}} \qquad (22.70)$$

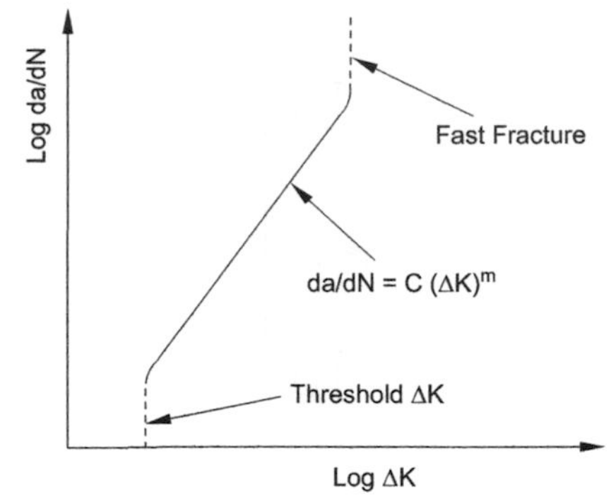

Figure 22.17 Threshold, steady and fast crack growth (log scales)

$$\text{and}\quad dN = \left(C^{-1}\left(\Delta\sigma\,\pi^{\frac{1}{2}}\alpha\right)^{-m} a^{-\frac{m}{2}}\right)da \qquad (22.71)$$

$$\text{Then}\quad \int_0^N dN = \int_{a_0}^{a_f}\left(C^{-1}\left(\Delta\sigma\,\pi^{\frac{1}{2}}\alpha\right)^{-m} a^{-\frac{m}{2}}\right)da \qquad (22.72)$$

where a_0 and a_f are the initial and final crack lengths, respectively.

Now $C^{-1}\left(\Delta\sigma\,\pi^{\frac{1}{2}}\alpha\right)^{-m}$ is a constant,

$$\text{Hence}\quad N = C^{-1}\left(\Delta\sigma\,\pi^{\frac{1}{2}}\alpha\right)^{-m}\int_{a_0}^{a_f} a^{-\frac{m}{2}}da \qquad (22.73)$$

$$= C^{-1}\left(\Delta\sigma\,\pi^{\frac{1}{2}}\alpha\right)^{-m}\left[\frac{a^{-\frac{m}{2}+1}}{-\frac{m}{2}+1}\right]_{a_0}^{a_f} \qquad (22.74)$$

$$= C^{-1}\left(\Delta\sigma\,\pi^{\frac{1}{2}}\alpha\right)^{-m}\left[a_f^{\left(-\frac{m}{2}+1\right)} - a_0^{\left(-\frac{m}{2}+1\right)}\right]\left(\frac{1}{-\frac{m}{2}+1}\right) \qquad (22.75)$$

Now $a_f >> a_0$ and m is usually in the range 3 to 4

This means that $a_f^{\left(-\frac{m}{2}+1\right)}$ is small compared to $a_0^{\left(-\frac{m}{2}+1\right)}$

$$\text{Hence,}\quad N = C^{-1}\left(\Delta\sigma\,\pi^{\frac{1}{2}}\alpha\right)^{-m}\left[-a_0^{\left(-\frac{m}{2}+1\right)}\right]\left(\frac{1}{-\frac{m}{2}+1}\right) \qquad (22.76)$$

Now C, $\pi^{\frac{1}{2}}$, α and m are constants

$$\text{Thus,}\quad N=\left(\frac{1}{\text{constant}}\right)\Delta\sigma^{-m}\left[a_0^{\left(-\frac{m}{2}+1\right)}\right]$$

$$=\left(\frac{1}{\text{constant}}\right)\left(\frac{1}{a_0^{\left(\frac{m}{2}-1\right)}\Delta\sigma^{m}}\right) \qquad (22.77)$$

Problem 8. A component within a chemical plant undergoes load cycles that cause maximum stresses of 1×10^9 Pa. The component is known to be defective in that it has a 6 mm long internal void. This crack is expected to cause failure if it is allowed to grow to 14 mm long. Determine the number of stress applications until failure is likely.

For the material, $\frac{da}{dN}=208.7\times10^{-27}(\Delta K)^{2.25}$

where $\frac{da}{dN}$ is in m/cycle and ΔK is in N/m$^{3/2}$.

It is an internal crack, so $a_0 = 3$ mm, $a_f = 7$ mm and maximum stress = 10^9 Pa

The following solution uses SI units.

$$\frac{da}{dN}=208.7\times10^{-27}(\Delta K)^{2.25}\ \text{m/cycle}$$

The stress range is from zero to the maximum stress, giving $\Delta\sigma = 10^9$ Pa

Substituting $\Delta K=\Delta\sigma\sqrt{\pi a}$ (from equation 22.57) gives:

$$\frac{da}{dN}=208.7\times10^{-27}\left(\Delta\sigma\sqrt{\pi a}\right)^{2.25}\ \text{m/cycle}$$

$$\text{i.e.}\quad \frac{da}{dN}=208.7\times10^{-27}\left(10^9\sqrt{\pi a}\right)^{2.25}$$

$$\text{i.e.}\quad \frac{da}{dN}=208.7\times10^{-27}\left(10^9\right)^{2.25}\pi^{2.25/2}a^{2.25/2}$$

$$\text{i.e.}\quad \frac{da}{dN}=208.7\times10^{-27}\left(10^9\right)^{2.25}\pi^{1.125}a^{1.125}$$

$$\text{i.e.}\quad \frac{da}{dN}=\left(134.5\times10^{-6}\right)\left(a^{1.125}\right)$$

Rearranging gives: $a^{-1.125}da=\left(134.5\times10^{-6}\right)dN$

$$\text{and}\quad \int_{a_0}^{a_f}a^{-1.125}da=\int_0^N\left(134.5\times10^{-6}\right)dN$$

where a_0 and a_f are the initial and final crack lengths of 3 mm and 7 mm respectively.

$$\text{Hence,}\quad \left[\frac{a^{-1.125+1}}{-1.125+1}\right]_{0.003}^{0.007}=\left[134.5\times10^{-6}N\right]_0^N$$

$$\text{and}\quad \frac{0.007^{-0.125}-0.003^{-0.125}}{-0.125}=134.5\times10^{-6}N$$

from which,

$$N=\left(\frac{0.007^{-0.125}-0.003^{\ 0.125}}{-0.125}\right)\left(\frac{1}{134.5\times10^{-6}}\right)$$

$$=\left(\frac{1.8594-2.0671}{-0.125}\right)\left(\frac{1}{134.5\times10^{-6}}\right)$$

i.e. **the number of stress applications until failure, $N = 12{,}350$ cycles**

Now try the following Practice exercise

Practice Exercise 76. Fracture mechanics applied to cyclic loading

1. A crankshaft of an internal combustion engine failed after several hundred hours of operation. Examination showed that failure was caused by a fatigue crack which originated at a forging lap about 2 mm deep, situated where the stress level alternated between +100 MPa and −100 MPa. The material used has a threshold value of ΔK for fatigue crack growth of 7.0 MN/m$^{3/2}$ at zero mean stress. Is this failure likely to be due to the producers of the crankshaft or is the failure more likely to have been due to maintenance and/or operation? Take the geometry factor, α, to be 1.
[$\Delta K = 15.85\,\text{MN/m}^{3/2}$; this is greater than the ΔK for fatigue crack growth of 7.0 MN/m$^{3/2}$ for this material. Therefore the 2 mm crack would be expected to have caused the failure]

2. After manufacture, ultrasonic examination of a machined part of a hydraulic system revealed that a plate had centre-line laminations (flaws). The equipment operates at pressures that load the plate to stresses of up to 50 MPa. The ΔK_C for the steel under zero tension loading in oil can be taken as 5.5 $MN/m^{3/2}$. The geometric factor, α, can be assumed to be unity. What size of flaw would be acceptable to the designers?
[Maximum acceptable crack size = 7.70 mm]

3. Non-destructive testing has identified a 5 mm internal flaw in a component that is subjected to a cyclic load of ±50 MPa. The rate of fatigue crack growth is given by:

$$\frac{da}{dN} = \left(6\times10^{-30}\right)\left(\Delta K\right)^3$$

where da/dN is in m/cycle and ΔK is in $N/m^{3/2}$. The number of cycles to failure, N, found by integrating the crack growth equation is:

$$N = \frac{1}{a_0^{(m/2)-1}\, C\Delta\sigma^m \pi^{m/2}\left(m/2-1\right)\alpha^m}$$

Assuming $\alpha = 1$ for the crack, estimate the number of cycles to failure of the component.

$$\left[N = 1.197\times10^6 \text{cycles}\right]$$

22.11 The J-Integral

The J-Integral, or J-contour Integral, was developed by ***James R Rice** in 1968. The technique is based on finding that for a two-dimensional crack, the sum of the strain energy density and the work done along an anticlockwise continuous path completely enclosing the crack tip, are independent of the path taken.

The J-Integral is defined as the line integral:

$$J = \int_p \left(W dy - T_i \frac{\partial u_i}{\partial x} ds\right) \quad (22.78)$$

where W is the strain energy density, T is the traction vector, u is the displacement vector (defined according to the outward normal n), along a path p and s is the arc length.

Figure 22.18 shows a path for a line integral around a crack.

This line integral appears strange on first sight, but it contains a considerable amount of information. A closer look at the component parts is required.

$$W = \int \sigma_{ij}\, d\varepsilon_{ij} \quad (22.79)$$

W is the energy density term and has SI units of J/m^3. In two dimensions:

$$W = \int_0^{\varepsilon_{11}} \sigma_{11}\, d\varepsilon_{11} + \int_0^{\varepsilon_{11}} \sigma_{12}\, d\varepsilon_{12} + \int_0^{\varepsilon_{11}} \sigma_{21}\, d\varepsilon_{21} + \int_0^{\varepsilon_{11}} \sigma_{22}\, d\varepsilon_{22} \quad (22.80)$$

*James Robert Rice (born 3 December 1940) is an American engineer, scientist, geophysicist, and Mallinckrodt Professor of Engineering Sciences and Geophysics at the Harvard John A. Paulson School of Engineering and Applied Sciences. Rice is known as a mechanician, who has made fundamental contributions to various aspects of solid mechanics. Two of his early contributions are the concept of the J-integral in fracture mechanics and an explanation of how plastic deformations localise in a narrow band. In recent years, Rice has focused on the mechanical processes involved in earthquakes. To find out more about Rice go to www.routledge.com/cw/bird

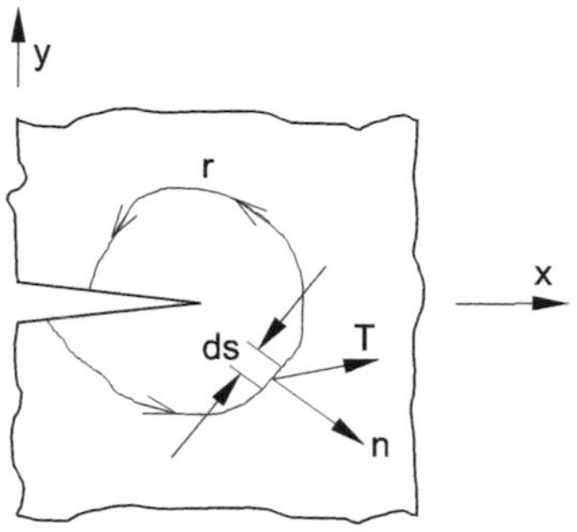

Figure 22.18 A path for a line integral around a crack

i.e. $W = \int_0^{\varepsilon_{11}} \sigma_{11}\, d\varepsilon_{11} + 2\int_0^{\varepsilon_{11}} \sigma_{12}\, d\varepsilon_{12} + \int_0^{\varepsilon_{11}} \sigma_{22}\, d\varepsilon_{22}$ (22.81)

The traction vector, T, is simply the force vector on a cross-section divided by that cross-section's area. Traction T_i at a point on path p is given by:

$$T_i = \sigma_{ij} n_{ij} \quad (22.82)$$

where $\boldsymbol{n}$ is the normal vector, pointing outwards. The *SI* units of $\boldsymbol{T}$ are N/m^2, but unlike stress, it is a vector.

$$T_{ij} = \begin{bmatrix} \sigma_{11} & \sigma_{12} \\ \sigma_{21} & \sigma_{22} \end{bmatrix} \begin{bmatrix} n_1 \\ n_2 \end{bmatrix} \quad (22.82)$$

Hence, $T_1 = \sigma_{11} n_1 + \sigma_{12} n_2$ (22.83)

and $T_2 = \sigma_{21} n_1 + \sigma_{22} n_2$ (22.84)

The second term in the J Integral can be expanded to give:

$$T_i \frac{\partial u_i}{\partial x} ds = \left(T_1 \frac{\partial u_1}{\partial x} + T_2 \frac{\partial u_2}{\partial x} \right) ds \quad (22.85)$$

The units of J are thus N/m.

Furthermore, it can be proven that:

- The J-Integral is path independent, i.e. p can be chosen arbitrarily within the body of a component.
- The J-Integral represents an energy release rate and is equal to G, for linear elastic bodies.

The use of the J-Integral is shown in the following problem.

Problem 9. Derive the J-integral for the double cantilever specimen shown in Figure 22.19, if the ends of each cantilever are pulled by a distributed load F.

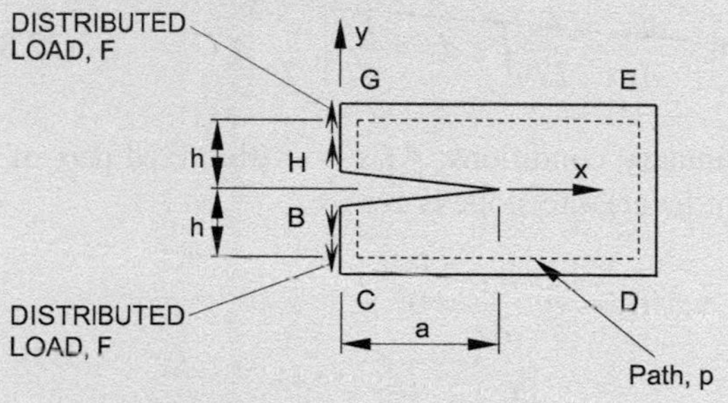

Figure 22.19 A rectangular path around a crack

The path is BCDEGH, following the contour of the specimen. The beauty of the chosen path here, is that many of the component parts of the equation become zero, or negligible.

CD and EG: These do not form part of the J-Integral because dy is negligible and $T = 0$ (as the slope is zero).

DE: The contribution is negligible, because stresses are very small. Consequently, W and T are negligible.

BC and GH: W is negligible. There is no bending stress (as there is no bending moment) and there is negligible shear stress.

The only non-trivial term is:

$$J = -2\int_0^h \left(T \frac{\partial u_i}{\partial x} ds \right) = -2\int_0^h \left(T_1 \frac{\partial u_1}{\partial x} + T_2 \frac{\partial u_2}{\partial x} \right) ds \quad (22.86)$$

However, $\frac{\partial u_1}{\partial x} = 0$ as the vertical lines have zero slope

Thus, $J = -2\int_0^h T_2 \frac{\partial u_2}{\partial x} ds$ (22.87)

$\frac{\partial u_2}{\partial x}$ is the slope and h is the depth of the cantilevered beam.

From conventional bending theory, $\frac{\partial^2 u_2}{\partial x^2} EI$ is equal to the bending moment M of a beam, where E is the Young's modulus and I is the second moment of area of the section.

For a cantilever, the bending moment $M = Fx$, so standard beam theory can be followed.

Thus, $M = \frac{\partial^2 u_2}{\partial x^2} EI = Fx_1$

from which, $\frac{\partial^2 u_2}{\partial x^2} = \frac{Fx_1}{EI}$

and $$\frac{\partial u_2}{\partial x} = \frac{F}{EI}\int x\,dx = \frac{F}{EI}\left(\frac{x^2}{2}\right) + C \qquad (22.88)$$

Boundary conditions: At $x = a$ (the fixed part of the cantilever), the slope is zero,

i.e. when $x = a$, $\frac{\partial u_2}{\partial x} = 0$

$$\text{and } 0 = \frac{F}{EI}\left(\frac{a^2}{2}\right) + C \text{ from which, } C = -\frac{Fa^2}{2EI} \qquad (22.89)$$

$$\text{From equation (22.88) } \frac{\partial u_2}{\partial x} = \frac{Fx^2}{2EI} - \frac{Fa^2}{2EI} \qquad (22.90)$$

Second moment of area, $I = \frac{Bh^3}{12}$

Substituting into equation (22.90) gives:

$$\frac{\partial u_2}{\partial x} = \frac{Fx^2}{2E\left(\frac{Bh^3}{12}\right)} - \frac{Fa^2}{2E\left(\frac{Bh^3}{12}\right)} = \frac{6Fx^2}{EBh^3} - \frac{6Fa^2}{EBh^3} = \frac{6F}{EBh^3}\left(x^2 - a^2\right) \qquad (22.91)$$

However, on segments BC and GH, $x = 0$, so:

$$\frac{\partial u_2}{\partial x} = \frac{6F}{EBh^3}\left(-a^2\right) \qquad (22.92)$$

From equation (22.87), the J-integral becomes:

$$J = -2\int_0^h T_2 \frac{\partial u_2}{\partial x} dy = -2\int_0^h T_2 \left\{\frac{6F}{EBh^3}\left(-a^2\right)\right\} dy$$

$$= 12\left(\frac{Fa^2}{EBh^3}\right)\int_0^h T_2 dy \qquad (22.93)$$

On segment BC and GH:

$$F = B\int_0^h T_2\,ds \text{ from which, } \int_0^h T_2\,ds = \frac{F}{B} \qquad (22.94)$$

From equations (22.93) and (22.94)

$$J = 12\left(\frac{Fa^2}{EBh^3}\right)\int_0^h T_2\,dy = 12\left(\frac{Fa^2}{EBh^3}\right)\left(\frac{F}{B}\right) = \frac{12F^2a^2}{EB^2h^3} \qquad (22.95)$$

It is worth pointing out that this same expression was obtained in Section 22.4, where G was derived by considering the change in energy (equation 22.30, page 452),

i.e. $$\boldsymbol{G = \frac{12F^2a^2}{EB^2h^3}} \qquad (22.96)$$

However, the definition of the J-Integral is more versatile than the definition of G, because the J-Integral can investigate a portion of the component and characterise the crack fully. Although the technique to determine J is quite different from that of G, both lead to the same result for linear elastic materials in the end. The parameter G is limited only to LEFM, whereas the J-Integral encompasses a much larger domain, as it can deal with both non-linear and elastic-plastic materials.

22.12 Crack extension resistance curves (*R*-curves)

The use of crack extension resistance curves or '*R*-curves' is another technique to enable LEFM to be applied to elastic-plastic fracture.

When considering energy and energy release thus far, the strain energy release rate, G, has been taken as a material-specific constant. However, this is only approximately true as size and geometry do have some effect, especially for plane stress conditions, and therefore for sheet materials. This was shown by Irwin, as long ago as 1959.

In such cases, the energy release rate is some function of the extension of the crack, i.e. a function of the change of crack length. This function is termed an *R*-curve. It is a plot of crack growth resistance rate, versus crack extension for a particular loading configuration.

$$R = R(\Delta a) = R(a - a_0)$$

where a_0 = initial crack length

a = current crack length

R_0 = initiation fracture energy (onset of nonlinearity)

R_F = fracture energy at steady state

A step function type R-curve describes the behaviour of a brittle material such as glass is shown in 22.20. Here, the crack grows immediately once a threshold level is reached, as $R_0 = R_F$

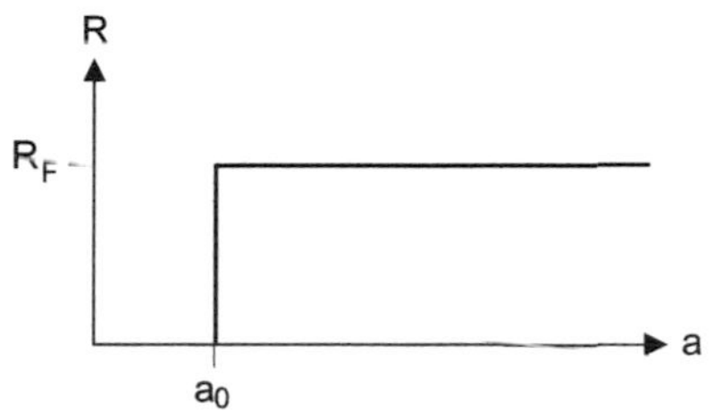

Figure 22.20 *R*-curve for a brittle material.

The R-curve for a more ductile material, such as mild steel, is shown in Figure 22.21, where the onset of crack growth is more gradual, with a curve beginning when $R = R_0$

The energy release rate from a Griffith crack can be

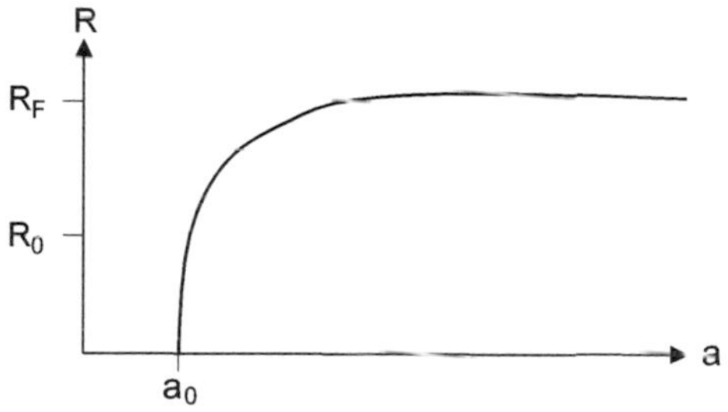

Figure 22.21 A curve begins when $R = R_0$, with a gradual on set of crack growth

obtained from equation (22.14), which accounts for the size of the plastic zone:

$$\left(EG_C\right)^{\frac{1}{2}} = \sigma\left(\pi a\right)^{\frac{1}{2}} \tag{22.97}$$

from which, $$G_C = \frac{\sigma^2 \pi a}{E} \tag{22.98}$$

This means that when different constant stress levels are plotted, a family of straight-line graphs result.

Figure 22.22 introduces a 2 variable approach, where the energy release rate depends on both the loading parameter and the length of the crack.

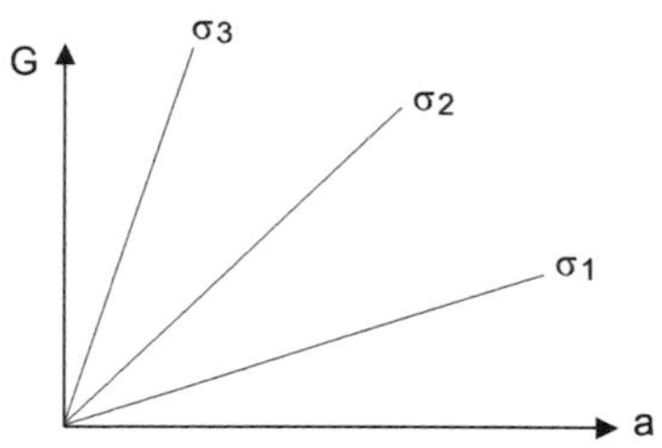

Figure 22.22 Family of straight line graphs for constant stress

This family of curves, accounting for the size of the plastic zone, can be thought of as loading curves, represented by the nonlinear function:

$$G = G(a, \lambda) \tag{22.99}$$

As K is a function of G, these loading curves are sometimes plotted from the function:

$$K = K\left(a, \lambda\right) \tag{22.100}$$

Under load, a crack could remain unchanged, grow a small amount (and then stop) or grow catastrophically. The outcome will depend upon:

(1) the initial length of the crack, a_0
(2) the resistance curve of the material, $R(a - a_0)$
(3) the loading curves of the specimen, $G(a, \lambda)$

Figure 22.23 shows an R-curve together with some loading curves.

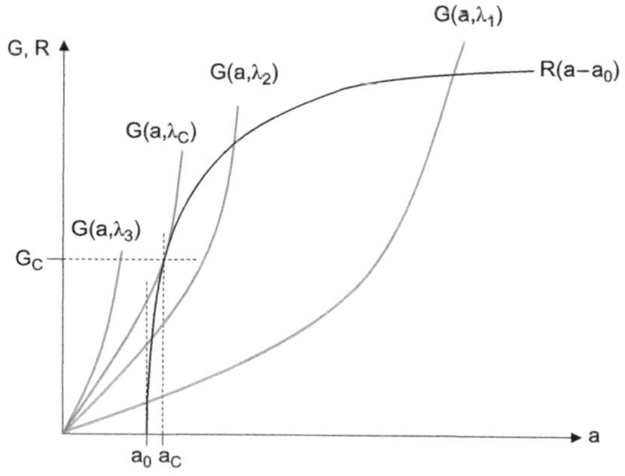

Figure 22.23 *R*-curve with loading lines

Whether a crack is stable depends on the way the two curves intersect.

With the loading parameter fixed at λ_1, the R-curve is above the loading curve for all cracks longer than a_0, and the crack will not grow.

When the loading parameter is fixed at λ_2, the crack will grow. This is because a is slightly larger than a_0, and so the loading curve is above the R-curve. However, when a is slightly larger again, the R-curve rises above the λ_2 loading curve and so crack growth will stop. This is easier to see on the more detailed Figure 22.24.

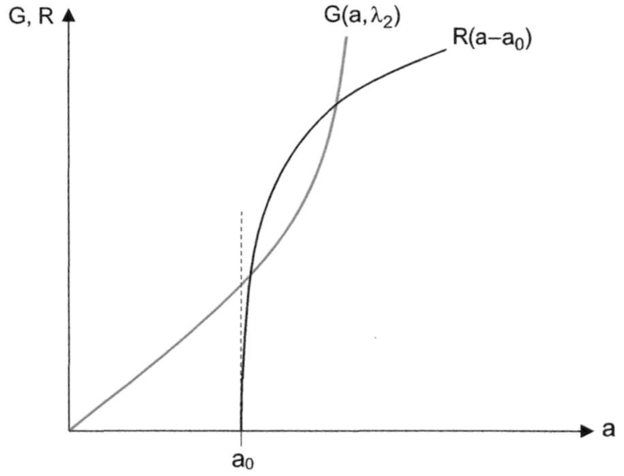

Figure 22.24 Detailed R-curve with a single loading line

When the loading parameter is fixed at λ_3 in Figure 22.23, the loading curve does not reach the initial crack length and so that loading cannot cause crack growth.

However, a critical load λ_C exists. When loaded above this level, the crack will propagate in an unstable manner and below this level, the crack will stop. Here, the intersection of the loading curve and the R-curve give two critical values, namely a_c and G_c. At this point, the R and G values are equal and tangent and have the same slope. This means that the following equations apply:

$$G(a,\lambda) = R(a - a_0) \qquad (22.101)$$

$$\frac{\partial G(a,\lambda)}{\partial a} = \frac{\partial R(a - a_0)}{\partial a} \qquad (22.102)$$

This means that the critical load and crack length can be obtained from the solution of these two equations.

For fully worked solutions to each of the problems in Exercise 74 to 76 in this chapter, go to the website:
www.routledge.com/cw/bird

Chapter 23

Material property considerations

Why is it important to understand: **Material property considerations**

Some loading conditions/environments invoke different modes of failure of structures. For example, a cyclical change of applied load severely reduces loading capacity compared to static loads or risk failure by fatigue. It is not surprising that different materials behave differently under variable loading; however, some steels exhibit significantly different behaviour. When involved with cyclic loading, designers use the concept of 'component life'. They also need a knowledge of the effects of stress concentrating features, as these adversely affect the life of components. Time and temperature significantly affect the mechanical performance of polymers, but these factors affect metals too. There are many cases where structures and mechanisms have to withstand some very high temperatures and here the predominant failure mode is creep. This is where the material stretches or ruptures under load when subjected to a high temperature environment. The creep of metals at lower temperatures is much smaller, but it can give rise to relaxation of components such as bolts. This directly affects the design of bolted joints where some form of sealing is required.

At the end of this chapter you should be able to:

- understand fatigue and the effects of cyclic loading and be able to design against fatigue
- understand mean stress and fatigue
- recognise applications of Goodman diagrams
- appreciate varying stress amplitudes and fatigue
- understand the effects of surface treatment and surface finish on fatigue
- understand corrosion and fatigue and recognise creep and the effects of high temperature
- apply creep testing and understand the extrapolation of creep data
- appreciate the effects of restraint and creep relaxation

Mathematical references

In order to understand the theory involved in this chapter on **material property considerations**, knowledge of the following mathematical topics are required: ***algebra, graphs, logarithms/exponentials and integration***. If help/revision is needed in these areas, see page 50 for some textbook references.

23.1 Introduction

The mechanical integrity of all structures within tall buildings, bridges, cars, aircraft, spacecraft or any machines depends upon the actual component that is installed. Great efforts are made to ensure that this is what the designers intended, but the history of failures shows that sometimes, this is not the case. That critical component, even if exactly within tolerance, will only perform safely and satisfactorily if the material is correctly specified. There are a number of areas where the characteristic properties of materials directly influence the performance under certain loading specific conditions. These loading conditions include potential problems from fatigue due to cyclic loading, and the phenomenon of creep, which affects performance at high temperature. Each of these topics has direct relevance to engineers and designers, who must recognise such cases, be aware of the issues and work accordingly.

23.2 Fatigue and the effects of cyclic loading

Fatigue failures occur after repeated application of a stress, such as the alternating load shown in Figure 23.1. This stress can be very much less than the ultimate tensile strength of the material. This means of failure is by far the most common cause of engineering failures in service.

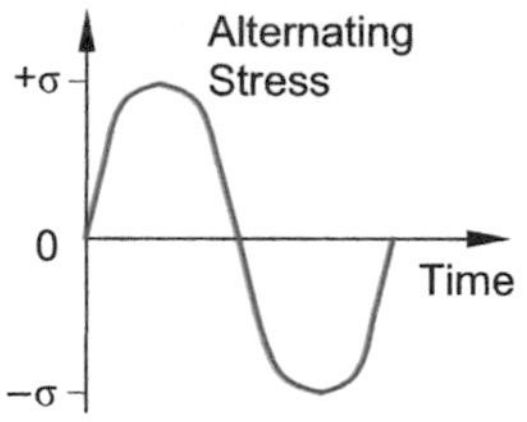

Figure 23.1 Stress due to an alternating load (1 cycle)

Specimens subjected to static loading display necking and then fail at the thinnest cross-section. When subjected to cyclic loading, failures are found to initiate from notches or local stress-raising features. This is shown diagrammatically in Figure 23.2.

Figure 23.2 Failures from static and from cyclic loading

Fatigue failures are due to the repetition of load and are therefore primarily dependent upon the load and the number of cycles. Components such as shafts tend to experience a large number of bending stress reversals and consequently, shaft failure by fatigue is common. Tests for fatigue have traditionally been carried out using a Wohler-type cantilever bending fatigue testing machine. Figure 23.3 shows a shaft driven through a cantilevered fixed bearing, with a vertical load which forms the free end of the cantilever. The direction of the vertical load remains constant as the shaft rotates putting the beam into tension and compression. This enables a large number of cycles to be accrued in a relatively short time.

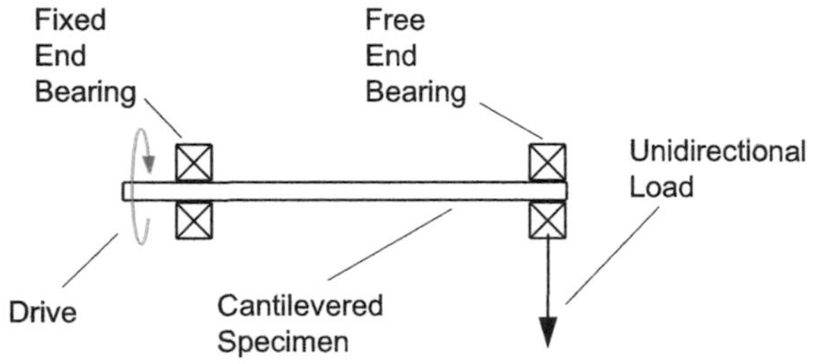

Figure 23.3 A Wohler cantilever fatigue testing machine

An illustration of a classic failure is given in Figure 23.4. It shows the initiation point on the surface - this is usually at a localised notch or some other imperfection that acts as a local stress raiser. As illustrated in Figure 23.4, the stress raiser is always extremely localised and at a microscopic level. Sweeping out from the initiation point, there is then a region of shell or wavelike steps, formed as the crack grows. The rest of the section is rougher. It results from a ductile failure due to insufficient material remaining to take the applied load, until failure is complete.

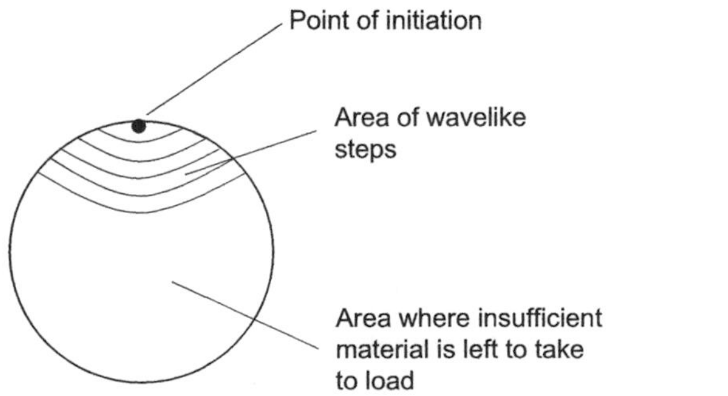

Figure 23.4 Classic appearance of the surface of a fatigue failure

The fatigue behaviour of materials is described by a graph plotting the amplitude of the cycled stress against the number of cycles, known as S-N curves. Unless otherwise stated, these graphs are for smooth, unnotched specimens in a noncorrosive environment. A diagrammatic S-N curve for most engineering materials is given in Figure 23.5.

This means that many components are design for a specified life, as is the case for many aerospace components. Replacement of stressed parts is widely used after a given number of flying hours or take-offs and landings.

As for most engineering materials, the S-N curve for low to medium carbon steels is also characterised by a great deal of scatter. However, there is a significant difference in that it exhibits a fatigue limit, S_F. This is hugely important, as it means that if the loads applied to components are known, then there is a value of stress that can be cycled indefinitely, without the risk of failure. This means that for most materials a value of S_E must be quoted, together with the number of cycles. As the majority of steels are low to medium carbon, for most steels, the S_F defines an infinite life. The fatigue limit occurs at approximately 10^6 cycles and the value of S_F is approximately 0.4 times the ultimate tensile strength, i.e. $S_F \approx 0.4\ S_{ULT}$ as shown in Figure 23.6.

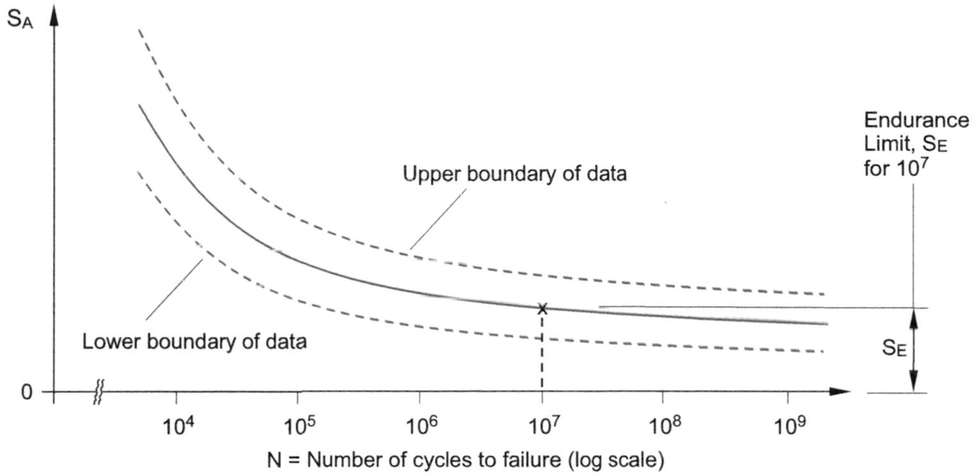

Figure 23.5 S-N curve for most engineering materials

All fatigue data is characterised by a great deal of scatter, and this is indicated by the upper and lower boundaries shown in Figure 23.5. The graph shows the continual decrease in the safe cycled stress level with increasing number of cycles (asymptotic to the x-axis). A fatigue strength is defined at a specific value of N, and this is termed the Endurance Limit, S_E; the value for N chosen is usually 10^7 or 10^8.

At 10^7 cycles, the endurance limit S_E is approximately 0.3 of the ultimate tensile strength of the material, i.e. $S_E \approx 0.3 S_{ULT}$ at $N = 10^7$.

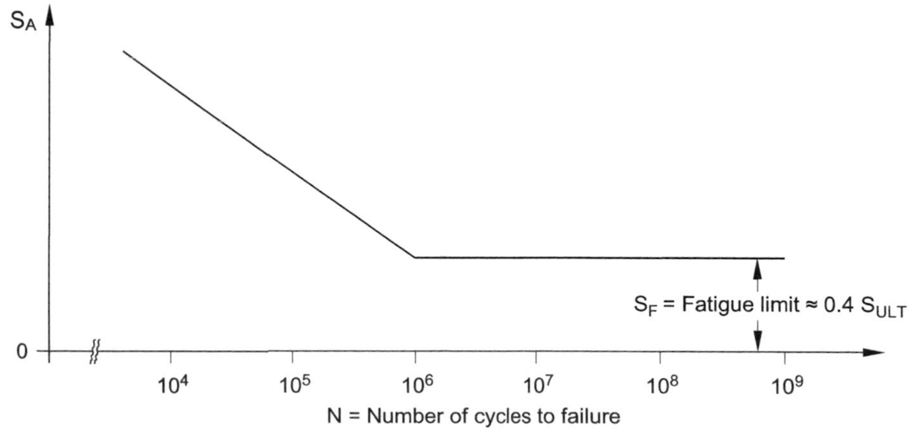

Figure 23.6 S-N curve for low to medium carbon steels

23.3 Design against fatigue

The following must be addressed and accounted for in order to ensure robust design in order to prevent fatigue:

(1) Notches due to unnecessarily sharp radii on shafts, or other unaccounted-for stress-raising feature. It is not rare to discover that one stress concentrating feature has been superimposed onto another.
(2) The effects of mean stress.
(3) The effects of varying stress amplitude.
(4) The environment
(5) Surface flaws such as inclusions near the surface, blow holes (resulting from the steel production itself), porosity (small holes in castings), inclusions in welds, as well as geometrical imperfections in welds (for example, undercuts). Surface flaws include damage due to misuse, for example, apparent minor impact damage and overloading.

The fatigue stress only needs to be reached in one place to cause a crack. Notches increase the stress locally, and so a stress concentration factor, K_T is defined in the usual way as:

$$K_T = \frac{\sigma_{Max}}{\sigma_{Min}}$$

Figure 23.7 shows the stress distribution on an unnotched and a notched specimen, each with a circular cross-section. The stress distribution across the notched specimen can be seen to be significantly higher around the notch itself.

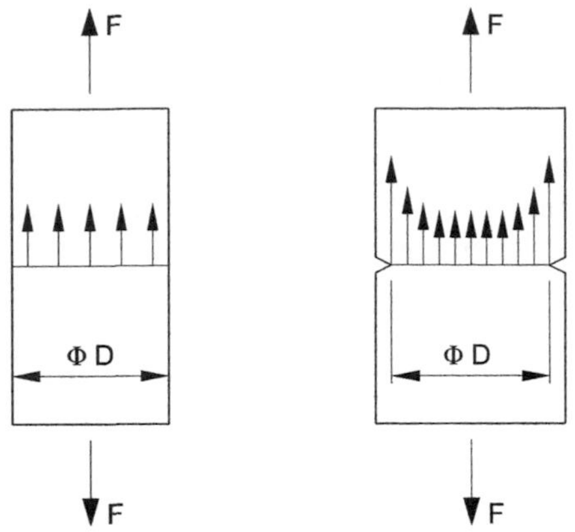

Figure 23.7 Stresses in a unnotched and notched specimen

The diameter (and hence the cross-sectional area) is the same for both specimens, even though the notched specimen has a larger cross-sectional area away from the notch.

$$\text{Since stress} = \frac{\text{force}}{\text{area}}, \text{ then force} = \text{area} \times \text{stress}$$

It may be shown that:

$$F = \pm A S_F \text{ for the unnotched specimen}$$

Since the fatigue stress needs only to be reached in one place, and initiate cracks:

$$F = \pm \frac{A S_F}{K_T} \text{ for the notched specimen}$$

where F = force, A = area, S_F = fatigue limit

The S-N curves for notched and unnotched specimens are shown in Figure 23.8.

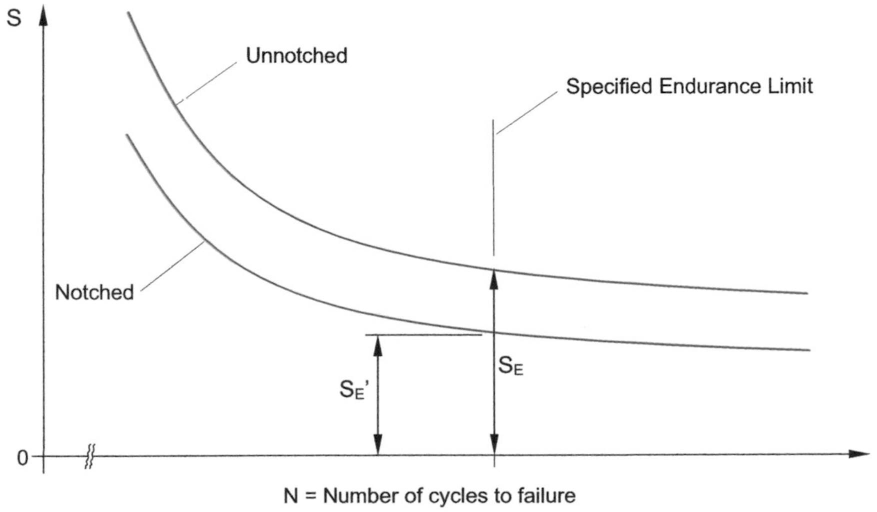

Figure 23.8 S-N curve for notched and unnotched specimens

S_E and S_E' are the stresses for a specified endurance limit, for the unnotched and the notched specimens respectively.

In theory, $\dfrac{S_E}{S_E'} = K_T$

In practice however, $$\frac{S_E}{S_E'} = K_F \qquad (23.1)$$

where K_F is the actual fatigue strength reduction factor and K_T is the stress concentration factor. K_F depends upon K_T but also on the notch sensitivity of the material. A notch **sensitivity factor, q**, is defined as:

$$q = \frac{K_F - 1}{K_T - 1} \qquad (23.2)$$

$K_F = K_T$ when $q = 1$ for a very notch sensitive material. $K_F = 1$ when $q = 0$ for a very notch insensitive material.

Problem 1. A threaded fastener is to carry a reversed axial load of 110 kN (i.e. ±110 kN). The stress concentration factor due to the threads is 2.0, the notch sensitivity factor is 0.78 and there is to be a factor of safety of 1.25 against fatigue. If the steel to be used has an ultimate tensile strength of 1.15 GPa, and the fatigue limit of steel can be taken to be 0.4 (S_{ULT}), estimate (a) the necessary cross-sectional area of the thread, (b) the safe core thread diameter

$$\text{Load} = \pm 110\,\text{kN} \quad K_T = 2.0 \quad q = 0.78 \quad S_F = 1.25$$

$$\text{UTS} = S_{ULT} = 1.15\,\text{GPa}$$

From equation (23.2), $q = \dfrac{K_F - 1}{K_T - 1}$ from which,

$$q(K_T - 1) = K_F - 1$$

so the actual fatique strength reduction factor,

$$K_F = q(K_T - 1) + 1$$

i.e. $$K_F = 0.78\,(2.0 - 1) + 1 = 1.78$$

(a) Assuming that the fatigue limit of steel $= 0.4 \times S_{ULT}$,

the working stress, $\sigma_{working} = \dfrac{0.4 \times S_{ULT}}{K_F \times S_F}$

$$= \frac{0.4 \times 1.15 \times 10^9}{1.78 \times 1.25} = 206.7 \times 10^6\,\text{Pa}$$

Cross-sectional area, $\text{A} = \dfrac{\text{load}}{\sigma_{working}}$

$$= \frac{110{,}000}{206.7 \times 10^6} = \mathbf{532.1 \times 10^{-6}\,m^2} \text{ or } \mathbf{532.1\,mm^2}$$

(b) Cross-sectional area, $A = \dfrac{\pi D^2}{4}$

from which, diameter, $D = \sqrt{\dfrac{4A}{\pi}}$

$$= \sqrt{\frac{4 \times 532.1 \times 10^{-6}}{\pi}} = 0.02603\,\text{m} = 26\,\text{mm}$$

Therefore, a safe core thread diameter must be greater than 26 mm

Problem 2. A shaft has a circular groove for which the theoretical stress concentration factor is 2.5. Similar shafts, but without this groove have been shown to have a fatigue strength of 200 MPa. If the notch sensitivity index (factor) is taken as 0.85, estimate the allowable working stress for the shaft, allowing a factor safety of 3.

$$K_T = 2.5 \quad S_F = 200\,\text{MPa} \quad q = 0.85$$

From equation (23.2), $q = \dfrac{K_F - 1}{K_T - 1}$ from which,

$$q(K_T - 1) = K_F - 1$$

and $$K_F = q(K_T - 1) + 1$$

i.e. $$K_F = 0.85(2.5 - 1) + 1 = 2.275$$

From equation (23.1), $K_F = \dfrac{S_E}{S_E'}$ so $S_E' = \dfrac{S_E}{K_F}$

i.e. $$S_E' = \frac{200 \times 10^6}{2.275} = 87.91 \times 10^6\,\text{Pa}$$

Applying a safety factor of 3, gives **allowable working stress** $= \dfrac{S_E'}{3}$

$$= \frac{87.91 \times 10^6}{3} = 29.3 \times 10^6\,\text{Pa} = \mathbf{29.3\,MPa}$$

Now try the following Practice exercise

Practice Exercise 77. Fatigue

1. Part of a sensing mechanism comprises a rectangular section cantilever beam 150 mm long and 25 mm wide. It is required to deflect ± 3.2 mm under cyclic loading, for its life of 10^7 cycles. One design alternative is to make the cantilever from steel and another to make it from an aluminium alloy. Determine the required thicknesses of the cantilever for each design. The relevant material data is given in the table below.
[4.78 mm, 9.24 mm]

	E- value (GPa)	UTS (MPa)	Stress for a life of 10^7 cycles
Proposed steel	200	510	$S_F = 0.4\ S_{ULT}$
Proposed Aluminium alloy	70	460	$S_E = 0.3\ S_{ULT}$

2. An aircraft's landing gear includes a short aluminium alloy bar which experiences a cyclic end loading of ± 35 MPa. The aluminium has an ultimate tensile strength of 265 MPa and a notch sensitivity factor of 0.8. Making reasonable assumptions for the endurance life for the material to achieve a life requirement of 10^7 cycles, estimate the maximum value of any stress concentration factor (K_T) in this part. [2.59]

cyclic load still has a detrimental effect on life, but not as great an effect as full tension to compression reversals. Thus far, a cyclic stress has been assumed to be a full sinewave (left hand graph of Figure 23.9, labelled 'Alternating or Reversed'). In reality, loads can be cyclic and fluctuate between two tensile values, never becoming compressive (centre graph of Figure 23.9, labelled 'Fluctuating'). Loads may be cyclic and fluctuate between a tensile value and zero (right hand graph of Figure 23.9, labelled 'Zero to Maximum').

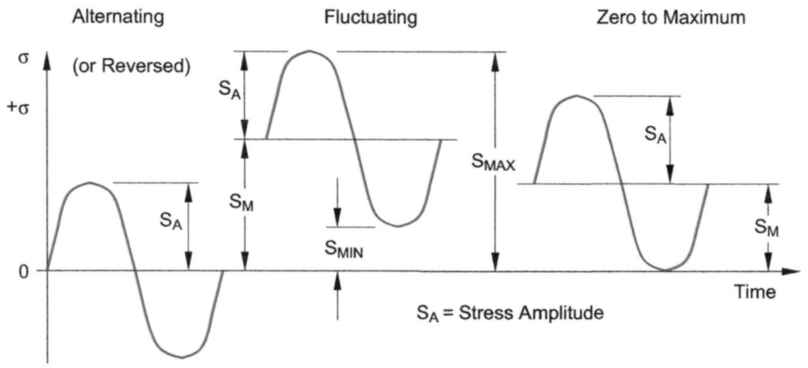

Figure 23.9 Types of cyclic stress

23.4 Mean stress and fatigue

Consideration of fatigue thus far has assumed that a cycle is approximately a sinewave, with a tensile maximum and a negative minimum value. However, this is not always the case. Figure 23.9 shows a full reversal together with other types of cyclic stress. Here the minimum stress is still tensile as the mean stress, S_M, is greater than the stress amplitude, S_A. Under these conditions, the

Figure 23.10 introduces a means to account for such situations. The left-hand graph presents S-N curves for increasing values of mean stress, S_M, starting from $S_M = 0$. Values of stress amplitude S_A for a given life of (for example) 10^7 cycles, are then transferred to the graph on the right-hand side. The $S_M = 0$ curve gives Point 1, which is on the S_A axis, and it represents the case of a conventional, equal tensile and compressive cycle. At the other extreme, when $S_A = 0$, there is no cycling of load. Failure occurs at the ultimate stress, S_{ULT} and this gives Point 2.

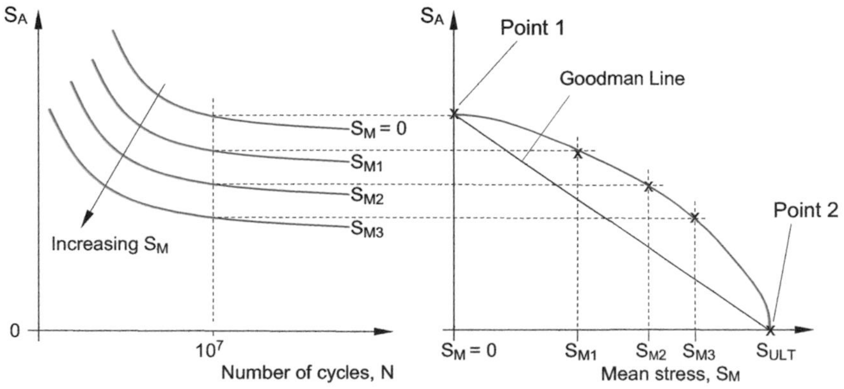

Figure 23.10 Graphical derivation of the Goodman Line

The parabola produced is known as the **Gerber parabola** and this can be formed for any stated number of cycles, N. Points above this parabola are unsafe for the value of N.
The **Goodman Line** is shown in Figure 23.10 as the line between points 1 and 2, for 10^7 cycles. This is known as a **Goodman Diagram**, and as the line is to the left of the parabola, it builds in an additional safety factor.

The Gerber parabola and the Goodman Line are shown in Figure 23.11, together with the **Soderburg Line**. The latter connects Point 1 on the vertical axis, where $S_M = 0$, to Point 3 on the horizontal axis, which coincides with the yield stress.

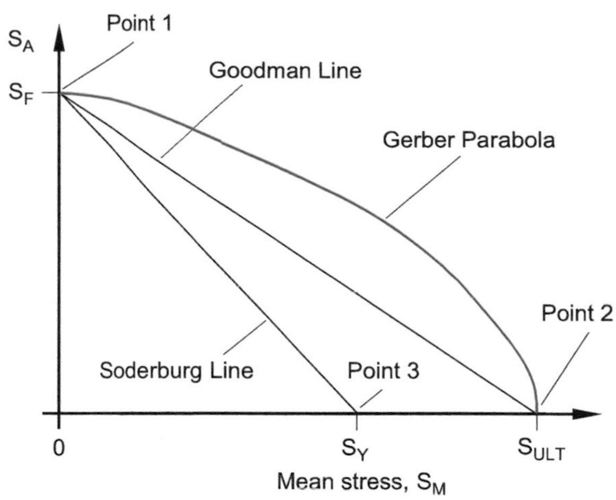

Figure 23.11 The Gerber Parabola, and the Goodman and Soderburg Lines

Problem 3. An automotive suspension component is to be made from a steel having an ultimate tensile stress of 900 MPa and a fatigue life, S_F, of 350 MPa. The loading consists of a steady load with a superimposed sinusoidal varying load of ± 255 MPa. Using a Goodman diagram, determine the maximum safe value of the base load that will allow an infinite life of this part.

$S_{ULT} = 900$ MPa $\quad S_F = 350$ MPa $\quad S_A = 255$ MPa

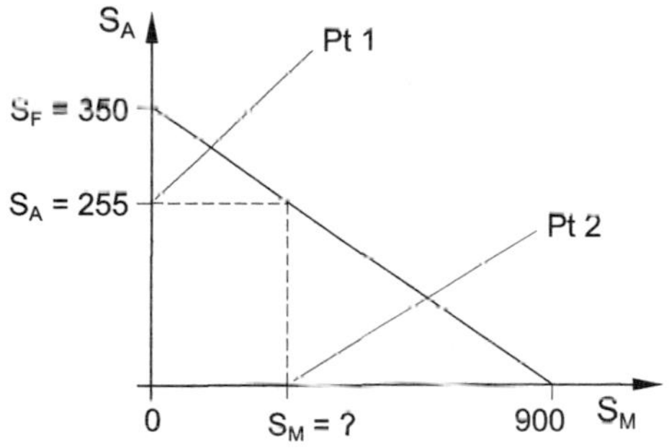

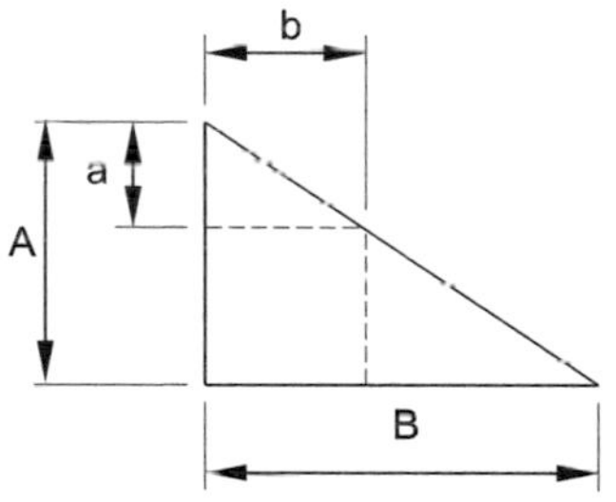

Figure 23.12

From Figure 23.12, and using similar triangles (see right hand side of figure):

$$\frac{A}{B} = \frac{a}{b} \quad \text{i.e.} \quad \frac{350}{900} = \frac{350 - 255}{S_M}$$

from which, **maximum safe value of the base load,**

$$S_M = \frac{900 \times 95}{350} = \mathbf{244.3\ MPa}$$

Problem 4. Critical parts of a railway bridge are to be made of structural steel having an ultimate tensile stress of 500 MPa. The applied load varies from 'dead load' to 'dead load plus live load'. The latter is four times as great as the dead load and sets up stresses in the same direction. A stress concentration factor of 2.2 is to be applied and the Goodman diagram may be used. Stating any assumptions made, determine the maximum stress likely to cause failure, and the working stress, without having to limit the number of trains that can cross during the life of the bridge.

$$S_{ULT} = 500\,\text{MPa}$$

$$\text{Live load, LL} = 4 \times \text{Dead load, DL} \quad K_T = 2.2$$

It is to be assumed that $S_F = 0.4\ S_{ULT}$

Figure 23.13 shows the alternating stress due to the loading. The variable alternating load is added to the constant dead load, so that the resultant is an alternating load between two positive values of stress.

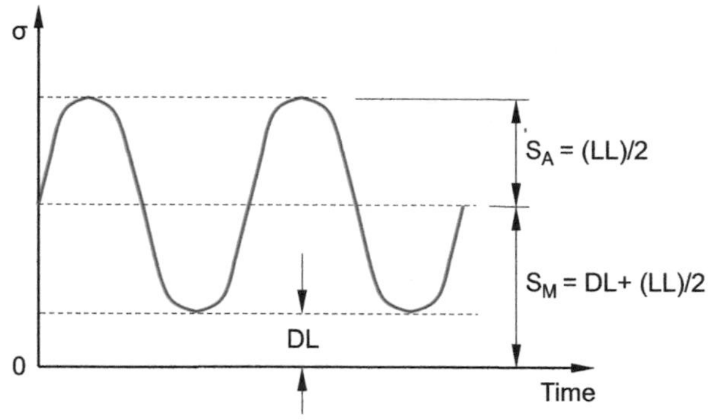

Figure 23.13

A Goodman diagram can be drawn, using 0.4 S_{ULT} on the vertical axis and S_{ULT} on the horizontal axis.
As S_M = 0.25 S and S_A = 0.75 S, similar triangles are produced as shown in Figure 23.14.

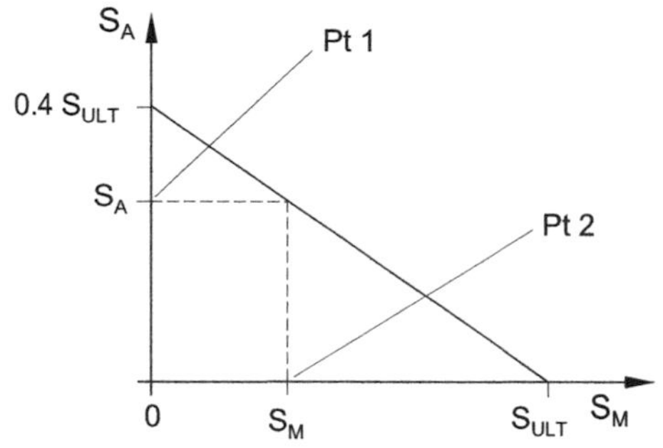

Figure 23.14

$S_A = (LL)/2$ i.e. Point 1 on Goodman Diagram
$S_M = DL + (LL)/2$ i.e. Point 2 on Goodman Diagram
Using similar triangles:

$$\frac{0.4 S_{ULT}}{S_{ULT}} = \frac{\left(0.4 S_{ULT} - (LL/2)\right)}{S_M} = \frac{\left(0.4 S_{ULT} - 2(DL)\right)}{3(DL)}$$

i.e. $$0.4 = \frac{\left(0.4 S_{ULT} - 2(DL)\right)}{3(DL)}$$

and $$0.4 \times 3(DL) = 0.4 \times 500 - 2 \times DL$$

i.e. $$1.2 \times DL = 200 - 2 \times DL$$

and 3.2 × DL = 200 and dead load,

$$DL = \frac{200}{3.2} = 62.5\,\text{MPa}$$

Live load, $LL = 4 \times DL = 4 \times 62.5 = 250$ MPa

Maximum stress likely to cause failure,

$$\sigma_{max} = 62.5 + 250 = \mathbf{312.5\ MPa}$$

$$\text{Working stress, } \sigma_{working} = \frac{\sigma_{max}}{K_T} = \frac{312.5}{2.2}$$

i.e. **the working stress = 142.0 MPa**

Now try the following Practice exercise

Practice Exercise 78. Mean stress and fatigue

1. A machine part can be assumed to be a 180 mm long, simply supported rectangular beam, made from steel. The centre of the beam is forced to oscillate vertically, through a maximum of 2 mm and minimum 1 mm under its cyclic loading. The Young's modulus and tensile strengths of the steel are 190 GPa and 375 MPa respectively. Using a Goodman Line, find the appropriate thickness (i.e. depth) of the beam. The vertical deflection can be taken to be $FL^3/48EI$
[3.88 mm]

2. Tests carried out on a certain aluminium showed an endurance limit, $S_E = 0.3\ S_{ULT}$ at 10^7 cycles. For a completely reversed bending this endurance limit was found to be 180 MPa. If the material is fatigue loaded under cycles of bending stress covering a stress range from S in tension to 0.5 S in compression, what is the maximum allowable value of S for 10^7 cycles?
[218.2 MPa]

23.5 Further applications of Goodman diagrams

The Goodman diagram can be modified to determine safe stresses in components where notches exist. Figure 23.15 shows how the stress concentration factor K_T is used to give the point on the Goodman line where there are notches. Again, the safe points are to the left of the corresponding Goodman Line.

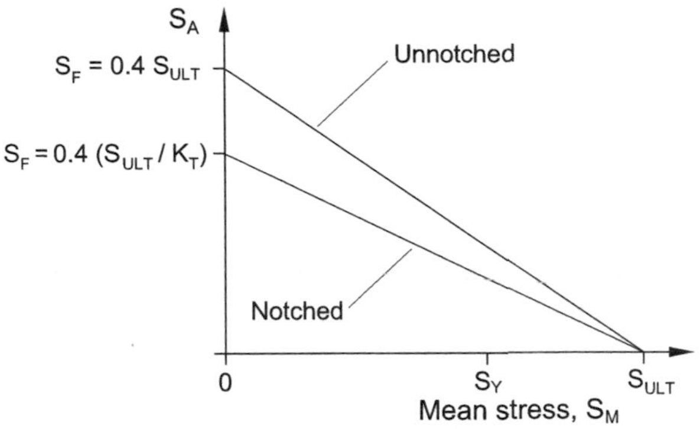

Figure 23.15 Goodman Lines for 10^7 cycles (say) for notched and unnotched specimens

Figure 23.16 shows the lines for different numbers of cycles, to emphasise that Goodman Lines are drawn for the specific number of life cycles required.

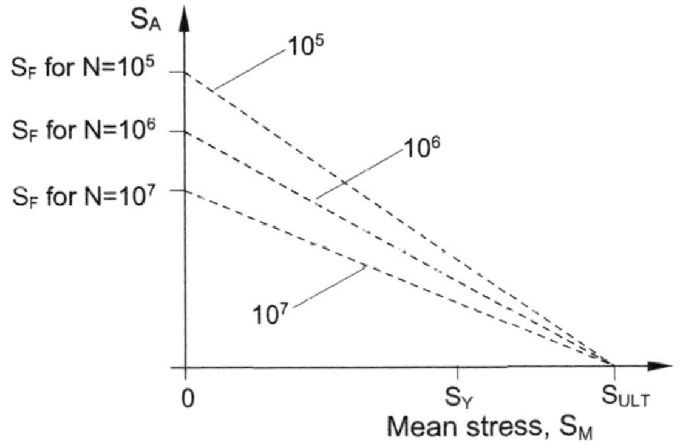

Figure 23.16 Goodman Lines for 10^5, 10^6 and 10^7 cycles

23.6 Varying stress amplitudes and fatigue

During service, a part is often subjected to not one, but a number of very different patterns of varying stress, quite different form the assumed single sinusoidal loading as often portrayed in the laboratory. The actual loading can be complex, and a comparison of some loading situations is illustrated in Figure 23.17.

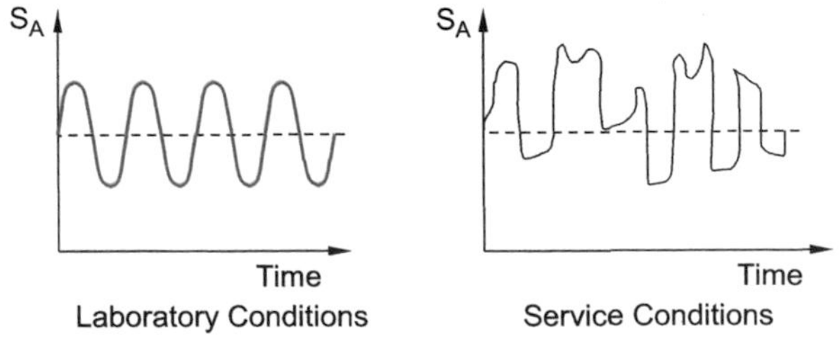

Figure 23.17 Assumed and real life service conditions

A simple approach to varying amplitude of stress, S_A is given by the **Linear Cumulative Damage Rule (Miner's Rule** – see Reference 39, page 498). It enables a number of sets of variable stress loading to be combined. A simple case is shown in Figure 23.18, where just two variable loads are shown, each of varying amplitude and frequency.

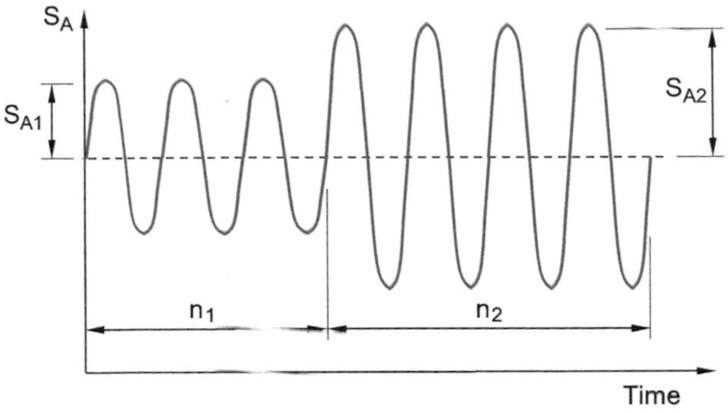

Figure 23.18 Attempts to model real life service conditions

The rule states that: $\frac{n_1}{N_1} + \frac{n_2}{N_2} + \frac{n_3}{N_3} + ... = 1$

For this case $\frac{n_1}{N_1} + \frac{n_2}{N_2} = 1$ (23.3)

where n_1 and n_2 are the number of cycles completed at stresses 1 and 2 respectively. N_1 and N_2 are the number of cycles to failure, at stresses 1 and 2 respectively. These can be obtained from S-N graphs such as the one shown in Figure 23.19.

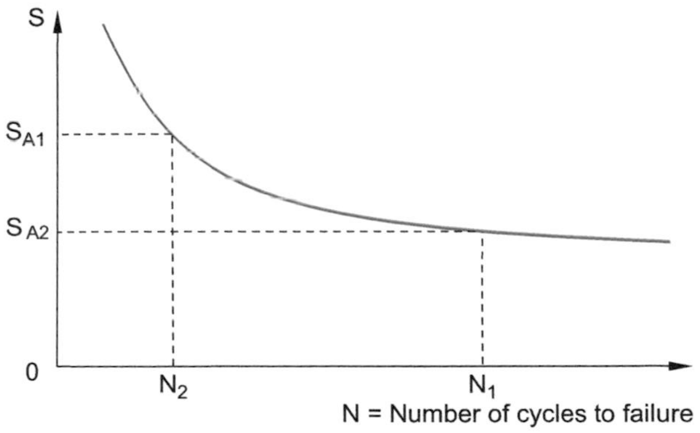

Figure 23.19 Change of loading and number of cycles used for a simple Miner's Rule

For multiple changes of stress amplitude:

$$\sum_{i}^{k} \frac{n_1}{N_1} = 1$$ where k is the number of changes of stress amplitude

If $\sum_{i}^{k} \frac{n_1}{N_1} < 1$ then the number of repetitions of the pattern to reach the fatigue life can be found

Results have shown the sum of the cycle ratios to differ from 1 to values from 0.5 to 2.0, and so appropriate safety factors need to be included. The great benefit of the Linear Cumulative Damage Rule is its relative simplicity. Other design models improve accuracy but introduce considerably greater complexity.

Problem 5. A certain material is known to fail after it has been subjected to a cyclic load of ± 250 MPa for 10^5 cycles. When the same material is tested at ± 160 MPa the life increases to 10^6 cycles. A part made from this material is known to operate successfully when its life is modelled as 1.5×10^4 cycles at 250 MPa followed by 10^5 cycles at 160 MPa. However, a proposed new application of the part requires three times this exact same loading pattern. Using the Linear Cumulative Damage Rule (Miner's Rule) determine whether this component can be used safely for the new application.

$$\frac{n_1}{N_1} + \frac{n_2}{N_2} + \ldots = \sum_{i}^{k} \frac{n_i}{N_i} = 1 \text{ for failure to occur}$$

$$\Sigma \frac{n_i}{N_i} = \frac{1.5 \times 10^4}{10^5} + \frac{1.0 \times 10^5}{10^6} = 0.15 + 0.10 = \mathbf{0.25}$$

As this is less than 1, the life can be increased.

Let n_T = number of repetitions of the pattern to reach the fatigue life

$$\text{Failure occurs when: } n_T \Sigma \frac{n_i}{N_i} = 1 = n_T (0.25)$$

This is using Miner's rule. If the right-hand side is less than 1, there is no failure.

Hence, $$n_T = \frac{1}{0.25} = \mathbf{4}$$

This means that the initial loading pattern could be multiplied by 4.

Hence multiplying the initial loading pattern by 3 means that the component is safe.

Now try the following Practice exercise

Practice Exercise 79. Stress amplitudes and fatigue

1. A machine tool component spends 15% of its life at an alternating stress of ±225 MPa, 35% at ±210 MPa, and the remainder at ±200 MPa. From the S-N diagram for this material the number of cycles to failure at ±225 MPa, ±210 MPa, and ±200 MPa are 370×10^3, 7.7×10^3 and 165×10^3 respectively. Using the Miner rule, how many cycles can the part undergo before failure?

   $\left[20.45 \times 10^3 \text{cycles}\right]$

2. The loading spectrum on an aircraft part is defined by the number of stress applications of different magnitudes in each 10,000 total applications as shown in the table below. The final column of the table gives the fatigue life at each stress as read from the ordinary S-N curve. Estimate the fatigue life of the part on the basis of the linear cumulative damage theory.

   $\left[1.016 \times 10^6 \text{cycles}\right]$

Stress Amplitude (MPa)	Frequency in each 10,000 cycles	Fatigue life on S-N curve (cycles)
310	500	70,000
230	1,000	650,000
180	3,000	3,000,000
160	5,500	35,000,000

23.7 The effects of surface treatment and surface finish on fatigue

(1) Surface hardening

Fatigue failures begin at the surface and so treatments such as carburising, nitriding, flame hardening and induction hardening are heat treatment processes used to harden the surface of steels. These are particularly beneficial in that they harden the surface, leaving a tough and shock-resistant core. However, these processes also increase fatigue lives by introducing compressive residual stresses on the outer surface. These stresses add to the surface loading, and thereby benefit fatigue life.

(2) Surface finish

Fatigue failures begin with defects on the surface and so surface finish has a profound effect on fatigue. This must be considered along with the cost of machining operations such as turning and milling. Coarse machining leads to more cost-effective manufacture, but finer machining, and even grinding, will lead to better surface finish and therefore better fatigue performance.

(3) Peening

This is a process where the surface layers of a component are bombarded with shot. This work hardens the outer surface and introduces compressive residual stresses locally, tending to close cracks. It is worth noting that this technique works for all metals. A warning needs to be flagged here as there is a potential problem with this process; if it is not done correctly, it is possible to roughen the surface to an extent that the fatigue properties are degraded.

(4) Cladding

The properties of pure aluminium are greatly improved by alloying and so pure aluminium is rarely used. An exception to this is when it is used as a cladding. Alclad, (a tradename of the Aluminium Company of America) has a thin layer of aluminium rolled onto a strong aluminium alloy core. This provides the exceptional corrosion resistance for just a slight increase in density, but this makes the fatigue properties slightly worse since the aluminium is softer. However, if there is a corrosion contribution to fatigue, as often happens in practice, then the corrosion fatigue resistance will improve.

(5) Plating

Plating provides smooth surfaces and helps to prevent corrosion. However, unlike the surface hardening processes discussed above, it results in tensile, rather than compressive residual stresses. Plating degrades fatigue performance, but it does provide protection from corrosion, and as corrosion will reduce fatigue resistance, the issue of plating components that are subjected to a large number of load reversals needs careful consideration.

23.8 Corrosion and fatigue

The combination of corrosion and fatigue is far worse than the sum of the two effects separately. Steels lose the fatigue limit in corrosive environments. The 'pits' formed by corrosive action will adversely affect the surface finish as well as act as imperfections. The dramatic effect of a corrosive fluid on the fatigue of steel is depicted in Figure 23.20.

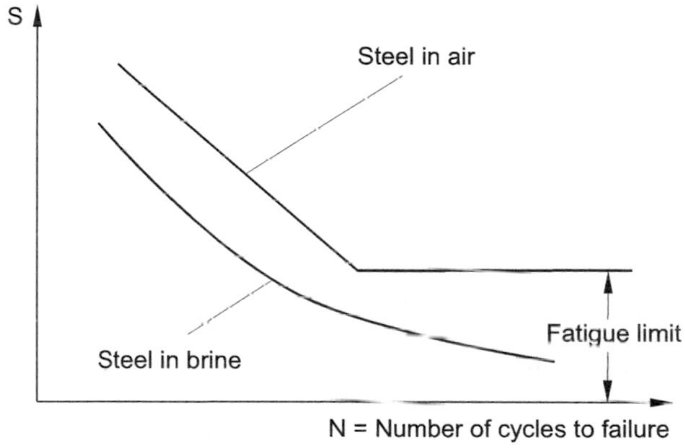

Figure 23.20

23.9 Creep - the effects of high temperature

Creep is a time-dependent deformation phenomenon which occurs under load. It is very much enhanced at elevated temperatures. At temperatures above approximately 0.4 of the melting point (in Kelvin), creep can become a problem, and at 0.5 of the melting point, creep is likely to be the most significant concern to the designer.

At the same time, thermodynamic aspects dictate that higher efficiencies can be obtained when operating at higher temperatures and so it is not surprising that creep is often the predominant failure mode of power plant, especially gas turbines.

23.10 Creep testing

To make successful predictions of the behaviour of materials under creep conditions requires good and plentiful test data. However, creep testing is by no means easy as throughout testing, it requires high temperature extensometry, and there is also a need to establish and maintain axiality, and all laboratory equipment must be kept free from external vibration. The necessity to maintain test conditions over long periods of time exacerbates each of these issues.

Figure 23.21 shows the three distinct stages that occur under creep conditions.

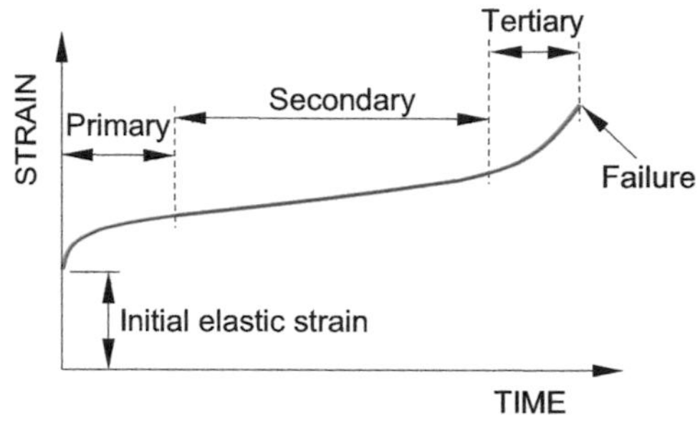

Figure 23.21 The three phases of creep, at constant stress and constant temperature

Primary creep (or transient creep) occurs rapidly initially and then decreases to a more constant value. Secondary creep (or steady-state creep) is steady but can go on for thousands of hours. The mechanism is mainly intergranular, i.e. within the grains themselves. Generally speaking, the primary and tertiary creep stages are much shorter than the secondary stage. Consequently, it is secondary creep which is of greatest importance to the design engineer. Tertiary creep was initially thought to be due to the change in cross-sectional area, causing the instantaneous stress to increase. However, it has been established that at constant stress, it is a distinct phase. Specimens exhibit necking and internal voids are formed which cause rapid increases in stress and so failure occurs relatively quickly. A creep failure is an intergranular process.

The effects of increasing stress and increasing temperature are shown diagrammatically in Figures 23.22 and 23.23.

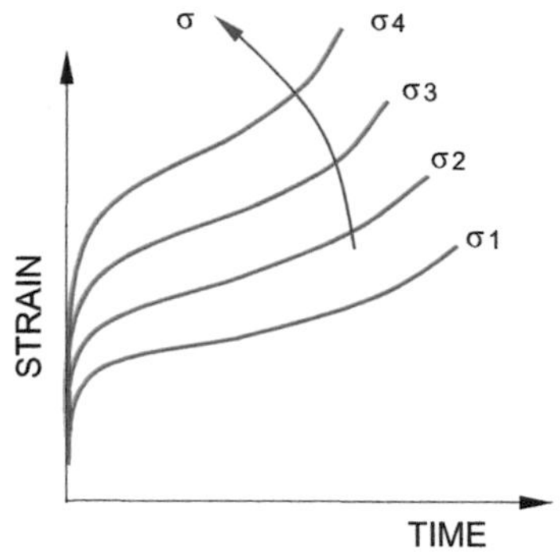

Figure 23.22 Graph of strain versus time for constant values of stress

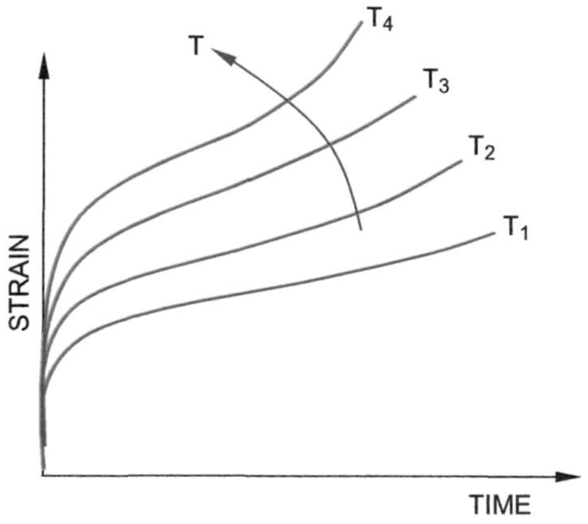

Figure 23.23 Graph of strain versus time for constant values of temperature

Cross-plotting of data allows the characteristics of specific materials to be readily displayed in graphical format as shown by the graphs in Figure 23.24 and Figure 23.25

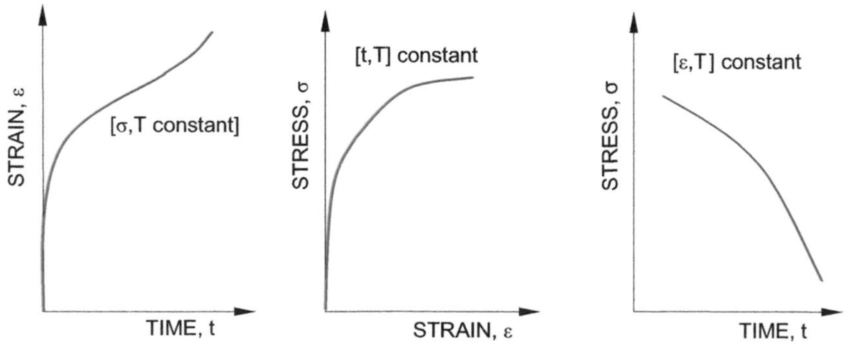

Figure 23.24 Graph of strain versus time, stress versus strain and stress versus time

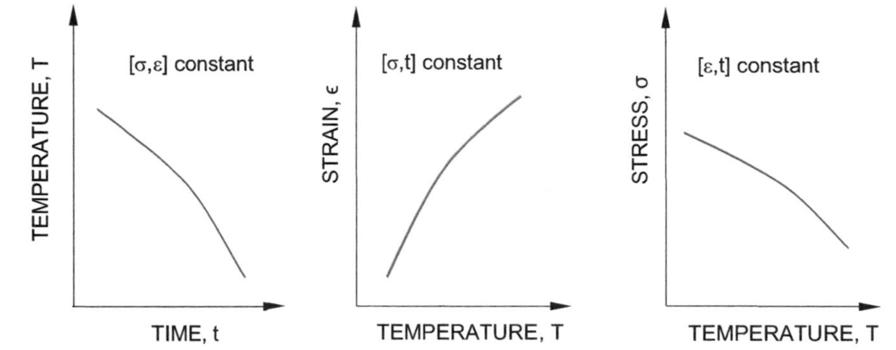

Figure 23.25 Graph of temperature versus time, strain versus temperature and stress versus temperature

23.11 Extrapolation of creep data

There are a number of techniques for the extrapolation of creep data. These include the parameters of **Dorn, Larson-Miller** and **Sherby-Dorn**. The shape of the creep curve for any material will depend on temperature and stress as these are the main factors controlling the work-hardening and recovery processes. As a result, an Arrhenius type diffusion equation is used to model creep. This equation is named after a Swedish scientist.

The Dorn Parameter

The equation used to derive the **Dorn Parameter** is:

$$\varepsilon = Ate^{-\Delta H/RT} \qquad (23.4)$$

A is dependent on stress, for example, $A = f\left(Be^{\sigma}\right)$ and ΔH is the activity energy.

Taking logarithms of both sides of equation (23.4) gives:

$$\ln\varepsilon = \ln\left(Ate^{-\Delta H/RT}\right) = \ln A + \ln t + \ln e^{-\Delta H/RT}$$

by the laws of logarithms

i.e. $$\ln\varepsilon = \ln A + \ln t - \frac{\Delta H}{RT} \qquad (23.5)$$

Now, $\ln\varepsilon = \ln A - \dfrac{\Delta H}{RT} = f(\sigma)$ for constant strain (23.6)

and $$\ln(t) = \left(\frac{\Delta H}{R}\right)\frac{1}{T} + \left[\ln(\varepsilon) - \ln(A)\right] \qquad (23.7)$$

If tests are carried out at set values of stress, then:

$$\boldsymbol{te^{-\Delta H/RT} = \text{a constant}} \qquad (23.8)$$

This is known as the **Dorn parameter**.

Taking logarithms of both sides of equation (23.8) gives:

$$\ln\left(te^{-\frac{\Delta H}{RT}}\right) = \ln k \text{ where } k \text{ is a constant}$$

i.e. $\ln(t) + \ln\left(e^{-\frac{\Delta H}{RT}}\right) = \ln k$ and $\ln(t) - \left(\dfrac{\Delta H}{R}\right)\dfrac{1}{T} = \ln k$

by the laws of logarithms

i.e. $\ln(t) = \left(\dfrac{\Delta H}{R}\right)\dfrac{1}{T} + \ln k$ which produces a straight-line graph when $\ln(t)$ is plotted vertically against $\dfrac{1}{T}$ horizontally, the gradient being $\dfrac{\Delta H}{R}$ for the material.

For set values of stress, equation (23.8) generates a set of straight lines as shown in 23.26. The average slope is ΔH/R and is specific to the material. These lines are characteristic of the Dorn parameter, making it appropriate where data forms parallel lines for the sets of stress values.

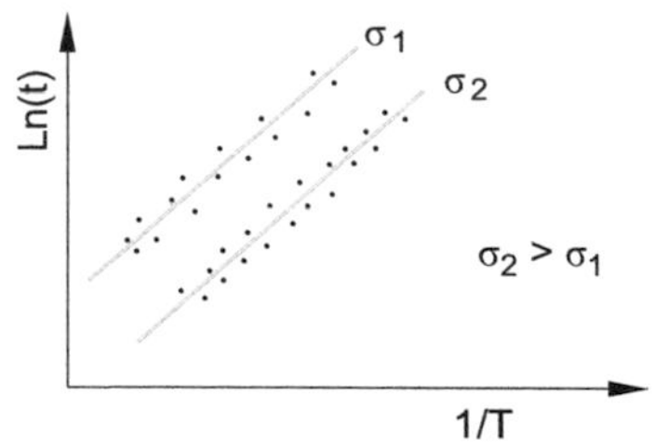

Figure 23.26 Typical data where the Dorn Parameter is appropriate

The ΔH/R term is known from the slope of the straight-line graphs (see Figure 23.26).
Figure 23.27 shows the Dorn Master curve.

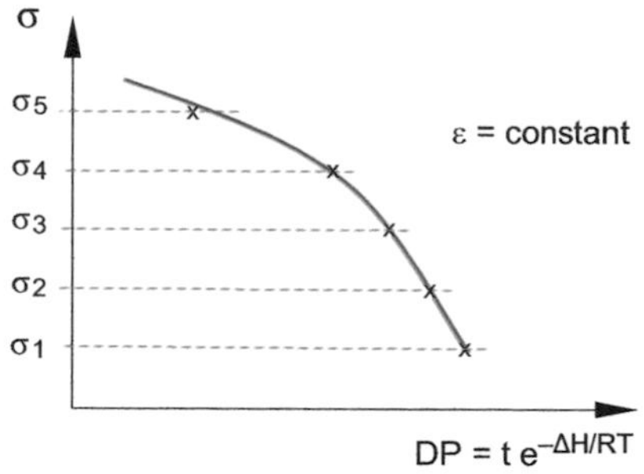

Figure 23.27 The Dorn Master curve

Any point corresponds to high temperature short times or low temperature long times. This enables data gathered from short tests to be used for long term creep design work.

Problem 6. The following readings were taken from a creep rupture test on an aluminium alloy used for aerospace applications.

T (°C)	100	125	150	200	255	255	140	125
t Hours)	2,700	85	4	240	2.4	26	1,200	900
Stress (MPa)	300	300	300	150	150	100	250	275

It is assumed the Dorn parameter, $te^{-\Delta H/RT}$ holds, where t is the time to rupture, T is in kelvin and ΔH/R is a constant for the material. Estimate the value of stress:

(a) that would cause rupture in 1500 hours at 120°C
(b) which would result in rupture at a service life of 30,000 hours at 120°C.

(a) Data that fits the Dorn parameter produces a series of straight lines when plotted on a graph of ln(t) against 1/T. Figure 23.28 shows two such lines, justifying the use of the Dorn parameter. The slope of each line is the value ΔH/R for the material.

An average value is taken, as neither is better than the other.
Therefore, average slope gives

$$\Delta H/R = (20557 + 20911)/2 = 20734$$

This enables the Dorn parameter (DP) master curve of stress against $te^{-\Delta H/RT}$ to be drawn. A log of the Dorn parameter has been plotted to make reading easier and more accurate.

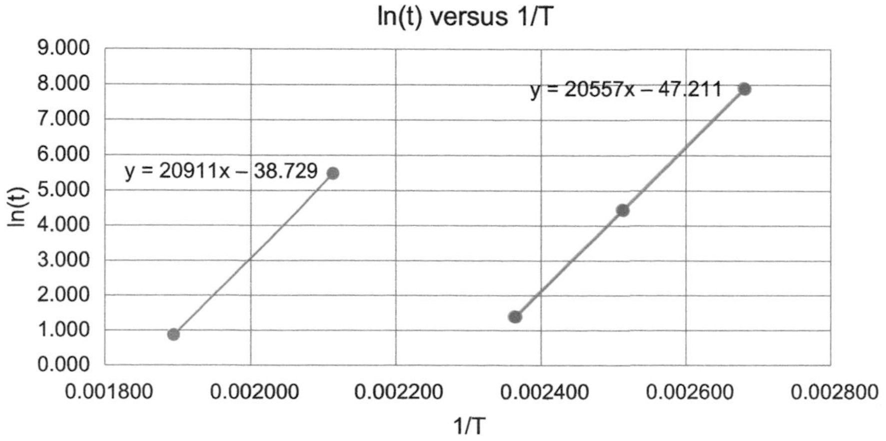

Figure 23.28 Graph of ln t versus 1/T

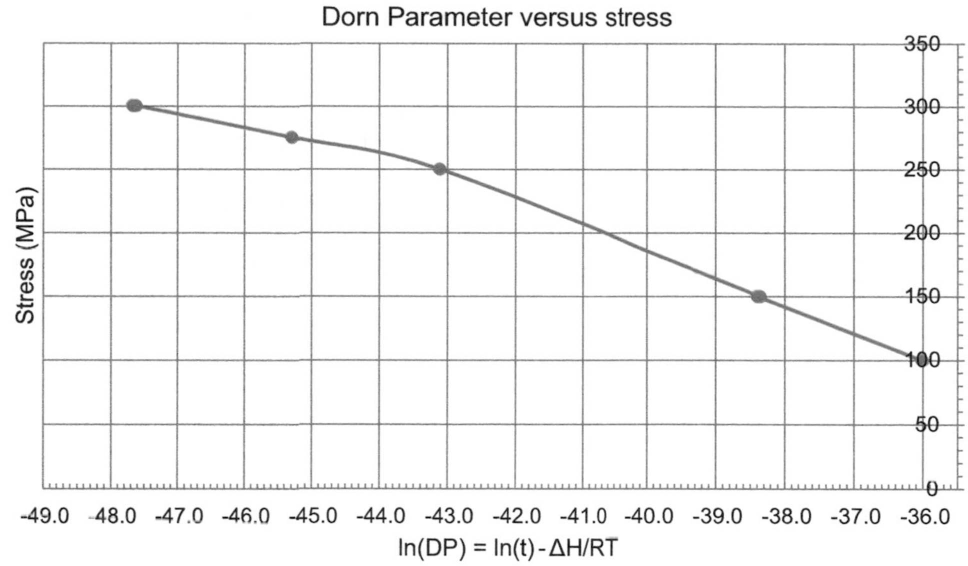

Figure 23.29 Graph of Dorn parameter versus stress

DP for rupture in 1500 hours at 120°C, i.e. (120 + 273)K = 393K:

$$DP = te^{-\Delta H/RT} = 1500e^{-\left(\frac{20734}{393}\right)}$$

$$= 1500\left(1.2229\times10^{-23}\right) = 1.834\times10^{-20}$$

$$\ln\left(DP\right) = \ln\left(1.834\times10^{-20}\right) = -45.45$$

From the graph of Figure 23.29, the **stress** corresponding to –45.45 can be seen to be approximately **278 MPa**

(b) DP for rupture in 30,000 hours at 120°C, i.e. 393K:

$$DP = te^{-\Delta H/RT} = 30{,}000e^{-\left(\frac{20734}{393}\right)} = 3.669\times10^{-19}$$

$$\ln\left(DP\right) = \ln\left(3.669\times10^{-19}\right) = -42.45$$

From the graph of Figure 23.29, the **stress** corresponding to –42.45 can be seen to be approximately **238 MPa**

The Larson-Miller Parameter

The equation used to derive the **Larson-Miller Parameter** (see reference 40, page 498) is:

$$\varepsilon = Be^{-\left(\frac{A+\Delta H}{RT}\right)}t \qquad (23.9)$$

ΔH is the activity energy, A is a function of stress, i.e. $A = f(\sigma)$ and B is a material constant.

Taking logarithms of both sides of equation (23.9) gives:

$$\ln\varepsilon = \ln\left\{Be^{-\left(\frac{A+\Delta H}{RT}\right)}t\right\} = \ln B - \left(\frac{A+\Delta H}{RT}\right) + \ln t$$

by the laws of logarithms (23.10)

and rearranging gives: $\left(\frac{A+\Delta H}{RT}\right) = \ln B - \ln\varepsilon + \ln t$

Multiplying each term by temperature T gives:

$$\left(\frac{A+\Delta H}{R}\right) = T\ln B - T\ln\varepsilon + T\ln t \qquad (23.11)$$

If strain is kept constant, then B and ε are both constants, so:

$$\ln B - \ln\varepsilon = K \quad \text{where } K \text{ is a constant} \qquad (23.12)$$

Hence, from equation (23.11),

$$\left(\frac{A+\Delta H}{R}\right) = T\left\{\ln t + K\right\} \qquad (23.13)$$

Now, A is a function of stress, and R is a constant, so:

$$\left(\frac{A+\Delta H}{R}\right) \text{is a function of stress} \qquad (23.14)$$

Equation (23.13) gives:

$$T\ln t = \left(\frac{A+\Delta H}{R}\right) - TK \qquad (23.15)$$

from which, $$\ln t = \left(\frac{A+\Delta H}{R}\right)\frac{1}{T} - K \qquad (23.16)$$

Equation (23.16) results in a series of straight lines for a given stress level, with a vertical axis intercept of –K, and a gradient of $\left(\frac{A+\Delta H}{R}\right)$.

The gradient, $\left(\frac{A+\Delta H}{R}\right)=T\{\ln t+K\}$ from equation (23.13), which is the Larson-Miller parameter, LMP.

Figure 23.30 shows typical data where the Larson-Miller parameter is appropriate.

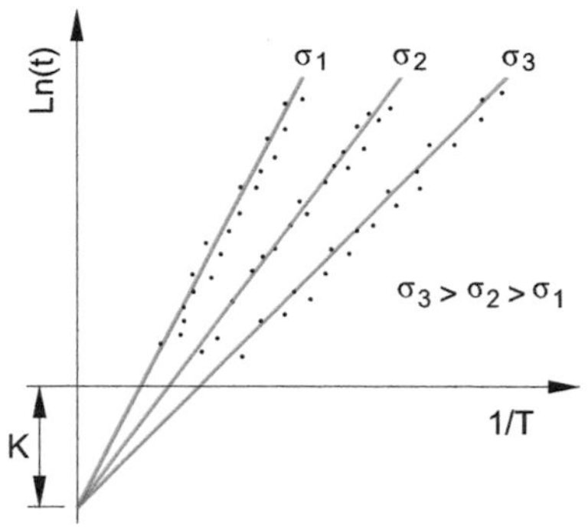

Figure 23.30 Typical data where the Larson-Miller parameter is appropriate

The Larson-Miller parameter is used when the data tends to form lines that meet at a single point, as shown in Figure 23.30. The Larson-Miller master curve is shown in Figure 23.31, which is produced from experimental data and so the life in hours can be determined once temperatures and stress levels are known.

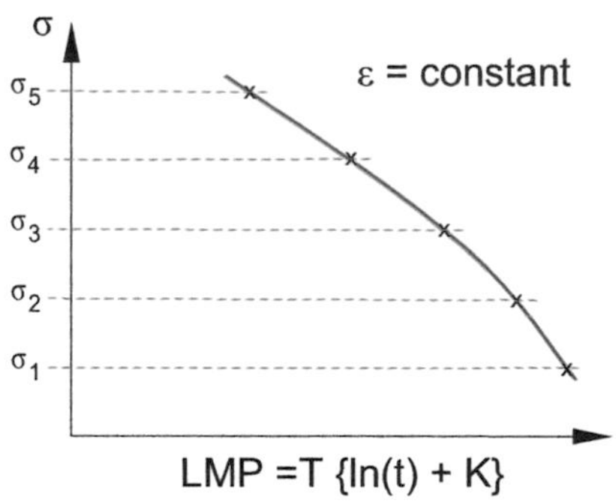

Figure 23.31 The Larson-Miller master curve

Natural logarithms are used in the development of the Larson-Miller parameter. However, most practitioners use logarithms to base 10.

This gives the Larson-Miller parameter,

$$LMP = T\,(\log(t) + C).$$

The change of base of the logarithm means that the constant used in the equation is multiplied by a factor of $\log_{10}(e)$ = 0.4343. The Larson-Miller constant, C, using this formula is around 20 for most metals.

The benefit of the master curves for both the Dorn and the Larson-Miller parameters is that they enable data to be interpolated rather than extrapolated. This is of great importance when designing equipment to operate at elevated temperatures.

Problem 7. The table below gives stress rupture data obtained from tests conducted at constant stress, and the Larson-Miller parameter is given by:

$LMP = T\left(C+\log(t)\right)$ where T is in Kelvin

Evaluate the constant C.

T (°C)	725	750	775	800	825
t (Hours)	21,000	8,400	3,300	1,300	500

There are several ways that this problem can be solved. Three methods are provided, all based on the fact that data collected at constant stress will give a straight-line graph of the type shown in Figure 23.32, where C is the y-intercept.

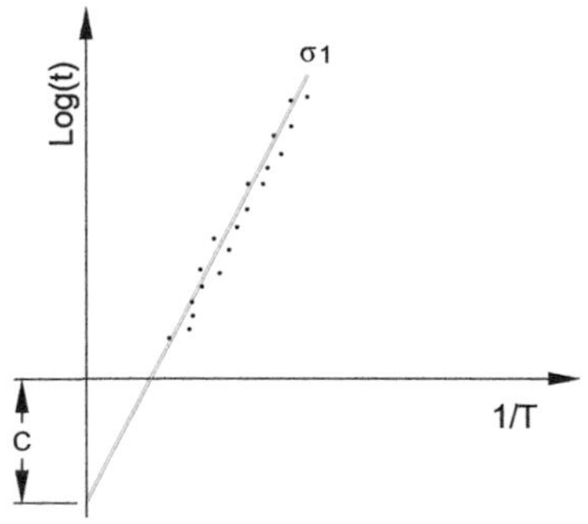

Figure 23.32

Given $LMP = T(C + \log t)$ then $\log(t)=\frac{LMP}{T}-C$ where T is in K (Kelvin = °C + 273)

T (°C)	t (Hours)	T (K)	$1/T$	$\log(t)$
725	21,000	998	0.001002	4.322
825	500	1098	0.000911	2.699

Method 1: Use an approximate solution to obtain a slope from two data points. Then substitute the x and y coordinates into the equation $y = mx + C$, where C is the y-intercept.

$$\text{Gradient}=\frac{4.322-2.699}{0.001002-0.000911}=\frac{1.623}{91\times10^{-6}}=17835$$

The linear equation is therefore: $y = 17835x + C$

Point (0.001002, 4.322) is on this line, so:

$$4.322 = (17835)(0.001002) + C$$

and $C = 4.322 - (17835)(0.001002)$

i.e. $\mathbf{C = -13.55}$

The accuracy could be improved if the gradient was obtained by drawing a graph and selecting a best fit.

Method 2: Use a spreadsheet to obtain the equation of the straight line.

The $1/T$ and $\log(t)$ values are required. These are generated from the spreadsheet below and a graph plotted as shown in Figure 32.33.

T(°C)	t (Hours)	$T(K)$	$1/T$	log t
725	21,000	998	0.001002	4.322
750	8,400	1023	0.000978	3.924
775	3,300	1048	0.000954	3.519
800	1,300	1073	0.000932	3.114
825	500	1098	0.000911	2.699

A spreadsheet can be used to give the equation of the line of best fit, and hence the equation of this line, i.e. $y = (17737)x - 13.454$

Note that the gradient is reasonably close to the value approximated by method 1,

i.e $\mathbf{C = -13.45}$

Method 3: An accurate and relatively quick method is to use the linear regression capability of a scientific calculator. After inputting the x and y coordinates, the y-intercept of the line of best fit can be obtained. In this case, $\mathbf{C = -13.46}$

The gradient obtained is 17764, which (as expected) is very close to the spreadsheet calculation of method 2.

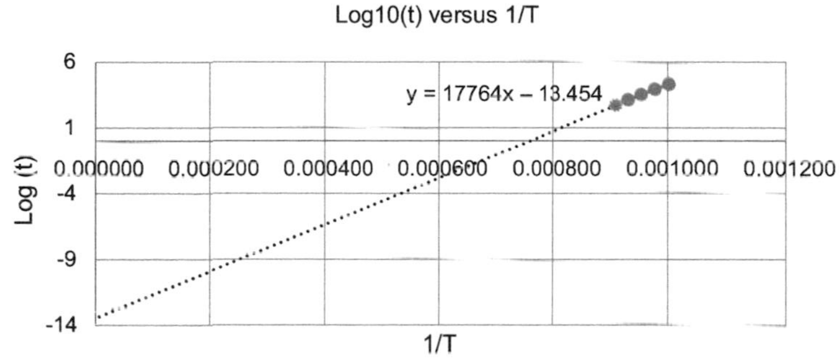

Figure 23.33

Now try the following Practice exercise

Practice Exercise 80. Extrapolation of creep data

1. The following stress rupture data was obtained for a nickel alloy:

T	(°C)	960	840	880	820	820	700	750	650	650
Stress	MPa	86	131	200	201	263	463	494	618	788
t	(Hours)	275	10,000	120	2,000	290	900	60	1,100	110

Plot the Larson-Miller parameter, $LMP = T(C + \text{Log}_{10}(t))$ against stress, where T is in K and the constant, C, may be taken to be 20. Find the temperature for a 100-hour rupture life at 300 MPa.

[$T = 827$°C]

2. In the design of a turbine blade, the creep life is dominated by the centrifugal force on the blade and the temperature. Both parameters vary along the length of the blade as does the stress. Values of temperature and stress are given in the following table for positions along the blade, from the blade root to the tip. The blade is manufactured from a nickel chrome alloy and the appropriate Larson-Miller creep curve for this alloy is shown in Figure 23.34 below, where $LMP = T(C + \log_{10}(t))$ and the constant $C = 20$.

	Position	x (mm)	T (°C)	T (K)	Sigma (MPa)
Root	1	0.00	310	583	135
	2	4.50	400	673	134
	3	15.50	600	873	117
	4	22.50	860	1133	103
	5	28.00	880	1153	93
	6	35.50	860	1133	80
	7	47.00	780	1053	48
Tip	8	48.5	705	978	47

Determine the time to rupture at each of the 8 positions along the blade and then plot a curve showing the position against time to rupture. Use this graph to predict the time to failure of the blade and the probable location of the rupture along its length, relative to the blade root.

[$t = 1585$ hours, failure site approximately 25 mm from blade root]

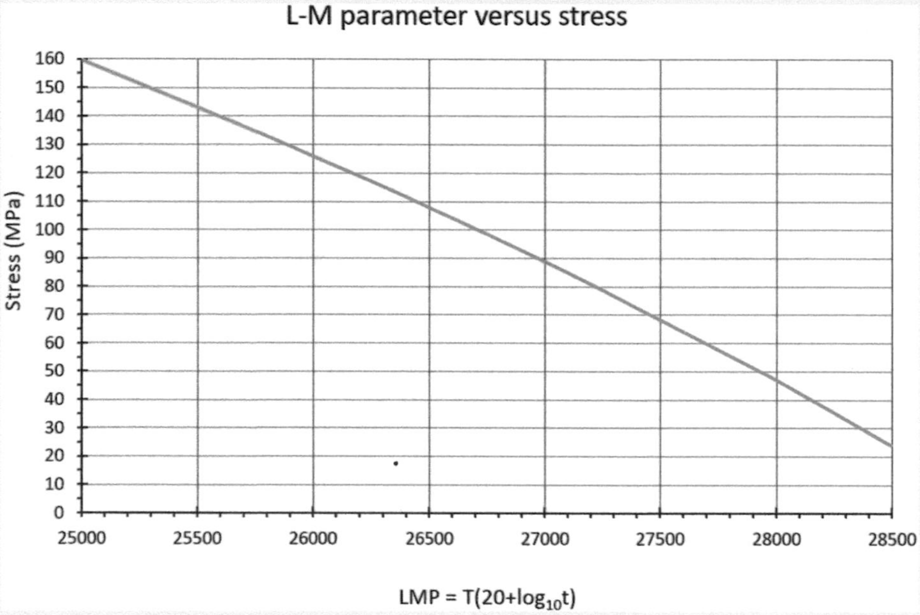

Figure 23.34 Larson-Miller parameter versus stress for blade material

Strain hardening and time hardening

In real world engineering, not all stages of operation of machinery occurs at the same stress. However, this too can be modelled. Slight modifications to methods described previously are required to deal with changes of stress, at either constant strain or at constant time.

A situation where there is a change of stress from σ_1 to σ_2 at constant strain is shown in Figure 23.35. This change is known as '**strain hardening**'. A similar situation is illustrated in Figure 23.36, but here the effects of changing stress from σ_1 to σ_2 at constant time are shown. This describes '**time hardening**'.

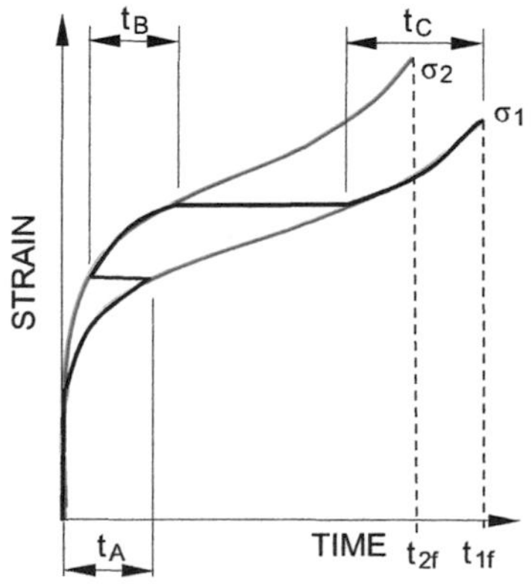

Figure 23.35 Change of stress curve at constant strain

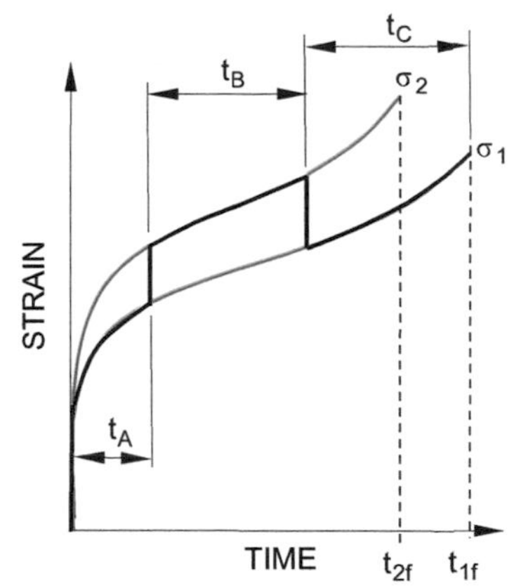

Figure 23.36 Change of stress curve at constant time

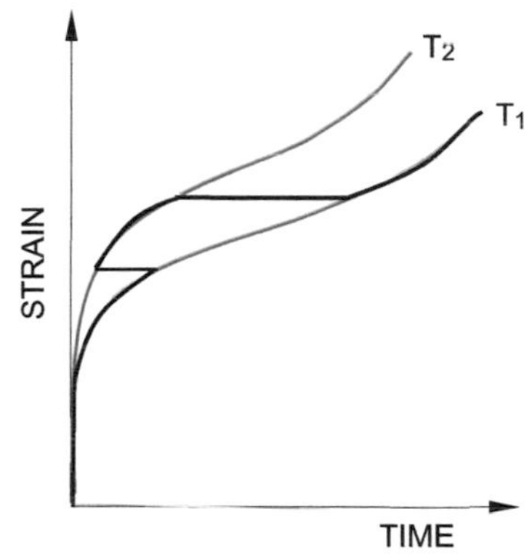

Figure 23.37 Change of temperature curve at constant strain

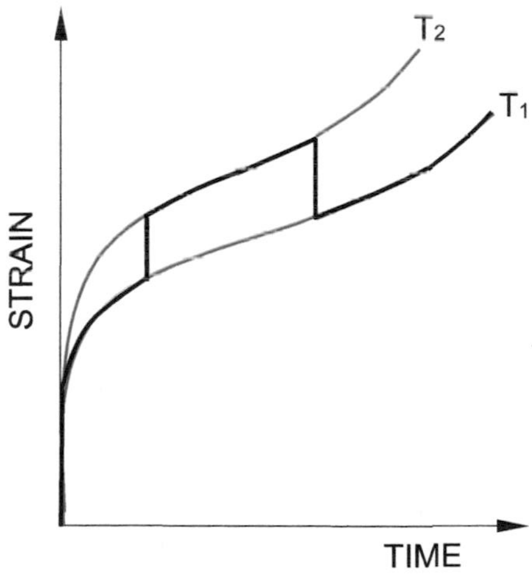

Figure 23.38 Change of temperature curve at constant time

The versatility of this method can be seen from Figure 23.37 and Figure 23.38, where it is shown to apply to changes from graphs of one temperature to another.

The **Life Fraction Rule** can be used to predict the times to failure, given the various known times.

$$\frac{t_1}{t_{1F}} + \frac{t_2}{t_{2F}} + \frac{t_3}{t_{3F}} + \ldots.. = 1$$

For the specific cases shown in Figure 23.35 and Figure 23.36, the Life Fraction Rule would be given by:

$$\frac{\boldsymbol{t_A}}{\boldsymbol{t_{1F}}} + \frac{\boldsymbol{t_B}}{\boldsymbol{t_{2F}}} + \frac{\boldsymbol{t_C}}{\boldsymbol{t_{3F}}} = \mathbf{1}$$

23.12 Effect of restraint – creep relaxation

The effects of creep are not limited to elevated temperatures. Bolted joints are renowned for exhibiting 'relaxation', where the tension in the bolt reduces with time, introducing the possibility of failure. Where this is a problem, certain techniques are employed to reduce the effect of this low temperature creep. These include:

1. the redesign of the bolted joint to minimise the use of soft gaskets,
2. tightening the fastener, waiting briefly, and then applying the torque again; this can be repeated a number of times,
3. using a low RPM power tool with a setting to apply the final torque.

A retightening sequence is shown in Figure 23.39, where the stress, and therefore the load in the bolt, relaxes with time.

Consider a bolted flange joining two pipes, as shown in Figure 23.40. The flange thickness, dimension A, can be considered to be rigid relative to the bolts, due to its size. In reality there would be some small amount of compression and there would also be washers to spread the load. The bolt length dimension B changes with time as it exhibits creep.
The original tightening strain will be:

$$\varepsilon_0 = \frac{\sigma_0}{E} \text{ where } \sigma_0 \text{ is the original stress.}$$

At time t, the instantaneous stress will be σ, and so the strain due to creep will be:

$$\varepsilon_C = \varepsilon_0 - \frac{\sigma}{E}$$

$$\frac{d\varepsilon_C}{d\sigma} = -\frac{1}{E}$$

Hence,
$$d\varepsilon_C = -\frac{d\sigma}{E} \qquad (23.17)$$

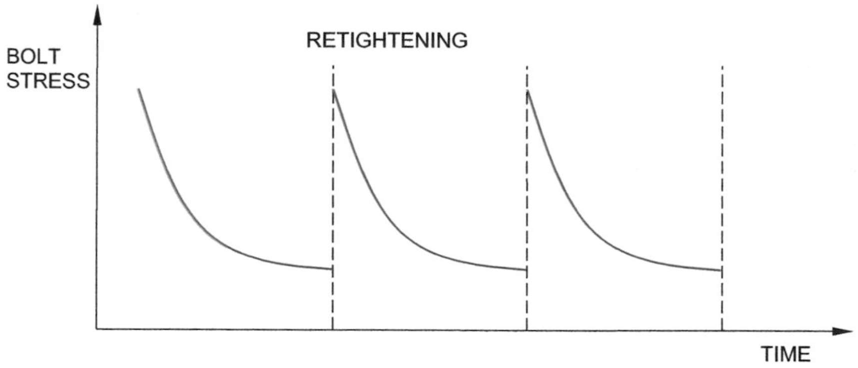

Figure 23.39 A theoretical retightening sequence

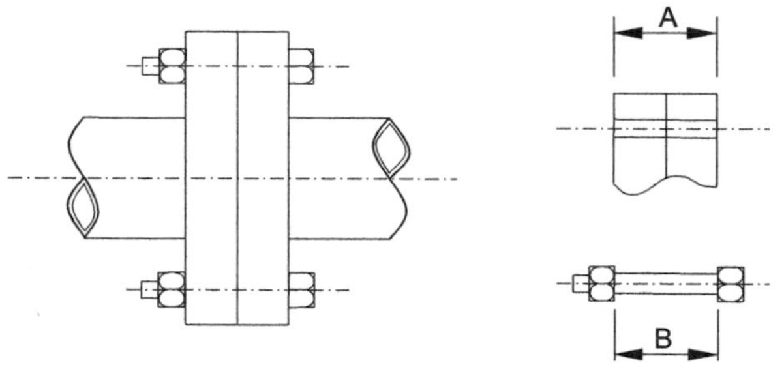

Figure 23.40 A bolted flange joining two pipes

Assuming a creep law of the Dorn or Larson-Miller type

$$\varepsilon_C = Be^{C\sigma}t \text{ where } B \text{ is a constant}$$

Differentiating with respect to time t gives:

$$\frac{d\varepsilon_C}{dt} = Be^{C\sigma} \tag{23.18}$$

and $$d\varepsilon_C = Be^{C\sigma}dt$$

Substituting from equation (23.17), gives:

$$-\frac{d\sigma}{E} = Be^{C\sigma}dt$$

$$\int_0^t dt = -\frac{1}{EB}\int_{\sigma_0}^{\sigma_1} e^{-C\sigma}\,d\sigma$$

Integrating by substitution, gives: $t = -\frac{1}{EB}\left[-\frac{1}{C}e^{-C\sigma}\right]_{\sigma_0}^{\sigma_1}$

i.e. $$t = \frac{1}{EBC}\left[e^{-C\sigma}\right]_{\sigma_0}^{\sigma_1}$$

and $$t = \frac{1}{EBC}\left[e^{-C\sigma_1} - e^{-C\sigma_0}\right]$$

Often, the strain rate of materials can be modelled by a power law approximation of equation (23.18).

$$\frac{d\varepsilon_C}{dt} = A\sigma^n \tag{23.19}$$

where A and n are known constants for the material.

$$d\varepsilon_C = A\sigma^n\,dt$$

Equation (23.17) gives:

$$d\varepsilon_C = -\frac{d\sigma}{E}$$

So, $$A\sigma^n = \left(-\frac{d\sigma}{E}\right)\frac{1}{dt}$$

$$dt = -\frac{\sigma^{-n}}{AE}d\sigma$$

$$\int_0^t dt = -\frac{1}{AE}\int_{\sigma_0}^{\sigma_1}\sigma^{-n}\,d\sigma$$

and $$t = -\frac{1}{AE}\left[\frac{\sigma^{1-n}}{1-n}\right]_{\sigma_0}^{\sigma_1} = \frac{1}{AE}\left[\frac{\sigma_1^{1-n} - \sigma_0^{1-n}}{n-1}\right]_{\sigma_0}^{\sigma_1}$$

$$= \left(\frac{1}{AE(n-1)}\right)\left[\frac{1}{\sigma_1^{n-1}} - \frac{1}{\sigma_0^{n-1}}\right]$$

$$(n-1)AEt = \frac{1}{\sigma_1^{n-1}} - \frac{1}{\sigma_0^{n-1}}$$

$$\boldsymbol{\frac{1}{\sigma_1^{n-1}} = \frac{1}{\sigma_0^{n-1}} + (n-1)AEt} \tag{23.20}$$

Problem 8. A bolted joint for aeroengine casings can be treated as two plates bolted togther. The bolts and the casings are made of a similar material. Laboratory tests on this material at service temperature show that the secondary creep strain rate is related to the stress by the relation:

$$\frac{d\varepsilon}{dt} = A\sigma^5 \quad \text{where } A \text{ is a constant}$$

At a stress of 40 MPa, it is found that:

$$\frac{d\varepsilon}{dt} = 3.0\times10^{-6} \quad \text{per hour}$$

Calculate the initial stress to which the bolts must be tightened if the stress in them must not be allowed to fall below 6 MPa over a 3 year duration. The Young's modulus at the service temperature can be taken to be 150 GPa. It can be assumed that the joint is rigid and that only secondary (steady) creep of the bolts is significant.

Equation (23.20) gives:

$$\frac{1}{\sigma_1^{n-1}} = \frac{1}{\sigma_0^{n-1}} + (n-1)AEt$$

The constant A can be found from:

$$\dot{\varepsilon}_c = 3.0\times10^{-6} \text{ per hour when } \sigma = 40 \text{ MPa}$$

So $\quad 3.0\times10^{-6} = A(40\times10^6)^5$

$$A = \frac{3\times10^{-6}}{(40\times10^6)^5} = 29.30\times10^{-45}$$

The initial stress in the bolts, σ_0, is to be found and the final stress, σ_1, is given as 6 MPa.

$$t = 3(365)(24) = 26282 \text{ hours}$$

$$\frac{1}{\sigma_0^{n-1}} = \frac{1}{\sigma_1^{n-1}} - (n-1)AEt$$

$$\frac{1}{\sigma_0^4} = \frac{1}{\sigma_1^4} - 4AEt$$

$$\sigma_0^{-4} = \sigma_1^{-4} - 4AEt$$

$$\sigma_0 = (\sigma_1^{-4} - 4tAE)^{-1/4}$$

$$= \left\{(6\times10^6)^{-4} - 4[3\times365\times24] (29.3\times10^{-45})(150\times10^9)\right\}^{-1/4}$$

$$= \left\{771.6\times10^{-30} - 462\times10^{-30}\right\}^{-1/4}$$

$$\sigma_0 = 7.539\times10^6 \text{ Pa} = \mathbf{7.54 \text{ MPa}}$$

Now try the following Practice exercise

Practice Exercise 81. Effect of restraint – creep relaxation

1. Part of a process plant uses a pressure vessel that operates continuously with fluids at a temperature of 425°C and a pressure of 1.2 MPa. There is a 300 mm diameter inspection cover which is secured to the vessel by means of 12 off, equi-spaced 25 mm diameter bolts. The Young's modulus for the bolt material is 210 GPa. Further tests on this material at 20 MPa and 425°C have shown the strain rate to be 800.0 × 10^{-12} per hour. It also follows a secondary creep rate given by the equation: $\frac{d\varepsilon}{dt} = A\sigma^n$ where A is a constant and $n = 4$.

   (a) Including a safety factor of 2, determine the initial tightening stress in the bolt, to ensure that there is no leakage due to relaxation of the bolts after 10,000 hours of operation.

   (b) For the design as in (a) above, what is the expected time before leakage occurs?

   [(a) $\sigma_0 = 45.86$ MPa
   (b) $t = 103\times10^3$ hours]

For fully worked solutions to each of the problems in Exercise 77 to 81 in this chapter, go to the website:
www.routledge.com/cw/bird

A revisionary list of formulae for Mechanics of Solids

Formula	Formula symbols	Units
Stress = $\dfrac{\text{applied force}}{\text{cross-sectional area}}$	$\sigma = \dfrac{F}{A}$	Pa
Strain = $\dfrac{\text{change in length}}{\text{original length}}$	$\varepsilon = \dfrac{x}{L}$	
Young's modulus of elasticity = $\dfrac{\text{stress}}{\text{strain}}$	$E = \dfrac{\sigma}{\varepsilon}$	Pa
Stiffness = $\dfrac{\text{force}}{\text{extension}}$	$k = \dfrac{F}{\delta}$	N/m
Modulus of rigidity = $\dfrac{\text{shear stress}}{\text{shear strain}}$	$G = \dfrac{\tau}{\gamma}$	Pa
Thermal strain = coefficient of linear expansion × temperature rise	$\varepsilon = \alpha T$	
Thermal stress in compound bar	$\sigma_1 = \dfrac{(\alpha_1 - \alpha_2) E_1 E_2 A_2 T}{(A_1 E_1 + A_2 E_2)}$	Pa
Ultimate tensile strength = $\dfrac{\text{maximum load}}{\text{original cross-sectional area}}$		Pa
Moment = force × perpendicular distance	$M = Fd$	N m
$\dfrac{\text{stress}}{\text{distance from neutral axis}} = \dfrac{\text{bending moment}}{\text{second moment of area}} = \dfrac{\text{Young's modulus}}{\text{radius of curvature}}$	$\dfrac{\sigma}{y} = \dfrac{M}{I} = \dfrac{E}{R}$	N/m^3
Torque = force × perpendicular distance	$T = Fd$	N m
Power = torque × angular velocity	$P = T\omega = 2\pi nT$	W
Horsepower	1 hp = 745.7 W	
Torque = moment of inertia × angular acceleration	$T = I\alpha$	N m
$\dfrac{\text{shear stress}}{\text{radius}} = \dfrac{\text{torque}}{\text{polar second moment of area}} = \dfrac{\text{(rigidity)(angle of twist)}}{\text{length}}$	$\dfrac{\tau}{r} = \dfrac{T}{J} = \dfrac{G\theta}{L}$	N/m^3
Average velocity = $\dfrac{\text{distance travelled}}{\text{time taken}}$	$v = \dfrac{s}{t}$	m/s
Acceleration = $\dfrac{\text{change in velocity}}{\text{time taken}}$	$a = \dfrac{v - u}{t}$	m/s^2

Formula	Formula symbols	Units
Linear velocity	$v = \omega r$	m/s
Angular velocity	$\omega = \dfrac{\theta}{t} = 2\pi n$	rad/s
Linear acceleration	$a = r\alpha$	m/s^2
Relationships between initial velocity u, final velocity v, displacement s, time t and constant acceleration a	$v_2 = v_1 + at$ $s = ut + \dfrac{1}{2}at^2$ $v^2 = u^2 + 2as$	m/s m $(m/s)^2$
Relationships between initial angular velocity ω_1, final angular velocity ω_2, angle θ, time t and angular acceleration α	$\omega_2 = \omega_1 + \alpha t$ $\theta = \omega_1 t + \dfrac{1}{2}\alpha t^2$ $\omega_1^2 = \omega_2^2 + 2\alpha\theta$	rad/s rad $(rad/s)^2$
Momentum = mass × velocity		kg m/s
Impulse = applied force × time = change in momentum		kg m/s
Force = mass × acceleration	$F = ma$	N
Weight = mass × gravitational field	$W = mg$	N
Centripetal acceleration	$a = \dfrac{v^2}{r}$	m/s^2
Centripetal force	$F = \dfrac{mv^2}{r}$	N
Density $= \dfrac{\text{mass}}{\text{volume}}$	$\rho = \dfrac{m}{V}$	kg/m^3
Work done = force × distance moved	$W = Fs$	J
Efficiency $= \dfrac{\text{useful output energy}}{\text{input energy}}$		
Power $= \dfrac{\text{energy used (or work done)}}{\text{time taken}}$ = force × velocity	$P = \dfrac{E}{t} = Fv$	W
Potential energy = weight × change in height Kinetic energy $= \dfrac{1}{2} \times \text{mass} \times (\text{speed})^2$	$E_p = mgh$ $E_k = \dfrac{1}{2}mv^2$	J J
Kinetic energy of rotation $= \dfrac{1}{2} \times \text{moment of inertia} \times (\text{angular velocity})^2$	$E_k = \dfrac{1}{2}I\omega^2$	J

Formula	Formula symbols	Units
Frictional force = coefficient of friction × normal force	$F = \mu N$	N
Angle of repose, θ, on an inclined plane	$\tan\theta = \mu$	
Efficiency of screw jack	$\eta = \dfrac{\tan\theta}{\tan(\lambda + \theta)}$	
SHM periodic time $T = 2\pi\sqrt{\dfrac{\text{displacement}}{\text{acceleration}}}$	$T = 2\pi\sqrt{\dfrac{y}{a}}$	s
$T = 2\pi\sqrt{\dfrac{\text{mass}}{\text{stiffness}}}$	$T = 2\pi\sqrt{\dfrac{m}{k}}$	s
simple pendulum	$T = 2\pi\sqrt{\dfrac{L}{g}}$	s
compound pendulum	$T = 2\pi\sqrt{\dfrac{(k_G^2 + h^2)}{gh}}$	s
Force ratio = $\dfrac{\text{load}}{\text{effort}}$		
Movement ratio = $\dfrac{\text{distance moved by effort}}{\text{distance moved by load}}$		
Efficiency = $\dfrac{\text{force ratio}}{\text{movement ratio}}$		
Kelvin temperature = degrees Celsius + 273		
Quantity of heat energy = mass × specific heat capacity × change in temperature	$Q = mc(t_2 - t_1)$	J
New length = original length + expansion	$L_2 = L_1\,[1 + \alpha(t_2 - t_1)]$	m
New surface area = original surface area + increase in area	$A_2 = A_1\,[1 + \beta(t_2 - t_1)]$	m^2
New volume = original volume + increase in volume	$V_2 = V_1\,[1 + \gamma(t_2 - t_1)]$	m^3
Pressure = $\dfrac{\text{force}}{\text{area}}$ = density × gravitational acceleration × height	$p = \dfrac{F}{A}$ $p = \rho g h$ 1 bar = 10^5 Pa	Pa
Absolute pressure = gauge pressure + atmospheric pressure		

Circular segment

In Figure F1, shaded area $= \dfrac{R^2}{2}(\alpha - \sin\alpha)$

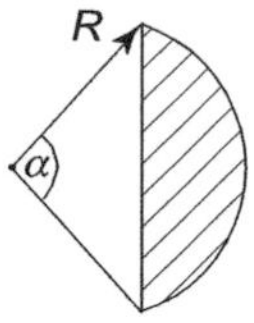

Figure F1

Summary of standard results of the second moments of areas of regular sections

Shape	Position of axis	Second moment of area, I
Rectangle		
length D breadth B	(1) Coinciding with B	$\dfrac{BD^3}{3}$
	(2) Coinciding with D	$\dfrac{DB^3}{3}$
	(3) Through centroid, parallel to B	$\dfrac{BD^3}{12}$
	(4) Through centroid, parallel to D	$\dfrac{DB^3}{12}$
Triangle		
Perpendicular height H base B	(1) Coinciding with B	$\dfrac{BH^3}{12}$
	(2) Through centroid, parallel to base	$\dfrac{BH^3}{36}$
	(3) Through vertex, parallel to base	$\dfrac{BH^3}{4}$
Circle		
radius R diameter D	(1) Through centre perpendicular to plane (i.e. polar axis)	$\dfrac{\pi R^4}{2}$ or $\dfrac{\pi D^4}{32}$
	(2) Coinciding with diameter	$\dfrac{\pi R^4}{4}$ or $\dfrac{\pi D^4}{64}$
	(3) About a tangent	$\dfrac{5\pi R^4}{4}$ or $\dfrac{5\pi D^4}{64}$
Semicircle		
radius R	Coinciding with diameter	$\dfrac{\pi R^4}{8}$

Bending stresses in beams

$$\frac{\sigma}{y} = \frac{M}{I} = \frac{E}{R} \qquad Z = \frac{I}{\bar{y}} \quad \hat{\sigma} = \frac{M}{Z}$$

Beam deflections due to bending

$$M = EI\frac{d^2y}{dx^2}$$

Torsion

$$\frac{\tau}{r} = \frac{T}{J} = \frac{G\theta}{l} \qquad P = T\omega$$

Complex stress and strain

$$\sigma_\theta = \frac{1}{2}\left(\sigma_x + \sigma_y\right) + \frac{1}{2}\left(\sigma_x - \sigma_y\right)\cos 2\theta + \tau_{xy}\sin 2\theta$$

$$\tau_\theta = \frac{1}{2}\left(\sigma_x - \sigma_y\right)\sin 2\theta - \tau_{xy}\cos 2\theta$$

$$\tan 2\theta = \frac{2\tau_{xy}}{\left(\sigma_x - \sigma_y\right)}$$

σ_1 = maximum principal stress

$$= \frac{1}{2}\left(\sigma_x + \sigma_y\right) + \frac{1}{2}\sqrt{\left(\sigma_x - \sigma_y\right)^2 + 4\tau_{xy}^2}$$

σ_2 = minimum principal stress

$$= \frac{1}{2}\left(\sigma_x + \sigma_y\right) - \frac{1}{2}\sqrt{\left(\sigma_x - \sigma_y\right)^2 + 4\tau_{xy}^2}$$

$$\tan 2\theta = \frac{\left(\sigma_y - \sigma_x\right)}{2\tau_{xy}}$$

$$\hat{\tau} = \pm\sqrt{\frac{1}{4}\left(\sigma_y - \sigma_x\right)^2 + \tau_{xy}^2}$$

$$\hat{\tau} = \frac{\left(\sigma_1 - \sigma_2\right)}{2}$$

$$\sigma_1 = \frac{16}{\pi d^3}\left(M + \sqrt{\left(M^2 + T^2\right)}\right)$$

$$\sigma_2 = \frac{16}{\pi d^3}\left(M - \sqrt{\left(M^2 + T^2\right)}\right)$$

$$\hat{\tau} = \pm\frac{16}{\pi d^3}\sqrt{M^2 + T^2}$$

$$\tan 2\theta = \frac{\gamma_{xy}}{\left(\varepsilon_x - \varepsilon_y\right)}$$

$$\varepsilon_1 = \frac{1}{2}\left(\varepsilon_x + \varepsilon_y\right) + \frac{1}{2}\sqrt{\left(\varepsilon_x - \varepsilon_y\right)^2 + \gamma_{xy}^2}$$

$$\varepsilon_2 = \frac{1}{2}\left(\varepsilon_x + \varepsilon_y\right) - \frac{1}{2}\sqrt{\left(\varepsilon_x - \varepsilon_y\right)^2 + \gamma_{xy}^2}$$

$$\sigma_x = \frac{E}{\left(1 - \nu^2\right)}\left(\varepsilon_x + \nu\varepsilon_y\right)$$

$$\sigma_y = \frac{E}{\left(1 - \nu^2\right)}\left(\varepsilon_y + \nu\varepsilon_x\right)$$

$$\sigma_x = \frac{E}{(1+\nu)(1-2\nu)}\left[(1-\nu)\varepsilon_x + \nu\varepsilon_y\right]$$

$$\sigma_y = \frac{E}{(1+\nu)(1-2\nu)}\left[(1-\nu)\varepsilon_y + \nu\varepsilon_x\right]$$

$$\tau_{xy} = G\gamma_{xy}$$

Membrane theory for thin-walled circular cylinders and spheres

$$\sigma_H = \frac{PR}{\eta_L t} \quad \sigma_L = \frac{PR}{2\eta_c t} \quad \sigma = \frac{RP}{2\eta t}$$

Energy methods

$$\frac{\partial U_e}{\partial P} = u \qquad \frac{\partial U_e}{\partial R} = \lambda$$

$$U_e = \frac{\sigma^2}{2E} \times \text{volume of rod}$$

$$\text{WD} = d\left(U_b\right) = \frac{M^2}{2EI} \times dx$$

$$U_T = \frac{T^2 \times dx}{2G \times J} \qquad U_s = \frac{\tau^2}{2G} \times \text{volume}$$

Theories of elastic failure

$$\sigma_1 = \sigma_{yp} \qquad \sigma_3 = \sigma_{ypc}$$

$$\sigma_1 - \nu\left(\sigma_2 + \sigma_3\right) = \sigma_{yp}$$

$$\sigma_3 - \nu\left(\sigma_1 + \sigma_2\right) = \sigma_{ypc}$$

$$\left(\sigma_1^2+\sigma_2^2+\sigma_3^2\right)-2v\left(\sigma_1\sigma_2+\sigma_1\sigma_3+\sigma_2\sigma_3\right)=\sigma_{yp}^{\ 2}$$

$$\sigma_1-\sigma_3=\sigma_{yp}$$

$$\left(\sigma_1-\sigma_2\right)^2+\left(\sigma_1-\sigma_3\right)^2+\left(\sigma_2-\sigma_3\right)^2=2\sigma_{yp}^{\ 2}$$

Thick cylinders and spheres

$$\sigma_r=A-\frac{B}{r^2}\qquad \sigma_\theta=A+\frac{B}{r^2}$$

$$\sigma=\frac{2}{3}A+\frac{B}{2r^3}\qquad \sigma_\theta=A+\frac{B}{r^2}-\left(1+3v\right)\left(\frac{\rho\omega^2r^2}{8}\right)$$

The buckling of struts

$$P=\frac{\pi^2EI}{l^2}$$

$$P_R=\frac{\sigma_{yc}\times A}{a\left(\frac{L_0}{k}\right)^2+1}\qquad \sigma_c=\frac{P}{A}+\frac{\Delta PP_e\bar{y}}{\left(P_e-P\right)I}$$

Effective lengths of struts (L_0)

Type of strut	Euler	BS449
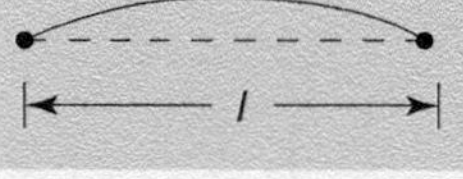	$L_0=l$	$L_0=l$
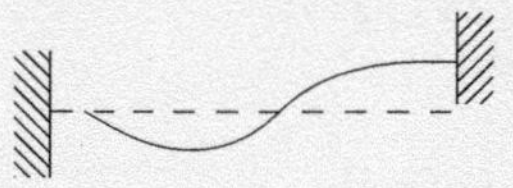	$L_0=l$	l
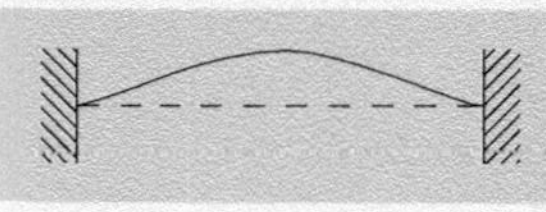	$L_0=0.5l$	$0.7l$
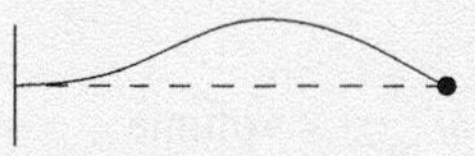	$L_0=0.7l$	$0.85l$
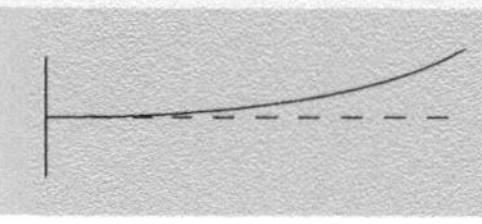	$L_0=2l$	$2l$

Asymmetrical bending of beams

$$\sigma=\frac{M(\cos\alpha)\bar{y}}{I_{xx}}+\frac{M(\sin\alpha)\bar{x}}{I_{yy}}$$

$$I_{xy}=A\bar{h}\bar{k}$$

$$\tan 2\theta=\frac{2I_{xy}}{\left(I_y-I_x\right)}$$

$$I_U=\frac{1}{2}\left(I_x+I_y\right)+\frac{1}{2}\left(I_x-I_y\right)\sec 2\theta$$

$$I_V=\frac{1}{2}\left(I_x+I_y\right)-\frac{1}{2}\left(I_x-I_y\right)\sec 2\theta$$

$$\sigma=\frac{M\cos\theta\, v}{I_{UU}}+\frac{M\sin\theta\, u}{I_{VV}}$$

$$\beta=\tan^{-1}\left(-\frac{I_{UU}\tan\theta}{I_{VV}}\right)$$

Shear stresses in bending and shear deflections

$$\tau=\frac{F}{bI}\int y\,dA\qquad \tau=\frac{F\left(B-Z\right)\bar{y}}{I}$$

Composites

$$\begin{pmatrix}\varepsilon_1\\ \varepsilon_2\\ \varepsilon_3\\ \varepsilon_4\\ \varepsilon_5\\ \varepsilon_6\end{pmatrix}=\begin{pmatrix}S_{11}&S_{12}&S_{13}&0&0&0\\ S_{12}&S_{22}&S_{23}&0&0&0\\ S_{13}&S_{23}&S_{33}&0&0&0\\ 0&0&0&S_{44}&0&0\\ 0&0&0&0&S_{55}&0\\ 0&0&0&0&0&S_{66}\end{pmatrix}\begin{pmatrix}\sigma_1\\ \sigma_2\\ \sigma_3\\ \sigma_4\\ \sigma_5\\ \sigma_6\end{pmatrix}$$

$$\left(S\right)=\begin{pmatrix}1/E_1&-\upsilon_{21}/E_2&0\\ -\upsilon_{12}/E_1&1/E_2&0\\ 0&0&1/G_{12}\end{pmatrix}$$

$$\frac{1}{|S|}=\left(\frac{1}{1-\upsilon_{12}\upsilon_{21}}\right)\left(E_1E_2G_{12}\right)$$

$$\left(R\right)=\begin{pmatrix}1&0&0\\ 0&1&0\\ 0&0&2\end{pmatrix}$$

$$\left(Q^*\right)=\left(T\right)^{-1}\left(Q\right)\left(R\right)\left(T\right)\left(R\right)^{-1}$$

$$\left(Q\right)=\left(S\right)^{-1}\qquad \begin{pmatrix}N\\ M\end{pmatrix}=\begin{pmatrix}A&B\\ B&D\end{pmatrix}\begin{pmatrix}\varepsilon^o\\ \kappa\end{pmatrix}$$

The Deviatoric strain energy (Tsai-Hill) failure criterion

$$\frac{\sigma_1^2}{F_1^2}+\frac{\sigma_2^2}{F_2^2}+\frac{\tau_{12}^2}{F_{12}^2}-\frac{\sigma_1\sigma_2}{F_1^2}=1$$

The Interactive tensor polynomial (Tsai-Wu) failure criterion

$$f_1\sigma_1+f_2\sigma_2+f_{11}\sigma_1^2+f_{22}\sigma_2^2+f_{66}\tau_{12}^2+2f_{12}\sigma_1\sigma_2^2=1$$

$$f_1=\frac{1}{F_{1T}}-\frac{1}{F_{1C}} \qquad f_{11}=\frac{1}{F_{1T}F_{1C}}$$

$$f_2=\frac{1}{F_{2T}}-\frac{1}{F_{2C}} \qquad f_{22}=\frac{1}{F_{2T}F_{2C}}$$

$$f_{66}=\frac{1}{F_{12}^2} \qquad f_{12}=-\frac{1}{2}\left(\frac{1}{F_{1T}F_{1C}F_{2T}F_{2C}}\right)^{\frac{1}{2}}$$

The matrix displacement method

$$\left(k^0\right)=\frac{AE}{l}\begin{pmatrix} C^2 & CS & -C^2 & -CS \\ CS & S^2 & -CS & -S^2 \\ -C^2 & -CS & C^2 & CS \\ -CS & -S^2 & CS & S^2 \end{pmatrix}$$

(columns: u_1^0, v_1^0, u_2^0, v_2^0; rows: u_1^0, v_1^0, u_2^0, v_2^0)

$$\begin{pmatrix} Y_i \\ M_i \\ Y_j \\ M_j \end{pmatrix}=EI\begin{pmatrix} 12/l^3 & -6/l^2 & -12/l^3 & -6/l^2 \\ -6/l^2 & 4/l & 6/l^2 & 2/l \\ -12/l^3 & 6/l^2 & 12/l^3 & 6/l^2 \\ -6/l^2 & 2/l & 6/l^2 & 4/l \end{pmatrix}\begin{pmatrix} v_i \\ \theta_i \\ v_j \\ \theta_j \end{pmatrix}$$

The finite element method

$$\left(k^o\right)=\int\frac{1}{4\Delta^2}\begin{pmatrix} b_1 & 0 & c_1 \\ b_2 & 0 & c_2 \\ b_3 & 0 & c_3 \\ 0 & c_1 & b_1 \\ 0 & c_2 & b_2 \\ 0 & c_3 & b_3 \end{pmatrix}E^1\begin{pmatrix} 1 & \mu & 0 \\ \mu & 1 & 0 \\ 0 & 0 & \gamma \end{pmatrix}\begin{pmatrix} b_1 & b_2 & b_3 & 0 & 0 & 0 \\ 0 & 0 & 0 & c_1 & c_2 & c_3 \\ c_1 & c_2 & c_3 & b_1 & b_2 & b_3 \end{pmatrix}t\,dA$$

Fracture mechanics

$$\left(E\,G_C\right)^{\frac{1}{2}}=\sigma\left(\pi a\right)^{\frac{1}{2}} \text{ for plane stress}$$

$$K=\sigma\sqrt{\pi a}\,\alpha \qquad \frac{da}{dN}=C\left(\Delta K\right)^m$$

$$J=\int_p\left(W\,dy-T_i\frac{\partial u_i}{\partial x}ds\right) \qquad G=\frac{12F^2a^2}{E\,B^2h^3}$$

Fatigue

$$K_T=\frac{\sigma_{MAX}}{\sigma_{MIN}} \qquad q=\frac{K_F-1}{K_T-1} \qquad \sum_i^k\frac{n_i}{N_i}=1$$

Answers to multiple-choice questions

Multiple-choice questions test 1 (Page 19)

1. (b)	**2.** (d)	**3.** (a)	**4.** (d)	**5.** (a)
6. (b)	**7.** (c)	**8.** (a)	**9.** (b)	**10.** (c)
11. (b)	**12.** (d)	**13.** (a)	**14.** (c)	**15.** (a)
16. (c)	**17.** (d)	**18.** (b)	**19.** (a)	**20.** (c)

Multiple-choice questions test 2 (Page 47)

1. (b)	**2.** (d)	**3.** (d)	**4.** (a)	**5.** (b)
6. (d)	**7.** (a)	**8.** (a)	**9.** (d)	**10.** (c)
11. (b)	**12.** (d)	**13.** (c)	**14.** (a)	**15.** (b)
16. (c)	**17.** (c)	**18.** (a)	**19.** (c)	**20.** (b)

Multiple-choice questions test 3 (Page 191)

1. (a)	**2.** (b)	**3.** (d)	**4.** (d)	**5.** (a)
6. (c)	**7.** (a)	**8.** (c)	**9.** (d)	**10.** (a)
11. (b)	**12.** (d)	**13.** (c)	**14.** (b)	**15.** (b)
16. (e)	**17.** (a)	**18.** (b)	**19.** (c)	**20.** (a)
21. (a)	**22.** (c)	**23.** (e)	**24.** (d)	**25.** (c)

Multiple-choice questions test 4 (Page 197)

1. (c)	**2.** (d)	**3.** (b)	**4.** (d)	**5.** (d)
6. (a)	**7.** (d)	**8.** (a)	**9.** (b)	**10.** (d)
11. (c)	**12.** (e)	**13.** (d)	**14.** (c)	**15.** (b)

Multiple-choice questions test 5 (Page 397)

1. (c)	**2.** (b)	**3.** (b)	**4.** (c)	**5.** (d)
6. (d)	**7.** (c)	**8.** (a)	**9.** (d)	**10.** (c)
11. (d)	**12.** (a)	**13.** (c)	**14.** (b)	**15.** (a)
16. (b)	**17.** (d)	**18.** (b)	**19.** (c)	**20.** (a)

References

1. Ross, C. T. F. *Computational Methods in Structural and Continuum Mechanics,* Ellis Horwood, Chichester, 1982.
2. Williams, J. G. *Stress Analysis of Polymers,* Longman, Harlow, 1973.
3. Macaulay, W. H. Note on the Deflection of Beams, *Messenger Math.,* 48, 129–130, 1919.
4. Stephens, R. C. *Strength of Materials,* Arnold, London, 1970.
5. Ross, C. T. F. *Advanced Applied Finite Element Methods,* Elsevier, Oxford, UK, 1998.
6. Johnson, D. *Advanced Structural Mechanics,* Collins, London, 1986.
7. Rockey, K. C., Evans, H. R., Griffiths, D. W. and Nethercot, D. A. *The Finite Element Method,* 2nd Edition, Collins, London, 1983.
8. Saada, A. S. *Elasticity – Theory and Applications,* J. Ross Publishing, Florida, 2009.
9. Den Hartog, J. P. *Advanced Strength of Materials,* McGraw-Hill, New York, 1952.
10. Timoshenko, S. P. and Goodier, J. N. *Theory of Elasticity,* 3rd Edition, McGraw-Hill, New York, 1984.
11. Ford, H. and Alexander, J. M. *Advanced Strength of Materials,* Ellis Horwood, Chichester, 1977.
12. Ross, C. T. F. *Finite Element Programs for Axisymmetric Problems in Engineering,* Ellis Horwood, Chichester, 1984.
13. Ross, C. T. F. *The Collapse of Ring-reinforced Circular Cylinders under Uniform External Pressure, Trans. RINA,* 107, 375–394, 1965.
14. Richards, T. H. *Energy Methods in Stress Analysis,* Ellis Horwood, Chichester, 1977.
15. Zienkiewicz, O. C. and Taylor, R. L. *The Finite Element Method,* 4th Edition, Volumes 1 to 3, McGraw-Hill, New York, 1989.
16. Cook, R. D., Malkus, S. S., Plesha, M. E. and Witt, R. D., *Concepts and Applications of Finite Element Analysis,* 4th Edition, Wiley, Chichester, 2001.
17. Segerlind, L. J. *Applied Finite Element Analysis,* 2nd Edition, Wiley, New York, 1984.
18. Irons, B. and Ahmad, S. *Techniques of Finite Elements,* Ellis Horwood, Chichester, 1980.
19. Fenner, R. T. *Finite Element Methods for Engineers,* Macmillan, London, 1996.
20. Davies, G. A. O. *Virtual Work in Structural Analysis,* Wiley, Chichester, 1982.
21. Smith, I. M. *Programming the Finite Element Method,* Wiley, Chichester, 1982.
22. Owen, D. R. J. and Hinton, E. *A Simple Guide to Finite Elements,* Pineridge, Swansea, 1980.
23. Coker, E. G. and Filon, L. N. G. *A Treatise on Photoelasticity,* Cambridge University Press, Cambridge, 1931.
24. Holister, G. S. *Experimental Stress Analysis – Principles and Methods,* Cambridge University Press, Cambridge, 1967.
25. Dally, J. W. and Riley, W. F. *Experimental Stress Analysis,* 2nd Edition, McGraw-Hill Kogakusha, New York, 1978.
26. Line, D. R. *Investigation into the Use of Some Commercial and Industrial Adhesives and Sealants for Electrical Resistance Strain Gauge Application*, Final Year Project, Dept. of Mech. Eng., Portsmouth Polytechnic, 1977–1978.
27. Milligan, R. V. The Effects of High Pressure on Foil Strain Gauges, *Exp. Mech.,* 4, 25–36, 1964.
28. Brace, W. F. Effect of Pressure on Electrical Resistance Strain Gauges, *Exp. Mech.,* 4, 212–216, 1964.
29. Megson, T. H. G. *Aircraft Structures for Engineering Students,* 5th Edition, Elsevier, Oxford, 2012.
30. Bolotin, V. V. *The Dynamic Stability of Elastic System,* Holden-Day, San Francisco, 1964.
31. NAFEMS = National Agency for Finite Element Methods and Standards (NAFEMS is a not for profit membership association, Hamilton, Lanarkshire, UK.)
32. Bird, J. O. and Ross, C. T. F. *Mechanical Engineering Principles*, 4th Edition, Routledge, Oxford, 2020.
33. Bird, J. O. *Bird's Higher Engineering Mathematics,* 9th Edition, Routledge, Oxford, 2021.

34. Bird, J. O. *Bird's Comprehensive Engineering Mathematics,* Routledge, Oxford, 2018.
35. Ross, C. T. F. *Finite Element Programs in Structural Engineering and Contiuum Mechanics*, Elsevier, Oxford, 1996.
36. Ross, C. T. F. *Pressure Vessels: External Pressure Technology,* 2nd Edition, Elsevier, Oxford, 2011.
37. Frye, C. D. *Microsoft Excel 2013 Step by Step*, Microsoft Press, Indianapolis, 2013.
38. Parker, A. P. *The Mechanics of Fracture and Fatigue*, E. & F. F. Spon, London, 1981.
39. M.A. Miner, Cumulative Damage in Fatigue, *J Appl Mech*, Vol. 12, 1945, pp. A159-64.
40. F. R. Larson & J. Miller, *Transactions ASME,* Vol.74, p.765–771, 1952.

Index

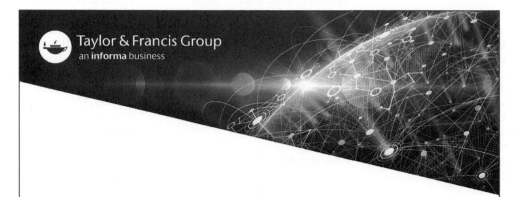

9780367651404